Tectonic Aspects of the Alpine-Dinaride-Carpathian System

The Geological Society of London
Books Editorial Committee

Chief Editor
BOB PANKHURST (UK)

Society Books Editors
JOHN GREGORY (UK)
JIM GRIFFITHS (UK)
JOHN HOWE (UK)
PHIL LEAT (UK)
NICK ROBINS (UK)
JONATHAN TURNER (UK)

Society Books Advisors
MIKE BROWN (USA)
ERIC BUFFETAUT (FRANCE)
JONATHAN CRAIG (ITALY)
RETO GIERÉ (GERMANY)
TOM MCCANN (GERMANY)
DOUG STEAD (CANADA)
RANDELL STEPHENSON (NETHERLANDS)

Geological Society books refereeing procedures

The Society makes every effort to ensure that the scientific and production quality of its books matches that of its journals. Since 1997, all book proposals have been refereed by specialist reviewers as well as by the Society's Books Editorial Committee. If the referees identify weaknesses in the proposal, these must be addressed before the proposal is accepted.

Once the book is accepted, the Society Book Editors ensure that the volume editors follow strict guidelines on refereeing and quality control. We insist that individual papers can only be accepted after satisfactory review by two independent referees. The questions on the review forms are similar to those for *Journal of the Geological Society*. The referees' forms and comments must be available to the Society's Book Editors on request.

Although many of the books result from meetings, the editors are expected to commission papers that were not presented at the meeting to ensure that the book provides a balanced coverage of the subject. Being accepted for presentation at the meeting does not guarantee inclusion in the book.

More information about submitting a proposal and producing a book for the Society can be found on its web site: www.geolsoc.org.uk.

It is recommended that reference to all or part of this book should be made in one of the following ways:

SIEGESMUND, S., FÜGENSCHUH, B. & FROITZHEIM, N. (eds) 2008. *Tectonic Aspects of the Alpine-Dinaride-Carpathian System.* Geological Society, London, Special Publications, **298**.

SCHULZ, B., STEENKEN, A. & SIEGESMUND, S. 2008. Geodynamic evolution of an Alpine terrane—the Austroalpine basement to the south of the Tauern Window as a part of the Adriatic Plate (eastern Alps). *In*: SIEGESMUND, S., FÜGENSCHUH, B. & FROITZHEIM, N. (eds) 2008. *Tectonic Aspects of the Alpine-Dinaride-Carpathian System.* Geological Society, London, Special Publications, **298**, 5–44.

GEOLOGICAL SOCIETY SPECIAL PUBLICATION NO. 298

Tectonic Aspects of the Alpine-Dinaride-Carpathian System

EDITED BY

S. SIEGESMUND
University of Göttingen, Germany

B. FÜGENSCHUH
University of Innsbruck, Austria

and

N. FROITZHEIM
University of Bonn, Germany

2008
Published by
The Geological Society
London

THE GEOLOGICAL SOCIETY

The Geological Society of London (GSL) was founded in 1807. It is the oldest national geological society in the world and the largest in Europe. It was incorporated under Royal Charter in 1825 and is Registered Charity 210161.

The Society is the UK national learned and professional society for geology with a worldwide Fellowship (FGS) of over 9000. The Society has the power to confer Chartered status on suitably qualified Fellows, and about 2000 of the Fellowship carry the title (CGeol). Chartered Geologists may also obtain the equivalent European title, European Geologist (EurGeol). One fifth of the Society's fellowship resides outside the UK. To find out more about the Society, log on to www.geolsoc.org.uk.

The Geological Society Publishing House (Bath, UK) produces the Society's international journals and books, and acts as European distributor for selected publications of the American Association of Petroleum Geologists (AAPG), the Indonesian Petroleum Association (IPA), the Geological Society of America (GSA), the Society for Sedimentary Geology (SEPM) and the Geologists' Association (GA). Joint marketing agreements ensure that GSL Fellows may purchase these societies' publications at a discount. The Society's online bookshop (accessible from www.geolsoc.org.uk) offers secure book purchasing with your credit or debit card.

To find out about joining the Society and benefiting from substantial discounts on publications of GSL and other societies worldwide, consult www.geolsoc.org.uk, or contact the Fellowship Department at: The Geological Society, Burlington House, Piccadilly, London W1J 0BG: Tel. +44 (0)20 7434 9944; Fax +44 (0)20 7439 8975; E-mail: enquiries@geolsoc.org.uk.

For information about the Society's meetings, consult *Events* on www.geolsoc.org.uk. To find out more about the Society's Corporate Affiliates Scheme, write to enquiries@geolsoc.org.uk.

Published by The Geological Society from:
The Geological Society Publishing House, Unit 7, Brassmill Enterprise Centre, Brassmill Lane, Bath BA1 3JN, UK

(*Orders*): Tel. +44 (0)1225 445046, Fax +44 (0)1225 442836)
Online bookshop: www.geolsoc.org.uk/bookshop

The publishers make no representation, express or implied, with regard to the accuracy of the information contained in this book and cannot accept any legal responsibility for any errors or omissions that may be made.

© The Geological Society of London 2008. All rights reserved. No reproduction, copy or transmission of this publication may be made without written permission. No paragraph of this publication may be reproduced, copied or transmitted save with the provisions of the Copyright Licensing Agency, 90 Tottenham Court Road, London W1P 9HE. Users registered with the Copyright Clearance Center, 27 Congress Street, Salem, MA 01970, USA: the item-fee code for this publication is 0305-8719/07/$15.00.

British Library Cataloguing in Publication Data

A catalogue record for this book is available from the British Library.

ISBN 978-1-86239-252-6

Typeset by Techset Composition Ltd., Salisbury, UK

Printed by MPG Books Ltd, Bodmin, UK

Distributors

North America
For trade and institutional orders:
The Geological Society, c/o AIDC, 82 Winter Sport Lane, Williston, VT 05495, USA
Orders: Tel +1 800-972-9892
Fax +1 802-864-7626
E-mail gsl.orders@aidcvt.com

For individual and corporate orders:
AAPG Bookstore, PO Box 979, Tulsa, OK 74101-0979, USA
Orders: Tel +1 918-584-2555
Fax +1 918-560-2652
E-mail bookstore@aapg.org
Website http://bookstore.aapg.org

India
Affiliated East-West Press Private Ltd, Marketing Division, G-1/16 Ansari Road, Darya Ganj, New Delhi 110 002, India
Orders: Tel +91 11 2327-9113/2326-4180
Fax +91 11 2326-0538
E-mail affiliat@vsnl.com

Contents

Foreword: Stefan M. Schmid	vii
SIEGESMUND, S., FÜGENSCHUH, B. & FROITZHEIM, N. Introduction: analysing orogeny—the Alpine approach	1
SCHULZ, B., STEENKEN, A. & SIEGESMUND, S. Geodynamic evolution of an Alpine terrane—the Austroalpine basement to the south of the Tauern Window as a part of the Adriatic Plate (eastern Alps)	5
SIEGESMUND, S., LAYER, P., DUNKL, I., VOLLBRECHT, A., STEENKEN, A., WEMMER, K. & AHRENDT, H. Exhumation and deformation history of the lower crustal section of the Valstrona di Omegna in the Ivrea Zone, southern Alps	45
FROITZHEIM, N., DERKS, J. F., WALTER, J. M. & SCIUNNACH, D. Evolution of an Early Permian extensional detachment fault from synintrusive, mylonitic flow to brittle faulting (Grassi Detachment Fault, Orobic Anticline, southern Alps, Italy)	69
VESELÁ, P., LAMMERER, B., WETZEL, A., SÖLLNER, F. & GERDES, A. Post-Variscan to Early Alpine sedimentary basins in the Tauern Window (eastern Alps)	83
DALLMEYER, R. D., NEUBAUER, F. & FRITZ, H. The Meliata suture in the Carpathians: regional significance and implications for the evolution of high-pressure wedges within collisional orogens	101
BERGER, A. & BOUSQUET, R. Subduction-related metamorphism in the Alps: review of isotopic ages based on petrology and their geodynamic consequences	117
TOMLJENOVIĆ, B., CSONTOS, L., MÁRTON, E. & MÁRTON, P. Tectonic evolution of the northwestern Internal Dinarides as constrained by structures and rotation of Medvednica Mountains, North Croatia	145
GRÖGER, H. R., FÜGENSCHUH, B., TISCHLER, M., SCHMID, S. M. & FOEKEN, J. P. T. Tertiary cooling and exhumation history in the Maramures area (internal eastern Carpathians, northern Romania): thermochronology and structural data	169
ROSENBERG, C. L. & SCHNEIDER, S. The western termination of the SEMP Fault (eastern Alps) and its bearing on the exhumation of the Tauern Window	197
LAMMERER, B., GEBRANDE, H., LÜSCHEN, E. & VESELÁ, P. A crustal-scale cross-section through the Tauern Window (eastern Alps) from geophysical and geological data	219
USTASZEWSKI, M. & PFIFFNER, O. A. Neotectonic faulting, uplift & seismicity in the central and western Swiss Alps	231
PLEUGER, J., NAGEL, T. J., WALTER, J. M., JANSEN, E. & FROITZHEIM, N. On the role and importance of orogen-parallel and -perpendicular extension, transcurrent shearing, and backthrusting in the Monte Rosa nappe and the Southern Steep Belt of the Alps (Penninic zone, Switzerland and Italy)	251
CIULAVU, M., FERREIRO MÄHLMANN, R., SCHMID, S. M., HOFMANN, H., SEGHEDI, A. & FREY, M. Metamorphic evolution of a very low- to low-grade metamorphic core complex (Danubian window) in the South Carpathians	281
TISCHLER, M., MATENCO, L., FILIPESCU, S., GRÖGER, H. R., WETZEL, A. & FÜGENSCHUH, B. Tectonics and sedimentation during convergence of the ALCAPA and Tisza–Dacia continental blocks: the Pienide nappe emplacement and its foredeep (N. Romania)	317
MIKES, T., BÁLDI-BEKE, M., KÁZMÉR, M., DUNKL, I. & VON EYNATTEN, H. Calcareous nannofossil age constraints on Miocene flysch sedimentation in the Outer Dinarides (Slovenia, Croatia, Bosnia-Herzegovina and Montenegro)	335
NAGEL, T. J. Tertiary subduction, collision and exhumation recorded in the Adula nappe, central Alps	365

BOUSQUET, R., OBERHÄNSLI, R., GOFFÉ, B., WIEDERKEHR, M., KOLLER, F., SCHMID, S. M., SCHUSTER, R., ENGI, M., BERGER, A. & MARTINOTTI, G. Metamorphism of metasediments at the scale of an orogen: a key to the Tertiary geodynamic evolution of the Alps — 393

MOLLI, G. Northern Apennine–Corsica orogenic system: an updated overview — 413

Index — 443

Foreword: Stefan M. Schmid

This Geological Society Special Publication is dedicated to Professor Stefan Schmid on the occasion of his 65th birthday in recognition of his merits in structural geology and tectonics. Stefan Schmid is without any doubt one of today's pre-eminent geologists, both as a scientist and teacher.

Born in 1943 in Wohlen, Switzerland, he first worked as a school teacher before starting his career as a geologist. In 1968, he finished his diploma thesis and in 1971 his PhD thesis, both at ETH Zürich under the guidance of Rudolf Trümpy. After six years as a postdoctoral fellow, that brought him first to Imperial College London (John Ramsay and Ernie Rutter) and afterwards to the Australian National University Canberra (with Mervyn Paterson), he then moved back to ETH Zürich as a lecturer. In 1989, he was appointed as full professor and head of Geologisch-Paläontologisches Institut at the University of Basel.

During his scientific career, he has contributed substantially to two main fields of Earth sciences, namely the structure and rheology of deformed rocks and the processes of mountain building, especially in Alpine-type orogens.

Together with colleagues at the Australian National University and the Center for Tectonophysics (Texas A & M University), Stefan carried out numerous experiments related to the deformation of calcite in the 1970s and 1980s and exploited deformation mechanisms in other minerals. Outstanding publications during this period include the description of 'superplasticity in fine-grained limestone' and the work on 'complete fabric analysis of quartz c-axis patterns'. Along with his experimental endeavours, Stefan has always kept contact with the 'field' and applied the experimental outcomes to field studies and *vice versa*. Cited over 500 times, his publication on 'shear sense criteria' with Carol Simpson clearly expresses this attitude. Application of these criteria for reconstruction of the kinematic history of mountain belts paved Stefan Schmid's future road from the microscale towards the macroscale in Earth sciences. Over the years, he has expanded and intensified his collaboration with colleagues from other branches of the Earth sciences and integrated this information with his sound basis as a structural geologist. Key publications include his work on the Insubric Line and his outstanding geological/geophysical transects through the Alps. In the last couple of years, Stefan has extended his area of interest both in the horizontal and vertical scale. Thanks to his close cooperation with geophysicists, he became fascinated by the recent outcomes of teleseismic tomography and their bearing on the Alps, Carpathians and Dinarides. On the horizontal scale, he moved (south)eastwards and became increasingly drawn to the Carpathians and Dinarides, first due to their challenging geology and secondly because of the attractiveness of eastern European countries.

Stefan's work in the fields of emplacement of granitic plutons, extensional tectonics and neotectonics found a broad readership, and he has lent his experience to social aspects of Earth sciences in the context of earthquake risk assessment and Alpine tunnels (NEAT).

During all this time, Stefan Schmid has incorporated and promoted young scientists. Hardly anybody can escape his overwhelming enthusiasm for good-natured debate, with students as well as non-specialists profiting immensely from his ability to extract the essentials from the geological chaos.

He clearly has had a great impact on the Earth science community, both as an outstanding scientist and as a fascinating personality, and he will, no doubt, do so for many years to come.

Happy 65th birthday, Stefan

Bernhard Fügenschuh, Niko Froitzheim and Siegfried Siegesmund

On-site structural geology lecture on the island of Elba. Stefan is explaining the kinematics of the detachment at Punte di Zuccale. His right hand is oriented parallel to sigma 1.

Introduction: analysing orogeny—the Alpine approach

S. SIEGESMUND[1], B. FÜGENSCHUH[2] & N. FROITZHEIM[3]

[1]*Gottinger Zentrum Geowissenschaften, University of Göttingen, Goldschmidtstrasse 03, D-37077, Göttingen (e-mail: ssieges@gwdg.de)*

[2]*Institut für Geologie & Paläontologie, Universität Innsbruck, Innrain 52, A–6020, Innsbruck*

[3]*Geologisches Institut, Universität Bonn, Nußalle 8, D–53115, Bonn*

The European Alps, the prototype collisional orogen and playground of geologists from all over the world, have been studied by generations of Earth scientists. The density of data is probably matched by no other mountain chain. Still, the Alpine chain is far from being over-studied, since many fundamental questions have not yet found a satisfactory and generally accepted answer, e.g. the formation of the Western Alpine arc. In recent years however, tectonic research on the Alpine mountain chains has made dramatic progress due to new findings (e.g. coesite), new methods (e.g. GPS), and new—or newly considered—concepts (e.g. subduction roll-back). Our picture of the Alpine orogeny has changed completely.

Extremely important for Alpine research, the opening of borders between western and eastern parts of Europe has opened new perspectives: seen from the east, the Alps are the result of the junction of the Dinarides and the Carpathians. Parts of the Alpine evolution, e.g. Jurassic tectonics in the Northern Calcareous Alps, can only be understood in the context of processes in the Internal Dinarides and Internal Carpathians. The exchange of information and ideas between Alpine, Carpathian, Pannonian and Dinaride Earth scientists—in which Stefan Schmid played and still plays a most important role—has been fruitful for all sides.

The present volume on the Alps, Carpathians and Dinarides (Fig. 1) includes articles that are related to key aspects of the tectonic evolution of these mountain chains. These articles are examples of the Alpine approach to orogeny, which combines careful fieldwork with a broad variety of laboratory methods, and integrates this into the extensive and detailed knowledge base that has been accumulated over a long history of geological research.

Key aspects of Alpine, Carpathian and Dinaride tectonics

Pre-Alpine heritage and Alpine reactivation

The Variscan continental basement in the Alps, Carpathians and Dinarides is strongly heterogeneous.

Schulz *et al.* present a compilation and review of geochronological, geochemical and structural data from the Austroalpine basement south of the Tauern Window and reconstruct the evolution of these units from a Neoproterozoic to Ordovician active margin setting, through a subsequent passive-margin setting at the northern periphery of Palaeo-Tethys, to Variscan collisional tectonics, Permian rifting, and Cretaceous collisional tectonics, and finally to Tertiary shear-zone development and intrusion of the Rieserferner Tonalite.

The Permian part of the history is the subject of three papers in this volume. During recent years, studies in the basement units have shown that Permian rifting and the related magmatism and metamorphism have strongly changed the structure and composition of the Alpine basement, leading to voluminous intrusions and widespread high-temperature/low-pressure metamorphism. These inherited features of the basement had a strong influence on Alpine tectonics. **Siegesmund *et al.*** analyse a cross-section of the Ivrea Zone, the most important tract of former lower continental crust exposed in the Alps, by means of structural geology and geochronology. In particular, they present data from a shear zone (Rosarolo Shear Zone) which formed in an extensional setting during the Permian, contemporaneously with magmatic underplating, and was rotated into its present subvertical orientation during the Alpine orogeny. The authors present evidence that the entire Ivrea Zone was verticalized during the Alpine events, and not during the Permian (as suggested by other researchers).

Froitzheim *et al.* present field, microstructural and textural evidence that the western part of the Orobic anticline in the southern Alps represents a Permian-age metamorphic core complex, and the basement-cover contact a top-to-the-SE, extensional detachment fault of Permian age as well. They demonstrate that normal shearing is coeval with both granitoid intrusion in the footwall of the detachment and volcanism in the hanging wall. This is the first description of a Permian

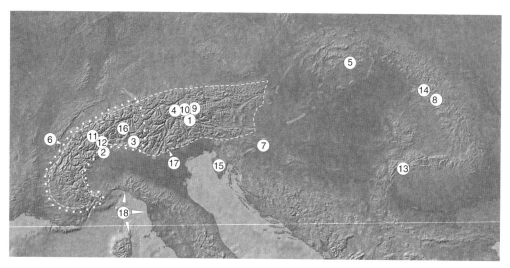

Fig. 1. Topography of the Alps, Carpathians and Dinarides (http://www.marine-geo.org/geomapapp; Carbotte et al. 2004). The study areas of papers in this volume are indicated: (1) Schulz et al.; (2) Siegesmund et al.; (3) Froitzheim et al.; (4) Veselá et al.; (5) Dallmeyer et al.; (6) Berger & Bousquet; (7) Tomljenovic et al.; (8) Gröger et al.; (9) Rosenberg & Schneider; (10) Lammerer et al.; (11) Ustaszewski & Pfiffner; (12) Pleuger et al.; (13) Ciulavu et al.; (14) Tischler et al.; (15) Mikes et al.; (16) Nagel; (17) Bousquet et al.; and (18) Molli.

detachment fault in the Alpine basement. It emphasizes the extensional character of Permian tectonics, in contrast to some earlier studies which rather suggested a transpressional setting. **Veselá et al.** have studied the remains of Post-Variscan sedimentary basins, strongly deformed by Alpine tectonics, in the western part of the Tauern Window of the Eastern Alps. They present two interesting new U–Pb zircon data and detailed sediment logs. Important conclusions are that the much-debated Kaserer Series is Upper Permian to Anisian, and that the Late Carboniferous and Permian clastic sediments were deposited in three elongate, normal-fault-bounded intramontane basins separated by horsts. They suggest that these basins were formed in a basin-and-range-like extensional tectonic scenario, very much in accordance with the findings of Froitzheim et al. in this volume. The arrangement of horsts and grabens predetermined the localization of major Alpine structures.

Mesozoic palaeogeography and Alpine subduction and collision processes

Early plate-tectonic models of the Alps and other collisional orogens suggested a rather simple evolution including a phase of mainly oceanic subduction followed by a phase of continental collision. However, structural, stratigraphic and geochronological studies have shown that in fact the Alps, Dinarides and Carpathians evolved in a series of subduction and collision events, dictated by a complex palaeogeography which involved microcontinents and ocean junctions. Changes in the tectonic style, e.g. from shortening to extension, may be induced by palaeogeography, e.g. renewed oceanic subduction after a continent collision. Pre-orogenic palaeogeography therefore influences or even controls processes as diverse as the development of foreland and intramontane basins, orogenic curvature, lateral extrusion, the formation of high- and ultrahigh-pressure metamorphic rocks and their exhumation.

Dallmeyer et al. focus on the Meliata zone in the western Carpathians, where some of the earliest stages of Alpine tectonics are recorded. The authors show that Middle to Late Jurassic subduction-related tectonics were followed, c. 50 Ma later, by Middle Cretaceous thrusting when the Meliata complex was emplaced on the Slovako-Carpathian units, and this was in turn followed by Late Cretaceous top-to-the-SE extension. This paper yields new data and interpretations for a still poorly understood part of the common history of the Carpathians, Dinarides and Alps.

Berger & Bousquet review age data for the blueschist- and eclogite-facies metamorphism in the Western and Central Alps. They conclude that the subduction-related metamorphism took place from 62 to 35 Ma (roughly 100 Ma later than the blueschist metamorphism in the Meliata unit), but at different times in different palaeogeographic/tectonic units. They assume that subduction was continuous in the Western and Central Alps but that the downgoing plate must have become fragmented on a large scale in order to explain the available pressure, temperature and age data.

Extrusion tectonics: from the Eastern Alps to the Carpathians and the Pannonian Basin

The concept of eastward extrusion has been successful in explaining the Tertiary to Recent kinematics of the Eastern Alps. The driving forces of this process, however, as well as the relative importance of N–S shortening and E–W extension, are still a matter of debate. Four articles in the volume are related to this complex.

Tomljenovic et al. decipher the tectonic evolution of Mount Medvednica in Croatia. This area records several stages of Cretaceous and Early Tertiary tectonics. The authors present palaeomagnetic evidence that during the Oligocene to earliest Miocene, the area was part of a tectonic block which rotated clockwise by 130°. This block is bounded to the north by the eastern extension of the Periadriatic Fault into the Mid-Hungarian Fault Zone, which was then active as a major dextral strike-slip fault accommodating eastward escape of the units to the north of it. Tomljenovic et al. suggest that the tectonic block broke away from the northern rim of the Dinarides, rotated, and was displaced eastward within the widening escape corridor. Whereas Tomljenovic et al. studied the southwestern end of the Mid-Hungarian Fault Zone, **Gröger et al.** investigated the northeastern end of this fault zone in the Maramures area of northern Romania. Combining structural fieldwork with thermochronology (zircon and apatite fission-track as well as apatite (U–Th)/He dating), they reconstruct a tectonic history including Cretaceous nappe stacking, Late Cretaceous exhumation, Palaeogene to Early Miocene burial and heating due to accumulation of sediments on top of the studied units, Mid-Miocene transpressional deformation (16 to 12 Ma), and finally Mid- to Late Miocene (12 to 10 Ma) exhumation related to transtension.

Rosenberg & Schneider describe an east–west-striking, ductile, sinistral strike-slip shear zone in the northern Tauern Window which they interpret as the westward and originally downward continuation of the Salzach-Ennstal-Mariazell-Puchberg Fault, a 300 km long sinistral strike-slip fault which accommodated eastward escape of the units to the south of it. This shear zone is a fine example of an exposed transition from brittle to ductile deformation. Moreover, shearing was kinematically linked to the formation of upright folds, showing that the shear zone developed in a transpressive framework during, and probably driven by, N–S convergence of Europe and Adria.

Lammerer et al. present a crustal-scale N–S cross-section through the Tauern Window, based on TRANSALP seismics and surface geology. They reconstruct how this cross-section evolved from an early stage of basement/cover duplex formation, to the uplifting of the duplex by the activity of north-directed thrust ramps at a deeper level. The authors emphasize the importance of N–S shortening in the evolution of the Tauern Window during Tertiary times. Instead of eastward tectonic escape, they rather suggest a kinematic framework of dextral transpression.

Orogen-parallel and orogen-perpendicular extension

Two papers deal with the western Central Alps, and, among other subjects, with synorogenic extensional tectonics. **Ustaszewski & Pfiffner** studied faults in the Penninic and Helvetic units of the Valais area (western Switzerland). The present-day deformation in this area, as known from earthquakes and geodesy, is characterized by orogen-perpendicular extension in the internal zone and shortening in the external zone. This strain field very probably results from the lateral density anomalies represented by the topography and the crustal root. The authors distinguish tectonic faults, gravitational faults, and composite faults—that is, tectonic faults with gravitational reactivation. Recent (postglacial) tectonic motion could only be documented for two faults, which suggests that the current strain is either predominantly aseismic or, alternatively, cumulated seismic moment is too low for producing surface rupture.

Pleuger et al. studied the Alpine deformation of the Monte Rosa Nappe and the Southern Steep Belt southwest and south of the Lepontine Dome. They suggest that the tectonic evolution of the area was dominated, during the Oligocene and Miocene (c. 35 to 15 Ma), by alternating phases of orogen-parallel extension, orogen-perpendicular extension, and orogen-perpendicular shortening. In this area, the Periadriatic (Insubric) Fault is not only a strike-slip fault and a backthrust, but it represents the boundary between a northern (Penninic) zone which underwent strong extension during the Tertiary, and a southern block (Southern Alps) where such extension did not take place. This resulted in complicated spatial and temporal variations of the kinematics of the Insubric Fault. The alternation of orogen-parallel and orogen-perpendicular extension probably reflects variations in the relative contributions of gravitational collapse and across-orogen convergence.

Like the Lepontine Dome, the Danubian Window in the Southern Carpathians is a metamorphic complex that was exhumed predominantly by orogen-parallel extension, in this case mainly during the Eocene. The main exhuming normal fault is the low-angle Getic detachment. **Ciulavu et al.**

present the results of a detailed study of metamorphism in this area. Low-temperature/low- to medium-pressure conditions during nappe stacking gave way to high-temperature/low-pressure conditions during the extensional exhumation of the window. Trends in cooling ages and in the thickness of crust excised by the Getic detachment suggest that the geometry of extension was strongly asymmetric. In this case, orogen-parallel extension probably resulted from the collision of the orogenic wedge with an irregularly shaped continental margin, including the Moesian Promontory and the Carpathian Embayment, in the downgoing European Plate.

Record of orogeny in foreland basins

Two articles investigate the sedimentary evolution of foreland basins located in two key areas of the Carpathians and Dinarides. **Tischler et al.** studied the area where the Mid-Hungarian Fault Zone at its northeastern termination meets the Carpathian arc. This area represents a major lateral discontinuity of the Carpathians, where several tectonic units of the Western Carpathians end (e.g. the Pieniny Klippen Belt). Tischler et al. describe the evolution of a foreland basin developing in Oligocene–Eocene times in front of the southeastward-advancing thrust system of the Pienide nappes, from Oligocene deep-marine clastics to Burdigalian molasse-type sediments. They interpret this basin as recording the convergence between and rotations of two continental blocks—ALCAPA and Tisza-Dacia.

Mikes et al. studied nannoplankton from deep-marine clastic (flysch) units associated with the thrust front of the Outer Dinarides, with important consequence for the tectonic evolution of the Dinarides. Like in other parts of the Alpine-Mediterranean orogens, nannoplankton evidence shows that the flysch units are significantly younger (up to Mid-Miocene in this case) than previously thought (Eocene). Previous age determinations were often based on foraminifers which now turn out to be redeposited. Hence, thrusting of the Outer Dinarides over Adria continued through much of Miocene times.

Tectonometamorphic evolution

Two papers deal with the relations between metamorphism and tectonics in the Alps. The paper by **Nagel** is a comprehensive review of the structural and metamorphic evolution of the Adula Nappe in the eastern Central Alps. The author discusses the question whether the high- and ultrahigh-pressure metamorphic rocks of the Adula Nappe were introduced into a lower-pressure mélange after their eclogite-facies metamorphism, or were subducted together with their country rocks. He comes to the conclusion that the latter is more likely. He emphasizes that significant internal deformation of the Adula Nappe occurred after peak-pressure conditions and that this deformation contributed to the exhumation of the high-pressure rocks. Two formerly published evolutionary models are confronted, featuring either one or two subducting slabs in the Penninic zone.

Bousquet et al. present a review of metamorphism in metasediments throughout the Alps. They suggest that the metamorphic evolution of the Western Alps is related to subduction followed by exhumation along a cool decompression path, that the metamorphic evolution of the Central Alps, in contrast, records collisional processes often with heating during decompression, and that the Eastern Alps show different stages and types of metamorphism because of two orogenic cycles which affected this part of the Alps.

Relations between Alps, Apennines and Corsica

Molli reviews the tectonics of the Northern Apennines and Corsica in relation to the Western Alps. He develops a picture including an early ('Alpine') stage of subduction of the Iberian Plate under the Adriatic Plate, which terminated in the Middle Eocene when Corsica, being part of the Iberian Plate, collided with the edge of the upper plate, leading to the emplacement of oceanic series on continental crust in northeast Corsica. In the Late Eocene, a new, 'Apenninic' subduction zone was created, in which the Adriatic Plate was subducted under the Iberian Plate, i.e. Corsica. This subduction zone soon began to retreat eastward, creating in its back the young extensional, partly oceanic basins of the Western Mediterranean.

The editors gratefully acknowledge the following colleagues for their reviews: P. Arkai, A. Berger, G. Bertotti, A. Bistacchi, R. Bousquet, D. Ciulavu, G. Dal Piaz, K. Decker, C. Doglioni, I. Dunkl, H. von Eynatten, G. Franz, U. Glasmacher, P. Gleissner, M. Handy, R. Kleinschrodt, J. Konzett, C. Krezsek, J. Kruhl, J. Malavieille, J. Mutterlose, F. Neubauer, H. Ortner, C. Passchier, D. Plasienka, M. Raab, J. von Raumer, M. Rockenschaub, S. Schmid, R. Schuster, J. Selverstone, S. Sinigoi, K. Stüwe, M. Wagreich, J. Wijbrans, W. Winkler.

Reference

CARBOTTE, S. M., ARKO, R., CHAYES, D. N. ET AL. 2004. New Integrated Data Management System for Ridge2000 and MARGINS Research. *EOS*, **85**, 51.

Geodynamic evolution of an Alpine terrane—the Austroalpine basement to the south of the Tauern Window as a part of the Adriatic Plate (eastern Alps)

BERNHARD SCHULZ[1], ANDRÉ STEENKEN[2] & SIEGFRIED SIEGESMUND[3]

[1]*Institut für Mineralogie der TU Bergakademie, Brennhausgasse 14, D-09596 Freiberg/ Sachsen, Germany (e-mail: Bernhard.Schulz@mineral.tu-freiberg.de)*

[2]*Instituto de Geosciêncas Universidade de Saõ Paulo, Rua do Lago 562, Cidade Universitária, 05508–080 Saõ Paulo, Brazil*

[3]*Geowissenschaftliches Zentrum Göttingen (GZG), Goldschmidtstraße 3, D–37077 Göttingen, Germany*

Abstract: The Austroalpine basement underwent a multistage Precambrian to Tertiary evolution. Meta-magmatic rocks occur in pre-Early Ordovician and post-Early Ordovician units. Protolith zircon ages and whole-rock trace element data define two magmatic evolution lines. An older trend with Th/Yb typical of subduction-related metamorphism, started by 590 Ma N-MORB-type and 550–530 Ma volcanic arc basalt-type basic suites which mainly involved depleted mantle sources, and was continued by mainly crustal-source 470–450 Ma acid magmatic suites. A presumably younger evolution by tholeiitic MORB-type and 430 Ma alkaline within-plate basalt-type suites is characterized by an intraplate mantle metasomatism and multicomponent sources. These magmatic trends can be related to a Neoproterozoic to Ordovician active margin and a subsequent Palaeo-Tethys passive margin along the north-Gondwanan periphery. During Variscan collision, the Austroalpine basement underwent multiphase deformation and metamorphism. Early deformation involved non-coaxial shearing with formation of sheath folds and calcsilicategneiss bodies in some regions. Syndeformational clockwise P–T paths in lower basement parts passed high-pressure and high-temperature amphibolite-facies stages and are interpreted by a Devonian to Carboniferous crustal stacking. A post-collisional Permian thermal event is documented by pegmatite intrusions, LP-HT assemblages and monazite ages. Ductile overprinting under greenschist-facies conditions during the Cretaceous is indicated by foliated pegmatites and monazite ages in samples with retrogressed garnet. The emplacement of the Oligocene Rieserferner pluton was controlled by sinistral shear zone deformation along the Defereggen–Antholz–Vals line. Shear zone activity ceased at 15 Ma and was superseded by brittle strike-slip movements along NW and SE trending faults.

The Alpine orogen resulted from the closure of the Penninic ocean and a subsequent multiphase collision of the southern Mediterranean microplates with the northern European Plate during Cretaceous to Oligocene times. During this collision, parts of the Adriatic microplate were thrusted in giant nappes upon the Penninic units exposed in the Tauern Window (Schmid *et al.* 2004). The Austroalpine basement complex in the Eastern Alps (Fig. 1) represents a major part of the Adriatic crust. Its erosional exposure to the south of the Tauern Window allows an insight into the complex Alpine (Cretaceous to Tertiary) and pre-Mesozoic magmatic and metamorphic evolution of the Adriatic–Apulian Plate.

The remnants of a pre-Mesozoic evolution turned out to be especially well preserved in this part of the Austroalpine basement, regardless of a more or less intense Alpine overprint. The Austroalpine basement represents one of the Intra-Alpine or Proto-Alpine terranes in the southern part of the European Variscan belt (Schätz *et al.* 2002). This is a huge collisional orogen assembled by crustal segments that were parts of the former microcontinents Avalonia, Cadomia, Armorica and the Intra-Alpine terranes. Since the Neoproterozoic, these microcontinents and terranes split off from the North-Gondwanan margin at various times, were moved in northern directions and successively collided with the southern margin of Laurussia (Stampfli 1996; von Raumer 1998; Stampfli & Borel 2002; Schätz *et al.* 2002; Stampfli *et al.* 2002; von Raumer *et al.* 2002, 2003, 2006). Rocks of the Austroalpine basement are increasingly important for the detailed reconstruction of the pre-Variscan evolution (Frisch *et al.* 1984; Frisch & Neubauer 1989; von Raumer & Neubauer 1993; Neubauer 2002*a*), as

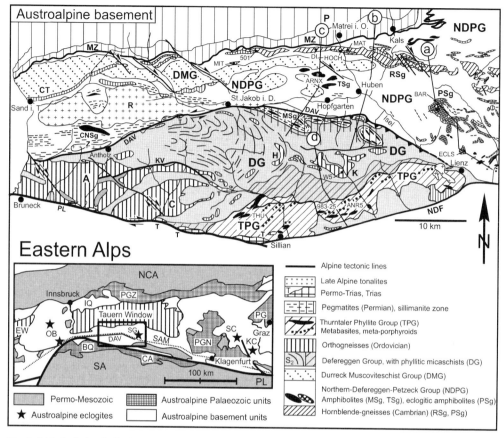

Fig. 1. Lithological units in the Austroalpine basement to the south of the central Tauern Window, Eastern Alps. Sampling locations of zircon dating and cross-sections in Figure 2 are marked. A, Antholz/Anterselva muscovite orthogneiss; BQ, Brixen Quartzphyllite; C, Casies/Gsies biotite orthogneiss; CA, Carnic Alps; CNSg, Croda Nera Subgroup (metabasites); CT, Campo Tures orthogneiss; DAV, Defereggen-Antholz-Vals line; DG, Defereggen Group (monotonous metapsammopelites); DMG, Durreck Muscoviteschist Group; EW, Engadine Window; H, Hochgrabe biotite orthogneiss; IQ, Innsbrucker Quartzphyllite; K, Kristeinertal biotite orthogneiss; KC, Koralpen Crystalline; KV, Kalkstein-Vallarga line; MSg, Michelbach Subgroup (amphibolites); MZ, Matreier Zone (Penninic); NDF, Northern Drauzug fault; NDPG, Northern-Defereggen-Petzeck Group; NCA, Northern Calcareous Alps; OB, Oetztal-Stubai Basement; P, Penninic Upper Schieferhuelle; PG, Palaeozoic of Graz; PGN, Palaeozoic of Gurktal Nappes; PGZ, Palaeozoic of Greywacke Zone; PL, Periadriatic Lineament, Pustertal line; PSg, Prijakt Subgroup (eclogitic amphibolites and hornblende-gneisses); R, Rieserferner tonalite; RSg, Rotenkogel Subgroup (hornblende-gneisses, orthogneisses); SA, Southern Alps; SAM, southern limit of Alpine metamorphism (Hoinkes *et al.* 1999); SC, Saualpen Crystalline; SG, Schobergruppe basement; T, Permo-Trias and Trias; TPG, Thurntaler Phyllite Group; TSg, Torkogel Subgroup (amphibolites); Z, Zinsnock tonalite.

geochemical and isotopic characteristics of pre-Carboniferous magmatic rocks and sediments can be recognized despite a considerable Variscan and Alpine overprint. The Austroalpine basement south of the Tauern Window involves the presumably post-Early Ordovician Thurntaler Phyllite Group, and lithologies appearing in several pre-Early Ordovician basement units (Schulz *et al.* 1993, 2001, 2004). They can be related to the Noric Composite terrane and the Celtic terrane, as established by Frisch & Neubauer (1989) in their listing of the Austroalpine pre-Alpine terranes.

Here we present insights to the complex geodynamics of a part of the Adriatic–Apulian Plate. The pre-Variscan evolution is illuminated by whole-rock geochemistry and zircon dating of meta-magmatic rocks and then interpreted within a frame of peri-Gondwanan plate tectonics. A time schedule of the Variscan collisional event and its Alpine overprint is detailed by structural

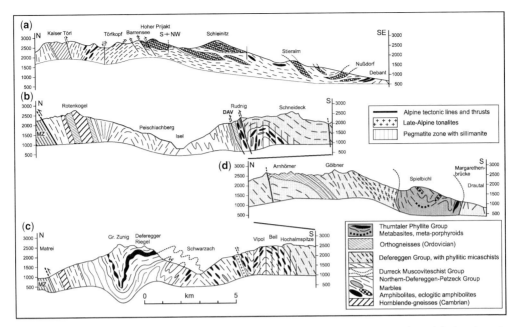

Fig. 2. Cross-sections of the Austroalpine basement to the south of the Tauern Window. See Figure 1 for locations of sections. (**a**) Schobergruppe with Prijakt eclogitic amphibolites. (**b**) From the Penninic Matrei Zone (MZ) across the Isel river to the south of Hopfgarten; an anticlinal structure of the pegmatite zone with fibrolitic sillimanite is exposed along the Michelbach valley (projected). (**c**) From the Matrei Zone to the south of Hopfgarten further to the south of the Defereggen-Antholz-Vals shear zone (DAV). (**d**) Southern part of the Defereggen Group and the Thurntaler Phyllite Group between Winkeltal and Kristeinertal.

and microstructural analysis, geothermobarometric reconstruction of P–T paths, the dating of metamorphic events and thermal modelling.

Basement units, tectonic lines and metamorphic ages

The Austroalpine crystalline basement between the southern margin of the Central Tauern Window and the Periadriatic Lineament is subdivided into several lithological groups (Figs 1 and 2). In part these can be traced to the west of the Tauferer valley (Sander 1925; Bianchi et al. 1930; Dal Piaz 1934; Senarclens-Grancy 1964; Hofmann et al. 1983; Schulz 1994a; Sassi et al. 2004). From north to south and the hanging wall (Figs 1 and 2) the following lithological units can be distinguished (Schulz & Bombach 2003): (1) NDPG—Northern-Defereggen-Petzeck Group; (2) DMG—Durreck Muscoviteschist Group; (3) DG—Defereggen Group; and (4) TPG—Thurntaler Phyllite Group. Their main characteristics are presented and were detailed in Behrmann (1990), Bücksteeg (1999), Schulz et al. (1993, 2001, 2004), Schulz (1995, 1997), Schulz & Bombach (2003) and Siegesmund et al. (2007).

Northern-Defereggen-Petzeck Group

The Northern-Defereggen-Petzeck Group was formerly labelled as the metapsammopelite-amphibolite-marble unit (Schulz et al. 1993) and is mainly composed of metapsammopelitic rocks with interlayered meta-quartzite and marble. Metabasic subgroups involving rock suites with different geochemical characteristics appear within the sequence (Figs 1 and 2):

(a) Prijakt Subgroup with enriched mid ocean ridge basalt-type (N-MORB) eclogitic amphibolites and volcanic arc basalt-type (VAB) hornblende-gneisses (Schulz 1995; Schulz & Bombach 2003).
(b) Rotenkogel Subgroup with volcanic arc basalt (VAB)-type hornblende-gneisses and mid ocean ridge basalt-type (MORB) amphibolites (Schulz 1997; Schönhofer 1999).
(c) Torkogel Subgroup with MORB- and within-plate basalt-type (WPB) amphibolites.
(d) Michelbach Subgroup with similar amphibolites as in the Torkogel Subgroup; the geochemical details are yet not known.
(e) Croda Nera Subgroup with MORB- and WPB-type metabasites (Steenken & Siegesmund 2000).

Durreck Muscoviteschist Group

The Durreck (Cima-Dura) Muscoviteschist Group is a separate unit within the Northern-Defereggen-Petzeck Group. It is characterized by monotonous garnet-muscovite schists and garnet-muscovite-quartzitic gneisses which appear to the west of Isel valley among the lower and upper parts of the Northern-Defereggen-Petzeck Group (Fig. 2a, b). Amphibolite lenses are rare and show within-plate basalt-type signatures (Godizart 1989; Schönhofer 1999). The outcrop of the Durreck Muscoviteschist Group is several kilometres along the Tauferer valley (Mazzoli et al. 1993) and fades out in a synformal position to the east of the Isel river (Figs 1 and 2).

Defereggen Group

The Defereggen Group is a several kilometres-thick monotonous metapsammopelitic sequence of banded quartzitic gneisses, paragneisses and micaschists with rare thin layers of amphibolite, marble, graphitic gneisses and quartz-feldspar gneisses (Figs 1b and 2). To the south, phyllitic micaschists occur in a lithological transition to the Thurntaler Phyllite Group (Schmidegg 1936; Senarclens-Grancy 1964; Schulz et al. 2001).

Thurntaler Phyllite Group

To the south, the Thurntaler Phyllite Group overlies the Defereggen Group along a partly overturned pre-Alpine foliation-parallel contact (Heinisch & Schmidt 1984; Schulz 1991; Kreutzer 1992). Quartzphyllites and phyllitic micaschists are interlayered by amphibolites, epidote-amphibolites, chlorite schists, quartzites, graphite phyllites and rare marbles (Fig. 2d). The metabasites have transitional MORB-type and WPB-type tholeiitic to alkaline compositions (Kreutzer 1992; Schulz & Bombach 2003).

Orthogneisses and meta-porphyroids

The Northern-Defereggen-Petzeck, Durreck Muscoviteschist and the Defereggen Groups are concordantly interlayered by numerous granitoid orthogneisses (Figs 1 and 2). They can be subdivided into several groups which occur without regional preference (Schulz et al. 2001; Schulz & Bombach 2003; Siegesmund et al. 2007): (a) tonalitic to granodioritic, rarely dioritic orthogneiss of Gsies/Casies; (b) fine-grained and strongly foliated biotite orthogneisses in the southern and eastern parts of the Defereggen Group at Hochgrabe and in the Kristeinertal; (c) leucocratic orthogneisses of Antholz/Anterselva and their lithological equivalents appear as former alkali feldspar granites; the orthogneiss of Sand in Taufers (Campo Tures) is as well a member of this suite (Hammerschmidt 1981; Mager 1981). In the Thurntaler Phyllite Group, horizons of meta-porphyroids were interpreted as metatuffites (Heinisch & Schmidt 1976, 1984; Schönlaub 1979).

Pegmatites

Between the Tauferer and Isel valleys, pegmatite rocks (Stöckhert 1987) with muscovite, tourmaline, garnet and sometimes spodumen crop out in an E–W striking zone along the border of the Northern-Defereggen-Petzeck Group and the Defereggen Group (Figs 1 and 2). To the south of Hopfgarten and St. Jakob in Defereggen, this zone is further characterized by the occurrence of fibrolitic sillimanite replacing garnet, and the crystallization of andalusite in the metapelitic host rocks (Schulz et al. 2001). To the east of the Isel river, these pegmatites are also abundant within the Northern-Defereggen-Petzeck Group of the Schobergruppe (Bücksteeg 1999). The Permian magmatic protolith and formation ages of the pegmatites are remarkably similar from the west at Uttenheim, Tauferer valley (262 ± 7 Ma Rb–Sr WR, Borsi et al. 1980a) via the southern Defereggen region (Michelbach valley, 253 ± 7 Ma Sm–Nd WR-garnet, Schuster et al. 2001) toward the east in the Kreuzeckgruppe (Strieden Complex, 261 ± 3 Ma Sm–Nd WR-garnet, Schuster et al. 2001 and references in Hoke 1990). As the pegmatites occur with variable structural habit (undeformed, with brittle deformation, foliated as pegmatite gneiss and pegmatite mylonite), they can be used for the discrimination of Alpine and pre-Alpine (pre-Permian) deformation.

Alpine tectonic lines

The northern border of the Austroalpine basement is marked by the Alpine Austroalpine-Penninic suture (Figs 1 and 2). The contact of the Austroalpine metamorphic rocks and the Penninic Matreier Zone is structurally concordant with similar orientations of steeply SSE-dipping foliation planes and slightly WSW-plunging lineations and fold axes within the transition zone (Schönhofer 1999; Schulz et al. 2001). Despite the structural transition, the lithological contact of the units is quite sharp. The Austroalpine lithological subunits as the hornblende-gneisses of the Rotenkogel Subgroup are cut off at acute angles along the Penninic-Austroalpine suture and they do not appear to the west (Fig. 1). Within the Austroalpine basement, the W–E striking Defereggen-Antholz-Vals line (DAV) represents a major tectonic discontinuity.

This sinistral wrench shear zone is characterized by a northern part with Alpine ductile deformation and a southern domain with brittle Alpine deformation; the latter is continued toward the east (Fig. 1) by a brittle thrust zone (Schulz 1989). Pseudotachylites occur in the southern cataclastic domain (Schulz 1989; Müller et al. 2000). Late-Alpine mica cooling ages from 15 to 28 Ma to the north of the Defereggen-Antholz-Vals line contrast late-Variscan mica ages to the south (Borsi et al. 1978; Steenken et al. 2002). This has been explained by a considerable component of vertical movement predating and/or accompanying sinistral strike-slip displacement along the shear zone (Borsi et al. 1978; Kleinschrodt 1987; Schulz 1989). Brittle Alpine deformation is observed along the Kalkstein-Vallarga fault system (Borsi et al. 1978; Guhl & Troll 1977, 1987; Schulz 1991) and the Northern Drauzug fault (Sprenger & Heinisch 1992; Bauer & Bauer 1993). Relicts of the anchito epizonal Permo-Mesozoic cover of the basement are found in the southern brittle part of the Defereggen-Antholz-Vals line (Trias of Staller Alm, Senarclens-Grancy 1964), along the Kalkstein-Vallarga line (KV, Permo-Trias of Kalkstein, Guhl & Troll 1977, 1987), along the sinistral wrenching SW–NE striking Zwischenbergen-Wöllatratten fault (Exner 1962), to the west of Sillian and immediately to the north of the Periadriatic Lineament (Fig. 1). The intrusion of the Oligocene (30 Ma, Borsi et al. 1979) Rieserferner tonalite (Romer & Siegesmund 2003 and the related Zinsnock pluton coincided and was lasted out by the movements along the Defereggen-Antholz-Vals line (Mager 1985a, b; Müller 1998; Müller et al. 2000; Steenken et al. 2000, 2002).

When interpreted to date cooling, radiometric ages of mica (K–Ar, Rb–Sr) give evidence of Alpine as well as pre-Alpine metamorphism and deformation of the basement south of the Tauern Window. To the west, the Defereggen-Antholz-Vals line with its vertical displacement component separates a northern domain with predominant Late-Alpine (28–15 Ma) mica ages from a southern domain with exclusively pre-Alpine (300–260 Ma) Rb–Sr mica ages (Borsi et al. 1978; Schuster et al. 2001). Toward the east and around the Isel river, this clear separation of mica cooling ages does not prevail. To the north of Hopfgarten in the Northern-Defereggen-Petzeck Group, the cooling ages increase from 28 to 59 Ma toward the east (Borsi et al. 1978). To the west of the Isel river in the Defereggen Group, Rb–Sr mica ages are at around 200 Ma (Borsi et al. 1978; Schuster et al. 2001; Steenken et al. 2002). In the Northern-Defereggen-Petzeck Group of the Schobergruppe to the east of the Isel river and in the northern parts of the Kreuzeck-Gruppe, mica cooling ages (90–60 Ma) from metapelites and Sm–Nd age data (80–110 Ma) from eclogitic amphibolites (Oxburgh et al. 1966; Troll 1978; Linner et al. 1996) signalize an Eo-Alpine (Cretaceous) metamorphic event. However, there exists a considerable record of 'mixed' Variscan-Alpine mica ages between 300 and 90 Ma in the basement to the east of the Isel river (Hoke 1990; Schuster et al. 2001). From the mica age data, a southern borderline of Alpine metamorphism (SAM, Fig. 1) has been defined by Hoinkes et al. (1999).

The magmatic evolution in a Peri-Gondwanan terrane

In their listing of Austroalpine pre-Alpine terranes, Frisch & Neubauer (1989) distinguished among a Noric Composite terrane, a Celtic terrane and the mainly ophiolitic Plankogel and Speik terranes. Following the definitions of terrane characteristics given by Frisch & Neubauer (1989), the Thurntaler Phyllite Group can be assigned to the Noric Composite terrane, as this unit involves presumably post-Early-Ordovician lithology, enclosing the 'porphyroid' meta-volcanosedimentary rocks (Heinisch 1981; Söllner et al. 1991, 1997; Meli & Klötzli 2001; Siegesmund et al. 2007). The greenschist- to epidote-amphibolite-facies metamorphic phyllitic units (e.g. Innsbrucker Quartzphyllite, Thurntaler Phyllite Group) are regarded as the higher metamorphic equivalents of the weakly metamorphosed and fossiliferous Early Palaeozoic units as Greywacke Zone, Gurktal Nappe Complex and Carnic Alps of which sedimentation started at the earliest in the Lower Ordovician and lasted until the Upper Devonian (Schönlaub 1979, 1993, 1997a, b; Heinisch et al. 1987; Loeschke 1989; Loeschke & Heinisch 1993; Neubauer & Sassi 1993; Läufer et al. 2001). The Noric Composite terrane as well involves units of a pre-Ordovician basement that has been intruded by numerous granitoids, as in the Defereggen Group and the Durreck Muscoviteschist Group (Peccerillo et al. 1979; Heinisch & Schmidt 1982; Sassi et al. 1985; Mazzoli & Sassi 1992). In contrast, the Celtic terrane was considered as a part of a Neoproterozoic to Early Palaeozoic or even older active continental margin. It encloses rock associations typical of arc magmatism, of which time period and nature has not been precisely defined when Frisch & Neubauer (1989) established the nomenclature of the Austroalpine terranes. The lithology in the Northern-Defereggen-Petzeck Group, especially the hornblende-gneisses of the Rotenkogel and Prijakt Subgroups, can be considered as part of the Celtic terrane. This is detailed below when geochemical characteristics and protolith ages of the Austroalpine meta-magmatic rocks are discussed.

Geochemistry of meta-magmatic rock suites

Whole-rock main and trace element data from the meta-magmatic rocks have been obtained through XRF and inductively coupled mass spectrometry (ICP-MS) analysis. Only samples with no signs of post-magmatic element mobility and with preserved igneous geochemical signatures were considered for interpretation.

It has been demonstrated that meta-magmatic rocks in the Austroalpine basement south of the Tauern Window cover considerable ranges of compositions (Peccerillo et al. 1979; Heinisch 1981; Godizart 1989; Schulz 1995, 1997; Schönhofer 1999; Steenken & Siegesmund 2000; Schulz et al. 2001; Schulz & Bombach 2003). When compiling geochemical whole-rock data from widely scattered metabasite outcrops, uncertainty remains if fractionates belong to a single magmatic process. Detailed study of larger metabasite outcrops as in the Croda Nera (Steenken & Siegesmund 2000), Torkogel, Rotenkogel (Schönhofer 1999; Steenken & Siegesmund 2000; Schulz & Bombach 2003) and Prijakt Subgroups (Schulz 1995) revealed wide and systematic correlations of element abundances. Whole-rock Zr and Ti inter-element characteristics of the metabasite layers in larger outcrops (Fig. 1) allowed distinguishing basaltic rock suites, following the method of Pearce (1982). Intra-suite element variations can be attributed to magmatic source variations and chemical fractionation. Further characterization of the suites was possible by rare earth and trace element data (Schulz 1995; Steenken & Siegesmund 2000; Schulz & Bombach 2003; Schulz et al. 2004):

N-MORB-type suite. Layered dykes or sills could have been the magmatic protoliths of the Prijakt eclogitic amphibolites as can be concluded from their appearance in the field, the lithological layering and the intra-suite element variations (Schulz 1995). Eclogitic amphibolites of the Prijakt Subgroup display N-MORB-type signatures of Zr–Ti (Fig. 3a), but with the large ion lithophile (LIL) elements and Ce slightly enriched compared to MORB. Chondrite-normalized (normalization values from McDonough & Sun 1995) rare earth element (REE) patterns are slightly enriched in the light REE (LREE) with $(La/Yb)_N$ ranging from 0.68 to 1.85. According to Thiéblemont et al. (1994), the chondrite-normalized $(Tb/Ta)_N$ 0.93–3.31 and $(Th/Ta)_N$ between 1.38 and 2.50 can be regarded as a sign of back-arc basin magmatism (Schulz & Bombach 2003; Schulz et al. 2004).

Volcanic arc basalt type suite. Major element characteristics and the presence of hornblendites interpreted as former autoliths support a plutonic origin for hornblende-gneisses in the Rotenkogel and Prijakt Subgroups (Schulz et al. 2004). The rocks have the Zr and Ti contents of volcanic arc basalt (VAB)-type magmatites (Fig. 3a). They follow a calc-alkaline fractionation trend with Zr/Y ratios from 1.3–13. They show enrichment of Sr, K_2O, Rb, Ba, Th and Ce relative to MORB and negative high field strength element (HFSE) anomalies; the elements Ta, Nb, Zr, Hf, Sm, Ti, Y, Yb and Cr appear as selectively depleted when compared to MORB (Schulz 1997; Schönhofer 1999; Schulz & Bombach 2003; Schulz et al. 2004). Both criteria are characteristic of magma generated in island arc and active continental margin settings (Wilson 1989). Subparallel pattern with systematically stronger enrichment of LREE are observed, the chondrite-normalised $(La/Yb)_N$ ranging from 2.6–8.1 (Schulz & Bombach 2003).

Within-plate basalt and MORB-type suites. The numerous isolated and m-scale amphibolites, and the metabasites of the Croda Nera, Torkogel and Michelbach Subgroups within the Northern-Daefereggen-Petzeck Group probably are former dykes or sills as can be concluded from their widely distributed appearance in the field (Godizart 1989; Schulz et al. 1993). They display a wide range of Zr and Ti compositions (Fig. 3a). This wide range cannot be explained by fractionation of a single protolith. A subdivision into MORB-type and within-plate basalt-type suites is possible when HFSE and REE are considered. Rocks with MORB-type signature display 2–10 times enrichment of large ion lithophile elements (LIL) compared to MORB, REE patterns with LREE depleted to flat at $(La/Yb)_N = 0.5–1.1$, and a marked negative Eu-anomaly, contrasting the N-MORB-type suite. A more primitive group with low Zr and Ti is distinguished from an evolved group, which are related by rocks of transitional compositions (Fig. 3a). Metabasites with the within-plate basalt-type Zr–Ti signatures have tholeiitic, transitional and alkaline compositions. The Nb (39.6–58.8 ppm) and Ta (3.4–8.2 ppm) contents are significantly higher than those of the N-MORB- and VAB-type rocks. There is significant enrichment in LIL, HFSE and REE, with $(La/Yb)_N = 14.6–19.6$ (Schulz et al. 1993, 2001; Steenken & Siegesmund 2000). A large range of initial $^{87}Sr/^{86}Sr$ in this suite (Fig. 3d, e) is interpreted to result from a variable contribution of enriched mantle and/or continental crust to a main depleted mantle source (Schulz et al. 2004). Among the mafic suites, the signatures of source

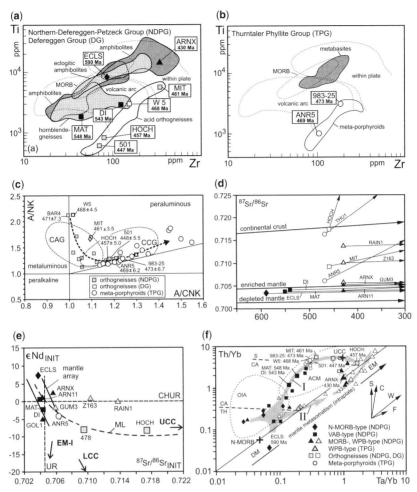

Fig. 3. Pre-Variscan magmatic rock suites in the Austroalpine basement to the south of the Tauern Window. Discrimination Zr–Ti after Pearce (1982). Samples with dated zircon are marked. Data compiled from Godizart (1989), Schulz (1995), Bücksteeg (1999), Schönhofer (1999), Steenken & Siegesmund (2000), Schulz et al. (2001, 2004), Schulz & Bombach (2003). (**a**) Northern-Defereggen-Petzeck Group and Defereggen Group with N-MORB-type eclogitic amphibolites of the Prijakt Subgroup, volcanic arc basalt-type hornblende-gneisses of Rotenkogel Subgroup, within-plate alkaline-basalt-type amphibolites of Torkogel Subgroup and Croda Nera Subgroup, MORB-type metabasites and acid orthogneisses. (**b**) MORB- to within-plate basalt-type metabasites and acid meta-porphyroids from Thurntaler Phyllite Group. (**c**) Discrimination diagram for granitoids after Maniar & Piccoli (1989) with data from acid orthogneisses in Northern-Defereggen-Petzeck Group and Defereggen Group, and meta-porphyroids from the Thurntaler Phyllite Group. The stippled arrow marks chemical and chronological evolution trend in the orthogneisses. (**d**) Evolution lines of $^{87}Sr/^{86}Sr$ in reference to depleted and enriched mantle, and continental crust (Rollinson 1993). Initial Sr calculated according to zircon protolith ages. (**e**) Initial $^{87}Sr/^{86}Sr$ vs εNd_{CHUR} for identification of magmatic sources. EM-I, enriched mantle; LCC, lower continental crust; ML, potential mixing line between depleted mantle and upper continental crust sources, as exemplified in I- and S-type Lachlan granitoids; UCC, upper continental crust. (**f**) Major evolution lines for pre-Variscan meta-magmatic rock suites in Ta/Yb vs. Th/Yb coordinates. Line I: subduction-related evolution involving N-MORB and volcanic arc basalt-type metabasites, orthogneisses and meta-porphyroids. Line II: trend of mantle-enrichment in MORB- and alkaline within-plate basalt-type metabasite suites. Vectors indicate the influence of subduction components (S), crustal contamination (C), within-plate enrichment (W), and fractional crystallization (F). ACM compositional field of active margin magmatites; CA, calcalkaline; OIA, compositional field of oceanic island arc magmatites; S, shoshonitic; TH, tholeiitic (Pearce 1983; Wilson 1989).

contamination by a crustal magma are most developed in the WPB-type rocks.

In the Thurntaler Phyllite Group, massive amphibolite layers with a thickness partly exceeding 10 m are abundant and could represent former dykes or sills (Schulz 1991; Kreutzer 1992). In the Zr–Ti coordinates (Fig. 3b), transitional MORB-type and within-plate basalt-type rocks with tholeiitic to alkaline compositions prevail. The transitional MORB-type rocks display a slight enrichment of LIL elements. Chondrite-normalized REE patterns show very slight enrichment of LREE with $(La/Yb)_N$ ranging between 1.5 and 2.3. Some of the alkaline within-plate basalt-type samples display stronger LREE enrichment with $(La/Yb)_N$ of 7.0 (Kreutzer 1992; Schulz & Bombach 2003).

Pre-Variscan acid magmatites. Among the orthogneisses (Zr–Ti is shown in Fig. 3a), one can distinguish metaluminous types (Fig. 3c) with a continental arc granite (CAG) affinity and peraluminous orthogneisses of a continental collision granite (CCG) S-type affinity at A/CNK-values of 1.05–1.35 and [Na/K] of 1.02–1.35 (Schulz et al. 2001; Schulz & Bombach 2003). The LREE are fractionated and the HREE are poorly fractionated in the CAG-type biotite and amphibole-bearing orthogneisses as from Gsies/Casies and Kristeinertal; whereas more-acidic CCG-type muscovite-bearing orthogneisses like from Antholz/Anterselva show higher HREE fractionation. A negative Eu anomaly occurs (Peccerillo et al. 1979; Mazzoli & Sassi 1992). The compositions of the meta-porphyroids in the Thurntaler Phyllite Group (Zr–Ti is shown in Fig. 3b) are restricted to the continental collision granite field (Fig. 3c) in the A/CNK–A/NK diagram of Maniar & Piccoli (1989). The REE patterns of meta-porphyroids perfectly match the data from the orthogneisses (Schulz & Bombach 2003) and demonstrate the geochemical similarity of both rock groups, which had already been stated by Heinisch (1981) and Heinisch & Schmidt (1982).

Apart from ages from the orthogneisses that turned out to be Ordovician metagranitoids (Borsi et al. 1973, 1978, 1980b; Satir 1975, 1976; Brack 1977; Cliff 1980; Hammerschmidt 1981; Sassi et al. 1985; Klötzli 1995, 1999a, b; Bücksteeg 1999), protolith age data of pre-Ordovician meta-magmatic rocks in the Austroalpine basement south of the Tauern Window was lacking (Thöni 1999). Therefore, selected samples out of the mafic meta-magmatic suites, the orthogneisses and the meta-porphyroids have been dated by the Pb–Pb single zircon evaporation method with thermal ionization mass spectrometry (TIMS) and by the sensitive high-resolution ion microprobe (SHRIMP) method (Schulz & Bombach 2003; Siegesmund et al. 2007).

Zircon ages from meta-magmatic rock suites

The zircon populations selected for further radiometric dating of the metabasites are homogeneous without distinct cores (Schulz & Bombach 2003) and display the typical oscillatory and sector zonation of CL generally attributed to growth in a melt (Hanchar & Miller 1993; Connelly 2000). Orthogneisses and meta-porphyroids show a comparably variable morphological range of zircon subpopulations. A discrimination among meta-porphyroids and orthogneisses based on zircon typology alone, as demonstrated by Mager (1981) and Kreutzer (1992), is hampered by more than 80% inherited crystals. Internal oscillatory and sector zoning of zircon was also observed in high-pressure metabasites (Corfu et al. 2003), thus further criteria for a distinction are necessary. Occurrence of zircon in mafic igneous rocks poses the question as to whether this mineral was assimilated from continental crustal rocks during magma formation, fractionation and ascent. It has been demonstrated that trace element characteristics allow distinguishing magmatic and metamorphic zircons, and are also indicators of the magmatic source (Hoskin & Schaltegger 2003). About 80 zircon grains from six metabasite samples were analysed with LA-ICP-MS (Schulz et al. 2006). Positive Ce and negative Eu anomalies and HREE enrichment in normalized zircon REE patterns are typical for an igneous origin. The Y in zircon is well correlated to HREE, Ce, Th, U, Nb and Ta and allows discrimination of compositional fields for each host rock type. Low Th/U ratios in zircon are correlated to low bulk Th/U in the host rocks and are likely a primary igneous characteristic that cannot be attributed to metamorphic recrystallization. Variations of zircon/host rock trace element ratios confirmed that ionic radii and charges control abundances and substitutions in zircon. Correlated host rock and zircon trace element concentrations further indicate that the metabasite zircons are not xenocrysts but crystallized from mafic melts, represented by the actual host rocks. As the actual host rocks represent the magmatic differentiates where the zircon populations crystallized, the metabasite zircon Pb–Pb ages should indicate the crystallization of the magmatic protoliths (Schulz et al. 2006).

Lead isotopes in selected zircons were analysed by the evaporation of single grains in a two-filament arrangement of a TIMS, as has been proposed by Kober (1986, 1987) and further applied in basement rocks (e.g. Klötzli 1997; Tichomirowa et al. 2001). The isotope data were corrected by contents of common lead derived from the $^{204}Pb/^{206}Pb$ ratios, following the model of Stacey & Kramers (1975).

A mean ratio of ^{207}Pb/^{206}Pb$_{CORR}$ for each analysed zircon has been gained. Apparent zircon ages were produced by iteration of the two equation system of the decay chains ^{238}U–^{206}Pb and ^{235}U–^{207}Pb; the error on ages was calculated as 2σ mean, using the mean ratio of ^{207}Pb/^{206}Pb$_{CORR}$ and the error of the measured ratios (Schulz & Bombach 2003; Schulz et al. 2004). When five to seven single zircon ages from one sample fall within a similar range, this has been considered to date a zircon magmatic crystallization event. Within error, weighted mean ages and isochron ages, calculated from a regression line in ^{204}Pb/^{206}Pb vs. ^{207}Pb/^{206}Pb coordinates (Ludwig 2001), are similar and quoted in Table 1.

Zircons from the N-MORB-type sample ECLS (Fig. 1) of eclogitic amphibolite display a weighted mean age of 590 ± 4 Ma. Weighted mean zircon ages from the hornblende-gneisses with the volcanic arc basalt-type geochemical signature are 533 ± 4 Ma, 543 ± 2 Ma, 548 ± 4 Ma and 550 ± 6 Ma. This time span is considered to reflect a longer-lasting magmatic crystallization process along several stages. A garnet-amphibolite (ARNX) from the Torkogel Subgroup is a Zr-rich late differentiate from the alkaline within-plate basalt-type metabasites and provided a weighted mean age of 430 ± 2 Ma. Data from the Kristeinertal orthogneiss (sample W5, 468 ± 5 Ma, Schulz & Bombach 2003) does not exactly coincide with the multigrain and discordant U–Pb ages of 427 Ma and 443 Ma reported by Cliff (1980), but are equivalent to a single zircon U–Pb age of 466 ± 10 Ma gained by Klötzli (1995) from the Gsies/Casies orthogneiss, which is another metaluminous member of the suite. The Pb–Pb zircon age from an orthogneiss sample (BAR4) in the Schobergruppe (478 ± 15 Ma) is confirmed by a U–Pb SHRIMP age of 470 ± 3 Ma from the same location; similarly the Pb–Pb zircon age from the meta-porphyroid (ANR5 469 ± 6 Ma) overlap within error to the U–Pb–SHRIMP age of 477 ± 4 Ma (Siegesmund et al. 2007). The Pb–Pb zircon ages from the biotite-bearing orthogneisses with the metaluminous continental arc granite-type character are slightly older with 478 ± 15 Ma (BAR4), 468 ± 5 Ma (W5), 463 ± 8 Ma (BAR1), 461 ± 4 (MIT) when compared to muscovite orthogneisses with peraluminous and continental collision granite-type compositions with 457 ± 5 Ma (HOCH) and 447 ± 6 Ma (501). From meta-porphyroids of the Thurntaler Phyllite Group, weighted mean ages are 473 ± 7 Ma (983–25) and 469 ± 6 Ma (ANR5). Further zircons from other samples of location ANR give 467 ± 5 Ma. The inherited zircons in the Ordovician orthogneisses and meta-porphyroids yield ages up to 2.0 Ga, around 800 Ma and at 600 Ma (Schulz & Bombach 2003; Siegesmund et al. 2007).

Although the meta-magmatic rock suites can be distinguished by their element abundances and protolith ages, their compositions show overlap and cover larger compositional fields as it could be explained by the fractionation of single primary magmas. The differences between the suites could be related to different sources. As zircons have not been found yet in the metabasites of the Thurntaler Phyllite Group, it is not clear if this suite is equivalent to and of similar protolith age as the corresponding rocks in the Northern-Defereggen-Petzeck Group, even when trace element characteristics coincide.

Whole-rock isotope characteristics

A contribution of various magmatic reservoirs to the mafic rock suites can be identified by oxygen, Nd and Sr isotope ratios, which should remain insensitive during subsequent magmatic fractionation events. Knowledge about magmatic sources of meta-porphyroids and orthogneisses would enable further interpretation in terms of a geodynamic evolution. Slopes and relative positions of whole-rock (WR) ^{87}Sr/^{86}Sr and εNd$_{CHUR}$ vs. time evolution lines of the meta-magmatic rocks in the frame of depleted mantle, enriched mantle and continental crust (Fig. 3d) allow to detail the aspect of the contribution of various magmatic sources when melts are generated during plate-tectonic processes. The ^{87}Sr/^{86}Sr vs. time evolution lines of the N-MORB-type and volcanic arc basalt-type rocks are subparallel to a depleted mantle growth curve and variably shifted toward the enriched mantle. However, even when low, the corresponding Sr$_{INIT}$ as well as the slightly lower εNd$_{CHUR}$ of the eclogitic amphibolite compared to the depleted mantle growth curve (Fig. 3d, e) signalize a minor contribution of an enriched mantle, a continental crust, or both. Furthermore, the N-MORB- and volcanic arc basalt-type metabasites display low δ^{18}O which range between 6.9 and 8.2‰, regardless of the SiO$_2$ content (Schulz et al. 2004). These values are slightly elevated compared to a mantle melt composition and signalize a differentiation from a mantle source, with only minor interaction or contribution of crustal material. However, this crustal component, low in the N-MORB-type eclogitic amphibolites and partly more prominent in the volcanic arc-type hornblende-gneisses (Fig. 3d, e) can be interpreted as a consequence of a subduction-related magmatism, as is also indicated by the trace element data (Schulz 1995; Schulz et al. 2001; Schulz & Bombach 2003).

Within the acid meta-magmatic suites, one can distinguish rocks with low and high Sr$_{INIT}$ and characterized by evolution paths with varying slopes (Fig. 3d). High Sr$_{INIT}$ and steep evolution

Table 1. *Summary of age data from the Austroalpine south of the Tauern Window. DG, Defereggen Group; DMG, Durreck Muscoviteschist Group; NDPG, Northern-Defereggen-Petzeck-Group; TPG, Thurntaler Phyllite Group*

Group	Sample	Name	Rock	Method	Age (Ma)	Event	Reference
NDPG	ECL	Oberlienz	eclam	Pb–Pb Zrn	590 ± 4	magprot	Schulz & Bombach 2003
NDPG	DI2	Arnitzalm	hbgn	Pb–Pb Zrn	543 ± 2	magprot	Schulz & Bombach 2003
NDPG	MAT2	Matrei	hbgn	Pb–Pb Zrn	548 ± 4	magprot	Schulz & Bombach 2003
NDPG	BAR99-7	Barrensee	hbgn	Pb–Pb Zrn	533 ± 4	magprot	Schulz & Bombach 2003
NDPG	ARNX	Arnitzalm	am	Pb–Pb Zrn	430 ± 2	magprot	Schulz & Bombach 2003
NDPG	BAR99-4	Barrensee	ms-ogn	Pb–Pb Zrn	471 ± 7	magprot	Schulz & Bombach 2003
NDPG	BAR99-1	Barrensee	ms-ogn	Pb–Pb Zrn	463 ± 8	magprot	Schulz & Bombach 2003
NDPG	MIT1-2	Mittagskogel	bt-ogn	Pb–Pb Zrn	461 ± 4	magprot	Schulz & Bombach 2003
NDPG	HOCH	S. Matrei	msogn	Pb–Pb Zrn	457 ± 5	magprot	Schulz & Bombach 2003
NDPG	501	Prägraten	bt-ogn	Pb–Pb Zrn	448 ± 6	magprot	Schulz & Bombach 2003
DG	W5	Kristeinertal	bt-ogn	Pb–Pb Zrn	468 ± 5	magprot	Schulz & Bombach 2003
TPG	ANR5	Anras	mporph	Pb–Pb Zrn	469 ± 6	magprot	Schulz & Bombach 2003
TPG	983-25	Abfaltern	Mporph	Pb–Pb Zrn	473 ± 7	magprot	Schulz & Bombach 2003
TPG	P1	Anras	Mporph	SHRIMP Zrn	477 ± 4	magprot	Siegesmund *et al.* 2007
NDPG	RS11-00	Tönig	mics	Rb–Sr Bt	204 ± 3	metam	Schuster *et al.* 2001
NDPG	RS11-00	Tönig	mics	Ar–Ar Ms	193 ± 2	cool < 400	Schuster *et al.* 2001
NDPG	RS13-00	Michelbachtal	pegm	Sm–Nd Grt	253 ± 7	magprot	Schuster *et al.* 2001
NDPG	RS13-00	Michelbachtal	pegm	Ar–Ar Ms	190 ± 3	cool < 400	Schuster *et al.* 2001
NDPG	AA74-21	Trojertal	pgn	Rb–Sr Bt	27 ± 1	cool	Borsi *et al.* 1978
NDPG	AA74-23	St. Veit	pgn	Rb–Sr Bt	48 ± 1	cool	Borsi *et al.* 1978
NDPG	AA74-26	Hopfgarten	pgn	Rb–Sr Bt	47 ± 1	cool	Borsi *et al.* 1978
NDPG	AA74-13	Huben	pgn	Rb–Sr Bt	59 ± 1	cool	Borsi *et al.* 1978
NDPG	AA73-85	St. Johann	pgn	Rb–Sr Bt	56 ± 2	cool	Borsi *et al.* 1978
NDPG	AA73-85	St. Johann	pgn	Rb–Sr Ms	201 ± 22	cool	Borsi *et al.* 1978
NDPG	AA74-08	Arntal	pgn	Rb–Sr Bt	297 ± 2	cool	Borsi *et al.* 1978
DG	AA73-92	Kristeinertal	pgn	Rb–Sr Bt	283 ± 2	cool	Borsi *et al.* 1978
DG	AA73-95	Villgratental	pgn	Rb–Sr Bt	294 ± 2	cool	Borsi *et al.* 1978
DG	AA74-19	Iseltal-West	pgn	Rb–Sr Bt	196 ± 2	cool	Borsi *et al.* 1978
DG		Gsiesertal	ms-ogn	Pb–Pb Zrn	460 ± 15	magprot	Klötzli 1995
DG		Gsiesertal	bt-ogn	Pb–Pb Zrn	466 ± 10	magprot	Klötzli 1995
DG		Winkeltal	ms-ogn	U–Pb Zrn	443 ± 13	magprot	Cliff 1980
DG		Kristeinertal	bt-ogn	U–Pb Zrn	427 ± 10	magprot	Cliff 1980
DG		Antholz	ms-ogn	Rb–Sr WR	434 ± 4	magprot	Borsi *et al.* 1973
DG		Campo Tures	ms-ogn	Rb–Sr WR	445 ± 24	magprot	Hammerschmidt 1981

NDPG		Schobergruppe	ms-ogn	Rb–Sr WR	440 ± 13	magprot	Brack 1977
NDPG		Schobergruppe	ms-ogn	U–Pb Zrn	455 ± 4	magprot	Bücksteeg 1999
NDPG		Barrenlesee	ogn	SHRIMP Zrn	470 ± 3	magproit	Siegesmund et al. 2007
NDPG	Alk8	Alkusersee	mics	EMP-Mnz	314 ± 20	metam	Schulz et al. 2005
NDPG	Alk2	Alkusersee	mics	EMP-Mnz	268 ± 8	metam	Schulz et al. 2005
NDPG	527	Mirschachsch	mics	EMP-Mnz	325 ± 75	metam	Schulz et al. 2005
NDPG	520	Mirschachsch	mics	EMP-Mnz	320 ± 28	metam	Schulz et al. 2005
NDPG	HPr10	Barrensee	mics	EMP-Mnz	95 ± 11	metam	Schulz et al. 2005
NDPG	P24	Bruggen	mics	EMP-Mnz	271 ± 15	metam	Schulz et al. 2005
DG	Sti14	Stierbichlsee	mics	EMP-Mnz	274 ± 12	metam	Siegesmund et al. 2007
DG	ST″-1	Stierbichlsee	staur	Pb–Pb St	281	metam	Siegesmund et al. 2007
DG	Kya-1	Stierbichlsee	kyan	Pb–Pb Ky	280	metam	Siegesmund et al. 2007
Rieserferner	Patsch	Patschertal	ton	U–Pb Allanite	32.4 ± 0.4	magprot	Romer & Siegesmund 2003
Rieserferner	Rain	Rainfälle	ton	U–Pb Allanite	31.8 ± 0.4	magprot	Romer & Siegesmund 2003
Rieserferner		Rieserferner	ton	Rb–Sr WR	31 ± 3	magproit	Müller et al. 2000 (recalc)
DG	AK9	Stemmeringer A.	peg	K-Ar Ms	193 ± 4.1	cool < 400	Steenken et al. 2002
DG	FT10	Sambock	peg	K-Ar Ms	97.4 ± 2.1	cool < 400	Steenken et al. 2002
DG	FT12	Bärentaler Alm	ms-ogn	K-Ar Ms	297.8 ± 6.7	cool < 400	Steenken et al. 2002
DG	FT22.1	Klosterfrauen Alm	peg	K-Ar Ms	189.2 ± 8.6	cool < 400	Steenken et al. 2002
DG	OG10	Gsies	bt-ogn	K-Ar Ms	309.8 ± 6.3	cool < 400	Steenken et al. 2002
DG	OG13a	Staller Sattel	ms-ogn	K-Ar Ms	294.7 ± 6	cool < 400	Steenken et al. 2002
DG	FT30	Weißbach Tal	pgn	K-Ar Ms	299.7 ± 8	cool < 400	Steenken et al. 2002
DG	FT1	Stemmeringer A.	pgn	K-Ar Bt	261.9 ± 5.6	cool	Steenken et al. 2002
NDPG	AK8	Staller Sattel	peg	K-Ar Ms	60.6 ± 1.3	cool < 400	Steenken et al. 2002
NDPG	FT14	Wengsee	ms-ogn	K-Ar Ms	100.8 ± 2.1	cool < 400	Steenken et al. 2002
NDPG	FT15	Mühlwald	ms-ogn	K-Ar Ms	58.4 ± 1.4	cool	Steenken et al. 2002
NDPG	FT20	Klosterfrauen A.	pgn	K-Ar Ms	319.5 ± 7.9	cool	Steenken et al. 2002
NDPG	BS23	Barrenle See	bt-ogn	K-Ar Bt	229.9 ± 4.9	cool	Steenken et al. 2002
NDPG	FT32	Patscher Tal	ms-ogn	K-Ar Bt	26.7 ± 0.7	cool	Steenken et al. 2002
Rieserferner	AST2	Jäger-Scharte	grd	K-Ar Bt	32.5 ± 0.9	cool	Steenken et al. 2002
Rieserferner	FT9	Osing	grd	K-Ar Bt	29.5 ± 0.9	cool	Steenken et al. 2002
Rieserferner	Patsch	Patscher Hütte	grd	K-Ar Bt	27.9 ± 0.7	cool	Steenken et al. 2002
Rieserferner	Rain	Rainfälle	ton	K-Ar Bt	27.3 ± 0.8	cool	Steenken et al. 2002
DG	FT1	Stemmeringer A.	pgn	Zrn FT	78 ± 6	Zrn PAZ	Steenken et al. 2002
DG	FT12	Bärentaler Alm	ms-ogn	Zrn FT	31.5 ± 1.9	Zrn PAZ	Steenken et al. 2002
DG	FT30	Weißbach Tal	pgn	Zrn FT	138 ± 9	Zrn PAZ	Steenken et al. 2002
DG	OG13a	Staller Sattel	ms-ogn	Zrn FT	91 ± 6	Zrn PAZ	Steenken et al. 2002

(Continued)

Table 1. Continued

Group	Sample	Name	Rock	Method	Age (Ma)	Event	Reference
NDPG	FT4	Speikboden-Fröz-A.	pgs	Zrn FT	25.1 ± 1.5	Zrn PAZ	Steenken et al. 2002
NDPG	FT6	Gritzen	pgs	Zrn FT	29 ± 1.9	Zrn PAZ	Steenken et al. 2002
NDPG	FT15	Mühlwald	ms-ogn	Zrn FT	19.1 ± 1.2	Zrn PAZ	Steenken et al. 2002
NDPG	FT32	Patscher Tal	ms-ogn	Zrn FT	23.5 ± 1.5	Zrn PAZ	Steenken et al. 2002
Rieserferner	ASt2	Jäger Scharte	grd	Zrn FT	27.7 ± 1.4	Zrn PAZ	Steenken et al. 2002
Rieserferner	FT9	Osing	grd	Zrn FT	27.6 ± 1.3	Zrn PAZ	Steenken et al. 2002
Rieserferner	FT24	Michelbach	ton	Zrn FT	27.6 ± 1.3	Zrn PAZ	Steenken et al. 2002
Rieserferner	FT26	Gossenbach	ton	Zrn FT	26.7 ± 1.3	Zrn PAZ	Steenken et al. 2002
DG	AK9	Stemmeringer A.	peg	Ap FT	13.9 ± 0.8	Ap PAZ	Steenken et al. 2002
DG	FT10	Sambock	peg	Ap FT	11.5 ± 0.7	Ap PAZ	Steenken et al. 2002
DG	FT12	Bärentaler Alm	ms-ogn	Ap FT	17.6 ± 1.7	Ap PAZ	Steenken et al. 2002
DG	FT22.1	Klosterfrauen A.	peg	Ap FT	21.7 ± 1.4	Ap PAZ	Steenken et al. 2002
DG	FT30	Weißbach Tal	pgn	Ap FT	18.8 ± 1.4	Ap PAZ	Steenken et al. 2002
DG	OG13a	Staller Sattel	ms-ogn	Ap FT	13.4 ± 0.9	Ap PAZ	Steenken et al. 2002
DG	AK8	Staller Sattel	peg	Ap FT	14.1 ± 1.1	Ap PAZ	Steenken et al. 2002
NDPG	FT4	Speikboden-Fröz-A.	pgn	Ap FT	17.8 ± 2.3	Ap PAZ	Steenken et al. 2002
NDPG	FT6	Gritzen	pgn	Ap FT	13 ± 1	Ap PAZ	Steenken et al. 2002
NDPG	FT8	Gsaritzer Alm	pgn	Ap FT	16.5 ± 2	Ap PAZ	Steenken et al. 2002
NDPG	FT14	Wengsee	ms-ogn	Ap FT	12.9 ± 0.8	Ap PAZ	Steenken et al. 2002
NDPG	FT15	Mühlwald	ms-ogn	Ap FT	14.1 ± 1	Ap PAZ	Steenken et al. 2002
NDPG	FT20	Göriacher Alm	pgn	Ap FT	9.6 ± 2	Ap PAZ	Steenken et al. 2002
NDPG	FT23.1	Wirtsalm	peg	Ap FT	16.1 ± 1.1	Ap PAZ	Steenken et al. 2002
NDPG	FT32	Patscher Tal	ms-ogn	Ap FT	13 ± 0.9	Ap PAZ	Steenken et al. 2002
NDPG	FT40	Jagdhaus Alm	mics	Ap FT	9 ± 0.7	Ap PAZ	Steenken et al. 2002
NDPG	FT41	Raintal	pgn	Ap FT	15.4 ± 2.5	Ap PAZ	Steenken et al. 2002
NDPG	FT42	Usprungtal	ms-ogn	Ap FT	10.1 ± 0.8	Ap PAZ	Steenken et al. 2002
Rieserferner	ASt2	Jäger Scharte	grd	Ap FT	17.7 ± 1.8	Ap PAZ	Steenken et al. 2002
Rieserferner	FT9	Osing	grd	Ap FT	15.6 ± 1.1	Ap PAZ	Steenken et al. 2002
Rieserferner	FT24	Michelbach	ton	Ap FT	18.5 ± 1.6	Ap PAZ	Steenken et al. 2002
Rieserferner	FT26	Gossenbach	ton	Ap FT	12.8 ± 1.8	Ap PAZ	Steenken et al. 2002
Rieserferner	Patsch	Patscher Hütte	grd	Ap FT	13.1 ± 1	Ap PAZ	Steenken et al. 2002
Rieserferner	Rain	Rainfälle	ton	Ap FT	12.7 ± 1.1	Ap PAZ	Steenken et al. 2002
NDPG				Zrn FT	21	Zrn PAZ	Fügenschuh 1995
NDPG				Zrn FT	24	Zrn PAZ	Fügenschuh 1995
Southern Alps				Zrn FT	81	Zrn PAZ	Fügenschuh 1995

Rensen		Rensen pluton	grd		Angelmaier et al. 2000
DG	FT25A	Kofl		128.4 + 6.9	Zrn FT	Zrn PAZ	Stöckhert et al. 1999
DG	FT25B	Kofl		119.2 + 5.5	Zrn FT	Zrn PAZ	Stöckhert et al. 1999
DG	FT26	Kofl		113.2 + 7.6	Zrn FT	Zrn PAZ	Stöckhert et al. 1999
DG	FT27	Kofl		121.8 + 7.9	Zrn FT	Zrn PAZ	Stöckhert et al. 1999
DG	FT29	Moar		40.7 + 3.3	Zrn FT	Zrn PAZ	Stöckhert et al. 1999
NDPG	GE01	Uttenheim		22.2 + 1.6	Zrn FT	Zrn PAZ	Stöckhert et al. 1999
NDPG	KAW2224	Schloss Land		27 + 1.2	Zrn FT	Zrn PAZ	Stöckhert et al. 1999
NDPG	KAW2227	Uttenheim		25.3 + 1.3	Zrn FT	Zrn PAZ	Stöckhert et al. 1999
DG	KAW2228	Spitzbach		33.8 + 2.4	Zrn FT	Zrn PAZ	Stöckhert et al. 1999
DG	KAW2232	Gasthof Sonne		94.9 + 6.8	Zrn FT	Zrn PAZ	Stöckhert et al. 1999
DG	KAW2233	Platten		45 + 2.3	Zrn FT	Zrn PAZ	Stöckhert et al. 1999
DG	KAW2234	Ried		113.2 + 7.1	Zrn FT	Zrn PAZ	Stöckhert et al. 1999
NDPG	RK81	Griesberg		21 + 1.4	Zrn FT	Zrn PAZ	Stöckhert et al. 1999
NDPG	RK92	Untergraber		25.2 + 4.8	Zrn FT	Zrn PAZ	Stöckhert et al. 1999
NDPG	RK106	Mühlen		21.7 + 1.2	Zrn FT	Zrn PAZ	Stöckhert et al. 1999

paths are typical of continental crust. Low Sr_{INIT} occur as well and indicate variable contribution of a mantle-derived material to the heterogeneous crustal sources of the former granitoids and volcanics (Faure 2001). Even when main and trace element signatures of the orthogneisses and meta-porphyroids are very similar, they are not co-magmatic rocks *sensu stricto*, as they involve sources with different geochemical characteristics. This difference among orthogneisses and meta-porphyroids is supported by WR $\delta^{18}O$ data. The orthogneisses are low in WR $\delta^{18}O$ (9.2–10.3‰) with a negative correlation to the SiO_2. In contrast, the Thurntaler meta-porphyroids have $\delta^{18}O$ of 13‰ at around 75 wt% SiO_2, as can be expected from siliceous crustal magmatites. Such large variations of $\delta^{18}O$ among the acid meta-magmatites cannot be explained by magmatic fractionation alone, but indicate variable contribution of mantle and/or further crustal components to the predominant crustal source.

The within-plate basalt-type rocks follow Sr-evolution lines parallel to, slightly beneath and in most cases considerably above the enriched mantle growth curve (Fig. 3d). They started at very different values of initial $^{87}Sr/^{86}Sr$, even when uncertainty about the protolith ages is considered. Strontium evolution lines situated between depleted and enriched mantle lines signalize contribution of both reservoirs. Furthermore, the Sr evolution lines above the enriched mantle are typical of a mantle source, which has been variably influenced by continental crust of different Sr_{INIT}. This gives the impression that the MORB- and WPB-type rocks cannot be derived from a homogeneous source. A contribution of several crustal components, each with a distinct and different Sr_{INIT}, appears more likely. In consequence, the initial $^{87}Sr/^{86}Sr$ in within-plate basalt-type samples do not correlate with the enrichment of large ion lithophile elements, Th, Nb and LREE. This is supported by WR $\delta^{18}O$ which range from 9.3 to 11.9‰ and display a negative correlation with SiO_2. A decrease of WR $\delta^{18}O$ of about >2‰ along the differentiation index exceeds a variation that could be explained by closed system crystal fractionation and contrasts a trend that could be expected by a possible interaction with continental crust enriched in $\delta^{18}O$ during a magma ascent (Hoefs 1997). Therefore, even the primitive members of the alkaline within-plate basalt-type suites are clearly influenced by batches of a continental crustal component variable in ^{18}O. This signalizes that the widespread within-plate basalt-type and the associated MORB-like rocks belong to several similar but independent fractionation lines. Different parent magmas that resulted from mixing of mantle reservoirs and crustal components fractionated in a similar way. Thus, it appears likely that the within-plate basalt-type rocks in the Northern-Dereggen-Petzeck Group and the MORB/WPB-type suite in the Thurntaler Phyllite Group have comparable ages and are related by a common magmatic process.

Early Palaeozoic magmatic evolution lines

In Ta/Yb–Th/Yb coordinates (Pearce 1983; Wilson 1989), whole-rock geochemical and isotopic data from meta-magmatic rocks in combination with single zircon protolith ages allows to identify two major pre-Variscan magmatic evolution lines. An apparently older evolution line-I comprises 590 Ma N-MORB-type basites, now appearing as eclogitic amphibolites, 550–530 Ma volcanic arc basalt-type magmatites, now hornblende-gneisses, and Ordovician 480–450 Ma acid plutonites and volcanics, now orthogneisses and meta-porphyroids. These rock suites follow a trend along a subduction-type or crustal contamination vector, then a fractional crystallization-type vector in the Ta/Yb–Th/Yb coordinates. The younger evolution line-II less well defined by radiometric data is composed of tholeiitic to transitional MORB- and alkaline within-plate basalt-type rock suites that follow an intraplate mantle metasomatism or within-plate enrichment trend. At low Ta/Yb–Th/Yb ratios, corresponding to a depleted mantle source, no further distinction among the suites and evolution lines is possible (Fig. 3f). Considering the trace element and isotope characteristics, the within-plate basalt-type rocks in the Northern-Dereggen-Petzeck Group and the MORB/WPB-type suite in the Thurntaler Phyllite Group are related by a common large-scale geodynamic process which generated multicomponent sources. Floyd et al. (2000) interpreted metabasites from the northeastern Bohemian Massif with a wide range of MORB- to WPB-like tholeiitic to alkaline compositions to have been generated through melting of a lithosphere on the one hand, and a depleted high-level asthenospheric mid-ocean ridge basalt (MORB)-type reservoir that mixed with an enriched mantle plume on the other hand. This could also explain the generation of the Austroalpine alkaline metabasites.

Structural and metamorphic evolution

Macro- and mesoscopic structures

Based on the Late Variscan Rb–Sr mica cooling ages and observations in the Permian pegmatites, structures of pre-Alpine ductile deformation can be expected to the south of the Dereggen-Antholz-Vals line. In the metasediments of the Dereggen Group, a predominant pre-Alpine foliation S_{V2} (monomineralic quartz layers are considered to represent an older foliation S_{V1}) is axial-planar to isoclinal F_{V2} folds (Fig. 4). Linear-planar fabrics in the Ordovician granitoids, now orthogneisses, are parallel to S_{V2} and the stretching lineation L_{V2} in the metasedimentary host rocks (Stöckhert 1985; Guhl & Troll 1987; Schulz 1994a). Feldspar porphyroclasts with sigma shape and predominant linear structures as well as planar structures indicate deformation of the orthogneisses by simple shear. A special type of F_{V2} folds are sheath folds (Quinquis et al. 1978), with strongly curved fold axes subparallel to the stretching lineation L_{V2}. Sheath folds in the cm- to dm-scale occur in quartzitic paragneisses (Fig. 5a), in calcsilicatic and graphitic layers within paragneisses, in marbles (Fig. 5c) and in biotite-plagioclase-paragneisses (Schulz 1988a, b, 1990; Schulz et al. 2001). Further F_{V2} structures are calcsilicategneiss layers and bodies that occur mostly within quartzitic paragneisses (Fig. 5b, d). Calcsilicategneiss-bodies, sheath folds and isoclinal folds F_{V2} and the corresponding axial plane foliation S_{V2} were refolded and overprinted at all scales by open to tight F_{V3}-folds and crenulations Lcr_{V3} whose axes are oriented mostly parallel to L_{V2} (Fig. 5a). F_{V3}-folds with steeply plunging axes are minor structures of large-scale syn- and antiforms (Figs 1 and 4b), labelled as 'Schlingen' structures (Schmidegg 1936). Toward the east and south of the region with 'Schlingen' in the Dereggen Group, the orientation of the F_{V3} fold axes grades from the steeply plunging direction into more gentle WSW or ENE plunging. In the Thurntaler Phyllite Group (Fig. 4c), a similar gentle plunging of L_{V2}–S_{V2} linear-planar fabrics and of later F_{V3} fold axes as in the adjacent Dereggen Group is observed (Schmidegg 1937; Heinisch & Schmidt 1984; Spaeth & Kreutzer 1989; Schulz 1991). Linear and linear-planar fabrics are found with similar orientations (Schulz 1988a, 1991) in the Upper Ordovician orthogneisses of the Dereggen Group and Northern-Dereggen-Petzeck Group units as well as in the Ordovician meta-porphyroid gneisses of the Thurntaler Phyllite Group and their metasedimentary host rocks. Therefore, these structures should be younger than the Ordovician meta-magmatic rocks and of a maximal Variscan age (Borsi et al. 1973; Stöckhert 1985; Schulz 1988a, 1990).

Zones of shear bands S_{V4} with a top-to-NE directed extensional movement can be mapped in phyllitic micaschists of the Dereggen Group adjacent to the Thurntaler Phyllite Group (Fig. 4c). Mylonitic shear bands S_{V5} with NW to west or south directed movement of the hanging walls

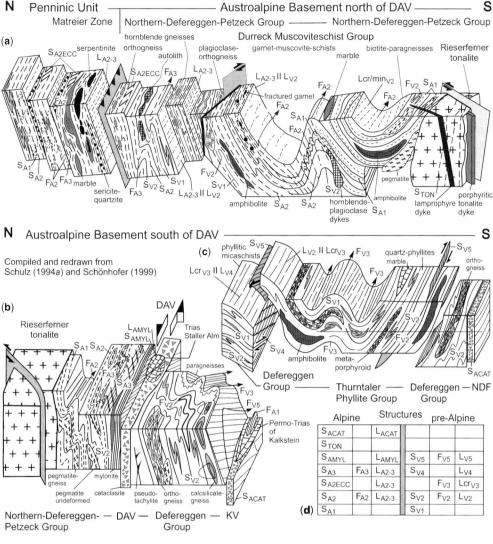

Fig. 4. Schematic structural cross-sections from the Matreier Zone (Penninic Unit) through the Austroalpine basement to the south of the Tauern Window. Compiled from observations reported in Schönhofer (1999), Schulz (1994b), Schulz et al. (2001). (**a**) Northern part from Matreier Zone (Virgental) to the Rieserferner Tonalite (St. Jakob). (**b**) Southern part from the Rieserferner across the Defereggen-Antholz-Vals line (Staller Sattel) to the Kalkstein-Vallarga line. (**c**) Thurntaler Phyllite Group in a synformal position. (**d**) Abbreviations of labels for structures: S_{V1} Variscan first foliation parallel to lithological banding and monomineralic quartz layers. S_{V2} Variscan main foliation dominant in orthogneisses and micaschists. S_{A1} first Alpine foliation, in Permian pegmatites. S_{A2} Alpine main foliation (subparallel to S_{A1} to the north); S_{A2ECC} shearbands within S_{A2}; S_{AMYL} L_{AML} mylonitic foliation and lineation in northern part of Defereggen-Antholz-Vals line. S_{A3} Alpine shearband foliation; S_{ATON} planar structures in tonalite; S_{ACAT} planar structures in fault zones; L_{ACAT} lineation in fault zones; F_{V2} Variscan isoclinal folds and sheath folds of S_{V1}; F_{V3} Variscan folds and Schlingen structures of S_{V2}; F_{A2} Alpine folds of S_{A1}; F_{A3} Alpine folds of S_{A2}; L, Lstr, Lcr, Lmin are lineation, stretching, crenulation and mineral elongation lineations with indices of relative age.

affected and overprinted the foliation-parallel lithological contact between both units (Heinisch & Schmidt 1984; Schulz 1991; Kreutzer 1992). Pervasive dynamic recrystallization of quartz is typical within the S_{V4} and S_{V5} shear bands. This quartz recrystallization has not been observed in the Alpine deformed Permo-Triassic sandstones of Kalkstein (Guhl & Troll 1987). In consequence,

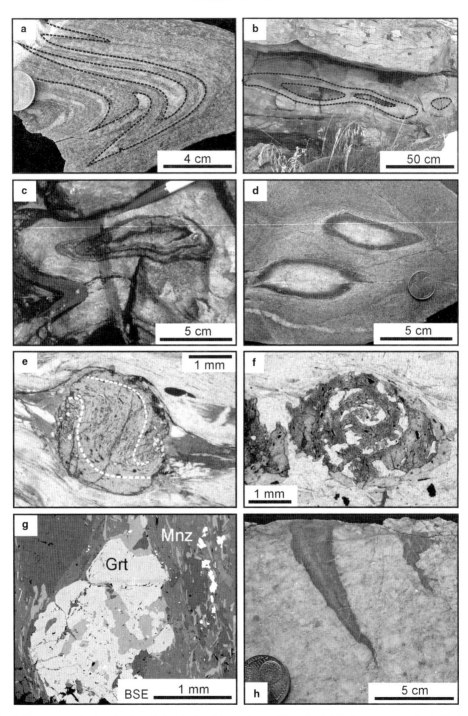

Fig. 5. Structures and microstructures in the Austroalpine metamorphic rocks. (**a**) Section perpendicular to Variscan fold axis F_{V3} with superposed open fold F_{V3} on isoclinal folds F_{V2}. Closed section of a F_{V2} sheath fold and isoclinal F_{V2} folds in quartz-rich layer. Defereggen Group. (**b**) Section of a zoned calcsilicategneiss-body in a matrix of quartzitic paragneiss, section perpendicular to the stretching lineation (Ackstallbach, Staller Sattel, Defereggen Group). (**c**) Closed oval section of a dm-scale sheath fold of a calcsilicategneiss layer in a marble (Storfenspitze, Defereggen Group). (**d**) Cm-scale cylindrical calcsilicategneiss-bodies with graphitic layers in a matrix of quartzitic

this can be taken as a further argument for a Variscan or pre-Alpine minimum age of the foliation-forming ductile deformation in the Defereggen and Thurntaler Phyllite Groups.

In the Schobergruppe to the east of the Isel river, strongly foliated orthogneisses, hornblende gneisses, paragneisses and micaschists concordantly overly, underly and are intercalated with the eclogitic amphibolites and related amphibolites (Figs 1 and 2). A dominant main foliation S_{V2} is parallel in all rock types and is deformed in an early generation of open to tight folds (Troll & Hölzl 1974; Troll et al. 1976, 1980; Behrmann 1990) with hinge lines parallel to the gently SE-plunging lineation. Later folds with NE–SW striking hinge lines deformed the lineation and are minor structures of kilometre-scale syn- and antiforms with gently dipping axial planes (Behrmann 1990; Bücksteeg 1999). In the Prijakt region of the Schobergruppe (Figs 1 and 2a), the lithological sequence is sporadically penetrated by metre-thick coarse-grained pegmatites. In most cases, the pegmatites are elongated parallel to the host rock foliation planes labelled S_{V2}. Deformed and foliated pegmatite gneisses as well as undeformed pegmatites with preserved textures of crystallization along the wall rocks are observed. When foliated, the pegmatites give evidence of a possible subliquidus deformation at still high temperatures, or alternatively, an Alpine overprint S_{A1} of the pre-Alpine structures at lower temperatures. Even when deformed and foliated, most of the large muscovites in the pegmatites display the typical magmatic mineral composition with low Si^{4+} at 6.1 p.f.u. and low Mg and Fe at <0.2 p.f.u. (Cipriani et al. 1971) and only in some muscovites the slightly elevated Si 6.1–6.4, Mg 0.2–0.4, Fe 0.2–0.6 p.f.u. (Schulz, unpublished data) of a metamorphic overprinting were recognized. Garnet in the pegmatites displays the typical Mn-rich magmatic composition (Schulz et al. 2005). In the hanging wall of the Penninic-Austroalpine boundary, Behrmann (1990) reported an Eo-Alpine penetrative ductile deformation at conditions of <450 °C with a formation of south-dipping shear planes of which quartz-c textures uniformly indicate a top-to-west and NW directed transport. Permian pegmatites are abundant in a distinct zone to the south of the Defereggen-Antholz-Vals line with metapelites bearing fibrolitic sillimanite and andalusite porphyroblasts (Figs 1 and 2b). Due to intercalated amphibolites, this zone is considered as the upper part of the Northern-Defereggen-Petzeck Group (Schulz et al. 2005). The pegmatites are parallel to, as well as cutting across, the SW- to south dipping foliation S_{V2}. To the south of this zone, the Defereggen Group is concordantly overlying along a foliation-parallel pre-Permian contact.

In the Northern-Defereggen-Petzeck Group to the north of the Defereggen-Antholz-Vals line and to the west of the Isel river, the pre-Alpine (Variscan) main foliation S_{2V} has been overprinted by an Alpine foliation S_{A1}, which also can be observed in Permian pegmatites (Fig. 4a). Further towards the north and in the structural transition to the underlying Penninic Unit, an Alpine foliation S_{A2} that is parallel to S_{A1} is increasingly dominant and pervasive. S_{A2} is the axial-plane foliation of Alpine folds F_{A2} (Fig. 4a). Foliation planes S_{A2}, fold axes F_{A2} as well as stretching lineations have similar orientation in the Austroalpine basement and the underlying Matreier Zone of the Penninic realm and both units are structurally concordant along the lithological transition to the south of the Virgen valley (Schönhofer 1999; Schulz et al. 2001). As further Alpine structures, a folding F_{3A} of S_{A2}, an ECC-type foliation S_{A3ECC} and a mylonitic foliation S_{AMYL} along the northern margin of the Defereggen-Antholz-Vals line and in distinct mylonite zones (Fig. 4a, b) have been described (Kleinschrodt 1987; Schönhofer 1999; Steenken et al. 2000).

Quartz-c-textures in calcsilicategneiss bodies

Calcsilicategneiss layers are widespread in the Defereggen Group. With a thickness of 1–7 cm, they are parallel to the foliation S_{V2} of the metapsammitic host rocks and display a symmetrical habit (Fig. 5b, d). Single calcsilicategneiss bodies have a cigar- or tongue-like shape. The long axis of the bodies is parallel to the stretching lineation L_{V2}. In the case of extremely cigar-shaped bodies, the extension parallel to the lineation can be several metres at a diameter of only several centimetres. It has been observed that several single calcsilicategneiss bodies develop from single or

Fig. 5. (Continued) paragneiss, cut perpendicular to stretching lineation (Hochgrabe, Defereggen Group). (**e**) Syncrystalline-rotated garnet with S-shaped inclusion trails of foliation in a micaschist, cut perpendicular to the stretching lineation (Degenhorn, Defereggen Group). (**f**) Syncrystalline-rotated garnet with double inclusion spiral in a micaschist, cut perpendicular to the stretching lineation (Langschneid, Defereggen Group). (**g**) Backscattered electron (BSE) image of a staurolite-garnet-micaschist with pre-Alpine monazite along the foliation plane (Alkus, Schobergruppe). (**h**) Cm-scale pseudotachylite in a foliated pegmatite in the southern cataclastic domain of the Defereggen-Antholz-Vals line (Bruggen, Defereggental).

parallel layers (Fig. 5d). When cut perpendicular to the long axes and lineation, the calcsilicategneiss bodies display flat to elliptical shapes. In detail, the calcsilicategneiss bodies are composed of several distinct zones. Centres of the bodies are whitish to grey with quartz, clinozoisite, plagioclase, calcite, garnet, green amphibole and sphene. The rims are darkish green with quartz, green amphibole, clinozoisite, plagioclase, garnet and sometimes graphite. Mineral assemblages in the rim layer can be explained by metasomatism along the contact zone of the carbonate-rich core and the psammitic host rock. Quartz-c-textures have been analysed by the universal stage in numerous calcsilicategneiss bodies (Schulz 1988a, b). In XZ sections of bodies that display no overprint by later folding F_{V3}, quartz grains are 0.1–1 mm in diameter. Surprisingly, although stable grain boundaries of static grain growth at amphibolite-facies temperatures prevail and the mode of quartz in the layers of the calcsilicategneiss bodies is mostly <50%, quartz-c fabrics display a high degree of preferred orientation. Layers in the quartzitic host rocks and the rims of the bodies display monocline cross girdles with transformations to inclined single girdles (Fig. 6b, d, e, i). Such fabrics are typical of a non-coaxial deformation involving simple shear when interpreted as outlined in Sander (1950), Bouchez (1977), Lister & Price (1978) and Lister et al. (1978). In contrast, layers in the centre of the bodies have the symmetric quartz-c fabrics of a coaxial prolate deformation, expressed in double large girdles considered to indicate axial constriction (Fig. 6b, h). Textures of coaxial oblate deformation as small girdles, typical of flattening, occur as well (Fig. 6f). The monocline quartz-c-textures in the marginal zones of the calcsilicategneiss-bodies in line with preferred orientation of grains and aggregates with their long axes parallel to the lineation signalize a

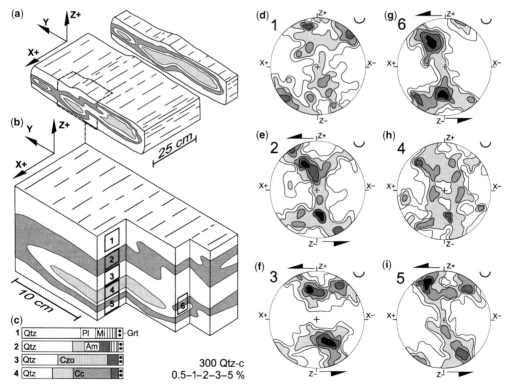

Fig. 6. Quartz-c-fabrics in a zoned calcsilicategneiss body (Defereggen Group, Weissenbachtal to the south of the Staller Sattel, 2100 m above sea level). The x-y-z-axes of finite strain are oriented according to the foliation plane (XY) and the stretching lineation (X). (**a**) Zoned calcsilicategneiss-body with overall oval shape in a matrix of paragneiss (block from Weissenbachtal, Staller Sattel). (**b**) Studied sample with layers characterized by variable modes. (**c**) Modes of layers: numbers refer to (b). Am, amphibole (green hornblende); Cc, calcite; Czo, clinozoisite; Grt, garnet; Mi, biotite and muscovite; Qtz, quartz. Note mode of <50% of quartz in layers 2–4. (**d–i**) Quartz-C fabrics in distinct layers (numbers refer to (b) and (c)) with C-axes in a lower hemisphere projection. Isolines of distribution 0.5–1–2–3–5% are based on 300 analysed quartz-c-axes for each diagram (see text for further explanation).

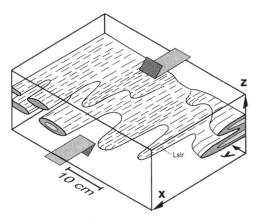

Fig. 7. Schematic 3D-model of calcsilicategneiss-bodies in a matrix of quartzitic paragneisses, with x-y-z axes of finite strain according to foliation (x-y) and the stretching lineation (x). Sheath-like elongation of bodies parallel to the stretching lineation. Oval-shaped zoned bodies in cuts x-z. Note occurrence of oval structures in parallel layers. Fold-like structures in cuts x-z. Model of the formation of zoned calcsilicategneiss-bodies out of single calcsilicate-layers by progressive simple shear with $\gamma > 10$ and a shearing direction parallel to the stretching lineation, according to Cobbold & Quinquis (1980).

formation of the bodies by progressive simple shear, as it has been proposed by Cobbold & Quinquis (1980) for the development of sheath folds in quartzitic layers (Fig. 7). However, the quartz-c-textures typical of coaxial deformation in the cores and centres of the bodies then could be explained by deformation partitioning. From the widespread occurrence of other structures like sigma-porphyroclasts in orthogneisses, syncrystalline-rotated garnets in micaschists (Fig. 5e, f) and F_{V2}-sheath folds in the Defereggen Group, one can conclude that progressive simple shear affected a several kilometre-thick pile of crustal rocks during the formation of the main foliation S_{V2}.

P–T paths of pre-Alpine metamorphism

The Austroalpine basement complex is a giant crystalline nappe that overlies the Penninic Unit in the Tauern Window (Fig. 1). Parts of the Austroalpine basement could have been involved in an Early-Ordovician ('Caledonian') high-temperature event, as it was reported from the northern Oetztal basement Winnebach area (Klötzli-Chowanetz et al. 1997) or has been deduced from the pervasive intrusion of granitoids (Purtscheller & Sassi 1975; Sassi et al. 2004). Devonian to Carboniferous metamorphism reaching eclogite facies conditions as a consequence of Variscan accretion and continental collision has been reported from the Northern Oetztal and other parts of the basement (Miller & Thöni 1995; Thöni 1999 and references therein). Furthermore, there is radiochronological and petrological evidence that Austroalpine metabasites in the Southern Oetztal basement (Hoinkes et al. 1991), as well as to the southeast, in the Saualpe and Koralpe (Thöni & Miller 1996), underwent an Eo-Alpine (Cretaceous) eclogite-facies metamorphism.

The Austroalpine basement to the south of the Tauern Window is situated between these Eo-Alpine eclogite locations. Domains with Alpine metamorphism to the north of the SAM (southern limit of Alpine metamorphism, Hoinkes et al. 1999, Fig. 1) appear as the potential locations of further Eo-Alpine high-pressure rocks. Cretaceous and younger mica ages in the Schobergruppe to the east of the Isel river (Troll 1978) and Sm–Nd garnet and Rb–Sr phengite data (Linner et al. 1996) from Prijakt eclogitic amphibolites and related rocks have been taken as argument that amphibolite-facies assemblages could be related to an Eo-Alpine high-pressure stage of metamorphism (Hoinkes et al. 1999), disregarding that pre-Alpine basement rocks were overprinted. However, Carboniferous and younger mica cooling ages to the south of the Defereggen-Antholz-Vals line and to the west of the Isel river (Borsi et al. 1978; Schuster et al. 2001; Siegesmund et al. 2007) signalize that amphibolite-facies metapelite assemblages should be related to a pre-Alpine metamorphism (Schulz 1990). The Alpine metamorphism did hardly exceed 300 °C in the Defereggen Group, as is demonstrated in Triassic carbonate rocks of Kalkstein and Staller Alm along the late-Alpine faults (Guhl & Troll 1987; Schulz 1991). The non-metamorphic slices of Permo-Triassic rocks along brittle faults as the Zwischenbergen–Wöllatratten line (Exner 1962) in the eastern parts of the basement also indicate that amphibolite-facies mineral assemblages are of pre-Alpine age.

In the Austroalpine metapelites, the garnets often display structures of a syndeformational crystallization (Passchier & Trouw 1996) and the porpyhroblasts are surrounded by the main foliation S_{V2}. Garnet continuous zonations mostly involve increasing XMg at decreasing Mn from cores to rims. The garnet chemical evolution can be correlated to the chemical variations in coexistent biotite and plagioclase by microstructural criteria and is interpreted to be the product of continuous reactions (St-Onge 1987; Schulz 1993a). For geothermobarometric calculations ('microstructurally-controlled geothermobarometry,' Schulz 1993a, b), mica and plagioclase enclosed in garnets were related to early stages, and mica and plagioclase aligned in the external foliation planes to late stages

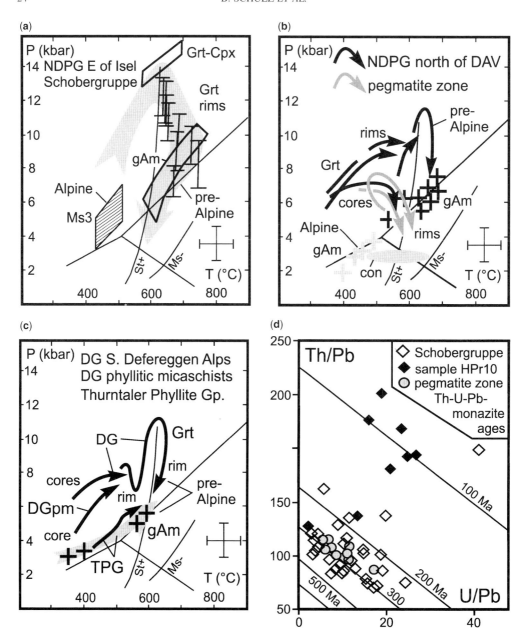

Fig. 8. P–T data and P–T paths from the Austroalpine basement south of the Tauern Window. Aluminosilicate and staurolite (St+) stability fields and muscovite-out (Ms−) univariant line after Spear (1993). (a) Prograde P–T paths reconstructed from zoned garnet in micaschists and paragneisses of the North-Deferegge-Petzeck Group to the west of the Isel river and the pegmatite zone. Crosses mark P–T conditions derived from amphiboles in metabasites from the basement and the Penninic unit (samples from Matrei Zone and Hinterbichl). Conditions for Late Alpine contact metamorphism (con) along the Rieserferner pluton are summarized from samples with different distance to the intrusion contact. Compiled from Mager (1985), Schönhofer (1999), Schulz (1994b), Schulz *et al.* (2001) and unpublished data. (b) Thermobarometry of metabasites and micaschists in the North-Deferggen-Petzeck Group to the east of the Isel river. Garnet-clinopyroxene (Grt-Cpx) and green amphibole (gAm) equilibria from Schobergruppe eclogitic amphibolites: see text for applied thermobarometers. Thermobarometric data from garnet micaschists, based on garnet–biotite and garnet–biotite–muscovite–plagioclase equilibria in garnet rims. Note superposition of pre-Alpine amphibolite-facies P–T conditions in metabasites and micaschists. Early-Alpine P–T conditions are derived

of garnet growth (Fig. 5e, f). Accordingly, P and T for crystallization of Mg-rich garnet rims have been calculated (Schulz 1990, 1993a; Schulz et al. 2005). P–T estimates from garnet cores are afflicted by comparably larger uncertainty, but may at least signalize an evolution with increasing temperature and pressure. The garnet–biotite thermometer of Battacharya et al. (1992) in combination with the garnet–muscovite–biotite–plagioclase barometer (Holland & Powell 1990), involving an internally consistent thermodynamic data set, the updated activity models for garnet (Ganguly & Saxena 1984; Ganguly et al. 1996) and for plagioclase (Powell & Holland 1993) were applied. Data from the Holland & Powell (1990) version of the garnet–aluminosilicate–plagioclase–quartz (GASP) barometer are intermediate between pressure estimates for biotite-Mg and biotite-Fe. Alternative calibrations (Holdaway 2001; Wu et al. 2004) yielded no substantially different results. The P–T conditions of the *garnet + clinopyroxene + plagioclase* bearing assemblages in eclogitic amphibolites were estimated by using the calibrations of the garnet–clinopyroxene Mg–Fe cation exchange thermobarometer by Ellis & Green (1979), Krogh (1988) and Ganguly et al. (1996), and the related geobarometers based on $Jd_{SS} + Qtz = Ab_{SS}$ (Holland 1980, 1983) or equilibria involving garnet, clinopyroxene and plagioclase (Newton & Perkins 1982; Perchuk 1991). Conditions of metamorphism in metabasites with the assemblage *Ca-amphibole + plagioclase (An >10) + quartz $\pm$ zoisite $\pm$ chlorite $\pm$ calcite* were determined by the thermobarometer of Zenk & Schulz (2004) which involves the experiments of Plyusnina (1982), the analytical expressions in Gerya et al. (1997) and an empirical modification for Fe^{3+} in amphibole by using the compositional data reported in Triboulet & Audren (1988) and Triboulet (1992). The error based on experimental estimates of amphibole equilibrium compositions has been reported as $\pm 37\,°C$ and ± 1.2 kbar (Gerya et al. 1997). These approaches allowed reconstructing the P–T evolution from numerous samples and the combination of thermobarometric data from adjacent garnet metapelites and metabasites (Fig. 8).

Micaschists of the Northern-Defereggen-Petzeck Group to the east of the Isel valley in the Schobergruppe and in the vicinity of the eclogitic amphibolites bear garnet, biotite, muscovite, plagioclase, quartz, staurolite and kyanite. Staurolite encloses and postdates garnet. The successive AFM assemblages are (with muscovite + quartz + plagioclase):

M1: garnet + biotite $\pm$ chlorite
M2: garnet + biotite $\pm$ staurolite $\pm$ kyanite
M3: staurolite + kyanite $\pm$ biotite.

Mineral assemblages M1 and M2 with garnet crystallized during formation of the main foliation S_{V2}; garnet is wrapped by the foliation. Sometimes, staurolite and kyanite crystallized after S_{V2} (Troll et al. 1976; Behrmann 1990; Schulz 1993b). Garnet core-to-rim zonations reveals either increase–decrease or simple decrease of Ca at increasing Mg. No garnet porphyroblasts with broad rims of a marked different composition in Mg and Ca, as were described by Purtscheller et al. (1987) from the Oetztal crystalline and interpreted as Eo-Alpine overgrowths on pre-Alpine garnet, were observed (Schulz 1993b; Schulz et al. 2005).

In the Prijakt region (Fig. 1, 2a), the metapelites are associated with eclogitic amphibolites. Coarse-grained eclogitic amphibolites show textural evidence of an early assemblage by garnet + clinopyroxene + plagioclase + quartz + rutile and a subsequent assemblage of Ca-amphibole + zoisite + plagioclase + quartz (Schulz 1993b; Schulz et al. 2005). Metabasite garnets enclose quartz, rutile, epidote, clinopyroxene and plagioclase. In many samples, clinopyroxenes are rimmed by fine-grained symplectites and show breakdown to green amphibole and plagioclase. Jadeite (Jd) contents of matrix omphacites range from 35 to 44 mole % from cores to rims; omphacites enclosed in garnet show the same range of compositions. Anorthite contents in plagioclase enclosed in garnet are at 7–9%, and 7–16% in xenoblasts in the matrix. Aligned large pargasite porphyroblasts partly enclose the garnets. Thermobarometry of the garnet–clinopyroxene assemblages yielded P–T conditions for the high-pressure event and range from 550–600 °C/14 kbar to 600–650 °C/15 kbar (Fig. 8a). Increase of Mg from cores to rims reflects increasing temperature during the high-pressure garnet crystallization. Amphiboles in the eclogitic amphibolites recorded a retrograde evolution from 730 °C/10 kbar to 600 °C/6 kbar. The P–T data from the

---

Fig. 8. (*Continued*) from Na and Si contents in decussate white mica (Ms_3). Data compiled from Schulz (1993b) and Schulz et al. (2005). (**c**) Prograde P–T paths (black arrows) reconstructed from zoned garnet in micaschists from the Defereggen Group (DG), from phyllitic micaschists of the Defereggen Group (DGpm) and the Thurntaler Phyllite Group (TPG) to the south of the Defereggen-Antholz-Vals shear zone. P–T data and P–T path from amphibolites in the TPG with zoned amphiboles are marked by crosses and a grey arrow. Data compiled from Schulz (1990, 1991, 1993a, 1997) and Schulz et al. (2001). (**d**) Monazite EMP ages from Schobergruppe and pegmatite zone metapelites in a histogram view (Schulz et al. 1995).

adjacent garnet metapelites matches and superposes the data from the metabasites and provides supplementary information to the pre-Pmax evolution (Fig. 8a). In some samples, muscovite (Ms_3) recrystallized with decussate orientation or overgrew the foliation planes and is interpreted to postdate the foliation-forming shearing. The Na and Si^{4+} contents in these muscovites define temperatures of maximal 500 °C (Cipriani et al. 1971) at pressures of around 4 kbar (Massonne & Szpurka 1997) of a post-S_{V2} metamorphism and Alpine overprint respectively (Behrmann 1990; Schulz 1993b).

Metapelite assemblages of the Northern-Defereggen-Petzeck Group to the west of the Isel river bear staurolite. However, staurolite is rare and disappears towards the north and does not occur in micaschists of the Durreck–Muscoviteschist Group. The increasing intensity of an Alpine structural overprint to the north of the Defereggen-Antholz-Vals line toward the Austroalpine-Penninic suture is not matched by any changes of the mineral chemistry in the garnet, biotite, muscovite, plagioclase and quartz-bearing metapelite assemblages or by changes within the garnet zoning profiles (Schönhofer 1999). Correspondingly, thermobarometry of garnet metapelites as well as of amphibolites with tschermakitic hornblende give no hints to metamorphic gradients related to this structural overprint. Maximal pressures range at 10–12 kbar and maximal temperatures are at 650 °C/7 kbar. Amphibole porphyroblasts coexisting with andesine in the metabasites are tschermakitic hornblendes and magnesio-hornblendes. They crystallized at decreasing T and P from 600 °C/6.5 kbar to 560 °C/5.0 kbar (Fig. 8b). In combination with the structural observations, it is concluded that a greenschist-facies Alpine overprint did not exceed the prevailing pre-Alpine amphibolite-facies metamorphic conditions (Schulz 1997; Schönhofer 1999). The maximal thermal conditions of the regional metamorphic Alpine overprint were 500 °C at 4 kbar, calculated from amphibolites in the Penninic Upper Schieferhuelle. The contact metamorphism in the vicinity of the Oligocene Rieserferner intrusion reached maximal conditions of 620 °C/3 kbar (Fig. 8b), with crystallization of sillimanite, staurolite, andalusite and K-feldspar in the presence of H_2O (Schulz 1994b).

A distinct zone with micaschists bearing fibrolitic sillimanite can be mapped to the south of the Defereggen-Antholz-Vals line in the vicinity of numerous pegmatite intrusions (Figs 1 and 2). This region represents the lithological transition zone between the Northern-Defereggen-Petzeck Group with abundant amphibolites and the overlying Defereggen Group. Staurolite encloses garnet and overgrew the S_{V2} foliation. Fibrolitic sillimanite crystallized in the foliation planes and replaced the garnet. Porphyroblasts of andalusite overgrew the foliation S_{V2}. The successive AFM assemblages (with muscovite + quartz + plagioclase) are:

M1: garnet + biotite ± chlorite
M2: garnet + biotite ± staurolite
M3: sillimanite ± biotite
M4: andalusite ± biotite

As fibrolitic sillimanite and biotite replace and pseudomorph the garnet porphyroblasts and K-feldspar does not appear, a garnet-consuming reaction $Grt + Ms + H_2O = Sil + Bt + Qtz$ can be derived (Schulz et al. 2001; Schuster et al. 2001). It is evident from the microstructures that the aluminosilicates are related to later stages of metamorphism and crystallized subsequent to garnet. Some of the garnet zonations in this zone are characterized by decreasing and then increasing Mg from cores to rims. The corresponding P–T evolution at medium pressures is first cooling, then heating and ends in the sillimanite stability field at 600 °C/5.0 kbar (Fig. 8b).

Micaschists from the central parts of the Defereggen Group bear garnet, biotite, muscovite, plagioclase, quartz, staurolite and kyanite. Staurolite encloses and postdates garnet (Schulz 1990). In the southern parts of the Defereggen Group, phyllitic micaschists with garnet + biotite + muscovite + chlorite + plagioclase + quartz prevail and staurolite as well kyanite are lacking, possibly due to inadequate bulk rock compositions. P–T data derived from zoned metapelite garnets are similar to the observations from the Northern-Defereggen-Petzeck Group (Fig. 8a, c); however, maximal pressures in the phyllitic micaschists to the south are comparably lower (Fig. 8c).

Assemblages with garnet, biotite, muscovite, oligoclase, quartz and lacking staurolite and aluminosilicates, are observed in garnet-bearing phyllitic micaschists belonging to the Thurntaler Phyllite Group (Schulz 1991). Maximal temperatures calculated from Mg-rich garnet rims are 560 °C/6 kbar (Fig. 8c). These conditions of an epidote-amphibolite facies metamorphism in the Thurntaler Phyllite Group are confirmed by assemblages with magnesio-hornblende/tschermakite, albite, oligoclase, epidote, chlorite, titanite and quartz in the metabasites (Schulz 1991, 1997; Schulz et al. 2001). The amphibole porphyroblasts are strongly zoned with actinolite in the core and tschermakite and magnesio-hornblende in the rim. They crystallized at increasing temperatures from 300 °C/2.5 kbar up to 600 °C/6 kbar (Schulz et al. 2001). Identical mineral assemblages, amphibole mineral chemical compositions and P–T conditions have been found in amphibolite samples from the western parts of the Kreuzeck Phyllite Group between the Möll and Drau valleys (Schulz, unpublished data).

Age constraints on Variscan-to-Permian metamorphism

In the Austroalpine metapelites, monazite is abundant in Ca-poor micaschists with <2 vol. % of epidote-group phases; whereas micaschists richer in Ca and with epidote-group phases bear no monazite. *In-situ* 'chemical' Th–U–Pb dating of monazite by electron microprobe analysis (EMP monazite dating, Montel *et al.* 1994, 1996; Suzuki *et al.* 1994; Dahl *et al.* 2005) is based on the observation that concentration of common lead in monazite (LREE, Th)PO$_4$ is negligible when compared to radiogenic lead resulting from decay of Th and U (Cocherie *et al.* 1998). In monazite, analysis of Th, U and Pb for calculation of model ages, as well as for Ca, Si, LREE and Y for corrections and evaluation of the mineral chemistry were carried out on a JEOL JX 8600 at 15 kV, 250 nA, at a beam size of *c.* 5 μm at the University of Salzburg. Mα1 lines were chosen for Th, Pb and U; Lα1 for La, Y, Ce; Lβ1 for Pr and Nd, and Kα1 for P, Si and Ca. The counting times for Pb, Th and U result in statistical errors (1σ) of typically 0.012 wt%, 0.057 wt% and 0.015 wt% for Pb, Th and U (Finger & Helmy 1998). The small Y interference on the Pb Mα line was corrected by linear extrapolation after measuring a Pb-free yttrium standard (Montel *et al.* 1996). A small Th interference on U Mα was also corrected. For each single analysis, a chemical age plus a respective error (mostly between ±20 and ±40 Ma, 1σ) were calculated using the equations of Montel *et al.* (1996). To control the quality of data, monazite with a concordant U–Pb age of 341 ± 2 Ma analysed by TIMS has been measured together with the specimen. Weighted average ages for monazite populations in the samples were calculated after Ludwig (2001). Different populations of monazite ages can occur in distinct samples.

All studied samples in the vicinity of the eclogitic amphibolites and in a zone with pegmatites to the south of the Defereggen-Antholz-Vals line bear monazite with pre-Alpine chemical model ages. Weighted mean ages calculated by avoiding outliers are 314 ± 20 Ma, 325 ± 75 Ma and 320 ± 28 Ma, 268 ± 8 Ma and 262 ± 20 Ma (Table 1). Samples situated immediately below the eclogitic amphibolites yielded exclusively pre-Alpine monazite ages. Especially samples located in the vicinity of a 30 cm-thick foliation-parallel pegmatite display a homogeneous population of Permian age. In a sample with strongly retrogressed garnet which is situated 50 m structurally below an eclogitic amphibolite, the majority of monazite grains is Eo-Alpine (95 ± 11 Ma, Table 1), but Permian grains occur as well. The data display a range between 90 and 360 Ma (Fig. 8d). The Variscan (350–320 Ma) and Permian (290–260 Ma) ages are arranged in a bimodal frequency distribution and are separated from the Eo-Alpine ages by some largely scattered ages ranging from 240–150 Ma. Apparently, the monazite ages do not correlate to the prograde P–T path sections recorded by the garnet. Furthermore, there is no correlation between the variable monazite Y contents and the garnet zonations with decrease of Y and HREE toward the rims (Schulz *et al.* 2005). Monazite should have crystallized after the thermal peak of metamorphism. In consequence, at best the oldest monazites may be related to the maximal temperatures during the amphibolite-facies metamorphism. Many studies have identified a major pulse of monazite growth in metapelites under amphibolite-facies conditions, which typically coincides with the appearance of staurolite and which could be linked to the breakdown of garnet to produce staurolite at high temperatures (Pyle & Spear 1999; Pyle *et al.* 2001; Fitzsimons *et al.* 2005). This could explain abundant pre-Alpine monazite in most of the samples where staurolite appears with kyanite and postdates garnet. Furthermore, when monazite crystallization was linked to such a garnet breakdown at high temperatures, this should indicate a pre-Alpine age of staurolite and kyanite, and in consequence a pre-Alpine age of the amphibolite-facies metamorphism. Presumably, a Mg-poor bulk rock composition in the sole sample with Eo-Alpine monazite ages allowed no crystallization of staurolite on the expense of garnet at high temperatures, as staurolite is not observed in this sample. In consequence, pre-Alpine monazite is scarce in this sample, as no Y was available from pre-Alpine garnet breakdown. On the other hand, during an Eo-Alpine retrograde stage of metamorphism at low temperatures (*c.* 500–400 °C), garnet in this sample underwent breakdown to chlorite, which could have initialized the formation of the new Eo-Alpine monazite.

It can be concluded that the amphibolite-facies and the precedent higher-pressure metamorphic stages in the Schobergruppe were pre-Alpine events. The rocks underwent an overprint by an Eo-Alpine metamorphic event at lower grade. Regional extent and distribution of the Eo-Alpine overprint and the corresponding monazite ages are not well known yet. Apparently, it has been recorded only in samples with a suitable bulk composition and a significant retrogressive decomposition of garnet at low temperatures. Many monazites in the Schobergruppe have Permian ages, which have not yet been known from this part of the Austroalpine basement. Therefore, monazite of homogeneous composition and exclusively Permian ages from micaschists with fibrolithic sillimanite and post-S_{V2} andalusite out of a distinct zone with numerous Permian pegmatites in the southern Defereggen Alps (Fig. 1) has been drawn for a comparison (Fig. 8d). Pb–Pb dating of a micaschist by staurolite stepwise leaching (Fig. 10a) confirmed the Variscan and Permian ages in the pegmatite zone (Siegesmund et al. 2007). The monazites provide an argument for a distinct Permian thermal event, which can be distinguished from a precedent Variscan metamorphism (Fig. 8a, b). The Permian thermal event at presumably low pressures due to the late appearance of andalusite (Schuster et al. 2001) also affected domains outside the distinct sillimanite-bearing zone, but apparently was related to the regional distribution of pegmatites.

Late Alpine intrusive history and tectonics

The Late Alpine (Oligocene) post-collisional stage in the Alps is characterized by the emplacement of calc-alkaline tonalitic to granodioritic intrusions with minor participation of gabbroic, dioritic and granitic melts. Besides smaller intrusions, the most well-known are the intrusions of Biella and Traversella, Adamello, Bergell and the Rieserferner. Their alignment mostly to the north of the Periadriatic Lineament led Davies & von Blanckenburg (1995) to the establishment of the slab-breakoff hypothesis. According to this model, oceanic crust will break apart in a depth between 50 km and 120 km due to the mineralogical phase changes at eclogite facies conditions in the subducting slab. Then the upwelling hot asthenosphere would induce melting in the overriding lithospheric mantle. Assimilation of crustal material is indicated by Nd and Sr isotope systematics that show mixing trends between enriched mantle sources and crustal material (Bellieni et al. 1981; von Blanckenburg & Davies 1995). Other models explained the calc-alkaline character of the Periadriatic intrusions by proposing their relation to the latest stage of Eocene subduction (Sassi et al. 1980; Bellieni et al. 1981; Kagami et al. 1991), melting of the crustal mountain root as a consequence of convective thinning of the lithospheric root (Dewey 1988) or extension-related melting (Laubscher 1983). However, these models did not explain the

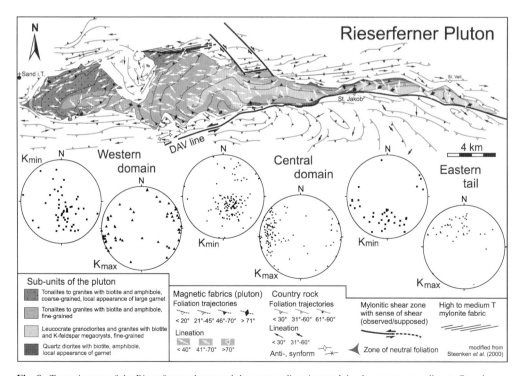

Fig. 9. Tectonic map of the Rieserferner pluton and the surrounding Austroalpine basement according to Steenken et al. (2000). Pole figures show the distribution of magnetic fabric axes (K_{min}–squares; K_{max}–triangles).

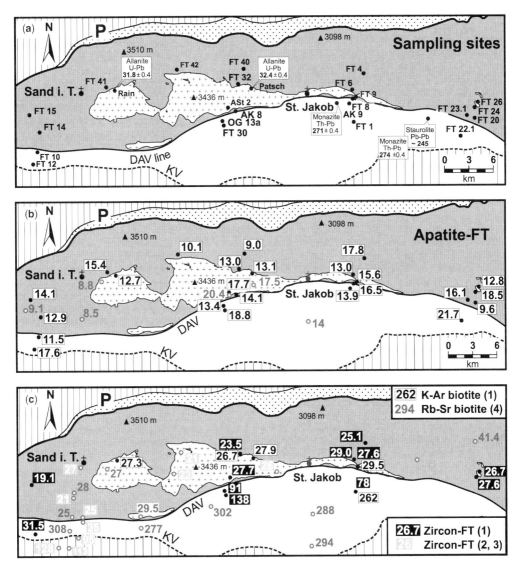

Fig. 10. Selected geochronological data from the Rieserferner pluton and the surrounding Austroalpine basement. (**a**) Apatite fission track sample localities (Table 1). Also shown are the U–Pb allanite ages of the pluton (Romer & Siegesmund 2003) as well as Pb–Pb staurolite and Th–Pb monazite ages from the southern block (Siegesmund *et al.* 2006; Schulz *et al.* 2005). (**b**) Apparent apatite fission track ages. Boxes show the data of Steenken *et al.* (2002). Shown in grey are representative data from Grundmann & Morteani (1985) and Coyle (1994). (**c**) Zircon fission track data and Rb–Sr and K–Ar biotite ages. Numbers in brackets label source of age data: (1) Steenken *et al.* (2002); (2) Stöckhert *et al.* (1999); (3) Angelmaier *et al.* (2000); (4) Borsi *et al.* (1978).

close proximity of the intrusions to the suture between the Penninic domain and the Apulian–Adriatic Plate.

The Rieserferner pluton with a length of c. 40 km and a maximum width of 6 km (Figs 1 and 9) is the largest Periadriatic intrusion to the south of the Tauern Window. The main body of the mainly tonalitic pluton is separated from the Defereggen-Antholz-Vals line by a few kilometres-wide slice of basement rocks that form the host of the small Zinsnock stock. Its eastern tail is in direct contact with the line. Further intrusions to the west are the Rensen and Altenberg plutons (Bellieni *et al.* 1984; Barth *et al.* 1989). The ambient conditions along the Rieserferner intrusion during its emplacement, inferred from contact metamorphic parageneses, indicate at least 600°–620° C in the immediate contact at pressures between 2.9 and 3.75 kbar (Cesare 1992, 1994), indicating an intrusion depth between 10.5 and 13.5 km. Time constraints on the crystallization of the Rieserferner pluton are based on an Rb–Sr whole rock errorchron (Borsi et al. 1973) that yielded an age of 31 ± 3 Ma after recalculation (Müller et al. 2000). This age was confirmed by near-concordant U–Pb allanite ages at 31.8 ± 0.4 Ma and 32.2 ± 0.4 Ma (Fig. 10a, Romer & Siegesmund 2003).

Pluton fabrics and their significance for the emplacement modus

Different emplacement models have been proposed for the Rieserferner pluton that take into account fabric elements in the pluton and its host as well as regional kinematics (Mager 1985b; Steenken et al. 2000; Rosenberg 2004; Wagner et al. 2006). The regional distribution of planar and linear fabrics in the pluton has been derived from analyses of the anisotropy of the magnetic susceptibility (AMS). Variations of the bulk susceptibility, geochemical characteristics and petrological field observations reflect a normal zoning of the pluton with partly preserved dioritic facies along the margin (Bellieni et al. 1981) and granitic facies in the centre (Steenken et al. 2000). Low bulk susceptibilities ($<380 \times 10^{-6}$ SI) go along with a low degree of eccentricity of the magnetic fabrics ($P' < 1.18$) and thereof magnetic fabrics have been directly related to the average orientation of its paramagnetic carriers (Hrouda 1982; Rochette et al. 1992). Additional measures, like the estimation of the biotite texture (Siegesmund et al. 1995) as well as magnetic fabric measurements at high-field conditions (Bergmüller et al. 1994; Bergmüller & Heller 1995; Siegesmund & Becker 2000; Hutton & Siegesmund 2001) demonstrate the validity of the relationship between magnetic and rock fabrics in case of the Rieserferner pluton. Combined results from AMS and direct measurements of planar fabrics in the field (Fig. 9) indicate that the Rieserferner pluton is constituted by two domes (Steenken et al. 2000; Wagner et al. 2006). These domes coincide with the two principal domains of the pluton that are separated by a remnant of the metapsammopelitic roof. To the east, limited fabric information still allows the interpretation of smaller domes that line up along the Defereggen-Antholz-Vals line. Only in the marginal areas of the central domain of the pluton, linear fabrics are penetrative due to an intense sub-solidus overprint at high to medium temperatures. In the large parts of the pluton where only minor sub-solidus deformation is observed, the information on the linear anisotropy is essentially based on the magmatic fabrics measured by the AMS. The linear magnetic fabric (Steenken et al. 2000) is generally subhorizontal or slightly to moderately plunging to the west. Lineations that plunge $>50°$ have been identified locally along the southern border close to the Defereggen-Anthols-Vals line and in the transition between the central and eastern parts of the pluton (Fig. 9).

A first model to explain the emplacement modus of the Rieserferner pluton invoked a lateral dislocation gradient along the Defereggen-Antholz-Vals line that enabled the ascent of the melts (Mager 1985b). In the most recent model of magma ascent and subsequent inflation (Wagner et al. 2006), the Defereggen-Antholz-Vals line acts as a feeder zone for the melts. Arguments for this feeder zone are provided by the steeply southward inclined planar fabrics as well as the local appearance of steep linear fabrics along the southern margin of the central part of the pluton and its eastern tail, mostly based on AMS analyses (Steenken et al. 2000). Those fabrics are considered to reflect the direction of magma ascent. Pluton-outward dipping foliations become moderately inclined toward the northern contact of the pluton and form at least two individual asymmetric domes that constitute the western and central domain (Fig. 9). The contact between the southern zone with steep fabrics and the northern part with shallow dipping fabrics supports the interpretation of two different fabric patterns: (i) belonging to the feeder zone; and (ii) to the main body of the intrusion. According to Wagner et al. (2006), inflation of the main body has followed the melt pressure enhanced fracturing of the country rock leading to an initial sub-horizontal sill-like intrusion in the σ_1–σ_2 plane. These authors further suggested that the feeder dyke along the Defereggen-Antholz-Vals line extends up to the top of the pluton. Consequently, the eastern tail of the pluton should represent the feeder of a sill-like part of the intrusion higher up in the crust that is now eroded. In a similar manner, the small Zinsnock stock could be interpreted as the feeder dyke for the western domain of the Rieserferner pluton.

However, this interpretation does not consider the contrasting geochemical characteristics of Zinsnock and Rieserferner (Bellieni 1980). The overall west to WNW plunging flow direction recorded by sub-solidus deformed granitoids is also unsatisfactorily explained by this model. Furthermore, the ongoing movement along the Defereggen-Antholz-Vals line during the intrusion is not considered. The emplacement model by Steenken et al. (2000) involves a west-directed expansion of the melt parallel to the σ_2 direction of the regional transpressional stress field. The locus of the magma ascent is indicated by steeply plunging lineations at the southeastern border of the central domain of the pluton in close contact to the Defereggen-Antholz-Vals line, where the latter shows a marked bending of its trend from c. east to west to a more WSW direction. This bending is responsible for the separation of the fault into two branches facilitating the formation of melt conduits in extension domains within a transpressive fault setting. Mylonitization of the country rock as well as of the granitoids is observed along the northern contact of the central domain (Kleinschrodt 1987) but cannot be traced further to the west (Wagner et al. 2006). The final upwelling of the intrusion is attributed to the buoyancy of the melts due to the marked contrast in density with the country rock. Structural evidence for this final ballooning stage is found in stretching and magnetic lineations in the immediate contact to the roof (Steenken 2002) that plunge parallel to the dip of the dome of the central domain of the pluton. Isoclinal folds in the roof of the pluton (Albertz et al. 1999; Rosenberg et al. 2000) and in the walls are structurally unrelated to the regional fold pattern but could be the result of the upwelling intrusion.

Cooling and exhumation of the basement

The cooling history of the Austroalpine basement and the incorporated Oligocene intrusions has been made accessible by Rb–Sr and K–Ar mica ages (Borsi et al. 1973, 1978; Hammerschmidt 1981; Prochaska 1981; Stöckhert 1984) as well as zircon and apatite fission track data (Grundmann & Morteani 1985; Coyle 1994; Stöckhert et al. 1999; Angelmaier et al. 2000). For a compilation of the entire data set, the reader is referred to Steenken et al. (2002). To the north of the Defereggen-Antholz-Vals line, Rb–Sr and K–Ar biotite data were fully reset to the Late Alpine ages when the area to the west of St. Jakob is considered (Fig. 10). In this western section between the Brenner fault and the Rieserferner pluton, a trend from younger ages at approximately 17 Ma around the Rensen intrusion towards c. 27 Ma in the western host rocks of the Rieserferner pluton is observed. This marked trend is smoothed along the strike of the Rieserferner pluton of which differential cooling history is manifested by a difference of only c. 3 Ma (Borsi et al. 1978; Steenken et al. 2002). The western domain of the pluton represents the deepest part of the exhumed intrusion; the vertical difference towards the eastern end depends on the geothermal gradient (e.g. 2 km at 28 °C/km). Towards lower temperatures, the cooling history can be retrieved by fission track data obtained from zircon and apatite (Fig. 10, Table 1). Apparent fission track ages are slightly younger than the K–Ar and Rb–Sr biotite ages, but document the same trend of differential cooling (Steenken et al. 2002). Along the Isel valley, zircon fission track samples have registered Late Alpine cooling, contrasting the Rb–Sr biotite data. This can be addressed to the interval of the zircon partial annealing zone at temperatures between 300 °C and 180 °C (Yamada et al. 1995) that is still below the generally accepted closure temperature of the Sr system in biotite at c. 300 °C (Jenkin et al. 1995). South of the Defereggen-Antholz-Vals line, the scarce zircon fission track ages between 34 and 128 Ma are difficult to interpret due to the lack of track length data and the corresponding annealing models. To the west, the distribution of zircon fission track ages appears to be controlled by the Kalkstein-Vallarga line (Borsi et al. 1973), which separates domains of Eocene cooling ages to the north from mostly Lower Cretaceous ages to the south (Stöckhert et al. 1999). Further to the east, this trend is not confirmed, as Cretaceous ages appear in close proximity to the Defereggen-Antholz-Vals line (Steenken et al. 2002).

The apatite fission track ages from 24 samples have been presented by Steenken et al. (2002). Thereby a group of 18 samples documents the final cooling history to the north of the Defereggen-Antholz-Vals line (Fig. 10b). These data are highly inhomogeneous. They fit the results of Coyle (1994) but frequently do not match the data from Grundmann & Morteani (1985) and Angelmaier et al. (2000). In part, this can be attributed to the altitude of the sampling locations (Steenken 2002). No clear correlation of apparent ages and altitude can be established. This may be explained by the wide spacing between individual sample localities and the existence of yet not recognized small-scale Late Alpine structures perturbing the temperature profile in the shallow crust. Additionally, a significant change of the morphology during the last 100 000 years due to tectonic and glacial effects (Stüwe et al. 1994; Mancktelow & Grasemann 1997) may have influenced the recorded information. A rather limited approach to a modelling by Steenken et al. (2002) is extended here to samples to the north of the Defereggen-Antholz-Vals line, where at least 30 confined horizontal tracks were measured. The modelling algorithm included by the AFTSolve software (Ketcham et al. 2000) was supplied by the track-length distribution and the counted spontaneous and induced tracks for each individual apatite grain. The procedure to find a best-fit T–t path for the track-length and individual grain age data is based on a Monte Carlo (MC) searching algorithm

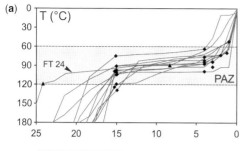

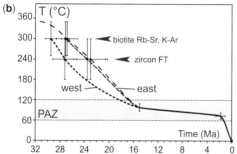

Fig. 11. (a) Summary of the modelled thermal histories obtained from track length data in samples with apatite (n = 12). Sample FT 24 does not fit the overall cooling trend. (b) Cooling path for the Rieserferner pluton and its surroundings starting with the closure of the Sr and Ar system in biotite (Steenken et al. 2002). The stippled line represents the western region of the pluton, whereas the dashed lines represent the cooling history of the more eastern parts of the pluton. PAZ partial annealing zone in apatite.

(Ketcham et al. 2000). The approach is handicapped by the lack of compositional data of the dated apatite crystals, especially the F/Cl that has an important effect on the fission track annealing behaviour. The chosen mono-kinetic annealing model (Laslett et al. 1987) is based on F-apatites. The apatite fission track ages obtained from the Rieserferner pluton and the surrounding basement in the north of the DAV fault range from 9.0 Ma to 18.5 Ma with an average of c. 12.9 Ma (Fig. 10b). They are slightly younger than those ages to the south of the fault that yield an average age of c. 14.3 Ma (Steenken et al. 2002). The track lengths of the entire sample set generally show uniform distribution with a mean track length between 11.7 μm and 13.7 μm, indicating some shortening while passing the partial annealing zone (PAZ). Only the sample FT 24 shows a bimodal track length distribution.

Simulations of the thermal history have been confined by the available apatite fission track ages from different parts of the pluton. Simulation runs of samples from the west were started at 19 Ma, those from the western and central part of the pluton at 24 Ma and from the east at 28 Ma. With the exception of the sample FT 24, the thermal history of all samples can be explained satisfactorily with an entrance into the PAZ at c. 15 Ma (Fig. 11a). Cooling rates, depending on the appropriate apatite fission track ages, are 10–15 °C/Ma. Subsequently, the samples remained in the central part of the PAZ where cooling is brought to a near standstill. Rapid final exhumation started during the last 4 Ma. The thermal history modelled from sample FT 24 from the Isel valley is contrasting this general scheme (Fig. 11a). This sample should have resided for a significantly longer time in the PAZ. Also, final cooling is less accelerated. However, this single observation should not be overrated. In general, apatite-FT ages to the north of the Defereggen-Antholz-Vals are not significantly younger in the western parts. In consequence, the west-upward rotation of the northern basement block as outlined in Borsi et al. (1978) and Kleinschrodt (1987) should be an older (pre-15 Ma) stage of the exhumation (Fig. 11).

Geodynamics in the Austroalpine to the south of the Tauern Window

Early-Palaeozoic active to passive margin evolution

The continental crust of the Austroalpine Adriatic Plate underwent a complex evolution. Sedimentation in the pre-Early Ordovician units (Northern-Defereggen-Petzeck Group, Durreck Muscovite-schist Group, Defereggen Group) started in the Precambrian and is dominated by monotonous clastic sequences of considerable thickness. Sedimentation ages and characteristics of the detrital provenance regions have not yet been constrained. The beginning of clastic sedimentation in the Thurntaler Phyllite Group is not precisely known (Rizzo et al. 1998); however, the occurrence of Ordovician volcanic rocks, now meta-porphyroids, provides an Early Palaeozoic time frame. Two major Early Palaeozoic magmatic evolution lines were identified within the Austroalpine basement south of the Tauern Window. An older evolution line comprises 590 Ma N-MORB-type basites, now eclogitic amphibolites, 550–530 Ma volcanic arc basalt-type magmatites, now hornblende-gneisses, and Ordovician 480–450 Ma acid plutonic and volcanic rocks, now orthogneisses and meta-porphyroids. The igneous geochemical signatures of these rock suites are more or less influenced by a subduction process and they can be assigned to an active margin setting. The younger evolution line is less well defined by radiometric data and composed of tholeiitic to transitional MORB- and alkaline within-plate basalt-type rock suites that follow an intraplate mantle metasomatism or within-plate enrichment trend. The geodynamic significance of these evolution lines and their magmatic suites will be discussed in the framework of the Austroalpine Gondwana-derived terranes. A sequence of tentative cross-sections characterizing a pre-Devonian evolution among Austroalpine microcontinental blocks is outlined in Fig. 12a–c. The labels of oceanic domains and terranes mainly follow Stampfli *et al.* (2002) and von Raumer *et al.* (2002, 2003).

The northern border of Gondwana was controlled by a Neoproterozoic to Cambrian active margin evolution, induced by subduction of a Proto-Tethys (Linnemann *et al.* 2000; von Raumer *et al.* 2006). Within the Austroalpine basement

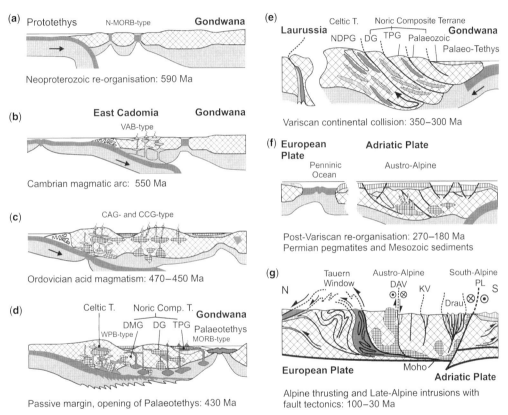

Fig. 12. Cross-sections model with stages a–g to illustrate the mainly documented stages of a plate tectonic magmatic and metamorphic evolution of the Austroalpine basement to the south of the Tauern Window. (**a–d**) Plate tectonic magmatic evolution of the Austroalpine northern margin of Gondwana during the Late Proterozoic and Early Palaeozoic (following von Raumer *et al.* 2002). The section is composed of an Eastern Cadomian Terrane with a Cambrian magmatic arc, an Ordovician terrane amalgamation assemblage and the Early Palaeozoic northern passive margin of the Palaeo-Tethys. DG, Defereggen Group; DMG, Durreck Muscoviteschist Group; NDPG, Northern-Defereggen-Petzeck Group; TPG, Thurntaler Phyllite Group. Note assignment of the lithological groups south of the Tauern Window to the Austroalpine Celtic and Noric Composite terranes of Frisch & Neubauer (1989). (**e**) Variscan continental collision with crustal thickening by tectonic stacking of lithotectonic units in the southern part of the Variscan orogen. High-pressure amphibolite-facies metamorphism occurs in structurally lower tectonic units as represented by the Northern-Defereggen-Petzeck Group. (**f**) The post-Variscan evolution is shown in two stages: to the right: ongoing subduction of Palaeotethys; uplift and extension with intrusion of Permian pegmatites in distinct basement units; sedimentation of Permian and Mesozoic cover rocks previous to opening of Meliata oceanic domain. To the left: Jurassic opening of Penninic ocean and separation of the Adriatic Plate. (**g**) Eo-Alpine thrusting of the Austroalpine basement upon the Penninic Units. Late stage of Alpine continental collision with further accretion of Penninic units at the base of the Austroalpine nappes. Subsequent Late Alpine intrusions (Rieserferner) are associated with shear zone activity along the Defereggen-Antholz-Vals line (DAV). The non-metamorphic Permo-Mesozoic sedimentary cover is preserved along the DAV, further wrench faults and in the Drauzug.

(Fig. 12a), the 590 ± 4 Ma N-MORB-type eclogitic amphibolites from the Prijakt Subgroup are a first and single observation of pre-Variscan ophiolites, as other occurrences (e.g. Plankogel and Speik complexes) have not yet been dated (Frisch & Neubauer 1989; Thöni 1999; Neubauer 2002a). According to their subduction-related trace element (Schulz 1995) and source signatures, the eclogitic amphibolites appear as part of an early or juvenile stage of an active margin evolution, presumably in a back-arc environment.

Much more prominent are the magmatites of an Early Cambrian active margin stage (Fig. 12b). Magmatic arc rocks belonging to this 'Cadomian' time span turn out to be more and more widespread in the peri-Gondwanan units (Linnemann et al. 2000; Tichomirowa et al. 2001), and have yet been reported from the Austroalpine basement in the Silvretta (Müller et al. 1995, 1996; Schaltegger et al. 1997). In the Ötztal basement, meta-gabbros have ages of 530–521 Ma (Miller & Thöni 1995). The hornblende-gneisses within the Rotenkogel and Prijakt Subgroups show geochemical signatures of former gabbros and diorites that were produced by magmatic processes in mantle wedge and arc crust above a subducted lithospheric slab. Protolith ages range between 550 and 530 Ma. This rock suite provides the main argument and evidence of an active continental margin setting during this period. It supports with precision the interpretation of Frisch et al. (1984, 1987) and Frisch & Neubauer (1989) who considered the Austroalpine gneiss-amphibolite association as an Early Palaeozoic active margin sequence within a Celtic terrane.

Indications supporting a subsequent Early Ordovician formation of oceanic crust through rifting (Schmidt & Söllner 1982) are lacking in the studied part of the Austroalpine, but arise from observations in basement units belonging to the Cadomian and Avalonian terranes to the west (Bernhard et al. 1996; Stampfli et al. 2002; von Raumer et al. 2002, 2003, 2006). Furthermore, direct observations on Early Ordovician collision and metamorphism (Peccerillo et al. 1979; Sassi et al. 1985; Söllner & Hansen 1987; Klötzli-Chowanetz et al. 1997) yet have not been reported from the region to the south of the Tauern Window. Therefore, these stages of the Early Palaeozoic evolution still remain hypothetical (Schulz et al. 2004). However, the distribution of magmatic suites among the lithological groups may provide a hint: Early Cambrian arc magmatites are restricted to the Northern-Defereggen-Petzeck Group, whereas the subsequent Ordovician acid magmatites are widespread in all the groups. This could be the consequence of an Early Ordovician juxtaposition of different basement blocks, not necessarily through continental collision, but alternatively by a strike-slip mode of amalgamation.

The interpretation of the Austroalpine Ordovician magmatism is still problematic. Protolith ages cover a wide time span from 420–494 Ma (von Raumer et al. 2002; Neubauer 2002a and references therein), enclosing data from Wildschönau metagabbro (477 ± 9 Ma, Loth et al. 2001) and a 484 ± 6 Ma felsic intrusion in the Oetztal (Bernhard et al. 1996). The former existence of a Cambrian magmatic arc has not been yet considered in the intense discussion, which circled around models of an extensional or initial rifting setting for a 'Caledonian' magmatic event (Schmidt & Söllner 1982; Bernhard et al. 1996). Ordovician magmatites involve suites of continental arc granitic magma, as crystallized in the Gsies/Casies-type tonalitic/granodioritic granitoids, to entirely anatectic continental collision granitic magmas, as crystallized in the Antholz/Anterselva-type granitoids. Apart from continental collision zones, both granitoid types can be expected in mature magmatic arcs, sitting on continental crust. In this light, the Ordovician acid magmatism (Fig. 12c) can be also interpreted as a consequence and culmination of a longer-lasting active margin setting since the Neoproterozoic and through Cambrian times. In such a scenario, the main heat source of crustal anatexis is the increasing amount of ascending melts, which enter a mature arc above a subducted slab (Wilson 1989). Stampfli (2000) suggested that the subduction of a mid-ocean ridge along the prevailing active margin could have provided additional heat to the root and the hinterland of the magmatic arc. The consequence was to increase magma ascent, which led to the volumetrically most important magmatic activity within the Austroalpine basement.

On one hand, the active margin setting could explain the Ordovician acid magmatism. On the other hand, there exists a widespread alkaline magmatism in the Austroalpine pre-Permian units, enclosing the Thurntaler Phyllite Group. In the Greywacke Zone and other less metamorphic Palaeozoic sequences, which allow biochronology, this alkaline magmatism has been found to cover a considerable time span from the Upper Ordovician to the Silurian/Devonian (Heinisch et al. 1987; Loeschke 1989; Loeschke & Heinisch 1993). The 430 ± 2 Ma zircon age demonstrates that within-plate basalt-type rocks in the Northern-Defereggen-Petzeck Group belong to this prominent Early Palaeozoic magmatism. Taking into account the trace element and isotopic data, the metabasites of the Thurntaler Phyllite Group should have similar protolith ages (Schulz et al. 2004). There is a general consensus that the Early Palaeozoic alkaline mafic magmatism accompanied the fragmentation of a northern passive margin of Gondwana by the development of intracontinental rifts and was associated with the opening of the Palaeotethys (Fig. 12d). The trace element and isotopic data from the within-plate basalt-type suites provide arguments for a progressive mantle metasomatism and a mixing of crustal material to sources from enriched and depleted mantle reservoirs. A large-scale process leading to such multicomponent sources could be expected above a deeply buried and partly assimilated oceanic slab in the hinterland of a subduction zone (Steenken & Siegesmund 2000). In this sense, the passive margin Palaeotethys opening event (Fig. 12d) can be seen as the final step of a precedent long-lasting Neoproterozoic to Ordovician active margin evolution.

Variscan-to-Permian metamorphic events

It has been outlined in palaeogeographic models (von Raumer & Neubauer 1993; Stampfli et al. 2002; von Raumer et al. 2006) that the Gondwana-derived terranes and then the Gondwana continent drifted to the north and successively collided with Laurussia during Devonian and Carboniferous times, forming the Variscan orogen. In the Austroalpine basement, the Variscan collisional event is documented by the post-Ordovician foliation S_{V2}, the associated structures of shearing and folding, partly under progressive simple shear, the refolding of these early structures, and an amphibolite-facies metamorphism. Detailed P–T path reconstruction from the various Austroalpine lithological units signalizes 'polymetamorphism' with distinct events—prograde stage, high-pressure stage, high-temperature amphibolite-facies stage at medium pressures, amphibolite-facies stage at low pressures—along complex P–T paths (Schulz 1990, 1993a, b, 1997; Schulz et al. 2001, 2005). The different shapes of syndeformational P–T paths and the variable P–T conditions in the basement units can be interpreted in models of successive stacking of crustal-scale nappes, as has been proposed from numerical thermal modelling (England & Thompson 1984; Davy & Gillet 1986). When compared to such models, the marked pressure variations in P–T paths from the Defereggen Group and Northern-Defereggen-Petzeck Group provide arguments for an overstacking of crustal units (Fig. 12e). Accordingly, during an early stage of a Variscan continental collision, the northern units of the pre-Early Ordovician crust were stacked and reworked by high-pressure metamorphism and coeval shearing. Later, in the course of the initial uplift of these rocks, the crustal stacking process with coeval shearing continuously progressed towards the south and caused a more temperature-dominated metamorphism of the Early Palaeozoic passive continental margin represented by the Thurntaler Phyllite Group. The subsequent common Late Variscan uplift of the welded units probably evolved in a dextral transpression system of NW-SE directed compression, leading to the Schlingen structures F_{V3} in the Defereggen Group and to later shearbands S_{V4} and S_{V5}.

There is increasing evidence for a distinct post-collisional Permian–Triassic thermal event in the Southalpine and Austroalpine units (Schuster et al. 2001 and references therein). To the south of the Tauern Window, the magmatic part of the Permian event is documented by the pegmatites. In a zone with numerous pegmatites, postdeformative crystallization of andalusite and fibrolitic sillimanite, as well as replacement of biotite and garnet by fibrolitic sillimanite, document a low-pressure stage of amphibolite-facies metamorphism. It can be discussed whether this low-pressure stage represents a segment of uplift and cooling along the Variscan P–T evolution, or a separate event of new heating, interrupting the Late Variscan cooling. Judging from the monazite data from metapelites, the latter interpretation can be preferred, as samples in the vicinity of the pegmatites display monazites with homogeneous Permian ages. The consequences of the Permian thermal event are not yet fully explored and understood. However, the elevated geothermal gradient signalizes an extensional tectonic setting with thinning of the continental crust (Fig. 12f). Such an extensional setting could have been possible by a slab roll back during the north-directed subduction of the Palaeotethys oceanic crust, as has been proposed by Stampfli (1996) and Stampfli & Mosar (1999).

Alpine deformation, metamorphism and uplift

In the Mesozoic, the Austroalpine basement and future Adriatic Plate has been fragmented by formation of the Meliata oceanic basin and separated from the northern European Plate (Fig. 12f) by the opening of the Penninic ocean (Neubauer 2002b). Alpine continental convergence started during the Cretaceous. Several stages of metamorphism and deformation can be distinguished in the course of the Alpine collision of the Austroalpine Adriatic Plate with the northern European Plate (Fig. 12g). One indication of the Eo-Alpine (Cretaceous) stage is the Alpine foliation of the Permian pegmatites (Stöckhert 1984). However, this first Alpine deformation appears inhomogeneous, as some pegmatites remained unfoliated between the domains of Eo-Alpine shearing. A coeval Alpine metamorphism in the Northern-Defereggen-Petzeck Group reached conditions of $< 500\,°C$ and caused a partial to complete resetting of the Variscan mica cooling ages. This

Eo-Alpine event is more pervasive in the eastern parts of the Austroalpine basement, as in the Saualpe and Koralpe (Thöni 1999; Hoinkes et al. 1999). The complex deep crustal structure of the Adriatic Plate with indication of a double Moho could be related to this Cretaceous (Eo-Alpine) event. The emplacement of the Austroalpine basement nappe upon the Penninic Unit occurred subsequent to the Eo-Alpine event, but before 30 Ma. The Late Alpine stage to the south of the Tauern Window is characterized by a regional transpressional tectonic regime (Kleinschrodt 1987; Steenken et al. 2000) and the emplacement of the Rieserferner intrusion. The c. 32 Ma Rieserferner pluton cuts across the already foliated pegmatites. Sinistral wrenching along the Defereggen-Antholz-Vals line occurred previous to, during and subsequent to the intrusion. A differential uplift of the northern block along the shear zone is documented by mica and zircon fission track ages and should be older as c. 15 Ma. According to apatite fission track data, the further uplift on both blocks along the shear zone occurred along cooling rates between $10-15\ °C/Ma$ and was accomplished before 15 Ma when the rocks entered the partial annealing zone. Rapid final exhumation commenced about 4 Ma ago.

The authors enjoyed the great hospitality of S. Michelitsch (Erlsbach) and other families in Eastern Tyrol during many years of long-term fieldwork and sampling. We acknowledge the contributions and cooperation of numerous colleagues in the projects on the Austroalpine basement, namely C. Triboulet (Paris), C. Audren (†2002; Rennes), K. Bombach (Freiberg), S. Pawlig (Goettingen), F. Finger, E. Krenn (Salzburg), B. Cesare (Padova), M. Raab (Melbourne) and B. Fügenschuh (Innsbruck). S. S. especially thanks T. Heinrichs (Göttingen) for the excellent and very fruitful cooperation during long-term joint fieldwork. J. F. von Raumer (Fribourg) and R. Kleinschrodt (Cologne) helped a lot with their constructive reviews. The German Science Foundation funded our studies in the Austroalpine basement for a long period and in numerous single projects. Most relevant projects for this publication were SCHU 676/1; 676/3; 676/8; 676/9. We received further support from Dr. W. Schnabel of the Austrian Geologische Bundesanstalt and from our hosting institutions, Institut für Geologie und Mineralogie at Erlangen, Göttinger Zentrum für Geowissenschaften and the Institut für Mineralogie at Freiberg. The contributions of around 50 students from Göttingen, Erlangen and Freiberg during several mapping projects is gratefully acknowledged.

References

ALBERTZ, M., ROSENBERG, C., MÖBUS, C. & HANDY, M. 1999. *The Roof of the Rieserferner Pluton: Implications for Syntectonic Magma Ascent and Emplacement along Crustal Scale Shear Zones.* The Origin of Granites and related rocks, IVth Hutton Symposium, **73**.

ANGELMAIER, P., DUNKL, I. & FRISCH, W. 2000. Altersprofile aus dem Zentralabschnitt der Transalp-Traverse (Ostalpen): neue K/Ar-, Zirkon- und Apatit-spaltspurendatierungen. *Terra Nostra, Schriften der Alfred-Wegener-Stiftung*, **5**, 4.

BARTH, S., OBERLI, F. & MEIER, M. 1989. U–Th–Pb systematics of morphologically characterized zircon and allanite; a high-resolution isotopic study of the Alpine Rensen Pluton (northern Italy). *Earth and Planetary Science Letters*, **95**, 235–254.

BAUER, W. & BAUER, P. 1993. Zur Geologie des metamorphen Kristallins zwischen Amlach und Lavant (Osttirol, Österreich). *Jahrbuch der Geologischen Bundesanstalt Wien*, **136**, 299–306, Wien.

BEHRMANN, J. H. 1990. Zur Kinematik der Kontinentkollision in den Ostalpen. *Geotektonische Forschungen*, **76**, 1–180.

BELLIENI, G. 1980. The Cima di Villa (Zinsnock) massif: geochemical features and comparisons with the Vedrette di Ries (Rieserferner) pluton (Eastern Alps–Italy). *Neues Jahrbuch für Mineralogie Abhandlungen*, **138**, 244–258.

BELLIENI, G., PECCERILLO, A. & POLI, G. 1981. The Vedrette di Ries (Rieserferner) Plutonic Complex: petrological and geochemical data bearing on its genesis. *Contributions to Mineralogy and Petrology*, **78**, 145–156.

BELLIENI, G., PECCERILLO, A., POLI, G. & FIORETTI, A. 1984. The genesis of late Alpine plutonic bodies of Rensen and Monte Alto (Eastern Alps); inferences from major and trace element data. *Neues Jahrbuch für Mineralogie Abhandlungen*, **149**, 209–224.

BERGMÜLLER, F. & HELLER, F. 1995. The field dependence of magnetic anisotropy parameters derived from high-field torque measurements. *Physics of the Earth and Planetary Interiors*, **96**, 61–76.

BERGMÜLLER, F., BÄRLOCHER, C., GRIEDER, M., HELLER, F. & ZWEIFEL, P. 1994. A torque magnetometer for measurement of high field anisotropy of rocks and crystals. *Measurement Science and Technology*, **5**, 1466–1470.

BATTACHARYA, A., MOHANTY, L., MAJI, A., SEN, S. K. & RAITH, M. 1992. Non-ideal mixing in the phlogopite-annite binary: constraints from experimental data on Fe-Mg partitioning and a reformulation of the garnet-biotite geothermometer. *Contributions to Mineralogy and Petrology*, **111**, 87–93.

BLANCKENBURG VON, F. & DAVIES, J. H. 1995. Slab breakoff: a model for syncollisional magmatism and tectonics in the Alps. *Tectonics*, **14**, 120–131.

BERNHARD, F., KLÖTZLI, U. S., THÖNI, M. & HOINKES, G. 1996. Age, origin and geodynamic significance of a polymetamorphic felsic intrusion in the Ötztal Crystalline Basement, Tirol, Austria. *Mineralogy and Petrology*, **58**, 171–196.

BIANCHI, A., DAL PIAZ, G. & MERLA, G. 1930. *Carta geologica delle Tre Venezie alla Scala di 1:100 000, Foglio Monguelfo 4b.* Ufficio Idrografico del R. Magistrato alle Aque; Venezia.

BORSI, S., DEL MORO, A., SASSI, F. P. & ZIRPOLI, G. 1973. Metamorphic evolution of the Austridic rocks to the south of the Tauern Window (Eastern Alps): radiometric and geopetrologic data. *Memorie Societa Geologica Italiana*, **12**, 549–571.

BORSI, S., DEL MORO, A., SASSI, F. P., ZANFERRARI, A. & ZIRPOLI, G. 1978. New geopetrologic and radiometric data on the Alpine history of the Austridic continental margin south of the Tauern Window. *Memorie dell Istituto Geologica e Mineralogica Universita Padova*, **32**, 1–17.

BORSI, S., DEL MORO, A., SASSI, F. P. & ZIRPOLI, G. 1979. On the age of the Vedrette de Ries (Rieserferner) massif and its geodynamic significance. *International Journal of Earth Sciences (Geologische Rundschau)*, **68**, 41–60.

BORSI, S., DEL MORO, A., SASSI, F. P., VISONA, D. & ZIRPOLI, G. 1980a. On the existence of Hercynian aplites and pegmatites in the lower Aurina Valley (Ahrntal, Austrides, Eastern Alps). *Neues Jahrbuch Mineralogie Monatshefte*, **1980**, 501–514.

BORSI, S., DEL MORO, A., SASSI, F. P. & ZIRPOLI, G. (1980b). New petrographic and radiometric data on the Ötztal and Stubai orthogneisses (Eastern Alps). *Neues Jahrbuch Mineralogie Monatshefte*, **1980**(2), 75–87.

BOUCHEZ, J.-L. 1977. Plastic deformation of quartzites at low temperature in an area of natural strain gradient. *Tectonophysics*, **39**, 25–50.

BRACK, W. 1977. *Geochronologische Untersuchungen an Gesteinen des Altkristallins in der Schobergruppe, Österreich*. PhD thesis, Ludwig-Maximilians-Universität München.

BÜCKSTEEG, A. 1999. Zur Geologie des Kristallins der Schobergruppe (Osttirol/Österreich). *Aachener Geowissenschaftliche Beiträge*, **33**, 1–206.

CESARE, B. 1992. Metamorfismo di contatto di rocce pelitiche nell' aureola di Vedrette di Ries (Alpi Orientali–Italia). *Atti Ticinensi di Science della Terra*, **35**, 1–7.

CESARE, B. 1994. Hercynite as the product of staurolite decomposition in the contact aureole of Vedrette di Ries, Eastern Alps, Italy. *Contributions to Mineralogy and Petrology*, **116**, 239–246.

CIPRIANI, C., SASSI, F. P. & SCOLARI, A. 1971. Metamorphic white micas: definition of paragenetic fields. *Swiss Bulletin of Mineralogy and Petrology*, **51**, 259–302.

CLIFF, R. A. 1980. U–Pb isotopic evidence from zircons for lower Palaeozoic tectonic activity in the Austroalpine nappe of the Eastern Alps. *Contributions to Mineralogy and Petrology*, **71**, 283–288.

COBBOLD, P. R. & QUINQUIS, H. 1980. Development of sheath folds in shear regimes. *Journal of Structural Geology*, **2**, 119–126.

COCHERIE, A., LEGENDRE, O., PEUCAT, J. J. & KOUMELAN, A. N. 1998. Geochronology of polygenetic monazites constrained by in-situ microprobe Th-U-total lead determination: implications for lead behaviour in monazite. *Geochimica et Cosmochimica Acta*, **62**, 2475–2497.

CONNELLY, J. N. 2000. Degree of preservation of igneous zonation in zircon as a signpost for concordancy in U/Pb geochronolgy. *Chemical Geology*, **172**, 25–39.

CORFU, F., HANCHAR, J. M., HOSKIN, P. W. O. & KINNY, P. 2003. Atlas of zircon textures. *In*: HANCHAR, J. M. & HOSKIN, P. W. O. (eds) *Zircon*. Reviews in Mineralogy and Geochemistry, **53**, 469–500.

COYLE, D. A. 1994. *The Application of Apatite Fission Track Analysis to Problems in Tectonics*. PhD thesis, La Trobe University Bundoora, Victoria, Australia.

DAHL, P. S., TERRY, M. P., JERCINOVIC, M. J. ET AL. 2005. Electron probe (Ultrachron) micronometry of metamorphic monazite: unraveling the timing of polyphase thermotectonism in the easternmost Wyoming Craton (Black Hills, South Dakota). *American Mineralogist*, **90**, 1712–1728.

DAL PIAZ, G. 1934. Studi geologica sull' Alto Adige e regione limitrofe. *Memorie dell Istituto Geologica e Mineralogica Universita Padova*, **10**, 1–238.

DAVIES, J. H. & VON BLANCKENBURG, F. 1995. Slab breakoff: a model of lithosphere detachment and its test in the magmatism and deformation of collisional orogens. *Earth and Planetary Science Letters*, **129**, 85–102.

DAVY, P. & GILLET, P. 1986. The stacking of thrust slices in collision zones and its thermal consequences. *Tectonics*, **5**, 913–929.

DEWEY, J. F. 1988. Extensional collapse of orogens. *Tectonics*, **7**, 1123–1139.

ELLIS, D. J. & GREEN, D. H. 1979. An experimental study of the effect of Ca upon garnet-clinopyroxene Fe–Mg exchange equilibria. *Contributions to Mineralogy and Petrology*, **71**, 13–22.

ENGLAND, P. C. & THOMPSON, A. B. 1984. Pressure-temperature-time paths of regional metamorphism I. Heat transfer during the evolution of regions of thickened continental crust. *Journal of Petrology*, **25**, 894–928.

EXNER, C. 1962. Die Perm-Trias-Mulde des Gödnachgrabens an der Störungslinie von Zwischenbergen (Kreuzeckgruppe, östlich Lienz). *Verhandlungen der Geologischen Bundesanstalt*, **1962**, 24–27.

FAURE, G. 2001. *Origin of Igneous Rocks. The Isotopic Evidence*. Springer, Berlin, Heidelberg, New York.

FINGER, F. & HELMY, H. M. 1998. Composition and total-Pb model ages of monazite from high-grade paragneisses in the Abu Swayel area, southern Eastern Desert, Egypt. *Mineralogy and Petrology*, **62**, 269–289.

FITZSIMMONS, J. C. W., KINNY, P. D., WETHERLEY, S. & HOLLINGSWORTH, D. A. 2005. Bulk chemical control on metamorphic monazite growth in pelitic schists and implications for U–Pb age data. *Journal of Metamorphic Geology*, **23**, 261–277.

FLOYD, P. A., WINCHESTER, R., SESTON, R., KRYZA, R. & CROWLEY, Q. G. 2000. Review of geochemical variation in Lower Palaeozoic metabasites from the NE Bohemian Massif: intracratonic rifting and plume-ridge interaction. *In*: FRANKE, W., HAAK, V. & ONCKEN, O. (eds) *Orogenic Processes: Quantification and Modelling in the Variscan Belt*. Geological Society, London, Special Publications, **179**, 155–174.

FRISCH, W. & NEUBAUER, F. 1989. Pre-Alpine terranes and tectonic zoning in the eastern Alps. *Geological Society of America Special Paper*, **230**, 91–99.

FRISCH, W., NEUBAUER, F. & SATIR, M. 1984. Concepts of the evolution of the Austroalpine basement complex (Eastern Alps) during the Caledonian-Variscan cycle. *International Journal of Earth Sciences (Geologische Rundschau)*, **73**, 47–68.

FRISCH, W., NEUBAUER, F., BRÖCKER, M., BRÜCKMANN, W. & HAISS, N. 1987. Interpretation of geochemical data from the Caledonian basement within the Austroalpine basement complex. In: FLÜGEL, H. W., SASSI, F. P. & GRECULA, P. (eds) *Pre-Variscan and Variscan Events in the Alpine-Mediterranean Mountain Belts.* Alfa, Bratislava, 209–226.

FÜGENSCHUH, B. 1995. *Thermal and kinematic history of the Brenner area (Eastern Alps, Tyrol).* PhD thesis, ETH Zürich.

GANGULY, J. & SAXENA, S. K. 1984. Mixing properties of aluminosilicate garnets: constraints from natural and experimental data, and applications to geothermobarometry. *American Mineralogist,* **69,** 88–97.

GANGULY, J., CHENG, W. & TIRONE, M. 1996. Thermodynamics of aluminosilicate garnet solid solution: new experimental data, an optimized model, and thermometric applications. *Contributions to Mineralogy and Petrology,* **123,** 137–151.

GERYA, T. V., PERCHUK, L. L., TRIBOULET, C., AUDREN, C. & SEŹKO, A. I. 1997. Petrology of the Tumanshet Zonal Metamorphic Complex, Eastern Sayan. *Petrology (Moscow),* **5,** 503–533.

GODIZART, G. 1989. *Gefüge, retrograde Metamorphose und Geochemie der Amphibolite im ostalpinen Altkristallin südlich des westlichen Tauernfensters (Südtirol, Italien).* PhD thesis, University of Erlangen-Nürnberg.

GUHL, M. & TROLL, G. 1977. Mehrphasige Faltengefüge in Altkristallin und Permotrias von Kalkstein in Osttirol, Österreich. *Verhandlungen der Geologischen Bundesanstalt,* **1977,** 45–52.

GUHL, M. & TROLL, G. 1987. Die Permotrias von Kalkstein im Altkristallin der südlichen Deferegger Alpen (Österreich). *Jahrbuch der Geologischen Bundesanstalt Wien,* **130,** 37–60.

GRUNDMANN, G. & MORTEANI, G. 1985. The young uplift and thermal history of the central Eastern Alps (Austria/Italy), evidence from apatite fission track ages. *Jahrbuch der Geologischen Bundesanstalt Wien,* **128,** 197–216.

HAMMERSCHMIDT, K. 1981. Isotopengeologische Untersuchungen am Augengneis vom Typ Campo Tures bei Rain in Taufers, Südtirol. *Memorie dell Istituto Geologica e Mineralogica Universita Padova,* **34,** 273–300.

HANCHAR, J. M. & MILLER, C. F. 1993. Zircon zonation patterns as revealed by cathodoluminescence and backscattered electron images: implications for interpretation of complex crustal histories. *Chemical Geology,* **110,** 1–13.

HEINISCH, H. 1981. Zum ordovizischen 'Porphyroid'-Vulkanismus der Ost- und Südalpen. *Jahrbuch der Geologischen Bundesanstalt Wien,* **124,** 1–109.

HEINISCH, H. & SCHMIDT, K. 1976. Zur kaledonischen Orogenese in den Ostalpen. *International Journal of Earth Sciences (Geologische Rundschau),* **65,** 459–482.

HEINISCH, H. & SCHMIDT, K. 1982. Zur Genese der Augengneise im Altkristallin der Ostalpen. *Neues Jahrbuch für Geologie und Paläontologie Monatshefte,* **1982,** 211–239.

HEINISCH, H. & SCHMIDT, K. 1984. Zur Geologie des Thurntaler Quarzphyllits und des Altkristallins südlich des Tauernfensters (Ostalpen, Südtirol). *International Journal of Earth Sciences (Geologische Rundschau),* **73,** 113–129.

HEINISCH, H., SPRENGER, W. & WEDDIGE, K. 1987. Neue Daten zur Altersstellung der Wildschönauer Schiefer und des Basaltvulkanismus im ostalpinen Paläozoikum der Kitzbüheler Grauwackenzone (Österreich). *Jahrbuch der Geologischen Bundesanstalt Wien,* **130,** 163–173.

HOEFS, J. 1997. *Stable Isotope Geochemistry.* Springer, Berlin, Heidelberg, New York, 201.

HOFMANN, K. H., KLEINSCHRODT, R., LIPPERT, R., MAGER, D. & STÖCKHERT, B. 1983. Geologische Karte des Altkristallins südlich des Tauernfensters zwischen Pfunderer Tal und Tauferer Tal (Südtirol). *Der Schlern,* **57,** 572–590.

HOINKES, G., KOSTNER, A. & THÖNI, M. 1991. Petrologic constraints for Eoalpine eclogite facies metamorphism in the Austroalpine Ötztal basement. *Mineralogy and Petrology,* **43,** 237–254.

HOINKES, G., KOLLER, F., RANTITSCH, G., DACHS, E., HÖCK, V., NEUBAUER, F. & SCHUSTER, R. 1999. Alpine metamorphism of the Eastern Alps. *Swiss Bulletin of Mineralogy and Petrology,* **79,** 155–181.

HOKE, L. 1990. The Altkristallin of the Kreuzeck Mountains, SE Tauern Window, Eastern Alps—basement crust in a convergent plate boundary zone. *Jahrbuch der Geologischen Bundesanstalt Wien,* **133,** 5–87.

HOLDAWAY, M. J. 2001. Recalibration of the GASP geobarometer in light of recent garnet and plagioclase activity models and versions of the garnet-biotite geothermometer. *American Mineralogist,* **86,** 1117–1129.

HOLLAND, T. J. B. 1980. The reaction albite = jadeite + quartz determined experimentally in the range 600–1200 °C. *American Mineralogist,* **65,** 129–134.

HOLLAND, T. J. B. 1983. The experimental determination of activities in disordered and short-range ordered jadeitic pyroxene. *Contributions to Mineralogy and Petrology,* **82,** 214–220.

HOLLAND, T. J. B. & POWELL, R. 1990. An enlarged and updated internally consistent thermodynamic dataset with uncertainties and correlations: the system $K_2O-Na_2O-CaO-MgO-MnO-FeO-Fe_2O_3-Al_2O_3-TiO_2-SiO_2-C-H_2-O_2$. *Journal of Metamorphic Geology,* **8,** 89–124.

HOSKIN, P. W. O. & SCHALTEGGER, U. 2003. The composition of zircon and igneous and metamorphic petrogenesis. In: HANCHAR, J. M. & HOSKIN, P. W. O. (eds) *Zircon.* Reviews in Mineralogy and Geochemistry, **53,** 27–62.

HUTTON, D. H. W. & SIEGESMUND, S. 2001. The Ardara Granite: reinflating the Balloon Hypothesis. *Zeitschrift der Deutschen Geologischen Gesellschaft,* **152,** 309–323.

HROUDA, F. 1982. Magnetic anisotropy of rocks and its application to geology and geophysics. *Geophysical Surveys,* **5,** 37–82.

JENKIN, G. R. T., ROGERS, G., FALLICK, A. E. & FARROW, C. M. 1995. Rb–Sr closure temperatures in bi-mineralic rocks: a mode effect and test for different diffusion models. *Chemical Geology,* **122,** 227–240.

KAGAMI, H., ULMER, P., HANSMANN, W., DIETRICH, V. & STEIGER, R. H. 1991. Nd–Sr isotopic and geochemical characteristics of the southern Adamello (northern Italy) intrusives: implications for crustal versus mantle origin. *Journal of Geophysical Research*, **96**, 14331–14346.

KETCHAM, R. A., DONELICK, R. A. & DONELICK, M. B. 2000. AFTSolve: a program for multi-kinetic modeling of apatite fission-track data. *Geological Materials Research*, **2**(1), 1–32.

KLEINSCHRODT, R. 1987. Quarzkorngefügeanalyse im Altkristallin südlich des westlichen Tauernfensters. *Erlanger geologische Abhandlungen*, **114**, 1–82.

KLÖTZLI, U. S. 1995. Geochronologische Untersuchungen an Metagranitoiden im ostalpinen Altkristallin W und S des Tauernfensters. Kurzfassungen Arbeitstagung Lienz Österreichische Geologische Bundesanstalt, 95–97.

KLÖTZLI, U. S. 1997. Zircon evaporation TIMS: method and procedures. *Analyst*, **122**, 1239–1248.

KLÖTZLI, U. S. 1999a. Resolving complex geological histories by zircon dating: a discussion of case studies from the Bohemian Massif and the Eastern Alps. *Mitteilungen der Österreichischen Geologischen Gesellschaft*, **90**, 31–41.

KLÖTZLI, U. S. 1999b. Th/U zonation in zircon derived from evaporation analysis: a model and its implications. *Chemical Geology*, **158**, 325–333.

KLÖTZLI-CHOWANETZ, E., KLÖTZLI, U. S. & KOLLER, F. 1997. Lower Ordovician migmatisation in the Ötztal crystalline basement (Eastern Alps, Austria): linking U–Pb and Pb–Pb dating with zircon morphology. *Swiss Bulletin of Mineralogy and Petrology*, **77**, 315–324.

KOBER, B. 1986. Whole-grain evaporation for $^{207}Pb/^{206}Pb$-age-investigations on single zircons using a double-filament thermal ion source. *Contributions to Mineralogy and Petrology*, **93**, 482–490.

KOBER, B. 1987. Single-zircon evaporation combined with Pb^+ emitter bedding for $^{207}Pb/^{206}Pb$-age investigations using thermal ion mass spectrometry, and implications to zirconology. *Contributions to Mineralogy and Petrology*, **96**, 63–71.

KREUTZER, S. 1992. *Zur Geologie des östlichen Thurntaler Quarzphyllitkomplexes und seiner tektonischen Einbindung in das Ostalpin der südöstlichen Deferegger Alpen, Osttirol*. PhD thesis, Rheinisch-Westfälische Technische Hochschule Aachen.

KROGH, E. J. 1988. The garnet-clinopyroxene Fe–Mg geothermometer—a reinterpretation of existing experimental data. *Contributions to Mineralogy and Petrology*, **99**, 44–48.

LASLETT, G. M., GREEN, P. F., DUDDY, I. R. & GLEADOW, A. J. W. 1987. Thermal annealing of fission tracks in apatite. *Chemical Geology, Isotope Geoscience Section*, **65**, 1–13.

LAUBSCHER, H. P. 1983. The late Alpine (Periadriatic) intrusions and the Insubric Line. *Memorie di Scienze Geologiche Italiana*, **26**, 21–30.

LÄUFER, A., HUBICH, D. & LOESCHKE, J. 2001. Variscan geodynamic evolution of the Carnic Alps (Austria/Italy). *International Journal of Earth Sciences (Geologische Rundschau)*, **90**, 855–870.

LINNEMANN, U., GEHMLICH, M. *ET AL.* 2000. From Cadomian subduction to Early Palaeozoic rifting: the evolution of Saxo-Thuringia at the margin of Gondwana in the light of single zircon geochronology and basin development (Central European Variscides, Germany). *In*: FRANKE, W., HAAK, V., ONKEN, O. & TANNER, D. (eds) *Orogenic Processes: Quantification and Modelling in the Variscan Belt*. Geological Society, London, Special Publications, **179**, 131–153.

LINNER, M., RICHTER, W. & THÖNI, M. 1996. Eo-Alpine eclogites in the Austroalpine basement S of the Tauern Window: geochemistry of eclogites and interlayered metasediments. *Journal of Conference Abstracts*, **1**, 363.

LISTER, G. S., PATERSON, M. S. & HOBBS, B. E. 1978. The simulation of fabric development in plastic deformation and its application to quartzite: the model. *Tectonophysics*, **49**, 107–158.

LISTER, G. S. & PRICE, G. P. 1978. Fabric development in a quartz-feldspar mylonite. *Tectonophysics*, **49**, 537–578.

LOESCHKE, J. 1989. Lower Palaeozoic volcanism of the Eastern Alps and its geodynamic implications. *International Journal of Earth Sciences (Geologische Rundschau)*, **78**, 566–616.

LOESCHKE, J. & HEINISCH, H. 1993. Palaeozoic volcanism of the Eastern Alps and its palaeotectonic significance. *In*: VON RAUMER, J. & NEUBAUER, F. (eds) *The Pre-Mesozoic Geology in the Alps*. Springer, Berlin, Heidelberg, New York, 441–455.

LOTH, G., EICHHORN, R., HÖLL, R., KENNEDY, A., SCHAUDER, P. & SÖLLNER, F. 2001. Cambro-Ordovician age of a metagabbro from the Wildschönau ophiolite complex, Greywacke Supergroup (Eastern Alps, Austria): a U–Pb SHRIMP study. *European Journal of Mineralogy*, **13**, 566–577.

LUDWIG, K. R. 2001. *Users manual for Isoplot/Ex rev. 2.49. A geochronological toolkit for Microsoft Excel*. Berkeley Geochronology Center Special Publication, **1a**, 1–55.

MAGER, D. 1981. Vergleichende morphologische Untersuchungen an Zirkonen des altkristallinen Augengneises von Sand in Taufers (Südtirol) und einiger benachbarter Gesteine. *Neues Jahrbuch Mineralogie Monatshefte*, **1981/9**, 385–397.

MAGER, D. 1985a. Geologische Karte der Rieserfernergruppe zwischen Magerstein und Windschar (Südtirol). *Der Schlern*, **6**, 358–379.

MAGER, D. 1985b. *Geologische und petrographische Untersuchungen am Südrand des Rieserferner-Plutons (Südtirol) unter Berücksichtigung des Intrusionsmechanismus*. PhD thesis, University of Erlangen-Nürnberg.

MANCKTELOW, N. S. & GRASEMANN, B. 1997. Time-dependent effects of heat advection and topography on cooling histories during erosion. *Tectonophysics*, **270**, 167–195.

MANIAR, P. D. & PICCOLI, P. M. 1989. Tectonic discrimination of granitoids. *Geological Society of America Bulletin*, **101**, 635–643.

MASSONNE, J.-H. & SZPURKA, Z. 1997. Thermodynamic properties of white micas on the basis of high-pressure experiments in the systems $K_2O-MgO-Al_2O_3-SiO_2-H_2O$ and $K_2O-FeO-Al_2O_3-SiO_2-H_2O$. *Lithos*, **41**, 229–250.

MAZZOLI, C. & SASSI, F. P. 1992. New chemical data on the Upper Ordovician acidic plutonism in the Austrides of the Eastern Alps. *In*: CARMIGNANI, L. & SASSI, F. P. (eds) Contributions to the geology of Italy with special regard to the Palaeozoic basements. *IGCP International Geological Correlation Program No. 276 Newsletter*, **5**, 263–277.

MAZZOLI, C., PERUZZO, L. & SASSI, R. 1993. An Austroalpine mylonite complex at the southern boundary of the Tauern Window: crystallization-deformation relationships in the Cima-Dura-Durreck Complex. *IGCP International Geological Correlation Program No. 276, Newsletter*, **6**, 22–25.

MCDONOUGH, W. F. & SUN, S. S. 1995. The composition of the earth. *Chemical Geology*, **120**, 223–253.

MELI, S. & KLÖTZLI, U. S. 2001. Evidence for Lower Palaeozoic magmatism in the Eastern Southalpine basement: zircon geochronology from Comelico porphyroids. *Swiss Bulletin of Mineralogy and Petrology*, **81**, 147–157.

MILLER, C. & THÖNI, M. 1995. Origin of eclogites from the Austroalpine Ötztal basement (Tirol, Austria): geochemistry and Sm–Nd vs. Rb–Sr isotope systematics. *Chemical Geology*, **122**, 199–225.

MONTEL, J.-M., FORET, S., VESCHAMBRE, M., NICOLLET, C. & PROVOST, A. 1996. A fast, reliable, inexpensive in-situ dating technique: electron microprobe ages on monazite. *Chemical Geology*, **131**, 37–53.

MONTEL, J.-M., VESCHAMBRE, M. & NICOLLET, C. 1994. Datation de la monazite à la microsonde électronique. *Comptes Rendues de l'Académie des Sciences Paris II*, **318**, 1489–1495.

MÜLLER, W. 1998. *Isotopic dating of deformation using microsampling techniques: the evolution of the Periadriatic fault system (Alps)*. PhD thesis, No. 12580, ETH Zürich.

MÜLLER, B., KLÖTZLI, U. S. & FLISCH, M. 1995. U–Pb and Pb–Pb dating of the older orthogneiss suite in the Silvretta nappe, eastern Alps: Cadomian magmatism in the upper Austro-Alpine realm. *International Journal of Earth Sciences (Geologische Rundschau)*, **84**, 457–465.

MÜLLER, B., KLÖTZLI, U. S., SCHALTEGGER, U. & FLISCH, M. 1996. Early Cambrian oceanic plagiogranite in the Silvretta Nappe, eastern Alps: geochemical, zircon U–Pb and Rb–Sr data from garnet-hornblende-plagioclase gneisses. *International Journal of Earth Sciences (Geologische Rundschau)*, **85**, 822–831.

MÜLLER, W., MANCKTELOW, N. S. & MEIER, M. 2000. Rb–Sr microchrons of synkinematic mica in mylonites; an example from the DAV Fault of the Eastern Alps. *Earth and Planetary Science Letters*, **180**, 385–397.

NEUBAUER, F. 2002a. Evolution of late Neoproterozoic to early Paleozoic tectonic elements in Central and Southeast European Alpine mountain belts: review and synthesis. *Tectonophysics*, **352**, 87–103.

NEUBAUER, F. 2002b. Tectonic evolution of the Eastern Alps: from Permian rifting to Cretaceous and Tertiary collision. Extended Abstract, TRANSALP Conference. *Memorie di Science Geologiche*, **54**, 175–178.

NEUBAUER, F. & SASSI, F. P. 1993. The Austroalpine quartzphyllites and related Palaeozoic Formations. *In*: VON RAUMER, J. & NEUBAUER, F. (eds) *The Pre-Mesozoic Geology in the Alps*. Springer, Berlin, Heidelberg, New York, 423–439.

NEWTON, R.C. & PERKINS, D. 1982. Thermodynamic calibration of geobarometers based on the assemblage garnet-plagioclase-orthopyroxene-clinopyroxene-quartz. *American Mineralogist*, **67**, 203–222.

OXBURGH, E. R., LAMBERT, R. S., BAADSGARD, H. & SIMONS, J. G. 1966. Potassium-Argon age studies across the southeast margin of the Tauern window in the Eastern Alps. *Verhandlungen der Geologischen Bundesanstalt*, **1966**, 17–33.

PASSCHIER, C. W. & TROUW, R. A. J., 1996. *Microtectonics*. Springer Verlag, Berlin, Heidelberg.

PEARCE, J. A. 1982. Trace element characteristics of lavas from destructive plate boundaries. *In*: THORPE, R. S. (ed.) *Andesites*. Wiley, London, 525–548.

PEARCE, J. A. 1983. Role of the sub-continental lithosphere in magma genesis at active continental margins. *In*: HAWKESWORTH, C. J. & NORRY, M. J. (eds) *Continental basalts and mantle xenoliths*. Shiva, Nantwich, 230–249.

PECCERILLO, A., POLI, G., SASSI, F. P., ZIRPOLI, G. & MEZZACASA, G. 1979. New data on the Upper Ordovician acid plutonism in the Eastern Alps. *Neues Jahrbuch Mineralogie Abhandlungen*, **137**, 162–183.

PERCHUK, L. L. 1991. Derivation of a thermodynamically consistent set of geothermometers and barometers for metamorphic and magmatic rocks. *In*: PERCHUK, L. L. (ed.) *Progress in Metamorphic and Magmatic Petrology*. University Press, Cambridge, 93–111.

PLYUSNINA, L. P. 1982. Geothermometry and geobarometry of plagioclase-hornblende assemblages. *Contributions to Mineralogy and Petrology*, **80**, 140–146.

POWELL, R. & HOLLAND, T. J. B. 1993. On the formulation of simple mixing models for complex phases. *American Mineralogist*, **78**, 1174–1180.

PROCHASKA, W. 1981. Einige Ganggesteine der Riesenfernerintrusion mit neuen radiometrischen Altersdaten. *Mitteilungen der Gesellschaft der Geologie- und Bergbaustudenten in Österreich*, **27**, 161–171.

PURTSCHELLER, F. & SASSI, F. P. 1975. Some thoughts on the pre-Alpine metamorphic history of the Austridic basement of the Eastern Alps. *Tschermaks Mineralogische und Petrographische Mitteilungen*, **22**, 175–199.

PURTSCHELLER, F., HAAS, R., HOINKES, G., MOGESSIE, A., TESSADRI, R. & VELTMAN, C. 1987. Eoalpine metamorphism in the crystalline basement. *In*: FLÜGEL, H. W. & FAUPL, P. (eds) *Geodynamics of the Eastern Alps*. Deuticke Verlag, Wien, 185–190.

PYLE, J. M. & SPEAR, F. S. 1999. Yttrium zoning in garnet: coupling of major and accessory phases during metamorphic reactions. *Geological Materials Research*, **1**(6), 1–49.

PYLE, J. M., SPEAR, F. S., RUDNICK, R. L. & MCDONOUGH, W. F. 2001. Monazite-xenotime-garnet equilibrium in metapelites and a new monazite-garnet thermometer. *Journal of Petrology*, **42**, 2083–2107.

QUINQUIS, H., AUDREN, C., BRUN, J. P. & COBBOLD, P. 1978. Intense progressive shear in Ile de Groix blueschists and compatibility with subduction or obduction. *Nature*, **273**, 43–45.

RIZZO, G., KLÖTZLI, U. S. & SPIESS, R. 1998. Single zircon Pb/Pb age constraints on maximum sedimentation ages for quartzphyllite complexes from the Eastern and Southern Alps. *Mitteilungen der Österreichischen Mineralogischen Gesellschaft*, **143**, 372–373.

ROCHETTE, P., JACKSON, M. & AUBOURG, C. 1992. Rock magnetism and the interpretation of anisotropy of magnetic susceptibility. *Reviews of Geophysics*, **30**, 209–226.

ROLLINSON, H. R. 1993. *Using Geochemical Data: Evaluation, Presentation, Interpretation*. Longman Scientific and Technical, London.

ROMER, R. L. & SIEGESMUND, S. 2003. Why allanite may swindle about its true age: *Contributions to Mineralogy and Petrology*, **146**, 297–307.

ROSENBERG, C. L. 2004. Shear zones and magma ascent: a model based on a review of the Tertiary magmatism in the Alps. *Tectonics*, **23**, 21.

ROSENBERG, C., WAGNER, R. & HANDY, M. R. 2000. *The roof of the Rieserferner Pluton: syntectonic magma ascent and emplacement along a crustal scale shear zone*. 2nd International TRANSALP-Colloquium–Programme and Abstracts, 20.

SANDER, B. 1925. *Note illustrative della carta geologica delle Tre Venezie, foglio Bressanone 1:100 000*. Ufficio Idrografico dell R. Magistrato, Sezione Geologiche, **56**, 1–60.

SANDER, B. 1950. *Einführung in die Gefügekunde geologischer Körper, Teil 2*. Springer Verlag, Wien.

SASSI, F. P., BELLIENI, G., PECCERILLO, A. & POLI, G. 1980. Some constraints on the geodynamic models in the Eastern Alps. *Neues Jahrbuch für Geologie und Palaeontologie Monatshefte*, **1980**, 541–548.

SASSI, F. P., CAVAZZINI, G., VISONA, D. & DEL MORO, A. 1985. Radiometric geochronology in the Eastern Alps: results and problems. *Rendiconti Societa Italiano de Mineralogia e Petrologia*, **40**, 187–224.

SASSI, F. P., CESARE, B., MAZZOLI, C., PRUZZO, L., SASSI, R. & SPIESS, R. 2004. The crystalline basement of the Italian eastern Alps: a review of the metamorphic features. *Periodico di Mineralogia*, **73**, 23–42.

SATIR, M. 1975. Die Entwicklungsgeschichte der westlichen Hohen Tauern und der südlichen Ötztalmasse auf Grund radiometrischer Altersbestimmungen. *Memorie dell Istituto Geologica e Mineralogica Universita Padova*, **30**, 1–84.

SATIR, M. 1976. Rb–Sr- und K–Ar-Altersbestimmungen an Gesteinen und Mineralien des südlichen Ötztalkristallins und der westlichen Hohen Tauern. *International Journal of Earth Sciences (Geologische Rundschau)*, **65**, 394–410.

SCHÄTZ, M., TAIT, J. & BACHTADSE, V. 2002. Palaeozoic geography of the Alpine realm, new palaeomagnetic data from the Northern Greywacke Zone, Eastern Alps. *International Journal of Earth Sciences (Geologische Rundschau)*, **91**, 979–992.

SCHALTEGGER, U., NÄGLER, T., CORFU, F., MAGGETTI, M., GALETTI, G. & STOSCH, H. G. 1997. A Cambrian island arc in the Silvretta nappe: constraints from geochemistry and geochronology study. *Swiss Bulletin of Mineralogy and Petrology*, **77**, 337–350.

SCHMID, S. M., FÜGENSCHUH, B., KISSLING, E. & SCHUSTER, R. 2004. Tectonic map and overall architecture of the Alpine orogen. *Eclogae Geologicae Helveticae*, **97**, 93–117.

SCHMIDEGG, O. 1936. Steilachsige Tektonik und Schlingenbau auf der Südseite der Tiroler Zentralalpen. *Jahrbuch der Geologischen Bundesanstalt Wien*, **86**, 115–149.

SCHMIDEGG, O. 1937. Der Triaszug von Kalkstein im Schlingengebiet der Villgrater Berge (Osttirol). *Jahrbuch der Geologischen Bundesanstalt Wien*, **87**, 111–132.

SCHMIDT, K. & SÖLLNER, F. 1982. Towards a geodynamic concept of the 'Caledonian event' in Central- and SW-Europe. *Verhandlungen der Geologischen Bundesanstalt*, **1982/3**, 251–268.

SCHÖNHOFER, R. 1999. Das ostalpine Altkristallin der westlichen Lasörlinggruppe (Osttirol, Österreich): Kartierung, Stoffbestand und tektonometamorphe Entwicklung. *Erlanger geologische Abhandlungen*, **130**, 1–128.

SCHÖNLAUB, H. P. 1979. Das Paläozoikum in Österreich. *Abhandlungen der Geologischen Bundesanstalt*, **33**, 1–124.

SCHÖNLAUB, H. P. 1993. Stratigraphy, biogeography and climatic relationships of the Alpine Palaeozoic. *In*: VON RAUMER, J. F. & NEUBAUER, F. (eds) *The Pre-Mesozoic Geology in the Alps*. Springer, Berlin, Heidelberg, New York, 65–92.

SCHÖNLAUB, H. P. 1997a. The biogeographic relationship of Ordovician strata and fossils of Austria. *Berichte der Geologischen Bundesanstalt*, **40**, 6–19.

SCHÖNLAUB, H. P. 1997b. The Silurian of Austria. *Berichte der Geologischen Bundesanstalt*, **40**, 20–41.

SCHULZ, B. 1988a. Deformation und Metamorphose im ostalpinen Altkristallin südlich des Tauernfensters (südliche Defereger Alpen, Österreich). *Swiss Bulletin of Mineralogy and Petrology*, **68**, 397–406.

SCHULZ, B. 1988b. Quarz- und Mikrogefüge zonierter Kalksilikatgneis-Körper im ostalpinen Altkristallin (südliche Defereger Alpen, Österreich). *Erlanger geologische Abhandlungen*, **116**, 117–122.

SCHULZ, B. 1989. Jungalpidische Gefügeentwicklung entlang der Defereggen-Antholz-Vals-Linie, Osttirol, Österreich. *Jahrbuch der Geologischen Bundesanstalt Wien*, **132**, 775–789.

SCHULZ, B. 1990. Prograde-retrograde P-T-t-deformation path of Austroalpine micaschists during Variscan continental collision (Eastern Alps). *Journal of Metamorphic Geology*, **8**, 629–643.

SCHULZ, B. 1991. Deformation und Metamorphose im Thurntaler Komplex (Ostalpen). *Jahrbuch der Geologischen Bundesanstalt Wien*, **134**, 369–391.

SCHULZ, B. 1993a. P-T-deformation paths of Variscan metamorphism in the Austroalpine basement: controls on geothermobarometry from microstructures in progressively deformed metapelites. *Swiss Bulletin of Mineralogy and Petrology*, **73**, 257–274.

SCHULZ, B. 1993b. Mineral chemistry, geothermobarometry and pre-Alpine high-pressure metamorphism of eclogitic amphibolites and mica schists from the

Schobergruppe, Austroalpine basement, Eastern Alps. *Mineralogical Magazine*, **57**, 189–202.

SCHULZ, B. 1994a. Geologische Karte 1:50 000 des Altkristallins östlich des Tauferer Tals (Südtirol). *Erlanger geologische Abhandlungen*, **124**, 3–30.

SCHULZ, B. 1994b. Microstructural evolution of metapelites from the Austroalpine basement north of the Staller Sattel during pre-Alpine and Alpine deformation and metamorphism (Eastern Tyrol, Austria). *Jahrbuch der Geologischen Bundesanstalt Wien*, **137**, 197–212.

SCHULZ, B. 1995. Geochemistry and REE magmatic fractionation patterns in the Prijakt amphibolitized eclogites of the Schobergruppe, Austroalpine basement (Eastern Alps). *Swiss Bulletin of Mineralogy and Petrology*, **75**, 225–239.

SCHULZ, B. 1997. Pre-Alpine tectonometamorphic evolution in the Austroalpine basement to the south of the central Tauern Window. *Swiss Bulletin of Mineralogy and Petrology*, **77**, 281–297.

SCHULZ, B. & BOMBACH, K. 2003. Single zircon Pb–Pb geochronology of the Early-Palaeozoic magmatic evolution in the Austroalpine basement to the south of the Tauern Window. *Jahrbuch der Geologischen Bundesanstalt Wien*, **143**, 303–321.

SCHULZ, B., NOLLAU, G., HEINISCH, H. & GODIZART, G. 1993. Austro-Alpine basement complex to the south of the Tauern Window. *In*: VON RAUMER, J. & NEUBAUER, F. (eds) *The Pre-Mesozoic Geology in the Alps*. Springer, Berlin, Heidelberg, New York, 493–512.

SCHULZ, B., SIEGESMUND, S., STEENKEN, A., SCHÖNHOFER, R. & HEINRICHS, T. 2001. Geologie des ostalpinen Kristallins südlich des Tauernfensters zwischen Virgental und Pustertal. *Zeitschrift der deutschen geologischen Gesellschaft*, **152**, 261–307.

SCHULZ, B., BOMBACH, K., PAWLIG, S. & BRÄTZ, H. 2004. Neoproterozoic to Early-Palaeozoic magmatic evolution in the Gondwana-derived Austroalpine basement to the south of the Tauern Window (Eastern Alps). *International Journal of Earth Sciences (Geologische Rundschau)*, **93**, 824–843.

SCHULZ, B., FINGER, F. & KRENN, E. 2005. Auflösung variskischer, permischer und alpidischer Ereignisse im polymetamorphen ostalpinen Kristallin südlich der Tauern mit EMS-Datierung von Monazit. Arbeitstagung 2005 der Geologischen Bundesanstalt Österreich, 141–153.

SCHULZ, B., BRÄTZ, H. & KLEMD, R. 2006. Host rock compositional controls on zircon trace-element signatures in metabasites from the Austroalpine basement. *Geochimica et Cosmochimica Acta*, **70**, 697–710.

SCHUSTER, R., SCHARBERT, S., ABART, R. & FRANK, W. 2001. Permo-Triassic extension and related HT/LP metamorphism in the Austroalpine—Southalpine realm. *Mitteilungen der Gesellschaft der Geologie und Bergbaustudenten in Österreich*, **45**, 111–141.

SENARCLENS-GRANCY, W. 1964. Zur Grundgebirgs- und Quartärgeologie der Defereger Alpen und ihrer Umgebung. *Zeitschrift der deutschen geologischen Gesellschaft*, **116**, 502–511.

SIEGESMUND, S., ULLEMEYER, K. & DAHMS, M. 1995. Control of magnetic rock fabrics by mica preferred orientation: a quantitative approach. *Journal of Structural Geology*, **17**, 1601–1613.

SIEGESMUND, S. & BECKER, J. 2000. The emplacement of the Ardara Pluton (Ireland): new constraints from magnetic fabrics, rock fabrics and age dating. *International Journal of Earth Sciences*, **89**, 307–327.

SIEGESMUND, S., HEINRICHS, T., ROMER, R. L. & DOMAN, D. 2007. Age constraints on the evolution of the Austroalpine basement to the south of the Tauern Window. *International Journal of Earth Sciences (Geologische Rundschau)*, **96**, 415–432.

SÖLLNER, F. & HANSEN, B. 1987. 'Panafrikanisches' und 'kaledonisches' Ereignis im Ötztal-Kristallin der Ostalpen: Rb–Sr- und U–Pb-Altersbestimmungen an Migmatiten und Metamorphiten. *Jahrbuch der Geologischen Bundesanstalt Wien*, **130**, 529–569.

SÖLLNER, F., HÖLL, R. & MILLER, H. 1991. U–Pb-Systematik der Zirkone in Meta-Vulkaniten ('Porphyroiden') aus der Nördlichen Grauwackenzone und dem Tauernfenster (Ostapen, Österreich). *Zeitschrift der deutschen geologischen Gesellschaft*, **142**, 285–299.

SÖLLNER, F., MILLER, H. & HÖLL, R. 1997. Alter und Genese rhyodazitischer Metavulkanite ('Porphyroide') der Nördlichen Grauwackenzone und der Karnischen Alpen (Österreich): Ergebnisse von U-Pb Zirkondatierungen. *Zeitschrift der deutschen geologischen Gesellschaft*, **148**, 499–522.

SPAETH, G. & KREUTZER, S. 1989. Bericht 1988 über geologische Aufnahmen im Thurntaler Quarzphyllit und Altkristallin auf Blatt 179 Lienz. *Jahrbuch der Geologischen Bundesanstalt*, **132**, 595–597.

SPEAR, F. S. 1993. Metamorphic Phase Equilibria and Pressure-Temperature-Time Paths. *Mineralogical Society of America Monograph Series*, **1**.

SPRENGER, W. & HEINISCH, H. 1992. Late Oligocene to Recent brittle transpressive deformation along the Periadriatic Lineament in the Lesach Valley (Eastern Alps); remote sensing and paleo-stress analysis. *Annales Tectonicae*, **6**, 134–149.

STACEY, J. S. & KRAMERS, J. D. 1975. Approximation of terrestrial lead isotope evolution by a two-stage model. *Earth and Planetary Science Letters*, **26**, 207–221.

STAMPFLI, G. 1996. The intra-Alpine terrain: a palaeotethyan remnant in the Alpine Variscides. *Eclogae Geologicae Helveticae*, **89**, 12–42.

STAMPFLI, G. 2000. Tethyan oceans. *In*: BOZKURT, F., WINCHESTER, J. A. & PIPER, I. D. A. (eds) *Tectonics and Magmatism in Turkey and the Surrounding Area*. Geological Society, London, Special Publications, **173**, 1–25.

STAMPFLI, G. M. & BOREL, G. D. 2002. A plate tectonic model for the Paleozoic and Mesozoic. *Earth and Planetary Science Letters*, **196**, 17–33.

STAMPFLI, G. M. & MOSAR, J. 1999. The making and becoming of Apulia. *Memorie di Science Geologiche*, **51**, 141–154.

STAMPFLI, G. M., VON RAUMER, J. F. & BOREL, G. D. 2002. Paleozoic evolution of pre-Variscan terranes: From Gondwana to the Variscan collision. *In*: MARTÍNEZ-CATALÁN, J. R., HATCHER, R. D., ARENAS, R. & DÍAZ GARCÍA, F. (eds) *Variscan-Appalachian Dynamics: The Building of the Late Paleozoic Basement*. Geological Society of America Special Paper, **364**, 263–280.

STEENKEN, A. 2002. The emplacement of the Rieserferner Pluton and its relation to the DAV Line as well as to the kinematic and thermal history of the

Austroalpine basement (Eastern Alps, Tyrol). *Geotektonische Forschungen*, **94**, 1–120.

STEENKEN, A. & SIEGESMUND, S. 2000. Evidence for an alkaline basaltic volcanism at the northern margin of Gondwana within the Austroalpine Basement Complex of the Eastern Alps (Austrian/Italian border). *Jahrbuch der Geologischen Bundesanstalt Wien*, **142**, 235–247.

STEENKEN, A., SIEGESMUND, S. & HEINRICHS, T. 2000. The emplacement of the Rieserferner pluton (eastern Alps, Tyrol): Constraints from field observations, magnetic fabrics and microstructures. *Journal of Structural Geology*, **22**, 1855–1873.

STEENKEN, A., SIEGESMUND, S., HEINRICHS, T. & FÜGENSCHUH, B. 2002. Cooling and exhumation of the Rieserferner Pluton (Eastern Alps, Italy/ Austria). *International Journal of Earth Sciences (Geologische Rundschau)*, **91**, 799–817.

STÖCKHERT, B. 1984. K–Ar determinations on muscovites and phengites and the minimum age of the Old Alpine deformation in the Austridic basement south of the western Tauern Window (Ahrn valley, Southern Tyrol, Eastern Alps). *Neues Jahrbuch für Mineralogie Abhandlungen*, **150**, 103–120.

STÖCKHERT, B. 1985. Pre-Alpine history of the Austridic basement to the south of the western Tauern Window (Southern Tyrol, Italy): Caledonian versus Hercynian event. *Neues Jahrbuch für Geologie und Paläontologie Monatshefte*, **1985**, 618–642.

STÖCKHERT, B. 1987. Das Uttenheimer Pegmatit-Feld (Ostalpines Altkristallin, Südtirol). Genese und alpine Überprägung. *Erlanger geologische Abhandlungen*, **114**, 83–106.

STÖCKHERT, B., BRIX, M. R., KLEINSCHRODT, R., HURFORD, A. J. & WIRTH, R. 1999. Thermochronometry and microstructures of quartz—a comparison with experimental flow laws and predictions on the temperature of the brittle-plastic transition. *Journal of Structural Geology*, **21**, 351–369.

ST-ONGE, M. R. 1987. Zoned poikiloblastic garnets: P–T paths and synmetamorphic uplift through 30 km of structural depth, Wopmay orogen, Canada. *Journal of Petrology*, **28**, 1–21.

STÜWE, K., WHITE, L. & BROWN, R. 1994. The influence of eroding topography on steady-state isotherms; application to fission track analysis. *Earth and Planetary Science Letters*, **124**, 63–74.

SUZUKI, K., ADACHI, M. & KAJIZUKA, I. 1994. Electron microprobe observations of Pb diffusion in metamorphosed detrital monazites. *Earth and Planetary Science Letters*, **128**, 391–405.

THIÉBLEMONT, D., CHÈVREMONT, P., CASTAING, C., TRIBOULET, C. & FEYBESSE, J. L. 1994. La discrimination géotectonique des roches magmatiques basiques par les éléments traces. Reélevation d'après une base de données et application à la chaîne panafricaine du Togo. *Geodinamica Acta*, **7/3**, 139–157.

THÖNI, M. 1999. A review of geochronological data from the Eastern Alps. *Swiss Bulletin of Mineralogy and Petrology*, **79**, 209–230.

THÖNI, M. & MILLER, Ch. 1996. Garnet Sm–Nd data from the Saualpe and the Koralpe (Eastern Alps, Austria): chronological and P-T constraints on the thermal and tectonic history. *Journal of Metamorphic Geology*, **14**, 453–466.

TICHOMIROWA, M., BERGER, H. J., KOCH, E. A. *ET AL.* 2001. Zircon ages of high-grade gneisses in the Eastern Erzgebirge (Central European Variscides)—constraints on origin of the rocks and Precambrian to Ordovician magmatic events in the Variscan foldbelt. *Lithos*, **56**, 303–332.

TRIBOULET, C. 1992. The (Na-Ca)amphibole-albite-chlorite-epidote-quartz geothermobarometer in the system S-A-F-M-C-N-H_2O. 1. An empirical calibration. *Journal of Metamorphic Geology*, **10**, 545–556.

TRIBOULET, C. & AUDREN, C. 1988. Controls on P–Tt–t-deformation path from amphibole zonation during progressive metamorphism of basic rocks (estuary of the River Vilaine, South Brittany, France). *Journal of Metamorphic Geology*, **6**, 117–133.

TROLL, G. 1978. The 'Altkristallin' of Eastern Tyrol between Tauern Window and Periadriatic Lineament. In: CLOSS, H., ROEDER, D. & SCHMIDT, K. (eds) *Alps, Apennines, Hellenides*. Inter-Union Commission on Geodynamics Scientific Report, No. 38, 149–154.

TROLL, G. & HÖLZL, E. 1974. Zum Gesteinsaufbau des Altkristallins der zentralen Schobergruppe, Osttirol. *Jahrbuch der Geologischen Bundesanstalt Wien*, **117**, 1–16.

TROLL, G., FORST, R., SÖLLNER, F., BRACK, W., KÖHLER, H. & MÜLLER-SOHNIUS, D. 1976. Über Bau, Alter und Metamorphose des Altkristallins der Schobergruppe, Osttirol. *International Journal of Earth Sciences (Geologische Rundschau)*, **65**, 483–511.

TROLL, G., BAUMGARTNER, S. & DAIMINGER, W. 1980. Zur Geologie der südwestlichen Schobergruppe (Osttirol, Österreich). *Mitteilungen der Gesellschaft der Geologie- und Bergbaustudenten in Österreich*, **26**, 277–295.

VON RAUMER, J. F. 1998. The Palaeozoic evolution in the Alps: from Gondwana to Pangea. *International Journal of Earth Sciences (Geologische Rundschau)*, **87**, 407–435.

VON RAUMER, J. F. & NEUBAUER, F. 1993. Late Precambrian and Palaeozoic evolution of the Alpine basement—an overview. In: VON RAUMER, J. F. & NEUBAUER, F. (eds) *The Pre-Mesozoic Geology in the Alps*. Springer, Berlin, Heidelberg, New York, 625–639.

VON RAUMER, J. F., STAMPFLI, G. M., BOREL, G. & BUSSY, F. 2002. Organization of pre-Variscan basement areas at the north-Gondwanan margin. *International Journal of Earth Sciences (Geologische Rundschau)*, **91**, 35–52.

VON RAUMER, J. F., STAMPFLI, G. M. & BUSSY, F. 2003. Gondwana-derived microcontinents—the constituents of the Variscan and Alpine collisional orogens. *Tectonophysics*, **365**, 7–22.

VON RAUMER, J. F., STAMPFLI, G., HOCHARD, C. & GUITÉRREZ-MARCO, J.-C. 2006. The Early Palaeozoic in Iberia—a plate-tectonic interpretation. *Zeitschrift der Deutschen Gesellschaft für Geowissenschaften*, **157**, 575–584.

WAGNER, R., ROSENBERG, C. L., HANDY, M. R., MÖBUS, C. & ALBERTZ, M. 2006. Fracture-driven intrusion and upwelling of a mid-crustal pluton fed from a transpressive shear zone—the Rieserferner

Pluton (Eastern Alps). *Geological Society of America Bulletin*, **118**, 219–237.

WILSON, M. 1989. *Igneous Petrogenesis. A Global Tectonic Approach*. Chapman and Hall, London.

WU, C.-M., ZHANG, J. & REN, L.-D. 2004. Empirical garnet-biotite-plagioclase-quartz (GBPQ) geobarometry in medium- to high-grade metapelites. *Journal of Petrology*, **45**, 1907–1921.

YAMADA, R., TAGAMI, T., NISHIMUA, S. & ITO, H. 1995. Annealing kinetics of fission tracks in zircon; an experimental study. *Chemical Geology*, **122**, 249–258.

ZENK, M. & SCHULZ, B. 2004. Zoned Ca-amphiboles and related P–T evolution in metabasites from the classical Barrovian metamorphic zones in Scotland. *Mineralogical Magazine*, **68**, 769–786.

Exhumation and deformation history of the lower crustal section of the Valstrona di Omegna in the Ivrea Zone, southern Alps

S. SIEGESMUND[1], P. LAYER[2], I. DUNKL[1], A. VOLLBRECHT[1], A. STEENKEN[3], K. WEMMER[1] & H. AHRENDT[1,†]

[1]*Gottinger Zentrum Geowissenschaften, University of Göttingen, Goldschmidtstrasse 03, D–37077, Göttingen (e-mail: ssieges@gwdg.de)*

[2]*Geophysical Institute, University of Alaska Fairbanks, Fairbanks, Alaska 99775, USA*

[3]*Instituto de Geociências, Universidade de São Paulo, Rua do Lago 562, Cidade Universitária, 05508-080 - São Paulo, SP-Brazil*

Abstract: The Ivrea Zone (southern Alps) is one of the key regions interpreted as exposing a section of the lower continental crust and was the subject of several review-type articles. The Ivrea–Verbano Zone was rotated into an upright position along the Insubric mylonite belt. In the southeast, this unit is in contact with the Strona Ceneri Zone, which is interpreted as upper continental crust crossing the Permian Cossato–Mergozzo–Brissagio Line (CMB Line). The CMB mylonites are locally overprinted by the mylonites and cataclasites of the Pogallo Line, which was active during the Jurassic. In addition, the sinistral, steeply inclined Rosarolo shear zone was active over a long time span from the ductile into the brittle field, i.e. from the Early Permian (high-temperature ultra-mylonites) to the Neo-Alpine basic dykes and pseudotachylites. The high-temperature mylonites accommodated crustal extension and may be related to normal faults generated by magmatic underplating. The reactivation at different crustal levels during exhumation and tilting is documented by strain increments at decreasing P/T conditions. Its present subvertical orientation was attained during the Neo-Alpine deformation. Constraints on its exhumation history are based on new ^{40}Ar/^{39}Ar hornblende ages, K–Ar biotite ages and zircon fission-track data along the NE–SW trending Valstrona section. A re-interpretation of existing U–Pb monazite ages is included, based on a higher closure temperature for monazite. The oldest monazite ages are observed in proximity to the Pogallo Line (c. 292 Ma). Heat input by mafic intrusions was sufficient to reset the U–Pb monazite system, as is evidenced by the youngest ages in the vicinity of the Insubric Line. The re-interpretation favours the hypothesis that the oldest monazite ages are the result of complete resetting by a Permian thermal event. The ^{40}Ar/^{39}Ar hornblende ages and K–Ar biotite ages document the cooling after Permian heating. Roughly parallel age progressions decrease from the Pogallo Line (hornblende: 271 Ma vs. biotite: 227 Ma) towards the Insubric Line (hornblende: 201 Ma vs. biotite: 156 Ma). Zircon fission-track ages run parallel to the biotite ages in the upper part of the profile, whereas towards the Insubric Line a significant deviation from the biotite age progression is attributed to tilting of the basement during the Oligocene. Zircon fission-track ages around 38 Ma are found close to the Insubric Line. No age offset, neither at the CMB nor at the Pogallo Line, is observed. This confirms the hypothesis that the Pogallo Line is an oblique normal fault, and that the CMB Line has accommodated only minor vertical displacement. The capture of the different cooling ages confirms the tilting of the Ivrea–Verbano Zone during the Neo-Alpine deformation and contradicts the tilting of the Ivrea–Verbano Zone during the Permian.

Introduction

The Ivrea–Verbano Zone (IVZ) is located between Locarno and Ivrea in the southern Alps of the Piedmont Region of Italy and the Swiss Canton of Ticino. Together with the adjacent Strona–Ceneri Zone, it was the first key region interpreted in terms of an exposed cross-section of the lower to middle continental crust (e.g. Berckhemer 1969; Mehnert 1975; Fountain 1976; Fountain & Salisbury 1981) (Fig. 1). However, the pre-Permian evolution of the region is still a matter of discussion (e.g. Hunziker & Zingg 1980; Schmid 1993; Handy et al. 1999; Vavra et al. 1996). Following its Palaeozoic tectonometamorphic history, the Ivrea crustal section has been exhumed to shallower crustal levels, tilted and emplaced into its present position. Over the last 20 years, a number of review articles were published on its metamorphic petrology and structural evolution (e.g. Zingg

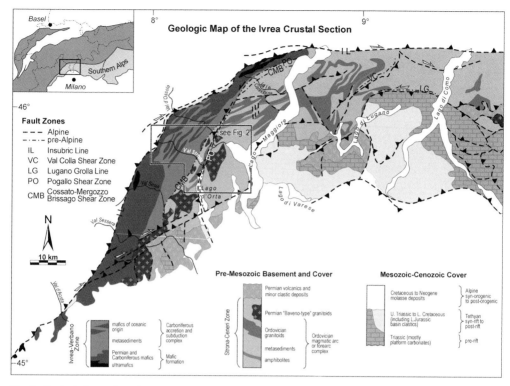

Fig. 1. Geological overview of the western part of the southern Alps (modified from Handy et al. 1999). The region consists of the metamorphic basement of the Ivrea–Verbano Zone (IVZ) and the Strona–Ceneri Zone (SCZ), and the Mesozoic to Cenozoic sedimentary cover. The outlined area around the Valstrona and Lago Maggiore was selected for structural and geochronological research (cf. Fig. 2).

et al. 1990; Boriani et al. 1990; Schmid 1993; Handy et al. 1999; Boriani & Giobbi 2004). In spite of this, the age, kinematics and geometry of the tectonic movements, which are responsible for exposing this almost complete crustal section at the Earth's surface, are the subject of ongoing debates (e.g. Handy & Zingg 1991; Schmid 1993; Handy et al. 1999).

Schmid et al. (1987) related the emplacement of the IVZ into its subvertical position adjacent to the Insubric Line to the Alpine orogeny, whereas Boriani et al. (1990) argued that vertical tilting took place prior to this, i.e. in the Permian. Handy et al. (1999) and Mulch et al. (2002) documented that the Strona Ceneri Zone (SCZ) was already moderately to steeply dipping in pre-Alpine times while the IVZ was tilted later, i.e. during the Late Oligocene to Early Miocene (Schmid et al. 1987).

Geochronological data are often used to constrain the timing and rates of geological processes, which are the major objectives in understanding the tectonic evolution of a region. In this study, the geochronology of a spectacular cross-section in the Valstrona di Omegna is presented (Fig. 2).

This crustal section reveals different levels of the continental crust including granulite facies rocks in the NW and amphibolite facies in middle/upper crustal levels in the SE. Late Palaeozoic magmatism and its impact on the host formations is coeval at all crustal levels. This event is characterized by the emplacement of large mafic intrusions in the lower crust, resulting in the partial melting of metasediments and by plutonic and volcanic activity at higher crustal levels (e.g. Schmid 1978, 1979; Fountain 1986, 1989; Schmid 1993; Voshage et al. 1990; Rivalenti et al. 1975, 1981; Quick et al. 1994; Snoke et al. 1999; Peressini et al. 2007).

In order to bracket the exhumation history of the Ivrea Zone, dense sampling was performed in several field campaigns along the Valstrona profile (Fig. 2) and various minerals were geochronologically investigated. In this paper, we present new $^{40}Ar/^{39}Ar$ ages for hornblende, K–Ar ages on biotite and fission-track ages for zircon. Published U–Pb monazite ages (Teufel & Schärer et al. 1989; Henk et al. 1997) from almost identical sample localities along the Valstrona, as well as

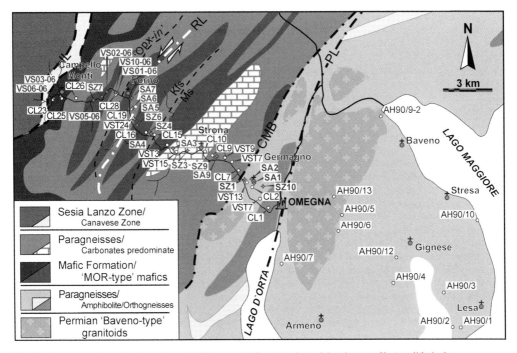

Fig. 2. Geology and sample sites along the Valstrona di Omegna–Lago Maggiore profile (modified after Bertolani 1969 and Handy *et al.* 1999). Sample sites for K–Ar Bt dating are indicated by open circles, whereas samples subjected to Ar–Ar Hbl and Zr fission-track dating are represented by grey stars.

published age data from other areas of the Ivrea Zone (Vavra *et al.* 1996; Boriani *et al.* 1990; Boriani and Villa 1997; Henk *et al.* 1997) are included to complete the dataset of the study. Together with available petrological data and structural field studies (Sills & Tarney 1984; Henk *et al.* 1997), we provide new constraints on the timing, kinematics and also to some extent on the displacement rates across distinct faults. The Rosarolo shear zone was selected to document the long-lasting time span and heterogeneity of deformation structures starting in the Permian.

Geological setting

The Ivrea Zone exposes a crustal section of the southern Alpine basement and is a SW–NE elongated body in the western Alps of Italy and Switzerland (Fig. 1). It is composed of the Ivrea–Verbano Zone (IVZ, 'formazione diorite-kinzigitica' by Novarese 1929) and the adjacent Strona–Ceneri Zone (SCZ, Serie dei Laghie of Boriani *et al.* 1990), and was the first region interpreted as an exposed cross-section of the entire continental crust (e.g. Berckhemer 1969; Mehnert 1975; Fountain 1976). The lowermost part of this section is often interpreted to represent the laminated lower crust, known from seismic sections all over the world. These are characterized by densely packed multiple sets of reflectors referred to as seismic lamellae (e.g. Fountain 1976; Burke & Fountain 1990; Rutter *et al.* 1993; Rabbel *et al.* 1998; Weiss *et al.* 1999).

The Ivrea Zone is separated from the Sesia Zone of the Central Alps by the greenschist facies Insubric mylonite belt (Fig. 1). The rocks tectonically incorporated into this mylonite belt are derived from the IVZ and the Sesia Zone, as well as from the Permo-Mesozoic sediments of the Canavese Zone. Shear criteria indicate that the mylonites accommodated back thrusting synchronous with back folding of the Central Alpine nappes. This was followed by a dextral strike-slip motion related to large transcurrent displacements in the Central Alps (Schmid *et al.* 1987).

The Pogallo Line, a 1–3 km-wide shear zone of mylonites, marks the subvertical contact between the SCZ and IVZ (Handy 1987). Hodges and Fountain (1984) first interpreted this shear zone in terms of a tilted normal fault of Mesozoic age. Boriani *et al.* (1990) identified the Cossato–Mergozzo–Brissagio Line (CMB-Line), which is closely associated with the Pogallo Line. The CMB Line (see Fig. 1) marks the transition between the IVZ and SCZ SW of the Val d'Ossola, where the Pogallo Line breaks away from the lithological contact between these two basement units (see discussion in Schmid 1993 and Boriani & Giobbi 2004). The CMB Line is characterized as a Late Variscan strike-slip zone showing a spatial and temporary relationship with mafic and granitic igneous rocks, i.e. the Appinite Suite. The continuity of the metamorphic pressure gradient across the CMB Line indicates that the vertical displacement was negligible (Handy et al. 1999).

While residing in the lower crust, the amphibolite to granulite facies paragneisses of the IVZ were intruded by huge volumes of mafic to intermediate plutonic rocks, known as the Mafic Complex (Rivalenti et al. 1981; Voshage et al. 1990). Most prominent in the south, the Mafic Complex dominates the IVZ along its limit with the Sesia Zone of the Central Alps. In the north, it is comprised of numerous sill-like intrusions up to several hundred of metres thick, whereas in the south, particularly in the Val Mastallone, mafic rocks up to 10 km thick occur. Along the northwestern border of the Mafic Complex, in the vicinity of the Insubric Line, several ultramafic bodies crop out. They extend from Baldissero in the south to Finero in the north. The geochemical composition and P–T conditions characterize them as derived from continental mantle material (Rivalenti et al. 1975, 1981; Garutti et al. 1978/1979; Shervais 1978/1979; Voshage et al. 1988; Hartmann & Wedepohl 1993).

Mafic rocks that contribute to the metamorphic history of the IVZ are separated into three groups (Zingg et al. 1990; Schmid 1993): (i) mafics of oceanic origin alternating with paragneisses (Sills & Tarney 1984); (ii) large bodies of 'Anzola gabbro'-type mafics; and (iii) parts of the banded mafics within the granulite facies region of the IVZ such as those within the layered complex (Rivalenti et al. 1981). The mafic rocks predating the 270 Ma magmatic event comprise the gabbro-diorites of the Mafic Complex in the Valsesia (see discussion in Voshage et al. 1990). Pin (1986) reported concordant 285 Ma U–Pb ages of magmatic zircons for these diorites. More recently, Peressini et al. (2007) found that the magmatic activity was bracketed between 290 Ma and 288 Ma.

At a higher structural level, the central part of the IVZ, known as the Kinzingite Formation (Novarese 1929), becomes dominant. In the Valstrona crustal profile, amphibolite-facies metapelites (the so-called kinzigites) and minor amphibolites constitute the uppermost part of the profile, whereas granulite-facies metapelites (known as stronalites) and mafic rocks make up the lowermost part in close proximity to the Mafic Complex. The appearance of subordinate silicate marbles, pegmatites and microgranites has also been described by Bertolani (1969). The degree of metamorphism of the Kinzigite Formation increases from amphibolite facies conditions in the SCZ and the lower part of the IVZ to granulite facies conditions close to the Insubric Line (Zingg 1983; Sills 1984; Henk et al. 1997). According to Sills (1984), granulite-facies metamorphism in the metapelites from the NW section of the Strona Valley reached maximum P–T conditions of $750 \pm 50\,°C$ and 6 ± 1 kbar. Henk et al. (1997) found peak metamorphic conditions of $810 \pm 50\,°C$ and 8.3 ± 2.0 kbar for a metagabbro from the base of the Mafic Complex. The lowest metamorphic conditions were obtained from alumosilicate-bearing gneisses near Omegna in the SCZ with P–T conditions of $580 \pm 30\,°C$ and 2.3 ± 0.5 kbar (Henk et al. 1997).

The metamorphic pressure gradient across the IVZ is about 0.41 kbar/km in the Valstrona section. This implies significant crustal thinning, particularly in the lowermost 5 km of the crust (Henk et al. 1997). These authors estimated that about 4 km of crustal attenuation occurred during the Early Permian. This estimate is in agreement with observations made by Sills and Tarney (1984), whereas Brodie and Rutter (1987) inferred only 2 km of lower crustal thinning in the IVZ. A higher pressure gradient of 1.7 kbar/km in the Valle Cannobina, located in the NW of the Valstrona (Fig. 1), suggests heterogeneous stretching subparallel to the lateral extent of the IVZ, which is attributed to the activity of the Pogallo shear zone (Handy et al. 1999). Normal faulting at the low angle Pogallo shear zone took place between 180 and 230 Ma (Hodges & Fountain 1984; Handy 1987).

The timing of peak metamorphism is still a matter of discussion. SHRIMP U–Pb data from a magmatic zircon population, with crystal shapes characteristic of calc-alkaline magmatites, yield a crystallization age of 355 ± 6 Ma (Varva et al. 1996). The oldest metamorphic zircon rims in the metasediments of the IVZ are Early Permian (c. 296 Ma). In contrast, U–Pb ages for monazite and zircon from the SCZ point to a mid-Palaeozoic metamorphic event at c. 400 Ma (Köppel & Grünfelder 1978/1979; Köppel 1974; Rigaletti et al. 1994). A numerical simulation (Henk et al. 1997) of the temperature development in the IVZ, in response to the emplacement of the Mafic Complex, suggests that the metamorphic peak occurred prior to 300 Ma. The range of U–Pb monazite ages (assuming a closure temperature of c. 600 °C) between 272 Ma and 292 Ma document the heating pulses related to the magmatic underplating, which had created the typical granoblastic structure in the Ivrea Zone. Additional geochronological data for the IVZ recorded the progressive cooling from the high temperatures prevailing during the Permo-Carboniferous metamorphic peak to temperatures of around 300 °C in the Jurassic (e.g. Zingg et al. 1990).

Results

Structural geology of the study area

Metamorphic foliation and banding in the IVZ are generally steeply inclined and trend NE–SW. An eastward plunge of stretching lineations is remarkably constant. Recently, Rutter et al. (2007) proposed that large-scale superimposed folding in the upper part of the IVZ in the Valstrona section occurred during regional migmatization, probably during the Hercynian orogeny. The IVZ in the Valstrona cross-section is bounded to the northwest by the Insubric mylonite belt. It is in contact with the adjacent SCZ to the southeast across the CMB Line (Boriani et al. 1990, see also Fig. 1). The CMB Line is considered to be a major subvertical tectonic discontinuity of Permian age (Boriani & Villa 1997; Mulch et al. 2002; or Boriani & Giobbi 2004). Starting in Permian times, the region was subjected to large-scale injection by basic magmas, accompanied by pervasive east–west stretching and crustal thinning of several kilometres. A network of conjugate high-temperature, low-angle shear zones in a layered lower crustal section related to this stretching has been identified (Brodie & Rutter 1987; Brodie et al. 1989, 1992; Rutter et al. 1993; Snoke et al. 1999). The corridor of shear zone outcrops, which passes through Anzola (Val d'Ossola) and Forno (Valstrona), can be traced for more than 20 km. The radiometric ages from the Anzola shear zone indicate that it relates to extension, which commenced prior to 280 Ma (Brodie et al. 1990), approximately coeval with the intrusion of the Mafic Complex. Metabasic rocks incorporated into inhomogeneous shear zones and transformed into mylonites and ultramylonites were deformed by dislocation creep accompanied by prograde dynamic recrystallization (Dornbusch & Skrotzki 2001).

The displacements along the CMB Line are clearly contemporaneous with the Permian mafic underplating events, and lead to the juxtaposition of the IVZ with the SCZ (Giobbi & Brodie 2004). The CMB mylonites are locally overprinted by a younger amphibolite to greenschist facies mylonite zone, the Pogallo Line. Mylonitization along the Pogallo Line clearly postdates the activity of the CMB in the Early Permian. The age of the Pogallo Line has been previously constrained to 160–240 Ma (Zing et al. 1990). In the Valstrona area, the Pogallo Line is only evidenced by brittle deformation due to the oblique low-angle normal fault geometry of the Pogallo Line, which has been reoriented in Alpine times (Handy 1987).

In the Valstrona cross-section, a long-term steeply inclined (present-day coordinates) shear zone is found, referred to as the Rosarolo shear zone (Fig. 2). The oldest high-temperature ultramylonites found in metabasic rocks can be related to the Anzola-type shear zones and incorporated mafic injections. At this time, the whole sequence was presumably in a horizontal position. The subsequent formation of pseudotachylites and other brittle deformation features took place within the upper crust. Shear sense criteria are principally sinistral, and antithetic shears only appear subordinate. The simplified sketches shown in Figure 3 illustrate that high-temperature mylonitization took place in the lower crust as part of an extensional fault system, while the evidence of frictional melting and brittle faulting indicates sinistral strike-slip movement in the middle and upper crust. In the following section, we describe the Rosarolo shear zone and argue in more detail to document the long time span and heterogeneity of deformation structures.

Fabrics of the Rosarolo shear zones

On the whole, the study area can be described as a high-strain shear zone. In its central parts, it consists of a network of sub-parallel anastomosing shear zones at the cm to m scale. Each of these minor mylonitic shear zones displays structures, which indicate that several strain increments occurred at different P/T conditions (Fig. 4). A large number of macro- and microstructures provide evidence that—with respect to present-day coordinates—sinistral shear operated over a long time span from the ductile to the brittle field (Ahrendt et al. 1990; Clausen 1990; see below). Only in a few cases are subordinate structures with an antithetic (dextral) sense observed. Because of distinct rheological contrasts between individual layers, which are mainly controlled by the ratio of mafic to felsic minerals, brittle and ductile deformation may also have operated more or less simultaneously. This resulted in a large variety of deformation structures and complex interactions of deformation processes. Macroscopically, the significance of rheological contrasts is best documented by ductile flow of felsic layers filling gaps produced by brittle faults affecting more mafic layers (Fig. 4b). Frequently, the common leucocratic segregations are cut by early formed mylonitic foliations, but were deformed and partly obliterated within the high strain cores of small-scaled shear zones (Fig. 4c).

Considering the shear zone as a whole, the mylonitic foliations are steeply inclined and strike NNW–SSE, grading into the NE–SW striking

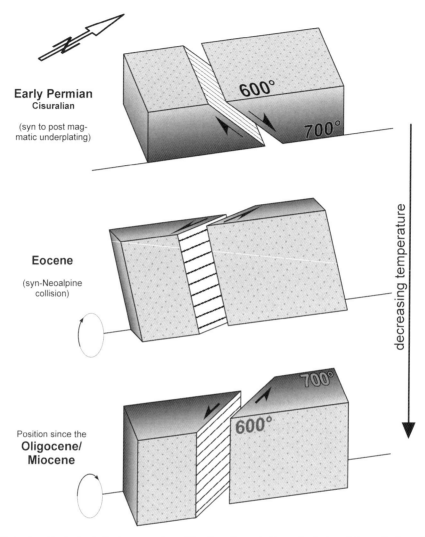

Fig. 3. Extensional tectonics in the lower crust resulting from the probable underplating of the mafic formation. The blocks were rotated during the Oligocene–Miocene uplift of the Ivrea basement and today appear as left lateral shear zones. Original vertical temperature profile now in a subhorizontal position.

metamorphic layering of the country rocks. This general sigmoidal trend of the foliations indicates a bulk sinistral sense of shear. In the central parts of the shear zone, the width of metapelitic layers is strongly reduced as compared to the more mafic lithologies. Locally, shear zones appear discontinuous because the main mylonitic foliation is transected by mylonitic horizons of different compositions. These probably represent former composite dykes (Fig. 5a) produced by crack-seal processes with alternating felsic and mafic input. Within the high strain area, the main brittle faults and the majority of associated pseudotachylites roughly follow the mylonitic foliation.

Boudin-shaped domains of comparatively low strain developed between the individual shear zones. Relics of early strain increments are also well preserved in these domains. Within the metabasic layers, a weak HT-mylonitization is indicated by fine-grained recrystallized rims around plagioclase and hornblende porphyroclasts. Around clinopyroxene, however, they are weakly developed. The orientation of the corresponding mylonitic foliation is strongly controlled by the metamorphic banding. Other often observed structures are boudins within mafic layers associated with tension or shear fractures, where the necking zones are filled with leucocratic segregations (Fig. 5b).

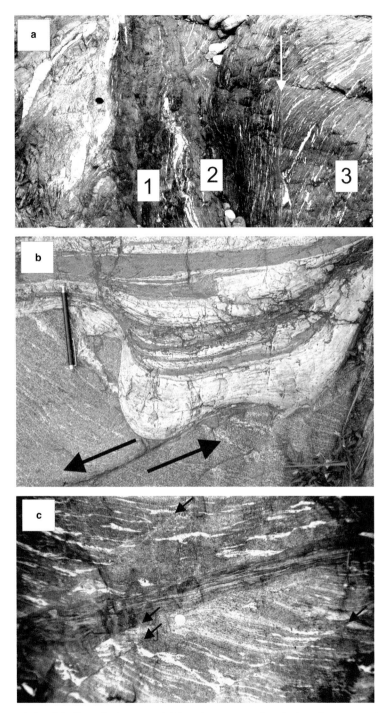

Fig. 4. Characteristics of the Rosarolo shear zone. (**a**) Complex structure of a polyphase shear zone at the metre scale. The dark ultramylonite (1) and cataclastic layers (2) form the core zone, with mylonitic country rock (3) in discontinuous contact to the shear zone core (arrow). (**b**) Ductile deformation of felsic layers, which probably originated from a composite dyke, filling a gap produced by a simultaneous shear fracture (see arrows). (**c**) Discontinuous shear zone at the dm scale. In the central high strain zone, the leucocratic injections are extremely stretched as compared to the mylonitic wall rock. Synthetic brittle–ductile shear planes (arrows) are developed at low angle to the central zone.

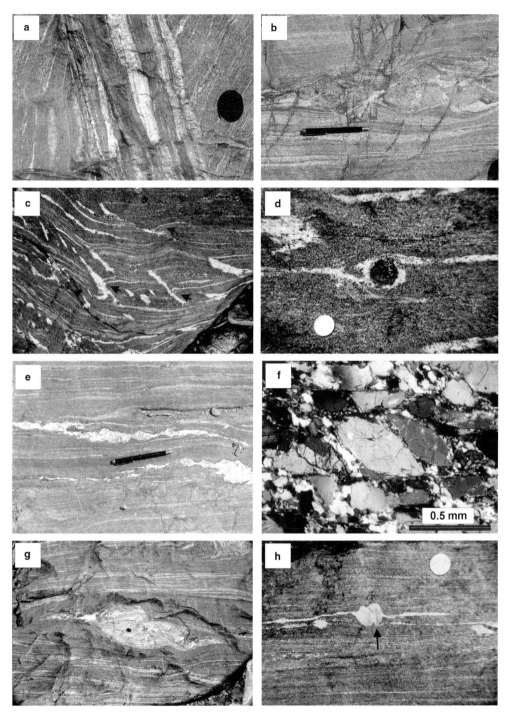

Fig. 5. Details of shear zone fabrics. (**a**) Finely laminated mylonite (left and right side) cut by a layered mylonite, which probably represents a former mylonitized composite dyke. (**b**) Synthetic sinistral shear fractures producing asymmetric boudinage of a mafic layer embedded in finely laminated mylonites; necking areas are filled with leucocratic injections. (**c**) Small-scale leucocratic veins within the mylonites, which are probably related to synthetic P-fractures, as indicated by dragging of the mylonitic foliation (arrows); locally the leucocratic veins are branched and form thin sills parallel to the mylonitic foliation.

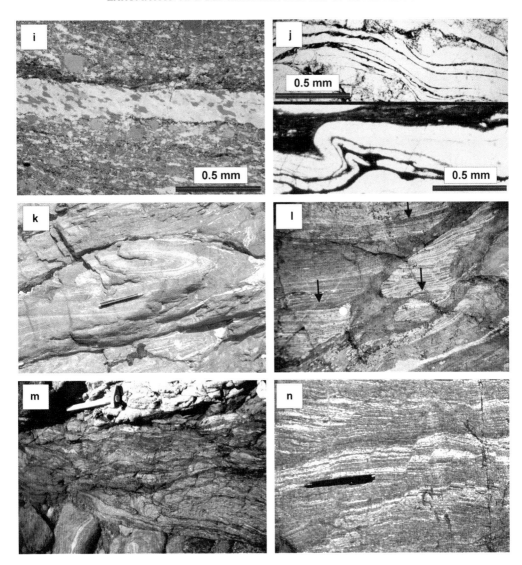

Fig. 5. (*Continued*) (**d**) Foliation-parallel leucocratic injection rotated together with a clinopyroxene porphyroblast/clast indicating a sinistral sense of shear. (**e**) Boudinage of a leucocratic segregation, which formed at a low angle to the mylonitic foliation; cores of boudins contain plagioclase porphyroclasts, while rims and necking zones consist of recrystallized aggregates. (**f**) SC-fabric displayed by hornblende 'fishes' and shape-preferred orientation of other constituents. (**g**) 'Litho-porphyroclast' (probably isolated boudin or remnant of a rootless intrafolial fold hinge); sigma-clast-like shape results from tails, which consist of stretched clast material and deformed segregations. (**h**) Brittle–ductile plagioclase porphyroclast with antithetic fragmentation and extremely long recrystallization tails. The tip of the right fragment (arrow) is affected by a continuing ductile deformation, which leads to the formation of a second generation recrystallization tail. (**i**) Mylonitic quartz layer displaying a distinct texture and shape-preferred orientation of grains, which indicate a sinistral sense of shear; crossed polars, with one quartz wave plate. (**j**) Mylonitic quartz layers separated by thin films of fine-grained material (top), which may indicate that the quartz layers were formed by repeated crack opening and healing parallel to the mylonitic foliation. Locally, these quartz layers also display asymmetric folds (bottom). (**k**) Elliptical cross-section of a sheath fold. (**l**) Mafic dyke, which cuts the mylonitic foliation or forms sills (arrows) parallel to it. (**m**) Cataclastic part of the shear zone containing fragments of the mylonitic host rock. (**n**) Set of planar sinistral fractures at the dm scale within the mylonites.

Typical structures along the less-deformed margin of the shear zone are synthetic P-faults (e.g. Logan *et al.* 1979), which are also filled with leucocratic injections (Fig. 5c). Dragging of adjacent layers gives evidence of sinistral shear along these planes. These veins are often connected with foliation-parallel segregations and occasionally were formed around larger porphyroblasts of clinopyroxene (Fig. 5d). Ductile deformation resulted in different kinds of boudinage, where other leucocratic veins formed both parallel or at low angle to the layering (Fig. 5e). The cores of these boudins often contain feldspar porphyroclasts, whereas recrystallized grains form the stretched areas. At the microscale, the sinistral sense of shear is best documented by SC-fabrics in mafic layers formed by 'hornblende fishes' together with corresponding oblique shape-preferred orientation of other components (Fig. 5f).

Individual layers contain small-scale lenses with complex fold structures (Fig. 5g), which may be interpreted as isolated relics of rootless hinge zones of intrafolial folds, which would imply localized high strain values. Along their boundaries, leucocratic segregations are emplaced, which partly are affected by ductile shear as indicated by asymmetric tails. Within these mylonites, isolated plagioclase porphyroclasts are embedded, which show both recrystallization tails and antithetic domino-like fracturing clearly indicating sinistral sense of shear (Fig. 5h). In many other comparable cases, however, the age relationships between ductile and brittle deformation remain questionable.

Within leucocratic horizons, quartz layers often display distinct textures associated with shape-preferred orientation obliquely inclined to the mylonitic foliation in a sinistral sense (Fig. 5i). These textures were qualitatively checked using a λ-plate pointed to the dominance of basal slip. Trails of solid opaque inclusions parallel to many of these quartz layers suggest that they originated by multiple crack healing parallel to the macroscopic layering (Fig. 5j).

Sheath folds appear at different dimensions ranging from the cm to m scale (Fig. 5k). They are related to an early HT deformation regime, because both mafic and felsic layers are homogeneously deformed.

During a late stage of the deformation history, irregular networks of mafic dykes intruded. These cut the mylonitic foliation or form short sills parallel to it (Fig. 5l). They are mainly composed of fine-grained recrystallized hornblende, partly containing relics of clinopyroxene, together with plagioclase and minor amounts of sphene. Locally, narrow bands of retrogressive biotite are interpreted as an incipient cleavage and represent the only macroscopic deformation feature.

Small cataclastic zones at the dm to m scale occur locally, which strike parallel to the central parts of the ductile shear zones. They mostly consist of lenticular fragments of the mylonitic country rock, which are embedded in an anastomosing network of narrow fractures filled with fine-grained gouge (Fig. 5m). The age relationship between these young cataclastic zones and the mafic dykes described above could not be determined, because cross-cutting structures of these two fabric elements were not observed.

Planar antithetic fractures at a high angle to the large-scale shear plane, which affect both felsic and mafic layers, can be attributed to late strain increments at low temperatures (Fig. 5n). In addition, brittle faults are developed which locally curve into the mylonitic foliation.

Several isotopic systems have been analyzed in order to constrain the age of metamorphism, the cooling history and the timing of the shearing.

Monazite U–Pb ages

Teufel & Schärer (1989) and Henk *et al.* (1997) presented 12 generally concordant U–Pb monazite ages from metasedimentary samples collected along the Valstrona cross-section (Fig. 2). From petrographic observations, it appears that the monazites are unzoned and oriented parallel to the main foliation. The monazite ages in the IVZ are all rather similar and ranging from 292 Ma to 276 Ma. Older (Caledonian) ages were detected in the SCZ (Fig. 6). In the IVZ, the ages become younger with increasing depth. This trend has been interpreted to reflect either cooling after regional metamorphism (Henk *et al.* 1997) or fluid circulation after the thermal peak of this metamorphism (Vavra *et al.* 1996).

$^{40}Ar/^{39}Ar$ hornblende dating

Eleven hornblende separates (Table 1, Fig. 7a, 7b) were dated at the Geochronology Laboratory of the Geophysical Institute of the University of Alaska Fairbanks (see Appendix 1 for analytical information). The samples were collected along the Valstrona di Omegna profile (Fig. 2).

With the exception of samples farthest from the Insubric Line (SA1, SA2, SA9; c. 270 Ma), showing concordance between isochron and plateau ages, all samples show considerable amounts of excess argon. This is especially true for samples (VS01-06, VS02-06, VS03-06 and VS05-06) in the vicinity (<4 km) of the Insubric Line, which have $^{40}Ar/^{36}Ar_i$ ratios between 1450 and 1900 showing the presence of considerable amounts of excess argon. None of the samples show a true plateau. The weighted average ages of

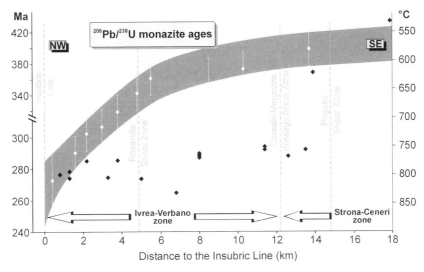

Fig. 6. Peak metamorphic temperatures (pale grey diamonds on grey trend) and U–Pb monazite ages (black diamonds) along the Valstrona section from the Insubric Line (0 km) to Omegna (c. 15 km). The data are taken from Henk et al. (1997) with one additional U–Pb monazite age from the SCZ of Grünfelder and Köppel (1971). Only the monazite ages in proximity to the Insubric Line can be explained by a complete reset of the U–Pb isotope system, whereas with distance the isotope system is only partly reset.

the youngest fractions averaging about 240 Ma, and the old integrated ages, are certainly biased by the excess argon. All four samples show well-defined isochrons with an average isochron age of 207.6 ± 2.3 Ma. The varying amounts of excess argon seen in these and other samples indicate open-system behaviour, and thus the isochron age is favoured in most cases (see Table 1). Isochron ages show a well-defined age progression becoming younger toward the Insubric Line (Fig. 7a).

K–Ar biotite ages

Another constraint on the cooling of the Ivrea Zone is gained from K–Ar biotite ages. Biotite separates were obtained from 30 metasedimentary rocks and measured at the K–Ar laboratory in Göttingen (see Appendix for analytical information). Of those samples, 19 correspond to the Valstrona whereas another 11 samples document the cooling history of the SCZ between the Lago d'Orta and Lago Maggiore (Table 2 and Fig. 8). Sample localities along the Valstrona largely correspond to the monazite samples of Henk et al. (1997) (Fig. 2). The K–Ar biotite ages continuously become younger from the SE to the NW, i.e. from the SCZ towards the IVZ. Lower Triassic ages (c. 245 Ma) were recorded in the SCZ near the Pogallo Line. Towards the Insubric Line, youngest ages are Late Jurassic, around 156 Ma, at the base of the IVZ. These data illustrate the migration of the zone of partial Ar retention in biotite, documenting the progressive cooling of the lower crust.

The trend of increasing K–Ar biotite ages advances in the direction of Lago Maggiore, where maximum ages of 335 Ma were recorded at the lakeside in the proximity of Lesa (Fig. 8). This profile also includes cooling ages of the Baveno granite scattering between 291 ± 7 Ma and 284 ± 7 Ma.

Zircon fission-track ages

The late stage of the cooling history is documented by zircon fission-track data for nine samples (see Appendix for analytical information). Analytical data are given in Table 3 and Figure 9. Only three of the nine dated samples passed the chi-square test, indicating that the single-grain ages do not form a normal distribution within the samples. The oldest, Lower Jurassic age of c. 178 Ma (SZ 10) is observed in the SCZ near the Pogallo Line, whereas the deepest part of the crustal section in the vicinity of the Insubric Line yielded Late Eocene ages of c. 38 Ma (VS 06-06, SZ 3).

Discussion

The new age data on the Valstrona cross-section together with the previously published work allow us to assemble an outline of the exhumation and deformation history of the region.

Table 1. *Hornblende data**

Sample	Distance to I.L. (km)	Integrated age ± 1σ (Ma)	Plateau (P) or weighted mean (W) age		Isochron (I) or errorchron (E) age		Ca/K
			Age ± 1σ (Ma)	Plateau information	Age ± 1σ (Ma)	Isochron information	
VS03-06	0.3	279.6 0.8	244.5 13.1	3 fractions (**W**) 71% ^{39}Ar release MSWD = 290.2	**201.7 13.0**	6 fractions (**E**) ^{40}Ar/^{36}Ar$_i$ = 1453 ±326 MSWD = 22.9	7.7
VS05-06	2.0	271.4 0.8	242.7 1.7	3 fractions (**W**) 56% ^{39}Ar release MSWD = 4.3	**210.3 4.4**	6 fractions (**I**) ^{40}Ar/^{36}Ar$_i$ = 1624 ±144 MSWD = 1.6	8.2
VS02-06	3.0	277.5 0.8	232.8 3.1	4 fractions (**W**) 73% ^{39}Ar release MSWD = 7.0	**206.1 3.0**	8 fractions (**I**) ^{40}Ar/^{36}Ar$_i$ = 1903 ±133 MSWD = 2.3	6.5
VS01-06	3.5	257.2 0.8	239.5 4.8	5 fractions (**W**) 78% ^{39}Ar release MSWD = 33.7	**211.1 7.0**	9 fractions (**E**) ^{40}Ar/^{36}Ar$_i$ = 1743 ±271 MSWD = 3.3	13.6
SA7	4.0	254.5 0.8	250.4 0.9	5 fractions (**P**) 73% ^{39}Ar release MSWD = 1.5	**222.7 12.5**	8 fractions (**I**) ^{40}Ar/^{36}Ar$_i$ = 1337 ±445 MSWD = 1.5	19.4
SA6	4.5	225.0 0.6	**221.9 0.6**	6 fractions (**W**) 94% ^{39}Ar release MSWD = 3.0	221.6 1.2	7 fractions (**E**) ^{40}Ar/^{36}Ar$_i$ = 327 ±60 MSWD = 4.8	6.5
SA5	5.5	294.8 0.8	261.9 0.9	3 fractions (**P**) 66% ^{39}Ar release MSWD = 0.1	**240.0 2.0**	5 fractions (**I**) ^{40}Ar/^{36}Ar$_i$ = 1177 ±52 MSWD = 2.1	9.9
SA4	6.5	284.6 0.8	246.7 1.5	1 fraction saddle (**W**)	**239.3 4.6**	5 fractions (**I**) ^{40}Ar/^{36}Ar$_i$ = 702 ±118 MSWD = 2.3	12.8
SA3	8.0	399.6 1.1	281.7 4.7	3 fractions (**W**) 44% ^{39}Ar release MSWD = 14.4	**250.4 3.7**	10 fractions (**E**) ^{40}Ar/^{36}Ar$_i$ = 777 ±23 MSWD = 7.2	15.4
SA9	10.5	264.2 0.9	**256.6 1.1**	5 fractions (**W**) 37% ^{39}Ar release MSWD = 2.8	261.2 9.9	5 fractions (**E**) ^{40}Ar/^{36}Ar$_i$ = 445 ±308 MSWD = 12.6	
SA2	12.5	265.0 0.8	**267.7 1.0**	3 fractions (**P**) 57% ^{39}Ar release MSWD = 1.3	269.5 2.0	6 fractions (**E**) ^{40}Ar/^{36}Ar$_i$ = 180 ±26 MSWD = 7.9	15.5
SA1	13.0	274.9 0.7	**271.1 0.7**	5 fractions (**P**) 84% ^{39}Ar release MSWD = 0.6	271.5 0.8	5 fractions (**I**) ^{40}Ar/^{36}Ar$_i$ = 266 ±18 MSWD = 0.6	13.2

***Bold**: Preferred age for each sample (ages reported at ±1 sigma).
Plateau: 3+ consecutive fractions, MSWD < c. 2.5, more than c. 50% ^{39}Ar release. If a group of fractions does not meet these criteria, a weighted average age is calculated. Isochron: MSWD < c. 2.5, otherwise it is considered an errorchron age.

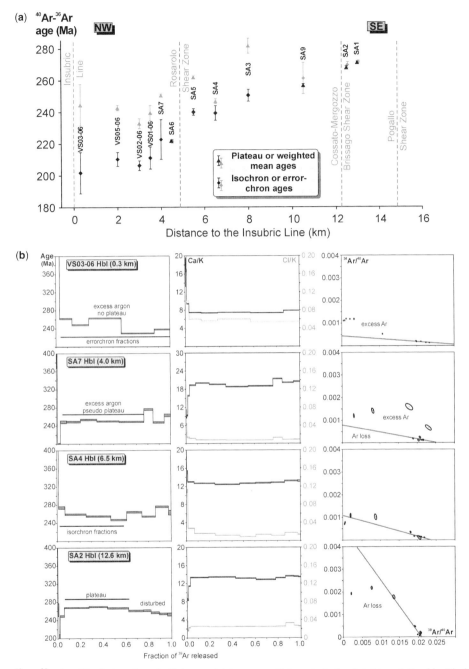

Fig. 7. ^{40}Ar–^{39}Ar hornblende data along the Valstrona. (**a**) Age versus distance to the Insubric Line relationship. Black symbols either corresponding to plateau and weighted mean ages (triangles) or isochron and errorchon ages (diamonds) are the preferred results of the data set. No age step is observed across the Cossato–Mergozzo–Bissago (CMB) shear zone suggesting a common cooling history of the IVZ and SCZ. (**b**) Age spectra of four representative hornblende analyses are shown. Besides the age information, i.e. heating step and isochron diagrams, the Ca/K and Cl/K ratios that might aid in the interpretation of the Ar analyses are presented. For the entire data set of the 12 analysed samples, the reader is referred to the online supplementary data at http://www.geolsoc.org.uk/SUP18311.

Table 2. *Compilation of the K/Ar biotite ages along the Valstrona–Lago Maggiore profile. The quoted errors comply with 95% of the confidence level (2σ). Ar-isotopic abundance:* ^{40}*Ar: 99.6000%;* ^{38}*Ar: 0.0630%;* ^{36}*Ar: 0.3370%. Spike–isotopic composition:* ^{40}*Ar: 0.009980%;* ^{38}*Ar: 99.9890000%;* ^{36}*Ar: 0.0009998%. Decay constants (1/a):* $\lambda\epsilon$*: 5.81 × 10*$^{-11}$*;* $\lambda\beta^-$*: 4.962 × 10*$^{-10}$*;* λ_{tot}*: 5.543 × 10*$^{-10}$*. Potassium:* ^{40}*K: 0.011670%;* K_2O/K*: 0.8302. Standard temperature pressure (STP): 0°C, 760 mmHg; Normal atmosphere (DIN 1343): 273.15 K, 1013.25 mbar. Molar volume (ml): 22413.8; Atomic weight (g/mol):* ^{tot}*Ar: 39.9477;* ^{40}*Ar: 39.9624;* ^{tot}*K: 39.1027.*

Sample	Mineral	Distance to I.L. (km)	K_2O (wt. %)	$^{40}Ar^*$ (nl/g)STP	$^{40}Ar^*$ (%)	K–Ar age ± 2σ (Ma)		± 2σ (%)
CL 23	bt	0.00	10.32	54.32	98.01	156.3	3.3	2.1
CL 25	bt	0.35	10.18	54.54	98.17	159.0	3.4	2.1
CL 26	bt	1.09	9.58	51.79	98.23	158.7	3.3	2.1
CL 28	bt	3.51	9.66	57.44	99.00	175.7	3.7	2.1
CL 19	bt	4.60	9.56	53.78	98.38	166.6	3.5	2.1
VST 24	bt	5.03	9.92	56.93	98.56	169.8	3.6	2.1
CL 16	bt	6.14	9.21	57.27	98.95	183.3	3.8	2.1
CL 15	bt	6.54	8.67	48.98	98.66	167.3	3.5	2.1
VST 3	bt	7.57	9.38	56.93	97.16	179.1	3.8	2.1
VST 15	bt	8.27	8.77	55.24	99.40	185.5	3.9	2.1
CL 10	bt	9.60	8.61	59.21	98.99	201.7	4.2	2.1
CL 9	bt	10.14	9.11	62.19	99.83	200.3	4.2	2.1
VST 9	bt	10.69	8.63	68.73	99.15	231.6	4.8	2.1
VST 7	bt	11.51	8.79	63.70	99.05	211.9	4.4	2.1
CL 7	bt	12.00	5.90	46.14	97.64	227.6	4.9	2.2
VST 13	bt	12.66	9.45	76.58	99.22	235.4	4.9	2.1
CL 2	bt	13.22	9.20	70.70	98.39	223.9	4.7	2.1
VST 5	bt	13.97	9.45	74.77	99.51	230.2	4.8	2.1
CL 1	bt	14.53	9.51	80.27	99.31	244.5	5.1	2.1
AH 90/7	bt	18.01	8.46	73.51	98.20	251.2	5.3	2.1
AH 90/13	bt	20.92	8.37	82.87	99.27	283.7	7.4	2.6
AH 90/6	bt	22.53	8.20	83.39	99.80	290.8	7.2	2.5
AH 90/9-2	bt	22.53	7.99	79.82	98.72	286.0	6.9	2.4
AH 90/5	bt	22.85	8.29	82.11	99.60	283.8	6.1	2.1
AH 90/12	bt	30.23	8.86	99.31	99.56	318.0	6.6	2.1
AH 90/4	bt	30.91	8.99	107.28	99.45	336.8	6.6	2.0
AH 90/3	bt	36.82	8.32	89.48	94.73	306.2	6.8	2.3
AH 90/10	bt	37.61	8.14	94.81	99.04	329.4	7.1	2.2
AH 90/2	bt	39.23	8.09	95.94	99.33	334.8	6.8	2.0
AH 90/1	bt	39.61	7.83	92.87	99.85	334.9	7.0	2.1

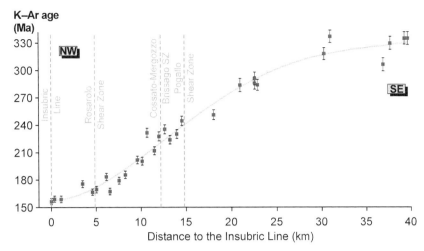

Fig. 8. K–Ar Bt ages between the Insubric Line (0 km) and Lago Maggiore (c. 40 km). Besides the IVZ, the profile also includes the K–Ar Bt cooling ages for the Strona Ceneri Zone (cf. Fig. 2 and Table 3). No abrupt steps of the cooling ages are observed across the two tectonic lines, i.e. the Cossato–Mergozzo–Bissago and Pogallo shear zones.

Significance of the geochronological data

Since the early work by Wagner et al. (1977), the estimated closure temperature interval of the U–Pb system in monazite has shifted significantly from 550 °C up to 650–725 °C (Copeland et al. 1988; Parrish 1990; Metzger et al. 1991). Kamber et al. (1998) even proposed a closure temperature in excess of 800 °C. The conditions leading to monazite growth in metamorphic rocks are poorly understood (Lanzirotti & Hanson 1996; Sawka et al. 1986; Bonn 1988; Seydoux-Guillaume 2001). Bonn (1988) and Milodowski & Hurst (1989) demonstrated that monazites may grow under greenschist facies or even during diagenetic conditions.

The peak of the granulite facies metamorphism in the IVZ was attained at c. 300 Ma (or earlier). This is based on the finite element approximation of thermal simulations by Henk et al. 1997, assuming a monazite closure temperature at 600 ± 50 °C (Smith & Barreiro 1990). In the amphibolite facies rocks (northwest and southeast of the CMB Line in Figs 1 and 6), the age of metamorphism is dated at c. 292 Ma (Henk et al. 1997). The high estimate of the monazite closure temperature by Kamber et al.

Table 3. Zircon fission-track ages obtained on gneiss samples of the Valstrona di Omegna profile across the Ivrea body. Cryst.: number of dated zircon crystals. Track densities (RHO) are as measured ($\times 10^5$ tr/cm^2); number of tracks counted (N) shown in brackets. Chi-sq P(%): probability obtaining Chi-square value for n degree of freedom (where n = no. crystals-1). Disp.: Dispersion, according to Galbraith & Laslett (1993).
*: Central ages calculated using dosimeter glass: CN2 with $\zeta_{CN2\text{-}zircon} = 127.8 \pm 1.6$

Sample	Distance to I.L. (km)	Cryst.	RhoS	(Ns)	RhoI	(Ni)	RhoD	(Nd)	Chi-sq. P (%)	Disp.	Central age ±1σ (Ma)	
VS 06-06	0.2	20	49.65	(859)	55.89	(967)	6.629	(4859)	4	0.16	37.7	2.3
SZ 3	3.0	18	47.90	(904)	52.56	(992)	6.631	(4859)	2	0.17	38.2	2.5
VS 10-06	4.5	20	80.51	(1195)	59.76	(887)	6.627	(4859)	0	0.23	56.8	4
SZ 4	6.0	20	83.73	(1251)	39.82	(595)	6.622	(4859)	53	0.01	88.4	4.7
SZ 6	6.5	15	95.19	(1048)	41.87	(461)	6.597	(4859)	0	0.27	97.1	9
SZ 7	8.5	20	123.51	(1979)	41.07	(658)	6.624	(4859)	84	0	126.1	6.2
SZ 9	10.5	15	126.33	(1515)	35.69	(428)	6.62	(4859)	0	0.29	151.5	14.4
SZ 1	13.0	20	159.80	(2372)	40.02	(594)	6.603	(4859)	54	0.02	166.3	8.3
SZ 10	13.8	21	217.20	(2199)	48.89	(495)	6.615	(4859)	0	0.24	180.1	13.4

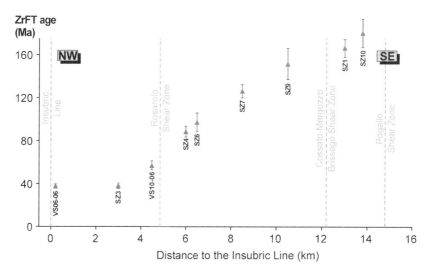

Fig. 9. Apparent zircon fission track data along the Valstrona profile (cf. Table 3). Note the distinct bend of the data trend at c. 3 km distance to the Insubric Line. A less-pronounced bend also occurs in the southeast of the Cossato–Mergozzo–Bissago (CMB) shear zone.

(1998) and the peak metamorphic temperatures (Sills & Tarney 1984; Henk et al. 1997) impede a straightforward interpretation of the monazite ages as either cooling or growth ages. The available U–Pb monazite ages (Henk et al. 1997) show a systematic decrease from 292 ± 2 Ma near the Pogallo Line to 276 ± 2 Ma at the base of the IVZ indicating younger ages for the granulite facies sequence. When assessing the role of multiple and long-lasting magmatic underplating with respect to the discussed P–T conditions (Zingg 1983; Sills & Tarney 1984; Henk et al. 1997), considering the above-discussed new closure temperature for U–Pb in monazite, the significance of the monazite ages may change. First, the elevated isotherms were only slightly affected (if at all) by the magmatic underplating. The SCZ provides an explanation for the ages (c. 292 Ma) in proximity to the Pogallo Line (see Fig. 7). The repeated heat input by the intrusion of mafic melts up to c. 285 Ma (Pin 1986; Peressini et al. 2007) or even longer (Voshage et al. 1990) was sufficient to reset the monazite U–Pb isotope equilibrium in the deeper parts of the IVZ explaining the younger ages. The long-lasting magmatic activity is also corroborated by 261 ± 4 Ma U–Pb SHRIMP ages on zircons from anatectic leucosomes in the IVZ metapelites (Vavra et al. 1996). However, the question arises whether the oldest monazite ages are already a result of reset due to the underplating.

The validity of the proposed 'closure temperature' of monazite of 800 °C or 725 °C can be easily constrained by the $^{40}Ar/^{39}Ar$ hornblende ages (see Figs 7 and 10). Following Dahl (1996) and Villa et al. (1996), the lower limit of the closure temperature of recrystallization-free amphiboles with values for $Z = 38.1\%$ (Z as defined by Dahl 1996) is ≥ 550 °C. Kamber et al. (1995) reported a temperature of 580 °C assuming a cooling rate of 1–2 °C/Ma. Boriani & Villa (1997) reported Z-values between 37.7% and 38.4% for amphiboles of the SCZ suggesting that at least the lower limit of the closure temperature interval is covered. The amphiboles of the SCZ and also within the amphibolite facies part of the IVZ have to be considered as formation ages dating the metamorphic peak at temperatures higher than 550 °C (e.g. Zingg 1983; Henk et al. 1997).

The $^{40}Ar/^{39}Ar$ hornblende ages and K–Ar biotite ages (Tables 1, 2 or Figs 7, 8 and 10) are interpreted to reflect the time of cooling of the crust after Permian heating caused by mafic intrusions. The closing temperature for hornblende should be around 500–600 °C (McDougall & Harrison 1999), while that for biotite can range between the extreme values of $300° \pm 50$ °C for simple cooling (Purdy & Jäger 1976; McDougall & Harrison 1999) and 450 °C for the total reset of the K–Ar system in recrystallization-free minerals (cf. Villa 1998). In this case, the lower closing temperature seems to be more appropriate assuming that the biotites in the vicinity of the Insubric Line behaved as open systems before and during the Permian heat flow.

The decrease of ages towards the rocks of a formerly deeper position simply reflects slow

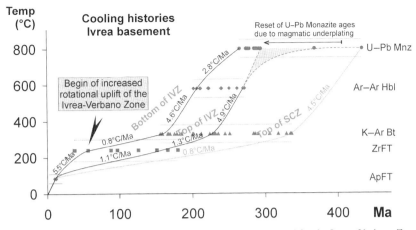

Fig. 10. Thermal history over the last 400 Ma for the Strona–Ceneri Zone and for the Ivrea–Verbano Zone.

exhumation. Permian or older hornblende and biotite cooling ages originate from rocks with a significantly higher crustal position before the late-stage tilting. Therefore, the whole transect represents a single trend. From the Valstrona section, we cannot find any indication for different cooling histories influenced by any movements along the Pogallo Line or the CMB Line. A comparable tendency may be assumed for the zircon FT ages which are only affected in the vicinity of the Insubric Line, indicating the beginning of an increased rotational uplift of the IVZ at around 40 Ma.

Thermal history–age of tilting

The obtained geochronological data from the Valstrona transect convey a certain amount of information on its tilting. Boriani et al. (1990) considered this process to be of Permian age, while Schmid et al. (1987) argued for an Eocene to Miocene age. Handy et al. (1999) found structural evidence indicating that the two crustal sections were rotated at different times: (1) the IVZ was tilted into its upright position during the Oligocene; and (2) the SCZ was already moderately steeply dipping.

The thermal history can be summarized as follows: the Permian underplating of mafic melts into the lower crust of the Ivrea section and the subsequent extension of the basement led to a pervasive heating, and partial or complete equilibration of, the U–Pb monazite isotope system between the Pogallo Line and Insubric Line in the Valstrona profile. Further to the SE, Caledonian ages are recorded, turning emphasis on the importance of the CMB Line during the Late Palaeozoic, when it acted as an extensional detachment facilitating the ascent of the mafic melts towards higher crustal levels

(e.g. Mulch et al. 2002). Peak metamorphic conditions gained during the emplacement of the mafic melts reached 810 °C and 8.3 kbar (Henk et al. 1997).

Thermal simulations by Henk et al. (1997), using an uplift rate of 0.4 mm/a for 10 Ma, indicate that the disturbance of isotherms will persist for slightly longer than 5 Ma. Only when considerably higher peak metamorphic temperatures (>1000 °C) are assumed, will the thermal anomaly fade away after around 20 Ma. In this way, the obtained Ar–Ar ages that are at least 20 Ma younger (when plateau ages are considered) than the corresponding U–Pb monazite ages could be perturbed by the thermal impact of the mafic intrusion as well. Moreover, the careful evaluation of the analytical results favours the Ar–Ar isochron ages for the samples in proximity to the Insubric Line at c. 210 Ma that are certainly beyond the time interval of the thermal impact. Therefore, the $^{40}Ar-^{36}Ar$ hornblende ages in the range between c. 270 Ma in proximity to the Pogallo and CMB Lines and c. 210 Ma at the deepest section of the profile are attributed to the regional cooling and exhumation of the Ivrea basement during Permian and Triassic times.

The majority of Ar and zircon FT ages of the profile are apparent ages for the time span between the Permian and Eocene. They formed in a relatively stable crust that experienced high heat flow during the Triassic to Jurassic rifting phases. At first glance, the shifts of Ar and zircon FT ages along the profile suggest a long-lasting, monotonous cooling process that started in the Permian and lasted through the entire Mesozoic era until the Palaeogene. However, this interpretation has to be rejected, because it contradicts the known evolutionary history of the southern Alps. The stratigraphic record suggests a calm, continental

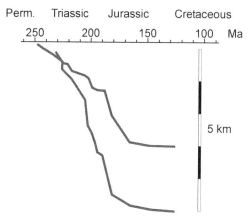

Fig. 11. Subsidence history of the southern Alps (from Stämpfli et al. 2002). The most intense extension happened during Early Jurassic time, between c. 200 and 175 Ma.

sedimentation for the post-Permian time (e.g. Stämpfli et al. 2002) until the mid-Triassic, and later an intense subsidence with alternating rift and basin facies development on the horst and graben structures formed by continental extension (see Fig. 11, Bertotti 1991). The Late Triassic and Jurassic crustal extension created a period of high heat flow in the southern Alps. Thus, the supposed exhumation for the western southern Alps during the Mesozoic is in sharp contradiction with the facts of sedimentation and structural records.

The hornblende $^{40}Ar/^{39}Ar$ ages (Fig. 7) show a monotonous trend of increasing isochron ages from 3.5 km towards higher crustal levels at the Pogallo Line. The oldest $^{40}Ar/^{39}Ar$ data, far from the Insubric Line, are practically identical to the mean of the monazite U–Pb ages. Thus, in that crustal level (the highest part of the profile), the post-Permian thermal pulses were not able to reset the Ar system of the hornblende. The very remarkable difference between the monazite ages and the youngest hornblende $^{40}Ar/^{39}Ar$ ages close to the Insubric Line indicates that the deepest part of the profile experienced a relatively high temperature of $>550\ °C$ at least until Triassic times. These hornblende ages, at c. 210 Ma, point to the beginning of extension-induced subsidence in the Lombardian basin.

The biotite K–Ar ages ranging between 230 and 156 Ma run parallel to the hornblende ages in the upper part of the profile. A minor shift to a shallower trend is located about 8 km from the Insubric Line (Fig. 12). No abrupt offset of the decreasing age profile is observed across the two tectonic lineaments located in the lower part of the profile, i.e. the CMB and the Pogallo shear zones. This suggests that Mesozoic tectonism affected the entire basement pile in a uniform way. This trend is typical in exhumed argon retention zones, as was documented by Kamp et al. (1989) for the southern Alps of New Zealand, where, similar to the Ivrea Zone, the former reset zones of the different chronometers were exhumed with different metamorphic mineral assemblages.

The slope recorded between 15 and 8 km of the profile forms the base of the preserved partial argon retention zone. Below this zone (between c. 8 and 0 km on the profile), the trend has a high angle, and we interpret this section as the former depth of zero argon ages. The partial retention zone was formed at the end of Mesozoic rifting. The c. 160 Ma age of the break-in-slope corresponds well to the end of the intense rifting documented by the subsidence curves of the southern Alps (see Fig. 11).

The oldest zircon FT ages (Fig. 12) were measured in the highest part of the profile, above the depth of the break-in-slope of the biotite K–Ar age trend. The oldest zircon ages correspond well to the supposed reset history of biotite ages. The partial annealing zone (PAZ, namely the oblique section of the zircon ages) occupies a deeper position than the partial retention zone of argon ages. This PAZ was formed later, in post-Jurassic (post-rift) times, probably during and after the sinking and stabilization of the perturbed isotherms. The lower break-in-slope (4 km on the profile, c. 50 Ma) is either the fossil depth of track instability, or the section with the low age gradient was formed during the rotation of the block.

The lower limit for the rotational exhumation of the Ivrea basement is confined by the apatite FT-thermochronometer, which indicates that cooling was below c. 100 °C and occurred around 12 Ma. The sporadic data published from the entire Ivrea complex do not reveal a trend (cf. Wagner et al. 1977; Hurford 1986; Hunziker et al. 1992). It is difficult to interpret these ages as providing kinematic evidence, because the young apatite FT data are mainly controlled by the Late Neogene exhumation of the western Alps (Fügenschuh & Schmid 2003), and the very rugged local relief also has an impact on the age pattern (see e.g. Stüve et al. 1994).

Summary and conclusions

Structural observations at the macro- to microscopic scale characterize the Rosarolo shear zone in the upper Valstrona as a long-term shear zone which initiated during the Early Permian magmatic underplating. It accommodated extension in the lower crust under high-temperature conditions. This sinistral sense of shear is consistent with the

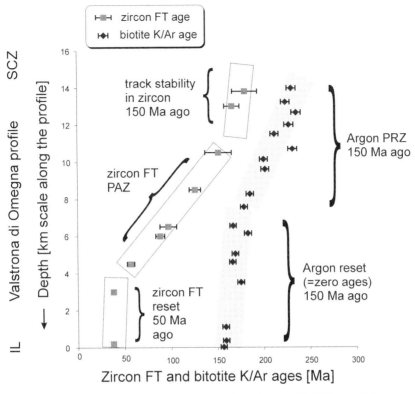

Fig. 12. Zircon FT and biotite K/Ar ages projected on an age/elevation plot. The Valstrona di Omegna profile is in the vertical position, as is hypothesized for the pre-40 Ma period. The steep intervals are the former zones of reset and a closed thermochronometer. The oblique intervals are exhumed partial retention and partial annealing zones (PRZ and PAZ respectively, for argon and FT chronometers).

gabbro-glacier growth of the Mafic Complex (see Quick et al. 1994). Its reactivation at different crustal levels during Mesozoic subsidence and final Eocene tilting is documented by structures indicating strain increments at different P/T conditions. Due to the late overall c. 90° tilt of this crustal section, the former extensional mylonites currently appear as a left-lateral transform shear zone. In order to confine the timing of this vertical rise, different isotopic systems along the Valstrona were analysed. They all show coherent age progression patterns related to the Late Variscan to Neo-Alpine geodynamic evolution of the Ivrea Zone's basement (Figs 12 and 13).

U–Pb monazite ages between 292 Ma and 274 Ma (Henk et al. 1997) are rather uniform. The younger ages in proximity to the Insubric Line, formerly the deepest part of the crust, are explained by the heat input by mafic melts leading to the complete equilibration of the U–Pb system in monazite. Recent high estimates on the closure temperature at 725 °C or even 800 °C of this isotope system point out that the monazite ages at 292 Ma at upper crustal levels of the IVZ reflect incomplete isotope reset and that the pre-Permian metamorphic peak is still an open question. Evidence for pre-Permian ages is only known from a few monazite ages in the SCZ. This contrasts with earlier interpretations based on numerical simulations on the temperature degradation with lower U–Pb closure temperatures for monazite (Henk et al. 1997), suggesting a complete obliteration of any pre-Permian geochronological record in the IVZ. However, considering the more recently reported SHRIMP-data by Peressini et al. (2007), who found out that the Mafic Complex grew from 290 (or earlier) to 288 Ma, the question arises whether the youngest monazite ages are simple cooling ages or the result of the latest events of underplating.

The age of the main metamorphic event in the SCZ is Early Palaeozoic (c. 450 Ma). $^{40}Ar/^{39}Ar$ hornblende data (Boriani & Villa 1997) point to a Variscan overprint at 320–350 Ma. These findings are in agreement with the oldest K–Ar biotite ages of c. 345 Ma in Lesa at the lakeside of Lago Maggiore. Hence, the K–Ar biotite ages for the Baveno granite at c. 291–284 Ma support the concept that

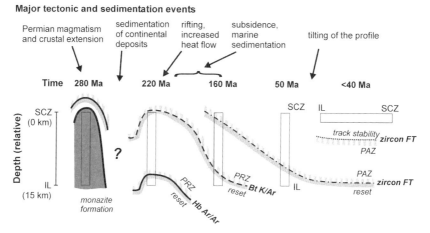

Fig. 13. Evolution of the Valstrona di Omegna profile through time (IL, Insubric Line; SCZ, Strona–Ceneri Zone). The long rectangles represent the profile, whereby it maintained a vertical position until c. 50 Ma, and then it was tilted to the present horizontal position at around 40 Ma ago. The curved lines express the approximate relation of the closure of the different isotope chronometers to the profile. The absolute depth of the profile cannot be determined. However, its position experienced a minor change, and the closure isotherms had a significant vertical oscillation driven by the basal heat flow and stretching of the lithosphere. PRZ, partial retention of argon; PAZ, partial annealing of fission tracks; reset, the decay products are not stable in the crystal, and the geochronometer shows zero age.

this pluton was emplaced in an already cooled basement with temperatures less than or equal to 300 °C. Its shallow emplacement level is also evident by the larger amount of miarolitic cavities.

The other applied chronometers, i.e. Ar-systematics in hornblende and biotite as well as the zircon fission-track data, document the cooling of the crust after the Permian underplating of mafic melts. Well-defined and roughly parallel age progressions of the hornblende and biotite data show a decrease from the Pogallo Line towards the Insubric Line cross-cutting the CMB Line, i.e. downwards from the middle continental crust to lower crustal levels. Carefully evaluated $^{40}Ar/^{39}Ar$ hornblende raw data yield ages from c. 271 Ma to c. 201 Ma. K–Ar biotite ages on the same distance range from c. 230 Ma to 156 Ma, whereas the zircon FT ages between 180 Ma to 38 Ma bracket a much broader time interval. No significant offsets of these ages are manifested, neither across the contact between the IVZ and the SCZ, as characterized by the CMB mylonites (Boriani & Giobbi 2004), the Pogallo Line, nor the Rosarolo shear zone. The mechanical juxtaposition of the IVZ against the SCZ along the CMB and/or Pogallo Lines is still under debate. Their juxtaposition exclusively during Early Mesozoic extensional shearing along the Pogallo Shear Zone is not documented in our data. It turns out that the CMB and Pogallo Lines do not disturb the post-Permian geochronology of the Ivrea Zone's basement. Missing age constraints on the displacement along the CMB Line is caused by the Late Permian thermal event that reset the monazite ages (see discussion above), or alternatively the total amount of displacement along the CMB was little or below the resolution of the applied methods. However, significant crustal attenuation between 2 and 4 km during the Early Permian to Middle Triassic has been proposed (Brodie & Rutter 1987; Henk et al. 1997). Three kilometres of crustal thinning some time during the Early Permian to Middle Triassic along the Pogallo Line was inferred from the cooling ages (Handy 1987). However, this is not confirmed by the $^{40}Ar/^{39}Ar$ hornblende and K–Ar biotite ages along the Valstrona, which show a roughly monotonous decrease towards the deeper part of the crust. The new data may confirm current interpretations of the Pogallo Line as a Mesozoic low-angle normal fault subsequently rotated into a steep orientation during Neo-Alpine times. The restoration of the IVZ–SCZ basement to a pre-Alpine configuration (Handy 1987; Schmid et al. 1987) explains the oblique low-angle normal fault. The deeper level is exposed in the Val Cannobino, whereas the upper level with a broader zone of cataclasis is exposed in the Valstrona.

Both the K–Ar biotite and the zircon FT age trends show a break-in-slope at roughly 160 Ma (Late Jurassic) after the end of the formation of Jurassic rift structures. This led to the formation of the Piedmont-Ligurian Ocean as part of the Tethys, probably during and after the sinking and stabilization of the perturbed isotherms. The intervals of gradual change of ages, interpreted as an exhumed partial argon retention zone and/or partial stability zone of zircon fission tracks, respectively, document the progressive cooling of the crust previously (K–Ar biotite data) and subsequently (zircon fission track data) to the Jurassic extension. The upper break-in-slope of the zircon fission track ages at c. 50 Ma, interpreted to be formed during, or immediately after the rotation of the IVZ, is consistent with the main Neo-Alpine collision during the Eocene. No evidence was found for any individual rotational uplift of the two basement slices or that tilting already took place earlier in the evolution.

With this paper, we wish to remember our colleague Hans Ahrendt who carried out most of the mapping in the Ivrea Zone, and almost all of the K–Ar analyses together with Petra Clausen and Doris Jaeger. Special thanks go to the staff of the nuclear reactors of Oregon State University and McMaster University for their careful neutron irradiations. For help with the sampling, we would like to thank Stefan Mosch and Pedro Oyhantcabal. For very helpful comments, we are most grateful to S. Sinigoi, B. Fügenschuh, J.H. Kruhl and S. Schmid.

Appendix: Analytical methods

$^{40}Ar/^{39}Ar$ hornblende dating

Eleven hornblende separates were dated at the Geochronology Laboratory of the Geophysical Institute of the University of Alaska Fairbanks. The monitor mineral MMhb-1 (Samson & Alexander 1987) with an age of 513.9 Ma (Lanphere & Dalrymple 2000) was used to check the neutron flux and calculate the irradiation parameter (J) for all samples. The samples and standards were wrapped in aluminium foil and loaded into aluminium cans of 2.5 cm diameter and 6 cm height. All samples were irradiated in position 5c of the uranium-enriched research reactor of McMaster University in Hamilton, Ontario, Canada for 30 megawatt-hours.

Upon their return from the reactor, the samples and monitors were loaded into 2 mm-diameter holes in a copper tray that was then loaded in an ultra-high vacuum extraction line. The monitors were fused, and samples heated, using a 6-watt argon-ion laser following the techniques described in York et al. (1981), Layer et al. (1987) and Layer (2000). Argon purification was achieved using a liquid nitrogen cold trap and a SAES Zr–Al getter at 400 °C. The samples were analysed in a VG-3600 mass spectrometer. The measured argon isotopes were corrected for system blank and mass discrimination, as well as calcium, potassium and chlorine interference reactions following procedures outlined in McDougall and Harrison (1999). System blanks generally were 2×10^{-16} mol ^{40}Ar and 2×10^{-18} mol ^{36}Ar that are 10 to 50 times smaller than fraction volumes. Mass discrimination was monitored by running both calibrated air shots and a zero-age glass sample. All ages are quoted to the ± 1 sigma level and were calculated using the constants of Steiger and Jäger (1977). The integrated ages is the age given by the total gas measured and is equivalent to a potassium–argon (K–Ar) age. The spectrum provides a plateau age if three or more consecutive gas fractions represent at least 50% of the total gas release and are within two standard deviations of each other (Mean Square Weighted Deviation less than c. 2.5).

K–Ar dating of biotite samples

Mica separation was performed by the standard techniques such as crushing, sieving, using the Frantz magnetic separator and selection by hand. The pure micas were ground in alcohol and sieved to remove altered rims which might have suffered argon loss.

The argon isotopic composition was measured in a Pyrex glass extraction and purification line coupled to a VG 1200 C noble gas mass spectrometer operating in static mode. The amount of radiogenic ^{40}Ar was determined by the isotopic dilution method using a highly enriched ^{38}Ar spike from Schumacher, Bern (Schumacher 1975). The spike is calibrated against the biotite standard HD-B1 (Fuhrmann et al. 1987). The age calculations are based on the constants recommended by the IUGS quoted in Steiger & Jäger (1977).

Potassium was determined in duplicate by flame photometry using an Eppendorf Elex 63/61. The samples were dissolved in a mixture of Hf and HNO_3 according to the technique of Heinrichs & Herrmann (1990). CsCl and LiCl were added as an ionization buffer and internal standard, respectively. The analytical error for the K–Ar age calculations has a 95% confidence level of 2σ. The procedural details for argon and potassium analyses at the laboratory in Göttingen are given in Wemmer (1991).

Zircon fission-track dating

The zircon crystals were collected by using a heavy liquid and magnetic separation, and then the crystals were embedded in PFA Teflon. To reveal the spontaneous tracks, the eutectic melt of NaOH-KOH was used at the temperature of 225 °C (Gleadow et al. 1976). The etching time varied from 23–54 hours. Neutron irradiations were made at the research reactor of Oregon Sate University. The external detector method was used (Gleadow 1981); after irradiation, the induced fission tracks in the mica detectors were revealed by etching in 40% HF acid for 30 min. Track counts were made with a Zeiss-Axioskop microscope (magnification of 1000) and a computer-controlled stage system (Dumitru 1993). The FT ages were determined by the ZETA method (Hurford & Green 1983) using age standards listed in Hurford (1998). The error was calculated using the classical procedure, i.e. by Poisson dispersion (Green 1981). Calculations and plots were made with the TRACKKEY program (Dunkl 2002).

References

AHRENDT, H., CLAUSEN, P. & VOLLBRECHT, A. 1990. Long-term shear zones in lower crustal rocks: an example from the Ivrea zone (Italian Alps). *Symposium Deformation Processes and the Structure of the Lithosphere*, Potsdam, May 3rd–10th, abstracts: p. 1.

BERCKHEMER, H. 1969. Direct evidence for the composition of the lower crust and Moho. *Tectonophysics*, **8**, 97–105.

BERTOLANI, M. 1969. La Petrografia delle Valle Strona (Alpi Occidentale Italiane). *Schweizerische Mineralogische und Petrographische Mitteilunngen*, **49**, 314–328.

BERTOTTI, G. 1991. Early Mesozoic extension and Alpine shortening in the western southern Alps: the geology of the area between Lugano and Menaggio (Lombardy, Northern Italy). *Memorie di Scienze Geologiche Padova*, **43**, 17–123.

BORIANI, A. & GIOBBI, E. 2004. Does the basement of western southern Alps display a tilted section through the continental crust? A review and discussion. *Periodico di Mineralogia*, **73**, 5–22.

BORIANI, A. & VILLA, I. 1997. Geochronology of regional metamorphism in the Ivrea–Verbano Zone and Serie dei Laghi, Italian Alps. *Schweizerische Mineralogische und Petrographische Mitteilungen*, **77**, 381–402.

BORIANI, A., ORIGONI GIOBBI, E., BORGHI, A. & CAIRONI, V. 1990. The evolution of the 'Serie dei Laghi' (Strona-Ceneri and Schisti dei Laghi): The upper component of the Ivrea–Verbano crustal section; southern Alps, north Italy and Ticino, Switzerland. *Tectonophysics*, **182**, 103–118.

BRODIE, K. & RUTTER, E. 1987. Deep crustal extensional faulting in the Ivrea zone of Northern Italy. *Tectonophysics*, **140**, 193–212.

BRODIE, K., REX, D. & RUTTER, E. 1989. On the age of deep crustal extensional faulting in the Ivrea Zone, northern Italy. *In*: COWARD, M. P., DIETTRICH, D. & PARK, R. G. (eds) *Alpine Tectonics*. Geological Society Special Publication, **45**, 203–210.

BRODIE, K., RUTTER, E. & EVANS, P. J. 1992. On the structure of the Ivrea–Verbano Zone (N. Italy) and its implication for present day lower continental crust geometry. *Terra Nova*, **4**, 34–40.

CLAUSEN, P. 1990. *Systematische K/Ar-Altersbestimmungen an Biotite in einem Querprofil durch die Ivrea-Verbano Zone (Val Strona di Omegna, Provinz Novara, Norditalien)*. Unpublished Diploma thesis, University Göttingen, 1–41.

COPELAND, P., PARRISH, R. R. & HARRISON, T. M. 1988. Identification of inherited radiogenic Pb in monazite and its implications for U–Pb systematics. *Nature*, **333**, 760–763.

DAHL, P. S. 1996. The effects of composition on retentivity of Ar and O in hornblende and related amphiboles: a field-tested empirical model. *Geochemica et Cosmochimica Acta*, **60**, 3687–3700.

DORNBUSCH, H.-J. & SKROTZKI, W. 2001. Microstructure and texture formation in a high-temperature shear-zone—with emphasis on amphibole. *In*: SIEGESMUND, S. (ed.) *Festschrift Klaus Weber*. Zeitschrift der Deutschen Geologischen Gesellschaft, **152**, 503–526.

DUMITRU, T. 1993. A new computer-automated microscope stage system for fission-track analysis. *Nuclear Tracks and Radiation Measurement*, **21**, 575–580.

DUNKL, I. 2002. TRACKKEY: a Windows program for calculation and graphical presentation of fission track data. *Computers & Geosciences*, **28**, 3–12.

FOUNTAIN, D. M. 1976. The Ivrea–Verbano and Strona-Ceneri zones, northern Italy: a cross section of the continental crust—new evidence from seismic velocities. *Tectonophysics*, **33**, 145–166.

FOUNTAIN, D. M. 1986. Implication of deep crustal evolution for seismic reflection seismology. *In*: BARAZANGI, M. & BROWN, L. (eds) *Reflection Seismology: The Continental Crust*. American Geophysical Union, Geodynamic Series, **14**, 1–7.

FOUNTAIN, D. M. 1989. Growth and modification of lower continental crust in extended terrains: the role of extension and magmatic underplating. *In*: MEREU, R. F., MUELLER, S. & FOUNTAIN, D. M. (eds) *Lower Crust Properties and Processes*. American Geophysical Union Monographs, **51**, 287–299.

FOUNTAIN, D. M. & SALISBURY, M. H. 1981. Exposed cross section through the continental crust: Implication for crustal structure, petrology and evolution. *Earth and Planetary Science Letters*, **5**, 263–277.

FÜGENSCHUH, B. & SCHMID, S. M. 2003. Late stage of deformation and exhumation of an orogen constrained by fission-track data: A case study of the Western Alps. *The Geological Society of America Bulletin*, **115**, 1425–1440.

FUHRMANN, U., LIPPOLT, H. J. & HESS, J. C. (1987) Examination of some proposed K–Ar standards: $^{40}Ar/^{39}Ar$ analyses and conventional K–Ar-Data. *Chemical Geology (Isotopic Geoscience Section)*, **66**, 41–51.

GARUTTI, G., RIVALENTI, G., ROSSI, A. & SINIGOI, S. 1978/1979. Mineral equilibria as geotectonic indicators in the ultramafics and related rocks of the Ivrea-Verbano Basic Complex (Italian Western Alps) Pyroxenes and Olivine. *Memorie di Scienze Geologiche*, **33**, 147–160.

GLEADOW, A. J. W. 1981. Fission-track dating methods: what are the real alternatives? *Nuclear Tracks*, **5**, 3–14.

GLEADOW, A. J. W., HURFORD, A. J. & QUAIFE, R. D. 1976. Fission track dating of zircon: improved etching techniques. *Earth and Planetary Research Letters*, **33**, 273–276.

GREEN, P. F. 1981. A new look at statistics in fission track dating. *Nuclear Tracks*, **5**, 77–86.

HANDY, M. R. 1987. The structure, age and kinematics of the Pogallo fault zone, southern Alps, northwestern Italy. *Eclogae Geologicae Helvetiae*, **80**, 593–632.

HANDY, M. R. & ZINGG, A. 1991. The tectonic and rheological evolution of an attenuated cross section of the continental crust: Ivrea crustal section, southern Alps, northwestern Italy and southern Switzerland. *Geological Society American Bulletin*, **103**, 236–253.

HANDY, M. R., FRANZ, L., HELLER, F., JANOTT, B. & ZURBRIGGEN, R. 1999. Multistage accretion and exhumation of the continental crust (Ivrea crustal section, Italy and Switzerland). *Tectonics*, **18**, 1154–1177.

HARTMANN, G. & WEDEPOHL, K. H. 1993. The composition of peridotite tectonites from the Ivrea Complex, northern Italy: residues from melt extraction. *Geochemica et Cosmochimica Acta*, **57**, 1761–1782.

HEINRICHS, H. & HERRMANN, A. G. 1990. *Praktikum der Analytischen Geochemie*. Springer Verlag.

HENK, A., FRANZ, L., TEUFEL, S. & ONCKEN, O. 1997. Magmatic underplating, extension and crustal reequilibration: insight from a cross section through the Ivrea-Verbano Zone and Serie del Laghi. N.W. Italy. *Journal of Geology*, **105**, 367–377.

HODGES, K. V. & FOUNTAIN, D. M. 1984. Pogallo Line, South Alps: an intermediate crustal level, low-angle normal fault. *Geology*, **12**, 151–155.

HUNZIKER, J. & ZINGG, A. 1980. Lower Paleozoic amphibolite to granulite facies metamorphism in the Ivrea Zone (southern Alps, northern Italy). *Schweizerische Mineralogische und Petrographische Mitteilungen*, **60**, 181–213.

HUNZIKER, J., DESMONS, J. & HURFORD, A. J. 1992. Thirty-two years of geochronological work in the Central and Western Alps: a review on seven maps. *Memoires de Geologie Lausanne*, **13**, 1–59.

HURFORD, A. J. 1986. Cooling and uplift patterns in the Lepontine Alps South Central Switzerland and an age of vertical movement on the Insubric fault line. *Contributions to Mineralogy and Petrology*, **93**, 413–427.

HURFORD, A. J. 1998. Zeta: the ultimate solution to fission-track analysis calibration or just an interim measure? *In*: VAN DEN HAUTE, P. & DE CORTE, F. (eds) *Advances in Fission-Track Geochronology*. Kluwer Academic Publishers, 19–32.

HURFORD, A. J. & GREEN, P. F. 1983. The zeta age calibration of fission-track dating. *Chemical Geology (Isotopic Geoscience Section)*, **41**, 285–312.

KAMP, P. J. J., GREEN, P. F. & WHITE, S. H. 1989. Fission track analysis reveals character of collisional tectonics in New Zealand. *Tectonics*, **8**, 169–195.

KAMBER, B. S., KRAMERS, J. D., NAPIER, R., CLIFF, R. A. & ROLLINSON, H. R. 1995. The Triangle Shear Zone, Zimbabwe: new data document an important event at 2.0 Ga in the Limpopo Belt. *Precambrian Research*, **70**, 191–213.

KAMBER, B. S., FREI, R. & GIBBS, A. J. 1998. Pitfalls and new approaches in granulite chronometry. An example from the Limpopo Belt. *Precambrian Research*, **91**, 269–285.

KÖPPEL, V. 1974. Isotopic U-Pb ages of monazite and zircons from the crust-mantle transition and adjacent units of the Ivrea and Strona-Ceneri Zones (southern Alps). *Schweizerische Mineralogische und Petrographische Mitteilungen*, **51**, 385–409.

KÖPPEL, V. & GRÜNFELDER, M. 1978/1979. Monazite and zircon U–Pb ages from the Ivrea and Ceneri Zones (southern Alps, Italy). *Memorie di Scienze Geologiche*, **33**, 55–70.

LANPHERE, M. A. & DALRYMPLE, G. B. 2000. First-principles calibration of ^{38}Ar tracers: Implications for the ages of $^{40}Ar/^{39}Ar$ fluence monitors. *U.S. Geological Survey Professional Paper*, **1621**, 10.

LANZIROTTI, A. & HANSON, G. N. 1996. Geochronology and geochemistry of multiple generations of monazite from the Wepawaug Schist, Connecticut, USA: implication for monazite stability in metamorphic rocks. *Contributions to Mineralogy and Petrology*, **125**, 332–340.

LAYER, P. W. 2000. Argon-40/Argon-39 age of the El'gygytgyn impact event, Chukotka, Russia. *Meteoritics and Planetary Science*, **35**, 591–599.

LAYER, P. W., HALL, C. M. & YORK, D. 1987. The derivation of $^{40}Ar/^{39}Ar$ age spectra of single grains of hornblende and biotite by laser step heating. *Geophysical Research Letters*, **14**, 757–760.

LOGAN, J. M., FRIEDMAN, M., HIGGS, N. G., DENGO, C. & SHIMAMOTO, T. 1979. Laboratory studies of simulated gouge and their application to studies of natural fault zones. *In*: Proc. Conf. VIII, Analysis of actual fault zones in bedrock. *U.S. Geological Survey Open File Rep.*, **79**, 305–343.

MCDOUGALL, I. & HARRISON, T. M. 1999. *Geochronology and thermochronology by the $^{40}Ar/^{39}Ar$ Method*. (2nd edn), Oxford University Press, New York.

MEHNERT, K. 1975. The Ivrea zone, a model of the deep crust. *Neues Jahrbuch Mineralogie Abhandlungen*, **125**, 156–199.

METZGER, K., RAWNSLEY, C. M., BOHLEN, S. R. & HANSON, G. N. 1991. U–Pb garnet, sphene, monazite and rutile ages; implication for the duration of high-grade metamorphism and cooling histories, Adirondack Mts., New York. *Journal of Geology*, **99**, 415–428.

MULCH, A., ROSENAU, M., DÖRR, W. & HANDY, M. R. 2002. The age and structure of dikes along the tectonic contact of the Ivrea–Verbano and Strona-Ceneri Zones (southern Alps, Northern Italy, Switzerland). *Schweizerische Mineralogische und Petrographische Mitteilungen*, **82**, 55–76.

NOVARESE, V. 1929. La Zona del Canavese e le formazioni adiacenti. *Memorie Descrittive Della Carta Geologia d'Italia*, **22**, 65–212.

PARRISH, R. R. 1990. U-Pb dating of monazite and its application to geological problems. *Canadian Journal of Earth Sciences*, **27**, 1431–1450.

PIN, C. 1986. Datation U-Pb sur zircons a 285 M.a. du complex gabbro-dioritique du Val Sesia-Val Mastallone et age tardi-hercyien du metamorphisme granitique de la zone Ivrea–Verbano (Italie), *Comptes Rendu de L'Academie des Sciences, Paris*, **303**, 827–830.

PERESSINI, G., QUICK, J. E., SINIGOI, S., HOFFMANN, A. W. & FANNING, M. 2007. Duration of a large mafic intrusion and heat transfer in the lower crust: a SHRIMP U–Pb zircon study in the Ivrea–Verbano Zone (Western Alps, Italy). *Journal of Petrology*, **50**, 1–34.

PURDY, J. W. & JAEGER, E. 1976. K-Ar ages on rock-forming minerals from the central Alps. *Memorie degli Istituti di Geologia e Mineralogia dell'Universita di Padova*, **30**, 1–321.

QUICK, J., SINIGOI, S. & MAYER, A. 1994. Emplacement of mantle peridotite in the lower continental crust, Ivrea Verbano Zone, Northwestern Italy. *Journal of Geophysical Research*, **99**, 21559–21573.

RABBEL, W., SIEGESMUND, S., WEISS, T., POHL, M. & BOHLEN, T. 1998. Shear wave anisotropy of the laminated lower crust beneath Urach (SW Germany) - a comparison with xenoliths and with exposed lower crustal sections. *Tectonophysics*, **298**, 337–356.

RIGALETTI, A. R., HEBEDA, E. H., SIGNER, P. & WIELER, R. 1994. Uranium–xenon chronology: precise determination of λ_{sf} $^{136}Y_{sf}$ for spontaneous fission of ^{238}U. *Earth and Planetary Science Letters*, **128**, 653–670.

RIVALENTI, G., GARUTI, G. & ROSSI, A. 1975. The origin of the Ivrea-Verbano basic formation (western Italian Alps): Whole rock geochemistry. *Bollettino della Società Geologica Italiana*, **94**, 1149–1186.

RIVALENTI, G., GARUTI, G., ROSSI, A., SIENA, F. & SINIGOI, S. 1981. Existence of different peridotite types and of a layered igneous complex in the Ivrea Zone of the western Alps. *Journal of Petrology*, **22**, 127–153.

RUTTER, E., BRODIE, K. & EVANS, P. 1993. Structural geometry, lower crustal magmatic underplating and lithospheric stretching in the Ivrea–Verbano Zone, N. Italy. *Journal of Structural Geology*, **15**, 647–552.

RUTTER, E., BRODIE, K., JAMES, T. & BURLINI, L. 2007. Large-scale folding in the upper part of the Ivrea–Verbano zone, NW Italy. *Journal of Structural Geology*, **29**, 1–17.

SAMSON, S. D. & ALEXANDER, E. C. 1987. Calibration of the interlaboratory $^{40}Ar/^{39}Ar$ dating standard, MMhb1. *Chemical Geology*, **66**, 27–34.

SAWKA, W. N., BANFIELD, J. F. & CHAPELL, B. W. 1986. A weathering-related origin of widespread monazite in S-type granites. *Geochemica et Cosmochimica Acta*, **50**, 171–175.

SCHMID, R. 1978/1979. Are the metapelites in the Ivrea Zone restites? *Memorie di Scienze Geologiche*, **33**, 67–69.

SCHMID, S. M. 1993. Ivrea zone and adjacent southern Alpine basement. *In*: VON RAUMER, J. F. & NEUBAUER, F. (eds) *Pre-Mesozoic Geology in the Alps*. Springer Verlag, New York, 567–583.

SCHMID, S. M., ZINGG, A. & HANDY, M. R. 1987. The kinematics and movements along the Insubric Line and the emplacement of the Ivrea Zone. *Tectonophysics*, **135**, 47–66.

SCHUMACHER, E. 1975. Herstellung von 99,9997% ^{38}Ar für die $^{40}K/^{40}Ar$ Geochronologie. *Geochronometria Chimia*, **24**, 441–442.

SEYDOUX-GUILLAUME, A.-M, 2001. Experimental determination of the incorporation of Th into orthophosphates and the resetting of geochronological systems of monazite, TU Berlin.

SHERVAIS, J. 1978/1979. Ultramafic and mafic layers in the Alpine-type Lherzolite massif at Balmuccia, N.W. Italy. *Memorie di Scienze Geologiche*, **33**, 135–145.

SILLS, J. D. 1984. Granulite facies metamorphism in the Ivrae zone, NW Italy. *Schweizerische Mineralogische und Petrographische Mitteilungen*, **64**, 169–191.

SILLS, J. D. & TARNEY, J. 1984. Petrogenesis and tectonic significance of amphibolites interlayered with metasedimentary gneisses in the Ivrea zone, S. Alps, N. Italy. *Tectonophysics*, **107**, 187–206.

SMITH, H. A. & BARREIRO, B. 1990. Monazite U-Pb dating of staurolite-grade metamorphism in pelitic schists. *Contributions to Mineralogy and Petrology*, **105**, 602–615.

SNOKE, A. W., KALAKAY, T. J., QUICK, J. E. & SINIGOI, S. 1999. Development of a deep crustal shear zone in response to tectonic intrusion of mantle magma into the lower crust, Ivrea-Verbano zone, Italy. *Earth and Planetary Science Letters*, **166**, 31–45.

STÄMPFLI, G. M., BOREL, G., MARCHANT, R. & MOSAR, J. 2002. Western Alps geological constraints on western Tethyan reconstructions. *Journal Virtual Explorer*, **8**, 77–106.

STEIGER, R. H. & JAEGER, E. 1977. Subcommission on geochronology: Convention on the use of decay constants in geo-and cosmochronology. *Earth and Planetary Science Letters*, **36**, 359–362.

TEUFEL, S. & SCHÄRER, U. 1989. Unravelling the age of high-grade metamorphism of the Ivrea Zone: A monazite single-grain and small fraction study. *Terra Abstract*, **1**, 350.

VAVRA, G., GEBAUER, D., SCHMID, R. & COMPSTON, W. 1996. Multiple zircon growth and recrystallisation during polyphase late Carboniferous to Triassic metamorphism in granulites of the Ivrea Zone (southern Alps): An ion microprobe (SHRIMP) study. *Contributions to Mineralogy and Petrology*, **122**, 337–358.

VILLA, I. 1998. Isotopic closure. *Terra Nova*, **10**, 42–47.

VILLA, I., GROBERTY, B. H., KELLEY, S. P., TRIGLIA, R. & WIELER, R. 1996. Assessing Ar transport paths and mechanisms in the McClure Mountains hornblende. *Contributions to Mineralogy and Petrology*, **126**, 67–80.

VOSHAGE, H., SINIGOI, S., MAZZUCHELLI, M. G., RIVALENTI, G. & HOFMANN, A. W. 1988. Isotopic constraints on the origin of ultramafic and mafic dikes in the Balmuccia peridotite (Ivrea Zone). *Contributions to Mineralogy and Petrology*, **100**, 261–267.

VOSHAGE, H., HOFMANN, A. W., MAZUCHELLI, G., RIVALENTI, G., SINIGOI, S., RACZEK, I. & DEMARCHI, G. 1990. Isotopic evidence from the Ivrea Zone for a hybrid lower crust formed by magmatic underplating. *Nature*, **347**, 731–736.

WAGNER, G. A., REIMER, G. M. & JÄGER, E. 1977. Cooling ages derived by apatite fission-track, mica Rb-Sr and K-Ar dating: the uplift and cooling history of the Central Alps. *Memorie Instituti di Geologia di Padova*, **XXX**, 1–28.

WEISS, T., SIEGESMUND, S., RABBEL, W., BOHLEN, T. & POHL, M. 1999. Physical properties of the Lower continental crust: an anisotropic perspective. *Pure and Applied Geophysics*, **156**, 97–122.

WEMMER, K. 1991. K/Ar-Altersdatierungsmöglichkeiten für retrograde Deformationsprozesse im spröden und duktilen Bereich–Beispiele aus der KTB-Vorbohrung (Oberpfalz) und dem Bereich der Insubrischen Linie (N-Italien). *Göttinger Arbeiten Geologie & Paläontologie*, **51**, 1–61.

YORK, D., HALL, C. M., YALANSE, Y., HANES, J. A. & KENYON, W. J. 1981. $^{40}Ar/^{39}Ar$ dating of terrestrial minerals with a continuous laser. *Geophysical Research Letters*, **8**, 1136–1138.

ZINGG, A. 1983. The Ivrea and Strona-Ceneri Zones (southern Alps, Ticino and N-Italy): A review. *Schweizerische Mineralogische und Petrographische Mitteilungen*, **77**, 361–380.

ZINGG, A., HANDY, M. R., HUNZIKER, J. & SCHMID, S. M. 1990. Tectonometamorphic history of the Ivrea Zone and its relationship to the crustal evolution of the southern Alps. *Tectonophysics*, **182**, 169–192.

Evolution of an Early Permian extensional detachment fault from synintrusive, mylonitic flow to brittle faulting (Grassi Detachment Fault, Orobic Anticline, southern Alps, Italy)

N. FROITZHEIM[1], J. F. DERKS[1], J. M. WALTER[2] & D. SCIUNNACH[3]

[1]*Geologisches Institut, Universität Bonn, Germany (e-mail: niko.froitzheim@uni-bonn.de)*

[2]*Mineralogisch-Petrologisches Institut, Universität Bonn, Germany*

[3]*Infrastruttura per l'Informazione Territoriale, Regione Lombardia, Milano, Italy*

Abstract: Lower Permian volcanic and sedimentary rocks of the Collio Formation in the Orobic Anticline do not rest with a depositional contact on the Variscan basement but are separated from it by the subhorizontal Grassi Detachment Fault, consisting of a cataclasite layer underlain by mylonite. Field relations indicate that both the cataclasite and the mylonite are Early Permian in age. The mylonite formed in a continuous process before, during, and after the intrusion of the Val Biandino Quartz Diorite in the footwall of the detachment fault. Microstructure and quartz texture of the mylonite indicate top-to-the-southeast displacement. Quartz textures of mylonite close to the intrusive bodies are characterized by c-axis single maxima near the Y-direction of the finite strain, indicating prism $<a>$ glide as the dominant gliding system and hence high temperatures (above $c.$ 500 °C) during mylonitization. This is explained by heat advection through the rising quartz diorite melt. During detachment faulting, the footwall of the Grassi Detachment Fault was bowed up to form a metamorphic core complex. The Ponteranica Conglomerate was deposited as a proximal, syntectonic fan-delta on the southeast side of the metamorphic core complex late in its evolution. The unconformity of the Verrucano Lombardo over the Collio Formation and the basement results from erosion of the topography created by detachment faulting, core complex updoming, and block tilting. These results indicate dramatic SE–NW stretching (in present-day coordinates) of the South-Alpine crust during the Early Permian. The return from the thickened, orogenic crust at the end of the Hercynian orogeny to the normal crustal thickness ($c.$ 30 km) of Late Permian and Early Triassic times was accommodated to a large extent by crustal extension, at least in this part of the southern Alps.

The Italian southern Alps are considered a good example of a thrust-and-fold belt, dominated by thin-skinned tectonics at shallow crustal levels (Laubscher 1985; Schönborn 1992). In the northern, internal part, basement rocks were also involved in thrusting. Much of the orogenic deformation was accommodated by rigid translation of relatively undeformed crustal slabs bounded by definite décollement horizons. Alpine (Cretaceous and Cenozoic) metamorphism nowhere exceeded very low-grade conditions (Crespi *et al.* 1982). Alpine deformation was sufficient to cause the surface exposure of different levels of the pre-Alpine continental crust, from the crust–mantle boundary and lower crust exposed in the Ivrea zone to the post-Hercynian sedimentary cover. This results in the favourable situation that the effects of Pre-Alpine tectonic processes can be studied in a complete crustal cross-section (Schmid *et al.* 1987; Bertotti *et al.* 1993; Handy *et al.* 1999). Such a study is particularly interesting in the case of magmatic, tectonic and sedimentary processes during the Early Permian. Intrusion of basaltic melts in the lower crust during the Early Permian (Ivrea zone, 285 Ma, Pin 1986), granite intrusion in the upper crust (e.g. Baveno granite, 277 ± 8 Ma; Pinarelli *et al.* 1988) and acidic to intermediate volcanism at the surface (e.g. Collio basin, 280–287 Ma; Cadel 1986) are approximately coeval parts of a magmatic system affecting different levels of the crust. This timing is confirmed by new zircon U–Pb ages and initial Hf isotopic compositions of volcanic rocks and shallow intrusions from a 250 km-long E–W transect including our study area, which showed that the volcanism was active between 275 and 285 Ma (Schaltegger & Brack 2007).

Clastic sediments accumulated together with the volcanic rocks in rapidly subsiding local basins. Especially during the last $c.$ 25 years, these have received considerable attention and been the subject of detailed stratigraphic and palaeotectonics studies (Wopfner 1984; Ori *et al.* 1986; Cassinis *et al.* 1988; Sciunnach 2001*a–c*, 2003; Muttoni *et al.* 2003). Probably in only a few Ma (late Sakmarian to middle Artinskian, $c.$ 287 to $c.$ 280 Ma) up to 1250 m thick volcanic and clastic sedimentary rocks of the Collio Formation were deposited in the area of the Orobic Anticline (Sciunnach 2001c). From the sedimentary facies architecture, Sciunnach (2001a) concluded that the Early Permian Collio basin represents a setting comparable with the present-day Salton Sea, a tectonic depression in the San Andreas strike-slip system, near the transition to oblique spreading in the Gulf of California.

An extensional to dextral strike-slip setting has generally been assumed for the deposition of the Early Permian succession in the southern Alps. Several authors have considered dominant dextral strike-slip faulting, with only subordinate normal faulting and crustal thinning (e.g. Handy et al. 1999). Bertoluzza & Perotti (1997), based on a comparison with finite-element modelling results, assumed that the Collio basin formed in the framework of strike-slip faulting without any extensional component. Normal faults were identified in the area of the Orobic Anticline (Blom & Passchier 1997; Sciunnach 2001b, 2003) but the extension accommodated by these appeared moderate. Different authors noticed the layer of cataclasite along the contact between the basement and the Collio Formation in the Orobic Anticline (e.g. Dozy 1935; Casati 1968; Sciunnach 2001c), but these cataclasites remained enigmatic. In the present paper, we will show that the contact between basement and cover rocks in the Orobic Anticline represents an important extensional detachment—the Grassi Detachment Fault—which evolved from a mylonitic shear zone active under high temperatures to a brittle fault, and that the extensional shearing was coeval with granitoid intrusion in the footwall and volcanism in the hanging wall. We will present evidence that the southern Alps were dramatically extended during the Early Permian.

Regional geological setting

The Orobic Anticline (Fig. 1) is an Alpine fault-bend fold (Schönborn 1992) formed during Cretaceous (?) to Cenozoic southward thrusting in the southern Alps. Hercynian basement is exposed in the core of the anticline. This includes a complex of mica schist and paragneiss with minor quartzite, amphibolite and marble (Morbegno Gneiss; described under the name 'Morbegnoschiefer' by Cornelius 1916). The characteristic paragenesis of the paragneiss is $Qz + Pl + Bi + Chl + Ms + Gar + Ky + St$. It was affected by widespread greenschist-facies retrogression. The Morbegno Gneiss is intruded by plutons, known as Val Biandino Quartz Diorite (Pasquaré 1967) and Valle Biagio Granite (De Sitter & De Sitter-Koomans 1949). Rb–Sr whole rock dating of the Val Biandino intrusion yielded 312 ± 48 Ma (Thöni et al. 1992). Taking only the samples with a granitic composition, the age is 286 ± 20 Ma. Six biotite concentrates separated from the same rock suite yielded Rb–Sr

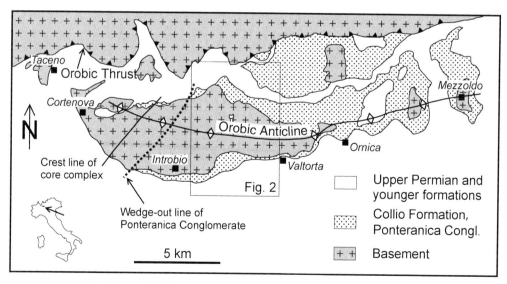

Fig. 1. Geological sketch map of the Orobic Anticline in the southern Alps east of Lago di Como. The anticline exposes Hercynian basement and the overlying volcanic and sedimentary rocks of Early Permian age (Collio Formation). It is located in the footwall of the south-directed Orobic Thrust of Alpine (Cretaceous or Tertiary) age. Frame: study area (see Figure 2).

ages between 277 ± 5 and 295 ± 6 Ma. The age is poorly constrained because of imperfect primary isotopic mixing and partial low-temperature alteration (Thöni et al. 1992). However, the age coincides with the better constrained (U–Pb zircon) age of the 'Volcanic Member' of the Collio Formation: 287 to 280 Ma (Cadel 1986; see below). Therefore, we assume that the intrusions represent the plutonic parts of the Collio magmatic system and are of the same age, that is, between c. 290 and 280 Ma, possibly with some older parts. This relationship is supported by the compositional similarity between plutons and volcanics (Cadel 1986). The lack of a physical continuity between plutons and volcanics does not contradict this assumption because, as we will show, the volcanites have been displaced away from the plutons by an extensional detachment fault.

Upper Carboniferous conglomerates ('Conglomerato basale'), which elsewhere in the southern Alps unconformably cover the basement, only occur in the easternmost part of the Orobic Anticline. In our study area in the western Orobic Anticline, the basement is directly overlain across a tectonic contact (see below) by the 'Volcanic Member' of the Collio Formation: welded tuffs, lapillistones, porphyric ignimbrites, massive volcanites, and volcanic breccias (Sciunnach 2001c). The age of the 'Volcanic Member' is 287 to 280 Ma (U/Pb on zircon, Cadel 1986). This is followed by the 'Arenaceous–volcanoclastic Member' and by the 'Upper arenaceous–pelitic Member' of the Collio Formation. In the western part of the anticline, these two members are partly replaced by the Ponteranica Conglomerate, poorly sorted, pebbly to cobbly conglomerates and coarse-grained sandstones with predominantly volcanic and subordinate basement clasts, representing a proximal fan-delta shed from the west or northwest. Across an angular unconformity, this succession is overlain by fluvial conglomerates, sandstones, and siltstones of the Verrucano Lombardo, of Late Permian age. West of the study area, the Verrucano Lombardo mostly rests directly on the basement (Morbegno Gneiss and Early Permian granitoids), without any intervening Collio Formation (Fig. 1). Especially in the area of Cortenova (Fig. 1), spectacular palaeosols developed over a reddened and deeply altered Valle Biagio Granite are stratigraphically overlain by Verrucano Lombardo conglomerates. The Verrucano Lombardo is followed by the siliciclastic to carbonatic Servino Formation (Lower Triassic), an evaporite horizon (Carniola di Bovegno), and various marine sedimentary rocks of Middle Triassic age.

Structural relations between Morbegno Gneiss, Val Biandino Quartz Diorite, and Collio Formation

In the study area, the 'Volcanic Member' of the Collio Formation is everywhere separated from the basement by a layer of black cataclasite, locally over four metres thick, except in the northwest corner of the area where the contact between the 'Volcanic Member' and the quartz diorite is an Alpine reverse fault (the Biandino Fault, Schönborn 1992; Figs 2 and 3). In some places, the 'Volcanic Member' is missing and the Ponteranica Conglomerate is directly on the basement, but again separated from the latter by the cataclasite. The cataclasite is often foliated and includes slivers and fragments of both volcanic and basement rocks in a fine-grained, black matrix mostly formed by illite. Basement inclusions predominate at the base and volcanite inclusions at the top of the cataclasite layer. We were unable to deduce a shear direction and sense from the structure of the cataclasite. Tourmaline which occurs in similar cataclasite farther east (Trabuchello–Cabianca anticline, Zhang et al. 1994) and farther west (Sciunnach 2001b) was not found here.

West of our study area, the cataclasite layer is sealed unconformably by the Verrucano Lombardo (Fig. 1: west of Introbio where the wedge-out line of the Ponteranica Conglomerate cuts the southern border of the basement outcrop; see Sciunnach 2001c). Therefore, it must be older than Late Permian. On the other hand, cataclasite formation affects the 'Volcanic Member', and in some places also the base of the Ponteranica Conglomerate. That is, the cataclasite zone formed between c. 287 to 280 Ma, the age of the 'Volcanic Member', and c. 267 Ma, when the deposition of the Verrucano Lombardo began (Sciunnach 2001c).

The Morbegno Gneiss underlying the cataclasite is a mylonite with a strong foliation and lineation. The foliation is parallel to the cataclasite layer and therefore also to the basement/Collio contact. North of the axial plane of the Orobic Anticline, the foliation dips north to northwest, in the hinge area it is shallow, and south of the axial plane, it dips mostly south to southeast. The stretching lineation (Figs 2 and 4) is oriented NW–SE (average trend 152°). The intensity of the mylonitic deformation is generally strong, but strongest near the top and decreasing downward. The mylonitic foliation is defined by alternating layers of dynamically recrystallized quartz, biotite- or chlorite-rich layers, and feldspar-rich layers. Shear-sense indicators are difficult to find in outcrops and hand specimens. This is somewhat surprising because they are extremely clear and consistent in thin

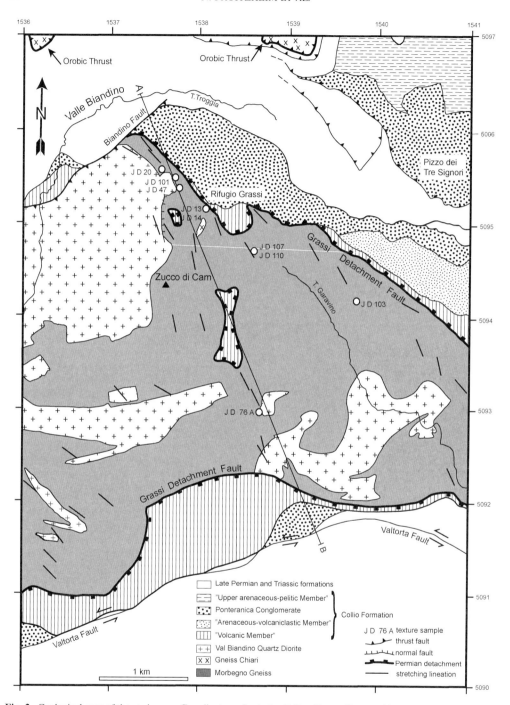

Fig. 2. Geological map of the study area. Coordinates refer to the Italian Gauss–Boaga grid.

sections (see below). Whereas the upper contact of the mylonite zone is sharp and defined by the cataclasite layer, the lower boundary is diffuse. The mylonite grades downward into 'normal' Morbegno Gneiss by an increasing spacing of the foliation, and by a decreasing intensity of the stretching lineation. Using these criteria, the thickness of the mylonitic zone can be estimated to be about 100 metres.

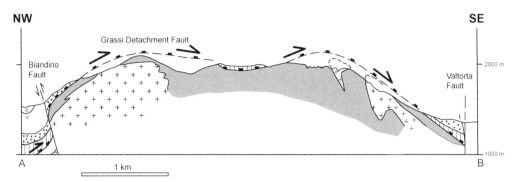

Fig. 3. Cross-section of the Orobic Anticline in the study area (trace in Fig. 2). The contact between basement and Collio Formation is formed by cataclasite representing the fault rock of the brittle Grassi Detachment Fault. Mylonite parallel to the contact in the uppermost part of the basement shows a transport direction parallel to the profile and shear sense towards the southeast.

The Val Biandino Quartz Diorite is mostly undeformed. In some outcrops near the borders of the intrusion west of Rifugio Grassi, xenoliths of mylonitic Morbegno Gneiss are enclosed in undeformed quartz diorite (Fig. 5), showing that part of the mylonitic shearing preceded quartz diorite intrusion. However, some parts of the quartz diorite intrusion close to the basement/Collio contact show a mylonitic structure themselves, with foliation and stretching lineation oriented just as in the Morbegno Gneiss. The thickness of mylonitized quartz diorite is only a few metres, much less than the thickness of mylonitized Morbegno Gneiss. Microstructural observations (see below) demonstrate the same kinematics for the mylonitized quartz diorite and Morbegno Gneiss. Therefore, part of the mylonitic shearing must have occurred after intrusion of the quartz diorite. Taken together, these observations indicate overall contemporaneity between mylonitization and quartz diorite intrusion. If our assumptions about the age of the quartz diorite are correct, the mylonites were formed around 290 to 280 Ma, that is, shortly before the cataclasite or overlapping in time with the formation of the cataclasite. The mylonites and the cataclasite layer hence represent the brittle and ductile part, respectively, of the same shear zone. The situation that mylonites are capped by cataclasite is typical for an extensional detachment fault where mylonites, formed at greater depth, are progressively juxtaposed to cataclasite formed at shallower depth (e.g. Lister & Davis 1989).

Microstructure and texture of Morbegno Gneiss mylonites

The mylonitized Morbegno Gneiss samples partly show fresh biotite in foliation-parallel layers, indicating deformation under conditions of at least the higher greenschist facies (Fig. 6a). This is the case

Fig. 4. Pronounced stretching lineation formed by quartz rods on a northwest-dipping foliation plane of gneiss mylonite below the Grassi Detachment Fault. Orientation of the stretching lineation: 327 (azimuth)/45 (dip). Footpath from Rifugio Pio X to Rifugio Grassi.

Fig. 5. Xenoliths of mylonitic Morbegno Gneiss enclosed in undeformed Val Biandino Quartz Diorite, indicating that part of the mylonitic shearing preceded quartz diorite intrusion. Thin lines: Foliation traces in Morbegno Gneiss. Footpath from Rifugio Pio X to Rifugio Grassi.

in the vicinity of the quartz diorite bodies. In samples away from the intrusions, biotite is syntectonically replaced by chlorite, documenting deformation under conditions of the lower greenschist facies (Fig. 6b). Garnet and staurolite are unstable and are replaced by biotite and/or chlorite in fractures, pressure shadows, and porphyroclast tails. Plagioclase is brittlely deformed, without any

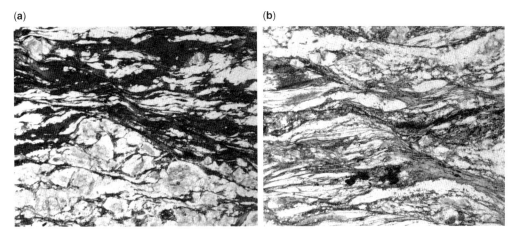

Fig. 6. Thin-section photographs of mylonitized Morbegno Gneiss underlying the Grassi Detachment Fault. (**a**) Relatively high-temperature, biotite-bearing mylonite close to quartz diorite pluton. Dark mineral is biotite. Sample JD 59, Gauss-Boaga coordinates 1537 952/5095 033. (**b**) Chlorite-bearing mylonite distant from quartz diorite pluton. Biotite has been completely replaced by chlorite (light grey). Sample JD 105, Gauss-Boaga coordinates 1539 600/5094 446. In both samples, shear bands indicate dextral (top-to-the-southeast) shear sense. Horizontal width is 3.44 mm in both photographs. Plane-polarized light.

indication of dynamic recrystallization. Quartz is always dynamically recrystallized. In some samples, quartz grain shapes suggest recrystallization by grain-boundary migration (e.g. sample JD 103, Fig. 7). Other samples have a uniform, very small grain size, suggesting subgrain rotation as the dominant recrystallization mechanism (e.g. sample JD 20, Fig. 7). In thin sections parallel to the X–Z plane of the finite strain, abundant shear-sense criteria indicate top-to-the-southeast shearing, both in the biotite-bearing and chlorite-bearing mylonites. The criteria include shear bands (Fig. 6), mica fish, sigma-type porphyroclasts, and oblique shape fabric of recrystallized quartz grains (Fig. 7). The quartz layers appear rather dark in X–Z thin sections under crossed polarizers (Fig. 7), indicating that a high proportion of grains has c-axes parallel to the Y-direction of the finite strain. This is confirmed by the results of neutron diffraction goniometry.

Textures of nine almost monomineralic quartz mylonites have been analysed using the texture diffractometer SV7-b operated at the Forschungszentrum Jülich (Jansen *et al.* 2000). The orientation distribution of the c-axes, which cannot be measured directly, was calculated from the orientation distribution function obtained from reflections of the first order prism $\{m\}$, the second order prism $\{a\}$, $\{111\}$ and the intrinsically overlapped positive and negative rhombs $\{r+z\}$. The results are shown in Figure 7. All textures except one (JD 103) are very similar and surprisingly clear and regular. $\{c\}$ $(=\{001\})$ forms a single maximum parallel to the foliation and perpendicular to the stretching lineation, that is, parallel to the strain Y-axis. This maximum is strong (up to 37 multiples of the random distribution) and partly round, partly slightly elongated in the plane perpendicular to the stretching lineation. From the $\{c\}$ pole figure, it would be impossible to extract the shear sense. This information is, however, given by the $\{m\}$ $(=\{100\})$ and $\{a\}$ $(=\{110\})$ pole figures. Both $\{m\}$ and $\{a\}$ pole figures show three maxima distributed on a great circle parallel to the X–Z plane, that is, on the periphery of the pole figures. These maxima are not symmetric with respect to the foliation trace, but rotated counterclockwise in the case of $\{m\}$ and clockwise in the case of $\{a\}$. The strongest of the $\{a\}$ maxima is rotated about 10° to 20° clockwise from the stretching direction. The textures are clear evidence for predominant prism $<a>$ slip. The strongest $\{a\}$ maximum represents the dominant slip direction. The obliquity indicates dextral (top-SE) shear sense.

Prism $<a>$ slip becomes predominant in quartz mylonites under temperatures above c. 500 °C (Stipp *et al.* 2002). At lower temperatures, basal $<a>$ and rhomb $<a>$ slip are usually the active glide systems. Temperatures above 500 °C during the mylonitization of the Morbegno Gneiss are in line with the observed stability of biotite. As demonstrated above, the mylonitic shearing was broadly synintrusive. Therefore, heat from the rising melt body is likely to have caused the elevated temperatures.

JD 103 is the only sample with a different texture. The thin section shows strongly sutured

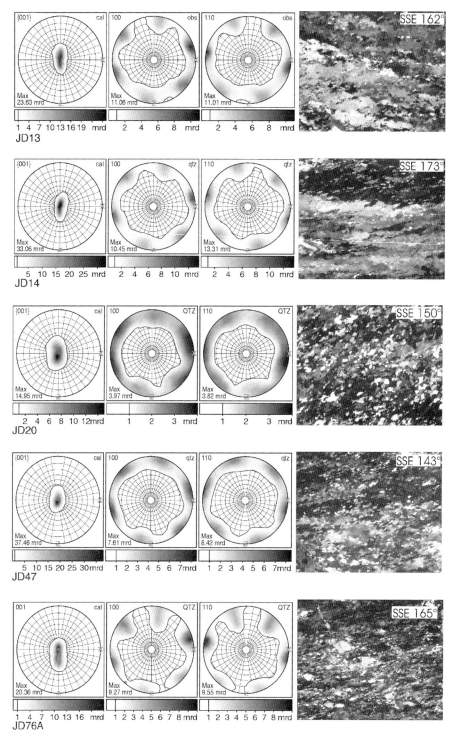

Fig. 7. Upper-hemisphere Wulff projections of quartz textures from Morbegno Gneiss mylonites, obtained from neutron diffraction. In the pole figures, the trace of the foliation is horizontal, the stretching lineation is at the left and right end of the foliation trace, and southeast is to the right. {c} (={001}) pole figures were calculated from the

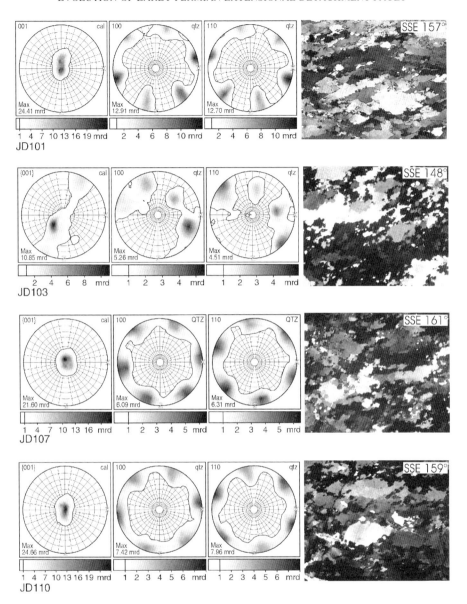

Fig. 7. (*Continued*) orientation distribution function; {m} (={100}) and {a} (={110}), pole figures were observed. All textures except one (JD 103) have c-axis single maxima close to Y, suggesting prism <a> slip as the dominant gliding system. Thin section photographs show X–Z sections perpendicular to the foliation and parallel to the stretching lineation. Photographs have same orientation as textures (foliation and lineation horizontal, southeast to the right). Horizontal width is 3.44 mm in all photographs; crossed polarizers. In all cases, dextral (top-to-the-southeast) shear sense is indicated by asymmetry of texture and of grain shape fabric.

and interlocking quartz grain boundaries suggestive of grain-boundary migration recrystallization. Biotite is completely replaced by chlorite in JD 103, whereas biotite was stable during deformation in the samples from which the other textures were determined. The {c} pole figure exhibits an oblique single girdle distribution suggesting top-to-the-southeast shearing. Two maxima are on this girdle, one near the periphery, and a stronger one about 40° away from the centre of the pole figure. Three {a} maxima are distributed on a great circle oblique to the foliation; the strongest of these is close to the periphery and rotated clockwise from the stretching lineation by c. 45°. This angle is much larger than for the other samples. The interpretation of this texture is difficult. One possibility is that it formed by superposition of the mylonitic shearing on some older texture of the Morbegno Gneiss. On the other hand, the weaker {c} maximum close to the periphery suggests the activity of basal <a> slip. In this case, the texture may represent deformation under lower temperatures than for the other samples, which is in line with the observed replacement of biotite by chlorite.

Mylonitized Val Biandino Quartz Diorite

In structurally high levels, Val Biandino Quartz Diorite shows signs of ductile deformation. In strained granodiorites with a weak foliation, the structure appears still magmatic but the quartz grains are completely dynamically recrystallized to a fine grain size by subgrain rotation recrystallization. Chlorite partly replaces magmatic biotite. Beginning mylonitization is represented by protomylonites with S-C fabrics. Quartz is dynamically recrystallized whereas feldspar grains remained largely undeformed. In higher-strain zones, recrystallized quartz layers form an interconnected shear zone network around feldspar porphyroclasts (Fig. 8). The asymmetry of the S-C fabric, σ-type porphyroclasts, mica fish, and oblique grain shape of quartz consistently indicate top-to-the-southeast shearing as observed in the Morbegno Gneiss mylonites. Together with the finding of mylonitized Morbegno Gneiss as xenoliths in undeformed quartz diorite (see above), these observations prove that the mylonitic shear zone underlying the Grassi Detachment Fault is synintrusive and hence, most probably, Early Permian in age.

Reconstruction of the Grassi Detachment Fault

The fault marked by the cataclasite layer at the basement/Collio contact accommodated strong exhumation of the footwall: it juxtaposed footwall mylonites that were deformed under high temperatures (above c. 500 °C) against hanging-wall volcanic rocks erupted at the Earth's surface. The hanging wall and footwall of the fault are today

Fig. 8. Mylonitized Val Biandino Quartz Diorite. Feldspar porphyroclasts (lower part of picture) were brittlely deformed; quartz (upper part of picture) was dynamically recrystallized. X–Z section perpendicular to the foliation and parallel to the stretching lineation. Orientation of stretching lineation: 328/18. Dextral (top-to-the-southeast) shear sense is indicated by oblique grain boundaries in quartz and by asymmetric feldspar porphyroclasts. Horizontal width is 3.44 mm; crossed polarizers. Gauss-Boaga coordinates of sample locality: 1537 294/5095 235.

completely separated. No rock type of the footwall occurs in the hanging wall, and *vice versa*. We therefore interpret the fault as an extensional detachment fault and call it 'Grassi Detachment Fault' after Rifugio Grassi, a mountain hut standing only a few metres away from the fault contact (Fig. 2). In our interpretation, the mylonites represent a shear zone which formed the downward continuation of the brittle Grassi Detachment Fault during an early stage of its activity. They were transported upward by the continuing tectonic unroofing of the footwall. Therefore, we assume that the brittle Grassi Detachment Fault had the same kinematics, top-to-the-southeast, as the mylonites.

The assumption that the Val Biandino Quartz Diorite and the Collio volcanites form the plutonic and volcanic parts, respectively, of one and the same magmatic system, is in line with the kinematics of the Grassi Detachment Fault. In the area where the plutons are exposed today, the 'Volcanic Member' is rather thin. It becomes thicker towards the east. Restoring a top-to-the-southeast displacement along the Grassi Detachment Fault would bring thick parts of the 'Volcanic Member' in a position above the Val Biandino Quartz Diorite (Fig. 9).

The Ponteranica Conglomerate represents a proximal, syntectonic fan-delta shed from west or northwest (Sciunnach 2003). It wedges out towards a structural high to the northwest. There are two wedge-out points, one on the northern and one on the southern flank of the Orobic Anticline. Connecting these two points results in a wedge-out line striking about 45° (NE–SW, Fig. 1). This is roughly perpendicular to the movement direction of the Grassi Detachment (155°). We assume that the Ponteranica Conglomerate was deposited during a late stage of detachment faulting when the fault had already been domed up to form a metamorphic core complex, the northwestern part of the fault was no longer active, and fault activity continued on the SE side (Fig. 9). The conglomerate was mainly fed by erosion of tilted blocks of the 'Volcanic Member'. Basement clasts are a subordinate component of the conglomerate. These were probably derived from tilted blocks of basement on the NW side of the core complex culmination, and/or

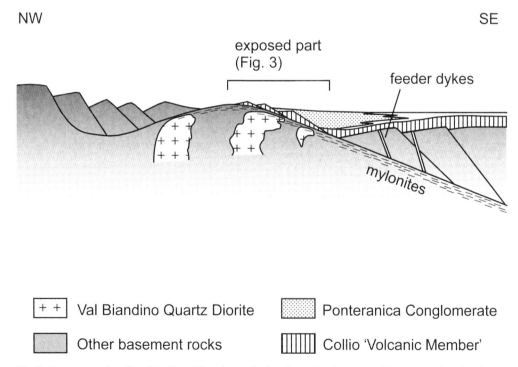

Fig. 9. Reconstructed profile of the Grassi Detachment Fault and associated metamorphic core complex, showing the situation at the end of fault activity and before the deposition of Verrucano Lombardo during the Late Permian. The Ponteranica Conglomerate has been deposited as a syntectonic fan-delta on the eastern flank of the metamorphic core complex. The volcanic rocks of the Collio Formation have been displaced towards the east relative to their deep-seated magma chambers represented by the Val Biandino intrusions.

local erosion of the core complex. The location of the crest line of the core complex (Fig. 1) is not well constrained. It was probably parallel to the wedge-out line of the Ponteranica Conglomerate and we assume that it was located only a few kilometres to the northwest of the latter (Fig. 9). In any case, it must have been oriented at a high angle to the crest line of the present-day Orobic Anticline. The Orobic Anticline formed in response to Alpine thrusting; it does not coincide precisely with the Permian metamorphic core complex. The antiformal shape of the Grassi Detachment Fault as seen in the profile of Figure 3 resulted from Alpine folding. Even at the NW end of the profile, the detachment probably dipped towards the SE before Alpine folding, because this area was still on the SE flank of the core complex (SE of the crest line). However, although the Orobic Anticline is an Alpine fold, the distribution of rock units within the anticline reflects Permian tectonics. The fact that basement rocks dominate in the western part of the anticline and Collio rocks in the eastern part (Fig. 1) is the result of Permian detachment faulting.

The surface break-away of the Grassi Detachment Fault is not exposed. Assuming that the mylonites were formed at a depth of only 10 km, and the fault initially cut through the brittle crust at an angle of 45° and was later rotated to shallow dip (rolling-hinge mechanism; Buck 1988; Wernicke & Axen 1988), the surface break-away would still be about 14 km NW of the Ponteranica Conglomerate wedge-out line, outside the outcrop area of the Orobic Anticline (Fig. 1). The area northwest of the wedge-out line is a structural high sealed by an unconformable blanket of Verrucano Lombardo. It represents the tectonically denuded and partly eroded part of the metamorphic core complex. We assume that the erosional unconformity of the Verrucano Lombardo over the Collio Formation and the basement results from the topography that was created by detachment faulting, block tilting and core complex updoming.

Conclusion

In this study, a Permian-age extensional detachment fault and associated metamorphic core complex have been identified for the first time in the Alps. This finding demonstrates that the return from the thickened, orogenic crust at the end of the Hercynian orogeny to the normal crustal thickness (c. 30 km) of Late Permian and Early Triassic times was accommodated to a large extent by crustal extension, at least in the middle part of the southern Alps. As discussed by Abers (2001), detachment faults form in the most rapidly extending rifts, at the present day, the Woodlark basin (10–40 mm/a opening rate, Taylor et al. 1999) and the Gulf of Corinth (15 mm/a, Clarke et al. 1998). Hence, we may assume that the Collio basin was such a rapidly extending rift, which is in line with the available age data suggesting rapid deposition. A metamorphic core complex formed on the northwestern flank of this basin. It is today exposed in the western part of the Orobic Anticline. In this core complex, footwall mylonites which had formed under temperatures of at least 500 °C were directly juxtaposed to hanging-wall volcanic rocks which had been erupted at the surface, indicating a large amount of exhumation. This is remarkably similar to the situation in metamorphic core complexes of the Colorado River extensional corridor in western North America, e.g. the Whipple Mountains, where Miocene footwall mylonites are overlain across the detachment fault by tilted blocks of volcanic rocks of the same age (Davis 1988).

The occurrence of Early Permian intrusive rocks (Val Biandino Quartz Diorite and associated rock types) closely coincides with the core complex. This suggests a genetic link between magmatism and core complex formation as suggested by Lister & Baldwin (1993). Thermal advection caused by rising magma leads to softening of the crust, localization of extension in the affected area, and thereby to a high local rate of extension. On the other hand, addition of magma enhances the updoming of the crust and differential uplift of the metamorphic core complex.

The metamorphic evolution of basement rocks in parts of the Austroalpine nappes is characterized by a Permian-age high-temperature/low-pressure stage which has been interpreted in terms of extensional tectonics (Schuster et al. 2001). Janák et al. (2004) hypothesized that a Permian rift within the Austroalpine became the site of intracontinental subduction during the Cretaceous. Veselá et al. (2008) present another example of Permian asymmetric rift basins in Penninic units of the western Tauern Window and show how the basin-bounding normal faults were inverted during Alpine shortening. Such studies complete the so-far rudimentary picture of Permian tectonics in the Alps and clarify the ways in which Permian structures influenced the kinematics of Alpine orogeny.

We thank Cees Passchier and an anonymous reviewer for their very helpful comments. The first author thanks Stefan Schmid who, in 1987, encouraged him to visit the area east of Lago di Como in search of pre-Alpine rift faults.

References

ABERS, G. A. 2001. Evidence for seismogenic normal faults at shallow dips in continental rifts.

In: WILSON, R. C. L., WHITMARSH, R. B., TAYLOR, B. & FROITZHEIM, N. (eds) *Non-volcanic Rifting of Continental Margins: A Comparison of Evidence from Land and Sea*. Geological Society, London, Special Publication, **187**, 305–318.

BERTOLUZZA, L. & PEROTTI, C. R. 1997. A finite-element model of the stress field in strike-slip basins: implications for the Permian tectonics of the Southern Alps (Italy). *Tectonophysics*, **280**, 185–197.

BERTOTTI, G., SILETTO, G. -B. & SPALLA, M. -I. 1993. Deformation and metamorphism associated with crustal rifting: the Permian to Liassic evolution of the Lake Lugano-Lake Como area (Southern Alps). *Tectonophysics*, **226**, 271–284.

BLOM, J. C. & PASSCHIER, C. W. 1997. Structures along the Orobic thrust, Central Orobic Alps, Italy. *Geologische Rundschau*, **86**, 627–636.

BUCK, W. R. 1988. Flexural rotation of normal faults. *Tectonics*, **7**, 959–973.

CADEL, G. 1986. Geology and uranium mineralization of the Collio basin (Central Southern Alps). *Uranium*, **2**, 215–240.

CASATI, P. 1968. Rapporti tra il basamento cristallino e le formazioni del Permiano presso Introbio in Valsassina (Lombardia). *Rendiconti dell'Istituto Lombardo di Scienze e Lettere*, **A101** (1967), 866–872.

CASSINIS, G., MASSARI, F., NERI, C. & VENTURINI, C. 1988. The continental Permian in the Southern Alps (Italy). A review. *Zeitschrift für geologische Wissenschaften*, **16**, 1117–1126.

CLARKE, P. J., DAVIES, R. R. & ENGLAND, P. C. ET AL. 1998. Crustal strain in central Greece from repeated GPS measurements in the interval 1989–1997. *Geophysical Journal International*, **135**, 195–214.

CORNELIUS, H. P. 1916. Zur Kenntnis der Wurzelregion im unteren Veltlin. *Neues Jahrbuch für Mineralogie, Geologie und Paläontologie*, **40**, 253–363.

CRESPI, R., LIBORIO, G. & MOTTANA, A. 1982. On a widespread occurrence of stilpnomelane to the South of the Insubric line, Central Alps, Italy. *Neues Jahrbuch für Mineralogie–Monatshefte*, **6**, 265–271.

DAVIS, G. A. 1988. Rapid upward transport of mid-crustal mylonitic gneisses in the footwall of a Miocene detachment fault, Whipple Mountains, southeastern California. *Geologische Rundschau*, **77**, 191–209.

DE SITTER, L. U. & DE SITTER-KOOMANS, C. M. 1949. Geology of the Bergamasc Alps, Lombardia, Italy. *Leidse Geologische Mededelingen*, **14**, 1–257.

DOZY, J. F. 1935. Die Geologie der Catena Orobica zwischen Corno Stella und Pizzo del Diavolo di Tenda. *Leidse Geologische Mededelingen*, **6**, 133–230.

HANDY, M. R., FRANZ, L., HELLER, F., JANOTT, B. & ZURBRIGGEN, R. 1999. Multistage accretion and exhumation of the continental crust (Ivrea crustal section, Italy and Switzerland). *Tectonics*, **18**, 1154–1177.

JANÀK, M., FROITZHEIM, N., LUPTÁK, B., VRABEC, M. & KROGH RAVNA, E. J. 2004. First evidence for ultrahigh-pressure metamorphism of eclogites in Pohorje, Slovenia: Tracing deep continental subduction in the Eastern Alps. *Tectonics*, **23**, TC5014, 10.

JANSEN, E., SCHÄFER, W. & KIRFEL, A. 2000. The Jülich neutron diffractometer and data processing in rock texture investigations. *Journal of Structural Geology*, **22**, 1559–1564.

LAUBSCHER, H. P. 1985. Large-scale, thin-skinned thrusting in the southern Alps: kinematic models. *Geological Society of America Bulletin*, **96**, 710–718.

LISTER, G. S. & BALDWIN, S. L. 1993. Plutonism and the origin of metamorphic core complexes. *Geology*, **21**, 607–610.

LISTER, G. S. & DAVIS, G. A. 1989. The origin of metamorphic core complexes and detachment faults formed during Tertiary continental extension in the northern Colorado River region, U.S.A. *Journal of Structural Geology*, **11**, 65–94.

MUTTONI, G., KENT, D. V., GARZANTI, E., BRACK, P., ABRAHAMSEN, N. & GAETANI, M. 2003. Early Permian Pangea 'B' to Late Permian Pangea 'A'. *Earth and Planetary Science Letters*, **215**, 379–394.

ORI, G. G., DALLA, S. & CASSINIS, G. 1986. Depositional history of the Permian continental sequence in the Val-Trompia—Passo Croce Domini area (Brescian Alps, Italy). *Memorie della Societá Geologica Italiana*, **34**, 141–154.

PASQUARÉ, G. 1967. Analisi geologico-strutturale del complesso intrusivo di Val Biandino (Alpe Orobie Occidentali). *Memorie della Società Geologica Italiana*, **6**, 343–357.

PIN, C. 1986. Datation U-Pb sur zircons à 285 Ma du complexe gabbro-dioritique du Val Sesia—Val Mastallone et age tardi-hercynien du métamorphisme granulitique de la zone Ivrea-Verbano (Italie). *Comptes Rendus de l'Académie des Sciences, Série II*, **303**, 827–830.

PINARELLI, L., BORIANI, A. & CONTICELLI, S. 1988. Rb–Sr geochronology of the Lower Permian plutonism in the Massiccio dei Laghi, Southern Alps (NW Italy). *Rendiconti della Società Italiana di Mineralogia e Petrografia*, **43**, 411–428.

SCHALTEGGER, U. & BRACK, P. 2007. Crustal-scale magmatic systems during intracontinental strike-slip tectonics, U, Pb and Hf isotopic constraints from Permian magmatic rocks of the Southern Alps. *International Journal of Earth Sciences*, **96**, 1131–1151.

SCHMID, S. M., ZINGG, A. & HANDY, M. 1987. The kinematics of movements along the Insubric Line and the emplacement of the Ivrea Zone. *Tectonophysics*, **135**, 47–66.

SCHÖNBORN, G. 1992. Alpine tectonics and kinematic models of the central Southern Alps. *Memorie di Scienze Geologiche (Padova)*, **44**, 229–393.

SCHUSTER, R., SCHARBERT, S., ABART, R. & FRANK, W. 2001. Permo-Triassic extension and related HT/LP metamorphism in the Austroalpine–Southalpine realm. *Mitteilungen der Gesellschaft der Geologie- und Bergbaustudenten in Österreich*, **45**, 111–141.

SCIUNNACH, D. 2001a. Benthic foraminifera from the upper Collio Formation (Lower Permian, Lombardy Southern Alps): implications for the palaeogeography of the peri-Tethyan area. *Terra Nova*, **13**, 150–155.

SCIUNNACH, D. 2001b. Early Permian palaeofaults at the western boundary of the Collio Basin (Valsassina, Lombardy). *Natura Bresciana. Annuario del Museo Civico di Scienze Naturali, Brescia, Monografia*, **25**, 37–43.

SCIUNNACH, D. 2001c. The Lower Permian in the Orobic Anticline (Southern Alps, Lombardy): a review based on new stratigraphic and petrographic data. *Rivista Italiana di Paleontologia e Stratigrafia*, **107**, 47–68.

SCIUNNACH, D. 2003. Fault-controlled stratigraphic architecture and magmatism in the Western Orobic Basin (Lower Permian, Lombardy Southern Alps). *Bollettino della Società Geologica Italiana, Volume Speciale*, **2**, 49–58.

STIPP, M., STÜNITZ, H., HEILBRONNER, R. & SCHMID, S. M. 2002. The eastern Tonale fault zone: a 'natural laboratory' for crystal plastic deformation of quartz over a temperature range from 250 to 700 °C. *Journal of Structural Geology*, **24**, 1861–1884.

TAYLOR, B., GOODLIFFE, A. M. & MARTINEZ, F. 1999. How continents break up: insights from Papua New Guinea. *Journal of Geophysical Research*, **104** (B4), 7497–7512.

THÖNI, M., MOTTANA, A., DELITALA, M. C., DE CAPITANI, L. & LIBORIO, G. 1992. The Val Biandino composite pluton: A late Hercynian intrusion into the South-Alpine metamorphic basement of the Alps (Italy). *Neues Jahrbuch für Mineralogie–Monatshefte*, **12**, 545–554.

VESELÁ, P., LAMMERER, B., WETZEL, A., SÖLLNER, F. & GERDES, A. 2008. Post-Variscan to Early Alpine sedimentary basins in the Tauern Window (eastern Alps). *In*: SIEGESMUND, S., FÜGENSCHUH, B. & FROITZHEIM, N. (eds) *Tectonic Aspects of the Alpine-Dinaride-Carpathian System*. Geological Society, London, Special Publications, **298**, 83–100.

WERNICKE, B & AXEN, G. J. 1988. On the role of isostasy in the evolution of normal fault systems. *Geology*, **16**, 848–851.

WOPFNER, H. 1984. Permian deposits of the Southern Alps as product of initial alpidic taphrogenesis. *Geologische Rundschau*, **73**, 259–277.

ZHANG, J. S., PASSCHIER, C. W., SLACK, J. F., FLIERVOET, T. F. & DE BOORDER, H. 1994. Cryptocrystalline Permian tourmalinites of possible metasomatic origin in the Orobic Alps, Northern Italy. *Economic Geology*, **89**, 391–396.

Post-Variscan to Early Alpine sedimentary basins in the Tauern Window (eastern Alps)

P. VESELÁ[1], B. LAMMERER[1], A. WETZEL[2], F. SÖLLNER[1] & A. GERDES[3]

[1]*Department of Earth and Environmental Sciences; Ludwig-Maximilians-Universität, Luisenstr. 37, D-80333 München, Germany (e-mail: petra.vesela@iaag.geo.uni-muenchen.de)*

[2]*Geologisch-Paläontologisches Institut, Universität Basel, Bernoullistrasse 32, CH-4056 Basel, Switzerland*

[3]*Institute of Geosciences, Petrology & Geochemistry, Johann Wolfgang Goethe-Universität, Senckenberganlage 28, D-60054 Frankfurt am Main, Germany*

Abstract: The crystalline basement of the Tauern Window is locally covered by Palaeozoic to Mesozoic sediments that experienced Alpine tectonometamorphism. The sedimentary cover has been subdivided into mappable lithological units. The correlation of these units, the use of some dated marker intervals and independent palinspastic restoration provide evidence that the depositional area was differentiated into basins and swells. At the end of the Variscan orogeny, during the Carboniferous and Permian, intermontane basins formed in basement rocks and mainly continental clastics accumulated in elongate troughs. Later, probably during the Triassic, there was levelling of the previous relief and subsidence of the basins, but continental sedimentation still prevailed although interrupted by some marine transgressions. Thereafter, probably during the Jurassic, the area was progressively flooded and the sedimentation became increasingly calcareous. The Upper Jurassic carbonates document complete submergence. In some areas, the Upper Jurassic carbonates directly rest on crystalline basement indicating renewed tectonic stretching. The sedimentary cover shows striking similarities with coeval deposits within the Germanic Basin and the study area is therefore considered to have been part of the southern European continental margin of the Tethys (the so-called Vindelician Land).

In addition to the classical ocean floor magmatic rocks and the overlying deep-marine deposits, mountain belts consist of continental margin and shelf deposits and their basement (e.g. Coward & Dietrich 1989). The latter especially store valuable information about the pre-orogenic palaeogeographical situation and the early history of an evolving continental margin. This is true for the Alps. However, where the continental crust and its sedimentary cover have experienced a strong tectonometamorphic overprint, deciphering the pre-orogenic history is far from straightforward.

Continental margins are characterized by deep-rooted faults and even the adjacent shelf areas may be affected by block-faulting. For instance, fault systems developed during the Late Palaeozoic (e.g. Arthaud & Matte 1977) became re-activated during the Mesozoic rifting of the central Atlantic and Tethys Oceans (e.g. Benammi & El Kochri 1998; Bouaziz *et al.* 1999). Such reactivated faults are known to influence the lithological development of the sedimentary cover (e.g. Faerseth 1996; Keeley 1996; Wetzel *et al.* 2003). It appears that pre-existing Late Palaeozoic faults and associated grabens strongly affected the tectonosedimentary development in that part of the Alps which is exposed today in the Tauern Window (Lammerer *et al.* 2008). It is the purpose of this paper to unravel the relationships between the Late Palaeozoic tectonic structures, including grabens, and the Mesozoic sediments.

Geological setting

Numerous small, elongated basins are known from the Alpine area and its northern foreland which formed at the end of the Variscan orogeny during the Late Palaeozoic as intermontane, fault-bounded basins (e.g. von Raumer 1998). Examples are the Permo-Carboniferous basins of northern Switzerland (Matter 1987), the Lake Constance and Landshut–Neuötting Basins (Lemcke 1988) continuing to basins within the Zentrale Schwellenzone in Austria (Kröll *et al.* 2006) which are only known from drilling and seismic imaging. Within the Alps, well-exposed examples are the Salvan-Dorénaz Basin in the Aiguille Rouge Massif (Capuzzo *et al.* 2003; Capuzzo & Wetzel 2004) and the basins within the Aar–Gotthard Massif (Franks 1966; Oberhänsli *et al.* 1988; Schaltegger & Corfu

From: SIEGESMUND, S., FÜGENSCHUH, B. & FROITZHEIM, N. (eds) *Tectonic Aspects of the Alpine-Dinaride-Carpathian System.* Geological Society, London, Special Publications, **298**, 83–100.
DOI: 10.1144/SP298.5   0305-8719/08/$15.00 © The Geological Society of London 2008.

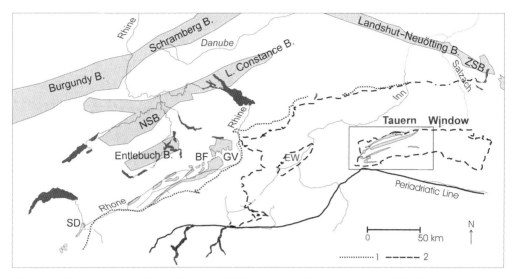

Fig. 1. Post-Variscan basins in Central Europe and Alpine realm (modified after Lemcke 1988; Ménard & Molnar 1988; Kröll *et al.* 2006; McCann *et al.* 2006). 1, Penninic-Helvetic thrust plane; 2, Austroalpine-Penninic thrust plane; SD, Salvan-Dorénaz Basin; NSB, Northern Swiss Permo-Carboniferous Basin; BF, Bifertengrätli Basin; GV, Glarner Verrucano Basin; EW, Engadine Window; ZSB, Zentrale Schwellenzone Basins. The inset frame shows the position of Figure 2.

1995). Besides the sediment fill with continental clastics and some volcaniclastic material, these basins share the similarity of being emplaced within basement rocks, mostly of Late Variscan age (Fig. 1).

In the eastern Alps, the European basement is exposed only in the Inner Tauern Window. Therefore, it represents an important link between the basement outcrops of central Europe and the Tisza Block in the Pannonian Basin (Haas & Péró 2004).

Post-Variscan sediments in the Tauern Window

The basement rocks of the Tauern Window consist of Variscan granitic plutons (locally called Zentralgneise) now metamorphosed and deformed which intruded into Lower Palaeozoic and older host rocks including amphibolites and graphite-bearing metasediments (e.g. Finger *et al.* 1993). Together they form a huge duplex structure, which has been uplifted along a deep-reaching ramp (Lammerer *et al.* 2008).

Within the Inner Tauern Window, four Late Palaeozoic to Mesozoic elongate, trough-like basins have been identified. Three of these, the Riffler–Schönach Basin, the Pfitsch–Mörchner Basin and the Maurerkees Basin, are separated by tectonic horsts consisting of basement rocks (Ahorngneiss Horst, Tux Gneiss Horst and Zillertal Gneiss Horst). The fourth basin, the Kaserer Basin, represents, in our interpretation, the southernmost rift-related trough, which developed during the Pangea break-up event that formed the Penninic Ocean. The sediments were detached and overthrust at the base of the Bündnerschiefer (Penninic) Nappes stack (Fig. 2).

The stratigraphically youngest post-Variscan unit that formed within the whole area is the Upper Jurassic Hochstegen Marble (Schönlaub *et al.* 1975; Kiessling 1992). It represents the eastern continuation of the Quinten Limestone of the Helvetic realm in Switzerland. The Hochstegen Marble covers both the former graben and horst areas and varies from between 20 and 400 m in thickness.

The age dating of metasediments which formed prior to the Hochstegen Marble is subject to some uncertainty, because metamorphism and deformation has destroyed almost all fossil material. A lower limit is given by the youngest age of the basement, which is 295 ± 3 Ma (Cesare *et al.* 2001) and an upper limit by the Hochstegen Marble, which began to be deposited in the Oxfordian at 160 Ma (Kiessling 1992). Lower Permian quartz porphyries (Söllner *et al.* 1991) and Anisian deposits with crinoids (Frisch 1975) represent further time markers. Plant fossils have been found so far only in the small Maurerkees Basin, proving a Late Carboniferous

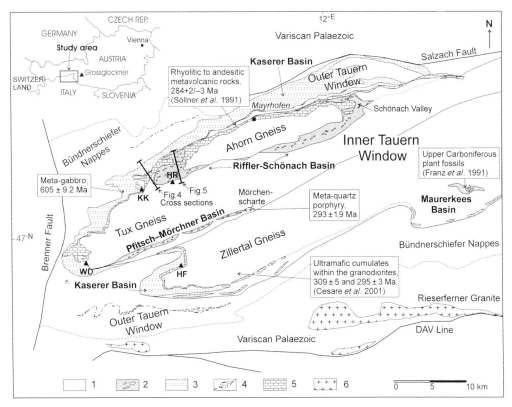

Fig. 2. Geological sketch map of the Tauern Window and the position of the post-Variscan basins. 1, Palaeozoic rocks and Variscan granites; 2, post-Variscan clastic sediments (Upper Carboniferous–Lower Jurassic); 3, Triassic clastic sediments and carbonates at the base of the Bündnerschiefer; 4, limestones, dolomites and cargneuls (Anisian); 5, Hochstegen Formation (Jurassic); 6, Alpine granites (Oligocene); HR, Hoher Riffler (3231 m); HF, Hochfeiler (3510 m); KK, Kleiner Kaserer (3039 m); WD, Wolfendorn (2776 m); DAV Line, Defereggen-Antholz-Vals Fault.

to Early Permian age (Franz *et al.* 1991; Pestal *et al.* 1999).

To unravel the post-Variscan history, the rock successions covering the Palaeozoic basement have been subdivided into mappable lithological units which can be correlated within the study area. Although these units fulfil the requirements of lithostratigraphical formations, their definition as such is beyond the scope of this paper. The most continuous succession is exposed in the Riffler–Schönach Basin and it is used as standard section within this paper. Various lithofacies associations have been distinguished on the basis of lithological changes and vertical succession, dominant grain size or grading. Due to the folding and metamorphism (which reached amphibolite facies), in most instances it has not been possible to classify the internal geometry of the beds, the detailed characteristics of any bounding surfaces or the palaeocurrent patterns. The mineral paragenesis reflects the composition of the protolith. In some instances, rock colour proved to be a useful criterion.

In view of these limitations, the reconstruction of the sedimentary environment and evolution of the basins is not an easy task, but field geology in the very well-exposed Alpine area has provided much new information and revealed some surprising findings. The purpose of this work is to define the lithostratigraphical units constituting the basin fill and to decipher the basin evolution. The stratigraphical scheme presented in this paper is based not just on the geochronological data and lithological changes but also on the correlation with units in the western Alps, southern Germanic Basin and the Bohemian Massif under the Bavarian and Austrian Molasse Basin.

Pfitsch–Mörchner Basin

The Pfitsch–Mörchner Basin extends from the Pfitsch Valley (Italy) in a SW–NE direction over

20 km to the Mörchenscharte (Austria). Its sediments are tightly folded into a syncline which plunges to the west and wedges out at the Mörchenscharte (2872 m) (Fig. 2). The clastic series of the basin fill starts with conglomerates and breccias (Fig. 3a) intercalated with volcaniclastics. The geochronological analysis of a volcanic extrusion from a locality near to the Mörchenscharte was carried out. As described below, zircons from meta-quartz porphyry cutting old basement rocks and covered

Fig. 3. (a) Metaconglomerates. Texturally immature and poorly sorted coarse-grained polymict matrix-supported metaconglomerates representing the proximal part of an alluvial fan (Unit I). Angular to subangular clasts and boulders up to 30 cm in size are predominantly granitic in origin, but also some amphibolites, graphite-bearing schists, marbles and very rare serpentinite clasts occur. The greyish matrix is of sand and silt size. Pebbles show relatively low prolate strain in a tectonically protected area at the Pfitscher Joch, Langsee, 2240 m (Austrian/Italian border). (b) Polished sections of a metaconglomerate from the Pfitscher Joch cut in x–z and y–z directions of the strain ellipsoid. Pebbles of aplites, granites and quartzites show strong flattening strain in an outcrop 500 m to the south of that shown in (a). Long axes of the specimens are 28 cm. (c) Same specimens as in (b) but the z-axis was stretched twice and the x-axis was shortened to $\frac{1}{2}$ by optical methods in order to partially remove the natural strain. The original angular shape of the more competent pebbles and the poor sorting is fairly good visible in the y–z cut. In the x–z direction, the natural strain is too high to provide a satisfactory result.

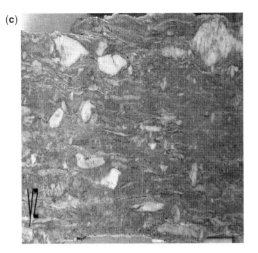

Fig. 3. (*Continued*)

by a meta-conglomerate provide a Late Carboniferous age (293 ± 1.9 Ma) of the magma extrusion. This documents a volcanic activity during initial phases of the basin formation along deep normal faults. The unconformably overlying meta-conglomerates are thus Early Permian or younger in age. Above this, the sediments grade into metapelites and quartzites. Although this fining-upwards succession is about 250 m thick, the geometric analysis of the stretched and flattened pebbles, however, suggests an original thickness in the range of 1 kilometre (Lammerer & Weger 1998) (Fig. 3b, c).

These clastic sediments are overlain by limestones, dolomites and cargneuls of the Aigerbach Formation, which is Middle to Late Triassic in age (Brandner *et al.* 2007). Later fine-grained clastic sediments were deposited; these are supposedly Late Triassic to Early Jurassic in age. To the west, the succession continues upward into the Upper Jurassic Hochstegen Marble. When tracing the Hochstegen Marble to the Wolfendorn area farther to the west, its substrate changes. There, the marble rests only on a thin veneer of Triassic rocks which in turn rest on the basement consisting of Variscan granitoids of the Tux Gneiss. This geometry suggests the onlapping of the basinal sediments onto an elevated area in the north which we interpret as a graben-horst geometry. The basement of the Pfitsch–Mörchner Basin consists of Early Variscan metamorphic rocks like amphibolites and graphite schists (Greiner Schists). These easily erodable rocks formed the graben floor, while the horst positions were made by more resistant granitoids.

The Kaserer Basin

We interpret the fill of the Kaserer Basin as being deposited in a basin that was situated to the south of the Tauern Window and which developed during the Pangean break-up into the Penninic Ocean. These beds form the base of the Penninic Bündnerschiefer and together they were thrust over the European basement as a nappe stack which forms the Outer Tauern Window. Kaserer Basin comprises of sediments of the so-called Kaserer Series, Seidlwinkel Formation and Aigerbach Formation (Middle to Upper Triassic marbles, dolomites and cargneuls) and the so-called Wustkogel Series.

We consider the complex structure of the Outer Tauern Window as an internal thrust and folded sediment stack. The evidence for this is the way that Middle to Upper Triassic carbonate rocks and evaporitic deposits (and other evaporite horizons of uncertain age) occur at several tectonic levels: at the base, within and on the top of the Kaserer and Wustkogel Series and also within the Bündnerschiefer Nappes. These evaporitic sediments presumably served as detachment horizons during the thrusting of the nappes.

Kaserer Series. The substrate of the Kaserer Series was initially formed by Palaeozoic and older basement rocks. During the Mesozoic, these rocks became extensively stretched and boudinaged and, finally, mantle-derived rocks have been emplaced at the basin floor during deposition of the sediments. This is documented by serpentinite bodies, which are incorporated into the Kaserer Series north of Mayrhofen (Thiele 1974) and by a meta-gabbro of Cambrian age (534 ± 9.4 Ma) exposed along a thrust plane in the Tuxer Joch area (Fig. 4). Geochronological data and dating methods of the

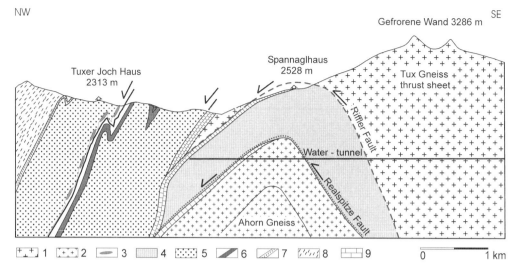

Fig. 4. Cross-section through the Tuxer Joch–Kaserer area. 1, Variscan granites, Ahorn Gneiss; 2, Variscan porphyric granite, Tux Gneiss; 3, Cambrian meta-gabbro within the thrust plane; 4, post-Variscan clastics of the Riffler–Schönach Basin; 5, Permo-Triassic Kaserer Series; 6, Triassic marbles; 7, cargneuls and quartzite, Upper Triassic; 8, Bündnerschiefer Series (Lower Jurassic–Upper Cretaceous); 9, Hochstegen Formation (Upper Jurassic).

meta-gabbro are presented in a separate section below (Figs 8a, b). Similar ages of metabasites occur in other positions within the Tauern Window basement and the Austroalpine basement more to the east (e.g. Kebede *et al.* 2005).

The basal deposits of the Kaserer Series consist of so far undated quartzites, meta-arkoses and mica- and graphite-bearing schists. Occasionally, these rocks exhibit graded bedding. They are well exposed at the Kleiner Kaserer (3091 m) where they have been thrust onto the Jurassic Hochstegen Marble. A succession, more than 100 m thick, with thin horizons of dolomitic marbles, marbles and cargneuls in the upper part of the Kaserer Series (Schöberspitzen), represents a correlative of the dated Anisian rocks (Frisch 1975) in the Wolfendorn area. The nature of the underlying strata and the correlation with the Wolfendorn deposits suggest that the Kaserer Series ranges in age from Late Permian to Early Triassic age, respectively.

Wustkogel Series. In the northern Tauern Window, the so-called Wustkogel Series overlies the Middle Triassic carbonates and form the immediate base of the Bündnerschiefer. The series consists of equigranular greenish quartzite, impure feldspar-rich meta-arenites and meta-arkoses and metaconglomerates showing thin-bedded variations in composition. The gradual transition to the Bündnerschiefer exhibits several coarsening-upward cycles with laterally persistent horizons of cargneuls and limy graphite-bearing schists intercalated into the quartzites. These beds are interpreted to represent fluvial- and delta-systems prograding into a progressively subsiding basin, where evaporitic conditions prevailed. This sequence is very well exposed in the Tuxer Joch and Gerlos area.

Stratigraphic Issues. The stratigraphy and the structure of the Outer Tauern Window is a still a matter of debate. Dietiker (1938), who mapped the area around Mayrhofen and Gerlos, proposed a Permo-Triassic age for the rocks of the Kaserer Series and saw a nappe contact with the Hochstegen Marble. On the contrary, Thiele (1974) presumed that the 'Kaserer Series' lay in sedimentary contact with the Hochstegen Marble and proposed an Early Cretaceous age, whilst noting that a definite stratigraphic position awaited fossil findings. Later workers accepted Thiele's opinion of a Cretaceous age due to an apparent conformable contact in the Wolfendorn area (Frisch 1974, 1980; Lammerer 1986; Rockenschaub *et al.* 2003).

Our more recent field observations, on the other hand, pose some strong arguments against a conformal position:

(1) At the type locality, the Kleiner Kaserer (3096 m), a tectonic horizon exists, as the Hochstegen Marble is parallel to the smooth Zentralgneiss surface, while the Kaserer Series beds are strongly folded. A tectonic horizon within the Kaserer Series is further documented by several bodies of serpentinite near Mayrhofen.

(2) There is a clear sedimentary oscillatory transition from the Kaserer Series to the overlying

Schöberspitzen limestones and dolomites which lie in the same horizon as the carbonates from the Kalkwandstange in which Frisch (1974) found Anisian crinoids. However, he supposed a tectonic contact with the underlying Kaserer Series, but outcrops are poor at that critical position.

(3) In drillings for the Brenner base tunnel, sheared anhydrite was encountered between Hochstegen Marble and the Kaserer Series (Brandner et al. 2007). This is uncommon in Cretaceous deposits of that area, but common for the Permo-Triassic series and for thrust horizons.

For all these reasons, we preliminarily suggest a ?Permo-Triassic age for the Kaserer Series.

A second debatable unit is the so-called Wustkogel Series. Here, we prefer to use a stratigraphically more neutral term, the 'Green Meta-arkoses Series' proposed by Thiele (1970). A Permo-Scythian age is generally assumed for this because of its affinity with the Gröden Sandstone Formation and the Buntsandstein. The term the Wustkogel Series refers to the Seidlwinkel Formation which is Middle Triassic in age, and its subjacent strata (Frasl 1958; Frisch 1968; Höck 1969). Thiele (1970) however doubted this stratigraphic position because the 'Green Meta-arkoses Series' lies on top of the Anisian carbonate rocks and grades into the Jurassic Bündnerschiefer' and so he proposed a Late Triassic age. We support this correlation but recognize its provisional nature. The Stubensandstein of Central Europe can be seen as an equivalent. Feldspar-rich sands were supplied to the Germanic Basin from uplifted areas in the south, the so-called Vindelician Land (Ziegler 1990). The same area might have delivered sediment at its southern side to the south European continental margin.

The Riffler–Schönach Basin

The Riffler–Schönach Basin is a well-exposed Variscan basin which today forms an elongate, SW–NE trending belt. This follows the general strike of the Ahorn Gneiss Horst and dips about $10°-16°$ to the southwest as well to the east under the Tux Gneiss thrust sheet (Fig. 2). The southeastern part is known as the 'Schönachmulde', (literally 'the Schönach syncline', see Thiele 1974; Miller et al. 1984; Sengl 1991). This term, with its suggestion of a syncline, is misleading, as the sediments dip uniformly to the south and are in upright position. The mesoscale asymmetric folds are all north-verging due to shearing and thrusting.

In the Hoher Riffler area, a section about 800 m thick is exposed; the lowermost part is internally thrust, doubling about 170 m of rocks (Fig. 5; Table 1). The clasts in the metaconglomerates are flattened and stretched in the E–W direction. The minimal original thickness of the basin fill is

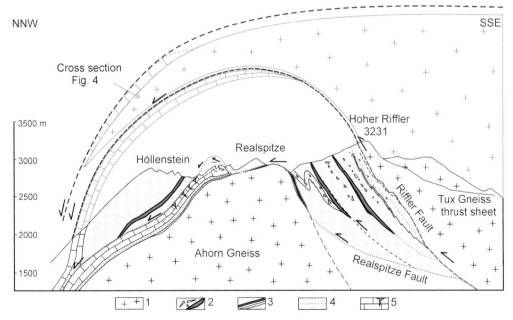

Fig. 5. Cross-section through the Hoher Riffler area. 1, Variscan granites and Palaeozoic basement rocks; 2, post-Variscan clastic sediments and meta-rhyodacite of the Riffler–Schönach Basin; 3, graphite-bearing schists and quartzites (? Lower Jurassic); 4, sandy marble (? Middle Jurassic); 5, Hochstegen Marble (Upper Jurassic) with karstification, in grey colour projected the situation from the Tuxer Joch, Kaserer area in the west (Fig. 4).

Table 1. *Lithofacies associations of the Riffler–Schönach Basin, W and NE from Hoher Riffler*

Lithofacies – short description	Interpreted original composition of the protolith	Interpretation of depositional processes and environment
"Zentralgneise", mylonitic	granites	
"Hochstegen Marble" – partially detached	lime, sandy lime and dolomite	gradual ingression of the sea
white and brown Fe-rich quartzite greyish schists and quartzite with marble layers	sand, partially pebbly laminae of sand, mud and marl	channel sandstone fining upward cycles, fan delta-near environment
greenish quartzite/ meta-arkose, micaceous lenses/metaconglomerate	feldspar-rich sand, partially pebbly mud, silt	stream reworked outer part of alluvial fans/ sheet flood and stream channel deposits
quartzites, marble nodules, marble layers calcareous quartzites mica-schists	calcareous sand marly and limy deposits sand mud	? periodical sea ingression events floodplain, crevasse splays
white and greyish quartzite/ metaconglomerate and -arkose	feldspar-rich fine to coarse sand, partially pebbly, mud clasts	meandering river
meta-rhyodacite graphite-bearing schists	rhyodacite carbonaceous mud, coal	subaerial lava flow plant and mud films, vegetated swamp deposits
greyish metaconglomerate/ meta-arkose/quartzite	massive, matrix-supported gravel, crudely bedded sand, pebbly	upper to middle part of wet alluvial fan, partially stream reworked
brownish quartzite meta-rhyodacite	rhyodacite	subaerial lava flow
greyish metaconglomerate/ meta-arkose/quartzite	massive, matrix-supported gravel, crudely bedded sand, pebbly	upper to middle part of wet alluvial fan, partially stream reworked
greyish quartzite/ meta-arkose brownish quartzite, strongly deformed	sand, partially pebbly	anastomosing river
graphite-bearing schists and quartzites- ? "Hochstegen Quartzite"	mud, sand carbonaceous mud	vegetated swamp deposits, crevasse splay
"Zentralgneise", mylonitic	granites	

estimated between two and three times of the present value—up to 1.9 km. The restoration of the geological section by use of balancing software (2DMove) gives a half-graben structure, about 2 km in depth and 7 km in width (Fig. 6). The true original extension cannot be reconstructed, because the basin is not fully exposed. The uniform lithology, except terrigenous clastic material at the base, and the wide extent of the Hochstegen Marble on the top suggest a rather low-relief landscape during Late Jurassic time.

The Ahorn Gneiss and Tux Gneiss initially formed granitic horsts bounding the half-graben of the Riffler–Schönach basin. The Hochstegen Marble covered the entire unit as a post-rift sediment. During Alpine convergence, the Tux Gneiss and its Hochstegen cover was thrust along the Realspitze Fault over the Riffler–Schönach basin which, in turn, was thrust along the Riffler Fault over the Ahorn gneiss forming now a duplex structure (Figs 5 and 6). By this stacking, the Hochstegen Marble occurs threefold within one section: twice

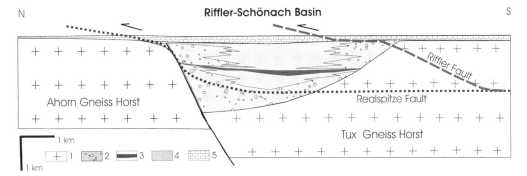

Fig. 6. Restoration of the original situation of the Riffler–Schönach Basin (W part). 1, Variscan granites and Palaeozoic basement rocks; 2, coarse post-Variscan clastics; 3, rhyodacite; 4, fine-grained post-Variscan clastics; 5, Hochstegen Formation (Jurassic).

covering granitic basement, once covering the metasediments of the Riffler–Schönach basin. The third Hochstegen Marble layer rests on top of the Tux Gneiss thrust sheet in the section further to the west (Fig. 4). The tip of the thrust sheets dips with about 50°–60° to the north. An analogous situation occurs in the eastern part of the Riffler–Schönach Basin, where in the Kirchspitze area the Ahorn Gneiss Horst is overthrust by basinal deposits (Thiele 1974, 1976).

The Hochstegen Formation in the Hoher Riffler area

The Hochstegen Formation covers the Tux Gneiss basement and the Riffler–Schönach basinal metasediments. At the base of the Formation, thin layers of graphite-bearing schists and graphite-bearing bedded or massive quartzites occur. An Early Jurassic age for this so-called 'Hochstegenquarzit' has been proposed (Frisch 1968). It is overlain by brownish sandy marbles (?Middle Jurassic). On the top, the main part of the formation is made up by bluish-grey, sulphide-bearing Hochstegen Marble which is locally at the base dolomitic (Oxfordian) and contains in higher horizons (Lower Tithonian) cherty nodules (Kiessling 1992).

Interpretation. The Hochstegen Formation in the Tauern Window overlies basinal sediments as well as the basement rocks (Figs 5 and 6). A notable feature is the shear deformation of the basement rocks (Zentralgneise) at the contact with the Hochstegen Marble, where the Zentralgneise locally become mylonites. This implies that the sediments rest on a tectonically exhumed and erosionally truncated basement. Extensional processes in formerly thickened continental crust could have caused a Basin-and-Range-like situation. The deposition of sand, silt, mud, coal and carbonates on the granitic basement implies considerably rapid tectonic subsidence and attendant erosion.

The lowest part of the Hochstegen Formation is composed of quartzite and graphite-bearing schists and displays characteristics of floodplain deposits with organic-rich mudstones formed in swamps, transected by low-gradient channels. The massive quartzites may represent fluvial sand bodies incised into the coastal plain deposits. The presence of overlying yellowish and brownish sandy marbles up to ten metres thick suggests a marine transgression followed abruptly. Above them, dolomites and greyish marbles document a marine environment as they contain various open-marine microfossils proving a Late Jurassic age (Kiessling 1992).

Lithostratigraphy of the Riffler–Schönach Basin fill

Six lithostratigraphical units constituting the basin fill have been defined. In the present paper, the description is restricted to the western part of the basin, which is very well exposed. The lithostratigraphy is however also applicable to the eastern part.

Unit I: Coarse clastics and volcanics

This unit is composed of greyish metaconglomerates, quartzites, meta-arkoses and meta-rhyodacites. The lowest part of the unit consists of very coarse matrix- to clast-supported polymictic metaconglomerates arranged in coarsening-upward successions. They are massive or crudely bedded, with weak grading. The matrix is sandy to pelitic; angular to subangular boulders of mostly granitic provenance are up to 30 cm in size. In the upper part, pebbles of vein-quartz clasts, gneisses, mudstones and graphite-bearing schists occur as stretched prolate bodies. In the Rötschneidkar (northwest of the Hoher Riffler), carbonate-rich rocks with greenish calc-silicate-rock pebbles occur in addition. In the upper part of Unit I, the amount of medium- to fine-grained sandy material increases. On top of the interval, very fine graphite-bearing micaschists, up to 8 m thick, are overlain by a dark-grey 15 to 30 m thick porphyritic meta-rhyodacite. The interval laterally extends for about 2 km and represents, thus, a good stratigraphical marker. The boundary to the underlying sandy and pebbly deposits is sharp.

Interpretation. The coarse clastics at the base suggest a wet alluvial fan depositional environment having predominantly plutonic and metamorphic rocks exposed in the catchment area. Coarse-grained, clast-supported beds (up to 2 m thick) may represent sieve deposits. Disorganized sediment gravity flows have been partially reworked through stream floods and in braided river systems in gravel and sandy bedforms in the middle and lower part of the fan. Lenses of fine sand and mud accumulated at decreasing flow velocities. The mud and coaly material accumulated in a marginal vegetated part of the fan away from the channels. The formation of alluvial fans implies considerable relative uplift (e.g. rising fault scarps) in tectonically active areas. Coarsening-upward conglomerates are overlain by unsorted debris-flow deposits while the fan prograded. The rhyodacitic lava flow documents subaerial volcanism during the initial phase of the basin formation. The volcanic layer was protected from erosion and preserved within the low-gradient fluvial system within a subsiding basin. The age of the volcanics is presumably Late Carboniferous–Early Permian, the age of many other volcanic and volcanoclastic deposits in the Alpine realm (e.g. Bonin *et al.* 1993; Capuzzo & Bussy 2000).

Unit II: Quartzites, metaconglomerates and mica-schists

This unit is about 150 m thick and comprises white and greyish Fe-rich quartzites, metaconglomerates and mica-schists. Unit II exhibits an overall fining-upward trend. Metaconglomerate lobes up to 3 m thick are intercalated into dm- to m-bedded quartzites. Pebbles, up to 10 cm in diameter, occur within a sandy, feldspar-rich matrix. The stretched pebbles are of vein-quartz, granitic and pelitic composition. Channel-like quartzite bodies with erosional lower bounding surfaces can be recognized. The quartzites contain < 2 cm thick, very fine magnetite-rich heavy mineral layers and lenses. Up sequence, the proportion of fine mica-schists significantly increases.

Interpretation. The laterally persistent, m-thick sands alternating with pelites indicate a meandering-river depositional environment. Channels filled with graded sands overlying pebbly basal lags typically occur. Unfortunately, internal structures like ripple- or cross-bedding have been obliterated by metamorphism. A red colour of the original sandstones is very likely and both the enrichment by iron and the deeply oxidized deposits point to a warm climate (e.g. Ollier 1969). Further up the sequence, fine sands interbedded with mudstones are suggestive of a wide, muddy flood plain. Several metres of quartzite with high amount of mudstone pebbles and mudstone lenses indicate vertical aggradation of the flood plain severely affected by crevasse splays. The stratigraphic position suggests an age in the range of Late Permian to Early Triassic. Several other Variscan basins in Europe display similar lithological characteristics and environmental setting during this time period (e.g. German Basin, Hauschke & Wilder 1999).

Unit III: Marbles and calcareous quartzites

This unit comprises quartzite, meta-arenites, calcareous quartzite and marble layers. It is about 60 m thick and extends laterally for about 1.5 km. The transition to underlying and overlying units is gradual. Metaconglomerate beds are intercalated into dm- to m-bedded quartzites. Feldspar-rich beds are common. Within the lower interval of this unit, layered marble nodules within the quartzite occur. Towards the top, they form continuous, 2 to 40 cm thick marble layers.

Interpretation. The increasing carbonate content, seen in nodules (due to boudinage of layers) and thin persistent layers first and an accumulation of carbonate in distinct beds later point to recurrent short periods of relative sea-level rise (e.g. Fürsich *et al.* 1991. A coastal plain setting is quite likely. During periods of lowered sea level, fluvial channels might have incised the plain. Unit III represents a laterally persistent horizon and is hence a useful stratigraphical marker.

Such a depositional system documents erosion of continental areas and considerable subsidence of the basins, which is typical for the post-Variscan period (Henk 1993; Ziegler 1982). We suggest a Middle Triassic age for the calcareous horizons and consider that they can be ascribed to the sea-level fluctuations of the Tethys which repeatedly flooded the southern part of the European continent at this time.

Unit IV: Impure quartzites

Unit IV is composed of 100 m thick quartzites to meta-arkoses and metaconglomerates with a specific light-green colour caused by phengite.

The pebbles are of vein-quartz, aplitic and granitic origin. Individual angular to subangular granitic clasts and boulders up to 10 cm in size within a sandy matrix are rare. The quartzites are well sorted. Thin micaceous lenses of 5 to 50 cm length are common. In the lower part of the unit, deposits with abundant feldspar clasts or mica-rich meta-arenites occur. The boundary to the underlying unit with calcareous nodules and marble layers is gradual.

Interpretation. Unit IV documents clastic sediment supply into the basin, often by sheet floods reworking the middle and distal parts of alluvial fans. Wide and shallow braided channels formed during high-energy floods due to heavy rain showers. Such supercritical high-density flows may have transported even individual large boulders over a considerable distance. During the peak flow, the adjacent areas were flooded and sheets of well-stratified sand or fine gravel with little silt and clay formed. Ripples with silty lee sides might be the origin of fine micaceous lenses which resemble flaser bedding. The high amount of feldspar clasts indicates a warm climate, at least seasonally dry, favouring the physical disintegration of granite rocks. Ephemeral floods transported debris throughout the basin. The sediments show striking lithological similarities with 'Green Meta-arkoses Series' of the Kaserer Basin, which is supposed to be Late Triassic in age. As palaeogeographic equivalents to these beds, we assign the fluviatile feldspar-rich sand deposits of the Vindelician Keuper (so-called Stubensandstein) in the German Basin (e.g. Ziegler 1990; Beutler *et al.* 1999), being aware that this is speculative.

Unit V: Quartzites, meta-arenites, schists and marbles

Unit V consists of greyish, ribbon-like, fine-grained, calcareous, micaceous schists and mica-quartzites and meta-arenites to metaconglomerates with intercalated, laterally persistent brownish marble horizons. It is about 40 m thick. The boundary with the underlying Unit IV is fairly sharp. All these rocks are characterized by a varying amount of finely disseminated ankerite. In the outcrop east of the Tuxer Ferner House, banded mica-schists and graded meta-arenites occur. Above, several fining-upward cycles, 1 to 5 m thick, with sharp lower surfaces are present. Polymictic, coarse conglomerates are overlain by fine conglomerate and graded quartzite beds. The amount of carbonate (up to 10%), fine-grained quartzites, meta-arenites and schists increases towards the top of a cycle. Thin impure marble beds, 5 to 10 cm thick occur repeatedly. The top of this unit consists of brownish calcareous and limonite-rich quartzites and fine conglomerates.

Interpretation. The depositional environment of Unit V is inferred to have been an alluvial fan within a coastal plain that temporarily developed into a fan-delta complex in a shallow bay depending on the relative change of sea level. The fan-delta complex is characterized by a cyclic succession of finely laminated mudstones and graded sandstones. They were transected by fluvial channels, depositing coarse basal lags, sand and mud in fining-upward cycles. The channel fills are overlain by mudstones and thin carbonate beds. The sediments contain about 5% of disseminated carbonate and, in the upper sandy part, limonite as well. Alluvial fans may prograde into the coastal setting. Calcareous horizons within the mudstones represent phases of relative high sea level. The greyish colour in the lower part suggests reducing conditions because of preserved organic matter while the brownish colour implies oxidation. A change in climate and/or groundwater level is likely. The stratigraphic position of Unit V, below the Upper Jurassic Hochstegen Marble and well above the probable Middle Triassic, suggests an age in the range of Early to Middle Jurassic. This is comparable to the sections north and southwest of the Vindelician High which document increasing flooding during this time (e.g. Trümpy 1980).

Unit VI: Marbles—(Hochstegen Formation)

The Hochstegen Formation consists of yellowish sandy marbles, dolomitic and pure greyish marbles. It occupies a significantly wider area than the older units. It rests on the rocks of Unit V and the boundary is fairly sharp. Unit VI starts with yellowish sandy marbles up to 10 m thick and dolomites. The upper part is made up of monotonous greyish marbles, 20 to 400 m thick. Synsedimentary normal faults occur exhibiting up to several metres of throw and convolute bedding. A palaeontological study on radiolarian associations has proved an Oxfordian and Tithonian age for the upper part of the Hochstegen Formation (Kiessling 1992). The main reason for the preservation of fossils was the early diagenetic pyritization. In the Riffler area, the Hochstegen Formation is about 40 m thick and is overthrust by the Tux Gneiss sheet, which again carries the Hochstegen Formation.

Interpretation. The lower part of the Hochstegen Formation documents the transition from coastal conditions to an open marine environment. Sandy marbles and dolomites represent sand bars and deposits close to the coast. The synsedimentary faulting, the sharp boundary to the underlying clastics and the pronounced decline in clastic material point to a clear increase in relative sea level. When compared to other Tethyan settings, rapid subsidence during the Middle Jurassic appears to be quite likely following the break-up of the Tethyan Ocean during the Bajocian (e.g. Borel 1995; Ziegler 2005). The diversity of radiolarian fauna within the upper part of the formation indicates an establishment of deeper water conditions during the Late Jurassic (Kiessling 1992).

Tectono-sedimentary evolution of the Riffler–Schönach Basin

At the end of the Variscan orogeny, numerous elongate intermontane basins formed in response to both consolidation of the crust and strike-slip movements (e.g. Ziegler et al. 2006). However, there is an ongoing debate on whether some of the basins in the Variscan Internides are of strike-slip or rift origin (Henk 1993; Capuzzo & Wetzel 2004; McCann et al. 2006).

The sedimentary record of the Riffler–Schönach basin starts with supply of continental clastics. In addition, rhyodacitic rocks of a presumed Late Carboniferous–Early Permian age document subaerial volcanism during the initial phases of basin formation and deep-reaching faulting. The basin fill consists of debris flow deposits that dominate in the western part of the basin and were formed on the proximal parts of alluvial fans. The coarse grain-size indicates high transport competence and, hence, a pronounced relief. The conglomerates contain granitic boulders implying rapid uplift and denudation of the basement. Comparison with other such basins suggest that the sedimentary cover had probably been removed prior to basin formation (Capuzzo et al. 2003; Capuzzo & Wetzel 2004). Abundant granite, feldspar- and vein-quartz clasts point to subordinate chemical weathering within the catchment area (e.g. Ollier 1969).

Dark mudstones accumulated under a humid climate from distal alluvial fans within a flood plain. The thickness of the floodplain mudstones increases significantly towards the eastern part of the basin. There, in the Schönach valley, fine-grained mudstones intercalated with quartzites and volcaniclastics are almost 300 m thick and display characteristics of distal parts of alluvial fan floodplain and playa-lake deposits. The thick mudstones are suggestive of rapid differential basin subsidence lowering the river gradient (see Capuzzo & Wetzel 2004). The overall basin fill geometry suggests a palaeoflow to the northeast during the formation of Unit I.

A warm, seasonally dry climate with a concomitant change in sediment composition to feldspar- and Fe-rich deposits characterized the depositional environment of Unit II for which a meandering river system is inferred. Within the Unit III, carbonates of presumed Anisian age are exposed. Accumulation of carbonates in distinct beds within a sandy coastal plain indicates periods of relative sea-level rise (e.g. Fürsich et al. 1991).

Unit IV comprises distal alluvial fans prograding into a coastal plain during ephemeral floods. Uplift and erosion of the hinterland may have been intensified. Braided rivers deposited well-sorted, but impure, sands. High amounts of feldspar clasts within the sands indicate the persistence of a warm and at least seasonally dry climate.

Units V and VI represent fluvio-deltaic depositional and coastal plain settings respectively, which have been rapidly flooded as open marine conditions were established on the top of Unit VI. Alluvial fans may have prograded into the coastal plain and carbonates represent phases of relative high sea level and marine incursions. Middle Jurassic sandy marbles form a sharp transition to the neritic environment and document progressive subsidence. In Late Jurassic times, deep marine conditions were established.

In Cretaceous time, the extensional regime changed progressively to compressional tectonics giving wrench faulting and stress-induced buckling of the lithosphere, causing a relative low-stand in sea level. During the earliest Cretaceous, large parts of western and central Europe were uplifted and subjected to erosion (Ziegler 2005). The Jurassic marine sediments in the Tauern Window may have been at least partially exhumed. The subsequent erosion possibly accounts for the strongly varying thickness of the Hochstegen Marble over the entire Tauern Window. In the southeast, the marble thins out or is completely missing (Fig. 2). This corresponds to the Cretaceous and Palaeocene evolution farther to the north. There the strong intraplate compressional deformation caused the partial destruction of the Mesozoic sedimentary cover of the Bohemian Massif in the foreland of the eastern Alps (e.g. Ziegler 1990, 2005).

Geochronological data and methods

The age of the meta-quartz porphyry from the Pfitsch-Mörchner Basin

West of the Mörchenscharte, a 5 m to 25 m thick meta-quartz porphyry cuts serpentinites, amphibolites and fine-grained or medium-grained clastic metasediments and is unconformably overlain by metaconglomerates. For a geochronological analysis, a medium-grained meta-quartz porphyry (MO1) was sampled 700 m to the west of the Mörchenscharte at an altitude of 2700 m. Zircons were separated into grain-size fractions. They form a homogeneous population, subhedral in shape with slightly corroded surfaces, but clear-cut edges and pyramidal apices. Crystal types according to Pupin (1980, 1985) are characterized as P3 to P5, as well as S9 to S22 which point to a subalkaline or calc-alkaline type of magma. Large zircon fractions 1 and 2 (150–180 μm) were assorted according to these criteria, respectively.

Conventional U–Pb analyses were performed in the laboratory at the Department of Earth and Environmental Sciences at the Ludwig-Maximillians-Universität (LMU). The data were plotted in a Tera & Wasserburg diagram, and calculations were made with the ISOPLOT program (Ludwig 2003). Data points of coarse grain-size fractions 1 and 2 are concordant; finer grain-size fractions show loss of radiogenic lead. The calculated discordia intersects the concordia at 293.0 ± 1.9 Ma and at the origin (Fig. 7). The age is interpreted as the time of zircon crystallization in the melt which is, within limits of error, identical with the extrusion age of the quartz porphyry magma.

The age of the meta-gabbro horizon from the Kaserer Basin

A fine-grained meta-gabbro (sample RAK) was sampled for age determination from the peak of the Rauhe Kopf (2150 m, Schmirntal, Brenner area). A MORB character was determined for this rock by Frisch (1984) using chemical characteristics. He assumed a Cretaceous age for the host rocks, the so-called Kaserer Series, and hence also for the intrusion. This age, however, was already questioned by field studies, as the Kaserer Series is in sedimentary contact with overlying Anisian carbonate rocks (Lammerer 2003).

Zircons of the meta-gabbro RAK were separated into two different types. Cathodo-luminescence investigations on type-1-zircons display predominantly only a single growth phase. An asymmetric inherited core is rarely developed. Zircon growth was uniform and displays irregularly cloudy luminescence. Type-2-zircons are of heterogeneous composition and display a three-phase growth history. A detrital core is surrounded by the volumetrically dominant main growth phase which contrasts in luminescence and is characterized by weakly developed oscillatory zoning. The anhedral

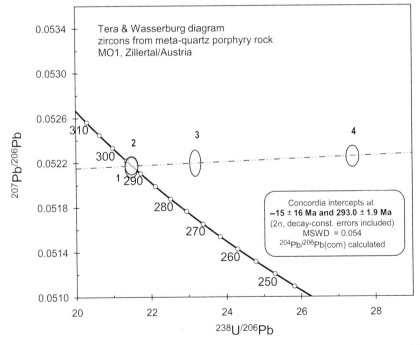

Fig. 7. Tera & Wasserburg diagram depicting zircon grain–size fractions from meta-quartz porphyry rock MO1 (Mörchenscharte/Zillertal, Austria). Large grain–size fractions 1 and 2 (150–180 μm) are, independent of their crystal type, concordant at 293 ± 1.9 Ma. Loss of radiogenic lead, possibly due to the enhancement of surface corrosion effects, is visible in smaller-sized zircon fractions 3 (80–100 μm) and 4 (42–53 μm). The age reflects the time of superficial extrusion of the precursor volcanic rocks.

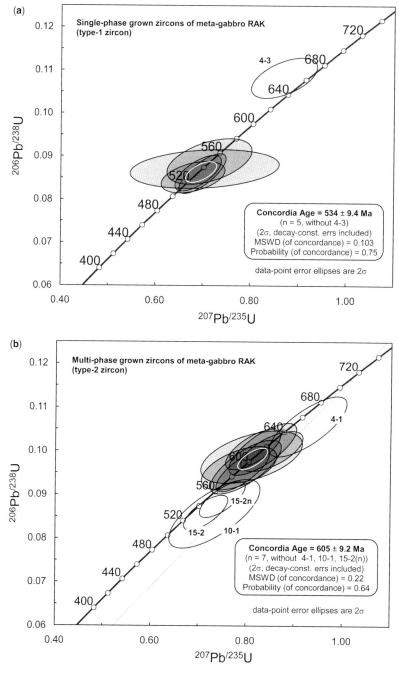

Fig. 8. (a) Concordia diagram of U–Pb single spot analyses with LA-ICP-MS on type-1-zircons from meta-gabbro RAK (Schmirntal/Brenner, Austria). Cathodoluminescence images reveal a simple growth history of this zircon type, which started occasionally at detrital cores. The main zircon growth has been dated at 534 ± 9.4 Ma. This age is interpreted as the time of zircon growth and suggests simultaneous crystallization of the gabbroic precursor rock.
(b) Concordia diagram of U–Pb single spot analyses with LA-ICP-MS on type-2-zircons from meta-gabbro RAK (Schmirntal/Brenner, Austria). Complex moulded zircons, indicated by cathodoluminescence images reveal a main zircon growth phase at 605 ± 9.2 Ma. Older inherited cores (about 716 Ma, upper intercept of discordia line) and an outer rim of subsequent zircon overgrowth are visible. Zircons of this type seem to be overtaken into the gabbroic melt and had experienced marginal zircon overgrowth.

outer shape of this zone can be explained by dissolution during a succeeding heating phase. In addition, a marginal domain of light and homogeneous luminescence is developed and may correspond to the similar-looking main growth zone in type-1 zircons. Unfortunately, this rim was too small to give reliable age results. As both zircon types originate from the same rock, a complex zircon growth history can be inferred.

Zircon grains were analysed for U, Th and Pb isotopes by Laser Ablation Inductive Coupled Plasma Mass Spectrometry (LA-ICP-MS) techniques at the Institute of Geosciences, Johann Wolfgang Goethe-University Frankfurt, using a Thermo-Finnigan Element II sector field ICP-MS coupled to a New Wave UP213 ultraviolet laser system. Data were acquired in peak jumping mode during 30 s ablation with a spot size of 20 and 30 μm, respectively. A common-Pb correction based on the interference- and background-corrected ^{204}Pb signal and a model Pb composition (Stacey & Kramers 1975) was carried out if necessary. The necessity of the correction is judged on whether the corrected ^{207}Pb/^{206}Pb lies outside of the internal errors of the measured ratios. Reported uncertainties (2σ) were propagated by quadratic addition of the external reproducibility (2 s.d.) obtained from the standard zircon GJ-1 during the analytical session and the within-run precision of each analysis (2 s.e.). Concordia diagrams (2σ error ellipses) and concordia ages with 2σ uncertainty were produced using Isoplot/Ex 2.49 (Ludwig 2003). For further details on the method, see Gerdes & Zeh (2006).

U–Pb analyses were made of inherited cores and the main growth phases of both zircon types. The age of the main zircon growth phase in type-1 zircons is dated at 534 ± 9.4 Ma (Fig. 8a) and can be interpreted as the phase of dominant zircon growth in the gabbroic melt. Discordant data point 4-3 suggests to have suffered uranium loss. The main growth phase in type-2 zircons is significantly older and of Precambrian age (605 ± 9.2 Ma; Fig. 8b). Two ages of inherited cores in type-2 zircons are discordant (4-1 and 10-1) because of the loss of radiogenic lead. A crystallization age of about 716 Ma (including the origin) can be inferred from the upper intercept of the regression line.

The different ages of the dominant growth phases in both zircon types are explained by their different growth history. The single-phase of type-1 zircons developed in the gabbroic melt, whereas type-2 zircons are considered to have been overtaken into the melt as older assimilated components. Only the outermost, small rim belongs perhaps to the stage of gabbroic melt formation at 534 ± 9.4 Ma.

Conclusions

Sedimentary basins exposed in the Tauern Window are part of the network of post-Variscan basins in western and central Europe. The Tauern Window exposes European continental crust which was drowned during the Late Jurassic, but which includes several small, elongate sedimentary basins which formed from the Late Palaeozoic as intermontane graben structures. The basins can be traced over 20 km or more wherein more than 1 to 2 km of sediments accumulated. As elongate belts, they run parallel to the Alpine tectonic strike and might have affected the orientation of Alpine compressive structures.

The onlap of sediments onto crystalline basement (well exposed in the Pfitsch–Mörchner Basin) demonstrates that the Late Palaeozoic basins were separated by highs. The sedimentary fill documents a rapid uplift, denudation of the Variscan orogenic belt and concomitant subsidence of the basins. The fault-bounded nature of these narrow basins is also evidenced by the existence of abrupt variations in the stratigraphic thickness. The facies pattern of the west Tauern Window suggests that three horsts experienced erosion for a prolonged period while the basins in between trapped sediments (Fig. 9). The sediments rest on a tectonically exhumed and erosionally truncated basement as the floor of the basin is made up of Variscan granitoids and Palaeozoic metamorphic rocks. Mylonitic zones occur within the basement rocks along steeply dipping normal faults. Basin-and-range-like extensional processes could have caused such a situation (Ménard & Molnar 1988).

In the representative, well-exposed basins like the Pfitsch–Mörchner or Riffler–Schönach Basins, no clear evidence for hiatus horizons within the sediments are seen. This absence is due to the Alpine folding and metamorphic overprint, as discordances between the strata are presumed to exist. Nevertheless, in some protected locations, major structures are still well preserved, so that lithostratigraphical correlation of the rock successions is feasible. The post-Palaeozoic history can be analysed, but only in part. The stratigraphic concept of more or less successive sedimentary evolution of the basins is, in part, contrary to previous interpretations. Our interpretation and the attempt to reconstruct the post-Variscan sedimentary evolution in the Tauern Window is based on the evidence that the area was exposed to prolonged erosion and experienced tectonic stretching until the Late Jurassic, when the area was flooded.

During the initial stages of the basin formation, subaerial volcanism was active. Major sediment delivery from continental sources, particularly by attendant erosion of the Variscan orogen, persisted

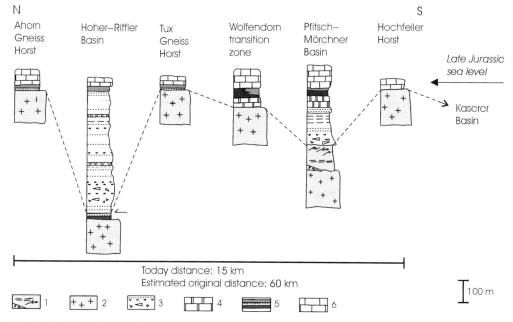

Fig. 9. Correlation of sections in the western Tauern Window. 1, pre-Variscan basement rocks; 2, Variscan granites; 3, post-Variscan clastic sediments; 4, Anisian carbonates; 5, graphitic schists and quartzites, sandy marbles (Early–Middle Jurassic); 6, Hochstegen Marble (Upper Jurassic).

until the Jurassic. The basin fill started with an overall fining-upward series of coarse clastic sediments which represent alluvial fans, fluvial sediments of meandering rivers and playa-lakes in pre-Middle Triassic times. Triassic marbles and cargneuls document that topography of the basin floors was close to sea level. Relative sea-level rise led to repeated flooding of the smooth surface of the southern European continental area. Above this calcareous horizon, fine-grained and well-sorted clastics were deposited by sheet floods and braided rivers until the Middle Jurassic. At this time, a transition from the coastal to deltaic conditions occurred. In response to continued crustal extension and relative sea-level rise, marine conditions were established from the Middle Jurassic, probably due to the break-up of the Tethys leading to rapid subsidence and the drowning of the continental margin. The Late Jurassic Hochstegen Formation was deposited mainly under deeper marine conditions when the entire area of the Tauern Window was submerged.

We thank Yvonne Legath (Munich) for laborious selection of the few zircons from the gabbro, Chris Walley (Swansea) for reading the manuscript and providing the final touches to the English, Horst Rass and Heinz Nyvelt from the Verbund Austria Hydropower in Mayrhofen for permission and support in visiting the water tunnels. This paper benefited from the engaged and encouraging reviews by Hilmar von Eynatten (Göttingen) and Wilfried Winkler (Zürich). The Deutscher Akademischer Austauschdienst (DAAD) financially supported Petra Veselá.

References

ARTHAUD, F. & MATTE, P. H. 1977. Late Paleozoic strike-slip faulting in southern Europe and northern Africa. *Geological Society of America Bulletin*, **88**, 1305–1320.

BENAMMI, M. & EL KOCHRI, A. 1998. Ouverture du rift atlasique: mise en évidence d'une paléofaille de transfer orientée N120. *Comptes Rendus de l'Académie des Sciences, Paris, Sciences de la terre et des planetès*, **327**, 845–850.

BEUTLER, G., HAUSCHKE, N. & NITSCH, E. 1999. Faziesentwicklung des Keupers im Germanischen Becken. *In*: HAUSCHKE, N. & WILDE, V. (eds) *Trias. Eine ganz andere Welt. Mitteleuropa im frühen Erdmittelalter*. Verlag Dr. Friedrich Pfeil, München, 129–174.

BONIN, B., BRÄNDLEIN, P., BUSSY, F. *ET AL.* 1993. Late Variscan magmatic evolution of the Alpine Basement. *In*: VON RAUMER, J. F. & NEUBAUER, F. (eds) *Pre-Mesozoic Geology in the Alps*, Springer-Verlag, Berlin, 171–201.

BOREL, G. 1995. Préalpes médianes romandes: courbes de subsidence et implications géodynamiques. *Bulletin de Société de Vaudois de Science Naturelle*, **83**, 293–315.

BOUAZIZ, S., BARRIER, E., TURKI, M. M. & TRICART, P. 1999. La tectonique permo-mésozoïque (anté-Vraconien) dans la marge sud téthysienne en Tunisie méridionale. *Bulletin de la Société géologique de France*, **170**, 45–56.

BRANDNER, R., REITER, F. & TÖCHTERLE, A. 2007. Geologische Prognose des Brenner Basis-tunnlesein Überblick. *In*: SCHNEIDER, E., JOHN, M. & BRANDNER, R. (eds) *BBT 2007, Internationales Symposium Brenner Basistunnel und Zulaufstrecken*. Innsbruck University Press, 13–23.

CAPUZZO, N. & BUSSY, F. 2000. High-precision dating and origin of synsedimentary volcanism in the Late Carboniferous Salvan Dorénaz basin (Aiguilles-Rouges Massif, Western Alps). *Schweizerische Mineralogische und Petrographische Mitteilungen*, **80**, 147–167.

CAPUZZO, N. & WETZEL, A. 2004. Facies and basin architecture of the Late Carboniferous Salvan-Dorenaz continental basin (Western Alps, Switzerland/France). *Sedimentology*, **51**, 675–697.

CAPUZZO, N., HANDLER, R., NEUBAUER, F. & WETZEL, A. 2003. Post-collisional rapid exhumation and erosion during continental sedimentation: the example of the Late Variscan Salvan-Dorenaz basin (Western Alps). *International Journal of Earth Sciences*, **92**, 364–379.

CESARE, B., RUBATTO, D., HERMANN, J. & BARZI, L. 2001. Evidence for Late Carboniferous subduction type magmatism in mafic – ultramafic cumulates of the Tauern window (Eastern Alps). *Contributions of Mineralogy and Petrology*, **142**, 449–464.

COWARD, M. & DIETRICH, D. 1989. Alpine tectonics — an overview. *In*: COWARD, M. P., DIETRICH, D. & PARK, R. G. (eds) *Alpine Tectonics*. Geological Society Special Publication, **45**, 1–29.

DIETIKER, H. 1938. *Der Nordrand der Hohen Tauern zwischen Mayrhofen und Krimml (Gerlostal, Tirol)*. PhD Thesis, ETH Zürich.

FAERSETH, R. B. 1996. Interaction of Permo-Triassic and Jurassic extensional fault-blocks during the development of the northern North Sea. *Journal of the Geological Society London*, **153**, 931–944.

FINGER, F., FRASL, G., HAUNSCHMID, B. *ET AL*. 1993. The Zentralgneise of the Tauern Window (Eastern Alps)—insight into an intra-Alpine Variscan batholite. *In*: NEUBAUER, J. F. & VON RAUMER, J. F. (eds) *The Variscan Basement in the Alps*. Springer Verlag, 375–391.

FRANKS, G. D. 1966. The development of the limnic Upper Carboniferous of the eastern Aar Massif. *Eclogae Geologicae Helvetiae*, **59**, 943–950.

FRANZ, G., MOSBRUGGER, V. & MENGE, R. 1991. Carbo-Permian pteridophyll leaf fragments from an amphibolite facies basement, Tauern Window, Austria. *Terra Nova*, **3**, 137–141.

FRASL, G. 1958. Zur Serientgliederung der Schieferhülle in den mittleren Hohen Tauern. *Jahrbuch der geologischen Bundesanstalt, Wien*, **101**, 323–472.

FRISCH, W. 1968. Geologie des Gebietes zwischen Tuxbach und Tuxer Hauptkamm bei Lanersbach (Zillertal, Tirol). *Mitteilungen der Geologie und Bergbaustudenten*, **18**, 287–336.

FRISCH, W. 1974. Die stratigraphisch-tektonische Gliederung der Schieferhülle und die Entwicklung des penninischen Raumes im westlichen Tauernfenster (Gebiet Brenner – Gerlospaß). *Mitteilungen der Österreichischen Geologischen Gesellschaft*, **66/67**, 9–20.

FRISCH, W. 1975. Ein Typ-Profil durch die Schieferhülle des Tauernfensters: Das Profil am Wolfendorn (westlicher Tuxer Hauptkamm, Tirol). *Verhandlungen der Geologischen Bundesanstalt*, Wien 1974/**2-3**, 201–221.

FRISCH, W. 1980. Post-Hercynian formations of the western Tauern window: sedimentological features, depositional environment, and age. *Mitteilungen der Österreichischen Geologischen Gesellschaft*, **71/72**, 49–63.

FRISCH, W. 1984. Metamorphic history and geochemistry of a low-grade amphibolite in the Kaserer Formation (marginal Buendner Schiefer of the western Tauern Window, the Eastern Alps). *Schweizerische Mineralogische und Petrographische Mitteilungen, Bulletin Suisse de Mineralogie et Petrographie*, **64**, 193–214.

FÜRSICH, F. T., OSCHMANN, W., JAITLY, A. K. & SINGH, I. B. 1991. Faunal response to transgressive-regressive cycles: example from the Jurassic of western India. *Palaeogeography, Palaeoclimatology, Palaeoecology*, **85**, 149–159.

GERDES, A. & ZEH, A. 2006. Combined U-Pb and Hf isotope LA-(MC-)ICP-MS analyses of detrital zircons: Comparison with SHRIMP and new constraints for the provenance and age of an Armorican metasediment in Central Germany. *Earth and Planetary Science Letters*, **249**, 47–62.

HAAS, J. & PÉRÓ, C. 2004. Mesozoic evolution of the Tisza Mega-unit. *International Journal of Earth Sciences (Geologische Rundschau)*, **93**, 297–313.

HAUSCHKE, N. & WILDE, V. (eds) 1999. *Trias, eine ganz andere Welt: Mitteleuropa im frühen Erdmittelalter*. Verlag Dr. Friedrich Pfeil, München.

HENK, A. 1993. Subsidenz und Tektonik des Saar-Nahe-Beckens (SW-Deutschland). *Geologische Rundschau*, **82**, 3–19.

HÖCK, V. 1969. Zur Geologie des Gebietes zwischen Tuxer Joch und Olperer (Zillertal, Tirol). *Jahrbuch der Geologischen Bundesanstalt, Wien*, **112**, 153–195.

KEBEDE, T., KLOETZLI, U., KOSLER, J. & SKIOLD, T. 2005. Understanding the pre-Variscan and Variscan basement components of the central Tauern Window, Eastern Alps, Austria; constraints from single zircon U-Pb geochronology. *International Journal of Earth Sciences*, **94**, 336–353.

KEELEY, M. L. 1996. The Irish Variscides: problems, perspectives and some solutions. *Terra Nova*, **8**, 259–269.

KIESSLING, W. 1992. Palaeontological and facial features of the Upper Jurassic Hochstegen Marble (Tauern Window, Eastern Alps). *Terra Nova*, **4**, 184–197.

KRÖLL, A., MEURERS, B., OBERLERCHER, G. *ET AL*. 2006. *Erläuterungen zu den Karten Molassezone Salzburg–Oberösterreich*. Geologische Bundesanstalt, Wien.

LAMMERER, B. 1986. Das Autochton im westlichen Tauernfenster. *Jahrbuch der Geologischen Bundesanstalt, Wien*, **129**, 51–67.

LAMMERER, B. & WEGER, M. 1998. Footwall uplift in an orogenic wedge; the Tauern Window in the Eastern Alps of Europe. *Tectonophysics*, **285**, 213–230.

LAMMERER, B. 2003. The Tauern Window, key to the understanding of the Eastern Alps. *Memorie di Scienze Geologiche, Padova*, **54**, 183–184.

LAMMERER, B., GEBRANDE, H., LÜSCHEN, E. & VESELÀ, P. 2008. A crustal-scale cross-section through the Tauern Window (eastern Alps) from geophysical and geological data. *In*: SIEGESMUND, S., FÜGENSCHUN, B. & FROITZHEIM, N. (eds) *Tectonic Aspects of the Alpine-Dinaride-Carpathian System*. Geogical Society, London, Special Publications, **298**, 219–229.

LEMCKE, K. 1988. *Geologie von Bayern 1: Das bayerische Alpenvorland vor der Eiszeit*. Schweizerbart'sche Verlagsbuchhandlung.

LUDWIG, K. R. 2003. *Users manual for isoplot/ex rev. 2.49: a geochronological toolkit for Microsoft Excel*. Berkeley Geochronology Center, Special Publication, **1a**, 1–56.

MATTER, A. 1987. Faziesanalyse und Ablagerungsmilieus des Permokarbons im Nordschweizer Trog. *Eclogae Geologicae Helvetiae*, **80**, 345–367.

MCCANN, T., PASCAL, C., TIMMERMAN, M. J. ET AL. 2006. Post-Variscan (End Carboniferous – Early Permian) basin evolution in Western and Central Europe. *In*: GEE, D. G. & STEPHENSON, R. A. (eds) *European Lithosphere Dynamics*. Geological Society London, Memoirs, **32**, 355–388.

MÉNARD, G. & MOLNAR, P. 1988. Collapse of Hercynian Tibetan Plateau into a Late Palaeozoic European Basin and Range province. *Nature*, **334**, 235–237.

MILLER, H., LEDOUX, H., BRINKMEIER, I. & BEIL, F. 1984. Der Nordwestrand des Tauernfensters—stratigraphische Zusammenhänge und tektonische Grenzen. *Zeitschrift der Deutschen Geologischen Gesellschaft*, **135**, 627–644.

OBERHÄNSLI, R., SCHENKER, F. & MERCOLLI, I. 1988. Indications of Variscan nappe tectonics in the Aar Massif. *Schweizerische Mineralogische und Petrographische Mitteilungen*, **68**, 509–520.

OLLIER, C. D. 1969. *Weathering*. Oliver and Boyd, Edinburgh.

PESTAL, G., BRUEGGEMANN-LEDOLTER, M., DRAXLER, I. ET AL. 1999. Ein Vorkommen von Oberkarbon in den mittleren Hohen Tauern. *Jahrbuch der Geologischen Bundesanstalt, Wien*, **141**, 491–502.

PUPIN, J. P. 1980. Zircon and granite petrology. *Contributions to Mineralogy and Petrology*, **73**, 207–220.

PUPIN, J. P. 1985. Magnatic zoning of Hercynian granitoids in France based on zircon typology. *Schweizerische Mineralogische und Petrographische Mitteilungen*, **65**, 29–56.

ROCKENSCHAUB, M., KOLENPRAT, B. & NOWOTNY, A. 2003. Das westliche Tauernfenster. *In*: ROCKENSCHAUB, M. (ed.) *Arbeitstagung der Geologische Bundesanstalt*. Wien, 7–38.

SCHALTEGGER, U. & CORFU, F. 1995. Late Variscan "Basin and Range" magmatism and tectonics in the Central Alps: evidence from U-Pb geochronology. *Geodinamica Acta*, **8**, 82–98.

SCHÖNLAUB, H. P., FRISCH, W. & FLAJS, G. 1975. Neue Fossilfunde aus dem Hochstegenmarmor (Tauernfenster, Österreich). *Neues Jahrbuch für Geologie und Paläontologie*, 1975/**2**, 111–128.

SENGL, F. 1991. *Geologie und Tektonik der Schönachmulde (Zillertaler Alpen, Tirol)*. PhD Thesis, Ludwig-Maximilians-Universität München.

SÖLLNER, F., HÖLL, R. & MILLER, H. 1991. U-Pb-Systematik der Zirkone in Meta-Vulkaniten ("Porphyroiden") aus der Nördlichen Grauwackenzone und dem Tauernfenster (Ostalpen, Österreich). *Zeitschrift der deutschen geologischen Gesellschaft*, **142**, 285–299.

STACEY, J. S. & KRAMERS, J. D. 1975. Approximation of terrestrial lead isotope evolution by a two-stage model. *Earth and Planetary Science Letters*, **26**, 207–221.

THIELE, O. 1970. Zur Stratigraphie und Tektonik der Schieferhülle der westlichen Hohen Tauern. *Verhandlungen der geologischen Bundesanstalt, Wien*, 230–244.

THIELE, O. 1974. Tektonische Gliederung der Tauernschieferhülle zwischen Krimml und Mayrhofen. *Jahrbuch der Geologischen Bundesanstalt, Wien*, **117**, 55–74.

THIELE, O. 1976. Der Nordrand des Tauernfensters zwischen Mayrhofen und Inner Schmirn (Tirol). *Geologische Rundschau, Stuttgart*, **65**, 410–421.

TRÜMPY, R. 1980. *Geology of Switzerland – a guide book*. Wepf, Basel, New York.

VON RAUMER, J. F. 1998. The Paleozoic evolution in the Alps: from Gondwana to Pangea. *Geologische Rundschau*, **87**, 407–435.

WETZEL, A., ALLENBACH, R. & ALLIA, V. 2003. Reactivated basement structures affecting the sedimentary facies in a tectonically "quiescent" epicontinental basin: an example from NW Switzerland. *Sedimentary Geology*, **157**, 153–172.

ZIEGLER, P. A. 1982. Triassic Rifts and Facies Patterns in Western and Central Europe. *Geologische Rundschau*, **71**, 747–771.

ZIEGLER, P. A. 1990. *Geological Atlas of Western and Central Europe*. 2nd edn. Shell Internationale Petroleum Maatschappij, The Hague.

ZIEGLER, P. A. 2005. Permian to Recent Evolution. *In*: SELLEY, R. C., COCKS, L. R. & PLIMER, I. R. (eds) *The Encyclopedia of Geology*. Elsevier, World Wide Web Address: http://www.encyclopediaofgeology.com/samples/europe2.pdf

ZIEGLER, P. A., CLOETINGH, S., CORNU, T. & BEEKMAN, F. 2006. Neotectonics and intraplate continental topography of the northern Alpine Foreland. *Earth-Science Reviews*, **74**, 127–196.

The Meliata suture in the Carpathians: regional significance and implications for the evolution of high-pressure wedges within collisional orogens

R. DAVID DALLMEYER[1], FRANZ NEUBAUER[2] & HARALD FRITZ[3]

[1]Dept. of Geology, University of Georgia, Athens/GA 30602, USA (e-mail: dallmeyr@uga.edu)

[2]Fachbereich Geographie und Geologie, Universität Salzburg, Hellbrunner Str. 34, A-5020 Salzburg, Austria

[3]Institut für Erdwissenschaften, Universität Graz, Heinrichstr. 26, A-8010 Graz, Austria

Abstract: The Meliata nappe of the Western Carpathian orogen is comprised of Triassic deep-sea metasedimentary rocks and fragments of blueschist-bearing ophiolite. These were structurally emplaced onto Permian/Triassic shelf sequences and its Variscan basement. The nappe records a succession of deformational events which formed under decreasing pressure conditions. Subduction-related burial and resulting blueschist metamorphism is dated by $^{40}Ar/^{39}Ar$ plateau ages recorded by four phengitic muscovite concentrates (160–150 Ma). These crystallized during ductile deformation characterized by predominantly coaxial NW–SE stretching. The structures were overprinted by semiductile, nonpenetrative fabric elements which formed under greenschist facies conditions and contemporaneously with fabrics which developed in footwall tectonic units. Kinematic indicators record top north to NW shear during Middle Cretaceous loading of the Meliata unit onto the Inner Carpathian nappe complex recorded by a $^{40}Ar/^{39}Ar$ whole rock phyllite age of 105.8 ± 1.5 Ma. A subsequent southeastern sense of shear is interpreted to have resulted from extension during Late Cretaceous uplift along hinterland margins of the tectonic wedge. The Meliata unit is part of a major Late Jurassic/Early Cretaceous suture which initially extended from the Alps to the Hellenides. It has been subsequently disrupted as a result of later strike-slip faulting following Tertiary collision of the Cretaceous orogen, and was transported onto extra-Alpine European units.

The resolution of chronology, mechanism of formation and subsequent exhumation of high-pressure rocks within collisional belts are of crucial importance in resolving the evolution of collisional orogenic belts (e.g. Ernst 1988; Schermer 1990; Wakabayashi & Unruh 1995; Froitzheim et al. 2003). Blueschist metamorphism has been generally interpreted to have resulted from burial within a subduction zone and may, therefore, represent a fossil suture zone during subsequent continent–continent collisional stages. The mode of structural transfer of high-pressure rocks from within a subduction zone to their final location within collisional belts has been controversial. Proposed models include (for a review, see Platt 1993; Ernst 2005): (1) break-up of a subduction zone and obduction onto continental elements; and (2) assembly of subducted-related imbricate sheets and transfer to the upper trailing edge of accretionary wedges by underplating (e.g. Platt 1993).

In this contribution, we present $^{40}Ar/^{39}Ar$ and structural data, which constrain the structural and temporal evolution of the Meliata unit, an important blueschist-bearing structural element within the Alpine–Carpathian system (Figs 1 and 2). Our data were never fully published although we used some data in an overview paper (Dallmeyer et al. 1996). This high-level structural unit has crucial importance because it contains high-pressure metamorphic sequences and occurs at a tectonostratigraphic level that has been classically accepted to be the suture within the Alpine–Carpathian collisional system. Small remnants of the Meliata suture are widespread in the Western Carpathians (Fig. 2), but also occur in Bükk Mountains within the Pannonian basin and the easternmost Alps (Fig. 1). Consequently, the new data have also significance for the entire Alpine–Carpathian system. Our data contribute, furthermore, to the general knowledge of tectonic processes within orogenic wedges.

Geological setting

The Western Carpathians represent the northeastern extension of the Alps, and are comprised of generally similar tectonic units (Fig. 1). Regional structural units include the Outer Carpathians, the Pienidic klippen belt and the Inner Carpathians

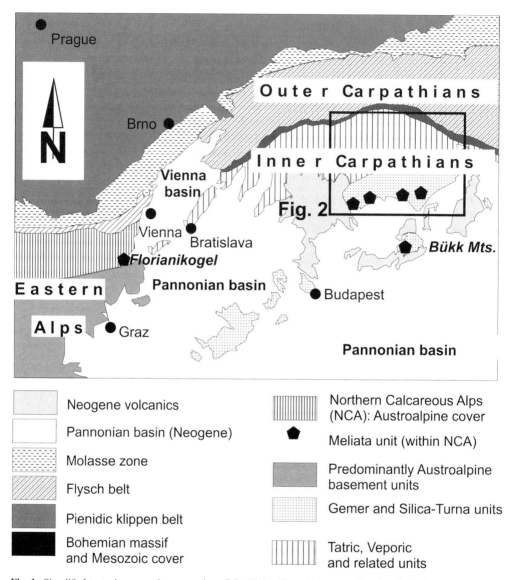

Fig. 1. Simplified tectonic map and cross-section of the Alpine-Carpathian orogen locating the Meliata suture. Note that Tatric, Veporic, Gemer and related units of the Carpathians correlate with Austroalpine units in the Eastern Alps, the Silica-Turňa unit with high structural levels within Northern Calcareous Alps. Small remnants of the Meliata unit are exposed at Florianikogel within Northern Calcareous Alps.

(Fig. 2). The Outer Carpathians include Cretaceous–Palaeogene flysch units, which represent an accretionary wedge derived from the European foreland (e.g. Tomek & Hall 1993; Tomek 1993). The Pienidic klippen belt is a zone of high transform displacement which separates the Outer and Inner Carpathians (Dal Piaz et al. 1995). The main, lower part of the Inner Carpathians comprise, from base to top, the Tatric, Veporic and Gemeric nappe complexes. These consist of an imbricated succession of very low-grade to medium-grade metamorphic nappe complexes comprising pre-Alpine basement and Late Palaeozoic–Cretaceous cover sequences (with shelf and passive continental margin affinities) and which are considered to represent the lateral extension of Austroalpine units of the eastern Alps (Fig. 1). Alpine metamorphism has been estimated to have reached c. 350 °C in Tatric and Gemer nappe complexes within medium pressures (e.g. Krist et al. 1992) and with

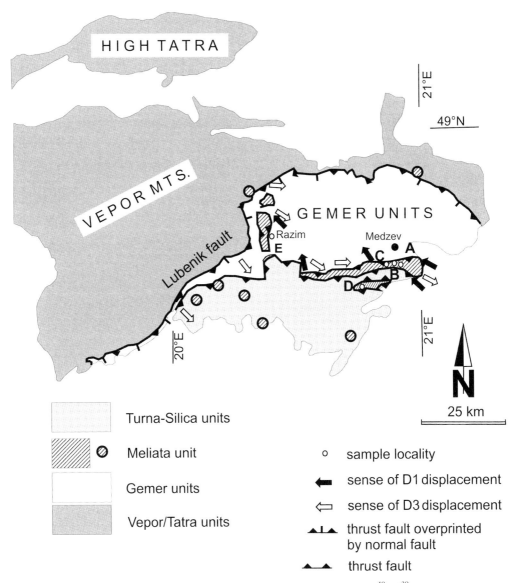

Fig. 2. Simplified structural map of the Inner Western Carpathians. Samples used for $^{40}Ar/^{39}Ar$ mineral dating are located (A–E).

associated deformation (e.g. Lexa et al. 2003). The basement of the Veporic unit reached in part medium-grade metamorphic overprint (c. 550 and 8–10 kbar, e.g. Lupták et al. 2000; Jának et al. 2001). Older K–Ar ages from all these units range between c. 400 and 250 respectively from 150 and 80 Ma and are compiled in Cambel & Král (1989) and Krist et al. (1992).

The Gemer units are structurally overlain by the blueschist-bearing Meliata tectonic element. The Meliata unit comprises a mixture of nearly unmetamorphic sedimentary units, as well as serpentinites and mafic rocks, the latter largely metamorphosed to blueschists. This represents a tectonic melange which includes fragments of mafic volcanic rocks of oceanic origin (e.g. Mello et al. 1998; Faryad 1999; Ivan 2002a, b; Faryad et al. 2005; Horvath & Árkai 2005), a widespread, fully recrystallized, coarse-grained, pure marble of Middle Triassic depositional age and Middle to Upper Triassic pelagic limestones and radiolarites (Kozur & Mock 1973; Mock et al. 1998; Kozur

1991; Channell & Kozur 1997; Velledits 2006 and references cited therein), the latter two rock types metamorphosed only in very low-grade metamorphic conditions (Árkai et al. 2003). This unit has been interpreted to have originated within a Middle Triassic oceanic basin situated along the southern margin of the Inner Carpathians. Conditions associated with blueschist metamorphism have been estimated at c. 380–460 °C and 10–12 kbar (Faryad 1995a, b; Faryad & Hoinkes 1999). The Turňa–Silica nappe system structurally overlies the Meliata unit. This includes Mesozoic sequences with palaeogeographic affinities similar to those of other higher structural units within the Eastern Alps. Characteristics of the Triassic and rare Jurassic sedimentary sequences suggest palaeogeographic correlations with the Northern Calcareous Alps of the Eastern Alps, particularly with Upper Juvavic and Tirolic nappe complexes (Kozur 1991; Rakús & Sýkora 2001). Furthermore, Middle–Upper Triassic Meliata-type radiolarites, pelagic limestones, serpentinites and white marbles are also found at the Florianikogel interlayered within units of Northern Calcareous Alps (Kozur 1991; Mandl & Ondreijokova 1991; Kozur & Mostler 1992).

Structural evolution

Peak conditions of blueschist metamorphism within the Meliata unit were reached prior to deformation as relatively coarse blue amphibole grains within internally massive blueschists indicate (Fig. 3a). These fabrics are overprinted by strained blueschist fabrics, which developed concomitantly with ductile deformation (D_{1a}) characterized by a penetrative foliation and crystallization of alkali amphiboles (Fig. 3b, c). This process resulted in formation of a stretching lineation which trends southeastward (Fig. 4a). East to NE-trending, subvertical tension gashes are common within mafic lithologies and are filled with blue amphibole, hematite, and phengite indicatle syn-metamorphic stretching of the less ductile deforming mafic lithologies. No preferred shear sense indicators have been found either in field or in thin sections. Both microfabrics and quartz-fabrics in quartz-rich phyllitic rocks record a predominant flattening strain. Hence, D_{1a} structures indicate c. NW–SE stretch. Structures formed during blueschist metamorphism have been overprinted by semiductile, non-penetrative shear bands which formed during maintenance of greenschist facies conditions (D_{1b}) within the same kinematic framework. These fabrics are locally overprinted by tight D_2 folds, which folded the S_1 foliation and partly formed a gently dipping axial plane foliation (Fig. 4b). The last ductile stage is particularly common in coarse-grained marbles, which display coarse, elongated calcite grains (Fig. 3d). Elongation is due to twinning, and twins are kinked, too. No recrystallization occurred after D_3 deformation. Consequently, these fabrics formed as latest, D_3, ductile event. The orientation of D_3 stretching lineation varies from SE to south (Fig. 4c). Overprinting criteria suggest an early ductile shear with top north to NW shear (D_{1b}) and a later top southeast sense of shear (D_3). D_3 is particularly well-developed within calcite textures of marbles (Fig. 3d) and also within radiolarites (Fig. 3e).

Blueschist metamorphic mineral parageneses do not occur within the structurally underlying Gemeric nappe complex. However, a similar low-grade metamorphism is recorded in both units. Structural studies generally suggest a two-phase deformation with early top-NW, ductile thrust faulting in Permian and Mesozoic cover sequences. Some thrust faults have been locally overprinted by later ductile normal faulting which was active during decreasing metamorphic temperature conditions, e.g. along the Lubeník fault (Fig. 2) and top ESE shear. The Pohorela fault separates the exhumed Vepor unit with Cretaceous-aged amphibolite-grade metamorphism from the overlying Gemer unit with very-low to low-grade metamorphism of Cretaceous age (e.g. Jának et al. 2001).

$^{40}Ar/^{39}Ar$ mineral dating

Four concentrates of phengitic muscovite for $^{40}Ar/^{39}Ar$ dating were separated from metasedimentary units (quartz-micaschist, quartz-carbonate-mica schist) within the Meliata zone to constrain the age of high-pressure metamorphism. The protolith ages of these rocks are unknown. A whole-rock metavolcanic phyllite of Permian cover sequences from footwall units (Gemer nappe: Fig. 2) was also analysed to constrain the age of greenschist facies deformation related to overthrusting of the Meliata unit. The whole rock approach was selected because of the small grain size of new-grown white mica. The sample locations and petrographic descriptions are given in Appendix 1. Analytical procedures were similar to those described in detail in Dallmeyer & Gil-Ibarguchi (1990) and Dallmeyer & Takasu (1992), and are described in detail in Appendix 2. The analytical data are presented in Tables 1 and 2 and are portrayed as age spectra in Figures 5 and 7, chemical mineral compositions in Table 3 and Figure 6.

The incremental age spectra of phengitic concentrates are internally discordant. However, 80–95% of the experimentally evolved gas

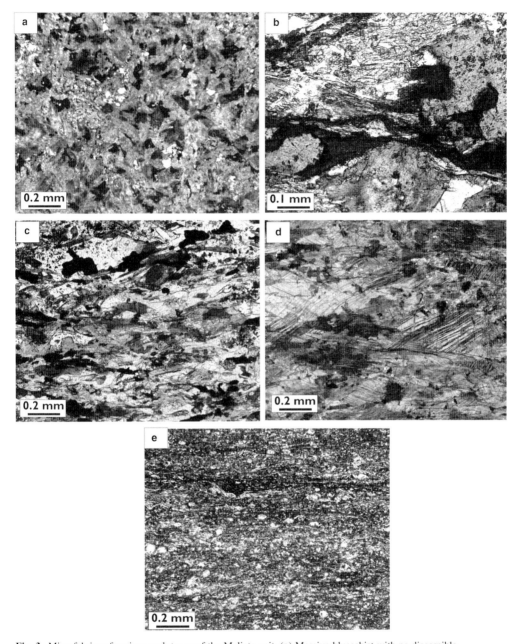

Fig. 3. Microfabrics of various rock types of the Meliata unit. (**a**) Massive blueschist with no discernible foliation. Note optically zoned blue amphibole. (**b**) Partly retrogressed blueschist. Note transformation of blue amphibole into chlorite and formation of a domainal fabric. (**c**) Strongly retrogressed blueschist and formation of a foliation. Note alignment of elongated blue amphibole parallel to foliation. (**d**) Coarse-grained marble with strongly deformed calcite grains with kinked twin lamellae. (**e**) Fine-grained Middle Triassic radiolarite with domainal fabric formed within very-grade metamorphic conditions.

records similar intra-sample ages which define plateaus between 160.0 ± 2.0 Ma and 150 ± 2.0 Ma. Variations in the chemical compositions of phengites (microprobe analyses) are presented in Figure 6. Samples A, C and D include grains with a relatively uniform phengite composition, variations of the atomic Si per formula unit between 3.35 and 3.50 following calculation procedures

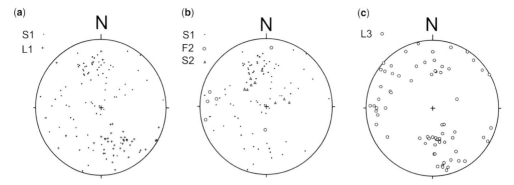

Fig. 4. Orientation data for Meliata unit. (**a**) Foliation S_1 and mineral lineation L_1 in mafic rocks. (**b**) Foliation S_1 and S_2, and F_2 fold axes in mafic rocks. (**c**) Stretching lineation L_3 in strongly strained, coarse-grained marble.

proposed by Massonne & Schreyer (1987). Sample B contains a subordinate population of zoned phengite grains with muscovite-rich rims, and muscovites with an atomic Si content of 3.01. Because of the relatively high temperatures of argon retention of phengite and of the closure temperature of phengite (c. 410 °C; e.g. Blankenburg et al. 1989; see also Hames & Bowring 1994 and Villa 1998 for the complications of the argon retention concept of white mica), we interpret the 160–150 Ma plateau ages to closely date phengite growth during blueschist metamorphism (Late Jurassic according to time-scale calibration of Gradstein et al. 2004).

The age spectrum displayed by the whole rock (sample E) is internally discordant with markedly varying ages recorded in low temperature increments (Fig. 7). This is matched by significant variations in apparent K/Ca ratios (Fig. 7). Anomalously low values are recorded in the four low temperature increments. Apparent K/Ca values increase throughout intermediate portions of the experiment. They systematically decrease in the three high-temperature increments. This systematic variation in apparent K/Ca ratios suggests that experimental evolution of gas occurred from several mineral phases. These included a subordinate, relatively low retention phase and a subordinate relatively high argon retention phase in addition to the predominant intermediate relatively high retention phase. Petrographic characteristics suggest these three phases are represented by very fine white mica, chlorite and detrital plagioclase, respectively. The seven intermediate temperature increments record apparent ages which define a plateau of 105.8 ± 1.5 Ma. This age is considered to be the geologically significant and is interpreted to date white mica formation during low-grade metamorphism and concomitant deformation.

Discussion

Our new structural data and the new ^{40}Ar/^{39}Ar age data together with published mineral age dating results of Maluski et al. (1993) and Faryad and Henjes-Kunst (1997) help to constrain the sequence of tectonothermal events within the Meliata and footwall units. The following succession of events is suggested (Fig. 6): (1) blueschist metamorphism and associated deformation within the Meliata unit; (2) semiductile to ductile, top to the north to NW shear; and (3) top to the SE shear.

The ^{40}Ar/^{39}Ar results suggest that blueschist deformation resulted in penetrative flattening and NW–SE stretching in association with Late Jurassic high-pressure metamorphism. Dominant coaxial stretching may indicate that pure shear was the predominant deformation regime maintained during blueschist metamorphism and likely occurred within the subduction channel because of preservation of high-pressure conditions during deformation stage D_{1a}.

A subsequent greenschist facies metamorphic deformation in both Meliata and footwall units likely was associated with tectonic assembly of the Meliata and footwall units. ^{40}Ar/^{39}Ar muscovite ages within Permo-Mesozoic cover of the Inner Western Carpathian nappe units decrease towards footwall units (Dallmeyer et al. 1996), and suggest assembly of nappes within this thick-skinned tectonic wedge occurred by footwall thrust propagation during the Early–Middle Cretaceous (e.g. Plašienka 1997). Generally, displacement along these thrusts was west and northwest. Loading of the Meliata unit onto the Inner Carpathian nappe complex (Gemer units) occurred during thrusting. As a result, the Meliata unit was then transported to progressively higher tectonic positions within the tectonic wedge. The new age

Table 1. $^{40}Ar/^{39}Ar$ analytical data for incremental-heating experiments on phengitic white mica concentrates from the Meliata nappe complex, Western Carpathians, Slovakia

Release temperature (°C)	$(^{40}Ar/^{39}Ar)^*$	$(^{36}Ar/^{39}Ar)^*$	$(^{37}Ar/^{39}Ar)^*$	^{39}Ar % of total	% ^{40}Ar nonatmospheric	$^{36}Ar_{Ca}$%	Apparent age (Ma)	Error (2σ, intra-laboratory, Ma)
Sample A, J = 0.010511								
480	14.00	0.03133	0.089	0.51	33.90	0.08	87.8	5.1
540	9.71	0.00314	0.038	5.89	90.43	0.33	159.3	0.3
575	8.89	0.00131	0.008	6.56	95.58	0.17	154.4	0.4
610	9.09	0.00092	0.012	5.89	96.94	0.34	159.8	0.2
645	9.06	0.00109	0.004	9.48	96.39	0.11	158.3	0.1
680	9.09	0.00069	0.003	14.73	97.69	0.13	160.9	0.3
710	9.10	0.00058	0.003	16.41	98.05	0.14	161.7	0.1
740	9.12	0.00060	0.004	13.93	98.01	0.18	162.0	0.1
770	9.04	0.00053	0.003	15.86	98.21	0.16	161.0	0.1
805	9.09	0.00053	0.004	5.15	98.23	0.22	161.8	0.2
840	9.86	0.00363	0.004	4.60	89.07	0.03	159.3	0.1
Fusion	11.06	0.00415	0.017	0.98	88.87	0.11	177.4	1.5
Weighted mean age	9.19	0.00119	0.007	100.00	96.36	0.17	160.1	0.2
Total without 480–575 °C and fusion				86.06			160.8	0.1
Sample B, J = 0.010325								
490	13.55	0.02261	0.601	2.05	51.03	0.72	124.5	1.2
550	10.52	0.00445	0.303	9.11	87.67	1.85	164.1	0.4
585	9.03	0.00046	0.008	8.97	98.44	0.44	158.5	0.4
620	8.99	0.00022	0.009	9.29	99.21	1.12	158.9	0.5
655	9.03	0.00014	0.004	9.16	99.43	0.84	159.9	0.2
680	9.03	0.00049	0.005	7.97	98.33	0.26	158.2	0.2
710	9.05	0.00039	0.005	15.13	98.65	0.33	159.1	0.3
740	8.96	0.00043	0.007	7.10	98.53	0.43	157.4	0.2
770	8.91	0.00029	0.006	11.54	98.99	0.60	157.3	0.2
800	9.18	0.00067	0.005	8.18	97.77	0.19	159.9	0.3
835	9.18	0.0018	0.18	8.74	96.15	0.41	157.3	0.3
Fusion	11.69	0.00240	0.063	2.75	93.91	0.71	193.6	1.2
Weighted mean age	9.34	0.00133	0.048	100.00	96.36	0.65	159.3	0.3
Total without 490–550 °C and fusion				86.09			158.5	0.2

(Continued)

Table 1. Continued

Release temperature (°C)	(^{40}Ar/^{39}Ar)*	(^{36}Ar/^{39}Ar)*	(^{37}Ar/^{39}Ar)*	^{39}Ar % of total	% ^{40}Ar nonatmospheric	^{36}Ar$_{Ca}$ %	Apparent age (Ma)	Error (2σ, intra-laboratory, Ma)
Sample C, J = 0.010305								
470	13.21	0.01812	0.105	2.60	59.48	0.16	140.5	0.4
530	9.89	0.00381	0.105	4.52	88.65	0.75	156.0	0.1
565	8.76	0.00079	0.013	4.90	97.29	0.46	151.9	0.2
600	8.68	0.00054	0.016	4.38	98.11	0.80	151.7	0.2
635	8.68	0.00055	0.010	5.74	98.07	0.52	151.7	0.1
670	8.66	0.00084	0.005	12.04	97.08	0.15	149.9	0.1
705	8.69	0.00091	0.003	14.51	96.83	0.09	150.0	0.1
740	8.68	0.00077	0.003	14.58	97.32	0.11	150.6	0.2
775	8.59	0.00060	0.003	16.95	97.87	0.15	149.9	0.1
810	8.57	0.00039	0.004	11.43	98.58	0.29	150.6	0.2
850	8.94	0.00147	0.021	6.53	95.10	0.38	151.6	0.3
Fusion	12.79	0.00914	0.150	1.82	78.93	0.45	178.6	0.6
Weighted mean age	8.92	0.00149	0.016	100.00	95.68	0.26	151.0	0.2
Total without 470–530 °C and fusion				91.06			150.5	0.1
Sample D, J = 0.010435								
490	9.84	0.00758	0.047	1.36	77.22	0.17	137.7	0.6
550	9.06	0.00215	0.015	6.01	92.95	0.19	151.9	0.1
580	8.43	0.00041	0.007	5.95	98.49	0.45	150.0	0.2
615	8.42	0.00021	0.005	5.16	99.19	0.70	150.7	0.2
650	8.49	0.00079	0.005	7.02	97.20	0.16	149.0	0.1
685	8.54	0.00057	0.003	11.82	97.97	0.13	151.1	0.2
720	8.59	0.00083	0.003	15.16	97.09	0.09	150.5	0.2
750	8.51	0.00047	0.003	14.87	98.30	0.17	150.9	0.2
780	8.42	0.00049	0.003	11.91	98.22	0.15	149.4	0.2
815	8.44	0.00034	0.004	8.63	98.73	0.30	150.3	0.1
850	8.75	0.00132	0.019	10.95	95.50	0.40	150.8	0.2
Fusion	9.31	0.00271	0.094	1.14	91.41	0.94	153.6	0.7
Weighted mean age	8.58	0.00085	0.008	100.00	97.09	0.24	150.3	0.2
Total without 490 °C and fusion				97.50			150.5	0.2

*Measured.
^{C}Corrected for post-irradiation decay of ^{37}Ar (35.1 day $\frac{1}{2}$ life).
*[^{40}Ar$_{tot.}$ – (^{36}Ar$_{atmos}$) (295.5)]/^{40}Ar$_{tot.}$
**Two sigma, intra-laboratory errors.

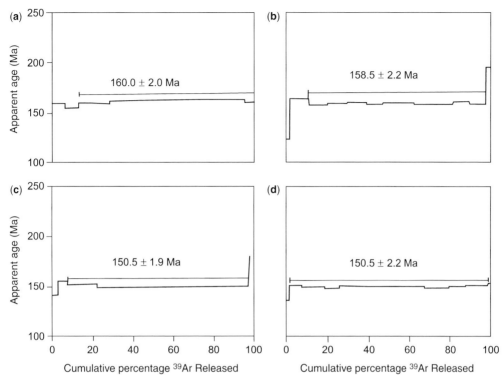

Fig. 5. ^{40}Ar/^{39}Ar incremental release spectra of phengitic muscovite concentrates from the Meliata unit. Analytical uncertainties (two sigma intra-laboratory) are represented by vertical width of scale bars. Experimental temperatures increase from left to right.

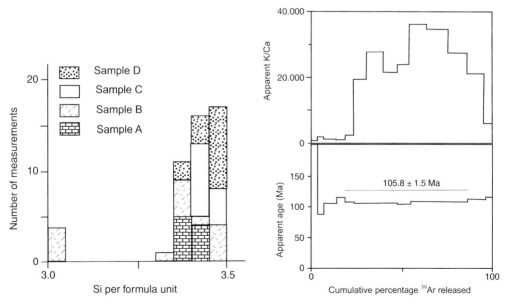

Fig. 6. Compositional variations of dated white mica concentrates from the Meliata unit. Plotted following methods of Massonne & Schreyer (1987).

Fig. 7. ^{40}Ar/^{39}Ar release spectrum and apparent K/Ca ration of a sample of a Permian cover sequence of the Gemer unit beneath the Meliata unit. Plotted as in Figure 5.

(105 Ma) of thrusting is in good agreement with similar $^{40}Ar/^{39}Ar$ white mica ages ranging from 105 to 95 Ma in similar structural levels (base of Northern Calcareous Alps, Greywacke zone) of eastern Alps (Dallmeyer et al. 1998; Frank & Schlager 2006). Consequently, a gap of c. 50 Ma exists between blueschist metamorphism recorded within Meliata units and thrust emplacement of these units onto Gemer units. It seems that thrusting emplaced a piece of Meliata oceanic crust, which was buried in an early stage of subduction and subsequently incorporated into an accretionary wedge.

Subsequently, Late Cretaceous southeastern shear developed as a result of extension along upper and trailing margins of the tectonic wedge. This is recorded by overprinting of ductile thrust fabrics within the Meliata unit by low-grade metamorphic deformation (especially within marble), and developed ductile normal faults which developed within upper portions of the footwall tectonic units, e.g. the Lúbenik fault. Final exhumation of the blueschist wedge could have related to a backstop developed along the southern margins of the western Carpathians. Kinematic characteristics of this final event appear to have been related to those postulated for the Eastern Alps (Neubauer et al. 1995).

These data presented herein and previously suggested palaeogeographic relationships provide important constraints in the tectonic resolution of the evolution of the entire Alpine–Carpathian belt. The existence of a Triassic oceanic domain comprising Meliata, Mureş and Vardar zones within the Alpine–Carpathian–Dinaric belt is important in all new tectonic models (e.g. Neubauer 1994, 2002; Stampfli & Mosar 1999; Csontos & Vörös 2004). This appears to have existed along southern margins of the Austroalpine domain and was part of the European continent in the early Mesozoic. This oceanic seaway separated European/Austroalpine units from interior Southalpine/Dinaric units. Subduction of that ocean occurred during the Late Jurassic (Fig. 8a), and thrust-loading of oceanic sequences onto the Austroalpine passive continental margin occurred during the late Early Cretaceous (Fig. 8b) similar to what occurred in the eastern Alps (Dallmeyer et al. 1998). These relationships suggest that the Alpine-Carpathian belt reflects two stages of Mesozoic subduction. The underlying Vepor unit and the Eclogite–Gneiss unit of the Austroalpine nappe complex were subducted during continent–continent collision between c. 95 and 90 Ma (Thöni & Jagoutz 1991; Thöni 1999). In the Eastern Alps, high-pressure metamorphism reached eclogite facies conditions (c. 22–30 kbar; Janák et al. 2005, 2006). Finally, the over-thickened orogenic wedge collapsed during the Late Cretaceous. This stage was associated with ductile top-southeast to east shear and formation of Gosau collapse basins (e.g. Willingshofer et al. 1999). Present tectonostratigraphic relationships resulted from Late Cretaceous together with Tertiary strike-slip dispersion of the previously assembled nappe succession (Ratschbacher et al. 1989; Csontos et al. 1992).

We gratefully acknowledge rapid and detailed reviews by Dušan Plašienka and Jan Wijbrans, and the patience of Siegfried Siegesmund as guest editor. We also acknowledge support and continuous discussion during fieldwork by Peter Reichwalder, Marian Jának and Marian Putiš, for help during microprobe analysis by Dan Topa (Salzburg), and recalculation of mineral formulas by Hans Genser (Salzburg). The work has been supported by grants from the Tectonics Program of the National Science Foundation (EAR-9316042) to RDD and Austrian Research Foundation (P8.652-GEO) to FN.

Appendix 1. Locations and descriptions of samples used for $^{40}Ar/^{39}Ar$ dating

Sample A (field no. FN-SLO-23): Sugov valley, c. 3 km SSW Medzew. The well-annealed and well-foliated mineral-rich quartzite comprises c. 80% quartz. c. 0.15–0.4 mm long white mica flakes, c. 0.5 mm long blue amphibole needles and chlorite show a well-developed foliation. Chlorite postdates formation of blue amphibole. Quartz grains are 0.1–0.3 mm large and have mostly straight grain boundaries.

Sample B (field no. FN-SLO-24): Sugov valley, hill south of abandoned marble quarry. Blue amphibole-bearing phengite–carbonate schist. c. 0.5 mm large blue amphibole grains are zoned with inclusion-rich centre and inclusion-free rims. They are embedded in a finer-grained matrix comprising twinned calcite (mostly 0.2–0.4 mm), chlorite, white mica, rare epidote, magnetite with leucoxene rims, K-feldspar, which are rich in various inclusions including blue amphibole, calcite, epidote and rutile and plagioclase (oligoclase).

Sample C (field no. FN-SLO-25): c. 2 km South of Medzew. Mica-carbonate schist. The rock is composed of calcite (c. 0.4 mm), Fe-rich carbonate, white mica (0.2–0.6 mm long), plagioclase, K-feldspar and quartz. The dominant calcite displays well-developed grain boundary migration.

Sample D (field no. FN-SLO-26): Hačava valley, c. 1 km NW Hačava village, WSW Vel Jelen vrch. The phengite–carbonate schist is exposed in close connection (N) to blueschist lenses. The rock is dominated by calcite, sericite (0.05–0.2 mm), chlorite and twinned plagioclase (oligoclase) and K-feldspar. Calcite is 0.3 to 0.5 m long, elongated and twinned. The rare K-feldspar is rich in various inclusions.

Sample E (field no. DA-SLO-22): Hnilec valley, road c. 2 km south of Helmanovce. Grey-greenish phyllite with more than 50% sericite, c. 20% chlorite, c. 20% quartz and less than 5% plagioclase. The grain size of sericite

(a) Late Jurassic

(b) Mid Cretaceous

(c) Late Cretaceous

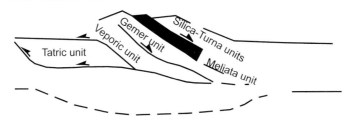

Fig. 8. Tectonic evolution of the Inner Western Carpathians as a model of the formation and emplacement of high-pressure units. (**a**) Late Jurassic blueschist metamorphism within the Meliata unit during subduction; (**b**) Middle Cretaceous loading of fragmented Meliata oceanic sequences onto the Inner Western Carpathians; (**c**) Late Cretaceous back-sliding along low-angle normal faults situated along the rear of the Carpathian tectonic wedge.

and chlorite is c. 0.004 mm, and these minerals display a well-developed foliation. Quartz and plagioclase (c. 0.005 mm in diameter), rutile, tourmaline, ilmenite respectively cubic opaque minerals are further minor constituents.

Appendix 2. $^{40}Ar/^{39}Ar$ analytical techniques

The techniques used during $^{40}Ar/^{39}Ar$ analysis of muscovite concentrates and whole rock phyllite were carried out at the Department of Geology, University of Georgia, Athens, GA, USA. The techniques generally followed those described by Dallmeyer & Gil-Ibarguchi (1990) respectively Dallmeyer & Takasu (1992). Optically pure (>99%) mineral concentrates were wrapped in aluminium foil packets, encapsulated in sealed quartz vials, and irradiated in the TRIGA Reactor at the US Geological Survey in Denver. Variations in the flux of neutrons along the length of the irradiation assembly were monitored with several mineral standards, including MMhb-1 (Sampson & Alexander 1987). The samples were incrementally heated until fusion in a double-vacuum, resistance-heated furnace following methods described by Dallmeyer & Gil-Ibarguchi (1990). Measured isotopic ratios were corrected for total system blanks and the effects of mass discrimination. Interfering isotopes produced during irradiation were corrected using factors reported by Dalrymple et al. (1981). Apparent $^{40}Ar/^{39}Ar$ ages were calculated from corrected isotopic ratios using the decay constants and isotopic abundance ratios listed by Steiger & Jäger (1977).

A variety of uncertainties is associated with ages determined by incremental release $^{40}Ar/^{39}Ar$ analyses. These include analytical uncertainties associated with

Table 2. Representative chemical composition of white mica of aliquots from samples of the Meliata unit used for $^{40}Ar/^{39}Ar$ dating

Sample	A	A	B-rim	B-core	C-rim	C	D-rim	D-core
SiO_2	50.10	46.70	43.80	50.60	50.10	49.20	50.70	50.6
TiO_2	0.15	0.13	0.04	0.16	0.06	0.1	0.16	0.16
Al_2O_3	25.70	25.10	35.60	26.20	23.20	23.60	25.90	26.20
Cr_2O_3	0.01	0.00	0.01	0.04	0.06	0.07	0.02	0.04
FeO	3.20	3.30	0.82	2.10	4.80	5.30	2.00	2.10
MnO	0.01	0.00	0.02	0.01	0.00	0.02	0.56	0.05
MgO	3.40	2.92	0.66	3.43	3.60	3.19	3.42	3.43
CaO	0.01	0.01	0.00	0.01	0.06	0.02	0.00	0.01
Na_2O	0.21	0.36	0.7	0.35	0.12	0.15	0.29	0.35
K_2O	10.9	10.1	10	10.7	11.1	10.9	10.8	10.7
BaO	0.08	0.12	1.40	0.13	0.01	0.01	0.19	0.13
Cl	0.01	0.05	0.00	0.00	0.02	0.00	0.00	0.00
Total (%)	93.78	88.79	93.05	93.73	93.13	92.56	94.04	93.77
Si	3.427	3.381	3.013	3.438	3.487	3.454	3.444	3.437
Ti	0.008	0.007	0.002	0.008	0.003	0.005	0.008	0.008
Al	2.072	2.142	2.886	2.098	1.903	1.953	2.073	2.097
Cr	0.001	0.000	0.001	0.002	0.003	0.004	0.001	0.002
Fe^{2+}	0.183	0.200	0.047	0.119	0.279	0.311	0.114	0.119
Mn	0.001	0.000	0.001	0.001	0.000	0.001	0.032	0.003
Mg	0.347	0.315	0.068	0.347	0.374	0.334	0.346	0.347
Ca	0.001	0.001	0.000	0.001	0.004	0.002	0.000	0.001
Na	0.028	0.051	0.093	0.046	0.016	0.02	0.038	0.046

Table 3. $^{40}Ar/^{39}Ar$ analytical data for incremental-heating experiment on whole-rock phyllite from the structural cover of the Gemeric nappe complex, Western Carpathians, Slovakia

Release temperature (°C)	$(^{40}Ar/^{39}Ar)^*$	$(^{36}Ar/^{39}Ar)^*$	$(^{37}Ar/^{39}Ar)^{*C}$	^{39}Ar % of total	% ^{40}Ar nonatmospheric	$^{36}Ar_{Ca}$ %	Apparent age (Ma)	Error (2σ, intra-laboratory, Ma)
400	24.81	0.00458	0.745	3.35	94.76	4.42	397.0	1.0
440	5.21	0.00162	0.263	3.12	91.11	4.42	87.4	0.9
460	6.37	0.00226	0.453	7.06	90.02	5.46	105.2	0.9
490	7.01	0.00245	0.550	4.56	90.22	6.11	115.6	0.5
520	6.34	0.00187	0.214	4.96	91.45	3.10	106.4	0.4
555	5.74	0.00009	0.025	6.72	99.45	7.33	104.6	0.2
590	5.72	0.00005	0.018	9.12	99.69	10.66	104.6	0.2
630	5.89	0.00046	0.023	7.95	97.63	1.35	105.5	0.1
665	5.87	0.00052	0.021	6.70	97.32	1.08	104.8	0.4
705	6.06	0.00080	0.014	9.83	96.02	0.46	106.7	0.5
745	6.16	0.00120	0.014	11.84	94.18	0.32	106.4	0.1
785	6.02	0.00069	0.018	11.10	96.55	0.71	106.6	0.2
825	6.18	0.00064	0.023	9.25	96.89	0.99	109.7	0.2
Fusion	6.81	0.00208	0.084	4.42	90.98	1.09	113.3	0.7
Weighted mean age	6.72	0.00110	0.118	100.00	95.39	3.05	116.1	0.4
Total without 400–490 °C, 825 °C and fusion				68.23			105.8	0.3

*Measured.
^{C}Corrected for post-irradiation decay of ^{37}Ar (35.1 day ½ life).
*$[^{40}Ar_{tot.} - (^{36}Ar_{atmos.})(295.5)]/^{40}Ar_{tot.}$
**Two sigma, intra-laboratory errors.

measurement of each isotopic ratio (inlet time extrapolation, mass discrimination correction, system blank, etc.) and overall uncertainties in age calculations (monitor age, atmospheric corrections, etc.). The latter are similar for each incremental age determination during analysis of one sample. Therefore, to evaluate potential differences in an incremental age analysis, only intra-laboratory uncertainties should be considered. These have been calculated by rigorous statistical propagation of uncertainties associated with measurements of each isotopic ratio (at two standard deviations of the mean) through the age equation. These intra-laboratory errors are listed in the data tables and corresponding figures. Comparison of Weighted mean ($\sim$total – gas), plateau and/or isotope correlation ages between different samples and between different laboratories requires consideration of the total inter-laboratory uncertainty. These are estimated at ± 1.25–1.5% of the quoted age, and are quoted in the text. Weighted mean ages ($\sim$total gas ages) have been computed for each sample by appropriate weighting of the age and percent ^{39}Ar released within each temperature increment. A 'plateau' is considered to be defined if the ages recorded by two or more contiguous gas fractions (with similar apparent K/Ca ratios) each representing >4% of the total ^{39}Ar evolved (and together constituting >50% of the total quantity of ^{39}Ar evolved) are mutually similar within a 1% intra-laboratory uncertainty. Analyses of the MMhb-1 monitor indicate that apparent K/Ca ratios may be calculated through the relationship $0.518 (\pm 0.0005) \times (^{39}Ar/^{37}Ar)_{corrected}$.

Plateau portions of the white mica and whole rock phyllite analyses have been plotted on $^{36}Ar/^{40}Ar$ vs. $^{39}Ar/^{40}Ar$ isotope correlation diagrams (not shown). Regression techniques followed methods described by York (1969). A mean square of the weighted deviates (MSWD) has been used to evaluate the isotopic correlation.

References

ÁRKAI, P., FARYAD, S. W., VIDAL, O. & BALOGH, K. 2003. Very low-grade metamorphism of sedimentary rocks of the Meliata unit, Western Carpathians, Slovakia: implications of phyllosilicate characteristics. *International Journal of Earth Sciences*, **92**, 68–85.

BLANCKENBURG VON, F., VILLA, I. M., BAUR, H., MORTEANI, G. & STEIGER, R. H. 1989. Time calibration of a P-T path from the western Tauern Window, eastern Alps: The problem of closure temperatures. *Contributions of Mineralogy and Petrology*, **101**, 1–11.

CAMBEL, B. & KRÁL, J. 1989. Isotope geochronology of the western Carpathian crystalline complex: the present state. *Geologica Carpathica*, **40**, 387–410.

CHANNELL, J. E. T. & KOZUR, H. W. 1997. How many oceans? Meliata, Vardar, and Pindos oceans in Mesozoic Alpine paleogeography. *Geology*, **25**, 183–186.

CSONTOS, L. & VÖRÖS, A. 2004. Mesozoic plate tectonic reconstruction of the Carpathian region. *Palaeogeography Palaeoclimatology Palaeoecology*, **210**, 1–56.

CSONTOS, L., NAGYMAROSY, A., HORVATH, F. & KOVAC, M. 1992. Tertiary evolution of the intra-Carpathian arc. *Tectonophysics*, **208**, 221–241.

DAL PIAZ, G., MARTIN, S., VILLA, I. M., GOSSO, G. & MARSCHALKO, R. 1995. Late Jurassic blueschist facies pebbles from the Western Carpathians orogenic wedge and paleostructural implications for Western Tethys evolution. *Tectonics*, **14**, 874–885.

DALLMEYER, R. D. & GIL-IBARGUCHI, J. L. 1990. Age of amphibolitic metamorphism in the ophiolitic unit of the Morais allochthon (Portugal): Implications for early Hercynian orogenesis in the Iberian Massif. *Journal of the Geological Society, London*, **147**, 873–878.

DALLMEYER, R. D. & TAKASU, A. 1992. $^{40}Ar/^{39}Ar$ of detrital muscovite and whole rock slate/phyllite, Narragansett Basin, RI-MA, USA: implications for rejuvenation during very low-grade metamorphism. *Contributions to Mineralogy and Petrology*, **110**, 515–527.

DALLMEYER, R. D., NEUBAUER, F., HANDLER, R., FRITZ, H., MÜLLER, W., PANA, D. & PUTIS, M. 1996. Tectonothermal evolution of the internal Alps and Carpathians: $^{40}Ar/^{39}Ar$ mineral and whole rock data. *Eclogae Geologicae Helvetiae*, **89**, 203–277.

DALLMEYER, R. D., HANDLER, R., NEUBAUER, F. & FRITZ, H. 1998. Sequence of thrusting within a thick-skinned tectonic wedge: Evidence from $^{40}Ar/^{39}Ar$ ages from the Austroalpine nappe complex of the Eastern Alps. *Journal of Geology*, **106**, 71–86.

DALRYMPLE, G. B., ALEXANDER, E. C., LANPHERE, M. A. & KRAKER, G. B. 1981. Irradiation of samples for $^{40}Ar/^{39}Ar$ dating using the Geological Survey TRIGA reactor. *U. S. Geological Survey Professional Paper*, **1176**, 1–55.

ERNST, G. 1988. Tectonic history of subduction zones inferred from retrograde blueschist P-T paths. *Geology*, **25**, 894–928.

ERNST, W. G. 2005. Alpine and Pacific styles of Phanerozoic mountain building: subduction-zone petrogenesis of continental crust. *Terra Nova*, **17**, 165–188.

FARYAD, S. W. 1995a. Petrology and phase relations of low-grade high-pressure metasediments from the Meliata Unit (West Carpathians, Slovakia). *European Journal of Mineralogy*, **7**, 71–87.

FARYAD, S. W. 1995b. Phase petrology and P-T conditions of mafic blueschists from the Meliata Unit, West Carpathians, Slovakia. *Journal of Metamorphic Geology*, **13**, 701–714.

FARYAD, S. W. 1999. Nature and geotectonic interpretation of the Meliata blueschists. *Geologica Carpathica*, **50**, 97–98.

FARYAD, S. W. & HENJES-KUNST, F. 1997. Petrological and K-Ar and $^{40}Ar/^{39}Ar$ age constraints for the tectonothermal evolution of the high-pressure Meliata unit, Western Carpathians (Slovakia). *Tectonophysics*, **280**, 141–156.

FARYAD, S. W. & HOINKES, G. 1999. Two contrasting mineral assemblages in the Meliata blueschists, Western Carpathians, Slovakia. *Mineralogical Magazine*, **63**, 489–501.

FARYAD, S. W., SPIŠIAK, J., HORVÁTH, P., HOVORKA, D., DIANIŠKA, I. & JÓZSA, S. 2005. Petrological and geochemical features of the Meliata mafic rocks from the sutured Triassic Oceanic Basin, Western Carpathians. *Ofioliti*, **30**, 27–35.

FRANK, W. & SCHLAGER, W. 2006. Jurassic strike slip versus subduction in the Eastern Alps. *International Journal of Earth Sciences*, **95**, 431–450.

FROITZHEIM, N., PLEUGER, J., ROLLER, S. & NAGEL, T. 2003. Exhumation of high- and ultrahigh-pressure metamorphic rocks by slab extraction. *Geology*, **31**, 925–928.

GRADSTEIN, F. M., OGG, J. G., SMITH, A. G., BLEEKER, W. & LOURENS, L.J. 2004. A New Geologic Time Scale, with special reference to Precambrian and Neogene. *Episodes*, **27**, 83–100.

HAMES, W. E. & BOWRING, S. A. 1994. An empirical evaluation of the argon diffusion geometry in muscovite. *Earth and Planetary Science Letters*, **124**, 357–167.

HORVÁTH, P. & ÁRKAI, P. 2005. Amphibole-bearing assemblages as indicators of microdomain-scale equilibrium conditions in metabasites: an example from Alpine ophiolites of the Meliata Unit, NE Hungary. *Mineralogy and Petrology*, **84**, 233–258.

IVAN, P. 2002a. Relics of the Meliata Ocean crust: Geodynamic implications of mineralogical, petrological and geochemical proxies. *Geologica Carpathica*, **53**, 245–256.

IVAN, P. 2002b. Relict magmatic minerals and textures in the HP/LT metamorphosed oceanic rocks of the Triassic-Jurassic Meliata ocean (Inner Western Carpathians). *Slovak Geological Magazine*, **8**, 109–122.

JÁNAK, M., PLAŠIENKA, D., FREY, M., COSCA, M., SCHMIDT, S., LUPTÁK, B. & MÉRES, S. 2001. Cretaceous evolution of a metamorphic core complex, the Veporic unit, Western Carpathians (Slovakia): P-T conditions and in situ Ar-40/Ar-39 UV laser probe dating of metapelites. *Journal of Metamorphic Geology*, **19**, 197–216.

JÁNAK, M., FROITZHEIM, N., LUPTÁK, B., VRBEC, M. & KROGH-RAVNA, E. J. 2005. First evidence for ultrahigh-pressure metamorphism of eclogites in Pohorje, Slovenia: tracing deep continental subduction in the eastern Alps. *Tectonics*, **23**, TC5014, doi:10.1029/2004TC001641.

JÁNAK, M., FROITZHEIM, N., VRBEC, M., KROGH-RAVNA, E. J. & DE HOOG, C. M. 2006. Ultrahigh-pressure metamorphism and exhumation of garnet peridotite in Pohorje, eastern Alps. *Journal of Metamorphic Geology*, **24**, 19–31.

KOZUR, H. 1991. The Evolution of the Meliata-Hallstatt ocean and its significance for the early evolution of the Eastern Alps and Western Carpathians. *Palaeogeography Palaeoclimatology Palaeoecology*, **87**, 109–135.

KOZUR, H. & MOCK, R. 1973. Zum Alter und zur tektonischen Stellung der Meliata-Serie des Slowakischen Karstes. *Geologický Zborník – Geologica Carpathica*, **24**, 365–374.

KOZUR, H. & MOSTLER, H. 1992. Erster paläontologischer Nachweis von Meliaticum und Süd-Rudabanyaicum in den Nördlichen Kalkalpen (Österreich) und ihre Beziehung zu den Abfolgen der Westkarpaten. *Geologisch-Paläontologische Mitteilungen der Universität Innsbruck*, **18**, 87–129.

KRIST, E., KORIKOVSIJ, S. P., PUTIS, M., JÁNAK, M. & FARYAD, S. W. 1992, *Geology and Petrology of metamorphic rocks of the Western Carpathian crystalline complexes*. Comenius University Press, Bratislava.

LEXA, O., SCHULMANN, K. & JEZEK, J. 2003. Cretaceous collision and indentation in the West Carpathians: View based on structural analysis and numerical modeling. *Tectonics*, **22**, Art. No. 1066.

LUPTÁK, B., JÁNAK, M., PLAŠIENKA, D., SCHIMDT, S. T. & FREY, M. 2000. Chloritoid-kyanite schists from the Veporic unit, Western Carpathians, Slovakia: implications for Alpine (Cretaceous) metamorphism. *Schweizerische Mineralogische und Petrographische Mitteilungen*, **80**, 213–223.

MALUSKI, H., RAJLICH, P. & MATTE, P. 1993. $^{40}Ar/^{39}Ar$ dating of the Inner Carpathians Variscan basement and Alpine mylonitic overprinting. *Tectonophysics*, **223**, 313–337.

MANDL, G. W. & ONDREIJOKOVA, A. 1991. Über eine triadische Teifasserfazies (Radiolarite, Tonschiefer) in den Nördlichen Kalkalpen – ein Vorbericht. *Jahrbuch der Geologischen Bundesanstalt (Wien)*, **134**, 309–318.

MASSONNE, H.-J. & SCHREYER, W. 1987. Phengite geobarometry based on the limiting assemblage with K-feldspar, phlogopite, and quartz. *Contributions to Mineralogy and Petrology*, **96**, 212–224.

MELLO, J., REICHWALDER, P. & VOZÁROVÁ, A. 1998. Bôrka nappe: high-pressure relic from the subduction-accretion prism of the Meliata ocean (Inner Western Carpathians, Slovakia). *Slovak Geological Magazine*, **4**, 261–273.

MOCK, R., SÝKORA, M., AUBRECHT, R., OŽVOLDOVÁ, L., KRONOME, B., REICHWALDER, P. & JABLONSKÝ, J. 1998. Petrology and stratigraphy of the Meliaticum near the Meliata and Jaklovce Villages, Slovakia. *Slovak Geological Magazine*, **4**, 223–260.

NEUBAUER, F. 1994. Kontinentkollision in den Ostalpen. *Geowissenschaften*, **12**, 136–140.

NEUBAUER, F. 2002. Contrasting Late Cretaceous to Neogene ore provinces in the Alpine-Balkan-Carpathian-Dinaride collision belt. *In*: BLUNDELL, D. J., NEUBAUER, F. & QUADT, A. (eds) *The timing and location of major ore deposits in an evolving orogen*. Geological Society (London) Special Publication, **204**, 81–102.

NEUBAUER, F., DALLMEYER, R. D., DUNKL, I. & SCHIRNIK, D. 1995. Late Cretaceous exhumation of the metamorphic Gleinalm dome, Eastern Alps: Kinematics, cooling history, and sedimentary response in sinistral wrench corridor. *Tectonophysics*, **242**, 79–89.

PLAŠIENKA, D. 1997. Cretaceous tectonochronology of the central western Carpathians, Slovakia. *Geologica Carpathica*, **48**, 99–111.

PLATT, J. P. 1993. Exhumation of high-pressure rocks: a review of concepts and processes. *Terra Nova*, **5**, 119–133.

RAKÚS, M. & SÝKORA, M. 2001. Jurassic of Silicicum. *Slovak Geological Magazine*, **7**, 53–84.

RATSCHBACHER, L., FRISCH, W., NEUBAUER, F., SCHMID, S. M. & NEUGEBAUER, J. 1989. Extension in compressional orogenic belts: the eastern Alps. *Geology*, **17**, 404–407.

SAMPSON, S. D. & ALEXANDER, E. C. 1987. Calibration of the interlaboratory $^{40}Ar/^{39}Ar$ dating standard, Mmhb-1. *Chemical Geology*, **66**, 27–34.

SCHERMER, E. R. 1990. Mechanism of blueschist creation and preservation in an A-type subduction zone, Mount Olympus region, Greece. *Geology*, **18**, 1130–1133.

STAMPFLI, C. M. & MOSAR, J. 1999. The making and becoming of Apulia. *Memorie di Scienze Geologiche*, **51**, 141–154.

STEIGER, R. H. & JÄGER, E. 1977. Subcommission of Geochronology: Convention on the use of decay constants in geo- and cosmochronology. *Earth and Planetary Science Letters*, **36**, 669–690.

THÖNI, M. 1999. A review of geochronological data from the Eastern Alps. *Schweizerische Mineralogische und Petrographische Mitteilungen*, **79**, 209–230.

THÖNI, M. & JAGOUTZ, E. 1991. Some new aspects of dating eclogites in metamorphic belts: Sm-Nd, Rb-Sr, and Pb-Pb isotopic results from the Austroalpine Saualpe and Koralpe type-locality (Carinthia/Styria, southeastern Austria). *Geochimica et Cosmochimica Acta*, **56**, 347–368.

TOMEK, Č 1993. Deep crustal structure beneath the central and inner West Carpathians. *Tectonophysics*, **226**, 417–431.

TOMEK, Č & HALL, J. 1993. Subducted continental margin imaged in the Carpathians of Czechoslovakia. *Geology*, **21**, 535–538.

VELLEDITS, F. 2006. Evolution of the Bükk Mountains (NE Hungary) during the Middle–Late Triassic asymmetric rifting of the Vardar–Meliata branch of the Neotethys Ocean. *International Journal of Earth Sciences*, **95**, 395–412.

VILLA, I. 1998. Isotopic closure. *Terra Nova*, **10**, 42–47.

WAKABAYASHI, J. & UNRUH, J. R. 1995. Tectonic wedging, blueschist metamorphism, and exposure of blueschists: are they compatible? *Geology*, **23**, 85–88.

WILLINGSHOFER, E., NEUBAUER, F. & CLOETINGH, S. 1999. Significance of Gosau basins for the upper Cretaceous geodynamic history of the Alpine-Carpathian belt. *Physics and Chemistry of Earth Part A: Solid Earth Geodesy*, **24**, 687–695.

YORK, D. 1969. Least squares fitting of a straight line with correlated errors. *Earth and Planetary Science Letters*, **5**, 320–324.

Subduction-related metamorphism in the Alps: review of isotopic ages based on petrology and their geodynamic consequences

ALFONS BERGER[1] & ROMAIN BOUSQUET[2]

[1]*Institute of Geological Sciences, University of Bern, Baltzerstrasse 1 & 3, 3012 Bern, Switzerland (e-mail: berger@geo.unibe.ch)*

[2]*Institut für Geowissenschaften, University of Potsdam, Karl-Liebknecht-Strasse 24, 14476 Golm, Germany*

Abstract: We summarize ages of the high-pressure/low-temperature (HP/LT) metamorphic evolution of the central and the western Alps. The individual isotopic mineral ages are interpreted to represent either: (1) early growth of metamorphic minerals on the prograde path; (2) timing close to peak metamorphism; or (3) retrograde resetting of the chronometers at still-elevated pressures. Therefore, each individual age cannot easily be transferred to a geodynamic setting at a certain time. These different data indicate a subduction-related metamorphism between 62 and 35 Ma in different units (e.g. Voltri Massif, Schistes Lustrés of the western Alps, Tauern Window). Oceanic and continental basement units show isotope ages related to eclogitic or blueschist facies metamorphism between 75 and 40 Ma. Most of these ages may record equilibration along the retrograde path, except of some Lu/Hf garnet ages and some zircon SHRIMP ages, which provide information on the prograde path. These different isotope ages are interpreted as different steps along pressure–time paths and so may provide some information on the geodynamic evolution. The data record a continuous subduction, which is ongoing for several tens of millions years. In a large-scale picture, we have to assume fragmentation of the downgoing plate in order to explain the available P–T and t data. This interpretation questions the ongoing driving force for subduction during the disappearance of the Alpine Tethys.

Introduction

The Alpine belt formed during a process of convergence, subduction and collision between the European and Adria continental plates during Mesozoic and Cenozoic times (e.g. Ernst 1971; Dal Piaz *et al.* 1972, 2003; Trümpy 1975; Le Pichon *et al.* 1988; Dewey *et al.* 1989). The intervening oceanic crust progressively deformed and partially accreted to the continental margins, which are now sandwiched between the overlying Austroalpine nappe units and the underlying rocks of the European domains. The occurrence of high-pressure–low-temperature metamorphism (HP/LT) of the oceanic and some adjacent continental units is now well known from many parts of the Alps (see for review Niggli 1978; Goffé & Chopin 1986; Frey *et al.* 1999; Oberhänsli *et al.* 2004). Nevertheless, there are different opinions about the age and palaeogeographic position of some HP/LT units (e.g. Cretaceous vs. Palaeogene, number of subduction sites; see for example Polino *et al.* 1990), which are partly due to conflicting determinations of the timing of metamorphism (e.g. Chopin & Maluski 1980; Hunziker *et al.* 1992; Gebauer 1999) and partly due to a fragmentary view of the Alpine belt.

The identification of ophiolites and their position in the Alpine edifice are crucial for the reconstructions of the Alps (e.g. Dal Piaz 1974; Dietrich 1980; Bigi *et al.* 1990; Polino *et al.* 1990; Höck & Koller 1989; Froitzheim *et al.* 1996). Furthermore, it is well accepted that high-pressure/low-temperature (HP-LT) metamorphism of such ophiolites is associated with subduction-related processes, whereas high-temperature/medium-pressure (HT–MP) metamorphism is associated with collision-related processes. A better understanding can be gained through integration of structural analysis at various scales, petrological estimates, and isotope dating (e.g. Compagnoni *et al.* 1977; Gosso *et al.* 1979; Polino *et al.* 1990; Oberhänsli 1994; Froitzheim *et al.* 1996; Handy & Oberhänsli 2004). The identification of ophiolites is well established in the Alps, whereas the metamorphic evolution in time and space is less clear. Another large difficulty arises by defining units (rock masses) which underwent the same geodynamic evolution (see discussion in Bousquet 2007). In other words, we have to combine information from local samples towards tectonic units or palaeo-plates to understand geodynamic processes in time and space.

Understanding geodynamic evolution during subduction requires the combination of different datasets. In this context, accurate constraints on pressure (P), temperature (T) and time (t) are key elements (e.g. Compagnoni et al. 1977; Ernst & Dal Piaz 1978). Recently, new isotopic ages and P–T conditions were published on the high-pressure/low-temperature metamorphic evolution using eclogites and blueschists (e.g. Gebauer 1999; Agard et al. 2002; Rubatto & Hermann 2003; Brouwer et al. 2005; Liati et al. 2005). Compiling all of the available data results in relatively high density of data, which makes the Alps suitable to investigate subduction processes in time and space. When compiled isotopic ages are combined with petrological information, the resulting information includes substantial uncertainties. These uncertainties will be summarized into two groups: (1) interpretation (geological meaning) of isotopic data ages per se; and (2) correlation of units in a geodynamic context.

With regard to point (1) individual isotopic mineral ages from high-pressure (HP) metamorphic samples may be interpreted as:

- Early metamorphic growth along a prograde path;
- Peak metamorphic conditions;
- Strong overprint along the retrograde path;
- Diffusional resetting of the chronometer; and
- Relict ages of the protolith.

With regard to point (2) the knowledge of a certain age has to be translated into a space-time movement of tectonic units or to fragments within mélange units. This includes the reconstruction of coherent unit for a certain part of the P–T–t evolution.

In this paper, we combine such different information and discuss possible models for the geodynamic evolution of the Alps. The Alps are starting to have enough information, in order to compare data from nature with numerical modelling. Such geodynamic modelling is available after several decades, and published models using different techniques and aims (e.g. Oxburgh & Turcotte 1974; Davy & Gillet 1986; Escher & Beaumont 1997). Modern geodynamic modelling proposed P–T–t paths, which can be compared with data from nature (e.g. Bousquet et al. 1997; Burov et al. 2001; Roselle et al. 2002; Goffé et al. 2003; Stöckert & Gerya 2005). Therefore, we have to understand the meaning of individual P, T and t data and link these into a geodynamic context, which is the aim of this contribution. The tectonic base, as used here, is summarized from the tectonic map of the Alps (Schmid et al. 2004a) and the geodynamic map of the Alps (Oberhänsli et al. 2004). Both the latter maps are based on the structural model of Italy (Bigi et al. 1990). In general, the tectonic units have to be subdivided into units belonging to the Adria margin, the Piemonte–Liguria Ocean, the Briançonnais microcontinent, the Valais Ocean and the European margin. In addition, smaller portions of the Meliata Ocean are preserved in the eastern part of the Alps. The main exhumed parts of the Western and Central Alps are related to the Late Cretaceous–Palaeogene orogen including subduction of the Alpine Tethys and collision of Adria and Europe. The tectonic evolution of the Alps has been already discussed from different points of view (e.g. Tapponier 1977; Frisch 1979; Trümpy 1982; Froitzheim et al. 1996; Stampfli et al. 1998; Dal Piaz 1999; Schmid & Kissling 2000; Dal Piaz et al. 2003; Schmid et al. 1996, 2004a, b).

This contribution summarizes data related to subduction and early evolution of the Alps. For simplicity, we do not deal with the latter and skip later important tectono-metamorphic evolution, although we are aware of the complex evolution following subduction, which includes the exhumation of the HP/LT rocks.

The pressure–temperature–time evolution during subduction

The main published results will be summarized in the following sections. This compilation does not give a complete summary of isotope ages and P–T data; rather it examines data that are directly relevant to the subduction history. For simplicity, data are mostly presented without the analytical error. The following data are organized into the major tectonic units from south(east) to north(west) (Sesia–Lanzo zone, Piemonte–Liguria Zone, Briançonnais and Valais).

Sesia Zone and Dent Blanche nappe

The Sesia Zone of the western Austroalpine is a portion of Adriatic continental crust recording Alpine eclogite-facies assemblages (Fig. 1). The recognition that granites underwent eclogite-facies conditions within this zone provided the first demonstration of subduction of continental crust (Dal Piaz et al. 1972; Compagnoni & Maffeo 1973; Compagnoni et al. 1977; Lardeaux et al. 1982; Oberhänsli et al. 1982, 1985; Lardeaux & Spalla 1991; Venturini 1995; Spalla et al. 1996). The Alpine evolution is characterized by relics of blueschist- to eclogite-facies prograde metamorphism, followed by a HT blueschist-facies re-equilibration during decompression (e.g. Castelli 1991; Pognante 1991 and references therein), and then by a greenschist facies overprint (Dal Piaz et al. 1972; Compagnoni et al. 1977;

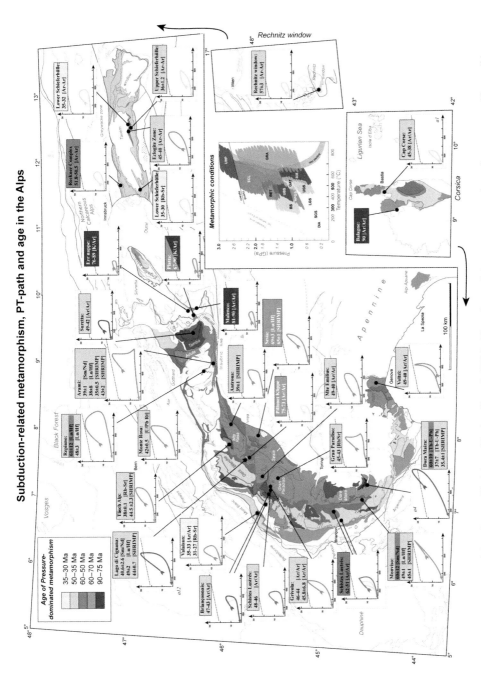

Fig. 1. Simplified metamorphic map of the Alps modified from Oberhänsli et al. (2004). The colours of the map represent the various metamorphic conditions preserved in the Alps (see colour code in the P/T diagram in the middle right). P–T paths are shown together with the isotope ages of subduction-related metamorphism for various localities across the Alps (the ages are colour coded as shown in the upper left). Data source and references for P/T and t data are given in Tables 1 and 2.

Oberhänsli et al. 1985), which is generally associated with mylonitic textures (Stünitz 1989; Spalla et al. 1991).

The more external Dent Blanche nappe s.l. is located at the same structural level as the Sesia–Lanzo Zone, namely on top of the entire ophiolitic Piemonte–Ligurian Ocean (Fig. 1). It consists of the Dent Blanche, M. Mary and Pillonet basement and cover units, which display relics of a blueschist facies imprint and a pervasive greenschist facies overprint (Ballevre et al. 1986; Dal Piaz 1999). Other basement slices (Mt. Emilius, Glacier-Rafray, Etirol-Levaz) preserve relics of an eclogitic imprint but are located at a lower structural level of the nappe stack (i.e. within the underlying ophiolitic Piemonte–Ligurian Ocean) and are reported as lower Austroalpine outliers (Dal Piaz 1999; Dal Piaz et al. 2001). More recently, the Sesia Zone and the Lower Austroalpine Outliers have been interpreted as one or more extensional allochthons within the Piemonte–Liguria Ocean (Froitzheim et al. 1996; Dal Piaz 1999; Dal Piaz et al. 2001; Schmid et al. 2004a). The original position of these units within an oceanic plate is of special importance for geodynamic and tectonic models (see discussion). The subduction direction of this unit is towards the SE, which is consistent with Tertiary subduction of the Piemonte–Ligurian Ocean.

SHRIMP zircon ages in the Sesia Zone indicate that the metamorphic evolution began at 76 Ma and subduction-related metamorphism at c. 65 Ma (Fig. 2; Rubatto et al. 1999). A Late Cretaceous subduction-related metamorphism in eclogite has been suggested based on Lu/Hf ages on garnet (69 Ma; Duchêne et al. 1997). While the Sesia evolution is well constrained between 60 and 80 Ma (see Frey et al. 1999; Handy & Oberhänsli 2004 for reviews), the similar eclogitic imprint of the lower Austroalpine outliers (Mt. Emilius, Glacier, Rafray, Etirol-Levaz units) are less clear. These units display Eocene Rb–Sr white mica ages (49–45 Ma; Dal Piaz et al. 2001).

Piemonte–Liguria Ocean

The Piemonte–Liguria Ocean (or Zone) is a structurally composite nappe system extending along the entire arc of the Western Alps (e.g. Bigi et al. 1990). In the northwestern Alps, it is classically divided into two principal units, the upper unit (e.g. Tsaté nappe or Combin zone in the north) and the lower unit (e.g. Zermatt-Saas Fee nappe in the north; e.g. Sartori & Thélin 1987; Ballèvre & Merle 1993; Dal Piaz 1999 and references therein). The distinction between these ophiolitic units was based both on lithostratigraphic and metamorphic differences (Bearth 1962, 1967; Kienast 1973; Dal Piaz 1965, 1974; Caby et al. 1978; Marthaler 1984; Marthaler & Stampfli 1989) and structural data (e.g. extensional faults; Ballèvre & Merle 1993; Reddy et al. 2003). These ophiolitic units extend south of the Aosta valley, through the Cottian to the Ligurian Alps; these are generally reported as external and internal units of the Piemonte–Liguria Zone, corresponding to the Combin and Zermatt-Sass units, respectively. The Zermatt-Saas Fee unit and southern homologues (Monviso, Voltri Group) are composed mainly of metamorphic ophiolites (metabasalt, metagabbro, serpentinite) with minor cover sequences. These units are well known for their high-pressure mineral assemblages. The discovery of coesite inclusions within garnet in Lago di Cignana suggests that some pieces of the Piemonte-Liguria Ocean were deeply subducted (c. 2.8 GPa; Reinecke 1991). The spatial distribution of possible ultra high-pressure rocks is not clear at the moment. The Combin zone (northwestern Alps) is characterized by blueschist facies relics scatteredly preserved below a pervasive greenschist facies overprint (Dal Piaz & Ernst 1978; Ayrton et al. 1982; Baldelli et al. 1983; Sperlich 1988; Martin et al. 1994) in the metabasites while garnet–chloritoid–phengite assemblages in surrounding metapelites indicate higher metamorphic conditions up to 1.4 GPa and 450 °C (Bousquet 2007). In the south, the Schistes Lustrés zone of the Cottian Alps is characterized by occurrences of Fe, Mg-carpholite bearing assemblages in the western part and chloritoid-bearing assemblages in the eastern part with Fe, Mg-carpholite only as relics in quartz veins (Goffé & Chopin 1986; Agard et al. 2001). On the basis of metapelite and metabasite mineralogy, P–T estimates for the uppermost unit at maximum depth increase from west to east from c. 1.0–1.2 GPa/300–350 °C to 1.4–1.5 GPa/450–500 °C (Agard et al. 2001; Messiga et al. 1999; Schwartz et al. 2000a).

Evidence of HP/LT metamorphism has also been found in the Avers nappe of the Central Alps (eastern Switzerland). Mineral assemblages formed of glaucophane–garnet ± chloritoid in metabasalts (Oberhänsli 1978), glaucophane–phengite in marbles, and garnet–chloritoid in calcschists (Wiederkehr 2004). The lower unit in the east is represented by the Platta, Lizun, Malenco and Arosa units (e.g. Schmid et al. 2004). The Platta Nappe shows no HP/LT metamorphism, but a very-low grade, Alpine metamorphism in the north and a greenschist facies metamorphism in the south (e.g. Müntener et al. 2000; Frey & Ferreiro-Mählmann 1999). The metamorphic evolution of the metasediments (e.g. Tsaté in the Western Alps and Avers nappe in eastern Switzerland) is in contrast to the greenschist facies metamorphism of some of the

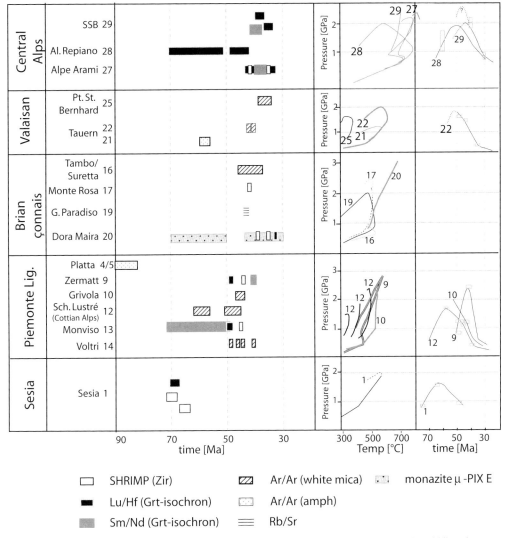

Fig. 2. Compiled isotope and petrological data from selected continental and oceanic units. In the middle column are exemplary P–T paths for some HP rocks Note that in the Central Alps a large variety of P–T paths are estimated and only a few are presented. The temperatures in the Central Alps are systematically higher compared to surrounding units. The right column presents the pressure–time paths using the isotope and P–T data (compare also Figs 1 and 3). Numbers refer to Table 1.

mentioned ophiolitic units. These are Balagne, Nebbio and Pineto-Tribbio units in the southwestern Alps and Corsica and the Platta-Lizun units in the eastern part of the Central Alps.

The eclogites of the Zermatt-Saas Fee unit show Lu/Hf and Sm/Nd ages of 49 and 41 Ma, respectively (Table 1, Fig. 2; Amato *et al.* 1999; Lapen *et al.* 2003). The Lu/Hf data are interpreted as the time of blueschist–eclogite transition (Lapen *et al.* 2003). The zircons record SHRIMP age of 44 Ma, which is interpreted as the HP/LT metamorphism (Rubatto *et al.* 1998). The Monviso eclogites show spread in Sm/Nd, Lu/Hf and SHRIMP ages, which may be related in part to analytical uncertainties. The Sm/Nd garnet isochrons result in ages between 60 ± 12 and 61 ± 9 Ma (Cliff *et al.* 1998), whereas the Lu/Hf garnet ages and SHRIMP zircon data are significantly younger (49 and 45 Ma; Duchêne *et al.* 1997; Rubatto & Hermann 2003; Table 1). These younger ages correspond somehow with Rb/Sr

Table 1. *Selected isotope age data relevant to the early tectonic history of the units. Units involved in the subduction processes, but without evidence of HP/LT metamorphism are shown in grey. The area-code can be found in Figures 2, 3 and 5. These data are presented graphically in Figure 1*

	Locality	Rock type	Method	Age	Err.	Interpretation	Reference	Area code
Austroalpine	**Reckner Complex**	Metasediments	Ar–Ar Phg	52–51		Blueschist	Dingeldey et al. 1997	**00**
	Err-nappe	Metaradiolarite	K/Ar Phg	89–76		PT max	Handy et al. 1996	
	Sesia	Eclogite	Zrn SH.	65	3	HP	Rubatto et al. 1999	**01**
		HP-Vein with Qtz, Jd-Ab-Ep	Zrn SH.	70	5	HP	Rubatto et al. 1999	
		Eclogite	Lu/Hf	69	3	HP	Duchêne et al. 1997	
		Metagranitoid	Rb/Sr	71	2	HP	Oberhänsli et al. 1985	
		Calcschists	Ar/Ar	65	2	HP	Venturini 1995	
		Metagabbros	Ar/Ar	74	0.2	HP	Venturini 1995	
		Marble	Rb/Sr	71	0.8	HP	Dal Piaz et al. 2001	
	Dt. Blanche (Pillonet Klippe)	Metagranitoid	Ar/Ar	75–73	0.7	HP	Cortiana et al. 1998	**02**
		Micaschists	Rb/Sr	75	0.8	HP	Cortiana et al. 1998	
		Micaschists	Rb/Sr	60.4	1.2	PT max	Reddy et al. 2003	
	Mt. Emilius	Eclogite	Rb/Sr	45	0.4	HP	Dal Piaz et al. 2001	**03**
		Gneiss	Rb/Sr	49–42	0.5	HP	Dal Piaz et al. 2001	
Piemonte-Liguria I	**Malenco**	Metagabbro	Ar/Ar Amph	83–71		P max	Villa et al. 2000	**04**
		Metagabbro	Ar/Ar Amph	73–67		T max	Villa et al. 2000	
	Platta	Basic dyke	K/Ar Rbk	84.3	1.3		Deutsch 1984	**05**
		Basic dyke	K/Ar Rbk	80.5	1.3		Deutsch 1984	
		Basic dyke	K/Ar Rbk	68.7	1.2		Deutsch 1984	
		Basic dyke	K/Ar Rbk	90.2	1.9		Deutsch 1984	
		Metaradiolarite	Ar/Ar Amph	80–67		PT max	Handy et al. 1996	
	Balagne	Metabasaltes	Ar/Ar Gln	90			Maluski 1977	**06**
Piemonte-Liguria II	**Rechnitz window**	Metabasaltes	Ar/Ar Amph	57	3	Pmax: Blueschists	Ratschbacher et al. 2004	**07**
	Zermatt (Täsch)	Metagabbro	Zrn SH.	49	3	HP	Rubatto et al. 1998	**08**
		Metapelites	Rb/Sr	38	0.1	500 °C isotherm	Amato et al. 1999	
	Zermatt (Lago di Cignana)	Eclogite	Zrn SH.	44.5	2.3	HP	Rubatto et al. 1998	**09**
		Metapelites	Zrn SH.	43.9	0.9	HP	Rubatto et al. 1998	
		Eclogite (without Cs)	Lu/Hf	48.8	2.1	Transition BS-EC	Lapen et al. 2003	
	Grivola (Urtier)	Eclogite	Sm/Nd	40.6	2.6	Final garnet growth	Amato et al. 1999	
		Metapelites	Rb/Sr	37.9	0.09	500 °C isotherm	Amato et al. 1999	
	Grivola	Eclogite	Rb/Sr	42–45		HP	Dal Piaz et al. 2001	**10**
		Vein in Eclogite	Ar/Ar Phg	46–44		Blueschist post eclogites	Bucher 2003	

Region	Rock type	Method	Age	Error	Interpretation	Reference	#
	Eclogites	Ar/Ar Phg	45.8	6.8	First part of the exhumation	Reddy et al. 2003	
Schistes Lustrés (Entrelor)	Metapelite	Ar/Ar	46–44		HP	Bucher 2003	11
Schistes Lustrés N' Cottian Alps	Metapelite	Ar/Ar Phg	51–45		HP?	Agard et al. 2002	12
Monviso	Metapelite	Ar/Ar Phg	62–55		HP	Agard et al. 2002	13
	Eclogite	Lu/Hf	49.1	1.2	Garnet growth	Dûchene et al. 1997	
	Eclogite	Zrn SH.	45	1	Eclogite stage	Rubatto & Hermann 2003	
Voltri	myl. Eclogite	Rb/Sr	41.6	0.4	?	Cliff et al. 1998	
	myl. Eclogite	Sm/Nd	60	12	?	Cliff et al. 1998	14
	Eclogite	Ar/Ar Phg	49	0.4	HP	Federico et al. 2005	
	Blueschist	Ar/Ar Phg	40	0.37	HP	Federico et al. 2005	
	Calcschist	Ar/Ar Phg	47.6	0.37	HP	Federico et al. 2005	
	Micaschist	Ar/Ar Pg	44.3	1.9	HP	Federico et al. 2005	
Tenda massif (Cap-Corse-Corsica)	Metapelite	Ar/Ar	45–38			Brunet et al. 2000	15
Briançonnais							
Tambo Suretta	Metagranite	Ar/Ar Phg	46	10	Pmax	Challendes et al. 2003	16
Monte Rosa (Furgg zone)	Qz–Cc–Phg–Rt Vein in eclogite	U/Pb Rt	42	0.6	HP	Lapen et al. 2007	17
Ruitor-Cogne	Quartzite	Ar/Ar Phg	47–43			Bucher 2003	18
Gran Paradiso	Micaschist	Rb/Sr	43	0.5	Phengite-apatite equilibr.	Meffan-Main et al. 2004	19
Dora Maira	Felsic-Eclogite	Lu/Hf	32.8	1.2	Garnet growth	Dûchene et al. 1997	20
	Felsic-Eclogite	μ-PIXE Mz	60	10	Prograde	Vagelli et al. 2006	
	Felsic-Eclogite	μ-PIXE Mz	37	7		Vagelli et al. 2006	
	Gneiss	U–Pb	39–35			Tilton et al. 1991	
	Calc-silicate	Zrn SH.	35.1	0.9		Rubatto & Hermann 2001	
Tauern (Valaisan?)							
Tauern (Eclogite Zone)	Eclogite	Ar/Ar Phg	40.9	0.13	HP	Ratschbacher et al. 2004	21
	Grt-micaschist	Ar/Ar Phg	40.1	0.22	HP	Ratschbacher et al. 2004	
	Ferrodiorite	Ar/Ar Amph	57	3	HP	Ratschbacher et al. 2004	
	Eclogite	Ar/Ar Phg	34.2	1.2	HP	Zimmermann et al. 1994	
Upper Schieferhülle	Micaschist	Ar/Ar Phg	36.0	1.2	Blueschist conditions	Zimmermann et al. 1994	22
	Micaschist	Rb–Sr	35.4–30		Garnet growth	Christensen et al. 1994	

(Continued)

Table 1. Continued

	Locality	Rock type	Method	Age	Err.	Interpretation	Reference	Area code
Valaisan	**Antrona (Mattone)**	Overprin. eclogite	Zrn SH.	38.5	0.7		Liati et al. 2005	23
	Balma	Overprin. eclogite	Zrn SH.	40.4	0.7		Liati & Froitzheim 2006	24
	Petit St Bernard	Metapelites	Ar/Ar	37–35		HP	Cannic et al. 1999	25
		Metapelites	Rb/Sr	31–27		HP?	Freeman et al. 1998	
Europe	**Lower Schieferhülle**	Quartzite	Ar/Ar Phg	34.9	1.4	Blueschist conditions	Zimmermann et al. 1994	26
Assimalied at the margin (Central Alps)	**Gruf**	Restitic granulite	Zrn SH.	32.7	0.5	T max at 1.2 GPa	Liati & Gebauer 2003	27
	Alpe Arami	Eclogite	Zrn SH.	35.8	2.8		Gebauer 1996	
		Peridotite	Zrn SH.	43	2		Gebauer 1996	
		Peridotite	Zrn SH.	35.4	0.5		Gebauer 1996	
		Eclogite	Sm/Nd	37.5	2.2	Cooling + decomp.	Becker 1993	
		Eclogite	Lu/Hf	36.6	8.9	Cooling + decomp.	Brouwer et al. 2005	
	Alpe Repiano	Eclogite	Lu/Hf	63	12	Transition blueschist-eclogite	Brouwer et al. 2005	28
		Eclogite	Lu/Hf	47.5	3.4	Somewhen during retrogression	Brouwer et al. 2005	
	M. Motti	Overprinted eclogite	Lu/Hf	35.8	1.6	Retrogression at c. 2 GPa	Brouwer et al. 2005	29
	Gorduno	Overprinted eclogite	Lu/Hf	38.1	2.9	Cooling + decomp.	Brouwer et al. 2005	29
	Cima di Gagnone	Grt-Lerzolithe	Sm/Nd	40.2	4.2	?	Becker 1993	29
	Mt. Duria	Peridotite	Zrn SH.	34.2	0.24	Retrogression at c. 2 GPa	Hermann et al. 2006	29

Abbreviations: Zrn SH = Zircon SHRIMP data, Amph = Amphibole, Rbk: Riebekite, Phg = Phengite, Pg = Paragonite, Rt = Rutile, Gln = Glaucophane, µ-PIXE Mz = chemical ages of monazite with PIXE. Lu/Hf and Sm/Nd are isochron data using garnet. HP = HP/LT metamorphism; BS =Blueschist facies, EC = Eclogite facies.

phengite isochron of Cliff *et al.* (1998). HP-phengites of Fe, Mg-carpholite-bearing metasediments of the external unit are dated between 60 and 55 Ma by Ar/Ar for peak pressure (Agard *et al.* 2002) while the exhumation down to greenschist facies conditions took place from 48 to 40 Ma (Hunziker *et al.* 1992; Markley *et al.* 1998; Agard *et al.* 2002; Bucher 2003). The greenschist facies metamorphism in the southern Platta and Malenco units is Cretaceous in age. These are recorded in K/Ar amphibole ages, which are 90–69 Ma (Deutsch 1984) and detail Ar/Ar data of amphiboles in the Malenco, which spread between 80–90 Ma (Table 2, Villa *et al.* 2000). Similar to these data, the data in the Balagne unit indicate a greenschist facies metamorphism at *c.* 80 Ma (Maluski 1977; see also discussion in Handy & Oberhänsli 2004).

Briançonnais

The palaeogeographic significance and the metamorphic evolution of the Briançonnais microcontinent have been under debate for a long time (e.g. Monié 1990; Stampfli 1993). Despite the relatively small size of this microcontinent, it displays a complex Alpine metamorphic pattern. According to the metamorphic history, three areas can be defined within the Briançonnais microcontinent.

- The so-called internal massifs (Dora Maira, Gran Paradiso, Monte Rosa units) show eclogite facies metamorphism. These are often metagranitoid rocks with interlayered thin layers of metasediments and/or metabasic rocks. These units are pervasively overprinted by alpine tectono-metamorphic reworking (eclogitic and greenschist to amphibolite facies; Dal Piaz & Lombardo 1986; Chopin *et al.* 1991; Frey *et al.* 1999; Dal Piaz 2001). The occurrence of coesite in the Dora Maira nappe gives a minimum pressure of *c.* 3 GPa (Chopin *et al.* 1991; Chopin & Schertl 1999; Simon & Chopin 2001). The estimated pressures in the Monte Rosa nappe differ between 1.8 to 2.4 GPa (Dal Piaz & Lombardo 1986; Le Bayon *et al.* 2006*a*) at temperatures around 500 °C. Maximum pressures are 1.8–2.0 GPa at 500 °C in the Gran Paradiso massif (cf. Benciolini *et al.* 1984; Brouwer *et al.* 2002; Le Bayon *et al.* 2006*b*). U/Pb of rutile in a vein of the Monte Rosa indicates HP/LT metamorphism at 42.6 Ma (Lapen *et al.* 2007). In addition, zircon SHRIMP data of 35 Ma in metasediments are published (Rubatto & Gebauer 1999), although the position of these metasediments inside the Monte Rosa unit has been questioned (see Dal Piaz 2001). Furthermore, Th/Pb monazite ages of 34–32 Ma in the Monte Rosa nappe indicate decompression down to *c.* 1.0 GPa (Engi *et al.* 2001*b*), which clearly are in conflict with an age for peak pressure at 35 Ma. The Gran Paradiso is dated by the Rb/Sr phengite–apatite isochron, which indicates an equilibration near maximum pressures at 43 Ma (Meffan-Main *et al.* 2004). The Dora Maira unit is dated by U/Pb in a range of minerals (Tilton *et al.* 1991), Lu/Hf in eclogite (Duchêne *et al.* 1997) and zircon SHRIMP age (Gebauer *et al.* 1997; Rubatto & Hermann 2001), Ar/Ar and Rb/Sr dating (Di Vincenzo *et al.* 2006). These data show a spread between 38 and 33 Ma, which has been correlated to subduction-related metamorphism. Chemical age data in monazite indicate two metamorphic events in the Dora Maira unit (Vagelli *et al.* 2006). Monazite cores record a metamorphic stage at 60 Ma, and a second monazite-forming event at 37 Ma. The latter age clearly overlaps with other data presented above (Table 1). The meaning of the 60 Ma monazite age is not clear at the moment.

- The internal Briançonnais (Ambin, Vanoise, Ruitor, Mt. Fort, Tambo/Suretta units) consists of combined basement and metasedimentary units, which shows a clear HP/LT imprint. Fe, Mg-carpholite (Goffé 1977, 1984; Goffé *et al.* 1973; Goffé & Chopin 1986), aragonite (Goffé & Velde 1984), jadeite (Cigolini 1995) in metasediments and occurrences of lawsonite and jadeite in metabasites (Lefèvre & Michard 1965; Schwartz *et al.* 2000*b*) indicate an evolution at the blueschist-eclogite transition (see Oberhänsli *et al.* 2004; Fig. 1). An increase in metamorphic grade from blueschist in the west towards blueschist–eclogite facies in the east has been documented in post-Hercynian metasediments (Bucher & Bousquet 2007). Paradoxically, in the northwestern Alps, only the uppermost units of the Briançonnais domain (the Mont Fort nappe) display HP metamorphic evolution (Schaer 1959; Bearth 1963; Bousquet *et al.* 2004).

The age of the HP metamorphism in the internal Briançonnais was debated for a while (see review in Caby 1996), despite the occurrences of glaucophane in Eocene sediments (Ellenberger 1958). In the Western Alps, Ar/Ar data on HP-phengites in post-Hercynian metasediments indicate that these units reached peak pressure between 50 and 43 Ma while exhumation and nappe stacking took place between 43 and 35 Ma (Bucher 2003). Ar/Ar ages in the Tambo/Suretta nappes (Central

Table 2. *Review of selected PT data relevant for subduction-related metamorphism in the Alps. For localities and paleogeographic subdivision, see Table 1*

	Locality	Rock type	P (GPa) max	T (°C) at P max	Reference
Austroalpine	Reckner Complex	Metasediments	1.05	350	Dingeldey et al. 1997
	Err-nappe	Metaradiolarite	0.7	325	Handy et al. 1996
	Sesia	Metapelites	1.7	600	Lardeaux et al. 1982; Vuichard & Ballèvre 1988
		Micaschistes	1.6	550	Pognante 1989
	Dt. Blanche (Pillonet Klippe)	Metagranitoid	0.8	400	Cortiana et al. 1998
	Mt. Emilius	Metagranitoid	1.8–1.5	530–560	Dal Piaz et al. 1983
		Metagabbros	1.6–1.5	550	Ballèvre et al. 1986
Piemonte-Liguria I	Malenco	Metagabbro	0.6	450	Bissig & Hermann 1999
	Platta	Metaradiolarite	0.8	450	Handy et al. 1996
Piemonte-Liguria II	Rechnitz window	Metabasalt	0.8	370	Koller 1985
	Zermatt (Täsch)	Eclogite	1.8	600	Barnicoat & Fry 1986; Oberhänsli 1980
	Lago di Cignana	Mn-rich quartzites	2.8	630	Reinecke 1991
	Grivola (Urtier)	Eclogites	2.3	550	Bousquet 2007
	Grivola	Vein in eclogites	1.5	530	Bousquet 2007
	Schistes Lustrés (Entrelor)	Metapelite	1.4	450	Bousquet 2007
	Schistes Lustrés	Metapelite	1.5	450	Agard et al. 2001
	N' Cottian Alps Monviso	Eclogite	1.9	580	Lombardo et al. 1978; Schwartz et al. 2000a
	Voltri	Eclogite	2.0	550	Vignaroli et al. 2005
	Corsica	Metapelites	1.2	350	Jolivet et al. 1998

Region	Unit	Rock type	P (GPa)	T (°C)	References
Briançonnais	Tambo Suretta	Metagranite (Tambo)	1.3	400	Boudin & Marquer 1993
		Metagranite (Suretta)	1.2	450	Challendes et al. 2003
	Monte Rosa	White-Schist	2.2	550	Le Bayon et al. 2006a
		Micaschist	1.5	500	Borghi et al. 1996
	Ruitor-Cogne	Quartzite – Val di Rhêmes	1.5	490	Bucher & Bousquet 2007
		Quartzite – Ruitor massif	1.2	460	Bucher & Bousquet 2007
	Gran Paradiso	Micaschist	2.0–1.8	540	Le Bayon et al. 2006b
		Micaschist	1.5	500	Borghi et al. 1996
		Felsic-Eclogite	3.0	750	Simon & Chopin 2001
	Dora Maira	Eclogite	2.0	590–560	Holland 1979
Tauern (Valaisan?)	Tauern (Eclogite Zone)	Metapelites	2.0	580–560	Frank et al. 1987
		Eclogite (Grossglockner)	1.7	570	Dachs & Proyer 2001
	Upper Schieferhülle	Micaschist	1.1–1.0	400	Holland & Ray 1985; Zimmermann et al. 1994
Valaisan	Antrona (Mattone)	Eclogite	1.5	500	Colombi & Pfeifer 1986
	Petit St Bernard	Metapelites	1.7	400	Goffé & Bousquet 1997; Bousquet et al. 2002
		Metapelites	1.5	470	(first decompression stage) Goffé & Bousquet 1997; Bousquet et al. 2002
		Eclogites	1.6–1.4	475–425	Cannic et al. 1996
Europa	Lower Schieferhülle	Micaschist	1.0	530	Selverstone et al. 1984
Assimilated at the margin	Alpe Arami	Eclogite	3.5	900	Ernst 1981; Dobrzhinetskaya et al. 2002
		Peridotite	6.0	1050	Paquin & Altherr 2001
	Alpe Repiano	Eclogite	2.8	650	Brouwer et al. 2005
	M. Motti	Eclogite	2.1	700–750	Brouwer et al. 2005
	Gorduno	Eclogite	2.3	750	Brouwer et al. 2005

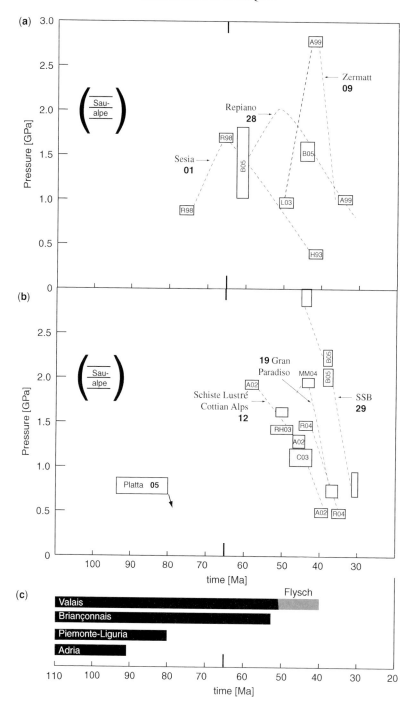

Fig. 3. Pressure–time 'path' for some selected samples/units. The key information includes the depth of subduction inferred from petrological estimates and related exhumation (compare also to Figure 2). (a) Pressure–time path, where dating of the prograde path is inferred. (b) Pressure–time path, where only the exhumation is documented. (c) Youngest sediments in the different units. Note that subduction must have been active, while sedimentation was occurring. (References: R98: Rubatto et al. 1998; R04: Ratschbacher et al. 2004; B05: Brouwer et al. 2005; A02: Agard et al. 2002; RH03: Rubatto & Hermann 2003; L03: Lapen et al. 2003; A99: Amato et al. 1999; MM04: Meffein Main et al. 2004 (for further information see Tables 1 and 2).)

Alps) indicate that the pressure-dominated low-temperature metamorphism occurred in the range of c. 45 Ma (Challandes et al. 2003). Similar ages (c. 45 Ma) have been obtained for the Tenda massif in Corsica (Brunet et al. 2000), which has a similar tectonic position.

The Briançonnais further to the west are composed only of sedimentary rocks (summary in Schmid et al. 2004a). These units display only greenschist metamorphic assemblages (see review in Goffé et al. 2004; Bousquet et al. 2004).

Valais Ocean

The rocks of the Valais Ocean can be followed from the Western Alps to the Engadine window (see Fig. 1 and Schmid et al. 2004a). The sedimentation in the Valais Ocean is mainly Cretaceous in age (e.g. Steinmann 1994), but continues also in the Palaeogene. The sediments are often stripped off from their substratum and were incorporated into an accretion wedge during subduction. Ophiolites and magmatic rocks associated with the metasediments are rare (Versoyen at the French/Italy boundary; Vals-Engadine and Chiavenna area in the Central and Eastern Alps). The metasediments as well as some of the magmatic rocks of the Valais Ocean show clear evidence for blueschist to eclogite facies metamorphism (e.g. Goffé & Oberhänsli 1992; Goffé & Bousquet 1997; Bousquet et al. 1998). Metamorphic pressures in the accretion wedge range from 1.2–1.4 GPa in the east and up to 1.7 GPa in the west at low temperatures (Bousquet et al. 2002). Eclogite facies rocks are also connected to the Valais Ocean (Antrona ophiolite; Oberhänsli 1994; Schmid et al. 2004a). In addition, some authors have recently related the Balma unit to the Valais Ocean (Pleuger et al. 2005; Liati & Froitzheim 2006) instead of the Zermatt-Saas Fee unit (Gosso et al. 1979).

Unfortunately, the timing of the HP metamorphic event of the Valais metapelites remains poorly resolved. An upper limit of the high-pressure metamorphism at 40–35 Ma is given by a Paleocene–Eocene Radiolaria in carpholite–chloritoid-bearing metasediments in the Central Alps (Bousquet et al. 2002). From isotope dating (Schürch 1987; Cannic et al. 1999) and the fossils in the HP-metasediment, it is reasonable to assume a high-pressure metamorphism around 40–35 Ma for the Valaisan domain. This age is in agreement with 39–40 Ma old zircon SHRIMP ages published for the Antrona and Balma units (Liati et al. 2005; Liati & Froitzheim 2006).

The oceanic part in the Eastern Alps is mainly visible in the Tauern Window (with several exceptions). The units in the Tauern Window are subdivided into a lower unit (Venediger nappe) and an upper unit (Glockner nappe). The lower unit is related to the European margin with its sedimentary cover (Schmid et al. 2004a). The upper unit includes magmatic rocks, schist and calcschists, which are correlated with the eastern continuation of the Valais Ocean (Schmid et al. 2004a). However, some sediments of the lower units have similarities to the Piemonte-Liguria and the Valais Ocean. However, these units are *not* separated by a microcontinent, and therefore, we will consider 'one' ocean in the Eastern Alps (see Stampfli et al. 2002). The magmatic and metasedimentary units of this ocean suffered blueschist to eclogite facies metamorphism (e.g. Miller 1974; Spear & Franz 1986; Zimmermann et al. 1994; Hoschek 2001). Maximum pressures reached in the eclogite zone are c. 2 GPa. These rocks underwent subduction-related metamorphism in the time between 55–45 Ma and exhumation from this time on (Zimmermann et al. 1994; Ratschbacher et al. 2004). The eclogite zone became coupled with the blueschist facies metasediments at around 35 Ma and both continued exhumation to 1.0 and 0.5 GPa (Table 1; Fig. 1; Ratschbacher et al. 2004).

Assimilated units at the European margin (Central Alps)

A former plate boundary is preserved as a Tectonic Accretion Channel in the Central Alps (TAC; Engi et al. 2001a). This represents a deep-seated part of the plate boundary including fragments of the mantle, metabasalts, metagabbros and metasediments. The continental basement units can be related to material above and below, but the source of the incorporated oceanic material is unclear. These units are treated as the trace either of the Valais Ocean (e.g. Froitzheim et al. 1996; Schmid et al. 1996, 2004a), or of the Piemonte-Liguria Ocean (Stucki et al. 2003). Studies in the mélange units of the Central Alps indicate that peak pressure of eclogites varies from c. 1.0 to c. 4.0 GPa at variable temperatures (e.g. Ernst 1981; Heinrich 1982; Pfiffner 1999; Paquin & Altherr 2001; Nimis & Trommsdorff 2001; Dale & Holland 2003; Brouwer et al. 2005; Fig. 2). The lowest pressure is reported from the northern part of the Adula nappe complex and the highest pressure from different lenses in the Southern Steep Belt (e.g. Heinrich 1982). The temperature of the eclogite stage is in the range of 700–800 °C, which is higher than in the subducted units of the Western Alps (e.g. Zermatt-Saas Fee unit; Fig. 2).

The available ages are concentrated on Alpe Arami and Cima di Gagnone bodies (Table 1; Fig. 1; Gebauer 1996, 1999; Becker 1993). The garnet Sm/Nd isochrons of these localities indicate ages between 38 and 42 Ma (Becker 1993). Zircon SHRIMP data indicate HP ages between 43 and 35 Ma (Gebauer 1996). Hermann *et al.* (2006) show that zircons developed at still-elevated pressure, but on the retrograde path at 34 Ma (Monte Duria body). The interpretations of the Sm/Nd ages in the Central Alps are under debate. Brouwer *et al.* (2005) conclude that these ages are reset by the high-temperature history of these bodies, which is supported by the same Lu/Hf ages in these localities (Fig. 2). In addition, Lu/Hf ages in one locality (Alpe Repiano) show growth of garnet at c. 61 Ma (Brouwer *et al.* 2005; Table 1). Using the Lu and Hf contents of the separates in combination with laser ablation ICP-MS data of the Lu and Hf content of garnet, it is possible to combine growth evolution with Lu concentrations in different garnet separates. The preserved zonation indicates that these ages represent garnet growth (see also discussion in Skora *et al.* 2006). The core of the garnets with high Lu content indicates growth at the garnet in position along the prograde path at c. 61 Ma.

Discussion

The presented ages show that the timing of subduction-related metamorphism is mainly spread between c. 75 and 40 Ma, with some local event at 33 Ma (Figs 2 and 3). Some of these ages represent the prograde path (i.e. Lu/Hf: Lapen *et al.* 2003; Ar/Ar in phengites: Agard *et al.* 2002) and others may be related to retrograde path at elevated pressures (e.g. Hermann *et al.* 2006). The isotopic HP age information, P–T data and geodynamic interpretation include a large number of uncertainties. These uncertainties and the consequences of combining the data can be summarized into the following questions:

- What is the meaning of each measured age *per se*?
- What is the palaeogeography before and during subduction?
- What is the type and amount of fragmentation during subduction?
- Can we reconstruct subduction in time and space?
- How many suture zones are hidden in the Central Alps?
- What is the palaeogeographic position of the Malenco-Platta units (and their southwestern equivalents)?

In the following discussion, we will mainly formulate these questions, rather than present ready solutions.

What is the meaning of each measured age?

Dating of rock forming minerals has the advantage that we directly combine petrological information with isotope dating. However, accessory phases (zircon, monazite, etc.) are often rich in U and Th, and they are robust in respect to isotopic resetting and hence are well suited for dating. In the case of accessory phases, the relation to pressure and temperature has to be evaluated. Datasets from accessory phases versus rock-forming minerals have both advantages and disadvantages.

In order to discuss the relation between isotopic age and P/T data, we have to discuss the 'closure-temperature' as introduced by Dodson (1973; see also Ganguely & Tirone 1999). However, the role of temperature-induced diffusion is of minor importance for most geochronometers at most grain sizes, and thermally induced diffusion is the key element of the 'closure-temperature' concept (see discussion in Villa 1998). In addition, other processes influence the 'resetting' of an isotope system in a certain mineral. These processes are mainly deformation, mineral reactions and interactions with fluids (see for example Federico *et al.* 2005). Unfortunately, the respective contributions from volume diffusion, mineral reactions, deformation or fluids on the resetting of a certain isotope system are unknown, so such processes cannot be generalized.

Despite these problems, the large number of individual data and several detailed studies on the meaning of the age indicate some processes are more important in one or the other mineral. For example, diffusional resetting is rarely a problem for U/Pb zircon ages. Therefore, many SHRIMP ages give important insights into the evolution of zircons. New zircon can be produced during HP events (e.g. Rubatto & Hermann 2003), but the relation to the P–T evolution is not always clear. The zircon SHRIMP ages are most useful if these data can be combined with trace element data and/or inclusion relationships (e.g. Rubatto 2002; Rubatto & Hermann 2003; Hermann *et al.* 2006). Ar/Ar mica ages in blueschist-facies units often record mineral growth along the prograde path (e.g. Agard *et al.* 2002; Federico *et al.* 2005). However, the role of excess Ar in high-pressure rocks has to be considered (e.g. Scaillet 1996; Brunet *et al.* 2000, but see also Gouzu *et al.* 2006). At higher temperatures and/or stronger deformation, resetting of this isotope system along the retrograde path often occurs, so in the higher temperature and strongly deformed units we exclude the Ar/Ar ages for our discussion.

Lu/Hf garnet ages are most useful because the distribution coefficient of the parent element is large between garnet and other minerals and diffusion of REE and Hf is slow in garnets (Scherer *et al.* 2000). Therefore, Lu/Hf in garnet has a huge potential for dating prograde metamorphic evolution (e.g. Lapen *et al.* 2003; Skora *et al.* 2006). In contrast to the accessory phases, garnet stability is well investigated. Diffusional resetting by volume diffusion in garnet is in most cases of minor importance, because of the low temperatures and slow element diffusion coefficients. Garnet isochron dating (especially Lu/Hf) has less problems with mineral reactions or volume-diffusion, but has other problems with REE-bearing mineral inclusions inside the garnet, which may disturb age information (Scherer *et al.* 2000). Another problem of porphyroblastic garnet is zonation of REE, which is either solely related to petrological processes or related to age zonation (Lapen *et al.* 2003; Skora *et al.* 2006).

In four localities, the zircon SHRIMP data are slightly younger than the Lu/Hf age (Zermatt-Saas Fee, Sesia, Arami, Monviso units), whereas the Sm/Nd garnet ages show no systematics with respect to other isotope ages. Lu-Hf ages record in most cases the early growth of garnet (e.g. Lapen *et al.* 2003; Skora *et al.* 2006), which occurs during the prograde path at the blueschist–eclogite transition. The meaning of the zircon U/Pb age in respect to P and T may be different, but the effect of fluids seems important in many cases. Taking this information into account, the Lu/Hf garnet ages may represent part of the prograde path except some HT samples in the Central Alps. Zircon SHRIMP data may relate to a range of events, depending on the metamorphic reactions and less availability of fluids; all of which influence the growth of new zircon shells.

The palaeogeographic situation

In order to combine P–T–t information with the geodynamic evolution of an orogen, we require a palinspastic reconstruction (e.g. Laubscher 1971*a*, *b*; Schmid & Kissling 2000; Schmid *et al.* 2004*a*, *b*). The Alpine tectonics can be subdivided into a Cretaceous and a Tertiary orogen, which are separated by the extensional Gosau event (e.g. Froitzheim *et al.* 1994, 1996). The Cretaceous orogen includes westward thrusting in the Austroalpine, which has been related to closing of the western Meliata Ocean. HP/LT rocks related to closing of the Meliata Ocean occur, which will not be discussed here (see Thöni & Miller 1996; Janak *et al.* 2004; Thöni 2006 for this theme). We discuss the Upper Cretaceous to Palaeogene subduction of the aforementioned oceans and continental fragments. The subducted units can subdivided into two basins and the related continental units. The basins are the Valais Ocean in the north and the Piemonte–Liguria Ocean in the south (Figs 1 and 4; e.g. Stampfli & Borel 2002). The northern

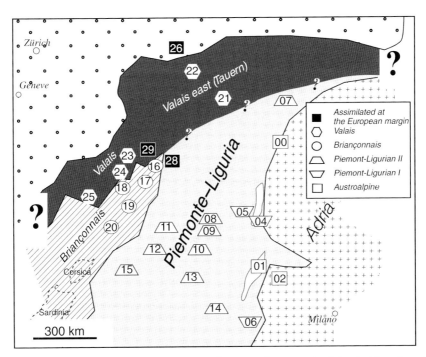

Fig. 4. Schematic palaeogeographical sketch for the Late Cretaceous (reconstruction is inspired from Stampfli & Marchand 1997; Stampfli *et al.* 2002; Stampfli & Borel 2002; Schmid *et al.* 2004*a*). Numbers of present-day units refer to Table 1.

continent is Europe and the southern one, as summarized in this contribution, is Adria. The forementioned oceans are separated by the Briançonnais unit, which represents a topographic high between the oceans, including a continental basement and its sedimentary cover. The oceanic metasediments in the eastern part (Tauern Window) show characteristics of the Piemonte–Liguria Ocean as well as the Valais Ocean, but these units are not separated by a continental domain (Frisch 1979; Stampfli & Borel 2002; Schmid et al. 2004a). Therefore, we will consider 'one' ocean in the Eastern Alps (Fig. 4, see also Stampfli et al. 2002). The different oceanic basins are part of the Alpine Tethys. This subdivision is accepted by most authors, but the palinspastic reconstruction and position of each tectonic unit are under debate. However, the inferred shape of these oceans requires a special palaeogeographic situation at the triple-point of the Valais and Piemonte–Liguria Oceans and the lateral end of the Briançonnais in the central part of the present-day Alps (Fig. 4).

It will be important for the discussion to constrain the size of these oceans in order to discuss subduction process (see below). Such data are obtained from palinspastic reconstructions and are in the range of 400–500 km for the combined Piemonte–Liguria and Valais Oceans in the west (Stampfli et al. 2002). The size of the ocean in the east (Tauern Window) must be somewhat smaller (Laubscher 1971a, 1975).

What is the type and scale of fragmentation during subduction?

In the former section, we concentrate on the prograde evolution during subduction. The difference in P–T path requires also differences in tectonic evolution, which most likely include off-scraping and fragmentation along the destructive plate boundary. However, the preserved information also includes the exhumation part of the investigated units, boudins or clasts. Compiling P–t evolution will give information on the time of fragmentation.

In general, subduction is considered to be a continuous process, which requires a coherent lithospheric plate (scale of kilometres to hundreds of kilometres) in order to maintain the driving force for subduction. In any scenario of a coherent subducting plate, HP/LT metamorphism occurs over a certain time interval (depending on the size of the ocean and the subduction rate). The present-day record of this long-lasting process and related metamorphism may be different, and two end-member scenarios can be discussed: (1) different exhumed parts of this process record the time interval of metamorphism; or (2) the record of this metamorphism is related to a single exhumation event. The latter scenario is not related to the time interval of metamorphism during subduction. However, both scenarios required fragmentation of the subducting plate at different scales (m and/or km). In this context, fragmentation means off-scraping of sections of the plate. In the case of large sheared-off proportions (km size) of a subducting plate, large ophiolitic sequences are transferred to the orogenic wedge (e.g. Zermatt Saas Fee, Platta units). In the example of small fragments (m-scale), we find complex tectonic mélange units, which contain HP/LT fragments of oceanic origin. The latter process was active in the Central Alps and possibly in the Tauern. The mechanism for separating pieces of the subducting oceanic crust and mantle slivers is not well understood. The understandings of such fragmentation processes are further obscured by later deformation related to exhumation. However, cutting out a large piece of the crust during subduction will change the geometry and driving force during subduction. In contrast, development of a tectonic accretion channel operates over long time intervals and subduction may be continued.

In terms of subduction and the record of HP metamorphic rocks, we may subdivide the units in the Alps into the following groups:

1. Coherent continental basement units (e.g. Dora Maira, Gran Paradiso, Monte Rosa and Sesia units);
2. Stripped-off metasediments (e.g. Schistes Lustrés);
3. Small HP fragments within mélange zones (e.g. Entrelor area, Central Alps); and
4. Coherent sections of oceanic lithosphere (e.g. Monviso, Zermatt-Saas Fee, Platta units).

Although some coherent continental basement units contain minor sedimentary cover units, most of the cover units were lost before reaching HP metamorphism. This is well visible in the Briançonnais of the Western Alps (e.g. Bucher et al. 2003).

The oceanic metasediments are related to different oceans (Valais and Piemonte–Liguria) and have different P–T–t evolution. Some of them are included in an accretion wedge along the upper part of the destructive plate boundary. These units show a strong internal deformation, which is evident from their structures and isotopic ages (Agard et al. 2002). Similar is true for tectonic mélange zones in the Central Alps, where different oceanic fragments show different P–T–t paths (Brouwer et al. 2005). The aforementioned metasediments and the deeper parts of the subduction channel are the possible locations of small-scale

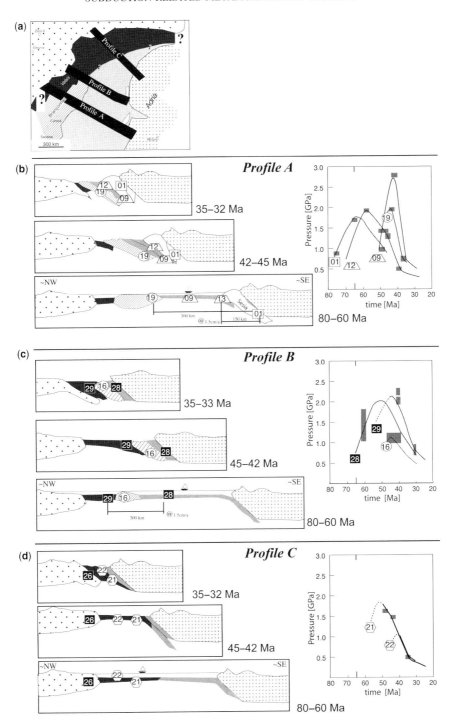

Fig. 5. Interpretation of the palaeogeographical reconstructions and available HP/LT data as given in Tables 1 and 2. (a) Palaeogeographical sketch map of the Alpine Tethys based on the supposed existence of two oceans and the Briançonnais microcontinent. The location of profiles in (b)–(d) are shown. (b) Schematic profile A through the Western Alps. (c) Schematic profile B through the Central Alps. (d) Schematic profile C through the area of the Tauern. Numbers in the profile and the pressure–time path correspond to units or fragments as used in Table 1 (see also Figs 1 and 2).

fragmentation. Small-scale fragmentation along a plate boundary may operate at times during ongoing subduction. In contrast, the separation and exhumation of complete ophiolite sequences is most likely to drastically change the geometry and driving force of subduction.

Can we reconstruct subduction in space and time?

As discussed in the last sections, there are many uncertainties related to the significance of age determinations and of the location of each unit in its former palaeogeographic position. Furthermore, there appears to be a large time span (c. 35 Ma) for HP/LT metamorphism. By combining the presented data (Figs 2 and 3), we have a tool to check the consistency of some tectonic models in terms of comparing map and section view (Fig. 5). In general, we have to consider coherent plates to maintain the driving force for subduction. One model indicates that subduction simply propagated to the northwest (e.g. Ernst 1971; Dal Piaz et al. 1972; Polino et al. 1990; Schmid & Kissling 2000 and literature therein). Some data fit such a propagation model well, but others do not. Such a model would also require similar ages along strike of the same palaeogeographic unit. For example, the basement of the Briançonnais (Dora Maira, Gran Paradiso, Monte Rosa) should have similar HP/LT ages at similar depth. In contrast to this prediction, there is significant variation in the HP/LT ages (35 Ma (60 Ma), 43 Ma, 45 Ma, see Table 1).

Testing possible scenarios, we may discuss ages perpendicular to the destructive plate boundary at different positions (Fig. 5). A distance between units before subduction can be calculated by combining the timing of HP metamorphism (at similar depth) and an assumed subduction rate (Fig. 5). In the last section, we present available P–T and t information for some units of interest (Figs 3 and 5). To examine space evolution, we employ pressure–time paths, using calculated metamorphic pressures and interpolation of available isotope ages (Fig. 3). In order to have comparable ages from burial to exhumation, we can calculate the distances of these units at a given subduction rate. Using the subduction-related metamorphism of the Sesia Zone as one point and calculating the distance to the metasediments in the Western Alps or the Zermatt-Saas Fee unit, we get distances of 150 km and 300 km respectively (using subduction rate of 1.5 cm/a; see plate-movement data in Dewey et al. 1989; Schmid et al. 1996). These distances fit roughly into the size of the oceans from independent plate reconstructions (Stampfli & Marchant 1997). Accepting the overall palaeogeographic configuration of the different subducted units, the differences in pressure–time path have a major consequence. It shows, that large-scale units (km-sized) are cut off from the subducting plate, but the overall subduction continues (see also section on fragmentation). These age-depth data can be also used the other way around. The timing and a certain subduction depth combined with a palinspastic reconstruction result in a certain subduction rate (1–2 cm/a). These subduction rates indicate a relatively slow subduction process compared with modern subduction rates worldwide.

How many oceans are hidden in the Central Alps?

In the Central Alps, we mainly find three units containing relics of oceanic crust: (1) the Tectonic Accretion Channel (TAC); (2) Chiavenna ophiolite; and (3) metasediments of the Northern Steep Belt. In the TAC units, there is one body indicating early HP metamorphism (c. 61 Ma), but most fragments record a pressure-dominated retrograde equilibration between 42 and 34 Ma (Tables 1 and 2; Gebauer 1996; Brouwer et al. 2005; Hermann et al. 2006). The rocks in the Southern Steep Belt are already incorporated in the nappe edifice and suffered Barrovian overprint at c. 32 Ma, which was contemporaneous with the ascent and emplacement of the Bergell pluton (Gebauer 1996; Berger et al. 1996; Oberli et al. 2004). Barrovian overprinting occurs much later in the northern part of the Lepontine. The metasediments of the Northern Steep Belt show relics of a blueschist facies metamorphism, although the age is not known. The Chiavenna ophiolite shows no evidence for any subduction-related metamorphism. The metamorphic ages (Barrovian?) in the Chiavenna ophiolite vary between 42–35 Ma (Talerino 2001; Liati et al. 2003). Geometrical consideration indicates that the main suture in the Central Alps is part of the Valais Ocean (Froitzheim et al. 1996; Schmid et al. 1996, 2004a). In this view, the trace of the Piemonte–Liguria Ocean is represented by the Avers and Platta nappes. This hypothesis is further supported by the occurrence of Briançonnais units in the collisional nappe stack between the Adula nappe complex and the Avers Nappe (Piemonte–Liguria Ocean). However, relics of the Piemonte–Liguria Ocean have also been proposed to occur in the Southern Steep Belt (Stucki et al. 2003). The Piemonte–Liguria Ocean would be missing in a cross-section through the Southern Steep Belt by assuming the Adula and other ophiolites are only the relics of the Valais Ocean. This situation can be explained by later deformation along the Periadriatic Lineament. Alternatively, the Adriatic Plate may have served as an upper plate for a long time, with fragments of different palaeogeographic units accreting to this upper plate. This would require strong tectonic transport to have occurred early in the tectonic history, whereas later tectonic movements were more coherent. This would allow juxtaposition of different fragments within the tectonic mélange units. In this view, the prograde Lu/Hf age at c. 61 Ma can be related to an early fragment along this upper plate.

What is the position of the Malenco–Platta system?

In eastern Switzerland, a part of the Piemonte–Liguria Ocean is preserved with its passive continental margin between the Adria continent and the ocean (e.g. Manatschal & Bernoulli 1999; Müntener et al. 2000). These passive continental margins and oceanic crust include the Platta, Malenco, Forno, Lizun and Arosa units. The Platta-Malenco system was dominated by west-directed thrusting, followed by greenschist facies overprinting during Cretaceous times (e.g. Deutsch 1984; Ring 1992; Handy et al. 1996; Villa et al. 2000), whereas other parts of the Piemonte–Liguria Ocean (e.g. Voltri, Monviso, Zermatt-Saas Fee units) were subducted during Early Tertiary times. The stacking of the Platta Nappe is towards the west, which is incompatible with a south-directed subduction in the Palaeogene (in the Central and Eastern Alps). All these data are in contrast to a joint evolution for the Tauern, Zermatt-Saas Fee and Platta units along the same plate boundary. Froitzheim et al. (1996) proposed that the Austroalpine units and the Platta unit were both subject to Cretaceous thrusting towards the west, with a migrating deformation front from east to west. However, the Platta–Malenco system is part of the Piemonte–Liguria Ocean and not part of the Meliata Ocean (as indicated by the age of the sediments). This interpretation includes major thrusting in Cretaceous times when the orogenic evolution is related to closing of the Meliata Ocean and the thickening of the Adriatic Plate. The thrusting of the Platta unit requires maintaining open oceans in order to deposit the preserved younger sediments. This requires an extraordinary shape and position of ocean–continent transition at that time and a large transfer zone somewhere in the northern part of the Eastern Alps (note the question marks in Fig. 6a). One possible solution would involve a large-scale plate boundary with curvilinear geometry (Fig. 6) or, alternatively, a series of large transform faults (Beccaluva et al. 1984; Stampfli et al. 1998, 2002). The first scenario would allow for west-directed thrusting of part of the Piemonte–Liguria Ocean and preservation of an open ocean with their magmatic crust, which were subsequently subducted elsewhere. Independent from such proposals, the Platta-Err and Malenco units are already in the Paleogene slice of the hanging plate (orogenic lid: Schmid et al. 1990). This does not exclude incorporation of metasediments of the Piemonte–Liguria Ocean at younger times (see relations of the Avers Nappe to the underlying Briançonnais units). At present, there appears no clear solution to this problem, but large changes in plate movement directions have to be assumed at the Cretaceous/Tertiary boundary (Le Pichon et al. 1988; Schmid & Kissling 2000).

Summary and conclusion

There are an increasing number of isotope ages on the HP/LT metamorphism in the Alps. Taking all ages into account, a 35 million-year history of HP/LT metamorphism is preserved in the Western and Central Alps (we exclude the Middle Cretaceous HP history in the Saualpe/Koralpe and related areas; see Thöni 2006). The compiled data show that preserved ages of HP/LT metamorphism are spread over the entire time interval, which is assumed for the subduction process. This is in contrast to the common view of single exhumation events for any one unit and the preservation of only 'one' HP event. The preserved stages of subduction in the Alps present a problem of possible ongoing subduction while exhumation is occurring. Examples of internal deformation in the HP/LT metamorphism lead to variation in ages depending on the individual position of the fragment (see for example Schiste Lustré of the Cottian Alps: Agard et al. 2002; Central Alps: Brouwer et al. 2005). In contrast, the mechanisms and dynamic of removal of large pieces of the oceanic crust during subduction is not understood at this point of research. The dilemma of the completely different P–T–t evolution of the Platta nappe and the Piemonte–Liguria units in the Western Alps requires a different geodynamic setting along the same plate boundary. We proposed a scenario which requires complex plate boundary geometries or a series of transfer faults.

The summary of HP/LT data indicates ongoing subduction and contemporaneus accretion of fragments along the destructive plate boundary. However, there are fundamental differences recorded in the size and type of fragmentation during subduction. Small-scale fragmentation along a plate boundary may be active during ongoing subduction. This process produced mélange zones, but the driving force for subduction remained. Therefore, it may be possible to find a number of different time steps of an ongoing subduction history preserved in a mélange. In contrast,

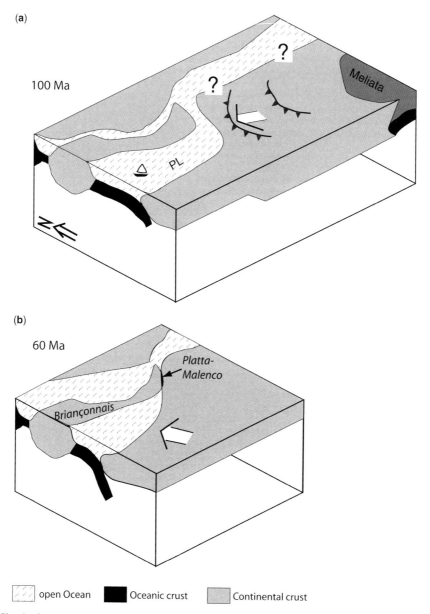

Fig. 6. Sketch of the possible position of the Platta–Malenco units in the Upper Cretaceous. (**a**) Situation before the Platta greenschist-facies metamorphism. Note the transfer zone located in the north of the Alps (located today at the north of the Northern Calcareous Alps). (**b**) Situation after nappe stacking of the Platta–Err system. Note the change in convergence direction.

the incorporation of large, oceanic slices in the nappe pile may predominantly occur during a change in plate arrangement.

We thank I. Mercolli and O. Müntener for comments on a first draft of the paper. Schweizerischer Nationalfonds has supported our research over several years (2000-055306.98, 20-63593.00, 20020-101826, and 200020-109637). Giorgio Dal Piaz provided an extended review and additional literature data and Gerhard Franz a careful review. Carl Spandler improved the English. We thank all of them very much.

References

AGARD, P., JOLIVET, L. & GOFFÉ, B. 2001. Tectonometamorphic evolution of the Schistes Lustrés complex: implications for the exhumation of HP and UHP rocks in the Western Alps. *Bulletin de la Société géologique de France*, **172**, 617–636.

AGARD, P., MONIE, P., JOLIVET, L. & GOFFE, B. 2002. Exhumation of the Schistes Lustrés complex: in situ laser probe $^{40}Ar/^{39}Ar$ constraints and implications for the Western Alps. *Journal of Metamorphic Geology*, **20**, 599–618.

AMATO, J. M., JOHNSON, C. M., BAUMGARTNER, L. P. & BEARD, B. L. 1999. Rapid exhumation of the Zermatt-Saas ophiolite deduced from high-precision Sm–Nd and Rb–Sr geochronology. *Earth and Planetary Science Letters*, **171**, 425–438.

AYRTON, S., BUGNON, C., HAARPAINTER, T., WEIDNMANN, M. & FRANK, E. 1982. Géologie du front de la nappe de la Dent Blanche dans la région des Monts-Dolin. *Eclogae Geologicae Helvetiae*, **75**, 269–286.

BALDELLI, C., DAL PIAZ, G. V. & POLINO, R. 1983. Le quarziti a manganese e cromo di varenche-St. Barthélémy, una sequenza di copertura oceanica della falda piemontese. *Ofioliti*, **8**, 207–221.

BALLÉVRE, M., KIENAST, J. R. & VUICHARD, J. P. 1986. La 'nappe de la Dent-Blanche' (Alpes occidentales); deux unites austroalpines independantes. *Eclogae Geologicae Helvetiae*, **79**, 57–74.

BALLÉVRE, M. & MERLE, O. 1993. The Combin Fault: compressional reactivation of a Late Cretaceous–Early Tertiary detachment fault in the Western Alps. *Schweizerische Mineralogische und Petrographische Mitteilungen*, **73**, 205–227.

BARNICOAT, A. C. & FRY, N. 1986. High-pressure metamorphism of the Zermatt-Saas ophiolite zone, Switzerland. *Journal of the Geological Society*, **143**, 607–618.

BEARTH, P. 1962. Versuch einer Gliederung alpinmetamorpher Serien der Westalpen. *Schweizerische Mineralogische und Petrographische Mitteilungen*, **42**, 127–137.

BEARTH, P. 1963. Contribution à la subdivision tectonique et stratigraphique du cristallin de la nappe du Grand Saint-Bernard dans le Valais (Suisse). *In*: DURAND, D. M. (ed.) *Livre à la mémoire du Professeur Fallot*. Paris.

BEARTH, P. 1967. Die Ophiolite der Zone von Zermatt-Saas Fee. *Beiträge zur geologischen Karte der Schweiz*, **132**, 1–130.

BECCALUVA, L., DAL PIAZ, G. V. & MACCIOTTA, G. 1984. Transitional to normal MORB affinities in ophiolitic metabasites from the Zermatt-Saas, Combin and Antrona units, Western Alps; implications for the paleogeographic evolution of the western Tethyan Basin. *Geologie en Mijnbouw*, **63**, 165–177.

BECKER, H. 1993. Garnet periodite and eclogite Sm–Nd mineral ages from the Lepontine dome (Swiss Alps): new evidence for Eocene high pressure metamorphism in the Central Alps. *Geology*, **21**, 599–602.

BENCIOLINI, L., MARTIN, S. & TARTAROTTI, P. 1984. Il metamorfismo eclogitico nel basamento del Gran Paradiso in unita piemontesi della Valle di Campiglia. *Memorie della Societa Geologica Italiana*, **29**, 127–151.

BERGER, A., ROSENBERG, C. & SCHMID, S. M. 1996. Ascent, emplacement and exhumation of the Bergell Pluton within the Southern Steep Belt of the Central Alps. *Schweizerische Mineralogische Petrographische Mitteilungen*, **76**, 357–382.

BIGI, G., CASTELLARIN, A., COLI, M., DAL PIAZ, G. V., SARTORI, R., SCANDONE, P. & VAI, G. B. 1990. *Structural Model of Italy, sheet 1.* C.N.R., Progetto Finalizzato Geodinamica, SELCA Firenze.

BISSIG, T. & HERMANN, J. 1999. From pre-Alpine extension to Alpine convergence: the example of the southwestern margin of the Margna nappe (Val Malenco, N. Italy). *Schweizerische Mineralogische Petrographische Mitteilungen*, **79**, 363–380.

BORGHI, A., COMPAGNONI, R. & SANDRONE, R. 1996. Composite P–T paths in the internal Pennine massifs of the western Alps: petrological constraints to their thermo-mechanical evolution. *Eclogae Geologicae Helvetiae*, **89**, 345–367.

BOUDIN, T. & MARQUIER, D. 1993. Metamorphisme et deformation dans la nappe de Tambo (Alpes centrales Suisses): evolution de la substitution phengitique au cors de la deformation alpine. *Schweizerische Mineralogische Petrographische Mitteilungen*, **73**, 285–299.

BOUSQUET, R. 2007. Metamorphic heterogeneities within a same HP unit: overprint effect or metamorphic mix? *Lithos*, doi:10.1016/j.lithos.2007.09.010.

BOUSQUET, R., GOFFÉ, B., HENRY, P., LE PICHON, X. & CHOPIN, C. 1997. Kinematic, thermal and petrological model of the central Alps: Lepontine metamorphism in the upper crust and eclogitisation of the lower crust. *Tectonophysics*, **273**, 105–127.

BOUSQUET, R., OBERHÄNSLI, R., GOFFÉ, B., JOLIVET, L. & VIDAL, O. 1998. High pressure–low temperature metamorphism and deformation in the Bündnerschiefer of the Engadine window: implications for the regional evolution of the eastern Central Alps. *Journal of Metamorphic Geology*, **16**, 657–674.

BOUSQUET, R., GOFFÉ, B., VIDAL, O., OBERHÄNSLI, R. & PATRIAT, M. 2002. The tectono-metamorphic history of the Valaisan domain from the Western to the Central Alps: new constraints for the evolution of the Alps. *Geological Society of America Bulletin*, **114**, 207–225.

BOUSQUET, R., ENGI, M., GOSSO, G. ET AL. 2004. Transition from the Western to the Central Alps. Explanatory note to the map 'Metamorphic structure of the Alps'. *Mitteilungen Österreichische Mineralogischen Gesellschaft*, **149**, 145–156.

BROUWER, F. M., VISSERS, R. L. M. & LAMB, W. M. 2002. Structure and metamorphism of the Gran Paradiso Massif, Western Alps, Italy. *Contribution of Mineralogy and Petrology*, **143**, 450–470.

BROUWER, F. M., BURRI, T., ENGI, M. & BERGER, A. 2005. Eclogite relics in the Central Alps: Regional distribution, metamorphic PT-evolution, Lu-Hf ages, and genetic implications on the formation of tectonic mélange zones. *Schweizerische Mineralogische Petrographische Mitteilungen*, **85**, 147–174.

BRUNET, C., MONIÉ, P., JOLIVET, L. & CADET, J.-P. 2000. Migration of compression and extension in the Tyrrhenian Sea, insights from $^{40}Ar/^{39}Ar$ ages on micas along a transect from Corsica to Tuscany. *Tectonophysics*, **321**, 127–155.

BUCHER, S. 2003. *The Briançonnais Units along the ECORS-CROP Transect (Italian-French Alps): Structures, Metamorphism and Geochronology*. PhD Thesis Universität Basel.

BUCHER, S. & BOUSQUET, R. 2007. Metamorphic evolution of the Briançonnais units along the ECORS-CROP profile (Western Alps): New data on metasedimentary rocks. *Swiss Journal of Geoscience*, **100**, 227–242.

BUCHER, S., SCHMID, S. M., BOUSQUET, R. & FÜGENSCHUH, B. 2003. Late-stage deformation in a collisional orogen (Western Alps): nappe refolding, back-thrusting or normal faulting? *Terra Nova*, **15**, 109–117.

BUROV, E., JOLIVET, L., LE POURHIET, L. & POLIAKOV, A. 2001. A thermomechanical model of exhumation of high pressure (HP) and ultra-high pressure (UHP) metamorphic rocks in Alpine-type collision belts. *Tectonophysics*, **342**, 113–136.

CABY, R. 1996. Low-angle extrusion of high-pressure rocks and the balance between outward and inward displacements of Middle Penninic units in the western Alps. *Eclogae Geologicae Helvetiae*, **89**, 229–267.

CABY, R., KIENAST, J.-R. & SALIOT, P. 1978. Structure, métamorphisme et modèle d'évolution tectonique des Alpes Occidentales. *Revue de Géographie physique et de Géologie dynamique*, **XX**, 307–322.

CANNIC, S., LARDEAUX, J.-M. & MUGNIER, J.-L. 1999. Neogene extension in the western Alps. GOSSO, G., JADOUL, F., SELLA, M. & SPALLA, M. I. (eds) *3rd Workshop on Alpine Geological Studies*, **55**.

CASTELLI, D. 1991. Eclogitic metamorphism in carbonate rocks: the example of impure marbles from the Sesia-Lanzo Zone, Italian Western Alps. *Journal of Metamorphic Geology*, **9**, 61–77.

CHALLANDES, N., MARQUER, D. & VILLA, I. M. 2003. Dating the evolution of C–S microstructures; a combined $^{40}Ar/^{39}Ar$ step-heating and UV laserprobe analysis of the Alpine Roffna shear zone. *Chemical Geology*, **197**, 3–19.

CHOPIN, C. & MALUSKI, H. 1980. Ar-40/Ar-39 dating of high pressure metamorphic micas from the Gran Paradiso area (Western Alps); evidence against the blocking temperature concept. *Contribution of Mineralogy and Petrology*, **74**, 109–122.

CHOPIN, C. & SCHERTL, H.-P. 1999. The UHP unit in the Dora-Maira massif, western Alps. *International Geologic Reviews*, **41**, 765–780.

CHOPIN, C., HENRY, C. & MICHARD, A. 1991. Geology and petrology of the coesite-bearing terrain, Dora-Maira massif, Western Alps. *European Journal of Mineralogy*, **3**, 263–291.

CHRISTENSEN, J. N., SELVERSTONE, J., ROSENFELD, J. L. & DEPAOLO, D. J. 1994. Correlation of Rb–Sr geochronology of garnet growth histories from different structural levels within the Tauern Window, Eastern Alps. *Contributions to Mineralogy and Petrology*, **118**, 1–12.

CIGOLINI, C. 1995. Geology of the Internal Zone of the Grand Saint Bernard Nappe: a metamorphic Late Paleozoic volcano-sedimentary sequence in South-Western Aosta Valley (Western Alps). *In*: LOMBARDO, B. (ed.) *Studies on Metamorphic Rocks and Minerals of the Western Alps. A Volume in Memory of Ugo Pognante*, **13**, 293–328.

CLIFF, R. A., BARNICOAT, A. C. & INGER, S. 1998. Early Tertiary eclogite facies metamorphism in the Monviso ophiolite. *Journal of Metamophic Geology*, **16**, 447–455.

COLOMBI, A. & PFEIFER, H.-R. 1986. Ferrogabbroic and basaltic metaeclogites from the Antrona mafic-ultramafic complex and the centovalli-Locarno region (Italy and Southern Switzerland)—first result. *Schweizerische Mineralogische Petrographische Mitteilungen*, **66**, 99–110.

COMPAGNONI, R. & MAFFEO, B. 1973. Jadeite-Bearing Metagranites and related Rocks in the Mount Mucrone Area (Sesia-Lanzo Zone, Western Italian Alps). *Schweizerische Mineralogische und Petrographische Mitteilungen*, **53**, 355–378.

COMPAGNONI, R., DAL PIAZ, G. V., HUNZIKER, J. C., GOSSO, G., LOMBARDO, B. & WILLIAMS, P. F. 1977. The Sesio-Lanzo Zone, a slice of continental crust with Alpine high pressure-low temperature assemblages in the western Italian Alps. *Rendiconti della Societa Italiana di Mineralogia e Petrologia*, **33**, 281–334.

CORTIANA, G., DAL PIAZ, G. V., DEL MORO, A., HUNZIKER, J. C. & MARTIN, S. 1998. ^{40}Ar-^{39}Ar and Rb–Sr dating on the Pillonet klippe less and Sesia-Lanzo basal slice in the Ayas Valley and evolution of the Austroalpine-Piedmont nappe stack. *Memorie di Scienze Geologiche*, **50**, 177–194.

DACHS, E. & PROYER, A. 2001. Relics of high-pressure metamorphism from the Grossglockner region, Hohe Tauern, Austria; paragenetic evolution and PT-paths of retrogressed eclogites. *European Journal of Mineralogy*, **13**, 67–86.

DAL PIAZ, G.-V. 1965. La formation mesozoica dei calescisti con pietre verdi fra la Valsesia e la Valtournanche ed i suoi rapporti strutturali con il recopimento Monte Rosa e con la Zona Sesia-Lanzo. *Bolletino della Società Geologica Italiana*, **84**, 67–104.

DAL PIAZ, G.-V. 1974. Le métamorphisme alpin de haute pression et basse température dans l'evolution structurale du basin ophiolitique alpino-appenninique. 1e partie: *Bolletino della Società Geologica Italiana*, **93**, 437–468; 2e partie Schweizerische Mineralogische Petrographische Mitteilungen, **54**, 399–424.

DAL PIAZ, G.-V. 1999. The Austroalpine–Piedmont nappe stack and the puzzle of Alpine Tethys. *Memorie di Scienze Geologiche*, **51**, 155–176.

DAL PIAZ, G.-V. 2001. Geology of the Monte Rosa massif: historical review and personal comments. *Schweizerische Mineralogische Petrographische Mitteilungen*, **81**, 275–303.

DAL PIAZ, G.-V. & ERNST, W. G. 1978. Areal geology and petrology of eclogites and associated metabasites of the piemontese ophiolite nappe, Breuil-St. Jacques area, Italian Western Alps. *Tectonophysics*, **51**, 99–136.

DAL PIAZ, G.-V. & LOMBARDO, B. 1986. Early Alpine eclogite metamorphism in the Penninic Monte Rosa-Gran Paradiso basement nappes of the northwestern Alps. *Memoir—Geological Society of America*, **164**, 249–265.

DAL PIAZ, G.-V., HUNZIKER, J. C. & MARTINOTTI, G. 1972. La Zona Sesia—Lanzo e l'evoluzione tettonico-metamorfica delle Alpi Nordoccidentali interne. *Memorie della Societa Geologica Italiana*, **11**, 433–460.

DAL PIAZ, G.-V., LOMBARDO, B. & GOSSO, G. 1983. Metamorphic evolution of the Mt. Emilius Klippe, Dent Blanche Nappe, Western Alps. *American Journal of Science, A*, **283**, 438–458.

DAL PIAZ, G.-V., CORTIANA, G., DEL MORO, A., MARTIN, S., PENNACHIONI, G. & TARTAROTTI, P. 2001. Tertiary age and paleostructural inferences of the eclogitic imprint in the Austroalpine outliers and Zermatt-Saas ophiolite, western Alps. *International Journal of Earth Sciences*, **90**, 668–684.

DAL PIAZ, G.-V., BISTACCHI, A. & MASSIRONI, M., 2003. Geological outline of the Alps. *Episodes*, **26**, 174–179.

DALE, J. & HOLLAND, T. J. B. 2003. Geothermobarometry, P–T paths and metamorphic field gradients of high-pressure rocks from the Adula Nappe, Central Alps. *Journal of Metamorphic Geology*, **21**, 813–829.

DAVY, P. & GILLET, P. 1986. The stacking of thrust slices in collision zones and its thermal consequences. *Tectonics*, **5**, 913–929.

DEUTSCH, A. 1984. Datierung an Alkaliamphibolen und Stilpnomelan der südlichen Platta-Decke. *Eclogae Geologicae Helvetiae*, **76**, 295–308.

DEWEY, J. F., HELMAN, M. L., TURCO, E., HUTTON, D. H. W. & KNOTT, S. D. 1989. Kinematics of the western Mediterranean. *In*: COWARD, D. & DIETRICH PARK, M. P. (eds) *Alpine Tectonics*. Geological Society Special Publication, **45**, 187–188.

DI VINCENZO, G., TONARINI, S., LOMBARDO, B., CASTELLI, D. & OTTOLINI, L. 2006. Comparison of $^{40}Ar-^{39}Ar$ and Rb–Sr data on phengites from the UHP Brossasco–Isasca Unit (Dora Maira Massif, Italy): implications for dating White Mica. *Journal of Petrology*, **47**, 1439–1465.

DIETRICH, V. J. 1980. The distribution of ophiolites in the Alps. *Ofioliti*, **1**, 7–51.

DINGELDEY, C., DALLMEYER, R. D., KOLLER, F. & MASSONNE, H.-J., 1997. P–T–t history of the Lower Austroalpine Nappe Complex in the 'Tarntaler Berge' NW of the Tauern Window: implications for the geotectonic evolution of the central Eastern Alps. *Contributions to Mineralogy and Petrology*, **129**, 1–19.

DOBRZHINETSKAYA, L. F., SCHWEINEHAGE, R., MASSONNE, H. J. & GREEN, H. W. I. 2002. Silica precipitates in omphacite from eclogite at Alpe Arami, Switzerland: evidence of deep subduction. *Journal of Metamorphic Geology*, **20**, 481–492.

DODSON, M. H. 1973. Closure temperature in cooling geochronological and petrological systems. *Contributions to Mineralogy and Petrology*, **40**, 259–274.

DUCHÊNE, S., BLICHERT-TOFT, J., LUAIS, B., TELOUK, P., LARDEAUX, J. M. & ALBAREDE, F. 1997. The Lu-Hf dating of garnets and the ages of the Alpine high-pressure metamorphism. *Nature*, **387**, 586–589.

ELLENBERGER, F. 1958. Étude géologique du pays de Vanoise. *Mémoire Service Explicite Carte géologie dét. France*, 561.

ENGI, M., BERGER, A. & ROSELLE, G. 2001a. The role of the tectonic accretion channel in collisional orogeny. *Geology*, **29**, 1143–1146.

ENGI, M., SCHERRER, N. C. & BURRI, T. 2001b. Metamorphic evolution of pelitic rocks of the Monte Rosa nappe: constraints from petrology and single grain monazite age data. *Schweizerische Mineralogische Petrographische Mitteilungen*, **81**, 305–328.

ERNST, W. G. 1971. Metamorphic Zonations on Presumably Subducted Lithospheric Plates from Japan, California and the Alps. *Contribution of Mineralogy and Petrology*, **34**, 43–59.

ERNST, W. G. 1981. Petrogenesis of eclogites and peridotites from the Western and Ligurian Alps. *American Mineralogist*, **66**, 443–472.

ERNST, W. G. & DAL PIAZ, G. V. 1978. Mineral parageneses of eclogitic rocks and related mafic schists of the Piemonte ophiolite nappe, Breuil-St. Jacques area, Italian Western Alps. *American Mineralogist*, **63**, 621–640.

ESCHER, A. & BEAUMONT, C. 1997. Formation, burial and exhumation of basement nappes at crustal scale: a geometric model based on the Western Swiss-Italian Alps. *Journal of Structural Geology*, **19**, 955–974.

FEDERICO, L., CAPPONI, G., CRISPINI, L., SCAMBELLURI, M. & VILLA, I. M. 2005. $^{39}Ar/^{40}Ar$ dating of high-pressure rocks from the Ligurian Alps: evidence for a continuous subduction–exhumation cycle. *Earth and Planetary Science Letters*, **240**, 668–680.

FRANK, W., HÖCK, V. & MILLER, C. 1987. Metamorphic and tectonic history of the central Tauern Window. *In*: FLUEGEL, H. W. & FAUPL, P. (eds) *Geodynamics of the Eastern Alps*. Deuticke, Vienna, 34–54.

FREEMAN, S. R., BUTLER, R. W. H., CLIFF, R. A., INGER, S. & BARNICOAT, T. A. C. 1998. Deformation migration in an orogen-scale shear zone array; an example from the basal Brianconnais thrust, internal Franco-Italian Alps. *Geological Magazine*, **135**, 349–367.

FREY, M. & FERREIRO-MÄHLMANN, R. 1999. Alpine metamorphism of the Central Alps. *Schweizerische Mineralogische Petrographische Mitteilungen*, **79**, 135–154.

FREY, M., DESMONS, J. & NEUBAUER, F. 1999. The new metamorphic map of the Alps. *Schweizerische Mineralogische Petrographische Mitteilungen*, **79**, 1–4.

FRISCH, W. 1979. Tectonic progradation and plate tectonics of the Alps. *Tectonophysics*, **60**, 121–139.

FROITZHEIM, N., SCHMID, S. & CONTI, P. 1994. Repeated change from crustal shortening to orogen-parallel extension in the Austroalpine units of Graubünden. *Eclogae Geologicae Helvetiae*, **87**, 559–612.

FROITZHEIM, N., SCHMID, S. & FREY, M. 1996. Mesozoic paleogeography and the timing of eclogite facies metamorphism in the Alps: a working hypothesis. *Eclogae Geologicae Helveticae*, **89**, 81–110.

GANGULY, J. & TIRONE, M. 1999. Diffusion closure temperature and age of a mineral with arbitrary extent of diffusion. *Earth and Planetary Science Letters*, **170**, 131–140.

GEBAUER, D. 1996. A P–T–t Path for a high pressure Ultramafic rock-association and their felsic country-rocks based on SHRIMP-Dating of magmatic and metamorphic Zircon domains. Example: Alpe Arami (Central Swiss Alps). *In*: HART, A. & BASU, S. R. (eds) *Earth Processes: Reading the Isotope Code*. AGU, DC, 307–328.

GEBAUER, D. 1999. Alpine Geochronology of the Central and Western Alps. *Schweizerische Mineralogische Petrographische Mitteilungen*, **79**, 191–208.

GEBAUER, D., SCHERTL, H. P., BRIX, M. & SCHREYER, W. 1997. 35 Ma old ultrahigh-pressure metamorphism and evidence for very rapid exhumation in the Dora Maira Massif, Western Alps. *Lithos*, **41**, 5–24.

GOFFÉ, B. 1977. Succession de subfacies métamorphiques en Vanoise méridionale (Savoie). *Contributions to Mineralogy and Petrology*, **62**, 23–41.

GOFFÉ, B. 1984. Le facies à carpholite-chloritoide dans la couverture briançonnaise des Alpes Ligures: un témoin de l'histoire tectono-metamorphique régionale. *Memorie della Societa Geologica Italiana*, **28**, 461–479.

GOFFÉ, B. & BOUSQUET, R. 1997. Ferrocarpholite, chloritoïde et lawsonite dans les métapelites des unités du Versoyen et du Petit St Bernard (Zone Valaisanne, Alpes occidentales). *Schweizerische Mineralogische und Petrographische Mitteilungen*, **77**, 137–147.

GOFFÉ, B. & CHOPIN, C. 1986. High-pressure metamorphism in the Western Alps: zoneography of metapelites, chronology and consequences. *Schweizerische Mineralogische und Petrographische Mitteilungen*, **66**, 41–52.

GOFFÉ, B. & OBERHÄNSLI, R. 1992. Ferro- and magnesiocarpholite in the 'Bündnerschiefer' of the eastern Central Alps (Grisons and Engadine Window). *European Journal of Mineralogy*, **4**, 835–838.

GOFFÉ, B. & VELDE, B. 1984. Contrasted metamorphic evolution in thrusted cover units of the Briançonnais zone (French Alps): a model for the conservation of HP–BT metamorphic mineral assemblages. *Earth and Planetary Science Letters*, **68**, 351–360.

GOFFÉ, B., GOFFÉ-URBANO, G. & SALIOT, P. 1973. Sur la présence d'une variété magnésienne de la ferrocarpholite en Vanoise (Alpes françaises): sa signification probable dans le métamorphisme alpin. *Comptes Rendus de l'Académie des Sciences Paris*, **277**, 1965–1968.

GOFFÉ, B., BOUSQUET, R., HENRY, P. & LE PICHON, X. 2003. Effect of the chemical composition of the crust on the metamorphic evolution of orogenic wedges. *Journal of Metamophic Geology*, **21**, 123–141.

GOFFE, B., SCHWARTZ, S., LARDEAUX, J.-M. & BOUSQUET, R. 2004. Explanatory notes to the maps: Metamorphic structure of the western Alps and Ligurian Alps. *Mitteilungen der Oesterreichen Mineralogischen Gesellschaft*, **149**, 125–144.

GOSSO, G., DAL PIAZ, G. V., PIOVANO, V. & POLINO, R. 1979. High pressure emplacement of early-alpine nappes, postnappe deformations and structural levels (internal northwestern Alps). *Memorie di Scienze Geologiche*, **32**, 1–16.

GOUZU, C., ITAYA, T., HYODO, H. & MATSUDA, T. 2006. Excess ^{40}Ar-free phengite in ultrahigh-pressure metamorphic rocks from the Lago di Cignana area, Western Alps. *Lithos*, **92**, 418–430.

HANDY, M., HERWEGH, M., KAMBER, B., TIETZ, R. & VILLA, I. M. 1996. Geochronologic, petrologic and kinematic constraints on the evolution of the Err-Platta boundary, part of a fossil continent-ocean suture in the Alps (Eastern Switzerland). *Schweizerische Mineralogische Petrographische Mitteilungen*, **76**, 453–474.

HANDY, M. & OBERHÄNSLI, R. 2004. Explanatory notes to the map: metamorphic structure of the Alps: Age map of the metamorphic structure of the Alps—tectonic interpretation and outstanding problems. *Mitteilungen der Oesterreichen Mineralogischen Gesellschaft*, **149**, 201–226.

HEINRICH, C. A. 1982. Kyanite-eclogite to amphibolite facies evolution of hydrous mafic and pelitic rocks, Adula nappe, Central Alps. *Contributions to Mineralogy and Petrology*, **81**, 30–38.

HERMANN, J., RUBATTO, D. & TROMMSDORFF, V. 2006. Sub-solidus Oligocene zircon formation in garnet peridotite during fast decompression and fluid infiltration (Duria, Central Alps). *Mineralogy and Petrology*, **88**, 181–206.

HÖCK, V. & KOLLER, F. 1989. Magmatic evolution of the Mesozoic ophiolites in Austria. *Chemical Geology*, **77**, 209–227.

HOLLAND, T. J. B. 1979. High water activities in the generation of high pressure kyanite eclogites of the Tauern Window, Austria. *Journal of Geology*, **87**, 1–27.

HOLLAND, T. J. B. & RAY, N. J. 1985. Glaucophane and pyroxene breakdown reactions in the Pennine units of the Eastern Alps. *Journal of Metamorphic Geology*, **3**, 417–438.

HOSCHEK, G. 2001. Thermobarometry of metasediments and metabasites from the Tauern Eclogite Zone of the Hohe Tauern, Eastern Alps, Austria. *Lithos*, **59**, 127–150.

HUNZIKER, J. C., DESMOS, J. & HURFORD, A. J. 1992. Thirty-two years of geochronological work in the Central and Western Alps: a review on seven maps. *Memoires de geologie Lausanne*, **13**, 1–59.

JANAK, M., FROITZHEIM, N., LUPTAK, B., VRABEC, M. & KROGH-RAVNA, E.-J. 2004. First evidence for ultrahigh-pressure metamorphism of eclogites in Pohorje, Slovenia; tracing deep continental subduction in the Eastern Alps. *Tectonics*, **23**, TC5014, doi:10.1029/2004TC001641.

JOLIVET, L., FACCENNA, C., GOFFÉ, B. ET AL. 1998. Midcrustal shear zones in postorogenic extension: Example from the northern Tyrrhenian Sea. *Journal of Geophysical Research*, **103**, 12123–12161.

JOLIVET, L., FACCENNA, C. & GOFFÉ, B. 2003. Subduction tectonics and exhumation of high-pressure metamorphic rocks in the Mediterranean orogens. *American Journal of Science*, **303**, 353–409.

KIENAST, J.-R. 1973. Sur l'existence de deux séries différentes au sein de l'ensemble des 'Schistes Lustrés-ophiolites' du Val d'Aoste, quelques arguments fondés sur les roches métalorphiques. *Comptes Rendus de l'Académie des Sciences Paris*, **276**, 2621–2624.

KOLLER, F. 1985. Petrologie und Geochemie des Penninikums am Alpenostrand. *Jahrbuch der geologischen Bundesanstalt Wien*, **128**, 83–150.

LAPEN, J. T., JOHNSON, C. M., BAUMGARTNER, L. P., MAHLEN, J. N., BEARD, B. L. & AMATO, J. M. 2003. Burial rates during prograde metamorphism of an ultra-high-pressure terrane: an example from Lago di Cignana, western Alps, Italy. *Earth and Planetary Science Letters*, **215**, 57–72.

LAPEN, J. T., JOHNSON, C. M., BAUMGARTNER, L. P., DAL PIAZ, G. V., SKORA, S. & BEARD, B. L. 2007. Coupling of oceanic and continental crust during Eocene eclogite-facies metamorphism: evidence from the Monte Rosa nappe, western Alps, Italy *Contributions to Mineralogy and Petrology*, **153**, 139–157.

LARDEAUX, J. M. & SPALLA, M. I. 1991. From Granulites to Eclogites in the Sesia Zone (Italian Western Alps)—a record of the opening and closure of the Piedmont Ocean. *Journal of Metamorphic Geology*, **9**, 35–59.

LARDEAUX, J.-M., GOSSO, G., KIENAST, J.-R. & LOMBARDO, B. 1982. Relations entre le métamorphisme et la déformation dans la zone Sésia-Lanzo (Alpes Occidentales) et le problème de l'éclogitisation de la croûte continentale. *Bulletin de la Société géologique de France*, **24**, 793–800.

LAUBSCHER, H. P. 1971a. Das Alpen-Dinariden-Problem und die Palinspastik der suedlichen Tethys. *Geologische Rundschau*, **60**, 813–833.

LAUBSCHER, H. P. 1971b. The large-scale kinematics of the Western Alps and the northern Apennines and its palinspastic implications. *American Journal of Science*, **271**, 193–226.

LAUBSCHER, H. P. 1975. Plate boundaries and microplates in Alpine history. *American Journal of Science*, **275**, 865–876.

LE BAYON, R., DE CAPITANI, C. & FREY, M. 2006a. Modelling phase-assemblage diagrams for magnesian metapelites in the system K_2O–FeO–MgO–Al_2O_3–SiO_2–H_2O: geodynamic consequences for the Monte Rosa nappe, Western Alps. *Contributions to Mineralogy and Petrology*, **151**, 395–412.

LE BAYON, B., PITRA, P., BALLÉVRE, M. & BOHN, M. 2006b. Reconstructing P–T paths during continental collision using multi-stage garnets (Gran Paradiso nappe, Western Alps). *Journal of Metamorphic Geology*, **24**, 477–496.

LE PICHON, X., BERGERAT, F. & ROULET, M.-J. 1988. Plate kinematics and tectonics leading to the Alpine belt formation: A new analysis. *In*: CLARK, S. P., BURCHFIEL, B. C. & SUPPE, J. (eds) *Processes in Continental Lithospheric Deformation*. Geological Society of America Special Paper, **218**, 111–131.

LEFÈVRE, R. & MICHARD, A. 1965. La jadéite dans le métamorphisme alpin, à propos dans les gisements de type nouveau, de la bande d'Acceglio (Alpes cottiennes, Italie). *Bulletin de la Société française de Minéralogie et de Cristallographie*, **LXXXVIII**, 664–677.

LIATI, A. & FROITZHEIM, N. 2006. Assessing the Valais Ocean, Western Alps: U–Pb SHRIMP zircon geochronology of eclogite in the Balma unit, on top of the Monte Rosa nappe. *European Journal of Mineralogy*, **18**, 299–308.

LIATI, A., FROITZHEIM, N. & FANNING, C. M. 2005. Jurassic ophiolites within the Valais Domain of the Western and Central Alps; geochronological evidence for re-rifting of oceanic crust. *Contributions to Mineralogy and Petrology*, **149**, 446–461.

LIATI, A., GEBAUER, D. & FANNING, C. M. 2003. The youngest basic oceanic magmatism in the Alps (Late Cretaceous; Chiavenna unit, Central Alps): geochronological constraints and geodynamic significance. *Contributions to Mineralogy and Petrology*, **146**, 144–158.

LIATI, A. & GEBAUER, D. 2003. Geochronological constraints for the time of metamorphism in the Gruf Complex (Central Alps) and implications for the Adula-Cima Lunga nappe system. *Schweizerische Mineralogische Petrographische Mitteilungen*, **83**, 159–172.

LOMBARDO, B., NERVO, R., COMPAGNONI, R. *ET AL.* 1978. Osservazioni preliminari sulle ofiolito metamorfiche del Monviso (Alpi Occidentali). *Rendiconti dell Società Italiana di Mineralogia e Petrologia*, **34**, 253–302.

MALUSKI, H. 1977. Application de la méthode $^{40}Ar/^{39}Ar$ aux minéraux des roches cristallines perturbées par les événements thermiques et tectoniques en Corse. *Bulletin de la Société géologique de France*, **19**, 849–855.

MANATSCHALL, G. & BERNOULLI, D. 1999. Architecture and tectonic evolution of nonvolcanic margins; present-day Galicia and ancient Adria. *Tectonics*, **18**, 1099–1119.

MARKLEY, M. J., TEYSSIER, C., COSCA, M. A., CABY, R., HUNZIKER, J.-C. & SARTORI, M. 1998. Alpine deformation and $^{40}Ar/^{39}Ar$ geochronological synkinematic white mica in the Siviez-Mischabel Nappe, western Pennine Alps, Switzerland. *Tectonics*, **17**, 407–425.

MARTHALER, M. 1984. Géologie des unités penniques entre le val d'Anniviers et le val de Tourtmagne. *Eclogae Geologicae Helvetiae*, **77**, 395–448.

MARTHALER, M. & STAMPFLI, G. M. 1989. Les schistes lustrés à ophiolites de la nappe du Tsaté: un ancien prisme d'accrétion issu de la marge active apulienne? *Schweizerische Mineralogische und Petrographische Mitteilungen*, **69**, 211–216.

MARTIN, S., TARTAROTTI, P. & DAL PIAZ, G. V. 1994. Alpine ophiolites: a review. *Bolletino Geofisica Teorica Applicatá*, **36**, 175–220.

MEFFAN-MAIN, S., CLIFF, R. A., BARNICOAT, A. C., LOMBARDO, B. & COMPAGNONI, R. 2004. A Tertiary age for Alpine high-pressure metamorphism in the Gran Paradiso massif, Western Alps: a Rb–Sr microsampling study. *Journal of Metamorphic Geology*, **22**, 267–281.

MESSIGA, B., KIENAST, J.-R., REBAY, G., RICCARDI, M. P. & TRIBUZIO, R. 1999. Cr-rich magnesiochloritoid eclogites from the Monviso ophiolites (Western Alps, Italy). *Journal of Metamorphic Geology*, **17**, 287–300.

MILLER, C. 1974. On the metamorphism of the eclogites and high-grade blueschists from the Pennine terrain of the Tauern window, Austria. *Schweizerische Mineralogische und Petrographische Mitteilungen*, **54**, 371–384.

MONIÉ, P. 1990. Preservation of Hercynian Ar-^{40}/Ar-^{39} Ages Through High-Pressure Low-Temperature Alpine Metamorphism in the Western Alps. *European Journal of Mineralogy*, **2**, 343–361.

MÜNTENER, O., HERMANN, J. & TROMMSDORFF, V. 2000. Cooling history and exhumation of lower crustal granulite and Upper Mantle (Malenco, Eastern Central Alps). *Journal of Petrology*, **41**, 175–200.

NIGGLI, E. 1978. *Metamorphic map of the Alps*. Commision of the Geological Map World.

NIMIS, P. & TROMMSDORFF, V. 2001. Revised thermobarometry of Alpe Arami and other garnet peridotites from the Central Alps. *Journal of Petrology*, **42**, 103–115.

OBERHÄNSLI, R. 1978. Chemische Untersuchungen an Glaukophan-führenden basischen Gesteinen aus den Bündnerschiefern Graubünden. *Schweizerische Mineralogische und Petrographische Mitteilungen*, **58**, 139–156.

OBERHÄNSLI, R. 1980. P-T Bestimmungen anhang von Mineralanalysen in Eklogiten und Glaukophaniten der Ophiolite von Zermatt. *Schweizerische Mineralogische und Petrographische Mitteilungen*, **60**, 215–235.

OBERHÄNSLI, R. 1986. Blue amphiboles in metamorphosed Mesozoic mafic rocks from the Central Alps. In: EVANS, B. W. & BROWN, E. H. (eds) *Blueschists, Eclogites*. Academic Press, London, 239–247.

OBERHÄNSLI, R. 1994. Subducted and obducted ophiolites of the Central Alps: Paleotectonic implications deducted by their distribution and metamorphism overprint. *Lithos*, **33**, 109–118.

OBERHÄNSLI, R., HUNZIKER, J. C., MARTINOTTI, G. & STERN, W. B. 1982. Monte Mucrone: Ein eoalpin eklogitsierter permischer Granit. *Schweizerische Mineralogische und Petrographische Mitteilungen*, **62**, 486.

OBERHÄNSLI, R., HUNZIKER, J.-C., MARTINOTTI, G. & STERN, W. B. 1985. Geochemistry, geochronology and petrology of Monte Mucrone: an example of Eo-alpine eclogitization of permian granitoids in the Sesia-Lanzo zone, Western Alps, Italy. *Chemical Geology*, **52**, 165–184.

OBERHÄNSLI, R., BOUSQUET, R., ENGI, M. *ET AL*. 2004. *Metamorphic structure of the Alps*. Commission of the Geological Map World.

OBERLI, F., MEIER, M., BERGER, A., ROSENBERG, C. & GIERE, R. 2004. ^{230}Th/^{238}U disequilibrium systematics in U-Th-Pb dating: Precise accessory mineral chronology and melt evolution tracing in the Alpine Bergell intrusion. *Geochimica Cosmochimica Acta*, **68**, 2543–2560.

OXBURGH, E. R. & TURCOTTE, D. L. 1974. Thermal Gradients and Regional Metamorphism in Overthrust Terrains with Special Reference to the Eastern Alps. *Schweizerische Mineralogische und Petrographische Mitteilungen*, **54**, 641–662.

PAQUIN, J. & ALTHERR, R. 2001. New constraints on the P-T evolution of the Alpe Arami garnet peridotite body (Central Alps, Switzerland). *Journal of Petrology*, **42**, 1119–1140.

PFIFFNER, M. A. 1999. *Genese der Hochdruckmetamorphen ozeanischen Abfolge der Cima Lunga Einheit (Zentralalpen)*. PhD ETH Zürich.

PLEUGER, J., FROITZHEIM, N. & JANSEN, E. 2005. Folded continental and oceanic nappes on the southern side of Monte Rosa (western Alps, Italy): Anatomy of a double collision suture. *Tectonics*, **24**, TC4013.

POGNANTE, U. 1989. Tectonic implications of lawsonite formation in the Sesia zone (Western Alps). *Tectonophysics*, **162**, 219–227.

POGNANTE, U. 1991. Petrological constraints on the eclogite- and blueschist-facies metamorphism and P-T-t paths in the Western Alps. *Journal of Metamorphic Geology*, **9**, 5–17.

POLINO, R., DAL PIAZ, G. V. & GOSSO, G. 1990. The Alpine Cretaceous orogeny: an accretionary wedge model based on integred statigraphic, petrologic and radiometric data. In: ROURE, F., HEITZMANN, P. & POLINO, R. (eds) *Deep Structure of the Alps*. Memoires de la Societé Géologique Suisse, **1**, 345–367.

RATSCHBACHER, L., DINGELDEYA, C., MILLER, C., HACKER, B. R. & MCWILLIAMS, M. O. 2004. Formation, subduction, and exhumation of Penninic oceanic crust in the Eastern Alps: time constraints from ^{40}Ar/^{39}Ar geochronology. *Tectonophysics*, **394**, 155–170.

REDDY, S., WHEELER, J., BUTLER, R. W. H. *ET AL*. 2003. Kinematic reworking and exhumation within the convergent Alpine Orogen. *Tectonophysics*, **365**, 77–102.

REINECKE, T. 1991. Very-high-pressure metamorphism and uplift of coesite-bearing metasediments from the Zermatt-Saas zone, Western Alps. *European Journal of Mineralogy*, **3**, 7–17.

RING, U. 1992. The Alpine geodynamic evolution of Penninic nappes in the eastern central Alps: geothermobarometric and kinematic data. *Journal of Metamophic Geology*, **10**, 33–53.

ROSELLE, G. T., THÜRING, M. & ENGI, M. 2002. MELONPIT: a finite element code for simulating tectonic mass movement and heat flow within subduction zones. *American Journal of Science*, **302**, 381–409.

RUBATTO, D. 2002. Zircon trace element geochemistry: partitioning with garnet and the link between U/Pb ages and metamorphism. *Chemical Geology*, **184**, 123–138.

RUBATTO, D. & GEBAUER, D. 1999. Eo/Oligocene (35 Ma) high-pressure metamorphism in the Gornergrat Zone (Monte Rosa, Western Alps): implications for paleogeography. *Schweizerische Mineralogische und Petrographische Mitteilungen*, **79**, 353–362.

RUBATTO, D. & HERMANN, J. 2001. Exhumation as fast as subduction? *Geology*, **29**, 3–6.

RUBATTO, D. & HERMANN, J. 2003. Zircon formation during fluid circulation in eclogites (Monviso, Western Alps); implications for Zr and Hf budget in subduction zones. *Geochimica et Cosmochimica Acta*, **67**, 2173–2187.

RUBATTO, D., GEBAUER, D. & FANNING, M. 1998. Jurassic formation and Eocene subduction of the Zermatt-Saas-Fee ophiolites: implications for the geodynamic evolution of the Central and Western Alps. *Contributions to Mineralogy and Petrology*, **132**, 269–287.

RUBATTO, D., GEBAUER, D. & COMPAGNONI, R. 1999. Dating of eclogite facies zircons: the age of Alpine metamorphism in the Sesia-lanzo Zone (Western Alps). *Earth and Planterary Science Letters*, **167**, 141–158.

SARTORI, M. & THÉLIN, P. 1987. Les schistes œillés albitiques de Barneuza (Nappe de Siviez-Michabel, Valais, Suisse). *Schweizerische Mineralogische und Petrographische Mitteilungen*, **87**, 229–256.

SCAILLET, S. 1996. Excess ^{40}Ar transport scale and mechanism in high pressure phengites: a case study from an eclogitized metabasite of the Dora Meira nappe, western Alps. *Geochemica Cosmochemcia Acta*, **60**, 1075–1090.

SCHAER, J.-P. 1959. Géologie de la partie septentrionale de l'éventail de Bagnes (entre le Val d'Hérémence et le Vaal de Bagnes, Valais, Suisse). *Archives des Sciences (Genève)*, **12**, 473–620.

SCHERER, E. E., CAMERON, K. L. & BLICHERT-TOFT, J. 2000. Lu-Hf garnet geochronology: closure temperature relative to the Sm–Nd system and the effects of trace mineral inclusions. *Geochemica Cosmochemcia Acta*, **64**, 3413–3432.

SCHMID, S. M., FUGENSCHUH, B., KISSLING, E. & SCHUSTER, R. 2004*a*. Tectonic map and overall architecture of the Alpine orogen. *Eclogae Geologicae Helvetiae*, **97**, 93–117.

SCHMID, S. M., FÜGENSCHUH, B., KISSLING, E. & SCHUSTER, R. 2004*b*. TRANSMED Transects IV, V and VI: Three lithospheric transects across the Alps and their forelands. *In*: CAVAZZA, W., ROURE, F., SPAKMAN, W., STAMPFLI, G. M. & ZIEGLER, P. A. (eds) *The TRANSMED Atlas: the Mediterranean Region from Crust to Mantle*. CD-ROM.

SCHMID, S. M. & KISSLING, E. 2000. The arc of the western Alps in the light of geophysical data on deep crustal structure. *Tectonics*, **19**, 62–85.

SCHMID, S. M., PFIFFNER, A., FROITZHEIM, N., SCHÖNBORN, G. & KISSLING, N. 1996. Geophysical-geological transect and tectonic evolution of the Swiss-Italian Alps. *Tectonics*, **15**, 1036–1064.

SCHMID, S. M., RÜCK, P. & SCHREURS, G. 1990. The significance of the Schams nappes for the reconstruction of the paleotectonic and orogenic evolution of the penninic zone along the NFP 20 East Traverse. *In*: ROURE, F., HEITZMANN, P. & POLINO, R. (eds) *Deep Structure of the Alps*, **1**, 263–287.

SCHÜRCH, M. L. 1987. *Les ophiolites de la zone du Versoyen: témoin d'un bassin à évolution métamorphique complexe*. PhD Université Geneve.

SCHWARTZ, S., LARDEAUX, J.-M., GUILLOT, S. & TRICART, P. 2000*a*. Diversité du métamorphisme éclogitique dans le massif ophiolitique du Monviso (Alpes occidentales, Italie). *Geodynamica Acta*, **13**, 169–188.

SCHWARTZ, S., LARDEAUX, J.-M. & TRICART, P. 2000*b*. La zone d'Acceglio (Alpes cottiennes): un nouvel exemple de croûte continentale éclogitisée dans les Alpes occidentales. *Comptes Rendus de l'Académie des Sciences Paris*, **320**, 859–866.

SELVERSTONE, J., SPEAR, F. S., FRANZ, G. & MORTEANI, G. 1984. High-pressure metamorphism in the SW Tauern Window, Austria; P–T paths from hornblende-kyanite-staurolite schists. *Journal of Petrology*, **25**, 501–531.

SIMON, G. & CHOPIN, C. 2001. Enstatite-sapphirine crack-related assemblages in ultrahigh-pressure pyrope megablasts, Dora-Maira Massif, Western Alps. *Contributions to Mineralogy and Petrology*, **140**, 422–440.

SKORA, S., BAUMGARTNER, L. P., MAHLEN, N. J., JOHNSON, C. M., PILET, S. & HELLEBRAND, E. 2006. Diffusion-limited REE uptake by eclogite garnets and its consequences for Lu–Hf and Sm–Nd geochronology. *Contribution of Mineralogy and Petrology*, **152**, 703–720.

SPALLA, M. I., LARDEAUX, J.-M., DAL PIAZ, G. V. & GOSSO, G. 1991. Métamorphisme et tectonique à la marge externe de la zone Sesia-Lanzo (Alpes occidentales). *Memorie di Scienze Geologiche*, **43**, 361–369.

SPALLA, I. M., LARDEAUX, J. M., DAL PIAZ, G. V., GOSSO, G. & MESSIGA, B. 1996. Tectonic significance of Alpine eclogites. *Journal of Geodynamics*, **21**, 257–285.

SPEAR, F. S. & FRANZ, G. 1986. P–T evolution of metasediments from the Tauern Eclogite Zone, south-central Tauern window, Austria. *Lithos*, **19**, 219–234.

SPERLICH, R. 1988. The transition from crossite to actinolite in metabasites of the Combin unit in Vallée St Barthélemy (Aosta, Italy). *Schweizerische Mineralogische und Petrographische Mitteilungen*, **68**, 215–224.

STAMPFLI, G. M. 1993. Le Briançonnais, terrain exotique dans les Alpes? *Eclogae Geologicae Helvetiae*, **86**, 1–45.

STAMPFLI, G. M. & BOREL, G. D. 2002. A plate tectonic model for the Paleozoic and Mesozoic constrained by dynamic plate boundaries and restored synthetic oceanic isochrons. *Earth and Planetary Science Letters*, **196**, 17–33.

STAMPFLI, G. M. & MARCHANT, R. H. 1997. Geodynamic evolution of the Tethyan margins of the Western Alps. *In*: PFIFFNER, O. A., LEHNER, P., HEITZMANN, P., MUELLER, S. & STECK, A. (eds) *Deep Structure of the Swiss Alps*. Birkhaüser AG, Basel, 223–239.

STAMPFLI, G. M., MOSAR, J., MARQUER, D., MARCHANT, R., BAUDIN, T. & BOREL, G. 1998. Subduction and obduction processes in the Swiss Alps. *Tectonophysics*, **296**, 159–204.

STAMPFLI, G. M., BOREL, G. D., MARCHANT, R. & MOSAR, J. 2002. Western Alps geological constraints on western Tethyan reconstructions. *Journal of the Virtual Explorer*, **8**, 77–106.

STEINMANN, M. C. 1994. *Die nordpenninischen Bündnerscheifer der Zentralalpen Graubündens: Tektonik, Stratigraphie und Beckenentwicklung*, PhD Thesis ETH Zürich.

STÖCKHERT, B. & GERYA, T. V. 2005. Pre-collisional high pressure metamorphism and nappe tectonics at active continental margins: a numerical simulation. *Terra Nova*, **17**, 102–110.

STUCKI, A., RUBATTO, D. & TROMMSDORFF, V. 2003. Mesozoic ophiolite relics in the Southern Steep Belt of the Alps. *Schweizerische Mineralogische und Petrographische Mitteilungen*, **83**, 285–300.

STÜNITZ, H. 1989. *Partitioning of metamorphism and deformation in the boundary region of the 'Seconda Zona Diorito-Kinzigitica', Sesia Zone, Western Alps*. PhD Thesis ETH Zürich.

TALERICO, C. 2001. *Petrological and chemical investigation of a metamorphosed oceanic crust-mantle section (Chiavenna, Bergell Alps)*. PhD Thesis ETH Zürich.

TAPPONNIER, P. 1977. Evolution tectonique du système alpin en Méditerranée: poinçonnement et écrasement rigide-plastique. *Bulletin de la Société géologique de France*, **29**, 437–460.

THÖNI, M. 2006. Dating eclogite-facies metamorphism in the Eastern Alps—approaches, results, interpretations: a review. *Mineralogy and Petrology*, **88**, 123–148.

THÖNI, M. & MILLER, C. F. 1996. Garnet Sm–Nd data from the Saualpe and the Koralpe (Eastern Alps, Austria): chronological and P–T constraints on the thermal and tectonic history. *Journal of Metamophic Geology*, **14**, 453–466.

TILTON, G. R., SCHREYER, W. & SCHERTL, H. P. 1991. Pb–Sr–Nd isotopic behavior of deeply subducted crustal rocks from the Dora Maira massif, Western Alps, Italy: II. What is the age of the ultrahigh pressure metamorphism? *Contribution of Mineralogy and Petrology*, **108**, 22–33.

TRÜMPY, R. 1975. On crustal subduction in the Alps. *In*: MAHEL, M. (ed.) *Tectonic Problems of the Alpine System*, 121–130.

TRÜMPY, R. 1982. Alpine paleogeography: a rappraisal. *In*: HSÜ, K. (ed.) *Mountain Building Processes*.

VAGGELLI, G., BORGHI, A., COSSIO, R. ET AL. 2006. Micro-PIXE Analysis of Monazite from the Dora Maira Massif, Western Italian Alps. *Microchimica Acta*, **155**, 305–311.

VENTURINI, G. 1995. Geology, geochemistry and geochronology of the inner central Sesia Zone (Western Alps, Italy). *Mémorie Geologic Lausanne*, **25**, 1–147.

VIGNAROLI, G., ROSSETTI, F., BOUYBAOUENE, M., MASSONNE, H.-J., THEYE, T., FACCENNA, C. & FUNICIELLO, R. 2005. A counter-clockwise P–T path for the Voltri Massif eclogites (Ligurian Alps, Italy). *Journal of Metamorphic Geology*, **23**, 533–555.

VILLA, I. M. 1998. Isotopic Closure. *Terra Nova*, **10**, 42–47.

VILLA, I. M., HERMANN, J., MÜNTENER, O. & TROMMSDORFF, V. 2000. $^{39}Ar/^{40}Ar$ dating of multiply zoned amphibole generations (Malenco, Italian Alps). *Contribution of Mineralogy and Petrology*, **140**, 363–381.

VUICHARD, J. P. & BALLÈVRE, M. 1988. Garnet-chloritoid equilibria in eclogitic pelitic rocks from the Sesia Zone (Western Alps); their bearing on phase relations in high pressure metapelites. *Journal of Metamorphic Geology*, **6**, 135–157.

WIEDERKEHR, M. 2004. *Prograde and retrograde evolution of HP-metasediments in relation with the deformation history (Avers, Switzerland)*. Unpublished Diploma thesis, Basel.

ZIMMERMANN, R., HAMMERSCHMIDT, K. & FRANZ, G. 1994. Eocene High Pressure Metamorphism in the Penninic Units of the Tauern Window (Eastern Alps)—Evidence from Ar^{-40}-Ar^{-39} Dating and Petrological Investigations. *Contributions to Mineralogy and Petrology*, **117**, 175–186.

Tectonic evolution of the northwestern Internal Dinarides as constrained by structures and rotation of Medvednica Mountains, North Croatia

BRUNO TOMLJENOVIĆ[1], LÁSZLÓ CSONTOS[2], EMŐ MÁRTON[3] & PÉTER MÁRTON[4]

[1]*University of Zagreb, Faculty of Mining, Geology and Petroleum Engineering, Pierottijeva 6, HR-10000 Zagreb, Croatia (e-mail: bruntom@rgn.hr)*

[2]*Eötvös Loránd University, Geology Department 1117 Budapest, Pázmány P s 1/c, Hungary/MOL PLC 1117 Budapest Október 23 u 18. Hungary*

[3]*Eötvös Loránd Geophysical Institute of Hungary, Palaeomagnetic Laboratory, 1145 Budapest, Columbus 17–23, Hungary*

[4]*Eötvös Loránd University, Geophysics Department, 1117 Budapest, Pázmány Péter sétány 1/c, Hungary*

Abstract: This paper attempts to explain the tectonic history and possible reasons for the change of trend of the northwestern part of the Internal Dinarides in a transitional area between the Southeastern Alps, central Dinarides and Tisia, north of Zagreb. Structural and palaeomagnetic data collected in pre-Neogene rocks at Medvednica Mountains, combined with palaeomagnetic data available from Neogene rocks in the surrounding area, point to the following conclusions:

(1) The reason for dramatic deflection in structural trend of the Internal Dinarides in the area north of Zagreb is a 130° clockwise rotation and eastward escape of a tectonic block comprising Medvednica Mountains and the surrounding inselbergs, bounded to the north by the easternmost tip of the Periadratic Lineament. In Medvednica Mountains, the main period of tectonic escape and associated clockwise rotation occurred in the Late Palaeogene, possibly in the Oligocene–earliest Miocene.

(2) When rotated into the original position, the trend of observed pre-Neogene structures of Medvednica Mountains becomes parallel to the major structural trend of the central Dinarides. In view of their original orientation, these structures are interpreted in the following way:

(a) The first D1 deformational event is attributed to the Aptian–Albian nappe stacking in the central–northern Dinarides that was accommodated by a top-to-the-north directed shearing and northward propagation of already obducted ophiolites of the Central Dinaridic ophiolite zone. This nappe stacking, which resulted in a weak regional metamorphism in tectonic units underlying the ophiolites, was orogen-parallel or at a very acute angle to known structural (and possibly palaeogeographic) trends. This implies a major left-lateral shear component along the former Adriatic margin and obducted Dinaridic ophiolite zone.

(b) This was followed by Early Albian orogen-perpendicular shortening (D2) that was accommodated by folding and top-to-the-west thrusting. This deformation resulted in gradual cooling of the metamorphic stack and also in uplift and erosion of the higher structural units.

(c) The D3 deformational event was driven by renewed E–W shortening that took place after the Paleocene, most probably during the Middle Eocene–Oligocene, i.e. synchronous with the main Dinaridic tectonic phase of the External Dinarides. This shortening was probably triggered by collision and thrusting of Tisia over the northern segment of the Internal Dinarides.

(d) This was finally followed by D4 pervasive, right-lateral N–S shearing that is tentatively interpreted as being related to the right-lateral shearing of the Sava zone during the Eocene–Oligocene.

(e) Following the main period of tectonic escape and induced clockwise rotation along the Periadriatic fault, possibly in the Oligocene–earliest Miocene, the Medvednica Mountains and the surrounding area were affected by repeated extensions and inversions since the Early Miocene to recent times. Palaeomagnetic data suggest that in the Early Miocene (but probably before the Karpatian) this area was part of a regional block that shifted northwards and rotated in a counter-clockwise sense. A second episode of counter-clockwise rotation occurred at the present latitude in post-Pontian times (since *c.* 5 Ma), driven by the counter-clockwise rotating Adriatic Plate.

From: SIEGESMUND, S., FÜGENSCHUH, B. & FROITZHEIM, N. (eds) *Tectonic Aspects of the Alpine-Dinaride-Carpathian System.* Geological Society, London, Special Publications, **298**, 145–167.
DOI: 10.1144/SP298.8   0305-8719/08/$15.00 © The Geological Society of London 2008.

The fold-thrust belt of the Dinarides in its northwestern and central parts (i.e. in the area of central–northern Bosnia and Herzegovina and northern Croatia) is classically subdivided into two tectonic domains of External and Internal Dinarides, bounded by the southeastern Alps and Tisia to the north and northeast, respectively (Fig. 1; see also Schmid et al. 2004).

The External Dinarides are composed of tectonic units derived from the eastern part of the Adriatic (or Apulian) microplate, which are largely composed of Mesozoic to Tertiary shallow-marine carbonate platform formations (e.g. Vlahović et al. 2005 and references therein). In this part of the belt, the main deformational phase (known as a 'Dinaridic phase') resulted in recently quite distinctive NW trend and SW vergence of km-scale compressional and imbricated structures (e.g. Tari-Kovačić & Mrinjek 1994; Blašković 1998). The age of this deformational phase is generally considered as Middle Eocene–Oligocene, being constrained by a transition from a carbonate platform into a flysch sedimentation that took place during the Middle Eocene and that locally continued up to the Early Oligocene, or even up to the Early Miocene (e.g. Marjanac 1990; Marjanac & Ćosović 2000; Vlahović et al. 2005). In the NW sector of the belt, these Dinaridic structures are strongly overprinted by Neogene to recent W-trending, S-verging folds and thrusts (e.g. Jamičić et al. 1995; Blašković 1998), which are the most distinctive deformational structures further north in the area of the adjacent Southern Alps (e.g. Doglioni & Bosellini 1987; Castellarin et al. 1992; Schönborn 1999). Thus, the boundary between the External Dinarides and the Southern Alps is not a sharp one and could be only arbitrarily placed within a transitional area in between, where prevalence of Neogene to recent top-to-the-south thrusting of the Southern Alps changes into Eocene–Oligocene top-to-the-southwest thrusting of the External Dinarides (e.g. Schmid et al. 2004).

The Internal Dinarides of central–northern Bosnia and Herzegovina and northern Croatia are comprised of the following tectonic zones: (1) the Bosnian flysch zone; (2) the zone composed of non- to low-grade metamorphic units derived from a distal Adriatic Plate margin involved in the Late Jurassic ophiolite obduction; (3) the Central Dinaridic ophiolite zone (CDOZ); and (4) the Sava zone or Sava-Vardar zone sensu Pamić (2002) located in between CDOZ and Tisia (Fig. 1). CDOZ basically represents a Middle–Late Jurassic to pre-Late Cretaceous tectonic assemblage that from base to top includes: (1) an ophiolitic mélange of Jurassic age (Pamić et al. 1998, 2002a; Babić et al. 2002); (2) slivers of ophiolites (peridotites and mafic extrusives) largely exposed in central-northern Bosnia and Herzegovina (e.g. Lugović et al. 1991; Pamić et al. 1998, 2002a); and (3) Upper Jurassic to Lower Cretaceous covering sequences. Slivers of ophiolites were derived from oceanic domain(s) neighbouring the Adriatic Plate in the east–northeast during Triassic–Jurassic times (i.e. from the Dinaridic branch of the Neotethys that presumably extended to the north into the Meliata Ocean; e.g. Dercourt et al. 1993; Halamić & Goričan 1995; Halamić et al. 1998; Babić et al. 2002; Schmid et al. 2004 with references therein).

The main tectonic phases of CDOZ are thought to be related to: (1) the Middle–Late Jurassic obduction of ophiolites over the mélange and the eastern Adriatic Plate margin as constrained by: (i) Early–Middle Jurassic (Bajocian) biostratigraphic age of ophiolitic mélange matrix (Babić et al. 2002), and (ii) by a Late Jurassic (Tithonian) shallow-marine carbonate sedimentation on top of already obducted peridotites (Strajin et al. 1977; Dimitrijević & Dimitrijević 1973); (2) the Early Cretaceous (120–110 Ma) low-grade to greenschist facies metamorphism in units underlying the ophiolites, possibly due to continued obduction over the eastern Adriatic margin (Tomljenović 2002); and (3) the Late Aptian–Early Albian compression documented by folding and thrusting of ophiolites, underlying metamorphics and their Jurassic–Lower Cretaceous covering sequences (Tomljenović 2002). This structural assemblage of CDOZ is unconformably overlain by Senonian Gosau-type succession of conglomerates, sandstones, shales and marls of Campanian age, followed by Maastrichtian and Paleocene–Eocene turbidite deposits largely exposed in inselbergs along and south of the Sava River valley between Zagreb in Croatia and Banja Luka in Bosnia and Herzegovina (e.g. Babić et al. 1976; Jelaska 1978; Crnjaković 1980, 1987). During the Eocene, these Upper Cretaceous–Palaeogene cover formations together with their underlying basement units of CDOZ were strongly affected by the so-called Dinaridic phase of shortening (Tomljenović 2002) that resulted in conspicuous NW–SE trend and SW-vergence of CDOZ structures in this part of the belt.

In the Sava zone (SZ), which is interpreted as a Cretaceous–Palaeogene suture zone between the Dinarides and Tisia (Pamić 2002), the Cretaceous–Palaeogene succession of conglomerates, sandstones and marls that grades into siliciclastic to carbonate flysch series is one of the most distinctive litostratigraphic units (see Pamić 2002 and Pamić et al. 2002b for a more extensive overview of SZ with references therein). Here, however, a part of this unit experienced a regional metamorphic overprint of very low- to medium-grade conditions (Pamić et al. 1992) of presumably Eocene age

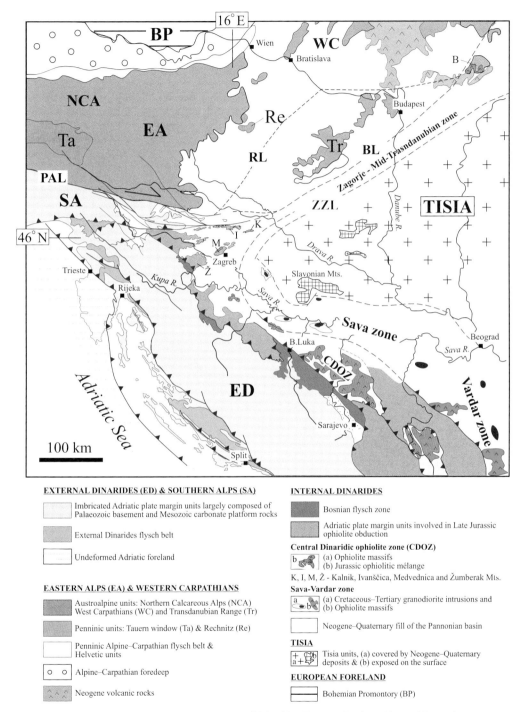

Fig. 1. Map showing major tectonic zones of the central Dinarides, Eastern–Southern Alps and Pannonian area (modified after Tomljenović 2002 and compiled from Aubouin et al. 1970; Csontos et al. 1992; Pamić et al. 1998; and Schmid et al. 2004). Major tectonic boundaries: PAL, Periadriatic Lineament; RL, Raba Line; BL, Balaton Lineament; ZZL, Zagreb–Zemplen Lineament.

(Lanphere & Pamić 1992). Most likely this metamorphism is related to thrusting of Tisia over the northern segment of the Internal Dinarides (Tari & Pamić 1998; Pamić et al. 2002b; Ustaszewski et al. 2005). According to Ustaszewski et al. (2005), this overthrusting generated N–S to NE–SW shortening that in turn resulted in NW-trending folds observed in Palaeogene flysch sediments of the SZ.

Hence, the well-pronounced NW trend of both External and Internal Dinarides in the central part of the belt is the result of Palaeogene shortening. However, in the area north of Zagreb (i.e. in Hrvatsko Zagorje area), the NW trend of CDOZ and SZ dramatically changes into NE–SW or even to E–W, with further eastward extension of these zones into Hungary (Fig. 1). There, distinctive units of CDOZ are well documented in boreholes and on the surface (e.g. Haas et al. 2000; Haas & Kovács 2001 and references therein), concentrated along the Zagorje–Mid-Transdanubian zone (ZMTZ in Fig. 1; Haas et al. 1995; Pamić & Tomljenović 1998). This zone is a major Oligo–Pliocene shear zone of the Intra-Carpathian area bounded to the north by the Periadriatic–Balaton fault system (Fodor et al. 1998; Csontos et al. 2005 and references therein) and to the SSE by the Zagreb–Zemplén or Mid-Hungarian Line (see Csontos & Nagymarosy 1998 and references therein). The sharp deviation of trend of the Internal Dinarides is also seen in the topography of inselbergs north of Zagreb, in their oblique to perpendicular orientation with respect to the rest of the Dinarides (Figs 1 and 2). These inselbergs, the largest of which are Medvednica, Kalnik and Ivanščica Mountains, are largely composed of distinctive CDOZ units (Babić et al. 2002 with references). They constitute a triangular-shaped block bounded from the north by the Periadriatic–Balaton fault system and to the SE by the Zagreb–Zemplén or

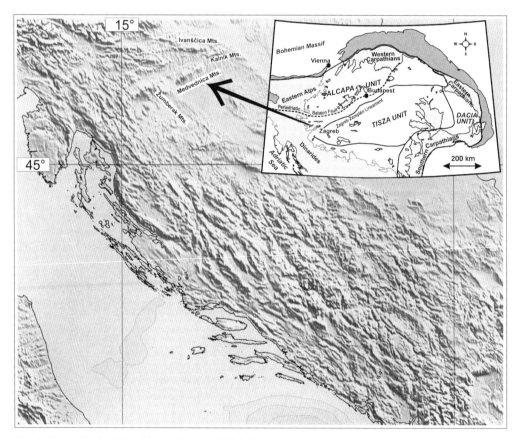

Fig. 2. Geographic location of the study area. The digital topographic model shows predominantly NW–SE orientated topographic trends in the Dinarides. Note the different NE–SW to E–W orientated trend of topography of Medvednica, Kalnik and Ivanščica Mountains in Hrvatsko Zagorje area. Insert shows location of the study area within a simplified Dinaridic–Carpathian tectonic framework. Black arrow points to Medvednica Mountains.

Mid-Hungarian Line, thus occupying the south-westernmost tip of ZMTZ.

The aim of this paper is to explain the tectonic history of the NW part of the Internal Dinarides and the possible reasons for the change of trend of Internal Dinaridic units in the Hrvatsko Zagorje area, exemplified by the structural evolution of Medvednica Mountains, the largest inselberg of this area. There, distinctive CDOZ units are the most completely preserved and exposed. Our study is based on: (1) structural analysis of deformation styles and tectonic transport/shear directions in CDOZ units and in their Senonian–Palaeogene cover exposed on Medvednica Mountains; and (2) palaeomagnetic measurements in Senonian–Palaeogene cover formations of Medvednica Mountains. We believe that the tectonic evolution proposed here for Medvednica Mountains is not only locally valid, but could be applied to the wider area of NW Internal Dinarides, in particular to the area of Hrvatsko Zagorje north of Zagreb. However, we are aware of the fact that the proposed kinematic scheme is only the first attempt for a more comprehensive tectonic scenario, and thus it has to be tested in other inselbergs of this area. After a short overview on major tectonic units of Medvednica Mountains, this paper presents a sequence of pre-Neogene deformational events (phases) constrained by documented sets of penetrative micro- to mesoscale structures that is followed by presentation of palaeomagnetic measurements and their results. These data are used for structural interpretation and reconstruction of the original position of Medvednica Mountains together with the surrounding inselbergs during the Palaeogene. Finally, this is followed by interpretation of deformational events and discussion about the possible cause and scale of clockwise rotation of Medvednica Mountains in the context of the tectonic evolution of the northwestern part of the Internal Dinarides.

Geology of Medvednica Mountains

The Medvednica Mountains expose regional-scale CDOZ tectonic units, unconformably overlain by Senonian–Paleocene and Neogene formations (Fig. 3). The following tectonic units can be differentiated, described below in their original structural order from bottom to top: (1) Eoalpine (122–110 Ma) anchi-epimetamorphic parautochthonous unit of Palaeozoic and Triassic clastic, carbonate, chert and basaltic protoliths (Belak et al. 1995; Judik et al. 2004; Lugović et al. 2006 and references therein); (2) non- to anchi-metamorphic (Judik et al. 2004) chaotic assemblage of ophiolite fragments (basalt, gabbro, serpentinite and diabase), greywackes, radiolarites and limestones, embedded in a sheared shaly-silty matrix of Early Jurassic to Bajocian age (Babić et al. 2002). This nappe stack is unconformably covered by (3) the Senonian–Paleocene (Gosau-type) sequence composed of coarse- to fine-grained clastics and sporadically rudist reef limestones at its base with proximal to distal turbidites higher upwards (Babić et al. 1973; Crnjaković 1987). (4) At the SW edge of Medvednica Mountains, a thick Triassic shallow marine succession dominated by platform carbonates of the Žumberak nappe is thrust upon the Senonian–Paleocene sequence, thus occupying the uppermost pre-Neogene structural position in this part of the mountains. (5) The Miocene (Ottnangian–Pontian) fill of the Pannonian basin unconformably overlies all the above described units (Šikić et al. 1977; Basch 1995). It is moderately tilted along the SE slope of the mountains. In contrast, along the NW foothills, the Late Miocene strata are frequently overturned and thrust by older units (Fig. 3). A seismic section roughly perpendicular to the topographic trend of the mountain indicates that even Plio–Quaternary strata might be incorporated in overthrusts (Tomljenović & Csontos 2001).

All the main structures of Medvednica Mountains are striking NE–SW (Fig. 3). A penetrative SE-dipping cleavage associated to NW-verging folding affected first the Palaeozoic–Mesozoic rocks of the parautochthonous and ophiolitic mélange units, and then their Senonian–Paleocene cover. Meso- to map-scale folds are parallel to a strong NE-trending stretching lineation in metamorphic rocks that pre-dates folding. Observed deformational styles, orientation of structures, inferred tectonic transport directions and proposed sequence of deformational events are described in detail in the following chapter.

Pre-Neogene structures, tectonic transport directions and timing

Based mostly on micro- to mesoscopic structural observations in Medvednica Mountains several structural events with different tectonic transport directions and metamorphic grades are differentiated (Fig. 4; Tomljenović 2002) and described below in chronological order. Tectonic transport directions were determined in metamorphic rocks as well as in brittle shear zones observed in Senonian–Paleocene rocks.

The oldest deformational event (D0), although not directly evident in Medvednica Mountains, is considered to be related to an ophiolite nappe emplacement over the Adriatic continental margin. This event, still not directly observed in microstructures, is possibly of Middle–Late

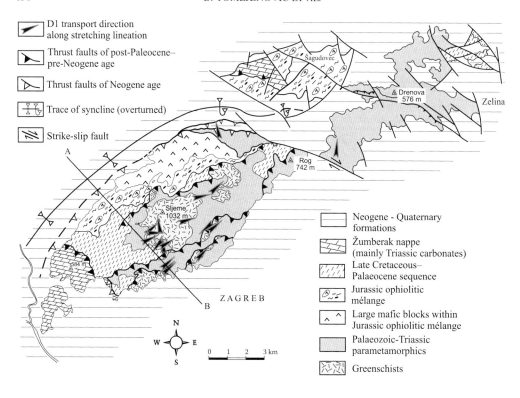

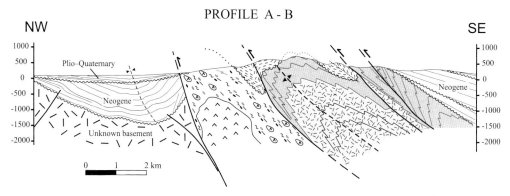

Fig. 3. Simplified geological map and cross-section of Medvednica Mountains (based on data from Šikić et al. 1977 and Basch 1995). Dark arrowheads indicate observed shear directions during the D1 deformational event. Cross-section is from Tomljenović (2002).

Jurassic age as suggested by Early–Middle Jurassic (Bajocian) biostratigraphic age of ophiolitic mélange matrix (Babić et al. 2002) and by Tithonian overstep sequences documented in the central part of the Internal Dinarides (Strajin et al. 1977; Dimitrijević & Dimitrijević 1973; Robertson & Karamata 1994; Pamić et al. 1998).

The D1 is characterized by a mylonitic shear fabric with a penetrative stretching lineation (L1) and boudinage of more competent 'layers' observed in the parautochthonous metamorphic unit of Medvednica Mountains. Observed shear-sense indicators (e.g. C'-type shear bands, σ-type porphyroclasts, mica-fish and displacement-controlled fibres in pressure shadows; Tomljenović 2002, 2005) point to a top-to-the-northeast tectonic transport direction in a modern coordinate system (Fig. 5). This mylonitization most probably occurred under peak metamorphic conditions as indicated by syn-kinematic growth of chlorite, epidote, zoisite and calcite

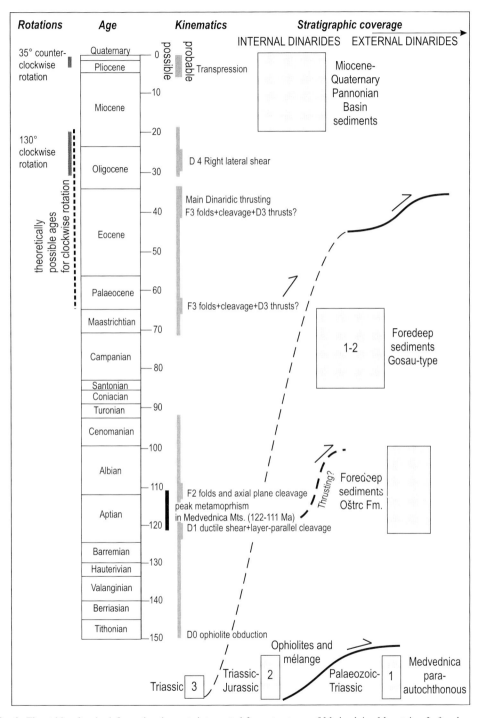

Fig. 4. Timetable of main deformational events interpreted from structures of Medvednica Mountains. Left column ('Kinematics') indicates possible ages of deformation; right column indicates probable ages. Numbering of deformational events refers to those listed in the text.

found in pressure-shadows (Fig. 5). Consequently, the possible age of this deformational event is Early Cretaceous as constrained by 122–110 Ma K–Ar ages on two whole rock samples from greenschists and three muscovite/phengite concentrates from metasediments (Belak *et al.* 1995).

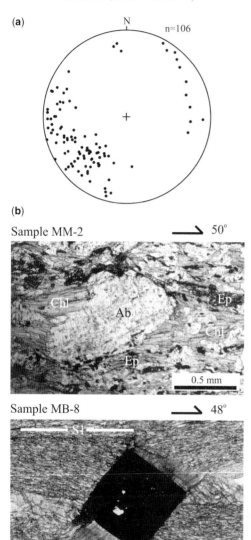

Fig. 5. (**a**) Equal-area, lower-hemisphere stereographic representation of L1 stretching lineation in the parautochthonous metamorphic unit of the central part of Medvednica Mountains. (**b**) Shear-sense indicators in greenschists and metacarbonates from the same unit indicating dextral (top-NE) shear direction. Top: Pressure-shadows of chlorite (Chl) around albite (Ab) porphyroclast in greenschist (Ep is epidote). Below: Pressure shadows of calcite fibrous overgrowths around pyrite. Both sections are normal to S1 and parallel to L1.

The next deformational event (D2) is characterized by asymmetric folding and a general penetrative cleavage. Micro- to mesoscale, tight to isoclinal folds with SE-dipping axial planes developed under relatively hot conditions, accompanied by axial plane cleavage (S2) and intersection lineation (L2) parallel to fold axes (F2) (Fig. 6). In the modern coordinate system, these ductile structures indicate a top-to-the-northwest directed shear. Since Senonian deposits unconformably cover different metamorphic rocks (Fig. 3), this folding must have preceded Senonian deposition. A likely age for this deformational event would be Late Aptian–Early Albian, i.e. following the peak metamorphic conditions, but still in temperature conditions favourable to ductile deformation.

D3 is best recorded in sediments of Senonian and Paleocene age (i.e. Gosau-type cover). It is characterized by NNE-trending and NW-vergent open to tight folds, sometimes associated with weak to penetrative axial plane cleavage (Fig. 7). The cleavage is best developed in marly or shaly lithologies and in places of more intense deformation such as in the footwall of the major D3 thrusts. Locally, these thrusts emplace the metamorphic basement above the Senonian–Paleocene succession (Fig. 8). Almost as a rule, the mesoscale D3 folds are found isolated and cut by reverse faults. Most frequently, these faults are developed in sets characterized by imbrication and sigmoidal drag folding of S_0 and S_3, local boudinage, Riedel shears and *en échelon* tension veins. All these structures clearly indicate a progressive non-coaxial shear with top-to-the-northwest tectonic transport direction (Fig. 8), i.e. generally perpendicular to F3 fold axes. Altogether, these structural elements are interpreted as an indication of progressive compressional deformation starting first with F3 flexural folding, followed by S3 cleavage and finally by NW-directed thrusting.

As described above, both the metamorphic and the (originally) overlying non-metamorphic Senonian–Paleocene succession are tightly folded with parallel fold axes and roughly parallel axial surfaces (compare Figs 6 and 7). Therefore it is debated whether these structures originate from one or several deformational events. The style of folding differs in metamorphic and Senonian sequences, being more ductile in metamorphic rocks. Moreover, a reconstruction of the Senonian depositional surface indicates that the Medvednica metamorphic rocks were already folded together with the originally overlying ophiolite mélange unit (see cross-section of Fig. 3). Comparing the stereoplots of S2 and S3 cleavages, a dispersion of the former is observed, while the latter are more grouped (Figs 6 and 7). This situation is anticipated in the case of an older deformation, refolded by a

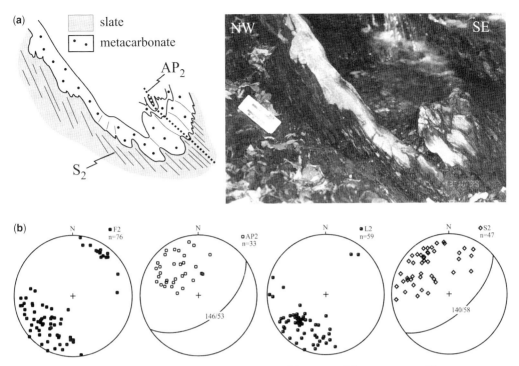

Fig. 6. (a) Mesoscale D2 folds and cleavage in metasediments of the parautochthonous metamorphic unit of the central part of Medvednica Mountains. (b) Equal-area, lower-hemisphere stereographic representation of F2 fold axes, AP2 axial planes, L2 cleavage/layering intersection lineation and S2 axial plane cleavage.

younger one. Thus, all these arguments suggest that there was a first phase of folding in pre-Senonian units, followed by a post-Paleocene folding of identical main shortening direction. The age of D3 deformational event is post-Paleocene–pre-Early Miocene, most probably Eocene as D3 thrusts are sealed by Lower Miocene strata.

The D4 event of Medvednica Mountains is characterized by a pervasive dextral strike-slip faulting with slickenlines on practically all previous, predominantly NE-striking surfaces, such as bedding, cleavage and D3 shear planes (Fig. 9). The stereoplots show that the slicks are somewhat oblique and not purely horizontal, and subvertical motion planes are missing. Nevertheless, all observed motions suggest right-lateral shear along pre-existing NE-striking surfaces weak enough to have reacted to strike-slip stresses without creating new (ideally vertical) surfaces. The possible age of this deformation event is post-Eocene–pre-Early Miocene (pre-Ottnangian) as it overprints shear planes of D3 deformation and does not affect the

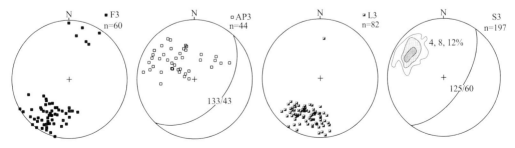

Fig. 7. Equal-area, lower-hemisphere stereographic representation of F3 fold axes, AP3 axial planes, L3 cleavage/bedding intersection lineation and S3 axial plane cleavage in Senonian–Paleocene sequence of the central part of Medvednica Mountains.

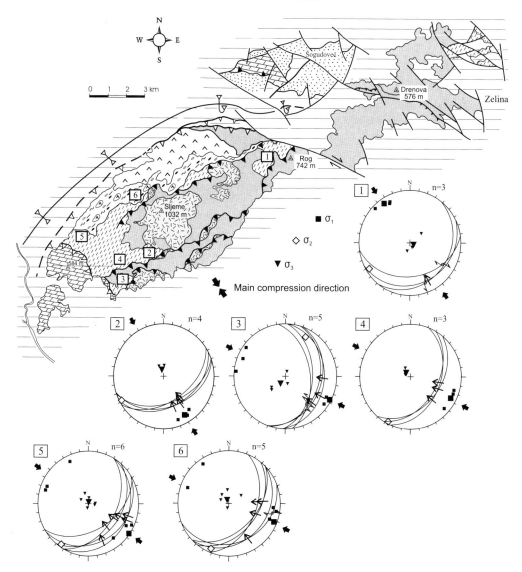

Fig. 8. Map showing orientation and location of major D3 thrusts (see Fig. 3 for legend). Stereoplots are equal-area, lower-hemisphere projections showing fault planes (traces) and slickenlines with observed character of offset indicated by arrows. Inward-pointing arrows indicate thrusts. Orientations of principal stress directions are calculated by numerical dynamic analysis method of Sperner (1996). Numbers in squares indicate location of measurements.

Lower Miocene (Ottnangian) or younger Miocene cover of Medvednica Mountains.

Summarizing the structural data, almost all the observed structural features fit well into the Dinaridic structural evolution. However, neither of the tectonic transport directions nor structural trends documented on Medvednica Mountains can be evaluated in their present positions: they strongly differ from their Dinaridic counterparts. It has to be mentioned here that in the neighbouring Žumberak Mountains, located just tens of kilometres SW of Medvednica Mountains (Figs 1 and 2), the ophiolitic mélange and the Upper Cretaceous and Palaeocene sequences are also exposed (Pleničar et al. 1975; Šikić et al. 1977; Devidé-Neděla et al. 1982). The latter are found there deformed in similar deformational styles as in Medvednica Mountains, characterized by open to tight folds dissected by thrusts, slip features including bedding-parallel slip and shear-related cleavage (Tomljenović 2002). However, in

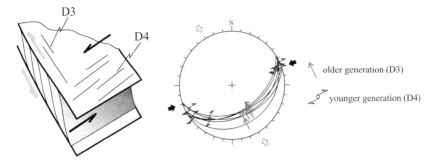

Fig. 9. Example to show D4 shear superimposed on earlier D3 shear surface. Stereoplot and palaeostress axes as on Figure 7. Double arrow indicates strike-slip shear.

contrast to structures observed on Medvednica Mountains, shear indicators in Upper Cretaceous strata in Žumberak Mountains clearly point to top-to-the-west-southwest tectonic transport direction (Prtoljan 2001; Tomljenović 2002), which is also suggested by the geometry and the NW trend of nappe boundaries and cut-offs (Pleničar et al. 1975; Šikić et al. 1977). The strong incongruity of orientation of D3 structures between Medvednica and Žumberak Mountains was the motivation for the palaeomagnetic study to determine existence, amount and timing of block rotation(s).

Palaeomagnetic measurements and results

In the surroundings of the Medvednica Mountains, Miocene sediments exhibit moderate counter-clockwise rotations (Fig. 10b; Márton et al. 2002a), which may have occurred during two tectonic events. The younger phase took place in post-Early Pontian times and is characterized by about 30° of counter-clockwise rotation. An older Miocene rotation (probably of Karpatian age) in the same sense is indicated by the declinations of the Ottnangian sediments which are around 325–315°. A similar situation is found SE of the Medvednica Mountains area, in the surroundings of the Slavonian Mountains of eastern Croatia (Márton et al. 1999), SW of Medvednica Mountains in the area of Žumberak Mountains. (Márton et al. 2006) and to the north, in Krško and Mura basins in Slovenia (Márton et al. 2002b, 2006). In the External Dinarides, the post-folding palaeomagnetic directions measured on Upper Cretaceous rocks are aligned with those observed on Miocene sediments of the above-mentioned areas (Fig. 10b; Márton et al. 2003). All rotations were probably driven by the motion of the Adriatic microplate (Márton et al. 2005; Márton 2006).

As suggested by structural observations, the Medvednica Mountains could have been in a different orientation before the Miocene. To test this hypothesis, palaeomagnetic samples were drilled from Senonian purple-grey marls and marly limestones at five localities (Fig. 10a), selected from a dozen exposures as the best candidates to obtain a good palaeomagnetic record. At four localities, the tilt of strata was monoclinal, while at the fifth, the samples were drilled from the two limbs of an E-plunging syncline (Table 1). All samples were oriented and field data such as bedding, cleavage and intersection lineation were recorded.

Standard-size specimens were cut from the cores, and palaeomagnetic and magnetic anisotropy measurements were carried out in the Palaeomagnetic Laboratory of Eötvös Loránd Geophysical Institute and the Geophysics Department of the Eötvös University. The palaeomagnetic measurements included measurements of the remanent magnetizations (NRM) in the natural state (followed by measurements of the anisotropy of the susceptibility, before demagnetization). The NRMs of selected specimens were demagnetized in increments with Alternating Field (AF) or thermal method. As AF demagnetization often could not eliminate the NRM signal, even in as high an AF field as 200 mT (Fig. 11), stepwise thermal demagnetization or the combination of the two methods was applied to treat the majority of samples. Susceptibility was remeasured after each heating step in order to detect possible changes in magnetic mineralogy on heating. The components of the NRM were identified as linear segments. However, in the case of locality 5, the NRM vector moved along a great circle without reaching a stable end point. For this locality, the intersections of the two generalized great circles, characterizing the two limbs of the syncline, were considered in tectonic interpretation, in combination with the locality mean palaeomagnetic directions for the others (see later).

Two components of NRM were found in most samples. Components with lower unblocking temperatures (component 'a') have positive

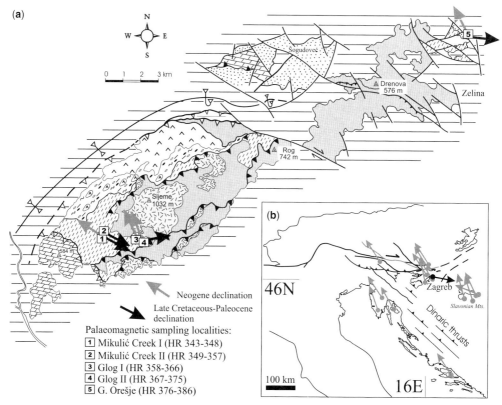

Fig. 10. Palaeomagnetic results from the Medvednica Mountains and from a wider surrounding. (**a**) Upper Cretaceous sampling localities from the Medvednica Mountains with declinations of pre-folding age (black arrows) and of post-folding age (grey arrows). See Fig. 3 for legend of the map. (**b**) Palaeomagnetic declinations from the wider surroundings of Medvednica Mountains (from Márton et al. 1999, 2000, 2002a, b, 2005; Fodor et al. 1998) recording a Late Neogene counter-clockwise rotation of c. 35°. Note the agreement between the declinations of post-folding age obtained from Upper Cretaceous rocks of the Medvednica Mountains and those signifying end-of-Miocene rotations in the wider surroundings.

inclinations: the declination for locality 2 is close to the stable European reference declination (around 8° in the Tertiary) while for the others it is counter-clockwise rotated (Table 1) and must have been imprinted before the final, post-Early Pontian counter-clockwise rotation of Medvednica Mountains and its wider surroundings (Figs 10b and 12).

The components with high unblocking temperature (component 'b') cluster in three groups (localities 1–3) and the great circles for the two limbs at G. Orešje (locality 5) are close to this cluster in the tectonic system. The overall-mean palaeomagnetic direction for the NRM of pre-folding age is calculated from the three stable end points and the two great circles. The rotation implied is about 100–110° with respect to the present North, in the clockwise sense (Fig. 12). This indicates a large clockwise rotation of Medvednica Mountains after the Late Cretaceous, most probably after the Paleocene. The upper age constraint for this rotation is provided by the palaeomagnetic direction for the Ottnangian sediments (Fig. 12; direction 'A').

As was earlier mentioned, the Senonian–Paleocene sediments of the Medvednica Mountains are deformed. At places, cleavage planes are well developed and the direction of intersection lineation can be measured in the field (e.g. localities Mikulić creek I & II, Glog I; for locations see Fig. 10). At other sampled localities, penetrative fabric was not observable macroscopically (localities Glog II, Gornje Orešje). Irrespective of the macroscopically observed fabric, the low field susceptibility anisotropy measurements revealed that the magnetic fabric was basically of sedimentary character, since the susceptibility minima were near vertical in the tectonic coordinate system, i.e. after full tectonic correction (Fig. 13) and the anisotropy degree

Table 1. *Summary of the palaeomagnetic directions obtained from Senonian rocks at Medvednica Mountains*

	Locality		n/no	D°	I°	k	$\alpha_{95}°$	$D_C°$	$I_C°$	k	$\alpha_{95}°$	dip	plunge	remark
1	Mikulić Creek I	a	5/6	302	+39	15	20					272/54, 258/60, 270/50	220/30	plunge + tilt corrected
	HR 343-348	b	6/6	111	−5	36	11	116	+42	43	10			
2	Mikulić Creek II	a	9/9	21	+75	58	7					250/54	220/30	plunge + tilt corrected
	HR 349-357	b	9/9	289	+6	95	5	295	−33	95	6			
3	Glog I	a	6/9	338	+53	103	7					268/52		tilt corrected
	HR 358-366	b	3/9	83	+1	85	13	80	+53	85	13			
4	Glog II	a	7/9	334	+69	137	5					170/50		
	HR 367-375													
5	Gornje Orešje S. limb	a	8/11	332	+47	16	14					46/35	90/30	plunge + tilt corrected
	HR 376-386 N. limb	b				great circle distribution						133/43		

Legend: n/no, number of used/collected samples; D°, I° (D_C, I_C), declination, inclination before (after) tilt correction; k and α_{95}, statistical parameters (Fisher 1953). Items 'a' are overprint components; items 'b' are components of pre-folding age. For locality 5, the pre-folding magnetization was constrained by the two great circles, representing the two limbs of the syncline.

was fairly low (1.7–6.8%). Higher values were observed both in cleaved (Mikulić creek I) and in visibly non-schistose rocks (Gornje Orešje). Nevertheless, there is a difference in the degree of magnetic lineation between more deformed rocks where it is 2.4–4.3%, while in the less deformed rocks it is below 1%. On full tectonic correction (plunge correction followed by bedding correction), maximum directions of susceptibility (and consequently, intermediate directions) become similar for all localities, i.e. magnetic lineations are NNE–SSW oriented (examples are shown in Fig. 13), and are aligned with the intersection lineation observed in the field (compare Figs 7 and 13). Thus, we conclude that magnetic lineation is tectonically induced and the magnetic fabric was imprinted while the strata were still horizontal.

An interesting aspect of the magnetic anisotropy results is that the orientation of the maxima in the presently studied Senonian sediments is different from those in the previously studied Tertiary sediments. In the latter, magnetic lineation maxima are NW–SE oriented, independent of the age of the rock (Egerian through Pannonian). The angular difference between the two generalized magnetic lineations is about 100° (counting through north), which agrees fairly well with the angular difference between the palaeomagnetic declinations for the same age groups (which is about 130°).

Complex structural interpretation

As described above, the External Dinaridic, Žumberak and the Medvednica–Hrvatsko Zagorje areas experienced almost the same degree of the Late Neogene counter-clockwise rotation (Figs 10 and 12), and, therefore, this late counter-clockwise rotation does not affect their relative orientation. However, if the original pre-Neogene structural trends and tectonic transport directions are to be reconstructed, retro-rotations should be applied to compensate for palaeomagnetically proven pre-Noegene rotation.

In the first step, a reconstruction of the Neogene rotations is achieved by retro-rotating the whole Dinaridic, south-Alpine area by 35° in a clockwise sense in order to compensate for a post-Ottnangian–recent counter-clockwise rotation described above (Fig. 14a, b). For this retro-rotation, we used a pole located at 45.36° N/9.10° E calculated by Calais *et al.* (2002) as a pole of recent counter-clockwise rotation of the Adriatic Plate. This retro-rotation affects also the region north of the Periadriatic fault system (Fig. 14b), because palaeomagnetic data from Austria suggest the same amount and sense of young rotation (Márton *et al.* 2000; Scholger & Stingl 2004).

In the next step, declination-differences measured in Senonian rocks are eliminated by

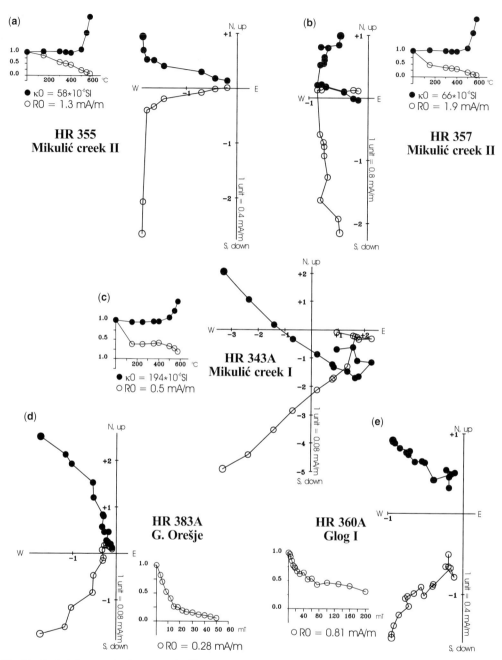

Fig. 11. Typical demagnetization curves for Upper Cretaceous sediments from the Medvednica Mountains Zijderveld plots plus NRM intensity(circles)/susceptibility(dots) versus temperature curves (**a**, **b**, **c**) or NRM intensity versus AF demagnetizing field curves (**d** and **e**). In the Zijderveld plots, the projection of the palaeomagnetic vector is shown in the horizontal (dots) and in a vertical (circles) plane.

retro-rotating the Medvednica Mountains and inselbergs of Hrvatsko Zagorje in counter-clockwise sense by 130° (Fig. 14b, c). This brings the Medvednica structural strike in alignment with the originally N–S striking Dinaridic Eocene structures (Fig. 14c). The applied 130° rotation should be post-Late Cretaceous (most probably post-Paleocene)–pre-Ottnangian, because the major

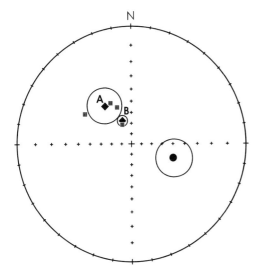

Fig. 12. Comparison between palaeomagnetic directions observed for Upper Cretaceous sediments from the Medvednica Mountains and for Cenozoic sediments from northern Croatia. Upper hemisphere, stereographic Schmidt projection, all vectors pointing downwards (positive inclination). Key: full circle with confidence circle: overall-mean palaeomagnetic direction calculated from components 'b' of the present study, after tectonic correction (interpreted as Cretaceous magnetizations); squares: locality mean directions calculated from overprint components 'a' of the Cretaceous sediments showing CCW rotation; diamonds: overall mean palaeomagnetic directions for Ottnangian sediments (A) and for Pannonian–Pontian sediments (B) measured around Medvednica Mountains (Márton et al. 2002a).

clockwise rotation was observed in Senonian and not in Miocene rocks. On the other hand, neither the D3 ductile, thrust- nor the D4 right-lateral shear-related structures observed in Senonian–Paleocene rocks were found in Ottnangian sediments. Therefore, the post-Paleocene D3 thrusting and the D4 right-lateral shear occurred in the same time interval or before the palaeomagnetically measured clockwise rotation (Fig. 4).

Reconstruction of the original D1–D4 tectonic transport directions

The main pre-Senonian nappe tectonics affected Medvednica Mountains in its original, Dinaridic position (Fig. 14d). When reconstructed, the original tectonic transport directions of D1, D2 and D3 events are top-to-the-north in the first, and top-to-the-west in the second and third events. Because of isotopic facies zones running parallel to the main Dinaridic structural boundaries

(Aubouin et al. 1970), it is generally assumed that former palaeogeographic boundaries, including ocean–continent margins, ran parallel to the Dinaridic structural trends (e.g. Pamić et al. 1998). In Medvednica Mountains the Middle–Late Jurassic obduction of the ophiolitic nappe(s) apparently did not leave any structurally measurable trace; this lack of data precludes a discussion of the obduction direction (D0 event) within the scope of this paper. Nevertheless, the first D1 shear event documented in Medvednica parautochthonous metamorphic unit indicates a major tectonic episode directed parallel to, or at a very acute angle to, the Dinaridic margin. In other words, the first major nappe stacking, resulting in a weak metamorphism of Aptian–Albian (120–110 Ma) age (Belak et al. 1995), was parallel to known structural (and possibly palaeogeographic) trends. This implies a major left-lateral shear component along the former Adriatic margin and obducted Dinaridic ophiolite zone.

This left-lateral D1 motion was followed by a more frontal D2 shortening almost perpendicular to the original margins/isopic facies zones. D2 folding and thrusting(?) created the now-known major tectonic boundaries and major folds within the nappe stack. Our data suggest that D2 episode occurred under still hot, ductile conditions within a relatively short amount of time after peak metamorphic conditions (Fig. 4). Thus, we propose that D2 structures of Medvednica Mountains also formed during the Early Albian, at c. 110 Ma. This deformation resulted in gradual cooling of the metamorphic pile and in exhumation and erosion of the higher structural units.

The post-Paleocene D3 deformational event was driven by E–W directed shortening, i.e. practically along the same direction as in the D2 event. The difference in D2 and D3 structural styles, the pre-Senonian erosion and the lack of difference in Senonian and Paleocene structures suggest that the D2 and D3 events were separated by a longer time of relative tectonic quiescence and that the similar orientation of D2 and D3 structural trends is coincidental. The D3 post-Paleocene event corresponds perfectly with the Middle Eocene–Oligocene Dinaridic phase of major folding and thrusting in the External Dinarides. Although the majority of deformation took place within the External Dinarides, Palaeogene tectonic activity is widely evident in the Sava zone of the Internal Dinarides as well, and most probably related to thrusting of Tisia over the northern segment of the Internal Dinarides (e.g. Tari & Pamić 1998; Pamić et al. 2002b; Ustaszewski et al. 2005).

In the original orientation, the D4 deformational event generated right-lateral shear surfaces parallel to the main (originally N–S) structural trends

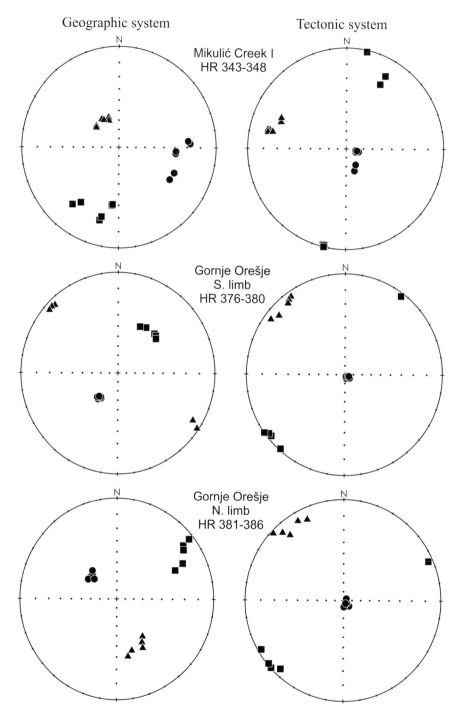

Fig. 13. Examples for the magnetic fabric of the Upper Cretaceous sediments from the Medvednica Mountains. Key: squares–maximum, triangles–intermediate, dots–minimum susceptibility directions. Stereographic Schmidt projection: all vectors point downward. Note the subvertical minima after full tectonic correction (evidence for basically sedimentary fabric) and maxima grouped along a NE–SW axis, due to a weak NW–SE shortening. It is particularly important that full tectonic corrections bring into alignment the maxima for the S and N limbs of the Orešje syncline, respectively.

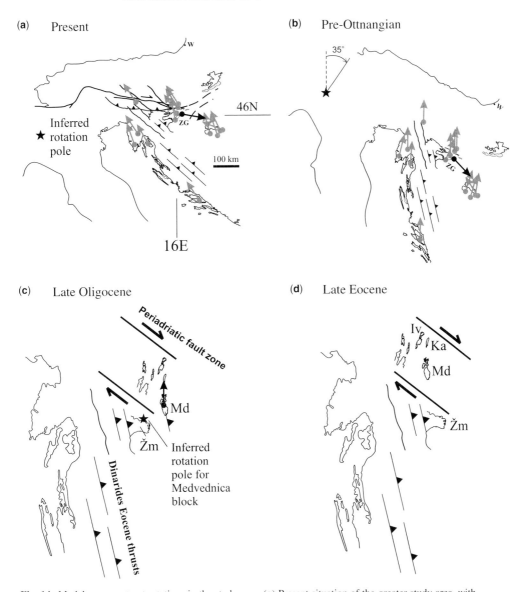

Fig. 14. Model to reconstruct rotations in the study area. (a) Present situation of the greater study area, with declination arrows. Light: Neogene declination directions. Dark: Late Cretaceous declination direction. (b) Position of the greater study area, reconstructed for Late Neogene counter-clockwise rotations. Rotation pole was fixed at the Adriatic Plate rotation pole calculated by Calais *et al.* (2002). (c) Blow-up of the study area. Note that Medvednica block (comprising the Hrvatsko Zagorje and inselbergs such as Ivanščica and Kalnik) is rotated backwards to reconstruct post-Late Cretaceous and pre-Ottnangian major clockwise rotation. The inferred age of this rotation is Late Oligocene, i.e. coinciding with the most intensive right-lateral shear along the Periadriatic fault zone. (d) Inferred Late Eocene position of studied blocks. Note the right-lateral shift of blocks between Late Eocene and Late Oligocene. Incipient right-lateral shear along the Periadriatic fault zone is not supposed to generate major rotation.

(Fig. 14c). Since timing of the main clockwise rotation is in the same time period, it is to be discussed whether the D4 event occurred prior to, during or after the 130° clockwise rotation. In the following text, we shall first discuss potential analogues or main dextral shear zones to explain these movements, and then propose alternative kinematic models to explain the D4 structures.

The Medvednica Mountains in their original and present position fall into the Central Dinaridic ophiolite zone of the Internal Dinarides (CDOZ), located at the boundary between the CDOZ and the more internal Sava zone (SZ) (Fig. 1; Pamić 2002). During the Palaeogene, the SZ is considered as a major mobile zone of the Internal Dinarides (Pamić et al. 2002a, b) with a probable eastward continuation into the Vardar zone of Serbia and further south into the Hellenides (Mercier 1968; Ricou et al. 1998). Pervasive right-lateral shearing is observed in both the Greek and Serbian sectors of the same zone (Mercier 1968; Grubić 2002), and it is suggested to be Palaeogene in age. On the other hand, the Medvednica Mountains are also in close proximity to the Periadriatic fault system (c. 30 km north of the mountain). This major right-lateral shear zone is thought to have been active during the Oligocene–Early Miocene (30 Ma peak activity) (Schmid et al. 1989; Ratschbacher et al. 1991; Fodor et al. 1998). It is also clear from structural and geochronological data that this fault system was already active during Eocene and Late Neogene–Quaternary times as a regional-scale transpressive fault system (Polinski & Eisbacher 1992; Fodor et al. 1998; Tomljenović & Csontos 2001). The ultimate cause of both regional dextral shear zones is the northward movement of the African/Apulian promontory during and after the Eocene (Channel & Horvath 1976; Besse & Courtillot 2002).

If the two alternatives for Sava-Vardar and Periadriatic shear systems are taken as the potential controlling factors, both pre- and post-rotational activity of D4 lateral shear bands might be plausible. Formation of right-lateral shear bands during rotation (i.e. in an intermittent, semi-rotated position) can be ruled out because right-lateral structures in Medvednica Mountains would be in unfavourable orientations for both main right-lateral shear zones to be activated. In a pre-rotational scenario, D4 right-lateral shear bands would be consistent with the synchronous right-lateral belt of the Sava-Vardar zone. While the precise timing of this tectonic event is unclear, D4 should have started in the Eocene, practically synchronous with W-directed D3 thrusting, assuming the northward movement of Africa as the primary tectonic driving force. In Medvednica Mountains, D4 dextral shear clearly post-dates D3 thrusting, as the lateral movement indicators overprint D3 thrust surfaces (Fig. 9). Thus, at least in Medvednica Mountains, the pervasive D4 right-lateral shear must be Oligocene or younger (Fig. 4).

In a post-rotational scenario, the D4 shear planes are not ideally oriented for the Periadriatic fault system: D4 shear planes are oblique to the Periadriatic fault by c. 25–30°. While some displacement could be accommodated along these structures, oblique striae parallel to the main Periadriatic motion would be expected. Lack of overprinting within a favourable orientation strongly suggests that the D4 shear planes were not active in their present orientation, and thus the pre-rotational scenario is more plausible.

Regional geology: a discussion

The reason for (possibly Late) Palaeogene clockwise rotation requires explanation. Major clockwise rotation could be easily caused by a right-lateral shear, and superficial rotations above a detachment plane (either in a shortening or stretching regime) can occur but these are not, in this case, supported by local or regional geology. On the other hand, there are two potential right-lateral shear zones which could be responsible for a major clockwise rotation. The Periadriatic fault system seems to be a probable candidate because it trends obliquely to the original Dinaridic trend (Fig. 14). Furthermore, the northern part of the rotated area exemplified by Ivanščica Mountains is clearly transposed and aligned parallel to the trend of the Periadriatic fault system (Figs 1 and 2). Several 10 (Schmid et al. 1989) to several 100 kilometres (Kázmér & Kovács 1985) of lateral offset have been proven or proposed for that fault system during the Late Palaeogene. Therefore, it is proposed that the right-lateral motion along the Periadriatic fault system tore off a part of the Dinaridic structural stack and rotated it in a clockwise sense along the shear zone. We propose that the bulk of the 130° clockwise rotation happened during peak activity of the Periadriatic fault system in Late Oligocene times.

Palaeomagnetic measurements in pre-Neogene rocks have not been carried out in other inselbergs surrounding Medvednica Mountains (Hrvatsko Zagorje area). However, the structural styles and orientations in pre-Neogene rocks of other inselbergs are similar in their incongruity and incompatibility as described for Medvednica Mountains (Figs 1 and 2). Therefore, most probably not only Medvednica but also a larger area suffered the same amount of Palaeogene rotations. For example, when reconstructed for 130–140° clockwise rotation, the Ivanščica Mountains, which are composed of similar and partly the same tectonic units as Medvednica Mountains (e.g. see in Babić et al. 2002), also become parallel to the main Dinaridic trend. Consequently, conspicuous CDOZ tectonic units exposed on Ivanščica Mountains like the Jurassic ophiolitic mélange become aligned with the Internal Dinaridic sector. It is therefore suggested that most, if not all, inselbergs of the Hrvatsko Zagorje area rotated away from their original directions, due to the major right-lateral shear along the Periadriatic fault system. Their rotation may have been accentuated later, due to Late Neogene transpression of the Hrvatsko Zagorje area (see Tomljenović & Csontos 2001).

While the Medvednica Mountains strike is dramatically different, the main structural trends in the adjacent Žumberak Mountains are parallel with the major structural trend of the central Dinarides (Pleničar et al. 1975; Šikić et al. 1977). Unfortunately, efforts to obtain palaeomagnetic data in pre-Neogene rocks of this area have proved to be unsuccessful so far, but we can postulate that, because of structural similarity, the Žumberak Mountains were not rotated with respect to the Dinarides. This implies that a major boundary between rotated and non-rotated parts should be found between Medvednica and Žumberak Mountains. As a possible candidate for this break, we supposed an Eocene–Oligocene precursor of the Sava fault found between the two massifs (Fig. 14c, d). This fault is known as a Tertiary strike-slip fault (e.g. Prelogović et al. 1998) and is thought to represent the SW margin of the Pannonian Basin (e.g. Horváth 1984). However, complex Neogene tectonics in the area of possible northern prolongation of this fault (including intensive Late Neogene–Recent shortening and strike-slip fault activity; e.g. Placer 1999; Tomljenović & Csontos 2001) does not enable a clear identification of a pre-Neogene northern segment of the Sava fault.

As proposed above, the North Croatian (Hrvatsko Zagorje) inselbergs were torn from the northern fringes of the Dinarides, and rotated in the Periadriatic dextral shear zone. However, the lithospheric or crustal scale of the process requires discussion. The high amount of rotation and the relatively ductile behaviour of the individual pieces speak in favour of crustal-scale deformation and rotation. On the other hand, the enduring proximity of these blocks is inconsistent with independent and unpredictable block translation within major shear zones. Furthermore, the lack of extensive volcanics during the Palaeogene also speaks against lithospheric-scale deformation. Oligocene magmatites, the direct equivalents of the Alpine Tonalite belt, can be only found at the Periadriatic line proper (e.g. Pamić & Palinkaš 2000) in the Mura depression, and not in the studied area (e.g. Pamić & Balen 2001). This suggests that the rotational deformation was not creating major fissures: neither in a lithospheric nor in a crustal scale. The north Croatian inselbergs were rotated possibly together with their lithospheric root, which should have behaved in a ductile manner. Eventually, softening due to an anomalous temperature gradient or to strain accumulation could have enabled this ductile behaviour.

Other rotations and major motions

In the Alpine–Pannonian–Dinaridic area, three major motions have been identified. Facies belt- and Palaeogene basin correlations (Kázmér & Kovács 1985; Csontos et al. 1992; Kázmér et al. 2003) as well as syn-sedimentary deformation phenomena in Palaeogene rocks (Fodor et al. 1992) suggest a tectonic escape of the Alcapa block from between the north and south Alpine sectors. Age constraints suggest a Late Eocene–Oligocene (Fodor et al. 1992) to Early Miocene (Csontos et al. 1992; Fodor et al. 1998) timing for this eastward-directed escape. Structural and geochronological data along the Periadriatic fault (Schmid et al. 1989, 1996; Schönborn 1992) indicate a dextral slip on the western portions of this fault in the time interval of 45–19 Ma. The second major motion is evidenced mainly by palaeomagnetic work (Márton 1987, 1993; Márton & Fodor 1995; Márton & Márton 1999) and suggests large-scale opposing rotations between the Alcapa and Tisza blocks of the Pannonian basin during the Ottnangian (between 19 and 17 Ma) (Márton 1987; Balla 1987; Csontos 1995), i.e. later than the main rotation of Medvednica Mountains. A third major motion in the sector occurred in the Late Miocene–Quaternary with a bulk 35° counter-clockwise rotation partly tied to dextral compressional reactivation along the Periadriatic fault system (Fodor et al. 1998; Tomljenović & Csontos 2001).

Conclusions

Four deformational events (D1–D4) of pre-Miocene age have been demonstrated based on structural analysis of the pre-Neogene tectonic and stratigraphic units of Medvednica Mountains in northern Croatia. All the main D1–D4 structures of Medvednica Mountains are striking NE–SW, which is almost orthogonal to the major NW trend of the central Dinarides. We propose that the reason for this dramatic deflection in structural trends is 130° clockwise rotation and eastward escape of the tectonic block comprising Medvednica Mountains and the surrounding inselbergs of the Hrvatsko Zagorje area bounded to the north by the easternmost tip of the Periadratic fault system. In Medvednica Mountains, the main period of tectonic escape and induced clockwise rotation occurred in the Late Palaeogene, possibly in the Oligocene–earliest Miocene.

The D1 deformational event is evidenced by a penetrative NE-trending stretching lineation (L1) observed in the parautochthonous low-grade to greenschist facies metamorphic unit. Observed shear-sense indicators point to a top-to-the-northeast tectonic transport direction in a recent coordinate system. When rotated into original position, D1 tectonic transport direction is top-to-the-north, attributed to the Aptian–Albian nappe stacking in the central-northern Dinarides and northward propagation of already obducted ophiolites of the Central Dinaridic ophiolite zone. This nappe stacking, which resulted in a weak regional metamorphism in tectonic units underlying the ophiolites, was orogen-parallel or at a very acute angle to known structural (and possibly palaeogeographic) trends. This implies a major left-lateral shear component along the former Adriatic margin.

This was followed by the D2 asymmetric folding with SE-dipping axial planes and axial plane cleavage (S2). The recent orientation of these ductile structures indicates a NW-directed shear. When rotated into original position, D2 shear direction is top-to-the-west, attributed to Early Albian, orogen-perpendicular shortening that took place after the peak metamorphic conditions, but still in temperature conditions favourable to ductile deformation. This deformation resulted in gradual cooling of the metamorphic stack and also in exhumation and erosion of the higher structural units.

D3 deformational event is characterized by NNE-trending and NW-vergent open to tight folds, sometimes associated with a weak to penetrative SE-dipping axial plane cleavage and SE-dipping thrusts. These structures clearly indicate a progressive non-coaxial shear with top-to-the-northwest tectonic transport direction in the present-day coordinate system. When rotated into original position, D3 shear direction is top-to-the-west, attributed to post-Palaeocene, most probably Middle Eocene–Oligocene E–W shortening, synchronous with the main Dinaridic tectonic phase of the External Dinarides. This shortening was probably triggered by collision and thrusting of Tisia over Palaeogene flysch series of the Sava zone, i.e. over the northern segment of the Internal Dinarides.

The D4 event is documented by pervasive right-lateral shearing on recently NE-striking surfaces of bedding, previously formed cleavage (S3) and D3 shears. D4 right-lateral shearing is tentatively interpreted as being related to the right-lateral shearing of the Sava zone during the Eocene–Oligocene.

Following the main period of tectonic escape and induced clockwise rotation in Late Palaeogene, possibly in Oligocene–earliest Miocene, the Medvednica Mountains and the surrounding area were affected by repeated extensions and inversions since the Early Miocene to recent times (Tomljenović & Csontos 2001). Palaeomagnetic data suggest that in the Early Miocene (but probably before the Karpatian) this area was part of a regional block that shifted northwards and rotated in a counter-clockwise sense. A second episode of counter-clockwise rotation occurred at the present latitude in post-Pontian times (since c. 5 Ma), driven by the counter-clockwise rotating Adriatic Plate (Márton et al. 2002a, b, 2003).

This work was financially supported by the Ministry of Science, Education and Sports of the Republic of Croatia (Project CROTEC, grant no. 195-1951293-3155) and by OTKA grants no. T043760 (LC) and T049616 (EM) of Hungary. The manuscript has benefited greatly from constructive reviews and suggestions by F. Neubauer and R. Schuster. Especially, we thank Stefan Schmid for his continuous moral and professional support; and for the initiation of many good debates and thoughts on the geology of the Dinarides.

References

AUBOUIN, J., BLANCHET, R., CADET, P. ET AL. 1970. Essai sur la geologie des Dinarides. *Bulletin de la Société Geoliquede France*, **XII** (6), 1060–1095.

BABIĆ, L. J., GUŠIĆ, I. & NEDĚLA-DEVIDÉ, D. 1973. Senonian breccias and overlying deposits on Mt. Medvednica (northern Croatia). *Geoloski Vjesnik, Zagreb*, **25**, 11–27.

BABIĆ, L. J., GUŠIĆ, I. & ZUPANIČ, J. 1976. Paleocene reef-limestone in the region of Banija, Central Croatia. *Geoloski Vjesnik, Zagreb*, **29**, 11–47.

BABIĆ, L. J., HOCHULI, A. P. & ZUPANIČ, J. 2002. The Jurassic ophiolitic mélange in the NE Dinarides: Dating, internal structure and geotectonic implications. *Eclogae Geolicae Helvetiae*, **95**, 263–275.

BALLA, Z. 1987. Tertiary paleomagnetic data for the Carpatho-Pannonian region in the light of Miocene rotation kinematics. *Tectonophysics*, **139**, 67–98.

BASCH, O. 1995. Geological map of Medvednica Mt. *In*: ŠIKIĆ, K. (ed.) *Geološki vodič Medvednice*. Croatian Geological Survey Zagreb.

BELAK, M., PAMIĆ, J., KOLAR-JURKOVŠEK, T., PECKAY, Z. & KARAN, D. 1995. Alpinski regionalnometamorfni kompleks Medvednice (sjeverozapadna Hrvatska). *In*: VLAHOVIĆ, I., VELIĆ, I. & ŠPARICA, M. (eds) *Zbornik radova 1, 1. Hrv. geol. kongres*. Croatian Geological Survey Zagreb, 67–70.

BESSE, J. & COURTILLOT, V. 2002. Revised and synthetic apparent polar wander path of the African, Eurasian, North American and Indian plates, and true polar wander since 200 Ma. *Journal of Geophysical Research*, **96**, 4029–4050.

BLAŠKOVIĆ, I. 1998. The Two Stages of Structural Formation of the Coastal Belt of the External Dinarides. *Geologia Croatica, Zagreb*, **51**, 75–89.

CALAIS, E., NOCQUET, J. M., JOUANNE, F. & TARDI, M. 2002. Current strain regime in the Western Alps from continuous Global Positioning System measurements, 1996–2001. *Geology*, **39**, 651–654.

CASTELLARIN, A., CANTELLI, L., FESCE, A. M. ET AL. 1992. Alpine compressional tectonics in the Southern Alps: Relationships with the N-Apennines. *Annales Tectonicae*, **6**, 62–94.

CHANNEL, J. E. T. & HORVÁTH, F. 1976. The African/Adriatic promontory as a palaeogeographical premise for alpine orogeny and plate movements in the Carpatho-Balkan region. *Tectonophysics*, **35**, 71–101.

CRNJAKOVIĆ, M. 1980. Sedimentation of transgressive Senonian in Southern Mountain Medvednica. *Geoloski Vjesnik, Zagreb*, **32**, 81–95.

CRNJAKOVIĆ, M. 1987. *Sedimentology of Cretaceous and Palaeogene clastics of Mountain Medvednica, Ivanšćica and Žumberak*. PhD thesis, University of Zagreb.

CSONTOS, L. 1995. Tertiary tectonic evolution of the Intra-Carpathian area: a review. *Acta Vulcanologica*, **7**, 1–13.

CSONTOS, L. & NAGYMAROSY, A. 1998. The Mid-Hungarian line: a zone of repeated tectonic inversions. *Tectonophysics*, **297**, 51–71.

CSONTOS, L., NAGYMAROSY, A., HORVÁTH, F. & KOVÁC, M. 1992. Tertiary evolution of the Intra-Carpathian area: a model. *Tectonophysics*, **208**, 221–241.

CSONTOS, L., MAGYARI, Á., VAN VLIET-LANOË, B. & MUSITZ, B. 2005. Neotectonics of the Somogy hills (Part II): Evidence from seismic sections. *Tectonophysics*, **410**, 63–80.

DERCOURT, J., RICOU, L. E. & VRIELYNCK, B. 1993 (eds) *Atlas Tethys Palaeoenvironmental Maps*. Gauthier-Villars, Paris.

DEVIDÉ-NEDĚLA, D., BABIĆ, L. J. & ZUPANIČ, J. 1982. Age maesstrichtien du Flysch de Vivodina dans le Žumberk et des environs d'Ozalj en Croatie occidentale (Yugoslavie). *Geoloski Vjesnik, Zagreb*, **35**, 21–36.

DIMITRIJEVIĆ, M. N. & DIMITRIJEVIĆ, M. D. 1973. Olistostrome mélange in the Yugoslavian Dinarides and the Late Mesozoic plate tectonics. *Journal of Geology*, **81**, 328–340.

DOGLIONI, C. & BOSELLINI, C. 1987. Eoalpine and Mesoalpine tectonics in the Southern Alps. *Geologische Rundschau*, **76**, 735–754.

FISHER, R. 1953. Dispersion on a sphere. *Proceedings of the Royal Society, London*, **217**, 295–305.

FODOR, L., KÁZMÉR, M., MAGYARI, A. & FOGARASI, A. 1992. Gravity-flow dominated sedimentation on the Buda paleoslope (Hungary): Record of Late Eocene continental escape of the Bakony Unit. *Geologische Rundschau*, **82**, 695–716.

FODOR, L., JELEN, B., MÁRTON, E., SKABERNE, D., ČAR, J. & VRABEC, M. 1998. Miocene-Pliocene tectonic evolution of the Slovenian Periadriatic fault: Implications for Alpine-Carpathian extrusion models. *Tectonics*, **17**, 690–709.

GRUBIĆ, A. 2002. Transpressive Periadriatic suture in Serbia and SE Europe. *Geologica Carpathica, Bratislava, Special Issue*, **53**, 141–142.

HAAS, J. & KOVÁCS, S. 2001. The Dinaridic-Alpine connection – as seen from Hungary. *Acta Geologica Hungarica, Budapest*, **44**, 345–362.

HAAS, J., KOVÁCS, S., KRYSTYN, L. & LEIN, R. 1995. Significance of Late Permian-Triassic facies zones in terrane reconstructions in the Alpine - North Pannonian domain. *Tectonophysics*, **242**, 19–40.

HAAS, J., MIOČ, P., PAMIĆ, J. *ET AL*. 2000. Complex structural pattern of the Alpine-Dinaridic-Pannonian triple junction. *International Jounal of Earth Sciences*, **89**, 377–389.

HALAMIĆ, J. & GORIČAN, Š. 1995. Triassic Radiolarites from Mountains. Kalnik and Medvednica (Northwestern Croatia). *Geologia Croatica, Zageb*, **48**, 129–146.

HALAMIĆ, J., SLOVENEC, D. & KOLAR-JURKOVŠEK, T. 1998. Triassic pelagic limestones in pillow lavas in the Orešje quarry near Gornja Bistra, Medvednica Mountain (Northwest Croatia). *Geologia Croatica, Zagreb*, **51**, 33–45.

HORVÁTH, F. 1984. Neotectonics of the Pannonian basin and the surrounding mountain belts: Alps, Carpathians and Dinarides. *Annals of Geophysics*, **2**, 147–154.

JAMIČIĆ, D., PRELOGOVIĆ, E. & TOMLJENOVIĆ, B. 1995. Folding and deformational style in overthrust structures on Krk Island (Croatia). *In*: ROSSMANITH, H. P. (ed.) *Mechanics of Jointed and Faulted Rocks – 2*. Balkema, Rotterdam.

JELASKA, V. 1978. Senonian-Paleogene flysch of the Mountain Trebovac area (north Bosnia): stratigraphy and sedimentology. *Geoloski Vjesnik, Zagreb*, **30**, 95–117.

JUDIK, K., ÁRKAI, P., HORVÁTH, P. *ET AL*. 2004. Diagenesis and low-temperature metamorphism of Mt. Medvednica, tCroatia: Mineral assemblages and phyllosilicate characteristics. *Acta Geologica Hungarica, Budapest*, **47**, 151–176.

KÁZMÉR, M. & KOVÁCS, S. 1985. Permian-Paleogene paleogeography along the eastern part of Insubric-Periadriatic Lineament system: evidence for continental escape of the Bakony-Drauzug unit. *Acta Geologica Hungarica, Budapest*, **28**, 71–84.

KÁZMÉR, M., DUNKL, I., FRISCH, W., KUHLEMANN, J. & OZSVÁRT, P. 2003. The Palaeogene forearc basin of the Eastern Alps and Western Carpathians: subduction erosion and basin evolution. *Journal of the Geological Society*, **160**, 413–428.

LANPHERE, M. & PAMIĆ, J. 1992. K-Ar and Rb-Sr ages of Alpine granite-metamorphic complexes in the Northwestern Dinarides and the southwestern part of the Pannonian basin in Northern Croatia. *Acta Geologia, Zagreb*, **22**, 97–123.

LUGOVIĆ, B., ALTHERR, R., RECZEK, I., HOFMANN, A. W. & MAJER, V. 1991. Geochemistry of peridotites and mafic igneous rocks from the Central Dinaric Ophiolite Belt, Yugoslavia. *Contributions to Mineralogy and Petrology*, **106**, 201–216.

LUGOVIĆ, B., ŠEGVIĆ, B. & ALTHERR, R. 2006. Petrology and tectonic significance of greenschists from the Medvednica Mountains. (Sava Unit, NW Croatia). *Ofioliti*, **31**, 39–50.

MARJANAC, T. 1990. Reflected sediment gravity flows and their deposits in flysch of Middle Dalmatia, Yugoslavia. *Sedimentology*, **37**, 921–929.

MARJANAC, T. & ĆOSOVIĆ, V. 2000. Tertiary Depositional History of Eastern Adriatic Realm. *In*: PAMIĆ, J. & TOMLJENOVIĆ, B. (eds) *Pancardi 2000, Vijesti 37/2*. Croatian Geological Society, Zagreb, 93–103.

MÁRTON, E. 1987. Paleomagnetism and tectonics in the Mediterranean region. *Journal of Geodynamics*, **7**, 33–57.

MÁRTON, E. 1993. The Itinerary of the Transdanubian Central Range: An assessment of relevant paleomagnetic observations. *Acta Geologica Hungarica, Budapest*, **37**, 135–151.

MÁRTON, E. 2006. Paleomagnetic evidence for Tertiary counterclockwise rotation of Adria with respect to Africa. In: PINTER, N., GRENERCZY, G., WEBER, J., STEIN, S. & MEDAK, D. (eds) *The Adria microplate: GSP Geodesy, Tectonics and Hazards*. NATO Science Series IV-61, 71–80.

MÁRTON, E. & FODOR, L. 1995. Combination of palaeomagnetic and stress data—a case study from North Hungary. *Tectonophysics*, **242**, 99–114.

MÁRTON, E. & MÁRTON, P. 1999. Tectonic aspects of a paleomagnetic study on the Neogene of the Mecsek Mountains. *Geophysical Transactions*, **42**, 159–180.

MÁRTON, E., PAVELIĆ, D, TOMLJENOVIĆ, B., PAMIĆ, J. & MÁRTON, P. 1999. First paleomagnetic results on Tertiary rocks from the Slavonian Mountains in the Southern Pannonian Basin, Croatia. *Geologica Carpathica, Bratislava*, **50**, 273–279.

MÁRTON, E., KUHLEMANN, J., FRISCH, W. & DUNKL, I. 2000. Miocene rotations in the Eastern Alps - Paleomagnetic results from intramontane basin sediments. *Tectonophysics*, **323**, 163–182.

MÁRTON, E., PAVELIĆ, D., TOMLJENOVIĆ, B., AVANIĆ, R., PAMIĆ, J. & MÁRTON, P. 2002a. In the wake of a counterclockwise rotating Adriatic microplate: Neogene paleomagnetic results from Northern Croatia. *International Journal of Earth Sciences*, **91**, 514–523.

MÁRTON, E., FODOR, L., JELEN, B., MÁRTON, P., RIFELJ, H. & KEVRIĆ, R. 2002b. Miocene to Quaternary deformation in NE Slovenia: complex paleomagnetic and structural study. *Journal of Geodynamics*, **34**, 627–651.

MÁRTON, E., DROBNE, K., ĆOSOVIĆ, V. & MORO, A. 2003. Palaeomagnetic evidence for Tertiary counterclockwise rotation of Adria. *Tectonophysics*, **377**, 143–156.

MÁRTON, E., PAVELIĆ, D., TOMLJENOVIĆ, B., MÁRTON, P. & AVANIĆ, R. 2005. Paleomagnetic investigations in the Croatian part of the Pannonian basin: a review. *Acta Geologica Hungarica, Budapest*, **48**, 225–233.

MÁRTON, E., JELEN, B., TOMLJENOVIĆ, B. *ET AL.* 2006. Late Neogene counterclockwise rotation in the SW part of the Pannonian Basin. *Geologica Carpathica, Bratislava*, **57**, 41–46.

MERCIER, J. L. 1968. Étude géologique des zones internes des Hellénides en Macédoine centrale (Grèce). Contribution à l'étude du métamorphisme et de l'évolution magmatique des zones internes des Hellénides. *Annales Géologiques des Pays Helléniques*, **20**, 1–792.

PAMIĆ, J. 2002. The Sava-Vardar Zone of the Dinarides and Hellenides versus the Vardar Ocean. *Eclogae Geolicae Helvetiae*, **95**, 99–113.

PAMIĆ, J. & BALEN, D. 2001. Tertiary magmatism of the Dinarides and the adjoining South Pannonian Basin. *Acta Vulcanologica*, **13**, 9–24.

PAMIĆ, J. & PALINKAŠ, L. 2000. Petrology and geochemistry of Paleogene tonalites from the easternmost parts of the Periadriatic Zone. *Mineralogy and Petrology*, **70**, 121–141.

PAMIĆ, J. & TOMLJENOVIĆ, B. 1998. Basic geologic data from the Croatian part of the Zagorje-Mid Transdanubian zone. *Acta Geologia Hungarica, Budapest*, **41**, 389–400.

PAMIĆ, J., ÁRKAI, P., O'NEIL, J. O. & LANTAI, C. 1992. Very low- and low-grade progressive metamorphism of Upper Cretaceous sediments in Mt. Motajica, northern Dinarides. In: VOZAR, J. (ed.) *Western Carpathians, Eastern Alps, Dinarides*. IGCP No 276, Bratislava.

PAMIĆ, J., GUŠIĆ, I. & JELASKA, V. 1998. Geodynamic evolution of the Central Dinarides. *Tectonophysics*, **297**, 251–268.

PAMIĆ, J., TOMLJENOVIĆ, B. & BALEN, D. 2002a. Geodynamic and petrogenetic evolution of Aline ophiolites from the central and NW Dinarides: an overview. *Lithos*, **65**, 113–142.

PAMIĆ, J., BALEN, D. & HERAK, M. 2002b. Origin and geodynamic evolution of Late Paleogene magmatic associations along the Periadriatic-Sava-Vardar magmatic belt. *Geodinamica Acta*, **15**, 209–231.

PLACER, L. 1999. Structural meaning of the Sava folds. *Geologija, Ljubljana*, **41**, 191–221.

PLENIČAR, M., PREMRU, U. & HERAK, M. 1975. Basic geological map of Yugoslavia M1:100 000, sheet Novo Mesto. *Institute of Geology Ljubljana, Federal Geological Institute Beograd*.

POLINSKI, R. K. & EISBACHER, G. H. 1992. Deformational partitioning during polyphase oblique convergence in the Karawanken Mountains, southeastern Alps. *Journal of Structural Geology*, **14**, 1203–1213.

PRELOGOVIĆ, E., SAFTIĆ, B., KUK, V., VELIĆ, J., DRAGAŠ, M. & LUČIĆ, D. 1998. Tectonic activity in the Croatian part of the Pannonian basin. *Tectonophysics*, **297**, 283–293.

PRTOLJAN, B. 2001. Relationscips of thrust-fold and horizontal mechanism of the Mountain Žumberak part of the Sava nappe in the northwestern Dinarides, West Croatia. *Acta Geologica Hungarica, Budapest*, **44**, 67–80.

RATSCHBACHER, L., FRISCH, W., LINZER, H.-G. & MERLE, O. 1991. Lateral extrusion in the Eastern Alps, part II: structural analysis. *Tectonics*, **10**, 257–271.

RICOU, L.-E., BURG, J.-P., GODFRIAUX, I. & IVANOV, Z. 1998. Rhodope and Vardar: the metamorphic and the olistostromic paired belts related to the Cretaceous subduction under Europe. *Geodinamica Acta*, **V 11**, 285–309.

ROBERTSON, A. H. F. & KARAMATA, S. 1994. The role of subduction-accretion processes in the tectonic evolution of the Mesozoic Tethys in Serbia. *Tectonophysics*, **234**, 73–94.

SCHMID, S. M., AEBLI, H. R., HELLER, F. & ZINGG, A. 1989. The role of the Periadriatic line in the tectonic evolution of the Alps. In: COWARD, M. P., DIETRICH, D. & PARK, R. G. (eds) *Alpine Tectonics*. Geological Society, London, Special Publications, **45**, 153–171.

SCHMID, S. M., PFIFFNER, O. A., FROITZHEIM, N., SCHÖNBORN, G. & KISSLING, E. 1996. Geophysical-geological transect and tectonic evolution of the Swiss-Italian Alps. *Tectonics*, **15**, 1036–1064.

SCHMID, S. M., FÜGENSCHUH, B., KISSLING, E. & SCHUSTER, R. 2004. Tectonic map and overall architecture of the Alpine orogen. *Eclogae Geolicae Helvetiae*, **97**, 93–117.

SCHOLGER, R. & STINGL, K. 2004. New paleomagnetic results from the Middle Miocene (Karpatian and

Badenian) in Northern Austria. *Geologica Carpathica, Bratislava,* **55**, 199–206.

SCHÖNBORN, G. 1992. Alpine tectonics and kinematic models of Central Southern Alps. *Memorie di Science Geologiche,* **44**, 229–393.

SCHÖNBORN, G. 1999. Balancing cross sections with kinematic constraints: The Dolomites (northern Italy). *Tectonics,* **18**, 527–545.

ŠIKIĆ, K., BASCH, O. & ŠIMUNIĆ, A. 1977. Basic geological map of Yugoslavia M 1:1000, sheet Zagreb. *Croatian Geological Survey, Federal Geological Institute Beograd.*

SPERNER, B. 1996. *Computer programs for the kinematic analysis of brittle deformation structures and the Tertiary tectonic evolution of the western Carpathians (Slovakia).* PhD thesis, University of Tübingen.

STRAJIN, V., MOJIĆEVIĆ, M., PAMIĆ, J. & SUNARIĆ-PAMIĆ, O. 1977. Basic geological map of Yugoslavia M1:100 000, sheet Vlasenica. *Geological Survey Sarajevo, Federal Geological Institute Beograd.*

TARI-KOVAČIĆ, V. & MRINJEK, E. 1994. The Role of Palaeogene Clastics in the Tectonic Interpretation of Northern Dalmatia (Southern Croatia). *Geologia Croatica, Zagreb,* **47**, 127–138.

TARI, V. & PAMIĆ, J. 1998. Geodynamic evolution of the northern Dinarides and the southern part of the Pannonian Basin. *Tectonophysics,* **297**, 269–281.

TOMLJENOVIĆ, B. 2002. *Structural characteristics of Medvednica and Samoborsko gorje Mts.* PhD thesis, University of Zagreb.

TOMLJENOVIĆ, B. 2005. Synmetamorphic Structural Fabric in Low-Grade Metamorphic Rocks from the Central Part of Mountain Medvednica. *In*: VELIĆ, I., VLAHOVIĆ, I. & BIONDIĆ, R. (eds) *Third Croatian Geological Congress.* Croatian Geological Survey, Zagreb.

TOMLJENOVIĆ, B. & CSONTOS, L. 2001. Neogene-Quaternary structures in the border zone between Alps, Dinarides and Pannonian Basin (Hrvatsko Zagorje and Karlovac Basins, Croatia). *International Journal of Earth Sciences,* **90**, 560–578.

USTASZEWSKI, K., RETTENMUND, S. & SCHMID, S. M. 2005. Investigating the Tisza-Dinarides Boundary: Structural and Petrological Features of the Inselbergs of Prosara, Motajica and Kozara (Northern Bosnia and Hercegovina). *In*: TOMLJENOVIĆ, B., BALEN, D. & VLAHOVIĆ, I. (eds) *7th Workshop on Alpine Geological Studies.* Croatian Geological Survey, Zagreb, 99–100.

VLAHOVIĆ, I., TIŠLJAR, J., VELIĆ, I. & MATIČEC, D. 2005. Evolution of the Adriatic Carbonate Platform: Palaeogeography, main events and depositional dynamics. *Palaeogeography, Palaeoclimatology, Palaeoecology,* **220**, 333–360.

Tertiary cooling and exhumation history in the Maramures area (internal eastern Carpathians, northern Romania): thermochronology and structural data

H. R. GRÖGER[1,2], B. FÜGENSCHUH[3], M. TISCHLER[1], S. M. SCHMID[1] & J. P. T. FOEKEN[4]

[1]*Geologisch Paläontologisches Institut, Universität Basel, Bernoullistrasse 32, 4056 Basel, Switzerland*

[2]*Present address: Statoil ASA, Forusbeen 50, 4035 Stavanger, Norway (e-mail: heigr@statoilhydro.com)*

[3]*Institut für Geologie und Paläontologie, Universität Innsbruck, Innrain 52, Bruno Sander Haus, 6020 Innsbruck, Austria*

[4]*Scottish Universities Environmental Research Centre (SUERC), Rankine Avenue, Scottish Enterprise Technology Park, East Kilbride, G75 0QF, Scotland*

Abstract: The Tertiary kinematic history of the Maramures area is constrained by integrating thermochronological (fission track and (U–Th)/He analysis) data with field-based structural investigations. This study focuses on the tectonic evolution of the northern rim of the Tisza–Dacia block during collision with the European margin. Cretaceous nappe stacking, related metamorphism as well as Late Cretaceous exhumation are evidenced by zircon fission track data. Subsequent Palaeogene to Early Miocene sedimentation led to burial heating and annealing of fission tracks in apatite. Final tectonic uplift was initiated during the convergence of Tisza–Dacia with the European margin, associated with transpressional deformation (16 to 12 Ma). This led to Mid-Miocene exhumation, recorded by apatite fission track cooling ages in the western part of the study area. Transtension between 12 and 10 Ma caused brittle deformation along E–W trending strike-slip faults and SW–NE trending normal faults, delimiting blocks that were tilted towards the SW. This fragmentation of the crust led to enhanced exhumation at rates of 1 mm/a in the central part of the study area, as is documented by Middle to Late Miocene cooling ages (13 to 7 Ma). The outside estimate for the total amount of exhumation since Middle Miocene times is 7 km.

This study addresses the cooling and exhumation history of basement units (Rodna horst and Preluca massif) in northern Romania (Maramures) by combining thermochronological analyses with structural field investigations. The study area is located in the NE prolongation of the Mid-Hungarian fault zone (Csontos & Nagymarosy 1998; Csontos & Vörös 2004), separating the ALCAPA and Tisza–Dacia mega-units (Fig. 1). This crustal-scale boundary plays a key role during the final emplacement of these two crustal-scale blocks (ALCAPA and Tisza–Dacia) in the Carpathian embayment during the Tertiary (Csontos & Nagymarosy 1998). Since the Mid-Hungarian fault zone is largely covered by Neogene sediments of the Pannonian basin, our study area represents one of the very few places where structures related to this important tectonic lineament can be studied in outcrop (Tischler et al. 2006).

Invasion of the ALCAPA and Tisza–Dacia blocks into the Carpathian embayment was triggered by a combination of lateral extrusion in the eastern Alps (Ratschbacher et al. 1991a, b) and retreat of the European lithospheric slab (e.g. Wortel & Spakman 2000; Sperner et al. 2005). Corner effects at the Moesian and Bohemian promontories led to opposed rotations of these continental blocks during their emplacement, well established by palaeomagnetic studies (e.g. Márton 2000; Márton & Fodor 1995, 2003; Márton et al. 2006). The Mid-Hungarian fault zone is of central importance, since it allows for differential movements between the ALCAPA and Tisza–Dacia blocks (Fodor et al. 1999; Csontos & Vörös 2004). Finally, soft collision of these blocks with the European continental margin resulted in the formation of the Miocene fold and thrust belt (Fig. 1; e.g. Royden 1988; Roure et al. 1993; Matenco & Bertotti 2000). A more detailed

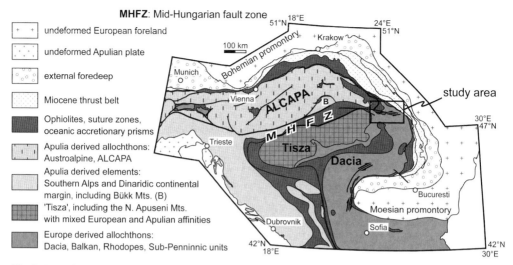

Fig. 1. Tectonic map of the Alpine–Carpathian–Pannonian area (simplified after Schmid et al. 2006). Three major continental blocks are located within the Carpathian embayment: ALCAPA, Tisza and Dacia. During emplacement in Tertiary times, a zone of repeated tectonic activity developed between ALCAPA and the previously consolidated Tisza–Dacia block: the Mid-Hungarian fault zone (MHFZ).

discussion of the Tertiary palaeogeographic and kinematic history of the ALCAPA and Tisza–Dacia blocks can be found in numerous earlier studies (e.g. Balla 1987; Royden & Baldi 1988; Săndulescu 1988; Csontos 1995; Csontos et al. 1992; Fodor et al. 1999; Fügenschuh and Schmid 2005).

The samples analysed in this study derive from pre-Mesozoic basement and the autochthonous sedimentary cover of the Tisza–Dacia block (Preluca massif, Rodna and Maramures Mountains, respectively) and from tectonic units that are part of an accretionary prism which is situated between the ALCAPA and Tisza–Dacia blocks, the so-called Pienides. According to a recent compilation of tectonic units by Schmid et al. (2006), the basement units are considered part of Dacia, but classically the Preluca massif, which is part of the Biharia nappe system, is attributed to Tisza (e.g. Haas & Péró 2004). We applied zircon fission track-, apatite fission track- and apatite (U–Th)/He thermochronological analyses. The combination of these methods constrains the thermal history of the samples in the temperature range between 300 °C and 40 °C, and is well suited to elucidate the thermal evolution in soft collisional regimes, which do not feature major exhumation and exposure of high grade metamorphic rocks (Royden 1993; Morley 2002).

Numerous other studies successfully addressed cooling and exhumation histories, either based on fission track analyses alone (e.g. Sobel & Dumitru 1997; Fügenschuh et al. 2000; Dunkl & Frisch 2002; Danišík et al. 2004; Fügenschuh & Schmid 2003, 2005) or on a combination of fission track and (U–Th)/He analyses (e.g. Foeken et al. 2003; Reiners et al. 2003; Persano et al. 2005; Stöckli 2005 and references therein). Within our working area, some apatite fission track data are already available from the work of Sanders (1998) and Sanders et al. 1999, who addressed the interplay between tectonics and erosion in the southern and eastern Carpathians and the Apuseni Mountains.

All chronostratigraphic ages will be given after Gradstein et al. (2004). Paratethys stages were correlated with Mediterranean stages according to Steininger & Wessely (2000).

Geological setting

The Precambrian to Palaeozoic basement units of the east Carpathians (so-called Bucovinian nappe stack of the Dacia block) mainly consist of metasediments and subordinately orthogneiss (Kräutner 1991; Voda & Balintoni 1994). The Alpine-age Bucovinian nappe stack consists, from bottom to top, of the Infrabucovinian, Subbucovinian and Bucovinian nappes, occasionally separated by Permian to Lower Cretaceous sediments (Săndulescu et al. 1981). Along most parts of the east Carpathians, late Early Cretaceous nappe stacking is generally considered to have taken place under sub-greenschist facies metamorphic conditions. In the so-called Rodna horst (Fig. 2), a part of the Bucovinian nappe stack located

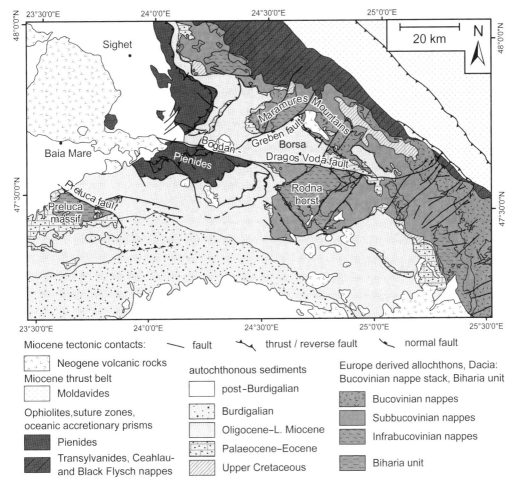

Fig. 2. Tectonic map of the study area. The Pienides, consisting of non-metamorphic flysch nappes, were emplaced in Early Burdigalian times (20.5 to 18.5 Ma, Tischler et al. 2006). Middle to Late Miocene brittle tectonics (16 to 10 Ma) led to the exhumation of the Rodna horst and the Preluca massif (Tischler et al. 2006). The most important Middle to Late Miocene structures are the Preluca fault, the Greben fault and the Bogdan–Dragos–Voda fault. To the west, the Bogdan–Voda fault is sealed by Neogene volcanic rocks. The map is compiled after Giusca & Radulescu (1967), Raileanu & Radulescu (1967), Raileanu & Saulea (1968), Ianovici & Dessila-Codarcea (1968), Ianovici & Radulescu (1968), Ianovici et al. (1968), Kräutner et al. (1978, 1982, 1983, 1989), Borcos et al. (1980), Dicea et al. (1980), Sandulescu (1980), Sandulescu & Russo-Sandulescu (1981), Sandulescu et al. (1981), Rusu et al. (1983), Sandulescu et al. (1991) and Aroldi (2001).

internally of the main east Carpathian chain, greenschist facies conditions during this event were postulated by Balintoni et al. (1997). The Preluca massif, consisting of metasediments (Rusu et al. 1983), is part of the Biharia nappe system of the North Apuseni Mountains. However, in contrast to most authors (e.g. Haas & Péró 2004), the Biharia nappe system was correlated with the Bucovinian nappe system (i.e. considered as part of Dacia) by Schmid et al. (2006), based on a re-interpretation of geophysical data from the Transylvanian basin. In any case, Tisza and Dacia became amalgamated during the Cretaceous and formed a consolidated Tisza–Dacia block ever since Early Tertiary times (Csontos 1995; Csontos & Vörös 2004).

Latest Early Cretaceous juxtaposition of the Bucovinian nappe stack against the Ceahlau and Black Flysch units (Săndulescu 1982) was followed by exhumation and erosion, leading to the sedimentation of the Late Cretaceous post-tectonic cover (Ianovici et al. 1968; Săndulescu 1994). Exhumation which followed the juxtaposition of the entire Alpine nappe pile against the internal

Moldavides (Săndulescu 1982) occurred during the latest Cretaceous (Săndulescu 1994) and led to erosion of large parts of these Upper Cretaceous cover units.

The deposition of a second stack of post-tectonic sediments during the Tertiary led to renewed burial. While the sedimentation already started in Palaeocene times in the Preluca massif (Rusu et al. 1983), non-deposition and/or erosion continues in the area of the Bucovinian nappe stack, as is indicated by a Palaeocene hiatus. Eocene conglomerates (Ypresian?, Kräutner et al. 1983) are the oldest Tertiary strata preserved on the previously exhumed Bucovinian nappe stack which they stratigraphically directly overlie except for a few places where the Late Cretaceous basins remained preserved (Fig. 2). Late Eocene (Late Lutetian–Priabonian) sedimentation is characterized by a general deepening of the depositional environment towards the NW. While platform carbonates indicating shallow water depths are found in large parts of the study area (main east Carpathian Chain and Rodna horst: Kräutner et al. 1978, 1982, 1983, 1989; Preluca massif: Rusu et al. 1983; de Broucker et al. 1998), sandy marls and marls dominate west of the Greben fault (Fig. 2; Săndulescu et al. 1991). Oligocene sediments overlying the basement units of the Rodna horst in direct stratigraphic contact document the Eocene palaeorelief (Kräutner et al. 1982).

Early Oligocene to Early Miocene (Aquitanian) sedimentation is dominated by siliciclastic turbidites, which show an overall coarsening-upward trend (Dicea et al. 1980; Săndulescu et al. 1991), thought to reflect the juxtaposition of ALCAPA and Tisza–Dacia (Tischler et al. 2008). A Burdigalian clastic wedge (Hida beds) testifies the overthrusting of the Pienides (Ciulavu et al. 2002), i.e. thrusting of ALCAPA onto Tisza–Dacia. The Pienides mainly consist of Eocene to Oligocene non-metamorphic flysch units (Aroldi 2001), which additionally contain phacoids of Pieniny-Klippen-type material embedded in Eocene flysch (Săndulescu 1980; Săndulescu et al. 1993).

Post-Burdigalian sedimentation starts with the deposition of the Middle Miocene (Badenian) Dej Tuff (Mason et al. 1998). Subduction-related calc-alkaline magmatism started during Middle Miocene times (13.5 Ma, Pécskay et al. 1995). Magmatic activity led to the formation of a linear volcanic chain along the inner side of the east Carpathians, with a general trend of decreasing ages towards the SE.

The most obvious structure in the study area is the E–W striking, predominantly left-lateral Bogdan–Dragos–Voda fault system (Tischler et al. 2006). The Bogdan–Voda fault to the west offsets the Palaeogene to Early Miocene cover of Tisza–Dacia, as well as the nappe pile of the Pienides. It is essentially sealed by Mid-Miocene volcanic rocks (Fig. 2). The Dragos–Voda fault delimits the Rodna horst to the north.

Methods

Fission track analysis: methodology and analytical procedure

For an overview of the methodology and applications of fission track (FT) analysis, the reader is referred to Wagner and van den Haute (1992), Andriessen (1995), Gallagher et al. (1998) and Reiners & Ehlers (2005). Regarding the fission track partial annealing zones, we used the temperature brackets given by Gleadow & Duddy (1981) and Green et al. (1989) for apatite (APAZ, 60 to 120 °C) and Hurford (1986) for zircon (ZPAZ, 190 to 290 °C), respectively.

After conventional crushing, sieving, magnetic and heavy liquid separation, apatite grains were mounted in epoxy resin, polished and etched for 40 seconds at room temperature in 6.5% HNO_3. Zircon grains were mounted in PFA® Teflon, polished and etched for 12 to 24 hours in a NaOH/KOH eutectic melt at about 225 °C. Irradiation was carried out at the High Flux Australian Reactor (HIFAR) with neutron fluxes monitored in CN5 for apatite and CN1 for zircon. All samples were analysed using the external detector method (Gleadow 1981) with muscovite as an external detector. Muscovite was etched for 40 minutes at room temperature in 40% HF.

Fission tracks were counted on a Zeiss® microscope with a computer-controlled scanning stage ('Langstage', Dumitru 1993) at magnifications of ×1250 for apatite and ×1600 for zircon (dry). Unless mentioned otherwise, all ages given are central ages (Galbraith & Laslett 1993). Ages were calculated using the zeta calibration method (Hurford & Green 1983) with a ξ value of 355.96 ± 9.39 (Durango standard, CN5) for apatite and 141.40 ± 6.33 (Fish Canyon tuff standard, CN1) for zircon. Calculations were done with the aid of the Windows software TrackKey (Dunkl 2002). For separation of subpopulations (i.e. the youngest population in partially annealed samples), the Windows software PopShare was used (Dunkl & Székely 2002).

Lengths of confined horizontal tracks in apatite were measured at ×1250 magnification. Track length distributions of confined horizontal tracks are diagnostic for different thermal histories (Crowley 1985; Gleadow et al. 1986a, b; Galbraith & Laslett 1993) and allow for thermal modelling. Thermal modelling of fission track parameters was carried out with the program AFTSolve (Ketcham et al. 2000) with the annealing model of Laslett et al. (1987).

Apart from temperature, chemical composition is known to have an effect on the annealing behaviour (Green et al. 1985, 1986; Crowley & Cameron 1987; OSullivan & Parrish 1995; Siddall & Hurford 1998). Thus, the long axis of etch pits on the polished surface (referred to as Dpar, Burtner et al. 1994) was measured (magnification ×2000) as an indicator of the annealing behaviour of apatite.

Apatite (U–Th)/He dating: methodology and analytical procedure

Apatite (U–Th)/He dating records the cooling of a sample between 80 and 40 °C (Wolf et al. 1998; Ehlers & Farley 2003 and references therein) and is complementary to apatite FT when reconstructing the latest exhumation stages. Apatite (U–Th)/He analyses were conducted on four samples: two from the Rodna horst and two from the Preluca massif. Suitability of samples for (U–Th)/He dating was mainly governed by sample quality (inclusions) and grain morphology. Between one to four apatite grains for each sample were hand-picked in ethanol at ×218 magnification using a binocular microscope. While the frosted nature of the apatite grains made the identification of inclusion-free grains difficult, small zircon inclusions were only observed in one replicate of sample P4 (II). To minimize grain size variation effects (Farley 2000), grains of similar radius were selected for each aliquot. ^{4}He, U and Th analyses were conducted following the procedures of Balestrieri et al. (2005). The total analytical uncertainty of He ages of each aliquot is approximately ±10% (2σ), governed largely by uncertainty in blank corrections and spike concentrations. Correction for He recoil loss was made using procedures described in Farley (2002). Analyses of a Durango apatite standard aliquot (two grains) yield 33.1 ± 0.7 Ma, which is indistinguishable from mean Durango ages measured at SUERC (32.8 ± 1.3 Ma, Foeken et al. 2006) and reported Durango ages (e.g. 32.1 ± 1.7 Ma, House et al. 2000).

Results of kinematic analyses

In the following, only a summary of the Middle to Late Miocene tectonic evolution of the study area will be given (see Tischler et al. 2006 for a detailed discussion). Two stages dominated by brittle sinistral strike-slip deformation are documented in the study area: Middle Miocene transpression (Fig. 3a), followed by Middle to Late Miocene transtension (Fig. 3b). During both stages, shortening remained NE–SW oriented. The most obvious structures related to these stages are the E–W striking Bogdan–Voda fault (BVF) and the Dragos–Voda fault (DVF). The left lateral activity of the Bogdan–Voda fault is already apparent in map view, as evidenced by the 25 km sinistral offset of the Burdigalian-age thrust contact of the Pienides (Fig. 3).

During the transpressional stage, the Bogdan–Voda fault was active as a sinistrally transpressive fault, terminating eastwards in a thrust splay geometry (Fig. 3a). Other major structures attributed to this first transpressional stage are NW–SE striking reverse faults (e.g. a back-thrust NE of Borsa), the Preluca fault (PF), as well as very open NW–SE to WNW–ESE striking folds.

During the later transtensional stage (Fig. 3b), the Bogdan–Voda and Dragos–Voda faults both acted together as one single continuous fault. Sinistral offset along the so-called Bogdan–Dragos–Voda fault system (Tischler et al. 2006) diminishes eastwards and terminates in an extensional horsetail splay geometry. Additional features attributed to this second stage are SW–NE-striking normal faults, such as the Greben fault and numerous faults within the Rodna horst (Fig. 3b).

Results of the thermochronological analysis

Sampling approach

All sample localities are depicted in Figure 4. The Preluca massif is covered by four samples (P1, P2, P3, P4) while three samples are from an immediately adjacent smaller basement body located further to the NE (P5, P6, P7). The Maramures Mountains further to the NE have been sampled along two orthogonal profiles, yielding a total of 10 samples. One profile is orogen-perpendicular (samples M01, M02, M03, M04, M05); the other one crosses the Greben fault and the horsetail splay of the Bogdan–Dragos–Voda fault (samples M06, M08, M07, M09, M13, M14). In the area of the Rodna horst, located S of the Dragos–Voda fault, four vertical profiles have been sampled, each of them within a distinct fault-bounded block. Each profile comprises five to six samples (see groups of samples listed under R1, R2, R3 and R4 in Fig. 4). Additional samples have been collected within the central block of the Rodna horst and adjacent to the Dragos–Voda fault (sample group R5). The basement samples comprise mainly paragneisses and rarely orthogneiss.

A last group of samples has been taken from sedimentary formations. The samples S2, S4 and

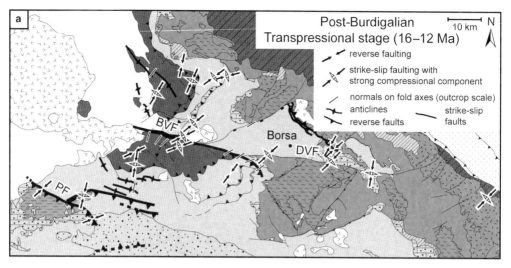
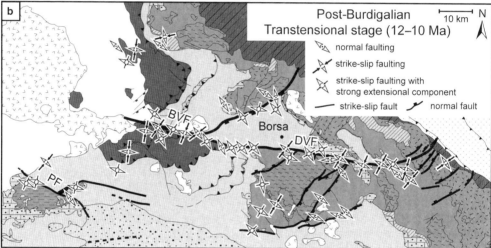

Fig. 3. Kinematics and structures related to the post-Burdigalian (16 to 10 Ma) tectonics in the study area (Tischler et al. 2006). The active structures are marked by thick lines (BVF = Bogdan–Voda fault, DVF = Dragos–Voda fault, PF = Preluca fault). A transpressional stage (**a**) precedes a transtensional stage (**b**), featuring constant SW–NE shortening. While the sinistral Bogdan–Voda fault is independently active during the transpressional stage (a), the Bogdan–Voda fault and Dragos–Voda fault are linked together during the transtensional stage (b).

S5 are from sedimentary units of the Pienides while S1 and S3 are from the post-tectonic Palaeogene to Early Miocene autochthonous cover of Tisza–Dacia (Late Oligocene Borsa sandstone). Samples have been taken from fine-grained sandstone horizons.

Zircon FT data

The zircon FT data are reported in Table 1. Zircon fission tracks were largely reset during the Cretaceous metamorphic overprint, associated with nappe stacking of the basement units.

The zircon fission track central ages of the Preluca massif and the Rodna horst range between 68.1 and 100.1 Ma (Fig. 4). Most samples pass the Chi-square test (Table 1, $\chi^2 > 5\%$). The zircon single grain age distributions show clusters (Fig. 4c, d) that indicate Late Cretaceous cooling (mainly Coniacian to Campanian) after a thermal event that led to full annealing of zircon (Fig. 4c, d), i.e. an event associated with temperatures exceeding 300 °C.

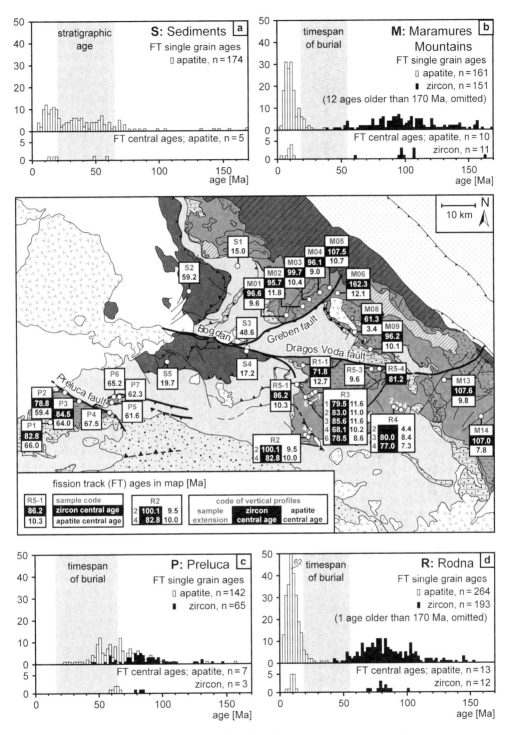

Fig. 4. Map indicating location of samples and results of the fission track analyses in terms of central ages, together with diagrams that summarize the single grain ages (a, b, c, d). In the sedimentary samples (a), the post-depositional thermal overprint caused strong annealing (S1, S4, S5), but also provenance single grain ages are preserved (S2, S3). In the Bucovinian nappe stack (b, d), apatite grains have been fully reset by the Palaeogene to Early Miocene burial, while zircon remained thermally undisturbed. Overlapping single grain ages of zircon and apatite from the Preluca massif reflect continuous exhumation through the ZPAZ and APAZ during the Late Cretaceous (c).

Table 1. Zircon fission track data. All samples have been analysed using the external detector method (Gleadow 1981) with a zeta value (Hurford & Green 1983) of 141.40 ± 6.33 (Fish Canyon Tuff standard, CN1). Code: sample code; Latitude: latitude in WGS84; Longitude: longitude in WGS84; Alt. [m]: altitude above sea level; N Grains: number of grains counted; Ps [$\times 10^5$ cm^{-2}]: spontaneous track density; Ns: number of spontaneous tracks counted; Pi [$\times 10^5$ cm^{-2}]: induced track density; Ni: number of induced tracks counted; Pd [$\times 10^5$ cm^{-2}]: standard track density; Nd: number of standard tracks counted; χ^2 [%]: Chi-square probability (Galbraith 1981); Central age $\pm 1\sigma$ [Ma]: zircon fission track central age (Galbraith & Laslett 1993)

Code	Latitude	Longitude	Alt. [m]	N Grains	Ps [$\times 10^5$ cm^{-2}]	Ns	Pi [$\times 10^5$ cm^{-2}]	Ni	Pd [$\times 10^5$ cm^{-2}]	Nd	χ^2 [%]	Central Age $\pm 1\sigma$ [Ma]
M01	24.496670	47.729790	540	10	176.37	1291	49.32	361	3.85	3065	34	96.6 ± 7.6
M02	24.560710	47.753720	580	4	109.39	367	30.10	101	3.77	3065	13	95.7 ± 13.5
M03	24.586640	47.772540	630	20	86.88	1468	35.45	599	5.80	3605	81	99.7 ± 6.8
M04	24.628100	47.791450	680	10	70.07	436	30.54	190	5.97	3605	84	96.1 ± 9.5
M05	24.667090	47.804300	745	20	98.44	2366	37.20	894	5.86	3605	<5	107.5 ± 7.4
M06	24.698590	47.790850	790	20	129.07	2543	33.35	657	5.91	3605	<5	162.3 ± 13.0
M08	24.770543	47.690263	820	20	123.09	2703	52.32	1149	3.69	3065	<5	61.3 ± 4.7
M09	24.833619	47.647505	1660	20	126.15	2501	32.18	638	3.50	3065	22	96.2 ± 6.5
M13	25.128220	47.571246	930	14	253.89	1812	58.71	419	3.54	3065	23	107.6 ± 8.5
M14	25.279510	47.478662	850	20	137.67	2830	30.60	629	3.38	3065	<5	107.0 ± 9.1
P1	23.574760	47.430957	315	14	79.60	1010	25.22	320	3.73	3065	55	82.8 ± 6.6
P2	23.628772	47.509842	215	16	155.74	1919	47.40	584	3.42	3065	38	78.8 ± 5.6
P3	23.686807	47.488712	610	35	128.38	3336	40.87	1062	3.81	3065	<5	84.5 ± 5.5
R1-1	24.559360	47.597860	1550	17	91.73	996	31.13	338	3.46	2967	25	71.8 ± 6.1
R2-2	24.597260	47.414920	1105	20	80.41	981	32.38	395	5.74	3605	78	100.1 ± 7.6
R2-4	24.590430	47.419300	705	26	115.20	2312	34.43	691	3.52	2967	12	82.8 ± 5.9
R3-1	24.620652	47.533782	2020	20	188.30	2550	55.68	754	3.35	3065	47	79.5 ± 5.1
R3-2	24.582850	47.423290	1465	20	74.15	1642	39.02	864	6.25	3605	62	83.0 ± 5.5
R3-3	24.608990	47.528930	1310	13	76.26	842	38.77	428	6.20	3605	37	85.6 ± 6.8
R3-4	24.604450	47.526330	1155	7	94.47	607	60.11	385	6.14	3605	50	68.1 ± 5.5
R3-6	24.587980	47.518690	945	20	92.24	934	49.77	504	6.03	3605	11	78.5 ± 6.4
R4-3	24.941010	47.494710	980	2	116.94	107	37.16	34	3.62	3065	86	80.0 ± 16.2
R4-4	24.960230	47.490430	700	20	134.72	1923	45.89	655	3.69	3065	<5	77.0 ± 5.8
R5-1	24.546451	47.552021	1150	8	204.11	635	57.54	179	3.46	3065	47	86.2 ± 8.4
R5-4	24.872870	47.596160	1270	20	161.18	2517	50.33	786	3.58	3065	10	81.2 ± 5.5

Some basement samples from the Maramures Mountains, however, yielded Cenomanian central ages (95.7 to 99.7 Ma; M01, M02, M03, M04, M09; Fig. 4b) which also pass the Chi-square test (Table 1, $\chi^2 > 5\%$). Hence they indicate earlier (Cenomanian) cooling after full annealing of zircon. In the most external samples (M05, M06, M13, M14), the coexistence of Late Cretaceous and Palaeozoic single grain ages (oldest single grain 310 Ma, M06; Appendix 1) indicates that annealing was only partial and was followed by Late Cenomanian cooling (Fig. 4b). Sample M08 with a younger Tertiary (61.3 Ma) central age forms an exception amongst this group.

No zircon FT single grain ages younger than Eocene have been found in samples from the basement units. Hence, it can be excluded that substantial heating, reaching temperatures within the ZPAZ, did result from Palaeogene to Early Miocene burial.

Apatite FT data

The apatite FT data are reported in Table 2. For the Preluca massif, an overlap between zircon and apatite single grain ages is observed (Fig. 4c). This overlap indicates continued slow cooling from temperatures above 200 °C down to temperatures below the lower limit of the APAZ (60 °C). The apatite central ages in the Preluca massif range from 59.4 to 67.5 Ma (Fig. 4c). A close look at the single grain age distribution reveals weak evidence for the existence of two clusters (Fig. 4c). Possibly the earlier cluster is related to Late Cretaceous cooling before renewed sedimentation. The second cluster could indicate partial annealing during Palaeogene to Early Miocene burial. Although etch pit long axes in the Preluca samples spread by at least 1 μm in each sample, suggesting compositional diversity, no clear correlation between etch pit long axis and single grain ages can be deduced (Appendix 2).

Three sedimentary samples (S1, autochthonous cover; S4 and S5, Pienides) show single grain ages that are younger than the stratigraphic age (Fig. 4a, Appendix 1). This indicates considerable post-depositional annealing. On the other hand, samples S2 (Pienides) and S3 (autochthonous cover) show single grain ages older than the stratigraphic age (Appendix 1).

In case of the samples from the Maramures Mountains and the Rodna horst, however, Palaeogene to Early Miocene burial caused total annealing of all apatite fission tracks (Fig. 4b, d). Central ages range between 7.3 and 12.7 Ma and thus indicate Middle to Late Miocene cooling. Two samples even yielded younger (Pliocene) central ages (M08: 3.4 Ma; R4-2: 4 Ma), best interpreted as indicating thermal overprint during Neogene volcanic activity (Pécskay *et al.* 1995). This interpretation is supported by the unusually young zircon FT central age of sample M08 (Fig. 4).

Confined track length measurements (Table 2) were used for thermal modelling of the apatite data that will be discussed below. In case of the sedimentary samples, the pre-depositional thermal history cannot be assumed to be identical for all constituent grains, hence confined track lengths were measured in dated grains only. In case of the basement samples, a common thermal history of individual samples allows for length measurements also in the non-dated grains (up to a maximum of 100 lengths). Only a limited number of samples provided sufficient track length data because of the low track densities. All basement samples show unimodal track length distributions, with mean values between 12.6 and 13.5 μm (Table 2). Sedimentary samples only provided a few track length data, which makes a characterization of the length distribution difficult.

Apatite (U–Th)/He data

The apatite (U–Th)/He data are reported in Table 3. In the Preluca massif, these apatite (U–Th)/He ages scatter and do not replicate within 2σ error. While the replicate of sample P4-II had small zircon inclusions and was therefore excluded, samples P3 and P4 both yielded apatite (U–Th)/He ages that are younger than the apatite FT single grain ages. Replicate ages from the Rodna massif yielded 12.0–13.3 Ma, which are, within 2σ error, indistinguishable from the accompanying apatite FT ages. This suggests fast cooling for the last stages of exhumation of the Rodna horst.

Thermal modelling of the apatite FT data

Thermal modelling used the apatite single grain ages and the confined track lengths as input data (Fig. 5). Additionally, time–temperature (t–T) constraints, independent from the apatite FT data, are incorporated. The t–T paths are modelled using the software AFTSolve (version 1.3.0, Ketcham *et al.* 2000) with the annealing model of Laslett *et al.* (1987). Modelling was done inverse monotonic with 10 000 runs for each model, using a Monte Carlo modelling scheme. Initial track length was 16.3 μm with a length reduction in age standard of 0.890. All models were run twice in order to identify reproducible trends.

In the case of the basement samples, such independent t–T constraints are given by the zircon FT central ages obtained on the same sample (or samples close by) and the age of the

Table 2. Apatite fission track data. All samples have been analysed using the external detector method (Gleadow 1981) with a zeta value (Hurford & Green 1983) of 355.96 ± 9.39 (Durango standard, CN5). Code: sample code; Latitude: latitude in WGS84; Longitude: longitude in WGS84; Alt. [m]: altitude above sea level; N Grains: number of grains counted; Ps [×10^5 cm^{-2}]: spontaneous track density; Ns: number of spontaneous tracks counted; Pi [×10^5 cm^{-2}]: induced track density; Ni: number of induced tracks counted; Pd [×10^5 cm^{-2}]: standard track density; Nd: number of standard tracks counted; χ^2 [%]: Chi-square probability (Galbraith 1981); Central age ± 1σ [Ma]: apatite fission track central age (Galbraith & Laslett 1993); N Length: number of confined track lengths measured; Mean Length ± 1σ [μm]

Code	Latitude	Longitude	Alt. [m]	N Grains	Ps [×10^5 cm^{-2}]	Ns	Pi [×10^5 cm^{-2}]	Ni	Pd [×10^5 cm^{-2}]	Nd	χ^2 [%]	Central Age ±1σ [Ma]	N Length	Mean Length ±1σ [μm]
M01	24.496670	47.729790	540	25	0.49	73	112.21	16596	119.59	6511	10	9.6 ± 1.2	–	–
M02	24.560710	47.753720	580	20	0.94	114	138.73	16842	98.14	6511	32	11.8 ± 1.2	–	–
M03	24.586640	47.772540	630	15	1.12	64	148.09	8471	77.35	3606	55	10.4 ± 1.3	–	–
M04	24.628100	47.791450	680	20	1.29	129	35.66	3580	14.10	4605	58	9.0 ± 0.9	–	–
M05	24.667090	47.804300	745	20	2.24	168	28.73	2155	7.74	4223	99	10.7 ± 0.9	80	12.94 ± 2.21
M06	24.698590	47.790850	790	3	0.51	10	87.59	1708	116.61	4196	39	12.1 ± 3.9	100	12.98 ± 3.20
M08	24.770543	47.690263	820	2	0.98	4	43.90	180	8.52	4451	60	3.4 ± 1.7	–	–
M09	24.833619	47.647505	1660	16	0.53	31	90.19	5276	96.19	5773	99	10.1 ± 1.8	–	–
M13	25.128220	47.571246	930	20	0.44	64	8.57	1261	10.81	4451	99	9.8 ± 1.3	–	–
M14	25.279510	47.478662	850	20	1.34	152	32.43	3668	10.50	4451	26	7.8 ± 0.7	43	13.35 ± 2.12
P1	23.574760	47.430957	315	20	3.06	297	8.40	815	10.23	4223	65	66.0 ± 4.9	100	13.01 ± 2.54
P2	23.628772	47.509842	215	20	6.33	713	16.79	1892	8.93	4223	9	59.4 ± 3.7	100	12.98 ± 1.92
P3	23.686807	47.488712	610	20	6.93	671	18.18	1760	9.37	4223	<5	64.0 ± 4.6	100	12.86 ± 1.70
P4	23.810507	47.465312	350	22	1.09	150	26.93	3724	94.57	6511	43	67.5 ± 6.0	–	–
P5	23.817260	47.488947	390	20	18.13	2489	44.93	6169	8.50	4223	<5	61.6 ± 3.0	100	12.75 ± 1.23
P6	23.824540	47.503686	430	20	14.12	1028	37.89	2758	9.80	4223	<5	65.2 ± 4.0	100	12.62 ± 2.08
P7	23.846548	47.503894	380	20	4.84	456	15.72	1482	11.44	4451	94	62.3 ± 3.8	100	12.60 ± 1.90
R1-1	24.559360	47.597860	1550	20	0.58	41	11.86	837	14.55	4766	97	12.7 ± 2.1	–	–
R2-2	24.597260	47.414920	1105	20	0.41	63	93.25	14333	121.63	4196	92	9.5 ± 1.2	–	–
R2-4	24.590430	47.419300	705	20	0.52	51	108.36	10717	118.28	4196	88	10.0 ± 1.4	100	12.80 ± 2.65
R3-1	24.620652	47.533782	2020	24	4.41	539	65.71	8030	9.78	4451	8	11.6 ± 0.7	59	13.12 ± 2.79
R3-2	24.582850	47.423290	1465	20	2.55	357	54.69	7646	13.04	4766	10	11.0 ± 0.8	100	13.50 ± 2.20
R3-3	24.608990	47.528930	1310	20	2.48	496	48.58	9716	12.79	4766	13	11.6 ± 0.7	100	13.25 ± 2.31
R3-4	24.604450	47.526330	1155	20	0.38	50	8.30	1096	12.54	4766	99	10.2 ± 1.5	–	–
R3-6	24.587980	47.518690	945	20	0.94	111	15.89	1873	8.17	4223	79	8.6 ± 0.9	–	–
R4-2	24.931050	47.498160	1305	20	0.72	85	38.19	4499	13.00	4451	46	4.4 ± 0.5	–	–
R4-3	24.941010	47.494710	980	20	2.02	216	53.45	5730	12.48	4451	84	8.4 ± 0.6	64	12.81 ± 2.48
R4-4	24.960230	47.490430	700	20	0.89	73	26.52	2185	12.27	4451	95	7.3 ± 0.9	–	–
R5-1	24.546451	47.552021	1150	20	1.31	93	20.94	1487	9.26	4451	93	10.3 ± 1.1	–	–
R5-3	24.951870	47.550260	1400	20	3.02	218	51.72	3734	9.22	4557	8	9.6 ± 0.9	100	13.32 ± 2.45
S1	24.296360	47.854000	550	38	2.69	514	41.23	7884	12.91	4605	29	15.0 ± 0.9	15	13.68 ± 2.01
S2	24.111950	47.782210	350	39	9.25	1787	33.57	6489	10.97	4605	<5	59.2 ± 5.0	61	12.14 ± 2.09
S3	24.351650	47.641020	530	40	7.21	1500	32.93	6850	12.27	4605	<5	48.6 ± 3.0	35	12.47 ± 2.33
S4	24.348370	47.635130	555	21	1.61	174	21.52	2320	12.59	4605	12	17.2 ± 1.6	–	–
S5	24.030170	47.599060	555	36	1.48	242	129.53	21178	94.83	4196	<5	19.7 ± 2.2	19	10.08 ± 2.71

Table 3. *Apatite (U–Th)/He data. Analyses of one Durango apatite standard aliquot (two grains) yield 33.1 ± 0.7 Ma. Code: sample code; ^{4}He [cc STP]: concentration of ^{4}He; U^{238} [ng]: concentration of U^{238}; Th^{232} [ng]: concentration of Th^{232}; Th/U: Th^{232}/U^{238} ratio; (U–Th)/He Age [Ma]; Ft: correction of α-recoil (Farley 2002); Corrected (U–Th)/He Age [Ma]; Error 2σ; FT Central age ±1σ[Ma]: apatite fission track central age (Galbraith & Laslett 1993)*

Code	4He [cc STP]	U^{238} [ng]	Th^{232} [ng]	Th/U	(U–Th)/He Age [Ma]	Ft	Corrected (U–Th)/He Age [Ma]	Error 2σ	FT Central Age ±1σ [Ma]
P3-I	4.29E-11	0.015	0.009	0.6	20.5	0.68	30.1	3.0	64.0 ± 4.6
P3-II	4.75E-11	0.025	0.018	0.7	13.4	0.69	19.3	1.9	
P4-I	5.79E-11	0.054	0.027	0.5	7.9	0.82	9.6	1.0	67.5 ± 6.0
P4-II	9.21E-11	0.014	0.045	3.2	30.6	0.80	38.3	3.8	
R2-4-I	1.54E-10	0.114	0.030	0.3	10.4	0.78	13.3	1.3	10.0 ± 1.4
R2-4-II	7.55E-11	0.058	0.027	0.5	9.6	0.80	12.0	1.2	
R3-2-1	8.00E-10	0.657	0.033	0.0	9.8	0.76	12.9	1.3	11.0 ± 0.8

unconformably overlying sediments. In the case of the Preluca massif, the time of maximum burial can also be constrained by assuming it to coincide with the end of Burdigalian sedimentation (Hida beds: 16 Ma). In the Bucovinian nappe stack, maximum burial temperatures are assumed to have been reached by a combination of sedimentary and tectonic burial during imbrication of the autochthonous sediments related to thrusting of the Pienides (18.5 Ma, Tischler et al. 2006).

Regarding the sedimentary samples, one t–T constraint is given by the stratigraphic age of the dated sediment (Săndulescu & Russo-Săndulescu 1981; Săndulescu et al. 1991; Aroldi 2001). Interestingly, a pre-Tertiary heating event has to be assumed in order to allow for the modelling of samples S2 and S3. The timing of maximum burial of the samples from the Pienide nappes (S2 and S5) is inferred to immediately pre-date nappe emplacement (20 Ma, Tischler et al. 2006), since erosion of the Pienides during thrusting is indicated by the deposition of a Burdigalian clastic wedge. In the case of the autochthonous cover (samples S1 and S3), the overthrusting of the Pienides is considered to have caused additional burial. Hence it was assumed that maximum temperatures were reached at the end of nappe emplacement (18.5 Ma, Tischler et al. 2006).

In the case of the Preluca massif, Palaeogene to Early Miocene burial caused partial annealing of apatite (Fig. 6). Late Cretaceous to Early Tertiary cooling is constrained by the zircon FT data and by the sedimentary unconformity. Thermal modelling suggests an increase of maximum temperatures from around 60 °C (P1) in the SW to around 80 °C in the NE (P5, P7), interpreted in terms of an increasing amount of burial towards the NE. The modelled maximum temperature near the Preluca fault (sample P3, 69 °C) in the upper part of the apatite helium partial retention zone is in good

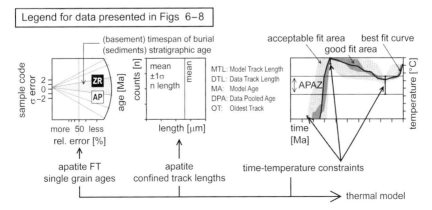

Fig. 5. Diagram explaining the thermal modelling of apatite FT data. This figure also serves as a legend for Figures 6–8. The single grain ages and the confined track lengths provide input data. Using additional independent constraints (e.g. zircon FT data, stratigraphic information), t–T paths fitting the data are modelled.

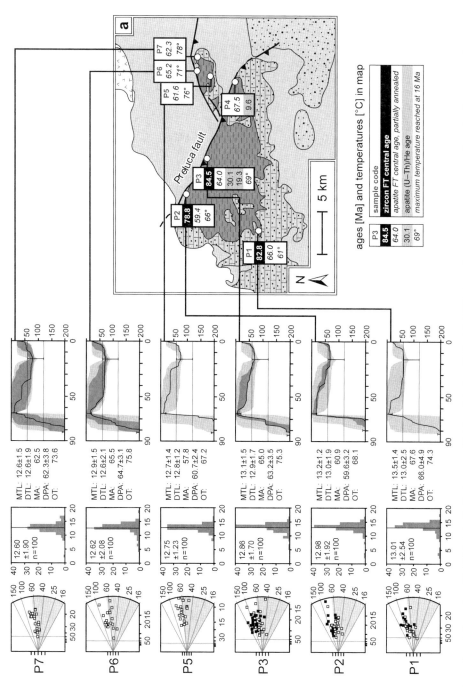

Fig. 6. Thermal modelling of apatite FT data in the Preluca massif. All radial plots (Galbraith 1990) show the same age spectrum to allow for direct comparison. Burial heating by the deposition of Palaeogene to Early Miocene sediments caused partial annealing of fission tracks in apatite. The models slightly indicate a deepening in burial by sedimentation from SW to NE, reflected by the increasing maximum temperatures modelled (a). The temperatures reached at the Preluca fault (69 °C) are confirmed by partial (P3) and full retention (P4) of He in apatite.

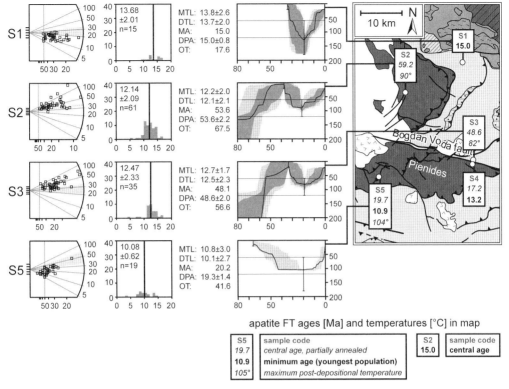

Fig. 7. Thermal modelling of apatite FT data from the sedimentary units. All radial plots (Galbraith 1990) show the same age spectrum to allow for direct comparison. All samples show at least post-depositional partial annealing of fission tracks in apatite; S1 has been fully annealed after deposition. Due to strong annealing in S4 and S5, minimum ages of the youngest populations were calculated. These minimum ages and the central age of S1 indicate Middle to Late Miocene cooling.

agreement with full (P4) and partial (P3) retention of helium, as is indicated by the (U–Th)/He data.

For samples S2 and S3, thermal modelling indicates heating to temperatures within the APAZ, followed by slow cooling (Fig. 7). The thermal history of S1, on the other hand, is characterized by fast heating to temperatures above the APAZ directly preceding fast cooling to temperatures within the APAZ. This short-lived thermal pulse is most likely related to the overthrusting by the Pienides. The thermal modelling of sample S5 shows that cooling from maximum temperatures occurred after 10 Ma (Fig. 7). The youngest grain populations in the strongly annealed samples south of the Bogdan–Voda fault indicate cooling during the Middle (S4: 13.2 Ma) and Late Miocene (S5: 10.9 Ma).

In the case of the Bucovinian nappe stack, the exhumation and burial history before 18.5 Ma was also modelled for all samples. However, due to the complete annealing of pre-Miocene apatite fission tracks, the track parameters are only sensitive regarding the final stages of exhumation (Fig. 8). The complete cooling paths, including the pre-Tertiary history, are only depicted for sample M14 (exemplary for the Maramures Mountains, Fig. 8a) and for R5-3 (exemplary for the Rodna horst, Fig. 8b). Both the zircon FT data and a sedimentary unconformity constrain Late Cretaceous to Eocene cooling in the Rodna horst. In the Maramures Mountains, however, the basement has already been exhumed in Cenomanian times. After maximum burial and heating at 18.5 Ma, most samples enter the upper limit of the APAZ between 15 and 12 Ma before present, and enhanced cooling occurred generally between 10 and 7 Ma.

Revealing the Miocene exhumation history of the Rodna horst

By integrating information provided by the available geological maps (Kräutner *et al.* 1978, 1982, 1983,

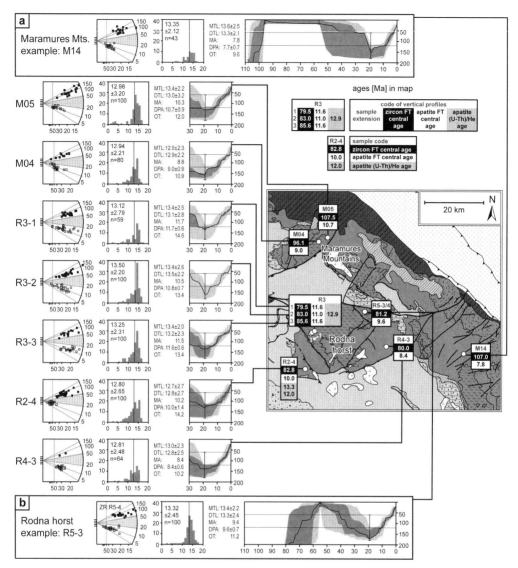

Fig. 8. Thermal modelling of samples from the Bucovinian nappe stack. All radial plots (Galbraith 1990) show the same age spectrum to allow for direct comparison. Because apatite has been fully annealed during Palaeogene to Early Miocene burial, the modelled t–T paths are tightly constrained only for the last stages of Miocene exhumation. Therefore, the results of the complete modelling, starting in the Cretaceous, are only given for the samples M14 (**a**) and R5-3 (**b**), which serve as examples for the Maramures Mountains and the Rodna horst, respectively.

1989) and by constructing schematic cross-sections (Fig. 9c, d), the progressive exhumation of a more internal part of the Bucovinian nappe stack—the Rodna horst—will now be discussed in four time slices (Fig. 10). The schematic cross-section 1 (Fig. 9c) depicts reverse faulting and open folding (5% profile-parallel shortening) related to the post-Burdigalian transpressional stage. In the perpendicular cross-section 2, however, normal faulting related to the subsequent transtensional stage is more evident (Fig. 9d, 7% profile-parallel extension). In section 1, the base Oligocene is tilted to the SW by some 3°, locally reaching up to 5° in the central block of the Rodna horst (Fig. 9b). Lines indicating 5° and 3° tilt are given for comparison in Fig. 9b. Two SW–NE striking normal faults

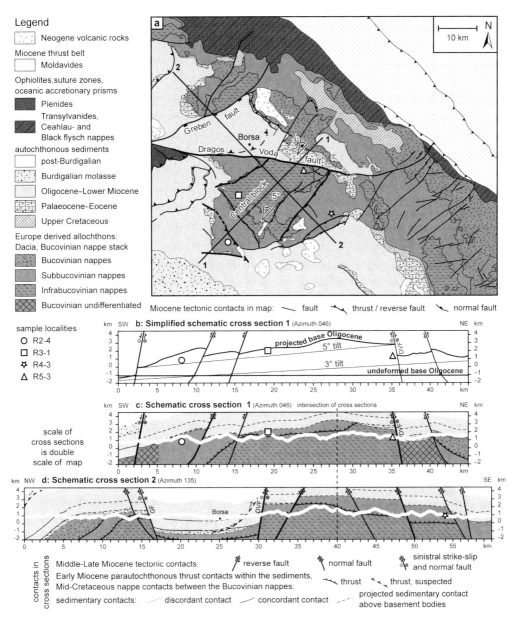

Fig. 9. Map (**a**) and schematic cross-sections (**b, c, d**) through the Bucovinian nappe stack in the study area. Schematic cross-section 1 (c) documents reverse faulting and open folding during the Middle Miocene transpressional stage, while section 2 (d) documents extension by normal faulting during the Middle to Late Miocene transtensional stage. The base of the Oligocene shows a general tilt towards the SW by approximately 3° (b). Along the central block of the Rodna horst, the tilt to the SW increases to 5°, realised along two normal faults delimiting this block by an increasing offset towards the Dragos–Voda fault (a). The suspected thrust at Borsa is drawn after Săndulescu pers. comm. 2002 ('Duplicature de Borsa').

(Fig. 9a), delimiting the central block, allow for this additional tilting. Offset across these normal faults increases towards the Dragos–Voda fault. This indicates that tilting occurred predominantly during the transtensional stage since deformation related to the transtensional stage dominates along the Dragos–Voda fault segment, delimiting the Rodna horst towards the north.

In the following, vertical movements predicted by structural observations are integrated with results from apatite FT thermal modelling in order to reconstruct the exhumation history of the Rodna horst. Relative uplift visible in cross-sections 1 and 2 (Fig. 9c, d) is estimated and correlated with Middle to Late Miocene brittle tectonics which occurred between 16 and 10 Ma ago (Tischler et al. 2006). Timing and amount of these vertical movements (Fig. 10, column B) are then directly compared to the modelled best-fit t–T paths inferred from apatite FT data obtained from the Rodna horst (Fig. 10, column C).

The undeformed base of the Oligocene deposits is used as a reference for differential tectonic uplift (Fig. 9b; Fig. 10, column B) and is defined as the null datum for the starting point at 20 Ma ago (Fig. 10Ba). All the subsequent time slices (Fig. 10Bb to Bd) show this reference plane ('undeformed base Oligocene') for comparison.

20 to 16 Ma time slice (Fig. 10, Row a)

This time slice shows the relative altitudes of the samples before the onset of Miocene brittle deformations 16 Ma ago (Tischler et al. 2006). The geological cross-sections (Fig. 9c, d) allow for an estimate of the original position of the samples relative to the base Oligocene (reference plane 'undeformed base Oligocene') at the following relative altitudes in Fig. 10Ba: R2-4 at −800 m, R3-1 at −200 m, R4-3 at −1000 m and R5-3 at −1400 m. The modelled t–T paths inferred from apatite FT data show that no significant cooling occurred before 16 Ma. All samples remain at temperatures above the APAZ (Fig. 10Ca).

16 to 12 Ma time slice: transpressional stage (Fig. 10, Row b)

During this transpressional stage, three tectonic features influencing tectonic uplift can be identified in cross-section 1 (Fig. 9c):

(1) Reverse faulting affects the SW corner of the Rodna horst. This reverse faulting influences only sample R2-4, situated in a hanging-wall position.

(2) Tilting is interpreted to play a subordinate role during the transpressional stage. Out of a total amount of 5°, only about 1° is assumed to have commenced between 16 and 12 Ma. The amount of relative uplift during tilting is estimated along cross-section 1, and is given by the vertical distance between a horizontal line and a line tilted 1° towards the SW, intersecting at the southern boundary fault of the Rodna horst (Fig. 9a). This vertical distance has been measured at the respective sample locations in section 1. The projected position of sample R4-3 is at the intersection of both sections.

(3) Additional relative tectonic uplift is realised by open folding in the cases of samples R3-2 and R4-3 in an anticlinal position, while R5-3 is situated in a synclinal position. The effect of open folding has been estimated by constructing the distance of the sample location to a tangent onto the synclines.

From the relative position of the specimens at the end of this stage (Fig. 10Bb), the following amounts of tectonically induced relative uplift can be deduced for this stage: +1300 m for R2-4, +1000 m for R3-1, +1000 m for R4-3 and +600 m for R5-3. Note that R2-4 and R3-1 are above the reference line at the end of this stage.

The uplift produced during the transpressional stage relative to the reference line base Oligocene causes a phase of slow cooling visible in the t–T paths (16 to 11 Ma interval in Fig. 10Cb). R3-1 and R2-4 enter the APAZ at very low cooling rates (cooling paths with a gradient of about 2.5 °C/Ma), while R4-3 and R5-3 remain at higher temperatures.

12 to 10 Ma time slice: transtensional stage (Fig. 10, Row c)

Normal and strike-slip faulting, together with 4° SW tilting led to variable amounts of relative tectonic uplift during this transtensional stage. First the amount of uplift produced by this tilting (4°) is estimated. Secondly, an additional uplift of the whole tilted block, as documented by an offset at its SW boundary fault, was applied. The amount of relative uplift for sample R4-3 is derived from its position projected into cross-section 1, corrected for normal faulting visible in cross-section 2. The tectonic uplift produced during this transtensional stage relative to the reference line (Fig. 10Bc) amounts to: +1400 m for R2-4, +2400 m for R3-1, +2200 m for R4-3 and +3400 m for R5-3. The relatively greater uplift values deduced for R3-1 and R5-3 result in their higher position relative to R2-4 and R4-3.

Relative tectonic uplift and erosion during the transtensional stage result in a phase of enhanced exhumation and cooling visible in the t–T paths (11 to 7 Ma interval of Fig. 10Cc). R5-3 enters and nearly crosses the APAZ at high cooling rates (the cooling path indicates 24 °C/Ma). At the end of this phase of enhanced cooling, R3-1 and R5-3 are at the lower limit of the APAZ, while R2-4 and R4-3 remain at higher temperatures. The relatively high cooling rates associated with this extensional stage resulted in Middle to Late Miocene apatite FT central ages in the Rodna horst.

10 to 0 Ma time slice (Fig. 10, Row d)

After cessation of the tectonic activity at 10 Ma (Tischler et al. 2006), the samples reached their present-day altitude relative to the 'undeformed base Oligocene' (Fig. 10Bd) at the following

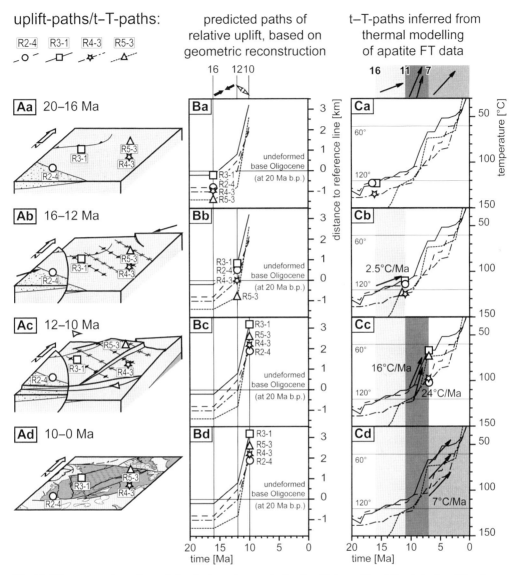

Fig. 10. Direct comparison of unscaled and schematic tectonic sketches (Column A), relative tectonic uplift paths (Column B) and cooling paths (best fit of 10 000 runs) inferred from thermal modelling of apatite FT data (Column C) in the Rodna horst, from top to bottom, four subsequent time intervals (see column A). Amounts of relative tectonic uplift are estimated in respect to the undeformed base Oligocene (Fig. 9b). Greyscale in column C denotes time intervals and associated changes in the rate of cooling (dark = fast cooling).

relative altitudes: +1900 m for R2-4, +2600 m for R5-3, +2200 m for R4-3 and +3200 m for R3-1. Final cooling to surface temperatures was associated with medium cooling rates of about 7 °C/Ma according to the cooling paths (Fig. 10Cd).

Summary of the Miocene exhumation history deduced for the Rodna horst

The Miocene cooling and exhumation observed in the Rodna horst is the result of a combination of tectonic uplift and erosion. Relative tectonic uplift during the transpressional stage (16 to 12 Ma), together with exhumation by erosion, allowed for the observed slow cooling (16 to 11 Ma). The transtensional stage (12 to 10 Ma), however, is reflected in a phase of enhanced cooling (11 to 7 Ma) related to a combination of tectonic and erosional exhumation. Although tectonic uplift and cooling are not contemporaneous, the tectonic stages are reflected in both relative uplift paths and cooling paths. During the transpressional stage (Fig. 10Ba

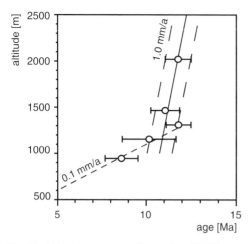

Fig. 11. Altitude vs. age plot of vertical profile R3 (Rodna horst), suggesting a two-stage exhumation history: enhanced exhumation of at least 1.0 mm/a between 12 and 11 Ma, followed by slower exhumation after 10 Ma. 1.0 mm/a and 0.1 mm/a lines are given as a reference.

to Bb) R3-1 and R2-4 are uplifted above the 'undeformed base Oligocene' during their cooling below the upper temperature limit of the APAZ (Fig. 10Ca to Cb). This implies that the reference plane 'undeformed base Oligocene' reflects a temperature interval close to the upper limit of the APAZ (i.e. 120 °C). Based on this observation, the base of the Oligocene sediments is estimated to have been subjected to temperatures around the upper limit of the APAZ by the beginning of Middle Miocene times. This estimation is corroborated by full annealing of apatite fission tracks, as is observed in Eocene sediments located east of Borsa (Sanders 1998).

An exhumation rate for the Rodna horst can be calculated via the altitude vs. age relation, extracted from the vertical profile R3 (Fig. 11). Enhanced exhumation rates are deduced for the time interval 12 to 11 Ma (≥ 1.0 mm/a), followed by overall slow exhumation after 10 Ma before present (the 0.1 mm/a-line is given as reference). Thus the altitude vs. age relationship corroborates the stratigraphic dating as well as the predominance of the transtensional stage along the Dragos–Voda fault.

Revealing the total amount of Miocene exhumation for the entire study area

The total amount of Miocene exhumation in the study area depicted in Fig. 12 has been estimated on the base of the apatite FT data. The maximum palaeodepth (D_{max}) is estimated on the basis of the maximum palaeotemperature (T_{max}), assuming a geothermal gradient of $\Delta T/\Delta Z = 20$ °C/km and an estimated surface temperature of 10 °C (T_s) on the basis of:

$$D_{max} = (T_{max} - T_s)/(\Delta T/\Delta Z)$$

A geothermal gradient of 20 °C/km has been derived by Sanders (1998), who balanced erosion along the Carpathian and Apuseni Mountains against the sedimentary infill of the Transylvanian basin. 20 °C/km is also in accordance with present-day heat-flow data (Veliciu & Visarion 1984; Demetrescu & Veliciu 1991). However, a palaeo-geothermal gradient is not easy to quantify. The area is affected by volcanism, which may provide additional heat sources. Therefore, the given amounts of exhumation represent maximum values.

In the case of samples displaying only partial annealing of apatite (samples P1, P2, P3, P5, P6, P7, S2, S3 and S5), thermal modelling indicates that T_{max} was reached before the onset of Miocene exhumation. In the case of fully annealed apatite (profiles R1, R2, R3, R4, R5, samples M01, M02, M03, M04, M05, M06, M08, M09, M13, M14 and S1), T_{max} has to have exceeded 120 °C. Thermally undisturbed zircon in all basement samples excludes temperatures above the lower limit of the ZPAZ ($T_{max} < 190$ °C) during Palaeogene to Early Miocene burial.

In the case of some samples (profiles R1, R2, R3, R4, R5, samples M01, M09), the estimation of T_{max} has been improved by assuming that the base of the Oligocene approximates the upper limit of the APAZ (120 °C, see above) before the onset of exhumation. In these cases, T_{max} can be derived by the vertical distance of the sample location to the base Oligocene based on maps and cross-sections (Fig. 9) and the estimated geothermal gradient of 20 °C/km. T_{max} for S4 has been estimated to 100 °C, in analogy to S5 since both samples show strong, but not full, annealing. All palaeodepth, and hence also all exhumation values, given in Figure 12 are rounded to 0.5 km intervals.

The Middle to Late Miocene cooling ages (central ages, youngest populations: 15.0 to 7.3 Ma) indicate that the entire amount of cooling and exhumation is the result of erosion that followed tectonic uplift during the strike-slip dominated post-Burdigalian tectonic activity in the study area. However, in the area directly north of the Bogdan–Voda fault, the total amount of exhumation (S2: 4.0 km, S3: 3.5 km) can hardly be exclusively of post-Burdigalian age, since the altitude of the base of the Middle Miocene sediments is found close to that of the collected samples. Hence, north of the Bogdan–Voda fault a significant part of the total amount of exhumation

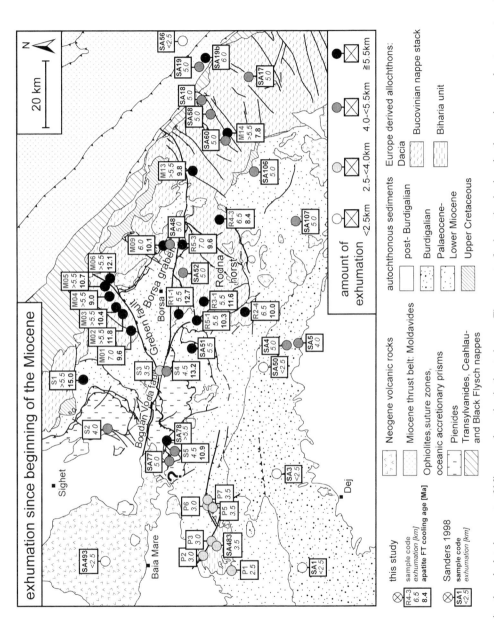

Fig. 12. Map showing the calculated amounts of Miocene exhumation in the study area. The amount of exhumation is given by the maximum palaeotemperatures inferred from apatite FT data, assuming a geothermal gradient of 20 °C/km. Sample code SA refers to additional data from Sanders (1998). For most samples, the entire amount of exhumation was realised after Middle Miocene times (i.e. after 16 Ma), as is indicated by Middle to Late Miocene cooling ages.

seems to be due to erosion that occurred earlier, i.e. during and directly following the intra-Burdigalian thrusting of the Pienides.

The strong thermal pulse found in the westernmost area of the southern part of the Pienide nappes (marked by a question mark, Fig. 12) occurred relatively late, as is indicated by young cooling ages: 10.9 Ma for the youngest population (Fig. 12), 11 Ma and 8 Ma according to Sanders (1998). Since the base of Middle Miocene sediments is in close neighbourhood, a thermal pulse caused by burial is rather unlikely. Hence, we interpret these Late Miocene cooling ages to be the result of a hydrothermal overprint. An intensive overprint of the sedimentary units of the Pienides close to the volcanic body at Baia Mare is documented by Săndulescu & Russo-Săndulescu (1981). Whether partial annealing is related to hydrothermal overprint or to heating by burial is difficult to judge.

Recapitulation of results

Burial by Palaeogene to Early Miocene sediments, locally combined with thrusting and frontal imbrication, caused full annealing of fission tracks in apatite in the case of the samples from the Bucovinian nappe stack. Locally, this is also the case for the sedimentary cover. Regarding the Preluca massif located in the SW part of the study area, however, the amount of burial due to the sedimentary overburden as well as the amount of exhumation were less substantial. Fission tracks in apatite from the Preluca massif remained relatively undisturbed during heating by burial, while (U–Th)/He data document considerable retention of He.

Exhumation is mainly the result of erosion triggered by uplift, accompanying and/or following post-Burdigalian strike-slip tectonic activity along the Bogdan–Dragos–Voda fault system. Middle Miocene cooling ages (15.0 to 13.2 Ma) document exhumation during a first transpressional stage in the eastern part of the study area (Fig. 12). In the central part of the study area, Middle to Late Miocene ages (12.7 to 7.3 Ma) indicate enhanced cooling and exhumation during and following a second transtensional stage. The vertical profile R3 indicates enhanced exhumation of at least 1 mm/a between 12 to 11 Ma (Fig. 11). Enhanced final exhumation is confirmed by apatite (U–Th)/He ages which coincide with the accompanying apatite FT ages within the error limits.

Discussion

Hydrothermal overprint

Beyond the Miocene cooling ages, a few Pliocene apatite FT central ages were also obtained in samples from the Bucovinian nappes (3.4 Ma for M08, 4.4 Ma for R4-2). Neogene volcanic activity is a likely explanation for these young ages. However, volcanism started in the Middle Miocene and activity only lasted until 9.0 Ma north of Baia Mare and 8.6 Ma in the area of the Rodna horst, respectively (Pécskay et al. 1995). Even though a second, younger phase of volcanism can be observed north of Baia Mare (8.1 to 6.9 Ma, Pécskay et al. 1995), Pliocene volcanism is only documented for areas located some 80 to 100 km south of our study area (Harghita Mountains). Thus we propose that the observed late annealing of apatite is due to hydrothermal activity that post-dates active volcanism. The present-day occurrence of hot springs at the southern margin of the Rodna horst also favours such an interpretation.

Burial and exhumation in the study area

The overall decrease in the amount of burial towards the SW, as deduced from the thermochronological data, is in accordance with the diminishing thickness of Oligocene sediments towards the SW (de Brouker et al. 1998). At the southern rim of the Preluca massif, apatite FT data indicate 2.5 km of burial (assuming a geothermal gradient of 20 °C; Fig. 12). About 0.8 km of Palaeocene to Early Miocene (Aquitanian) sediments are still preserved in this area (Rusu et al. 1983). Values of 2500 km for the total sedimentary overburden are in good accordance with the thickness of the Burdigalian strata which reaches about 2 km in the Transylvanian basin directly to the south, as is documented in seismic sections (de Broucker et al. 1998; Ciulavu et al. 2002; Tischler et al. 2008).

Concerning the area of the Rodna horst, our apatite FT data suggest that 5.5 km of sediments accumulated above the base Oligocene. A minimum thickness of Oligocene deposits of 2.8 km is documented in the area of Borsa (Kräutner et al. 1982). Geometric projections of the Oligocene deposits in the profiles of Figure 9c and 9d, based on available maps (Ianovici et al. 1968; Kräutner et al. 1989; Săndulescu et al. 1991), indicate a minimum thickness which diminishes from 4.4 km NW of the Greben fault to 2.7 km SW of the Rodna horst. Higher burial depths in the NW part of the study area are directly documented by the fully annealed apatite FT sample S1 from the Oligocene deposits of this area. Yet the observed or inferred thickness of the post-Eocene deposits is still insufficient for explaining the burial depth inferred from our FT data. Excess burial can be explained by additional tectonic loading related to the emplacement of the Pienides during the Burdigalian and accompanying imbrications of the autochthonous cover (Tischler et al. 2006). Especially in the NW parts of the

Rodna horst, additional tectonic burial appears to be indicated by tectonic considerations. In the SE corner of the Rodna horst, however, additional Burdigalian deposits could also allow for the necessary total overburden. However, it has to be kept in mind that a higher geothermal gradient than the assumed 20 °C/km can also explain apparent discrepancies between palaeotemperatures and amounts of overburden. The apparent volcanic activity in the area may have increased the geothermal gradient. Therefore, the mentioned values for post-Miocene exhumation (Fig. 12) should be considered as maximum estimates.

Although the emplacement of the Pienide nappes probably led to tectonic burial in most parts of the study area, certainly in parts of the autochthonous sediments adjacent to the frontal thrust (sample S1), exhumation that accompanied or postdated the emplacement of the Pienides is indicated by the truncation of tectonic contacts by an unconformity at the base Middle Miocene (Tischler et al. 2006). This truncation could either be explained by exhumation by erosion in the internal part of an accretionary wedge or by exhumation which postdates nappe emplacement.

Regarding the age and the amounts of exhumation, our data are generally in good accordance with earlier apatite FT data (Sanders 1998; Sanders et al. 1999; see Fig. 12). However, in contrast to Sanders (1998), our data indicate full annealing of apatite in all the basement units of the Bucovinian nappe stack. Hence we infer slightly higher values of exhumation.

Relationships between uplift and cooling

The age vs. altitude relation (Fig. 11) indicates fast exhumation between 12 and 11 Ma in the Rodna horst, contemporaneous with the stratigraphically dated tectonic transtensional stage (12 to 10 Ma). On the other hand, thermal modelling of FT data indicates fast cooling between 11 and 7 Ma, i.e. slightly later. However, given high exhumation rates, the observed cooling history of a sample is not only influenced by the exhumation rate alone (Brown 1991; Mancktelow & Grasemann 1997). Starting at rates exceeding 1 mm/a, advective heat transport will cause a delay of cooling with respect to exhumation (Mancktelow & Grasemann 1997). Advective heat transport not only causes a decrease in the spacing of the isotherms (i.e. an increase in geothermal gradient) but can also lead to formation of non-planar isotherms in areas with rugged relief (Stüwe et al. 1994; Mancktelow & Grasemann 1997; Braun 2002). Following these authors, we interpret the delay between fast exhumation and fast cooling, as documented by Middle to Late Miocene cooling ages, as an effect of advective heat transport during a limited phase of fast tectonic exhumation. Sanders (1998) determined mean erosion rates of 0.5 mm/a in the East Carpathians for about the same time interval, i.e. from Middle Miocene to Pliocene times (15 to 5 Ma).

Uplift and exhumation at a regional scale

Middle Miocene ages in the western part of the study area reflect a first stage of exhumation related to the transpressional stage of deformation. The stratigraphically dated Middle Miocene transpressional stage (16 to 12 Ma) is interpreted as a result of the NE-directed perpendicular convergence during soft collision of Tisza–Dacia with the NW–SE striking European margin.

The enhanced Middle to Late Miocene exhumation rates deduced from thermochronological analyses are contemporaneous with the transtensional stage of deformation documented in the study area. The transtensional stage of deformation is thought to be the result of oblique convergence of Tisza–Dacia with the NW–SE striking European margin, evidenced by eastward thrusting in the external Miocene thrust belt (Matenco & Bertotti 2000). Blocking of eastward movement of Tisza–Dacia in the north led to sinistral strike-slip activity along E–W trending faults and coeval normal faulting along NE–SW directed faults (Tischler et al. 2006). This fragmentation of the crust by normal and sinistral strike-slip faulting into SW down-tilted blocks led to the development of triangle-shaped graben (Borsa graben) and corresponding horst structures (Rodna horst; Fig. 12). Exhumation of internal units during foreland propagating thrusting is also known from the westernmost Carpathians. Danišík et al. (2004) report enhanced Middle Miocene exhumation in the northern Danube basin, documented by apatite FT ages (16 to 13 Ma), contemporaneous with north-directed emplacement and thrusting in the Miocene thrust belt further to the N (16 to 14 Ma, e.g. Jiricek 1979).

Although the main activity of the Bogdan–Dragos–Voda fault system terminated at 10 Ma, extensional veins within the volcanic body of Baia Mare, featuring hydrothermal ore deposits (7 to 8 Ma, Lang et al. 1994), suggest minor extensional activity after 10 Ma. Possible Late Miocene normal and strike-slip faulting may explain the extraordinarily young apatite FT ages (7.8 Ma, Fig. 12) in the SE part of the study area. Such a Late Miocene exhumation would be contemporaneous with strike-slip deformation in the external East Carpathian Miocene thrust belt, dated at 9 Ma (Matenco & Bertotti 2000).

Conclusions

The basement units of the Maramures Mountains, the Rodna horst and the Preluca massif suffered a last stage of metamorphism under sub-greenschist to lowermost greenschist facies conditions during the later Early Cretaceous, followed by Late Cretaceous cooling and exhumation. Palaeogene to Early Miocene sedimentation and thrusting caused renewed heating by burial, which led to full annealing of fission tracks in apatite in parts of the study area.

Soft collision of Tisza–Dacia with the European margin initiated a first transpressional tectonic stage (16 to 12 Ma) which caused exhumation by folding and reverse faulting in the western part of the study area. Middle Miocene cooling ages (15.0 to 13.2 Ma) are related to this tectonic stage.

Transpression was followed by transtension (12 to 10 Ma) during ongoing NE–SW shortening. The late-stage evolution of the Rodna horst is dominated by combined normal and strike-slip faulting, which led to its fragmentation into tilted and fault-bounded blocks. Fast exhumation in the central part of the Rodna area resulted in Middle to Late Miocene cooling ages (12.7 to 7.3 Ma). Exhumation rates exceeded 1 mm/a during this stage and caused advective heat transport leading to delayed cooling.

We are most grateful for the excellent introduction into the study area and its geology provided by M. Săndulescu and L. Matenco and their ongoing support. L. Matenco also critically reviewed a first version of the manuscript. Fruitful discussions with D. Badescu, M. Marin, I. Balintoni and D. Radu are also highly appreciated. H.R.G. is very grateful to F. Stuart for his introduction to the lab. procedures and methods of (U–Th)/He analysis. Furthermore, we thank I. Dunkl and M. Raab for their careful reviews and constructive remarks. Financial support by the Swiss National Science foundation (NF-project Nr. 21-64979.01, granted to B.F.) is gratefully acknowledged.

References

ANDRIESSEN, P. A. M. 1995. Fission track analysis: principles, methodology, and implications for tectono-thermal histories of sediment basins, orogenic belts and continental margins. *Geologie en Mijnbouw*, **74**, 1–12.

AROLDI, C. 2001. *The Pienides in Maramures—Sedimentation, Tectonics and Paleogeography*. Cluj, PhD thesis, Universitaria Babes—Bolyai Cluj.

BALESTRIERI, M. L., STUART, F. M., PERSANO, C., ABBATE, E. & BIGAZZI, G. 2005. Geomorphic development of the escarpment of the Eritrean margin, southern Red Sea from combined apatite fission-track and (U–Th)/He thermochronometry. *Earth and Planetary Science Letters*, **231**, 97–110.

BALINTONI, I., MOSONYI, E. & PUSTE, A. 1997. Informatii si interpretari litostratigrafice, metamorfice si structurale privitoare la masivul Rodna (Carpatii orientali). *Studia Universitates Babes—Bolyai. Geologia XLII*, 51–66.

BALLA, Z. 1987. Tertiary paleomagnetic data for the Carpatho-Pannonian region in the light of Miocene rotation kinematics. *Tectonophysics*, **139**, 67–98.

BORCOS, M., SANDULESCU, M., STAN, N., PELTZ, S., MARINESCU, F. & TICLEANU, N. 1980. *Geological Map 1:50 000 Cavnic*. Institutul de Geologie şi Geofizica, Bucharest.

BRAUN, J. 2002. Quantifying the effect of recent relief changes on age-elevation relationships. *Earth and Planetary Science Letters*, **200**, 331–343.

BROUCKER, G. DE, MELLIN, A. & DUINDAM, P. 1998. Tectonostratigraphic evolution of the Transylvanian Basin, Pre-Salt sequence, Romania. *In:* DINU, C. (ed.) *Bucharest Geoscience Forum*. Special volume, **1**, 36–70.

BROWN, R. W. 1991. Backstacking apatite fission track 'stratigraphy': a method for resolving the erosional and isostatic rebound components of tectonic uplift histories. *Geology*, **19**, 74–77.

BURTNER, R. L., NIGRINI, A. & DONELICK, R. A. 1994. Thermochronology of the Lower Cretaceous source rocks in the Idaho-Wyoming thrust belt. *AAPG Bulletin*, **78**, 1613–1636.

CIULAVU, D., DINU, C. & CLOETINGH, S. A. P. L. 2002. Late Cenozoic tectonic evolution of the Transylvanian basin and north-eastern part of the Pannonian basin (Romania): constraints from seismic profiling and numerical modeling. *EGU Stephan Mueller Special Publication Series*, **3**, 105–120.

CROWLEY, K. D. 1985. Thermal significance of fission-track length distributions. *Nuclear Tracks*, **10**, 311–322.

CROWLEY, K. D. & CAMERON, M. 1987. Annealing of etchable fission-track damage in apatite: effects of anion chemistry. *Geological Society of America Abstract Program*, **19**, 631–632.

CSONTOS, L. 1995. Tertiary tectonic evolution of the Intra-Carpathian area: a review. *Acta Vulcanologica*, **7**, 1–13.

CSONTOS, L. & NAGYMAROSY, A. 1998. The Mid-Hungarian line: a zone of repeated tectonic inversions. *Tectonophysics*, **297**, 51–71.

CSONTOS, L. & VÖRÖS, A. 2004. Mesozoic plate tectonic reconstruction of the Carpathian region. *Palaeocgeography, Palaeoclimatology, Palaeoecology*, **210**, 1–56.

CSONTOS, L., NAGYMAROSY, A., HORVÁTH, F. & KOVÁČ, M. 1992. Cenozoic evolution of the Intra-Carpathian area: a model. *Tectonophysics*, **208**, 221–241.

DANIŠÍK, M., DUNKL, I., PUTIS, M., FRISCH, W. & KRAL, J. 2004. Tertiary burial and exhumation history of basement highs along the NW margin of the Pannonian basin an apatite fission track study. *Austrian Journal of Earth Sciences*, **95/96**, 60–70.

DEMETRESCU, C. & VELICIU, S. 1991. Heat flow and lithosphere structure in Romania. *In:* CERMAK, V. & RYBACH, L. (eds) *Terrestrial Heat Flow and the Lithosphere Structure*. Springer-Verlag, Berlin, New York, 187–205.

DICEA, O., DUTESCU, P., ANTONESCU, F. *ET AL.* 1980. Contributii la cunoasterea stratigrafiei zonei transcarpatice din maramures. *Dari de Seama Institutul Geologie și Geofizica*, **LXV**, 21–85.

DUMITRU, T. 1993. A new computer-automated microscope stage system for fission-track analysis. *Nuclear Tracks and Radiation Measurements*, **21**, 575–580.

DUNKL, I. 2002. TrackKey: a Windows program for calculation and graphical presentation of fission track data. *Computers & Geosciences*, **28**, 3–12.

DUNKL, I. & FRISCH, W. 2002. Thermochronological constraints on the Late Cenozoic exhumation along the Alpine and Western Carpathian margins of the Pannonian basin. *EGU Stephan Mueller Special Publications Series*, **3**, 135–147.

DUNKL, I. & SZÉKELY, B. 2002. Component analysis with visualization of fitting PopShare, a Windows program for data analysis. Goldschmidt Conference Abstracts 2002. *Geochimica et Cosmochimica Acta*, **66/15A**, 201.

EHLERS, T. A. & FARLEY, K. A. 2003. Apatite (U–Th)/He thermochronometry: methods and applications to problems in tectonic and surface processes. *Earth and Planetary Science Letters*, **206**, 1–14.

FARLEY, K. A. 2000. Helium diffusion from apatite: general behaviour as illustrated by Durango fluorapatite. *Journal of Geophysical Research*, **105**, 2903–2914.

FARLEY, K. A. 2002. (U/Th)/He dating: techniques, calibrations, and applications. *In*: PORCELLI, P. D., BALLENTINE, C. J. & WIELER, R. (eds) *Noble Gas Geochemistry*. Reviews in Minerology and Geochemistry, **47**, 819–843.

FODOR, L., CSONTOS, L., BADA, G., GYÖRFI, I. & BENKOVICS, L. 1999. Tertiary tectonic evolution of the Pannonian Basin system and neighbouring orogens: a new synthesis of paleostress data. *In*: DURAND, B., JOLIVET, L., HORVÁTH, E. & SÉRANNE, M. (eds) *The Mediterranean Basins: Tertiary Extension within the Alpine Orogen*. London, Geological Society, **156**, 295–334.

FOEKEN, J. P. T., DUNAI, T. J., BETOTTI, G. & ANDRIESSEN, P. A. M. 2003. Late Miocene to present exhumation in the Ligurian Alps (southwest Alps) with evidence for accelerated denudation during the Messinian salinity crisis. *Geology*, **31**, 797–800.

FOEKEN, J. P. T., STUART, F. M., DOBSON, K. J., PERSANO, C. & VILBERT, D. 2006. A diode laser system for heating minerals for (U–Th)/He chronometry. *Geochemistry Geophysics and Geosystems*, **7**, Q04015, doi: 10.1029/2005GC001190.

FÜGENSCHUH, B. & SCHMID, S. M. 2003. Late stages of deformation and exhumation of an orogen constrained by fission-track data: a case study in the Western Alps. *GSA Bulletin*, **115**, 1425–1440.

FÜGENSCHUH, B. & SCHMID, S. M. 2005. Age and significance of core complex formation in a highly bent orogen: evidence from fission track studies in the South Carpathians (Romania). *Tectonophysics*, **404**, 33–35.

FÜGENSCHUH, B., MANCKTELOW, N. S. & SEWARD, D. 2000. Cretaceous to Neogene cooling and exhumation history of the Oetztal-Stubai basement complex, Eastern Alps: a structural and fission track study. *Tectonics*, **19**, 905–918.

GALBRAITH, R. F. 1981. On statistical models for fission track counts. *Mathematical Geology*, **13**, 471–478.

GALBRAITH, R. F. 1990. The radial plot: graphical assessment of spread in ages. *Nuclear Tracks in Radiation Measurements*, **17**, 207–214.

GALBRAITH, R. F. & LASLETT, G. M. 1993. Statistical models for mixed fission track ages. *Nuclear Tracks in Radiation Measurements*, **21**, 459–470.

GALLAGHER, K., BROWN, R. & JOHNSON, C. 1998. Fission track analysis and its applications to geological problems. *Annual Reviews in Earth and Planetary Science Letters*, **26**, 519–571.

GIUSCA, D. & RADULESCU, D. 1967. *Geological Map 1:200 000 Baia Mare*. Institutul de Geologie și Geofizica, Bucharest.

GLEADOW, A. J. W. 1981. Fission-track dating methods: what are the real alternatives? *Nuclear Tracks*, **5**, 3–14.

GLEADOW, A. J. W. & DUDDY, I. R. 1981. A natural long-term track annealing experiment for apatite. *Nuclear Tracks*, **5**, 169–174.

GLEADOW, A. J. W., DUDDY, I. R., GREEN, P. F. & LOVERING, J. F. 1986a. Confined track lengths in apatite: a diagnostic tool for thermal history analysis. *Contributions to Mineralogy and Petrology*, **94**, 405–415.

GLEADOW, A. J. W., DUDDY, I. R., GREEN, P. F. & HEGARTY, K. A. 1986b. Fission track lengths in the apatite annealing zone and the interpretation of mixed ages. *Earth and Planetary Science Letters*, **78**, 245–254.

GRADSTEIN, F., OGG, J. & SMITH, A. 2004. *A Geologic Time Scale*. Cambridge University Press, Cambridge.

GREEN, P. F., DUDDY, I. R., GLEADOW, A. J. W. & TINGATE, P. R. 1985. Fission track annealing in apatite: Track length measurements and the form of the Arrhenius plot. *Nuclear Tracks*, **10**, 323–328.

GREEN, P. F., DUDDY, I. R., GLEADOW, A. J. W., TINGATE, P. R. & LASLETT, G. M. 1986. Thermal annealing of fission tracks in apatite, 1. A qualitative description. *Chemical Geology*, **59**, 237–253.

GREEN, P. F., DUDDY, I. R., LASLETT, G. M., HEGARTY, K. A., GLEADOW, A. J. W. & LOVERING, J. F. 1989. Thermal annealing of fission tracks in apatite, 4. Quantitative modelling techniques and extension to geological timescales. *Chemical Geology*, **79**, 155–182.

HAAS, J. & PÉRÓ, S. 2004. Mesozoic evolution of the Tisza Mega-unit. *International Journal of Earth Science*, **93**, 297–313.

HOUSE, M. A., FARLEY, K. A. & STÖCKLI, D. F. 2000. Helium chronometry of apatite and titanite using Nd: YAG laser heating. *Earth and Planetary Science Letters*, **183**, 365–368.

HURFORD, A. J. 1986. Cooling and uplift patterns in the Lepontine Alps South Central Switzerland and an age of vertical movement on the Insubric fault line. *Contributions to Mineralogy and Petrology*, **92**, 413–427.

HURFORD, A. J. & GREEN, P. F. 1983. The zeta age calibration of fission-track dating. *Isotope Geoscience*, **1**, 285–317.

IANOVICI, V. & DESSILA-CODARCEA, M. 1968. Geological Map 1:200 000 Radauti. Institutul de Geologie şi Geofizica, Bucharest.

IANOVICI, V. & RADULESCU, D. 1968. Geological Map 1:200 000 Toplita. Institutul de Geologie şi Geofizica, Bucharest.

IANOVICI, V., RADULESCU, D. & PATRULIUS, D. 1968. Geological Map 1:200 000 Viseu. Institutul de Geologie şi Geofizica, Bucharest.

JIRICEK, R. 1979. Tectonic development of the Capathian arc in the Oligocene and Neogene. In: MAHEL, M. (ed.) Tectonic Profiles Through the Western Carpathians. Geological Institute Dionyz Stur, Bratislava, 205–214.

KETCHAM, R. A., DONELICK, R. A., & DONELICK, M. B. 2000. AFTSolve: a program for multi-kinetic modeling of apatite fission-track data. Geological Materials Research, 2, (electronic).

KRÄUTNER, H. G. 1991. Pre-Alpine geological evolution of the East Carpathian metamorphics. Some common trends with the West Carpathians. Geologica Carpathica, 42, 209–217.

KRÄUTNER, H. G., KRÄUTNER, F., SZASZ, L., UDUBASA, G. & ISTRATE, G. 1978. Geological Map 1:50 000 Rodna Veche. Institutul de Geologie şi Geofizica, Bucharest.

KRÄUTNER, H. G., KRÄUTNER, F. & SZASZ, L. 1982. Geological Map 1:50 000 Pietrosul Rodnei. Institutul de Geologie şi Geofizica, Bucharest.

KRÄUTNER, H. G., KRÄUTNER, F. & SZASZ, L. 1983. Geological Map 1:50 000 Ineu. Institutul de Geologie şi Geofizica, Bucharest.

KRÄUTNER, H. G., KRÄUTNER, F., SZASZ, L. & SEGHEDI, I. 1989. Geological Map 1:50 000 Rebra. Institutul de Geologie şi Geofizica, Bucharest.

LANG, B., EDELSTEIN, O., STEINITS, G., KOVACS, M. & HALGA, S. 1994. Ar-Ar dating of adulari—a tool in understanding genetic relations between volcanism and mineralization: Baia Mare (Gutii Mountains), northwestern Romania. Economic Geology, 89, 174–180.

LASLETT, G. M., GREEN, P. F., DUDDY, I. R. & GLEADOW, A. J. W. 1987. Thermal annealing of fission tracks in apatite, 1. A quantitative analysis. Chemical Geology, 65, 1–13.

MANCKTELOW, N. S. & GRASEMANN, B. 1997. Time-dependent effects of heat advection and topography on cooling histories during erosion. Tectonophysics, 270, 167–195.

MÁRTON, E. 2000. The Tisza Megatectonic Unit in the light of paleomagnetic data. Acta Geologica Hungary, 43, 329–343.

MÁRTON, E. & FODOR, L. 1995. Combination of palaeomagnetic and stress data—a case study from North Hungary. Tectonophysics, 242, 99–114.

MÁRTON, E. & FODOR, L. 2003. Tertiary paleomagnetic results and structural analysis from the Transdanubian Range (Hungary): rotational disintegration of the ALCAPA unit. Tectonophysics, 363, 201–224.

MÁRTON, E., TISCHLER, M., CSONTOS, L., FÜGENSCHUH, B. & SCHMID, S. M. 2006. The contact zone between the ALCAPA and Tisza–Dacia mega-tectonic units of Northern Romania in the light of new paleomagnetic data. Swiss Journal of Geoscience, 100, 109–124.

MASON, P. R. D., SEGHEDI, I., SZAKASC, A. & DOWNES, H. 1998. Magmatic constraints on geodynamic models of subduction in the Eastern Carpathians, Romania. Tectonophysics, 297, 157–176.

MATENCO, L. & BERTOTTI, G. 2000. Tertiary tectonic evolution of the external East Carpathians (Romania). Tectonophysics, 316, 255–286.

MORLEY, C. K. 2002. Tectonic settings of continental extensional provinces and their impact on sedimentation and hydrocarbon prospectivity. In: RENAUT, R. W. & ASHLEY, G. M. (eds) Sedimentation in Continental Rifts. Society of Sedimentary Geology, Special publications, 73, 25–55.

ÒSULLIVAN, P. B. & PARRISH, R. R. 1995. The importance of apatite composition and single-grain ages when interpreting fission track data from plutonic rocks: a case study from the Coast Ranges, British Columbia. Earth and Planetary Science Letters, 132, 213–224.

PÉCSKAY, Z., EDELSTEIN, O., SEGHEDI, I., SZAKÁCS, A., KOVACS, M., CRIHAN, M. & BERNÁD, A. 1995. K-Ar datings of Neogene-Quaternary calc-alkaline volcanic rocks in Romania. In: DOWNES, H. & VASELLI, O. (eds) Neogene and related magmatism in the Carpatho-Pannonian Region. Acta Vulcanologica, 7, 53–61.

PERSANO, C., STUART, F. M., BISHOP, P. & DEMPSTER, T. J. 2005. Deciphering continental breakup in eastern Australia using low temperature thermochronometers. Journal of Geophysical Research, 110, doi: 10.1029/2004JB003325.

RAILEANU, G. & RADULESCU, D. 1967. Geological Map 1:200 000 Bistrita. Institutul de Geologie şi Geofizica, Bucharest.

RAILEANU, G. & SAULEA, E. 1968. Geological Map 1:200 000 Cluj. Institutul de Geologie şi Geofizica, Bucharest.

RATSCHBACHER, L., MERLE, O., DAVY, P. & COBBOLD, P. 1991a. Lateral extrusion in the Eastern Alps; Part 1, Boundary conditions and experiments scaled for gravity. Tectonics, 10, 245–256.

RATSCHBACHER, L., FRISCH, W., LINZER, H. G. & MERLE, O. 1991b. Lateral extrusion in the Eastern Alps; Part 2, Structural analysis. Tectonics, 10, 257–271.

REINERS, P. W. & EHLERS, T. A. (eds) 2005. Low-temperature thermochronology: techniques, interpretations, and applications. Reviews in Mineralogy and Geochemistry, 58.

REINERS, P. W., ZHOU, Z., EHLERS, T., XU, C., BRANDON, M. T., DONELICK, R. A. & NICOLESCU, S. 2003. Post-orogenic evolution of the Dabie Shan, eastern China, from (U–Th)/He and fission track thermochronology. American Journal of Science, 303, 489–518.

ROURE, F., BESSEREAU, G., KOTARBA, M., KUSMIEREK, J. & STRZETELSKI, W. 1993. Structure and hydrocarbon habitats of the Polish Carpathian Province. American Association of Petroleum Geologists Bulletin, 77, 1660.

ROYDEN, L. H. 1988. Late Cenozoic Tectonics of the Pannonian Basin System. In: ROYDON, L. H. & HORVÁTH, F. (eds) The Pannonian Basin: A Study In Basin Evolution. Tulsa, Oklahoma, The American Association of Petroleum Geologists, 45, 27–48.

ROYDEN, L. H. 1993. The tectonic expression of slab pull at continental convergent boundaries. *Tectonics*, **12**, 303–325.

ROYDEN, L. H. & BÁLDI, T. 1988. Early Cenozoic tectonics and paleogeography of the Pannonian and surounding regions. *In*: ROYDON, L. H. & HORVÁTH, F. (eds) *The Pannonian Basin: A Study In Basin Evolution*. Tulsa, Oklahoma, The American Association of Petroleum Geologists, **45**, 1–16.

RUSU, A., BALINTONI, I., BOMBITA, G. & POPESCU, G. 1983. *Geological Map 1:50 000 Preluca*. Institutul de Geologie şi Geofizica, Bucharest.

SANDERS, C. 1998. *Tectonics and Erosion-Competitive Forces in a Compressive Orogen: A Fission Track Study of the Romanian Carpathians*. PhD thesis, Vrije Universiteit Amsterdam, 1–204.

SANDERS, C. A. E., ANDRIESSEN, P. A. M. & CLOETINGH, S. A. P. L. 1999. Life cycle of the East Carpathian orogen: erosion history of a double vergent critical wedge assessed by fission track thermochronology. *Journal of Geophysical Research*, **104**, 29095–29112.

SĂNDULESCU, M. 1980. Sur certain problèmes de la corrélation des Carpathes orientales Roumaines avec les Carpathes Ucrainiennes. *Dari de Seama Institutul Geologie şi Geofizica*, **LXV**, 163–180.

SĂNDULESCU, M. 1982. Contributions à la connaissance de nappes Crétacées de Monts du Maramures (Carpathes Orientales). *Dari de Seama Institutul Geologie şi Geofizica*, **LXIX**, 83–96.

SĂNDULESCU, M. 1988. Cenozoic Tectonic History of the Carpathians. *In*: ROYDEN, L. H. & HORVÁTH, F. (eds) *The Pannonian Basin; a study in basin evolution*. Tulsa, Oklahoma, The American Association of Petroleum Geologists, **45**, 17–26.

SĂNDULESCU, M. 1994. Overview on Romanian Geology. *In*: 2nd Alcapa Congress, Field guidebook, *Romanian Journal of Tectonics and Regional Geology*, **75**, 3–15.

SĂNDULESCU, M. & RUSSO-SĂNDULESCU, D. 1981. *Geological Map 1:50 000 Poiana Botizii*. Institutul de Geologie şi Geofizica, Bucharest.

SĂNDULESCU, M., KRÄUTNER, H. G., BALINTONI, I., RUSSO-SĂNDULESCU, D. & MICU, M. 1981. *The Structure of the East Carpathians, (Guide Book B1)*. Carpathian Balkan Geological Association, 12th Congress, Bucharest.

SĂNDULESCU, M., SZASZ, L., BALINTONI, I., RUSSO-SĂNDULESCU, D. & BADESCU, D. 1991. *Geological Map 1:50 000 No 8d Viseu*. Institutul de Geologie şi Geofizica, Bucharest.

SĂNDULESCU, M., VISARION, M., STANICA, D., STANICA, M. & ATANASIU, L. 1993. Deep structure of the inner Carpathians in the Maramures-Tisa zone (East Carpathians). *Romanian Journal of Geophysics*, **16**, 67–76.

SCHMID, S. M., FÜGENSCHUH, B., MATENCO, L., SCHUSTER, R., TISCHLER, M. & USTASZEWSKI, K. 2006. The Alps-Carpathians-Dinarides-connection: a compilation of tectonic units. 18th Congress of the Carpathian-Balkan Geological Association, Belgrade, Serbia, Sept. 3–6, 2006. *Conference Proceedings*, 535–538.

SIDDALL, R. & HURFORD, A. J. 1998. Semi-quantitative determination of apatite anion composition for fission-track analysis using infra-red microspectroscopy. *Chemical Geology*, **150**, 181–190.

SOBEL, E. R. & DUMITRU, T. A. 1997. Thrusting and exhumation around the margins of the western Tarim basin during the India-Asia collision. *Journal of Geophysical Research*, **102**, 5043–5063.

SPERNER, B., CRC 461 TEAM 2005. Monitoring of Slab Detachment in the Carpathians. *In*: WENZEL, F. (ed.) Perspectives in modern seismology. *Lecture Notes in Earth Sciences*, **105**, 187–202.

STEININGER, F. F. & WESSELY, G. 2000. From the Tethyan ocean to the Paratethys Sea: Olicocene to Neogene stratigraphy, paleogeography and paleobio-geography of the circum-Mediterranean region and the Oligocene to Neogene basin evolution in Austria. *Mitteilungen der Österreichischen Geologischen Gesellschaft*, **92**, 95–116.

STÖCKLI, D. F. 2005. Application of low-temperature thermochronometry to extensional tectonic settings. *In*: REINERS, P. W. & EHLERS, T. A. (eds) *Low-Temperature Thermochronology: Techniques, Interpretations, and Applications. Reviews in Mineralogy and Geochemistry*, **58**, 411–448.

STÜWE, K., WHITE, L. & BROWN, R. 1994. The influence of eroding topography on steady-state isotherms. Applications to fission-track analysis. *Earth and Planetary Sciene Letters*, **124**, 63–74.

TISCHLER, M., GRÖGER, H. R., FÜGENSCHUH, B. & SCHMID, S. M. 2006. Miocene tectonics of the Maramures area (Northern Romania)—implications for the Mid-Hungarian fault zone. doi: 10.1007/s00531-006-0110-x. *International Journal of Earth Sciences*, **96**, 473–496.

TISCHLER, M., MATENCO, L., FILIPESCU, S., GRÖGER, H. R., WETZEL, A. & FÜGENSCHUH, B. 2008. Tectonics and sedimentation during convergence of the ALCAPA and Tisza–Dacia: continental blocks the Pienide nappe emplacement and its foredeep (N. Romania) *In*: SIEGESMUND, S., FÜGENSCHUH, B. & FROITZHEIM, N. (eds) *Tectonic Aspects of the Alpine-Dinaride-Carpathian Systems*, **298**, 317–334.

VELICIU, S. & VISARION, M. 1984. Geothermal models for the East Carpathians. *Tectonophysics*, **103**, 157–165.

VODA, A. & BALINTONI, I. 1994. Corelari lithostratigrafice în cristalinul Carpatilor Orientali. *Studia Universitates Babes—Bolyai, Geologia*, **XXXIX**, 61–66.

WAGNER, G. A. & VAN DEN HAUTE, P. 1992. *Fission-Track Dating*. Stuttgart, Enke Verlag.

WOLF, R. A., FARLEY, K. A. & KASS, D. M. 1998. Modeling of the temperature sensitivity of the apatite (U–Th)/He thermochronometer. *Chemical Geology*, **148**, 105–114.

WORTEL, M. J. R. & SPAKMAN, W. 2000. Subduction and slab detachment in the Mediterranean-Carpathian region. *Science*, **290**, 1910–1917.

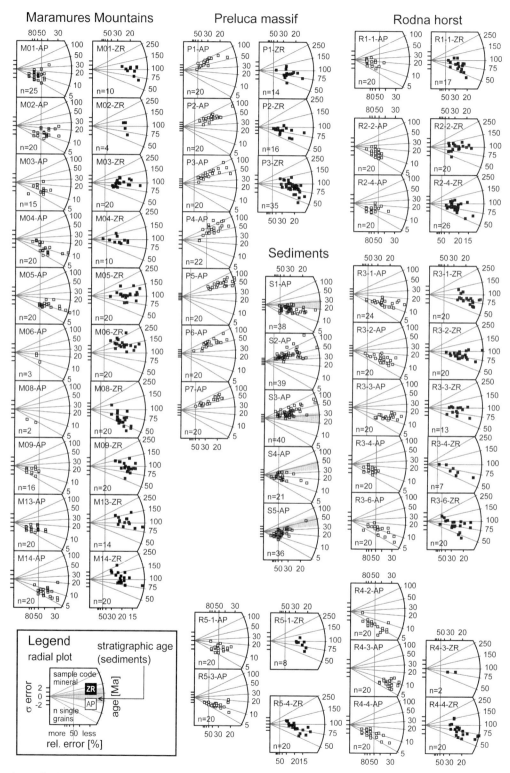

Appendix 1a. Results of fission track analyses in the study area: radial plots. All radial plots (Galbraith 1990) of the same data group (Ap/Zr) show the same age spectrum to allow for direct comparison.

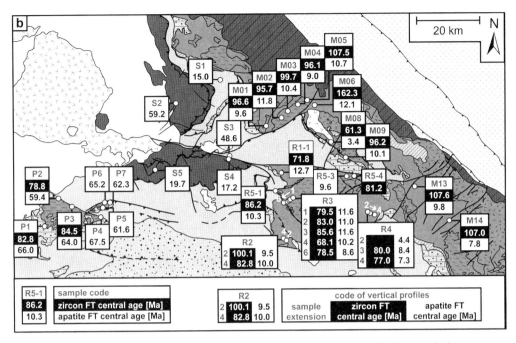

Appendix 1b. Results of fission track analyses in the study area: ages in map view. All given ages in the map are central ages (Galbraith & Laslett 1993).

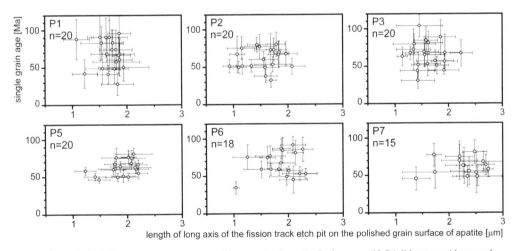

Appendix 2. Etch pit long axes measurements of the samples from the Preluca massif. P4 did not provide enough spontaneous tracks. Generally, as many as possible etch pit long axes per grain were measured, up to a maximum of ten. The grain ages are plotted with their single grain error and the mean of the etch pit long axes is plotted with one standard deviation.

The western termination of the SEMP Fault (eastern Alps) and its bearing on the exhumation of the Tauern Window

CLAUDIO L. ROSENBERG & SUSANNE SCHNEIDER

Freie Universität Berlin, Institut für Geologie, Malteserstr. 74–100, 12249 Berlin Germany (e-mail: cla@zedat.fu-berlin.de)

Abstract: The SEMP (Salzach–Ennstal–Mariazell–Puchberg) Fault strikes along more than 300 km from the southern margin of the Vienna Basin to the northern Tauern Window accommodating a sinistral displacement of 60 km during Tertiary time. We present new structural data, showing that the SEMP Fault continues into the Tauern Window within a 50 km long mylonitic belt of approximately 2 km width, which we term the Ahorn shear zone. This sinistral shear zone, which marks the northern boundary of the Zentral Gneiss, strikes E to ENE, dips subvertically, and is characterized by gently W-dipping to subhorizontal stretching lineations. S-side-up kinematic indicators in the Y–Z fabric plane and a pronounced southward increase in the inferred temperature of sinistral shearing are observed within the shear zone. Microstructural observations indicate that deformation of quartz at the northernmost boundary of the Zentral Gneiss occurred by dislocation glide with only incipient dynamic recrystallization, suggesting a temperature of approximately 300 °C. Further south, temperatures greater than 300 °C are inferred because all samples are affected by dynamic recrystallization of quartz, and dynamic recrystallization of feldspars also occurred in the southernmost part of the shear zone. These findings point to transpressive deformation accommodating a significant component of south-side-up displacement in addition to sinistral shearing. The sinistral mylonitic foliation forms the axial-plane foliation of the large-scale, ENE-striking upright folds of the western Tauern Window. From east to west, deformation becomes increasingly distributed, passing from an area of interconnected shear zones in the east to a homogeneously deformed mylonitic belt in the west, which terminates into a belt of WNW-striking, upright folds. From the above, we suggest the following: (1) the SEMP Fault extended beyond the brittle-ductile transition to a depth where temperatures exceeded 500 °C (>20 km depth?). These mylonites should be included in the seismic interpretation profiles as a major crustal discontinuity; (2) the large-amplitude, upright folds of the Tauern Window formed at the same time as the sinistral mylonites, and hence during south-side-up differential displacement; and (3) part of the 60 km lateral displacement of the SEMP fault is transferred into a vertical displacement at the western end of the Ahorn shear zone and into a fold belt accommodating NNE-oriented shortening, west of the Ahorn shear zone.

The Tauern Window represents a large (160 × 30 km) exposure of intensively folded lower (European) Plate of the Tertiary Alpine orogeny (Fig. 1c) in the eastern Alps. The uplift and exhumation of this deep structural level occurred mainly in the Miocene, at a time in which the remaining parts of the presently exposed eastern Alps had already been exhumed to a structural level close to the surface. The differential exhumation of this axial zone of the eastern Alps throughout the Miocene was largely accommodated by a series of faults and shear zones bounding the Tauern Window (Fig. 1a, b). These comprise the extensional Brenner and Katschberg faults, marking the western and eastern boundaries of the window respectively, and the Mölltal and SEMP faults, marking the southeastern and northern boundaries of the window, respectively (Fig. 1a, b). The present paper investigates the structure of the SEMP fault along and across its strike direction and also the relationship between this fault and the internal deformation and exhumation of the Tauern Window.

The SEMP (Salzach–Ennstal–Mariazell–Puchberg) Fault (Ratschbacher *et al.* 1991; Decker *et al.* 1994) strikes along more than 300 km from the southern margin of the Vienna Basin to the northern Tauern Window (Fig. 1a), accommodating a sinistral displacement of 60 km (Linzer *et al.* 1997, 2002) during Tertiary time. Strike-slip deformation along the Inntal Fault (Fig. 1a) is inferred to have been active since the Late Rupelian based on dated sediments deposited along the paleo-Inntal Fault (Ortner & Stingl 2001; Ortner *et al.* 2003). Considering that the SEMP and the Inntal Faults belong to the same sinistral system, which disrupted the North Calcareous Alps into lozenge-shaped blocks during the Tertiary (Frisch *et al.* 1998), one could argue by analogy that the activity of the SEMP Fault also started in the Oligocene. The sinistral DAV Fault (Fig. 1b), marking the southern border of the Tauern Window, is also inferred to be Oligocene in age (Mancktelow *et al.* 2001), and according to some authors it continued to be active during the Miocene (Wagner *et al.* 2006).

From: SIEGESMUND, S., FÜGENSCHUH, B. & FROITZHEIM, N. (eds) *Tectonic Aspects of the Alpine-Dinaride-Carpathian System.* Geological Society, London, Special Publications, **298**, 197–218.
DOI: 10.1144/SP298.10   0305-8719/08/$15.00 © The Geological Society of London 2008.

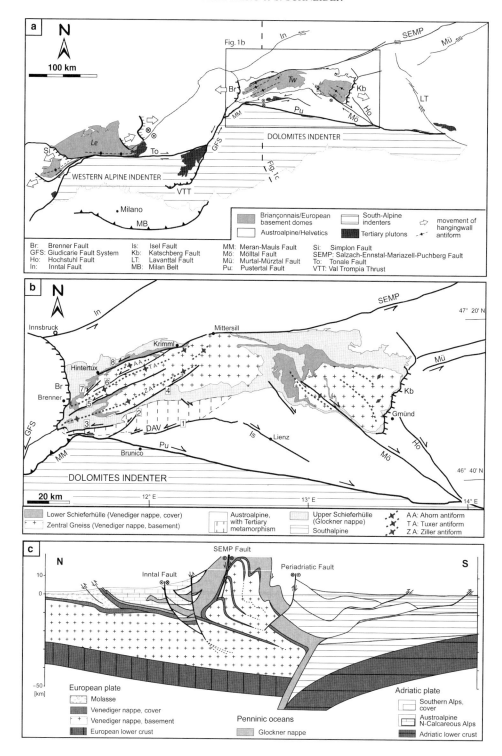

Fig. 1. Tectonic maps and cross-section of the eastern Alps. (**a**) Simplified tectonic map of the Alps, after Handy *et al.* (2005), showing the major Tertiary fault systems. *Le*, Lepontine dome; *Tw*, Tauern Window. Stippled line

In addition, earthquakes recorded throughout the 20th century show a concentration of seismic activity along a linear structure, which roughly corresponds to the fault trace of the SEMP (Reinecker & Lehnhardt 1999; their Fig. 2), suggesting that this fault system may still be active today.

The amount of sinistral displacement and the location of individual fault segments of the SEMP are soundly constrained in the North Calcareous Alps (e.g. Linzer et al. 2002), whereas the western termination of this fault in the Tauern Window is not well known. The SEMP Fault becomes ductile as it enters the Tauern Window (Wang & Neubauer 1998), close to Mittersill (Fig. 1). Further west, it was suggested that the SEMP Fault splays into three distinct shear zones (Linzer et al. 2002): the Ahrntal Fault in the South (No. 4 in Fig. 1b), the Greiner shear zone (No. 5 in Fig. 1) in the central part of the Zentral Gneiss, and a third shear zone (No. 8 in Fig. 1) at the northern margin of the Zentral Gneiss (Fig. 1). We term this shear zone the Ahorn shear zone, because it affects most of the 'Ahorn Kern' (Ahorn antiform of Zentral Gneiss, Fig. 1b). As shown in Figure 1b, at least two more sinistral shear zones subparallel to the latter three have been mapped in the western Tauern Window (Lammerer & Weger 1998; Mancktelow et al. 2001; Kurz et al. 2001). Taken together, these closely spaced sinistral shear zones, which strike parallel to upright antiforms, show an 'en-echelon' structure in which the western termination of each shear zone is progressively displaced eastward from south to north (Fig. 1b), except for the Ahrntal Fault (no. 4 in Fig. 1b), whose western termination is however not well known yet. This geometry suggests that the entire western Tauern Window represents a large-scale zone of sinistral, transpressive displacements. Each of these sinistral fault segments is discussed in more detail below.

Sinistral shearing along the Greiner shear zone (Behrmann 1988, 1990) was suggested to be Cretaceous or Eocene, because quartz microstructures in these mylonites were inferred to be statically annealed during the peak of regional metamorphism (30 Ma; Christensen et al. 1994) of the Tauern Window (Behrmann & Frisch 1990). In addition, dextral shear zones overprinting the Greiner shear zone were inferred to be c. 28 Ma old, suggesting that sinistral shearing must be older than 28 Ma (Barnes et al. 2004).

The Ahrntal Fault shows an apparent sinistral displacement of the Zentral Gneiss margin in map view (Fig. 1b). Except for one study, however, in which outcrop-scale kinematic indicators were described from a single locality of the Ahrntal Fault (Reicherter et al. 1993), no structural or geochronological investigations of this shear zone exist yet. The spatial continuity between this shear zone and the SEMP Fault is questionable, because a continuous zone of steeply dipping foliations does not occur between the Ahrntal Fault (no. 4 in Fig. 1b) and Mittersill (Fig. 1b), i.e. the northern boundary of the Zentral Gneiss.

The Ahorn shear zone (no. 8 in Fig. 1b) also lacks any structural and geochronological investigations. It strikes subparallel to, and in direct continuation of, the latter, hence being the most obvious continuation of the SEMP Fault within the Tauern Window. A structural and microstructural investigation of this shear zone is the prime subject of the present work.

In the following, we present new structural and microstructural data, which allow us to constrain: (1) the kinematics of the SEMP Fault at deep structural levels; (2) the relationship between this fault and the Brenner fault; (3) the significance of the western termination of the SEMP fault for the exhumation of the Tauern Window; and (4) the anatomy of an orogen-scale strike-slip fault system from a deep-seated structural level, to the upper, brittle crust.

The 'Ahorn' shear zone

The existence of a sinistral shear zone overprinting the northwestern margin of the Zentral Gneiss has been inferred in previous studies. Wang & Neubauer (1998) showed that the SEMP Fault passes into the ductile field in the area west of Mittersill (Fig. 1b). Lammerer & Weger (1988; their Fig. 3a) mapped a sinistral fault along the northwestern margin of the Tuxer Antiform (no. 7 in Fig. 1b). Linzer et al. (2002; their Fig. 1) extended the latter fault eastward, along the

Fig. 1. (*Continued*) indicates the trace of cross-section of Fig. 1c. (**b**) Simplified tectonic map of the Tauern Window and surrounding areas. Modified from Rosenberg et al. 2007. Fault-name abbreviations are as in Fig. 1a.
Numbers indicate the references below, used to compile sinistral faults in the western Tauern Window. 1, Borsi et al. (1978); Kleinschrodt (1987); 2, Mancktelow et al. (2001); 3, Kurz et al. (2001); Mancktelow et al. (2001); 4, Linzer et al. (2002); 5, Behrmann (1988); Barnes et al. (2004); 6, Lammerer & Weger (1998), 7, Lammerer & Weger (1998); 8, Linzer et al. (2002) and own work (Fig. 2). (**c**) N–S cross-section of the Alps, striking through the western Tauern Window (stippled line in Fig. 1a), based on surface and seismic data (TRANSALP Line). Modified from Schmid et al. (2004). The SEMP Fault is missing in the original cross-section.

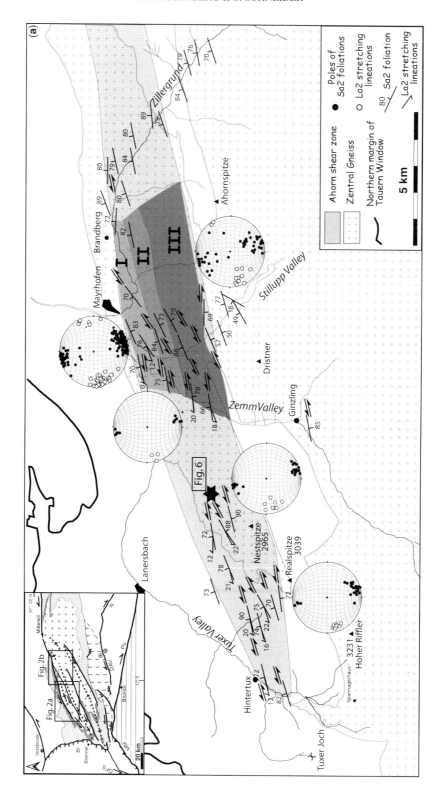

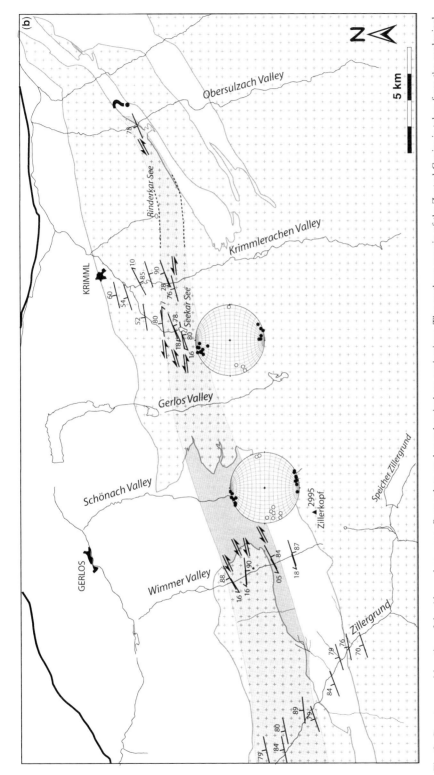

Fig. 2. Structures and location of the Ahorn shear zone. Stereoplots are lower hemisphere projections. The northern margin of the Zentral Gneiss is taken from the geological maps of the Geologische Bundesanstalt (in press), numbers 149, 150 and 151. Note the southward-directed transition from northward dip to southward dip within the Ahorn shear zone. (a) Western part of the Ahorn shear zone. The area of the Stillupp and Zemm Valleys is characterized by sinistral mylonites within a biotite-free orthogneiss. Zone III is characterized by sinistral mylonites within an orthogneiss where biotite is stable. Zone II is intermediate between zone I and zone III. Black star indicates the location of Figure 6. (b) Eastern part of the Ahorn shear zone. See Figure 2a for legend. Stippled boundaries indicate the area mapped by Cole et al. (2007). Boundaries of the shear zone are taken from their description. Question mark indicates area east of which the Ahorn shear zone has not yet been mapped.

northern margin of the Ahorn antiform (no. 8 in Fig. 1b) to the Salzach Fault, and westward, almost to the Brenner Fault (Fig. 1b). However, no field data supporting the existence of these shear zones and describing the type of deformation in terms of kinematics, microstructures, and inferred deformation temperatures were documented by the latter investigations.

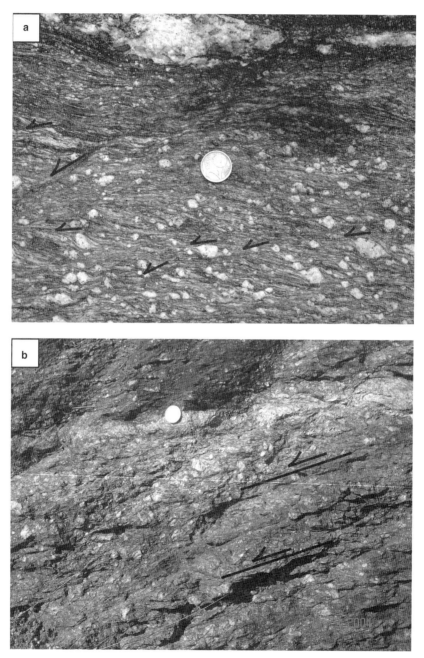

Fig. 3. Sinistral shear bands in the Zentral Gneiss. (**a**) Sinistral shear bands and sinistral sigma clasts in the Sa2 foliation of the Ahorn Orthogneiss from area II (Fig. 2a) (Zemm Valley). (**b**) Discrete C-C' structures, from the northern most margin of the Ahorn Kern (area I in Fig. 2), Zemm Valley, 1 km north of Figure 3a.

Figure 2 shows the location of the Ahorn shear zone based on our structural mapping. The criterion used to delimit the shear zone was the occurrence of a steeply dipping mylonitic foliation (Sa2), overprinting the main Alpine foliation (Sa1) and pervasively associated with sinistral kinematic indicators (Fig. 3a, b). This second Alpine foliation is rarely observed in the western Tauern Window, outside of this shear zone. As summarized in Table 1, the Sa1 schistosity is generally interpreted to have formed during Early Tertiary, N-directed nappe stacking, and the Sa2 schistosity, during (Miocene?) upright folding.

The Ahorn shear zone has an average width of nearly 2 km, and strikes along the northern margin of the Zentral Gneiss, between Hintertux in the west and Mittersill in the east (Fig. 1). In some areas, the shear zone affects both the Zentral Gneiss and the neighbouring schists of the Schieferhülle, as observed west of the Zemmtal (Fig. 2a). In other areas, as to the south of Krimml (Fig. 2b), the shear zone crosscuts the Zentral Gneiss, without overprinting its northernmost border. Close to its western termination, the shear zone abandons the northern border of the Zentral Gneiss and enters into the Schieferhülle, where its width rapidly decreases in the area of Tuxertal (Fig. 2a), before terminating approximately 2 km south of Hintertux (Figs 1b, 2a). This is indicated by the disappearance of the wide mylonitic belt with its Sa2 mylonitic foliation.

To the west of the shear zone termination (Fig. 2a), sinistral shear senses on E- to NE-striking foliations are not uncommon and generally occur where the S1 foliation is locally rotated into a subvertical orientation. In these outcrops, it is difficult to distinguish whether the foliation is an Sa1 or Sa2, or alternatively a composite foliation. However, the widespread and continuous occurrence of Sa2 foliations clearly ceases west of Hintertux and the main structural grain is formed by Sa1 foliations folded by the upright F2 folds. At its eastern end, the Ahorn shear zone appears to be continuous with and subparallel to the SEMP Fault, suggesting that they are part of one and the same structure.

Structure of the shear zone and relationship to its country rocks

Internal structure

Profiles across the shear zone west of Krimml (Figs 1b, 2b) show a similar structural trend. The Sa2 mylonitic foliation systematically strikes ENE, but the dip direction changes across the shear zone. Foliations dip steeply to the NNW at the northern margin of the shear zone, they become vertical further south, and finally steeply SSE-dipping at the southern margin of the shear zone (Fig. 2a). In contrast, lineations systematically plunge to the WSW with angles varying between 0–25°. Kinematic indicators in the form of pervasive C′ shear bands, indicating a sinistral shear sense are very common.

The average axial ratio of feldspar clasts in the X–Z plane of the deformation ellipsoid progressively increases from north to south (Fig. 4). However, strain analyses performed on the mylonitic augengneiss by the 'Fry' analysis (Fry 1979) on feldspar clasts do not show a strain gradient across the shear zone (Fig. 4). Therefore, the increase in the axial ratio of feldspars is interpreted to be the result of a temperature gradient. In the north of the shear zone, the temperature of deformation was not sufficient to allow the intracrystalline plastic deformation of feldspar, whereas in the southern part of the shear zone this temperature was attained. Based on a review of a large number of natural investigations, Fitz Gerald & Stünitz (1993) showed that the minimum temperature needed for the dynamic recrystallization of feldspar is higher than 450 °C and generally even higher than 500 °C, unless recrystallization occurs by nucleation of new grains and migration of their boundaries. In addition, feldspar ductility (qualitatively expressed by the increase in the axial ratio of feldspar aggregates) increases with increasing temperature (Rosenberg & Stünitz 2003).

In areas dominated by the same lithology, e.g. the porphyritic facies of the Zentral Gneiss as in the Zemmtal or Stillupptal (Fig. 2a), a comparison between Sa2 fabrics across the shear zone shows that the style of the pervasive C′ planes, changes from north to south. These planes are more sharply defined in the north (Fig. 3b), where they look like brittle-ductile structures. Further south, they are characterized by larger widths (Fig. 3a), suggesting a less-localized displacement, as expected for higher temperature conditions.

Relationship between sinistral shearing and upright folding

The structural evolution across the shear zone is similar in all investigated sections. Below, we describe this evolution from south to north.

South of the Ahorn shear zone (Fig. 2), the Zentral Gneiss was affected by two phases of folding. F1 folds are tight to isoclinal, generally recumbent (Fig. 5a–c), with an axial plane schistosity, which forms the main foliation of the western Tauern Window. The axial plane schistosity of F1 represents the first Alpine schistosity (Sa1) in the Zentral Gneiss as previously suggested by most structural investigations in the Tauern Window (e.g. Lammerer & Weger 1998; Table 1).

Table 1. *Compilation of inferred deformation phases in the Tauern Window. Terminology is as found in the literature*

Phase	Structures and tectonic significance	Fabric	Metamorphism/age
	Norris et al. 1971		
F_v	Folding with subhorizontal axes		Pre-alpine.
F_A^1	Subisoclinal folds in the Peripheral Schieferhülle. More open folds in the Inner Schieferhülle. Fold axes subparallel to F_v. NE-directed tectonic transport. During late stages of F_A^1 and after F_A^1 formation of Hochalm antiform and Reisseck synform.		Alpine
	Steepening of W-limb of Hochalm antiform between F_A^1 and after F_A^2		Alpine
F_A^2	Flexures, joints and faults, related to vertical movements.		Alpine
	De Vecchi & Baggio 1982		
I	Thrusting and formation of the Vizze syncline (Greiner Zone)		Upper Cretaceous
II	Initial thrusting of the Glockner nappes		Tauern Crystallization. Paleocene-Eocene
III	Final thrusting of the Glockner nappes		Tauern Crystallization. Eocene-Lower Oligocene
IV	Upright folding and uplift of the Tuxer block with respect to the southern part of the Tauern Window	Retrograde metamorphism	Upper Oligocene-Miocene
	Miller et al. 1984		
F1	Folding related to nappe stacking	S1 parallel to bedding	Alpine
F2	Local and minor refolding, with N–S striking fold axes.	No schistosity	
F3	Prominent upright folding	No schistosity	
F4	Minor backthrusting	No schistosity	
	Selverstone 1985		
F1	Folding related to formation of nappes with extensive shearing of fold limbs	S1	**USH**: synkinematic growth of porphyro-blasts (grt and Pl) started during F1.
F2	Refolding of previous folds and associated nappes. Upright folds.	Local formation of S2 by crenulation of Bt and Phe.	**LSH**: growth of porphyro-blasts post-dates F1 and is more or less static.
	Lammerer 1988		
D1	Thrusting of Austroalpine on top of Penninc. Top to the N.		Eocene
D2	Isoclinal recumbent folds, involving the uppermost Zentral Gneiss.	Intense schistosity and Lstr., at high angle to fold axes.	
D3	Upright to S-dipping Tux and Zillertal antiforms		
D4	Differential uplift of TW along the Salzach fault		
	Behrmann 1990		
D1	Thrusting (possibly towards 330°, see p. 107)	S1 under prograde conditions	Older than 70–55 My
D2	Tectonic significance unclear.	Formation of S2 or intensification of S1. E–W trending Lstr.	Between 70 and 55 My
D3	Upright folding, N–S shortening during E–W extension or extension in all directions. Sinistral shearing (Greiner shear zone, Knuttenalm, DAV and Speikboden).	Local formation of S3	Between 70–55 and 20 Ma
D4	Early Tertiary Extension of the western Tauern dome		Static thermal peak (Tauern Cristallisation) at the end of D3 D4 between 20 and 15 Ma
	Kupferschmied 1993		
D1	Isoclinal recumbent fold with amplitudes of ca. 5 km	S1 generally sub-parallel to lithological boundaries.	'Early' Alpine age.

(Continued)

Table 1. *Continued*

Phase	Structures and tectonic significance	Fabric	Metamorphism/age
D2	Large-scale and nearly coaxial (to D1 ?) folds with similar amplitude of D1.	Folding of S1 and formation of S2 in mica-rich rocks.	Alpine
	Inger & Cliff 1994		
D2	F2 folds (Sonnblick antiform and Mallnitz synform).	Folding of S1. Formation of S2	White mica ages of 28–29 Ma.
D3	Extensional unroofing	Brittle-ductile shear bands, down-to-the-SE shear sense	Alpine
	Kurz et al. 1996		
	Southeastern TW		
D0	NNE-directed nappe stacking along brittle thrust planes.		
D1	N-directed ductile nappe stacking. Continuous transition from brittle- to ductile nappe stacking between D0 and D1	S1 parallel to the thrust surfaces and L1, S- to SSE-dipping	S1: Hbl + cpx + bio + ep + Ab, suggesting 6 kb and 500°C
D2		S2 and L2 completely obliterate S1. S1 and S2 form a composite foliation.	Epidote-amphibolite facies
D3	Refolding of the nappe pile during exhumation of the Hochalm dome and development of the Sonnblick dome		
	Northeastern TW		
D0	as above		
D0	as above		
D1	as above	Venediger Nappe: S1 is pre-peak metamorphism and overgrown by Ky, Cld, Hbl and wM.	
D2		Subhorizontal S2 and W- to NW-trending L2 obliterate S1 in the Glockner nappe. S1 and S2 form composite foliation.	
	Lammerer & Weger 1998		
D1	Stacking of Ahorn-, Tux-, and Zillertal gneisses.		Early Tertiary (62 Ma)
D2	Formation of Ahorn, Ziller, and Tux upright antiforms		
D3	Strike-slip faults disrupted the folded structure		
	Kurz et al. 2001		
D1	Underplating and top-to-the-N stacking	Penetrative S1 and N–S trending Lstr 1.	
			D1 and D2 are separated by the 'Tauern-Crystallization', D2 is locally contemporaneous to Tauern Crystallization.
D2	Emplacement of penninic nappes onto the European foreland	S2 penetrative foliation and W to NW/SE trending stretching lineation L2.	
D3	Formation of the dome structure. Shear localization along the margins of the dome.	S3 and L3 along shear zones, boundng the Tauern domes.	
	Steffen & Selverstone 2006		
F2	Tight, recumbent folds with N or S plunging axes		
F3	Upright, ENE-striking axial planes and shallowly W-dipping fold axes		
	Present Study		
F1	Tight, recumbent folds	Sa1	Early Tertiary
F2	Upright, ENE-striking folds coeval to sinistral shearing along the SEMP and exhumation of the Tauern Dome.	Sa2 foliation in the Ahorn shear zone	Miocene

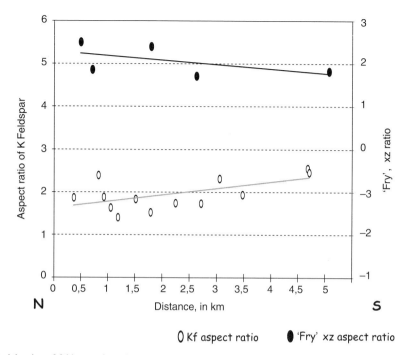

Fig. 4. Axial ratios of feldspars clasts (open ovals) plotted again N–S distance in the Stillupptal (see Fig. 2 for location). Filled ovals: Aspect ratio of deformation ellipses measured in the X–Z section of the inferred deformation ellipsoid on the basis of the Fry analysis. Each aspect ratio is the result of a Fry analysis performed on one sample with 30 to 50 feldspar clasts as centre points.

The F1 folds and the Sa1 foliation are refolded by open to tight F2 folds (Fig. 5a–c), which do not form an axial plane schistosity in this area. Their axial plane is steep to subvertical (Fig. 5b, c) and strikes ENE.

Further north, within the southern part of the Ahorn shear zone (area III in Fig. 2a), F2 folds become tighter and associated with a steeply dipping axial-plane schistosity (Sa2; Fig. 5d) in addition to the folded Sa1. Locally, the folded Sa1 is overprinted by sinistral shear zones, whose length varies from metres to hundreds of metres. These shear zones are sub-parallel to the axial planes of F2.

Still further north within the Ahorn shear zone (area II in Fig. 2a), F1 folds are no longer observed and F2 folds are also very rare. Where found, they occur in the form of dismembered hinges. The Sa2 foliation is very pronounced and it is associated with a pervasive C-C' fabric (Fig. 3a, b), systematically indicating a sinistral sense of shear. Locally, shear bands and/or shear zones indicating south-side-up displacements (Fig. 5e) also occur within the Y–Z planes, pointing to a transpressive type of deformation. The Sa1 foliation cannot be distinguished anymore because it is completely overprinted by the pervasive Sa2 foliation.

The F2 folds described above (Figs 5b, c and 6a) are parasitic folds of the large-scale, upright, ENE-striking antiforms, which form the structural grain of the western Tauern Window (Fig. 1a, b). The structural sequence described above points to a N-directed increase in the intensity of F2 shortening within our study area, culminating at the northernmost contact of the Zentral Gneiss, which is pervasively overprinted by a mylonitic Sa2 foliation. The parallelism between mylonitic Sa2 schistosity and the axial plane of the F2 folds suggests contemporaneous sinistral shearing and folding in the western Tauern Window. This interpretation is also supported by the direct observation of the Sa2 foliation with associated sinistral kinematic indicators (Fig. 6c), forming the axial plane schistosity of the F2 upright folds of the Ahorn antiform (Fig. 6a).

Microstructural changes across the shear zone

The microstructures of the mylonitic Zentral Gneiss vary across the Ahorn shear zone. We describe them from S to N below.

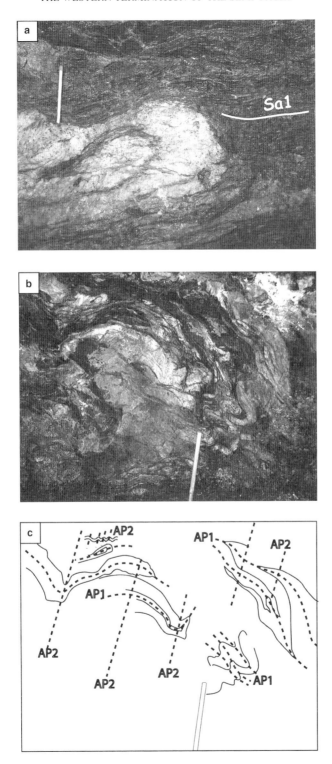

Fig. 5. *Continued.*

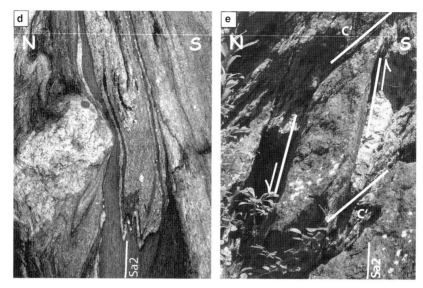

Fig. 5. (*Continued*) Field photographs from the Stillupp and Zemm valleys (Fig. 2). (**a**) Fa1 recumbent folds, within migmatitic gneisses, showing an Sa1 axial plane foliation within the dark, mica-rich layers. Upper Stillupp valley. (**b**) Open, upright Fa2 folds within migmatitic gneisses, refolding Fa1 folds, in the upper Stillupp valley. Hammer of 80 cm length for scale. (**c**) Line drawing of Figure 3c, showing the traces of the axial planes of the first phase (AP 1) and of the second phase (AP 2) of folding. (**d**) Isoclinal, upright Fa2 folds, with pronounced Sa2 axial plane foliation. This foliation is characterized by systematic and pervasively distributed sinistral kinematic indicators (mostly shear bands). Zemm Valley. (**e**) South-side-up shear sense in the Y–Z plane. Stillupp valley. Shorter side of the photographs is 60 cm long.

At the southernmost boundary of the Ahorn shear zone (area III in Fig. 2a), where the Zentral Gneiss is only locally overprinted by the Sa2 foliation within metre-scale sinistral shear zones (Fig. 7a), quartz grains have lobate boundaries (Fig. 7b), with lobe sizes of 50 to 100 μm, indicating recrystallization by grain boundary migration (Regime III of Hirth & Tullis 1992). Feldspars also show new grains with lobate boundaries within elongate tails of feldspar porphyroclasts (Fig. 7b). These observations point to a deformation temperature probably above 450–500 °C (for review, see Fitz Gerald & Stünitz 1993).

Further north (area II in Fig. 2a), the microstructures of quartz and feldspar are significantly different. Quartz grains have lobate grain boundaries (Fig. 7c), but the size of the lobes and of the grains is smaller compared to the samples located southward. No evidence for dynamic recrystallization of feldspar grains is found within these samples. Assuming similar strain-rate conditions between this area and the one described above, these microstructural differences indicate that sinistral shearing in the latter sample occurred under lower temperature conditions. This assumption is reasonable in the light of the distributed character of deformation within all parts of the shear zone.

In the northernmost area (area I in Fig. 2a), quartz grains are more elongate than in the previous samples (Fig. 7e, f), but they are not recrystallized, or only very locally. In contrast, they show strong undulose extinction (Fig. 7e), eventually passing into deformation bands. Where present, recrystallized grains have small sizes (10 to 20 μm) and serrated boundaries, which point to recrystallization by bulge nucleation. Feldspars form competent clasts in these mylonites and their occasional internal deformation only occurs by cataclasis (Fig. 7f) or shearing along retrograded, saussuritic domains. Flame perthites and exsolution to albite and plagioclase are common at the rims of the clasts. Sinistral shearing within these rocks was mainly partitioned into an interconnected weak layer (Handy 1990), consisting of fine-grained aggregates of white mica (Fig. 7e, f), locally containing minor amounts of quartz and albite. These fine-grained aggregates flow around the elongate and partly boudinaged quartz grains (Fig. 7e, f), indicating that quartz was more competent under these temperature conditions. This competence contrast does not persist in the southern part of the shear zone, where deformation is equally partitioned within the quartz and the mica aggregates (Fig. 7c, d). Assuming a similar strain rate as in the samples

Fig. 6. Structures and microstructural relationship between F2 folding and Sa2 foliation in the Ahorn dome, Inner Elskar (Fig. 2a for location). (**a**) F2-Folded contact of Zentral Gneiss and Schieferhülle on the northern limb of the Ahorn antiform (Fig. 1), Inneres Elskar, Ziller Valley. Black line marks the contact between Zentral Gneiss and Lower Schieferhülle. (**b**) Detail of Figure 6a, showing the axial plane foliation Sa2, cross-cutting the boundary between Zentral Gneiss (below; ZG) and Triassic quartzites above. (**c**) Micrograph, with crossed-polarizers, indicating sinistral shear bands in the Zentral Gneiss sampled in the outcrop of Figure 6b.

of Figure 7(c) and (e), the microstructural difference described above can be attributed to a northward decrease in the temperature of deformation.

The aforementioned microstructural changes go together with a change in the modal composition of the samples. From south to north, the content of biotite progressively decreases, becoming replaced by fine-grained white micas (compare Fig. 7c, e), oxides and/or chlorite within grain fractures. At the northern margin of the Zentral Gneiss, biotite is almost completely absent. These findings support the interpretation of the northward transition of recrystallization mechanisms in terms of a decrease in deformation temperature. The northward decrease in the axial ratio of the feldspar clasts at relatively constant bulk strain (Fig. 4) is also consistent with lower temperatures in the north: the higher axial ratios in the south are likely due to increased feldspar ductility compared with brittle feldspar behaviour in the north.

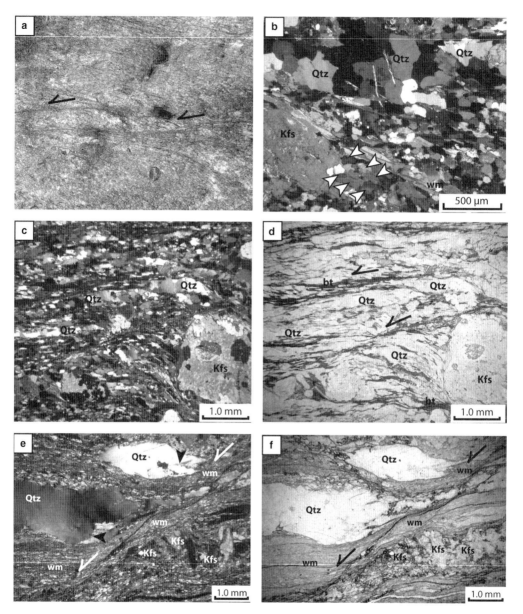

Fig. 7. Structure and microstrucures of Sa2 foliation along a S–N traverse through the Zemmtal. (**a**) High-temperature, sinistral shear zones at the southernmost margin of the Ahorn Shear zone, Zemm Valley. In this area, the mylonitic Sa2 foliation only occurs within discrete, metre-long sinistral shear zones, which overprint the Sa1 foliation.
(**b**) Photomicrograph with cross-polarized light showing the microstructures within a shear zone of Figure 7a. Note the large and lobate grains of quartz (qtz) and the recrystallized grains of feldspar (white arrowheads) in the pressure shadow of a larger clast (Kfs). wm: white mica. (**c**) Microphotograph with cross-polarized light. Sa2 foliation from the southern part of the Ahorn shear zone, Zemm Valley. Note the dynamically recrystallized aggregates of quartz.
(**d**) Microphotograph with plane light. Same sample as in Figure 7c. Note the localization of deformation into sinistral shear bands, consisting of quartz, white mica and biotite. (**e**) Microphotograph with cross-polarized light. Sa2 foliation from the northernmost part of the Ahorn shear zone, Zemm Valley. Note the elongate and boudinaged quartz grains, with deformation bands and undulose extinction. Recrystallization of quartz is very limited (black arrowheads).
(**f**) Microphotograph with plane light. Same sample as in Figure 7e. Note that deformation does not localize into quartz aggregates, but into sinistral shear bands, consisting of very fine-grained white mica.

Structure of the country rocks west of the Ahorn shear zone

The area located immediately west of the termination of the Ahorn mylonitic belt, i.e., west of Hintertux (Figs 1b and 2b) is affected by intense shortening accommodated by upright folds striking WNW–ESE (Fig. 8). These folds fold the Sa1 schistosity and do not form a new axial plane foliation. Therefore, they can probably be attributed to the F2 phase. However, the orientation of the axial planes of these folds differs from that of the F2 axial planes measured further east, which strike WSW–ENE, as do most structures of the western Tauern Window (Fig. 1b). At present, we do not have more data to constrain the northern, southern and western termination of the area characterized by upright folds with WNW-striking axial planes. However, we emphasize the spatial coincidence between the rotation of the F2 axial planes from an ENE to a WNW strike and the termination of the Ahorn shear zone (Fig. 8).

Discussion

Alpine deformation phases and fabrics

The Ahorn shear zone is characterized by a mylonitic Sa2 foliation, which forms the axial plane schistosity of the northernmost upright antiforms of the western Tauern Window, i.e. of the Ahorn antiform (Fig. 1b). This relationship, which points to contemporaneous upright folding and sinistral shearing of the Ahorn core, is in contrast to previous work (Kurz et al. 2001), which suggested that deformation in this area pre-dated the formation of the Tauern dome and the formation of shear zones bordering the dome.

The fact that the mylonitic Sa2 foliation is not folded and is axial planar to the F2 upright folds suggests that it formed in a steep orientation, probably similar to the present one (Fig. 2), because no significant deformation phases younger than F2 are observed in this study area (Table 1). A similar structural relationship between sinistral shearing and upright folding has been suggested by Kleinschrodt (1987) and Wagner et al. (2006) for the DAV Shear Zone (Fig. 1b), which marks the southern border of the Tauern structural and thermal dome (Frisch et al. 2000).

A static recrystallization event associated with the Tertiary metamorphic peak has often been invoked in the Venediger Nappe Complex of the Tauern Window (e.g. Behrmann 1990; Kurz et al. 2001). This static event has been interpreted as the result of the 'Tauern crystallization' of Sander (1920), although Sander originally described it as a dynamic metamorphic event. We found no microstructural evidence pointing to static recrystallization in the Ahorn Kern, neither within the Ahorn shear zone nor outside of the shear zone (Fig. 7). Even the high-temperature deformation fabrics of the southern border of the shear zone and the Sa1 fabrics outside of the shear zone (Fig. 9) indicate the preservation of dynamic fabrics characterized by lobate grain boundaries and subgrains within the quartz aggregates. Therefore, both the peak of Tertiary metamorphism and the retrograde metamorphic overprint occurred under dynamic conditions in our study area. Interestingly, Steffen et al. (2001) and Steffen & Selverstone (2006) showed that even some apparently 'post-kinematic' Garbenschiefer fabrics south of our study area are syndeformational. This may suggest that the 'Tauern crystallization' was syntectonic throughout the Tauern Window.

Age of folding and sinistral shearing

F2 upright folds were suggested to be late and mainly post-peak of Tertiary metamorphism, i.e. upper Oligocene to Miocene, because the F2 folds fold the isograds of Tertiary metamorphism (de Vecchi and Baggio 1982). Behrmann (1990; Table 1) suggested that F2 folding occurred between 70–55 and 20 Ma. Wagner et al. (2006) interpreted the steep orientation of the Austroalpine basement south of the western Tauern Window as due to F2 folding, which was inferred to be Late Oligocene based on cross-cutting relationships with the 30 Ma old Rieserferner intrusion.

In the present study, we showed that the upright folds of the Tauern Window probably formed at the same time as sinistral shearing along the western termination of the SEMP Fault. Therefore, the age of displacement along the SEMP Fault may be used to date F2 folding in the western Tauern Window. Sinistral shearing along the SEMP is inferred to be of Karpatian age (17 Ma; Peresson & Decker 1997) on the base of deformed, dated conglomerates (Steininger et al. 1989). The major phase of lateral escape and hence of activity of the SEMP Fault was suggested to be between 23 Ma and 12–13 Ma (Frisch et al. 1999). As a consequence, sinistral shearing along the Ahorn shear zone and upright folding in the western Tauern Window may also be lower to Middle Miocene in age. However, Most et al. (2003) showed a very pronounced southward younging of apatite fission track ages across the Ahorn shear zone (Fig. 10), from 12 Ma at the northern shear zone boundary to 7 Ma at the southern boundary. Therefore, significant south-side-up displacements were accommodated by the Ahorn shear zone in the upper Miocene. The spatial coincidence of this pronounced age gradient

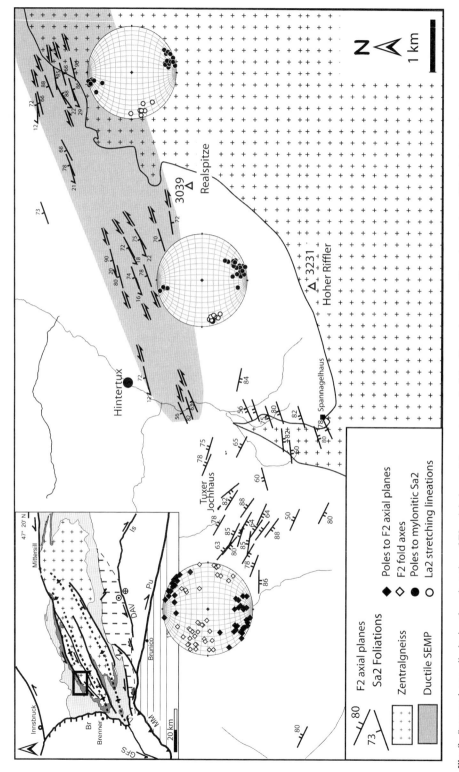

Fig. 8. Structural map, displaying the orientation of F2 axial planes at the western end of the Ahorn shear zone. Note the rotation of F2 axial planes from ENE-striking in the east, to WNW-striking in the west.

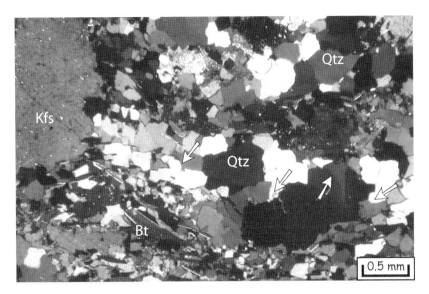

Fig. 9. Microstructures of Sa1 schistosity. White arrows directed to the bottom show large lobate boundaries of quartz grains, indicating dynamic recrystallization by grain boundary migration. White arrow directed to the top shows the occurrence of deformation bands within a quartz grain. These microstructures suggest that this Early Alpine schistosity was not statically annealed.

(Fig. 10) and the location of the transpressive mylonitic belt of the Ahorn shear zone, suggest that the south-side-up displacement indicated by the FT ages of apatites (Most *et al.* 2003) may be contemporaneous with sinistral displacements. In this case, sinistral shearing in the Ahorn shear zone would also have been active until the upper Miocene.

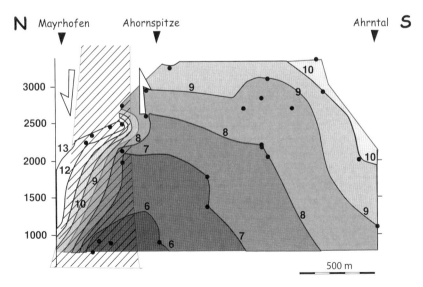

Fig. 10. Fission track apatite ages, modified after Most *et al.* (2003). Note the very rapid younging of fission track ages located exactly along the Ahorn shear zone, indicating a pronounced south-side-up displacement, probably younger than 7 Ma. Numbers on curves are ages in millions of years.

Kinematics of the Ahorn shear zone

The occurrence of a southward increase in the temperature of deformation within the Ahorn shear zone, and the fact that the Ahorn shear zone separates an area with Tertiary amphibolite-facies metamorphism in the south (e.g. Hörnes & Friedrichsen 1978) from an area in the lowest greenschist facies (just above the brittle-ductile transition in quartz) in the north points to a south-side-up component of displacement acting along the shear zone. It is difficult to quantify the absolute temperature difference between the southern and northern margins of the Ahorn shear zone only on the base of quartz recrystallization mechanisms, because the transition temperatures from one mechanism to the other are temperature-, but also strain-rate dependent (Stipp et al. 2006). However, a large number of field studies (Fitz Gerald & Stünitz 1993, for a review) suggest that dynamic recrystallization of feldspar initiates at temperatures above 450–500 °C. Therefore, this temperature may be considered as a minimum estimate of the maximum temperature during shearing at the southern margin of the Ahorn shear zone. The lowest temperature of shearing, at the northern margin of the Ahorn shear zone, can be constrained on the basis of quartz grains, showing microstructural evidence for dislocation glide (Fig. 7e), but no evidence for dislocation creep. This transition is inferred to be at temperatures ≥ 280 °C for rocks deforming at strain rates within the commonly inferred range of 10^{-11} s^{-1} to 10^{-14} s^{-1} (Stipp et al. 2002).

These temperature estimates point to a difference of approximately 200 °C between the southern and northern boundaries of the Ahorn shear zone, in the area of the Stillupp and Zemm valleys (Fig. 2a). Considering a geotherm of c. 30 °C/km in order to describe a simplified crust for illustrative purposes, the aforementioned temperature difference may correspond to a vertical offset of c. 7 km. This offset could result from sinistral shearing parallel to the west-plunging stretching lineations (Fig. 2). However, given the transpressive character of the shear zone, the transport direction may have been steeper than the stretching lineation (e.g. Robin & Cruden 1994).

As a consequence, the lateral offset of 60 km, which affected the SEMP Fault east of the Tauern Window (Linzer et al. 2002), is partly transferred into a vertical one in the Ahorn shear zone. This conclusion is consistent with the western termination of the shear zone, which is observed to pass into a zone of upright folds, probably accommodating a component of vertical extension.

Kinematic link between the SEMP and the Brenner Faults

The Ahorn shear zone terminates in the area east of the Tuxer Joch (Fig. 2a), approximately 15 km east of the Brenner Fault (Fig. 2). Therefore, in contrast to previous interpretations (Linzer et al. 2002), the SEMP Fault and the Brenner Fault are not in spatial continuity, and the Brenner Fault does not form the lateral ramp of the SEMP Fault. The lateral displacement of the Ahorn shear zone passes into a WNW–striking folded structure which accommodates shortening in an approximately NNE–SSW directed orientation (Fig. 11), and not into an E–W directed extension.

The fact that folds with anomalous orientations (WNW-striking axial planes, Fig. 8) occur exactly in the spatial continuation of the Ahorn shear zone termination suggests that the latter structures are genetically related. We envisage that NNE-oriented shortening becomes partitioned into a sinistral displacement within the Ahorn shear zone and a NNW-oriented shortening component to the south of the shear zone (Fig. 11). West of the shear zone termination, NNE-directed shortening is not partitioned into a lateral, sinistral displacement, hence resulting in folds with axial planes approximately perpendicular to the shortening direction. As a consequence, the sinistral displacements of the SEMP Fault and the Ahorn shear zone are transferred into an area of NNE-directed shortening (Fig. 11), not in an E–W extensional deformation.

The other sinistral shear zones previously suggested to be splays of the SEMP Fault (nos. 4 and 5 in Figure 1b; Linzer et al. 2002) also lack a kinematic continuity with the Brenner Fault. The spatial relationship between Brenner Fault and Greiner shear zone (no. 5 in Figure 1b) cannot be satisfyingly solved due to the lack of outcrops in the critical area, where overprinting structures would be expected (Behrmann 1988). We note however, that if the Greiner shear zone continued westward until the Brenner Fault without a marked change of strike, it would reach the Brenner Fault at its southern end (Fig. 1b). In this area, the kinematics of the west-dipping Brenner extensional fault would predict a dextral shear zone, associated with exhumation of the footwall of the Brenner Fault (e.g. Fügenschuh et al. 1997) and not a sinistral shear zone. Therefore, a direct kinematic link between the Greiner shear zone and the Brenner Fault is unlikely. The same line of arguments precludes a kinematic link between the sinistral Ahrntal Fault (Fig. 1) and the Brenner Fault. The westward continuation of the Ahrntal Fault would also reach the Brenner Fault at its southernmost margin.

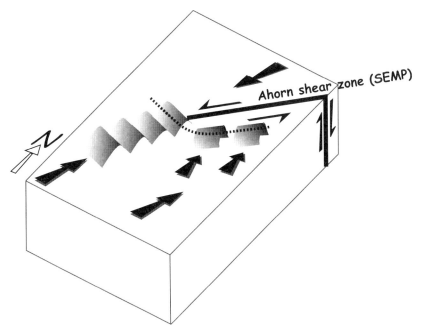

Fig. 11. Schematic block diagram indicating the relationship between inferred direction of shortening and the resulting first order structures at the western termination of the Ahorn shear zone. West of the termination of the Ahorn shear zone, NNE-oriented shortening leads to the formation of WNW-striking axial planes. South of the termination of the Ahorn shear zone, NNE shortening is partitioned into a lateral ENE-striking sinistral shear component and a pure shear component oriented perpendicular to the shear zone, i.e. NNW.

Anatomy of the SEMP Fault from the upper crust to the middle/lower crust

An along-strike, brittle-ductile transition within the SEMP Fault has been described in the area of Mittersill (Wang & Neubauer 1998; Fig. 1). This transition is described within metapelites and carbonatic schists containing calcite mylonites. The brittle-ductile transition in these rocks is controlled by the onset of intracrystalline plastic deformation of micas and calcite, which is constrained to occur at temperatures <250 °C (e.g. Burkhard 1993), whereas the same transition in the quartz-dominated Zentral Gneiss probably occurred at temperatures ≥280 °C as constrained by Stipp et al. (2002) for quartz veins in the contact aureole of the Adamello batholith. In the area of Rinderkarsee (Fig. 2b), located approximately 20 km west of Mittersill, Cole et al. (2007) described a wide zone (1300 m) of sinistral deformation within the Zentral Gneiss, in which shearing is heterogeneously distributed into individual shear zones. Further west, at Seekarsee (Fig. 2b), sinistral shear zones of tens to hundreds of metres thickness alternate with areas of hundreds of metres thickness where the Zentral Gneiss is largely undeformed, locally preserving its prealpine magmatic fabric. West of Krimml (Figs 1 and 2b), sinistral deformation becomes homogeneously distributed within a zone of approximately 2 km thickness. Undeformed areas within this zone do not occur anymore and the Zentral Gneiss consists everywhere of an S-C type mylonite (Fig. 3).

These changes in the spatial pattern of deformation partitioning are often suggested to characterize the depth-dependent brittle-ductile transition of large-scale faults (e.g. Twiss & Moores 1992). The fact that the microstructures observed in the areas of Zemmtal and Stillupptal (Fig. 7) indicate sinistral shearing at temperatures much higher than those described further east by Cole et al. (2007) and Wang & Neubauer (1998) suggests a continuous westward increase in the deformation temperature and hence a deeper exposure level of the SEMP Fault, from Mittersill to Rinderkarsee, and to the Ahorn Kern (Fig. 2). Taken together, these interpretations suggest that the SEMP Fault consists of discrete brittle faults in the upper crust (presently exposed east of Mittersill; Linzer et al.

2002) of an anastomozing network of shear zones close to the brittle-ductile transition, as presently exposed south of Krimml (Cole et al. 2007), and of a distributed deformation as shown by the structures exposed in the area west of Krimml (Fig. 2a, b). In contrast to Wang & Neubauer (1998), we find no coaxial flattening in the ductile part of the SEMP Fault, and no decrease in the shear zone width with increasing depth.

Conclusions

The SEMP Fault extended beyond the brittle-ductile transition to a depth where temperatures probably exceeded 500 °C (>20 km depth?). The present-day surface exposure of this fault shows a transition from brittle faulting in the east to an area of anastomozing, ductile shear zones and finally to a homogeneously deformed mylonitic belt further west.

The large-amplitude, upright folds of the Tauern Window formed at the same time as the sinistral mylonites during the activity of the SEMP Fault, which may have started in the Oligo(?)-Miocene. Vertical, differential displacements indicated by pronounced younging of apatite and zircon fission track ages (Most et al. 2003) exactly across the trace of the Ahorn shear zone, suggest that the shear zone was still active in the uppermost Miocene. Part of the pronounced lateral displacement of the SEMP Fault is transferred into a vertical displacement at its western end, and into NNE-directed shortening, accommodated by WNW-striking, upright folds.

The sharp increase of metamorphic grade across the shear zone contrasts to the gradually increasing grade observed along the southern margin of the western Tauern Window. These features are similar but symmetrically opposite to the central Alps, where the sharpest metamorphic gradient and vertical displacement are localized along the southern side of the Lepontine dome (Fig. 1a).

Although we reject the idea that the SEMP Fault continuously splays into the three major shear zones of the western Tauern Window (the Ahorn, Greiner and Ahrntal shear zones), these three shear zones were probably coeval and belonged to the same sinistral system of the SEMP Fault. The en-echelon structure of these shear zones (Fig. 1b) and their parallelism to the upright folds suggest that the entire western Tauern Window may be regarded as a restraining bend connecting the sinistral displacements of the Giudicarie Fault System with the sinistral displacements of the SEMP Fault. Laboratory experiments (Ratschbacher et al. 1991a, EXP 1–23; Rosenberg et al. 2004; Rosenberg et al. 2007) show that the strike of the SEMP Fault may have been approximately NE, as the fault nucleated, i.e. much closer to the orientation of the Giudicarie Fault System. Subsequent deformation rotated the SEMP Fault into the present ENE strike, whereas the Giudicarie Fault, which marks the boundary of a nearly rigid indenter, maintained its initial orientation.

Due to its vertical orientation, this mylonitic belt was not imaged as a reflector along the recent TRANSALP seismic line, hence it was not shown in the geological interpretations of the seismic lines (e.g. Lüschen et al. 2004, 2006; Schmid et al. 2004). One exception is the interpretation of Ortner et al. (2006), who associated the northern termination of several gently south-dipping reflectors to two steep faults with south-side-up displacement, bounding the upper Schieferhülle. Our field data indicate that the Ahorn shear zone should be included in the cross-sections of the western Tauern Window, as a wide (2 km) subvertical structure, mainly located within the northern boundary of the Zentral Gneiss.

Field work has been supported by the Deutsche Forschungsgemeinschaft (Ro 2177/4-1). The high quality of the thin sections of W. Michaelis is much appreciated. Lothar Ratschbacher was extremely helpful, providing literature and suggesting localities where one could or could not possibly follow the ductile terminations of the SEMP Fault. We acknowledge discussions with Jochen Babist and Konrad Hammerschmidt who shared with us their ideas and field observations. Jane Selverstone and Hugo Ortner provided very constructive reviews of the manuscript. Oliver Kettler collected some of the structural data in Zillergrund. C. R. is indebted to Stefan Schmid for an excellent training in structural geology.

References

BARNES, J. D., SELVERSTONE, J. & SHARP, Z. D. 2004. Interactions between serpentinite devolatilization, metasomatism and strike-slip strain localization during deep-crustal shearing in the Eastern Alps. *Journal of Metamorphic Geology*, **22**, 283–300.

BEHRMANN, J. H. 1988. Crustal-scale extension in a convergent orogen: the Sterzing-Steinach mylonite zone in the Eastern Alps. *Geodynamica Acta*, **2**, 63–73.

BEHRMANN, J. H. 1990. Zur Kinematik der Kontinentalkollision in den Ostalpen. *Geotektonische Forschungen*, **76**, 1–180.

BEHRMANN, J. & FRISCH, W. 1990. Sinistral ductile shearing associated with metamorphic decompression in the Tauern Window, Eastern Alps. *Jahrbuch der Geologischen Bundesanstalt*, **133**, 135–146.

BORSI, S., DEL MORO, A., SASSI, F. P., ZANFERRARI, A. & ZIRPOLI, G. 1978. New geopetrologic and radometric data on the alpine history of the austridic continental margin south of the Tauern Window (Eastern Alps). *Consiglio Nazionale della Ricerca*, **32**, 3–17.

BURKHARD, M. 1993. Calcite twins, their geometry, appearance and significance as stress-strain markers and indicators of tectonic regime: a review. *Journal of Structural Geology*, **15**, 351–368.

CHRISTENSEN, J. N., SELVERSTONE, J., ROSENFELD, J. L. & DEPAOLO, D. J. 1994. Correlation by Rb-Sr geochronology of garnet growth histories from different structural levels within the Tauern Window, Eastern Alps. *Contributions to Mineralogy and Petrology*, **118**, 1–12.

COLE, J. N., HACKER, B. R., RATSCHBACHER, L., DOLAN, J. F., SEWARD, G., FROST, E. & FRANK, W. 2007. Localized ductile shear below the seismogenic zone: structural analysis of the exhumed SEMP strike-slip fault, Austrian Alps. *Journal of Geophysical Research*, **112**, doi: 10.1029/2007JB004975.

DE VECCHI, G. & BAGGIO, P. 1982. The Pennine zone of the Vizze region, in the Western Tauern Window (Italian Eastern Alps). *Bollettino della Società Geologica Italiana*, **101**, 89–116.

DECKER, K., PERESSON, H. & FAUPL, P. 1994. Die miocäne Tektonik der östlichen Kalkaplen: Kinematik, Paläospannungen und Deformationsaufteilung während der 'lateralen Extrusion' der Zentralalpen. *Jahrbuch der Geologischen Bundesanstalt*, **137**, 5–18.

FITZ GERALD, J. D. & STÜNITZ, H. 1993, Deformation of granitoids at low metamorphic grades. I. Reactions and grainsize reduction. *Tectonophysics*, **221**, 269–297.

FRISCH, W., KUHLEMANN, A., DUNKL, I. & BRÜGEL, A. 1998. Palinspastic reconstruction and topographic evolution of the Eastern Alps during late Tertiary tectonic extrusion. *Tectonophysics*, **297**, 1–15.

FRISCH, W., BRÜGEL, A. J., DUNKL, I., KUHLEMANN, J. & SATIR, M. 1999. Post-collisional large-scale extension and mountain uplift in the Eastern Alps. *Memorie di Scienze Geologiche (Padova)*, **51**, 3–23.

FRISCH, W., DUNKL, I. & KUHLEMANN, J. 2000. Post-collisional orogen-parallel large-scale extension in the Eastern Alps. *Tectonophysics*, **327**, 239–265.

FRY, N. 1979. Random point distributions and strain measurement in rocks. *Tectonophysics*, **60**, 89–105.

FÜGENSCHUH, B., SEWARD, D. & MANTCKELOW, N. 1997. Exhumation in a convergent orogen: the western Tauern Window. *Terra Nova*, **9**, 213–217.

HANDY, M. R. 1990. The solid-state flow of polymineralic rocks. *Journal of Geophysical Research*, **96**, 8647–8661.

HANDY, M. R., BABIST, J., WAGNER, R., ROSENBERG, C. L., KONRAD, M. 2005. Decoupling and its relation to strain partitioning in continental lithosphere: insight from the Periadriatic fault system (European Alps). *In*: GAPAIS, D., BRUN, J. P. & COBBOLD, P. R. (eds) *Deformation Mechanisms, Rheology and Tectonics: from Minerals to the Lithosphere*. Geological Society of London, Special Publication, **243**, 249–276.

HIRTH, G. & TULLIS, J. 1992. Dislocation creep regimes in quartz aggregates. *Journal of Structural Geology*, **14**, 145–159.

HÖRNES, S. & FRIEDRICHSEN, H. 1978. Oxygen isotope studies of the Austroalpine and Pennine units of the eastern Alps. *In*: CLOOS, H., ROEDER, D. & SCHMIDT, K. (eds) *Alps, Appennines, and Hellenides*. Inter-Union Committee on geodynamics Science Report, **38**, 127–131.

INGER, S. & CLIFF, R. A. 1994. Timing of metamorphism in the Tauern Window, Eastern Alps: Rb-Sr ages and fabric formation. *Journal of Metamorphic Geology*, **12**, 695–707.

KLEINSCHRODT, R. 1987. Quarzkorngefügeanalyse im Altkristallin südlich des westlichen Tauernfensters (Südtirol/Italien). Erlanger geologische Abhandlungen, **114**, 1–82.

KUPFERSCHMIED, M. 1993. Structural studies in the Western Habach Group (Tauern Window, Salzburg, Austria). *Abhandlungen der Geologische Bundesanstalt*, **49**, 67–78.

KURZ, W. & NEUBAUER, F. 1996. Deformation partitioning during updoming of the Sonnblick area in the Tauern Window (Eastern Alps, Austria). *Journal of Structural Geology*, **18**, 1327–1343.

KURZ, W., UNZOG, W., NEUBAUER, F. & GENSER, J. 2001. Evolution of quartz microstructures and textures during polyphase deformation within the Tauern Window (Eastern Alps). *International Journal of Earth Sciences*, **90**, 361–378.

LAMMERER, B. 1988. Thrust-regime and transpression-regime tectonics in the Tauern Window (Eastern Alps). *Geologische Rundschau*, **77**, 143–156.

LAMMERER, B. & WEGER, M. 1998. Footwall uplift in an orogenic wedge: the Tauern Window in the Eastern Alps of Europe. *Tectonophysics*, **285**, 213–230.

LINZER, H.-G., MOSER, F., NEMES, F., RATSCHBACHER, L. & SPERNER, B. 1997. Build-up and dismembering of a classical fold-thrust belt: from non-cylindircal stacking to lateral extrusion in the eastern Northern Calcareous Alps. *Tectonophysics*, **272**, 97–124.

LINZER, H.-G., DECKER, K., PERESSON, H., DELL' MOUR, R. & FRISCH, W. 2002. Balancing orogenic float of the Eastern Alps. *Tectonophysics*, **354**, 211–237.

LÜSCHEN, E., LAMMERER, B., GEBRANDE, H., MILLHAN, K. NICOLICH, R. & TRANSALP WORKING GROUP 2004. Orogenic structure of the Eastern Alps, Europe from TRANSALP deep seismic reflection profiling. *Tectonophysics*, **388**, 85–102.

LÜSCHEN, E., BORRINI, D., GEBRANDE, H., LAMMERER, B., MILLAHN, K., NEUBAUER, F. & NICOLICH, R. 2006. TRANSALP—deep crustal Vibroseis and explosive seismic profiling in the Eastern Alps. *Tectonophysics*, **414**, 9–38.

MANCKTELOW, N. S., STÖCKLI, D. F., GROLLIMUND, B. ET AL. 2001. The DAV and the Periadriatic fault system in the Eastern Alps south of the Tauern Window. *International Journal of Earth Sciences*, **90**, 593–622.

MILLER, H., LEDOUX, H., BRINKMEIER, I. & BEIL, F. 1984. Der Nordwestrand des Tauernfenstersstratigraphische Zusammenhänge und tektonische Grenzen. *Zeitschrift der deutschen geologischen Gesellschaft*, **135**, 627–644.

MOST, P., DUNKL, I. & FRISCH, W. 2003. Fission track tomography of the Tauern Window along the TRANSALP profile. *Memorie di Scienze Geologiche, Padova*, **54** (special vol.), 225–226.

NORRIS, J., OXBURGH, E. R., CLIFF, R. A. & WRIGHT, R. C. 1971. Structural, Metamorphic, and Geochronological Studies in the Reisseck and Southern Ankogel Groups, the Eastern Alps, Part IV, Structure. *Jahrbuch der geologischen Bundesanstalt*, **114**, 198–234.

ORTNER, H. & STINGL, V. 2001. Facies and basin development of the Oligocene in the lower Inn Valley, Tyrol Bavaria. *In*: PILLER, W. E. & RASSER, M. W. (eds) Palaeogene of the Eastern Alps. *Österreichische Akademie der Wissenschaften Schriftenreiche der Erdwissenschaftlichen Kommissionen*, **14**, 153–196.

ORTNER, H., REITER, F. & BRANDNER, R. 2003. Kinematics of the Inntal shear zone and its relation to other major faults crossing the northern TRANSALP seismic section. *Memorie di Scienze Geologiche, Padova*, **54**, 189–192.

ORTNER, H., REITER, F. & BRANDNER, R. 2006. Kinematics of the Inntal shear zone–sub-Tauern ramp fault system and the interpretation of the TRANSALP seismic section, Eastern Alps, Austria. *Tectonophysics*, **414**, 241–258.

PERESSON, H. & DECKER, K. 1997. Far-field effects of late Miocene subduction in the Eastern Carpathians: E-W compression and inversion of structures in the Alpine-Carpathian-Pannonian region. *Tectonics*, **16**, 38–56.

RATSCHBACHER, L., MERLE, O., DAVY, P. & COBBOLD, P. 1991a. Lateral extrusion in the Eastern Alps, Part 1: Boundary conditions and experiments scaled for gravity. *Tectonics*, **10**, 245–256.

RATSCHBACHER, L., FRISCH, W. & LINZER, H.-G. 1991b. Lateral extrusion in the Eastern Alps: Part II. Structural analysis. *Tectonics*, **10**, 257–271.

REICHERTER, K., FIMMEL, R. & FRISCH, W. 1993. Sinistral strike-slip faults in the Central Tauern Window (Eastern Alps, Austria). A short note. *Jahrbuch der Geologischen Bundesanstalt*, **136**, 495–502.

REINECKER, J. & LEHNHARDT, W. A. 1999. Present-day stress field and deformation in eastern Austria. *International Journal of Earth Sciences*, **88**, 532–550.

ROBIN, P.-Y. F. & CRUDEN, A. R. 1994. Strain and vorticity patterns in ideally ductile transpression zones. *Journal of Structural Geology*, **16**, 447–466.

ROSENBERG, C. L. & STÜNITZ, H. 2003. Deformation and recrystallization of plagioclase along a temperature gradient. The example of the Bergell tonalite. *Journal of Structural Geology*, **25**, 391–410.

ROSENBERG, C. L., BRUN, J.-P. & GAPAIS, D. 2004. Indentation model of the Eastern Alps and the origin of the Tauern Window. *Geology*, **32**, 997–1000.

ROSENBERG, C. L., BRUN, J.-P., CAGNARD, F. & GAPAIS, D. 2007. Oblique indentation in the Eastern Alps: insights from laboratory experiments. *Tectonics*, **26**, TC2003; doi: 10.1029/2006TC001960.

SANDER, B. 1920. Geologische Studien am Westende der Hohen Tauern. II Bericht. *Jahrbuch der Geologische Staatsanstalt*, **70**, 273–296.

SCHMID, S. M., FÜGENSCHUH, B., KISSLING, E. & SCHUSTER, R. 2004. Tectonic map and overall architecture of the Alpine orogen. *Eclogae Geologicae Helvetiae*, **97**, 93–117.

SELVERSTONE, J. 1985. Petrologic constraints on imbrication, metamorphism, and uplift in the Tauern Window, Eastern Alps. *Tectonics*, **4**, 687–704.

STEFFEN, K. J., SELVERSTONE, J. & BREARLEY, A. 2001. Episodic weakening and strengthening during synmetamorphic deformation of a deep crustal shear zone in the Alps. *In*: HOLDSWORTH, R., STRACHAN, R., MAGLOUGHLIN, J. & KNIPE, R. (eds) *The Nature and Tectonic Significance of Fault Zone Weakening*, Geological Society of London, Special Publication, **186**, 141–156.

STEFFEN, K. J. & SELVERSTONE, J. 2006. Retrieval of P-T information from shear zones: thermobarometric consequences of changes in plagioclase deformation mechanisms. *Contributions to Mineralogy and Petrology*, **151**, 600–614.

STEININGER, F., RÖGL, F., HOCHULI, P. & MÜLLER, C. 1989. Lignite deposition and marine cycles. The Austrian Tertiary lignite deposits. A case history. *Sitzungsbericht der Akademie der Wissenschaften, Wien, mathematisch- naturwissenschaftliche Klasse*, **197**, 309–332.

STIPP, M., STÜNITZ, H., HEILBRONNER, R. & SCHMID, S. M. 2002. The eastern Tonale fault zone: a 'natural laboratory' for crystal plastic deformation of quartz over a temperature range from 250 to 700 °C. *Journal of Structural Geology*, **24**, 1861–1884.

STIPP, M., TULLIS, J. & BEHRENS, H. 2006. Effect of water on the dilocation creep microstructure and flow stress of quartz and implications for the recrystallized grain size piezometer. *Journal of Geophysical Research*, **111**, B04201; doi: 10.1029/2005JB003852.

TWISS, R. J. & MOORES, E. M. 1992. *Structural Geology*. W. H. Freeman & Co.

WAGNER, R., ROSENBERG, C. L., HANDY, M. R., MÖBUS, C. & ALBERTZ, M. 2006. Fracture-driven intrusion and upwelling of a mid-crustal pluton fed from a transpressive shear zone: the Rieserferner pluton (Eastern Alps). *Geological Society of America, Bulletin*, **118**, 219–237; doi: 10.1130/B25842.1.

WANG, X. & NEUBAUER, F. 1998. Orogen-parallel strike-slip faults bordering metamorphic core complexes: the Salzach-Enns fault zone in the Eastern Alps. *Journal of Structural Geology*, **20**, 799–818.

A crustal-scale cross-section through the Tauern Window (eastern Alps) from geophysical and geological data

B. LAMMERER[1], H. GEBRANDE[1], E. LÜSCHEN[2] & P. VESELÁ[1]

[1]*Dept. of Earth and Environmental Sciences, Ludwig-Maximilians-University Munich Luisenstr. 37, D-80333 München, Germany (e-mail: Lammerer@lmu.de)*

[2]*Federal Institute of Geosciences and Natural Resources (BGR), Stilleweg 2, D-30655 Hanover, Germany*

Abstract: A restorable geological cross-section through the entire crust of the Tauern Window is presented. It is drawn from surface geology and seismic data of the TRANSALP vibroseis section using balancing software. The architecture of the window is characterized by three horses in a large duplex structure and folded granitic sills. The duplex was later uplifted along two large faults at its northern rim. The first is a blind fault along the deep-reaching sub-Tauern ramp with a displacement of 17 km. The tip of the hanging wall block wedged underneath the Austroalpine and Penninic nappes and caused a triangle structure. This led to backthrusting and backfolding within the marginal rocks of the window. At the second one, the Tauern North Boundary Fault occurred in our retrodeformation, a throw of $c.$ 3 km. A total shortening of the crust or parts of the crust of $c.$ 60 km in north–south direction led to uplift of the Tauern Window.

The Tauern Window is a structural key element within the edifice of the eastern Alps (Fig. 1). The whole nappe stack and the deepest tectonic units are here exposed due to an Oligocene to Recent uplift of about 30 km (Selverstone *et al.* 1984; Selverstone 1985; Blanckenburg *et al.* 1989; Fügenschuh *et al.* 1997; Frisch *et al.* 1998). Its uplift led to a major re-deformation of the entire orogenic wedge.

The western Tauern Window is crossed by the TRANSALP deep seismic profile. TRANSALP was a multidisciplinary research program with partner institutions from Austria, Germany, Italy and Switzerland. Vibroseis and explosion seismic measurements and receiver function studies were carried out in 1998–2001 along a more than 300 km long profile between Munich and the plain of Venice (TRANSALP Working Group 2002; Kummerow *et al.* 2004; Lüschen *et al.* 2004, 2006). It offered the opportunity to correlate the deeper structures of the western Tauern Window with surface geological data.

The Tauern Window is a tectonic double window within the Austroalpine crystalline nappes (Fig. 2). The inner or tectonic lower part belongs to the former European continental margin which had formed as a consequence of the breakup of Pangaea in Middle Jurassic time. In the outer or higher Tauern Window, a Penninic nappe system (Upper Schieferhülle or Bündnerschiefer nappes) is exposed, which originated partly from the continental–oceanic transition and, to the main part, from the Penninic–Liguric oceanic basin (Alpine Tethys). The window is surrounded by Austroalpine nappes. To the north of the Tauern Window, they are mainly composed of low-grade metamorphic quartzphyllites (Lower Austroalpine nappe) and the very low-grade metamorphic greywacke zone which carried the non-metamorphic Northern Calcareous Alps to the north (Upper Austroalpine nappe). To the south of the Tauern Window, the Austroalpine nappes are composed mainly from high to medium-grade ortho- and paragneisses and amphibolites (Schulz *et al.* 2001).

In the inner Tauern Window, late Hercynian granites, granodiorites and tonalites are exposed over vast areas. Minor gabbros and some ultramafic cumulates are also present. The magmatic suite took place between 309 ± 5 Ma to 298 ± 3 Ma (Cesare *et al.* 2001). During Alpine metamorphism, they were deformed into orthogneisses ('Zentral Gneiss'). Three major Zentral Gneiss bodies are separated by fault zones. The Ahorn gneiss to the north is a porphyric biotite–granite gneiss with large K-feldspars. The Tux gneiss is a monotonous granodioritic orthogneiss and the Zillertal gneiss to the south contains the whole suite of magmatic rocks from leucogranite to gabbros but with predominance of tonalitic gneisses. All Zentral Gneiss bodies are folded with wavelengths of 5–10 km. This indicates an initial strongly anisotropic crust, which was characterized by large plutonic sills between layered host rocks. The Ahorn gneiss is folded into a narrow brachyanticline which plunges gently to the southwest and northeast, respectively (Fig. 3). The anticlines of the

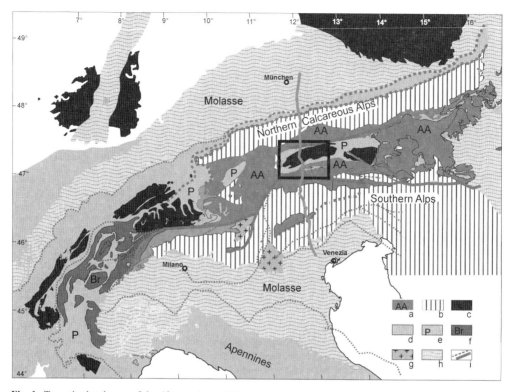

Fig. 1. Tectonic sketch map of the Alps. a, Austroalpine and Southalpine basement; b, Austroalpine and Southalpine Mesozoic cover; c, European basement; d, European cover; e, Valais and Ligurian oceanic sediments and ophiolites; f, Briançonnais terrane; g, Tertiary magmatites; h, Tertiary sediments of the Molasse and Rhinegraben; i, dotted line: thrusts, bold line: faults with mainly strike-slip movements. Bold line between München and Venezia, TRANSALP seismic section; Apennine, undifferentiated. The inset frame marks the position of Figure 2 in the western Tauern Window. After Schmid et al. 2004.

Tux and Zillertal gneisses plunge to the west near the Brenner Pass.

Host rocks to these intrusions are made from graphite-bearing metapelites, quartzites, banded gneisses, amphibolites and serpentinites or meta-ophicalcites (Greiner schists). In part, a deformed and metamorphosed coloured mélange can be inferred. The complex is interpreted as being derived from a Cadomian island arc and marginal basin tectonic setting along the margin of Gondwana (Frisch & Neubauer 1989; von Raumer et al. 2002). Newer single zircon U–Pb dating, however, points to Devonian and Early Carboniferous intrusion and sedimentation ages and hence these rocks might have also formed, at least partially, during early stages of the Variscan orogeny (Kebede et al. 2005).

The cover rocks of the inner Tauern Window show a clear affinity to the Germanic facies realm and are thus very similar to those of the Helvetic and some of the Penninic crystalline massifs of the Swiss Alps. Post Hercynian sedimentation started shortly after the emplacement of the plutons. ?Upper Carboniferous or Lower Permian plant fossils in graphite schists are reported from the southern Tauern Window (Franz et al. 1991; Pestal et al. 1999). Clastic sediments filled topographic depressions or tectonic grabens until Lower Jurassic times, interrupted by a short marine ingression during the Anisian, which is documented by carbonate horizons (see Veselá et al. 2008). A graben-horst or a basin-and-range topography is presumed. The Upper Jurassic Hochstegen marble is the youngest exposed sediment of the inner Tauern Window. It was deposited as a deeper marine platform carbonate that covered the entire region (Kiessling 1992). Proven Cretaceous rocks are unknown until now.

The outer or upper Tauern Window is formed by nappes which originate from the European continental slope and the Penninic Ocean basin (P in Fig. 1). Its rock successions are very similar to those of the North Penninic Bündnerschiefer nappes of eastern Switzerland (Engadin window)

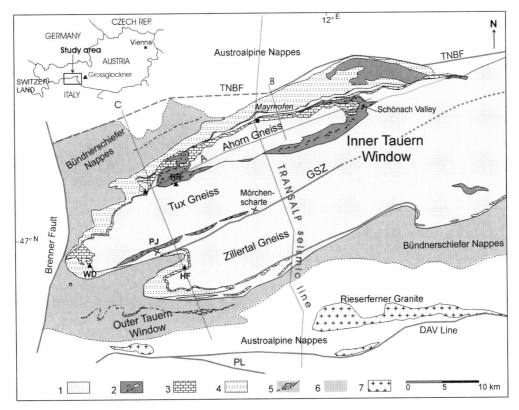

Fig. 2. Geological sketch map of the western Tauern Window. Inner Tauern Window: 1, Hercynian orthogneisses and old European basement rocks; 2, clastic carbonaceous metasediments of the European basement (?Late Carboniferous to ?Early Jurassic); 3, Hochstegen marble (Late Jurassic). Outer Tauern Window: 4, clastic and carbonaceous metasediments of the continental margin and slope (?Permian and Triassic); 5, thrust horizon within the Penninic nappes, decorated with lenses of serpentinite, quartzites and dolomites; 6, limy, marly and pelitic metasediments of the Penninic ocean basin (Bündnerschiefer, Jurassic and Cretaceous); 7, Alpine granites (Oligocene); TNBF, Tauern North Boundary Fault; GSZ, Greiner shear zone; DAV–Line, Defereggen–Antholz–Vals–Line; PL, Pustertal line, a segment of the Periadriatic line; HR, Hoher Riffler, 3228 m; PJ, Pfitscher Joch, 2230 m; WD, Wolfendorn, 2775 m; HF, Hochfeiler, 3510 m; A, section of Fig. 3; B, section of Fig. 4; C, section of Figure 5.

and are considered to be its continuation. Its base consists of ?Permo-Triassic clastic sediments and Anisian and ?Ladinian carbonates. Slices of serpentinite and sheared Palaeozoic microgabbros (Veselá *et al.* 2008) prove internal shear zones within this basal series. The main mass of the Penninic nappes is made of Bündnerschiefer-type phyllites and calcphyllites. Again, a tectonic horizon

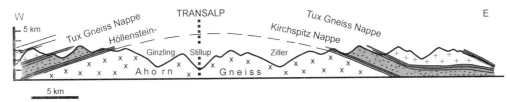

Fig. 3. West–east section along the strike of the Ahorn granite anticline. The folded sill of the Ahorn gneiss plunges gently to the west and to the east respectively. It culminates near the Stillup valley, where the extrapolated top of the gneiss is some 5 km higher than the exposed margins. The TRANSALP seismic section crosses the structure near the axial culmination in a north–south direction (dotted).

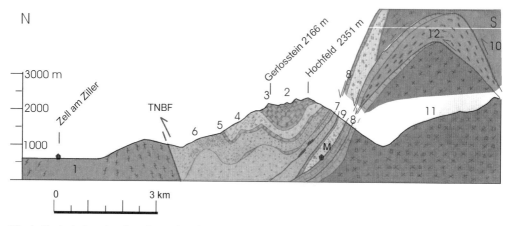

Fig. 4. Geological section along the northern boundary of the Tauern Window along the Ziller Valley and up-structure projection of the Ahorn gneiss structure. *Lower Austroalpine nappes*: 1, quartzphyllite. *Penninic nappes*: 2, clastic metasediments (Late Triassic or ??early Triassic); 3, carbonates of the Gerlosstein (Middle Triassic); 4, dolomites and cargneuls (Middle Triassic); 5, dolomites (Middle Triassic); 6, Kaserer Series (?early Triassic; ??Cretaceous); 7, thrust horizon with lenses of serpentinite. *Inner Tauern Window duplex:* 8, Hochstegen marble (late Jurassic); 9, porphyry (early Permian); 10, Tux orthogneiss; 11, Ahorn porphyric granite; 12, metaconglomerates of the Höllenstein nappe (?Permian to Middle Jurassic). TNBF, Tauern North Boundary Fault; M, Mayrhofen.

divides the Bündnerschiefer. It is decorated with lenses of serpentinites, quartzites, dolomites and even gypsum which are sometimes interpreted as olistholites (Miller *et al.* 1984; Thiele 1974), but we assume a thrust horizon with lenses of the substratum which divides the complex into a lower and upper Bündnerschiefer nappe. The lower nappe originated at the continental margin and hence has coarser-grained clastic rocks at its base; the upper nappe originated from a more distal part within the Tethys basin and contains more ophiolitic material. Up to 1000 m of prasinites or amphibolites can be found in this upper Bündnerschiefer nappe in the southwestern part of the Tauern window (Pfunderer Berge). This, however, was not a target of our studies here.

The tectonic sections

Cover rocks are exposed only marginally in the Ziller valley. Along the northern margin of the Tauern Window, the Ahorn gneiss and the covering Late Jurassic Hochstegen marble dip with 45° to 65° to the north (Fig. 4). Two small nappes follow which again carry a thin veneer of Hochstegen marble. The lower one consists mainly of metaconglomerates, the higher one of meta-quartz porphyries which is derived from the Tux gneiss domain, where in places a Permian quartz porphyry covers the basement. This nappe is named Porphyrmaterialschieferschuppe by local workers (Thiele 1976). The Penninic nappes show only the Kaserer Series and middle Triassic carbonates at the Gerlosstein which are topped by ?Late Triassic metasediments. The Bündnerschiefer are not present here: they are found mainly to the west of the Zillertal. In the meridian of the TRANSALP section, they are cut by the Tauern North Boundary Fault (TNBF).

The quartzphyllites close to the fault are overturned and dip steeply to the south indicating a similar orientation of the TNBF which is not directly exposed. A better understanding of the internal window architecture comes from areas more to the west, where the axes plunge with 12°–15° towards the Brenner Pass area and where cover rocks are exposed (Fig. 5).

Hochstegen marble covers the Ahornkern only in its northern part. To the south, however, it lies on coarse and fine clastic sediments which filled a Post Variscan depression, the Riffler Schönach Basin. It is, presumably, tectonic in origin, as some volcanic rocks are intercalated and because the granite beneath the Hochstegen marble is locally mylonitic (during the Middle Jurassic extension?), while the weaker marble itself is only slightly deformed. The same situation arises in the Tux gneiss area. In the Hochfeiler area is the Zillertal gneiss deformed into a narrow syncline and a wider anticline, which plunge both with 12–15° westwards to the Brenner Pass. An early fault-propagation fold in connection with the thrusting or a large parasitic fold in the course of the duplex development has formed here. The tonalite gneiss is directly covered by Hochstegen marble

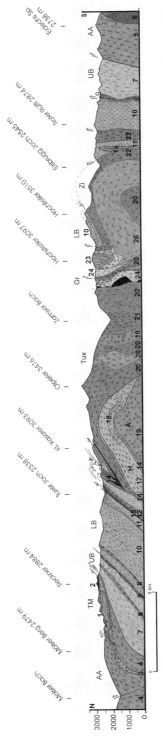

Fig. 5. Geological section through the western Tauern Window. **Austroalpine nappes**: 1, Rensen granite and dykes, Oligocene; 2, Jurassic shales and cherts; 3, serpentinite; 4, Triassic carbonates and cargneuls; 5, quartzphyllite (mainly ?Ordovician); 6, gneisses south of the Tauern Window. **Penninic nappes**: 7, phyllites and calcphyllites of the higher Bündnerschiefer nappe; 8, amphibolites and prasinites; 9, thrust horizon with lenses of serpentinites and Triassic quartzites, dolomites, gypsum and breccias; 10, phyllites of the lower Bündnerschiefer nappe; 11, ?Permo–Triassic clastic metasediments and cargneuls (Wustkogl and Kaserer Series); 12, dolomite marbles (Middle Triassic); 13, tectonic horizon with lenses of Cambrian microgabbro. **Inner Tauern Window duplex system**: *Post Variscan metasediments*: 14, Hochstegen marble (Upper Jurassic); 15, blackshists (± cyanite) and quartzites (?Liassic) and brown sandy limestones (?Dogger); 16, Triassic limestone or dolomite marbles, white hematite or magnetite-bearing quartzites; 17, clastic sediments, metaconglomerates, metarkoses (Pre Upper Jurassic); 18, dazitic porphyry; *Late Variscan plutonites*: 19, Ahorn porphyric biotitegranite; 20, Tux granodiorite; 21, migmatic rocks and injection gneisses; 22, Zillertal granites, granodiorites, tonalites and gabbros; *Pre Variscan and early Variscan rocks*: 23, black graphite schists; 24, amphibolites and garbenschiefer; 25, serpentinites and meta-ophicalcites; 26, injected gneisses and amphibolites. AA, Austroalpine; TM, Tarntal Mesozoic; LB, Lower Bündnerschiefer nappes; UB, Upper Bündnerschiefer nappes; Tux, Tux gneiss; Gr, Greiner Series; Zi, Zillertal gneiss; A, Ahorn gneiss; H, Höllenstein nappe with clastic metasediments of the Riffler–Schönach Basin.

showing again a topographic high position during Triassic and Lower Jurassic. The Penninic nappes are thrust directly over the Hochstegen marble. The Zillertal gneiss itself is thrust over the folded early Variscan Greiner Serie Upper Carboniferous to Jurassic sediments, indicating a low topographic position throughout the entire time span. To the east, in the meridian of the TRANSALP line, the Greiner Serie is not anymore present.

A repeated horst and graben succession or a basin-and-range type extension would best explain this situation. In the Alpine compressive phase, the stretched crust was inverted and the horsts of the Zillertal gneiss and the Tux gneiss were thrust over the graben sediments in between which, in turn, are also dislocated (Höllenstein and Greiner area).

The TRANSALP section and the structural evolution

The deep structure of the Tauern Window is inferred from the seismic image (Fig. 6). In the depth-migrated vibroseis section dips a broad band of reflectors under the northern rim of the Tauern Window down to the lower crust. We calculate along the northern end of the Tauern Window with two independent but sub-parallel faults in the depth—the deeper Sub Tauern Ramp and the Tauern North Boundary Fault. A second prominent feature is a band of sub-horizontal reflectors around 5 km under the centre of the Tauern Window. We interpret it as a ductile detachment horizon within pre- or early-Variscan layered metamorphic rocks where a granitic sill was detached and folded on top of this horizon to form the anticline of the Ahorngneiss (Fig. 7). In the southern part of the Tauern Window, all major reflectors dip with 35–40° to the south and hence were correlated with faults and anisotropic layering within the Tauern Window rocks.

The resulting cross-section is restorable and was line length and area balanced by use of the 2DMove software from Midland Valley Inc. (Glasgow). Figure 8 gives a redrawn and simplified model of the gross structural evolution of the Tauern Window. The structures follow the classical rules of thrust tectonics.

The original situation is given in Figure 8a. It shows a smooth surface of exposed basement rocks and two Permo-Carboniferous or younger basins (shown in black) covered with the Jurassic Hochstegen limestone. Possible younger rocks are not drawn. If they had been there, they were scraped off by the Penninic and Austroalpine nappes and transported to the northern margin of the eastern Alps where they form the Helvetic and Ultrahelvetic nappes along the Alpine front of Bavaria. The Helvetic rock succession lacks any Jurassic strata but starts with Cretaceous sediments, indicating a top-Jura detachment horizon.

A large duplex with three horses (Ahorn gneiss, Tux gneiss and Zillertal gneiss) developed beneath the Penninic and Austroalpine nappes (Fig. 8a–d). The sequence of stacking of the horses proceeded from south to north and started with a thrusting of the Zillertal gneiss onto the Tux gneiss (Fig. 8b).

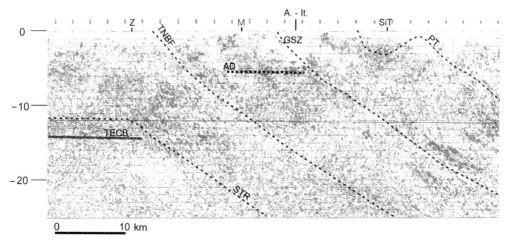

Fig. 6. Depth-migrated vibroseis data of the TRANSALP line between CDP 4500 and 7000. CDP distance is 25 metres, depth scale is in kilometres. Some important structural elements are included. Explanation: A.-It. Austrian–Italian border; Z, Zell am Ziller; M, Mayrhofen; SiT, Sand in Taufers; PT, Penninic thrust; TECB, top of European crystalline basement; STR, Sub Tauern Ramp; TNBF, Tauern North Boundary Fault; GSZ, Greiner shear zone; AD, Ahorngneiss detachment.

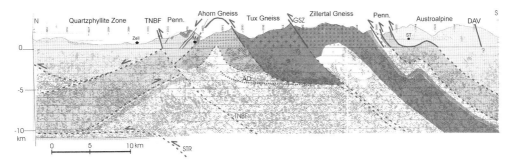

Fig. 7. Upper 12 km of the TRANSALP line between CDP 4400 and 6800 (depth migrated). Depth scale in kilometres, surface geology extrapolated to depth. M, Mayrhofen; ST, Sand in Taufers; STR, Sub Tauern Ramp; TNBF, Tauern North Boundary Fault; GF, Greiner Fault; AD, Ahorngneiss detachment.

The Greiner schist zone which occurs in between shows that a strongly anisotropic crustal layering was present. As a low-friction horizon, thick graphite schists may have facilitated the detachment. In addition, the tip of the horse is folded, where the more than one kilometre thick granite sill thins out to several metres when approaching the Pfitsch valley (Fig. 5). The southern sedimentary basin rocks were tightly folded together with the Greiner schists. Its strongly stretched or flattened pebbles are best visible in the Pfitscher Joch–Mörchner Scharte area (for localities, see Fig. 2).

In a next step, the Tux gneiss was thrust over the Ahorngneiss and its sedimentary basin (Fig. 8c. In the course of these movements, the metasediments of this northern basin were thrown over the Ahorngneiss horst area (see Figs 2 and 5). The Ahorngneiss was then detached and folded into a tight anticline (Fig. 8d). The folding of granites supports the idea of already sheeted granitic intrusions as

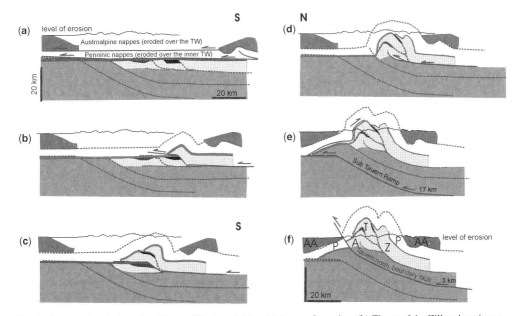

Fig. 8. Structural evolution of the Tauern Window. (**a**) Pre Alpine configuration. (**b**) Thrust of the Zillertal gneiss on the Greiner basin and the Tux gneiss. (**c**) Thrust of the Tux gneiss on the Riffler–Schönach basin and the Ahorn gneiss. (**d**) Detachment and folding of the Ahorn gneiss—duplex formation completed. (**e**) Movements along the Sub Tauern Ramp. Backthrusting along the Tauern north rim. (**f**) Movement along the Tauern North Boundary Fault—present situation. AA, Austroalpine nappes; P, Penninic Bündnerschiefer nappes; A, Ahorn gneiss; T, Tux gneiss; Z, Zillertal gneiss.

sills or lakkoliths, which intruded into a former sub-horizontally, layered crust.

The duplex formation led to a first stage of rapid uplift of the Tauern Window–but with different uplift histories of the three blocks. A second stage of deformation affected the whole Tauern Window as the entire duplex was uplifted along the Sub Tauern Ramp with a displacement of about 17 km (Fig. 8e). The Tauern block wedged under the Penninic–Austroalpine nappe stack. As a consequence, south-vergent backthrusting and backfolding occurred all along the north rim (Rossner & Schwan 1982).

In the third stage, the sub Tauern Ramp became inactive and movements shifted to the Tauern North Boundary Fault (Fig. 8f). This fault shows a throw of about 3 kilometres and merges into the Salzachtal–Ennstal strike-slip fault. The reverse fault movements overprint Oligocene–Miocene sinistral strike-slip movements at the Salzachtal–Ennstal fault (Ratschbacher et al. 1991). It cuts the Penninic nappes and the Tauern duplex obliquely.

Discussion

The Tauern Window is clearly a strongly compressive structure. On the other hand, ENE–WSW directed stretching lineations are widespread and many workers include the Tauern Window into an escape model (Ratschbacher et al. 1991; Selverstone et al. 1995; Frisch et al. 1998). As we find several structurally isolated horses within the window with different structural histories, one should refine this story. As the TNBF cuts the structures obliquely and the movement direction of the Adria Plate south of the Insubric line is also oblique to the Tauern duplex, the possibility of stacking versus a northwest direction and a subsequent rotation under transpressive movements should be taken into consideration. This could explain at least part of the ductile stretching within the inner Tauern Window. By this, a sinistral differential movement between the rotating blocks would occur, which is present along the northern rim and in the Greiner zone (Behrmann & Frisch 1990). An east–west directed extension and tectonic unroofing of the window could, on the other hand, have facilitated the nappe stacking and folding within the window.

Three phases of uplift are inferred from the above-described structural evolution–duplex formation, movements along the Sub Tauern Ramp and reverse faulting along the Tauern North Boundary Fault. Three phases of uplift were also found by modelling the uplift history in the Austroalpine to the immediate south of the Tauern Window (Steenken et al. 2002). The first is a relatively rapid exhumation of nearly 1 mm/a north of the Defereggen–Vals–Antholz line (DAV) between the intrusion of the Rieserferner pluton at 31 ± 3 Ma and about 23 Ma. Afterwards, exhumation rates slowed down to 0.4 mm/a at 13 Ma and to 0.2 mm/a during the final exhumation.

A similar path but with higher rates was calculated for the western Tauern Window by Blanckenburg et al. (1989), Selverstone et al. (1995) and Fügenschuh et al. (1997). Uplift rates up to 4 mm/a were calculated between 30 Ma and 20 Ma. From 20 Ma to 10 Ma, the uplift slowed down to 1.0 mm/a (Fügenschuh et al. 1997) and low uplift rates of 0.2 mm/a characterize the last 10 Ma.

Even when there is no proof, we argue that the major changes of the cooling rates correlate with the changing structural phases. The rapid initial uplift could correspond to the duplex formation and ductile folding of the Ahorngranite. A shortening of the Tauern crust of 35 km between 30 Ma and 20 Ma was necessary to produce the duplex which reached a height of 13 km (isostatic effects and tectonic unroofing of the cover by lateral extrusion not calculated). With other words: ten million years of convergence of the Adria Plate versus the European Plate with an average velocity of 3.5 mm per year would have produced an uplift of 1.3 mm per year in the Tauern Window if the movement was absorbed by the Tauern Window only. In the southern part, the Bündnerschiefer could have reached the surface during this phase.

After blocking of the duplex due to geometrical reasons, displacement continued along the Sub Tauern Ramp with a shortening of 17 km during the following 10 million years. This means not necessarily that the velocity of convergence has slowed down to 1.7 mm/a, because other parts of the Alps start to be involved into the movements at that time. Along the Valsugana Fault the movements starts around 12 Ma ago (Castellarin & Cantelli 2000). The thrust along the Sub Tauern Ramp led to a further uplift of eight kilometres and the top of the window was exposed to erosion. In the sector of the TRANSALP line, an erosion of three to four kilometres of Bündnerschiefer and resistant Tauern window granites occurred. At least since that time, the Tauern Window should have been a high mountain area. The upthrusting of the whole Tauern Window along the Sub Tauern Ramp should have affected the Molasse basin. Revived or accelerated flexural down-bending by the northward migration of the load centre could have caused the transgression of the Upper Marine Molasse, which occurred between 21 and 17 Ma. Afterwards, the clastic sediments overwhelmed the marine phase and the final freshwater phase started.

During the last 10 Ma, the uplift shifted to the Tauern North Boundary Fault where the final exposition of around 2–3 km took place. All of this uplift must have been accommodated by erosion of granites, which slowed down the process to an average of 0.2–0.3 mm/a.

Conclusions

Combined field studies and seismic imaging allowed to draw a depth-extrapolated cross-section through the entire crust of the Tauern Window. The section can be balanced and a sequence of movements can be inferred. Duplex formation with three large horses and folding of kilometre-thick granite sills characterize the early structural evolution. A uniform thickness of the horses of about 10 km suggests a detachment along the brittle–ductile transition in a sub-horizontally stratified crust.

In a second stage, uplift occurred along the 23 km high Sub Tauern Ramp which dips with an angle of 30° to the southeast and cuts through the entire upper and middle crust. The northern tip of the Tauern Window was wedged under the Austroalpine nappes and formed a triangle zone which caused widespread backthrusting and backfolding which is visible at many places along the northern part of the Tauern Window (Rossner & Schwan 1982).

A ramp beneath the Tauern was suggested earlier by Lammerer & Weger 1998, as the internal structures of the Tauern Window could otherwise not explain its high position. Differences to that model occur at the northern end of the Tauern Window by the recognition of the importance of the North Boundary Fault for the uplift history. In the southern part of the Tauern Window the deep structures dip much more gently than was expected from surface studies in this sector. This is due to a large parasitic syncline–anticline pair. Further to the east, where these parasitic folds have died out into the air, much gentler dips to the south are common.

A further conclusion is that the Tauern Window interacts strongly with the south Alpine structures which becomes evident, when regarding the large-scale structures, given by receiver function (Kummerow et al. 2004), explosion- and vibroseismic studies (Lüschen et al. 2006). The Val Sugana Fault system and the south-directed movements in the southern Alps seem to be a consequence of the deep wedging of the Tauern Window between the south Alpine upper crust, which was thrust over the Tauern Window, and the lower crust, which has pushed the Tauern Window to the north. The Alpine edifice is thus controlled by a thin-skinned wedge of Austroalpine nappes and off-scraped nappes from the European crust (Helvetic nappes and molasse nappes). Thick-skinned tectonics in the southern Alps is due to the deep-reaching European crust southwards under the Dolomite Mountains (Fig. 9).

The geometry and the role of the Pustertal line remains enigmatic and is in general interpreted as part of the Periadriatic lineament system with large dextral strike-slip components. The Pustertal line is displaced to the north around 60 km and rotated clockwise (10–20°) in comparison to the Insubric–Tonale Line of the western Alps. In addition, there is no remarkable uplift along the Pustertal line—which is different in the western Alps. This means that the Pustertal line is sheared off in depth and transported horizontally. There is, until now, no good model to show the decollement level. As an attempt, we propose a disrupted geometry of this fault in depth, which also might gave an

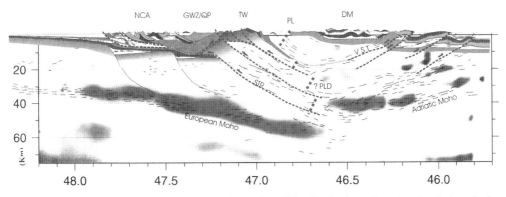

Fig. 9. Deep structure of the Alps from receiver function studies and line drawing from vibroseis and explosion seismic studies. NCA, Northern Calcareous Alps; GWZ/QP, greywacke zone and quartzphyllite zone; TW, Tauern Window; DM, Dolomite Mountains; PL, Pustertal line; PLD, Pustertal line, disrupted?; VST, Val Sugana thrust.

explanation as to why it is not visible in the seismic section. As so many factors are still unclear, we avoided the inclusion of the Insubric line into our model. As a pure strike-slip fault, it would not much affect the nappe stacking within the Tauern Window.

We thank the reviewer M. Rockenschaub and C. Doglioni for helpful hints and critical remarks.

References

BEHRMANN, J. H. & FRISCH, W. 1990. Sinistral ductile shearing associated with metamorphic decompression in the Tauern Window, Eastern Alps. *Jahrbuch der Geologischen Bundesanstalt Wien*, **133**, 135–146.

BLANCKENBURG, F. V., VILLA, I., BAUR, H., MORTEANI, G. & STEIGER, R. H. 1989. Time calibration of a PT-path in the Western Tauern Window, Eastern Alps: The problem of closure temperatures. *Contributions to Mineralogy and Petrology*, **101**, 1–11.

CASTELLARIN, A. & CANTELLI, L. 2000. Neo-Alpine evolution of the Southern Eastern Alps. *Journal of Geodynamics*, **30**, 251–274.

CESARE, B., RUBATTO, D., HERMANN, J. & BARZI, L. 2001. Evidence for Late Carboniferous subduction type magmatism in mafic—ultramafic cumulates of the Tauern window (Eastern Alps). *Contributions to Mineralogy and Petrology*, **142**, 449–464.

FRANZ, G., MOSBRUGGER, V. & MENGE, R. (1991). Carbo-Permian pteridophyll leaf fragments from an amphibolite facies basement, Tauern Window, Austria. *Terra Nova*, **3**, 137–141.

FRISCH, W. & NEUBAUER, F. 1989. Pre-Alpine terranes and tectonic zoning in the eastern Alps. *Geological Society of America, Special Paper*, **230**, 91–100.

FRISCH, W., KUHLEMANN, J., DUNKL, I. & BRÜGEL, A. 1998. Palinspastic reconstruction and topographic evolution of the Eastern Alps during late Tertiary tectonic extrusion. *Tectonophysics*, **297**, 1–15.

FÜGENSCHUH, B., SEWARD, D. & MANCKTELOW, N. 1997. Exhumation in a convergent orogen: the western Tauern window. *Terra Nova*, **9**, 213–217.

KEBEDE, T., KLÖTZLI, U., KOSLER, J. & SKIÖLD, T. 2005. Understanding the Pre Variscan and Variscan basement components of the central Tauern Window, Eastern Alps (Austria): constraints from single zircon U-Pb geochronology. *International Journal of Earth Sciences (Geologische Rundschau)*, **94**, 336–353.

KIESSLING, W. 1992. Palaeontological and facial features of the Upper Jurassic Hochstegen marble (Tauern Window, Eastern Alps). *Terra Nova*, **4**, 184–197.

KUMMEROW, J., KIND, R., ONCKEN, O., GIESE, P., RYBERG, T., WYLEGALLA, K. & SCHERBAUM, F. 2004. A natural and controlled source seismic profile through the eastern Alps: TRANSALP. *Earth and Planetary Science Letters*, **225**, 115–129.

LAMMERER, B. & WEGER, M. 1998. Footwall uplift in an orogenic wedge: the Tauern Window in the Eastern Alps of Europe. *Tectonophysics*, **285**, 213–230.

LÜSCHEN, E., LAMMERER, B., GEBRANDE, H., MILLAHN, K., NICOLICH, R. & TRANSALP Working Group. 2004. Orogenic structure of the Eastern Alps, Europe, from TRANSALP deep seismic reflection profiling. *Tectonophysics*, **388**, 85–102.

LÜSCHEN, E., BORRINI, D., GEBRANDE, H., LAMMERER, B., MILLAHN, K., NEUBAUER, F., NICOLICH, R. & TRANSALP Working Group. 2006. TRANSALP— deep crustal Vibroseis and explosive seismic profiling in the Eastern Alps. *Tectonophysics*, **414**, 9–38.

MILLER, H., LEDOUX, H., BRINKMEIER, I. & BEIL, F. 1984. Der Nordwestrand des Tauernfensters– stratigraphische Zusammenhänge und tektonische Grenzen. *Zeitschrift der Deutschen Geologischen Gesellschaft*, **135**, 627–644.

PESTAL, G., BRÜGGEMANN-LEDOLTER, M., DRAXLER, I., EIBINGER, D., EICHBERGER, H., REITER, C. H. & SCEVIK, F. 1999. Ein Vorkommen von Oberkarbon in den mittleren Hohen Tauern. *Jahrbuch der Geologischen Bundesanstalt Wien*, **141**, 491–502.

RATSCHBACHER, L., FRISCH, W., LINZER, H.-G. & MERLE, O. 1991. Lateral extrusion in the Eastern Alps: Part 2. Structural analysis. *Tectonics*, **10**, 257–271.

RAUMER, J. VON, STAMPFLI, G. M., BOREL, G. & BUSSY, F. 2002. Organisation of pre-Variscan basement areas at the north-Gondwana margin. *International Journal of Earth Sciences (Geologische Rundschau)*, **91**, 35–52.

ROSSNER, R. & SCHWAN, W. 1982. Zur Natur der südvergenten Deformationsstrukturen im NW-Teil des Tauernfensters (Tirol, Österreich). *Mitteilungen der Gesellschaft der Geologie und Bergbaustudenten Österreichs*, **28**, 35–54.

SCHMID, S. M., FÜGENSCHUH, B., KISSLING, E. & SCHUSTER, R. 2004. Tectonic map and overall architecture of the Alpine orogen. *Eclogae Geologicae Helvetiae*, **97**, 93–117.

SCHULZ, B., SIEGESMUND, S., STEENKEN, A., SCHÖNHOFER, R. & HEINRICHS, T. 2001. Geologie des ostalpinen Kristallins südlich des Tauernfensters zwischen Virgental und Pustertal. *Zeitschrift der Deutschen Geologischen Gesellschaft*, **152**, 161–307.

SELVERSTONE, J. 1985. Petrologic constraints on imbrication, metamorphism and uplift in the SW Tauern window, Eastern Alps. *Tectonics*, **4**, 687–704.

SELVERSTONE, J., SPEAR, F. S., FRANZ, G. & MORTEANI, G. 1984. High-pressure metamorphism in the SW Tauern Window, Austria: P-T paths from hornblende-kyanite-staurolite schists. *Journal of Petrology*, **25**, 501–531.

SELVERSTONE, J., AXEN, G. J. & BARTLEY, J. M. 1995. Fluid inclusion constraints on the kinematics of footwall uplift beneath the Brenner Line normal fault, Eastern Alps. *Tectonics*, **14**, 264–278.

STEENKEN, A., SIEGESMUND, S., HEINRICHS, M. T. & FÜGENSCHUH, B. 2002. Cooling and exhumation of the Rieserferner pluton (Eastern Alps, Italy/Austria). *International Journal of Earth Sciences (Geologische Rundschau)*, **91**, 799–817.

THIELE, O. 1974. Tektonische Gliederung der Tauernschieferhülle zwischen Krimml und Mayrhofen. *Jahrbuch der Geologischen Bundesanstalt Wien*, **117**, 55–74.

THIELE, O. 1976. Der Nordrand des Tauernfensters zwischen Mayrhofen und Inner Schmirn (Tirol). *Geologische Rundschau*, **65**, 410–421.

TRANSALP WORKING GROUP. 2002. First deep seismic reflection images of the Eastern Alps reveal giant crustal wedges and transcrustal ramps. *Geophysical Research Letters*, **29**; doi: 10.1029/2002GL014911.

VESELÁ, P., LAMMERER, B., WETZEL, A., SÖLLNER, F. & GERDES, A. 2008. Post-Variscan to Early Alpine sedimentary basins in the Tauern Window (Eastern Alps). *In*: SIEGEMUND, S., FÜGENSCHUH, B. & FROITZHEIM, N. (eds) *Tectonic Aspects of the Alpine-Dinaride-Carpathian System*. Geological Society, London, Special Publications, **298**, 83–100.

Neotectonic faulting, uplift and seismicity in the central and western Swiss Alps

MICHAELA USTASZEWSKI[1,2] & O. ADRIAN PFIFFNER[1]

[1]*Institute of Geological Sciences, University of Bern, Baltzerstrasse 1-3, 3012 Bern, Switzerland*

[2]*Now at: Department of Geosciences, National Taiwan University, No. 1, Sec. 4, Roosevelt Road, Taipei 10617, Taiwan (e-mail: michaela@ntu.edu.tw)*

Abstract: Our study aims to characterize the post-glacial neotectonic activity by finding surface expressions of recently active tectonic faults. The central and western Swiss Alps were chosen as the study area because surface uplift rates are very high, indicating ongoing uplift of the external basement massifs. Moreover, the Valais area coincides with enhanced seismic activity. Active faults were searched by mapping lineaments on aerial photographs and subsequent field studies. Three main types of faults could be distinguished: gravitational faults (i.e. faults related to mass movements); tectonic faults; and composite faults (i.e. tectonic faults with a component of gravitational and post-glacial rebound-related reactivation). A large number of tectonic faults were found (over 1700), but only two unequivocally post-glacially active tectonic faults could be distinguished. Indications for their post-glacial (re-)activation are displaced Quaternary landforms or sediments. Large gravitational faults, as well as composite faults often correlate with deep-seated gravitational slope deformations (DSGSD). The latter occur mainly along valley slopes, particularly where a pervasive foliation strikes parallel to the valley. Fault orientations show correlations either with the regional main foliation (e.g. Aar and Gotthard massif), the orientation of valleys (e.g. Bedretto and Urseren valley), or pre-existing tectonic structures (e.g. faults parallel to joints that are perpendicular to the strike of major structures in the Helvetic nappes). Comparisons of fault orientations with orientations of nodal planes of earthquake focal mechanisms of the last 20 years show a poor indicative correlation. The central and western Swiss Alps host a large number of faults prone for reactivation in today's stress field. However, for most of these faults, no indications of their last phase of activity exist. The low number of unambiguously active tectonic faults suggests that the current strain is either predominantly aseismic or, alternatively, cumulated seismic moment is too low for producing surface rupture.

The rate of the active convergence between the European and the African Plates is known from various geodetic studies (e.g. Global Positioning System (GPS) surveys) and plate motion models. DeMets *et al.* (1990, 1994) give a value of 3 mm/a for the converging plate motion in a NW–SE direction at the longitude of the Strait of Gibraltar, increasing eastward to 8 mm/a near Sicily according to the NUVEL-1A global kinematic model for plate motions. This is in congruence with values determined by Calais *et al.* (2000, 2002), Ferhat *et al.* (1998), Serpelloni *et al.* (2005) and Vigny *et al.* (2002). Grenerczy *et al.* (2005) report N–S convergence of the Adriatic and European Plates at rates of 2–3 mm/a. Tesauro *et al.* (2005) compiled GPS-data from 53 permanent stations from four different networks in Western Europe, with observation periods of up to seven years between 1996 and 2003, that allowed a determination of horizontal displacement rates relative to Eurasia. Stations located between 4° and 16 °E, comprising the domain of the Alps and the northern Alpine foreland, show horizontal displacement rates between 0.1 and 2.9 mm/a relative to Eurasia (Tesauro *et al.* 2005, fig. 2). Because convergence and deformation rates may differ on a local scale from the plate motion, Ferhat *et al.* (1998) and Vigny *et al.* (2002) argue that a large proportion of the geodetically observed convergence occurs aseismically or outside the Alpine chain, respectively.

In the Alps, the regional stress field varies along and across the strike of the orogen. Various studies were conducted in the western and central Alps showing east–west extension and north–south compression (e.g. Calais *et al.* 2002; Nocquet & Calais 2003, 2004) or radial extension (Champagnac *et al.* 2003, 2004; Kastrup 2002). Results of Maurer *et al.* (1997), Eva & Solarino (1998), Eva *et al.* (1998) and Bistacchi *et al.* (2000) in the Penninic realm of the Valais and surrounding areas support those geodynamic models that predict extensional deformation perpendicular to the crest of a mountain range, while flanks and lowlands continue to undergo crustal shortening. This observation is corroborated by Delacou *et al.* (2004). Furthermore, numerical modelling studies by Selzer (2006) and Pfiffner *et al.* (2000) suggested that horizontal extension occurs in the shallow parts of the core of a contracting orogen due to gravitational instability of the overthickened lithosphere.

Seismicity and surface uplift rates are further hints supporting active deformation in the Alps. A compilation of all instrumentally recorded earthquakes in Switzerland since 1975 indicates a strong alignment of seismic events in the Valais (Fig. 1). Gubler (1991), Kahle *et al.* (1997) and Schlatter *et al.* (2005) published surface uplift data from the Swiss Nivellement Net LHN 95, which shows a strong uplift maximum of about 1.5 mm/a in Switzerland near Brig and Sierre in the Valais (Fig. 2). The conspicuous surface uplift pattern and enhanced seismicity in the Valais and surrounding areas raised the question of whether these regions host any topographic expressions that can be attributed to ongoing motion along potentially seismogenic faults. These areas were therefore chosen for the investigation of post-glacially active tectonic faults (Fig. 1). The aim of this study is a better understanding of the very recent tectonic activity of the central and western Swiss Alps. The applicability of such an approach has been successfully demonstrated in a precursory study by Persaud (2002) and Persaud & Pfiffner (2004), which showed that tectonic surface faults are parallel to nodal fault planes of earthquakes in the eastern Swiss Alps.

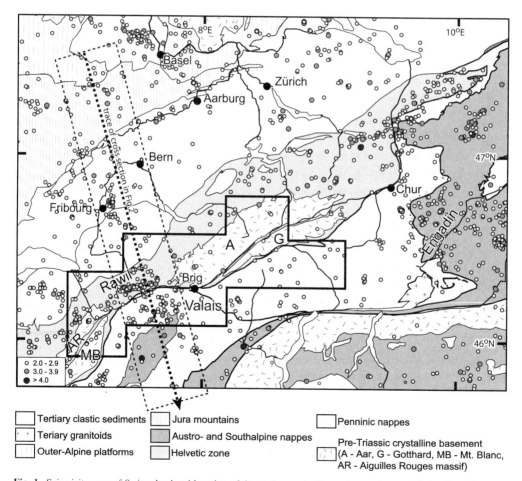

Fig. 1. Seismicity map of Switzerland and location of the study area in the western and central Swiss Alps. The study area is indicated by the black polygon. All earthquakes between 1975–2005 with $M_w \geq 2$ are plotted, data are taken from the Swiss Seismological Service. Trace of cross-section of Figure 2b indicated. Dashed rectangle corresponds to swath used for projection of earthquakes onto cross-section shown in Figure 2b.

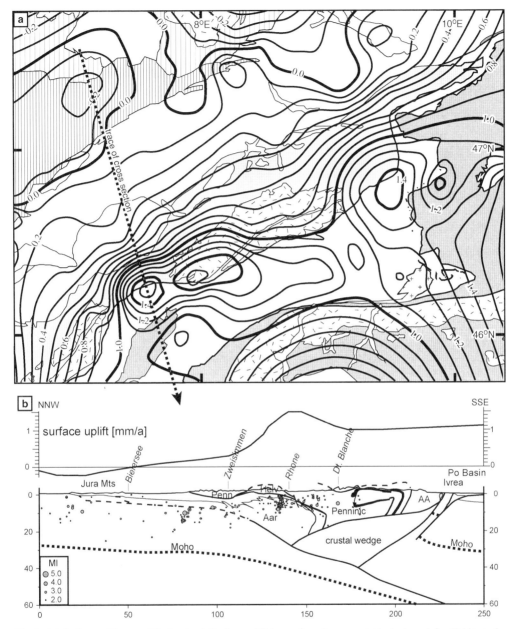

Fig. 2. (**a**) Surface uplift map of Switzerland. Surface uplift in mm/a, reference point is Aarburg (after Kahle *et al.* 1997). Schlatter *et al.* (2005) report slightly lower maxima. For legend, see Figure 1. (**b**) Comparison of surface uplift and seismicity along a cross-section. Earthquakes are plotted from a 40 km broad swath shown in Figure 1 onto the profile. (AA, Austro- and Southalpine units; Helv, Helvetic units; thick black lines, oceanic suture.)

We used a combined approach involving earthquake catalogue, surface uplift data, aerial photographs and field studies to evaluate neotectonic activity in the Swiss Alps. We attribute the term 'neotectonic' only to Holocene tectonic activity, i.e. spanning the time since the Last Glacial Maximum (LGM).

First, aerial photographs from the entire area were investigated for linear geological features. Subsequently, selected displacement-controlled lineaments were visited in the field to study their origin. This allowed to differentiate between faults of gravitational, tectonic and composite (tectonic and gravitational) origin. Two unequivocally post-glacially active tectonic faults are described in greater detail in this paper.

Data

Seismicity and surface uplift

Two strong indicators for neotectonic activity in Switzerland are seismicity and surface uplift (Figs 1 and 2). The Swiss Seismological Service publishes annual reports of all seismic events in Switzerland and surrounding regions (e.g. Deichmann et al. 2004; Baer et al. 2005; Deichmann et al. 2006). Pronounced seismic activity is recorded in the area of Basel, in the greater area of Chur and the Engadin, and along a linear alignment north of the Rhône valley in the Valais. Surface uplift rates (after Gubler 1991; Kahle et al. 1997) are highest in the area of Chur and the Engadin (1.4–1.5 mm/a) and Brig and Sierre (1.5 mm/a). Schlatter et al. (2005) give slightly lower values for Chur and the Engadin (1.4 mm/a) as well as around Brig (1.3 mm/a). The steepest gradient of uplift rates, however, is observed in the area of the Rawil depression between the external basement uplifts of the Aar and the Aiguilles Rouges/Mont Blanc massif.

A projection of earthquake foci from a 40 km broad corridor onto a profile running across western Switzerland compared with the surface uplift rates shows a strong correlation of these two features (Fig. 2b). The steep gradient of the surface uplift rates is found along the front of the Aar massif and coincides with an earthquake cluster in the Aar and Aiguilles Rouges/Mont Blanc massifs. These earthquakes reveal predominantly strike-slip and reverse-faulting mechanisms (Kastrup 2002). We assume that the seismic activity is linked to the recent tectonic uplift, i.e. the external massifs are still bulging upwards by ongoing thrusting. Earthquakes in the Molasse foreland and at the northern front of the Penninic nappes (Préalpes) are linked to the Fribourg fault and are strike-slip (Kastrup 2002). Fault plane mechanisms of earthquakes in the Penninic units south of the Rhône River show predominantly strike-slip and normal faulting motions (Kastrup 2002). From the inversion of focal mechanisms, Kastrup (2002) and Kastrup et al. (2004) calculated the recent stress field. According to them, the Helvetic zone in our transect is characterized by a strike-slip to normal faulting regime with NNE–SSW extensional component and WNW–ESE compression.

The Helvetic zone in eastern Switzerland displays a pure strike-slip regime with NW–SE compression and NE–SW extension.

Lineaments and faults

Recent to sub-recent faults can be marked by very subtle morphological characteristics in the Alpine landscape, often unnoticed even by a geologist's eye. An efficient tool to detect such structures is the analysis of aerial photographs and the subsequent visit in the field. Sub-vertical to vertical faults are easier to detect than inclined normal or reverse faults, which demand additional advertence for their detection (Bistacchi & Massironi 2000; Bistacchi et al. 2000).

In a first step, aerial photographs were investigated for geological linear features. Comparison with geological maps of different scales allowed a first classification of the lineaments into displacement-controlled lineaments (i.e. proper faults) and lithology-, fabric- or anisotropy-controlled lineaments (i.e. lineaments caused by pre-existing foliation, joints, lithology contrasts or differences in rock erodibility). In a subsequent step, displacement-controlled lineaments were separated into three types: gravitational faults, tectonic faults and composite faults. The latter are formed by a complex interplay of gravitational, tectonic and post-glacial rebound processes (Ustaszewski et al. in press). Ground truthing helped to identify the fault types. Based on our experience from fieldwork, we were also able to interpret and classify lineaments not visited in the field. Main distinction criteria were the morphological expression of the lineaments, their continuity and their large-scale geological and geomorphological setting. A crucial step was the determination of the age of a fault's activity (pre-glacial or post-glacial), for which detailed structural and geomorphological field studies were indispensable. An interpretation was only possible if the fault displaced glacial and post-glacial deposits and landforms. Figure 3 shows the mapped faults. The fault map is not complete for the following reasons. First, the use of aerial photographs for lineament detection has the drawback that lineaments with moderate or weak morphological expression are not visible in forested areas. Secondly, glaciers cover a considerable portion of the study area, thus prohibiting the detection of possible lineaments. Thirdly, anthropogenic modification of the landscape may also have concealed a certain number of lineaments.

The morphological expression of a fault is largely controlled by the fault type (normal, reverse or strike-slip) and the lithology of the bedrock. In the Penninic Bündnerschiefer (schistes lustrés), faults are weakly preserved and exhibit

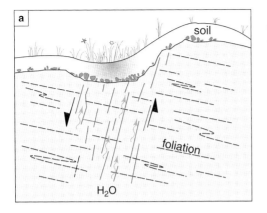

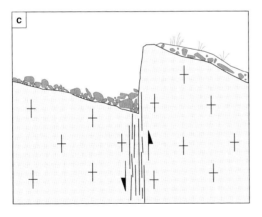

Fig. 4. Lithology-dependent fault morphology (modified after Persaud 2002; Persaud & Pfiffner 2004). (**a**) Faults in schists are characterized by small offsets, subtle depressions and vegetation cover. (**b**) Faults in carbonates show generally a pronounced morphology with fault scarp (often not covered by vegetation) and a clear surface offset or incision. Faults in carbonates often display karstification features. (**c**) Faults in granites and gneisses have a pronounced morphology and—if parallel to the foliation—a clear fault scarp.

a rounded morphology (Fig. 4a). They tend to be covered by vegetation. Limestone-dominated lithologies are marked by a high abundance of parallel, sub-vertical joints and parallel running faults (Fig. 4b). These faults are typically associated with gullies and incisions in mountain ridges. The faults can be highly karstified as they form pathways for water circulation. Crystalline rocks may host schistosity-parallel faults (Fig. 4c), which may be reactivated by gravitational movements. This is particularly true if the schistosity is steeply dipping and parallel to the thalweg (valley line). The preservation potential of faults in crystalline rocks is high. For a more detailed description of lithology-dependent lineament and fault morphology, the reader is referred to Persaud & Pfiffner (2004).

Gravitational faults. For this study, we only took into account large gravitational faults that develop mostly in higher parts of extended mass movements, as is the case in sagging mountain slopes ('Talzuschub'). These gravitational slope processes are known in the literature as 'deep-seated gravitational slope deformations' (DSGSD). Typical morphological characteristics of DSGSD and related faults are as follows (after Radbruch-Hall *et al.* 1976; Mahr 1977; Mahr & Nemcok 1977; Radbruch-Hall 1978; Savage & Varnes 1987; Dramis & Sorriso-Valvo 1994; Agliardi *et al.* 2001; Dramis *et al.* 2002; Gutiérrez-Santolalla *et al.* 2005 and therein; Hippolyte *et al.* 2006):

- The areal extension of DSGSD covers mostly at least 1 km^2 in size, but is limited and does not cross a thalweg or deep incisions in a slope.
- The thickness of the involved masses amounts to several tens to hundreds of metres.
- A continuous shear horizon at the base of the mass movement may be present or not.
- The velocity of the (creeping) movement of DSGSD is extremely slow. Longer inactive time spans alternate with phases of higher activity, often as a consequence of intense rainfalls or earthquakes.
- Pre-existing geological structures have greater influence on the development of DSGSD than topographical proportions.
- Mountain slopes affected by DSGSD often show near-surface slope movements that cause a very uneven morphology. Associated features include: bulging, sackung (in the definition of Zischinsky 1967, 1969; Savage & Varnes 1987; Radbruch-Hall 1976), landslides, open tension cracks, toppling, double (or twin) ridges and ridge-top depressions.
- Faults developing in the upper part of mass movements can have both a linear or curved form.

- The faults run parallel to the mountain ridges and often exhibit a relative uplift of the valley-block, thus forming uphill-facing scarps (also called antislope scarps, counter-scarps, or antithetic scarps).

Examples of gravitational faults can be found southwest of Sierre (VS) at the Mont Noble, where the western and northern mountain slopes are strongly affected by gravitational processes (for location see Fig. 3). Double ridges with several metres deep ridge-top depressions developed near the summit. A large sackung with a ≤ 5m high curved downhill-facing scarp representing the head scar of a landslide can be observed west of the summit. Trees growing in the affected area display curved trunks ('Säbelwuchs') or inclined boles.

In the following paragraphs, we present two examples of gravitationally reactivated pre-existing structural features. In the first case, two tectonic faults are reactivated, and in the second case, the foliation serves as a gliding plane for gravitational movements.

The mass movement at Lac de Fully: two faults near Lac de Fully, NW of Martigny in the Rhône valley, display fresh fault scarps (Fig. 5a, b). These faults, which are located in Permo-Carboniferous sediments, strike NNE–SSW (Fig. 5e), and have been active for a longer time since at least one of them, i.e. the northwestern fault (f2 in Fig. 5), displays long and large quartz-slickenfibre lineations of at least three different generations (Fig. 5f). These fibres must have formed deeper beneath the land surface by precipitation from hot fluids. The fibres are preserved only in the lower 0.5 to 1 m of the scarp-wall. The youngest quartz fibres indicate a relative upward movement of the northwestern block. The scarp height varies between 0.5 m and 5 m. Pervasive open joints perpendicular to the faults are observed in the bedrock (Fig. 5f), causing a high degree of disintegration of the bedrock. To both sides of the fault, the landscape shows glacial overprints, glacially polished rock surfaces and large boulders.

In order to verify post-glacial movements along the faults, a 2 m long and up to 1.5 m wide trench was dug across the southeastern fault (f1 in Fig. 5), where the fault crosses a depression (Fig. 5c). The trench (Fig. 5d) reached the bedrock on its eastern side, revealing a step in the black Carboniferous shales. The shales dip steeply to the WNW, are heavily weathered with white and grey alteration areas, where shales turn into clay. A thick layer of grey-ochre sandy clay is draped over the black shales. Several steps in the bedrock represent fault offsets. A regolith with a minimum thickness of 0.5 m and imbricated pebbles dips towards the fault scarp. Thin layers of regolith and slope-wash deposits parallel the lowermost thick regolith, but are dragged towards the fault. A 70 to 80 cm thick sequence of slope-wash deposits and a 30 cm thick soil layer level out any displacement morphology at the trenching site.

No suitable material for ^{14}C-dating was found in the trench dug across the southeastern fault (f1 in Fig. 5). Nevertheless, several indications for post-glacial movements could be observed. The most obvious are the very fresh scarps offsetting a glacially modified landscape. An additional hint at a post-glacial displacement is the preservation of quartz fibres only in the lower part of the fault scarp of f2. We interpret that the fibres have been exposed at the surface by the last, post-glacial motion along the fault. The height at which the fibres are still preserved coincides with the height of the lowermost scarp section that bars the water run-off of the basin NW of the scarp. Glacially polished surfaces are found on both sides of the scarp. The throw of the fault, as observed today (western block down), contradicts the indication of the quartz fibres, thus suggesting an inversion of the fault movement. Last, but not least, the drag of the Holocene sediments in the trench across f1 also indicates a lowering of the western block. Several indications point to a post-glacial reactivation of the faults induced by gravitational movements. (1) The scarp height varies considerably along strike of the faults. (2) The large displacements (1 to 2 m) of both faults would indicate differential displacement rates of about 0.1 to 0.2 mm/a; such rates are well compatible with DSGSD rates (e.g. Gutiérrez-Santolalla et al. 2005; Hippolyte et al. 2006). (3) A high degree of disintegration of the rock caused by open fissures. (4) Double ridges at the Col du Demècre. The gravitational movement seems to affect the whole ridge at the Col de Demècre, leading to extensional spreading of the ridge and its lowering towards the NW. Only the central parts of the described faults are thought to act as primary detachment faults for this gravitational movement. Lateral flanking of the mass movement is enabled by the NW–SE striking open fissures.

The mass movements of Cari: a deep-seated gravitational deformation is observed along the mountain ridge of Pizzo del Sole between the northern Leventina Valley (Quinto to Faido) and the Piora Valley (Fig. 6). The entire mountain ridge pertains to an overturned north-dipping limb of a backfold of granite gneiss and mica gneiss (Lucomagno basement block, Fig. 6d). To the north, a thick zone consisting of Triassic dolomites, cargneules, evaporites and slates forms a steep belt reaching down at least 2 km (Piora zone, Herwegh & Pfiffner 1999; Pfiffner 2000). From inspection

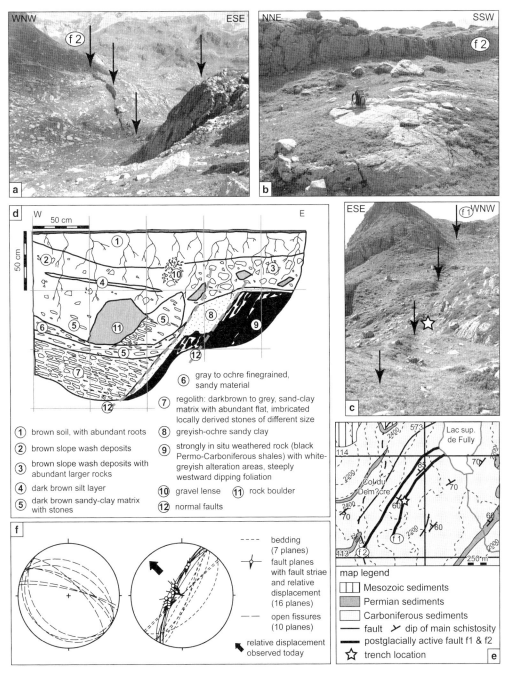

Fig. 5. Gravitationally reactivated tectonic faults at Lac de Fully (location indicated in Fig. 3). (a) Pronounced uphill-facing scarp of fault f2. (b) Glacially polished rock surface proving the existence of a Late Glacial glacier at the fault location. (c) Fault trace of fault f1, white star indicates trench location. (d) Trench log of the trench across fault f1. (e) Geological map of the area of Lac de Fully. Coordinate numbers correspond to the Swiss national km grid; geology after Badoux et al. (1971). (f) Structural data of the fault f2 and the surrounding geology (lower-hemisphere equal-area projections).

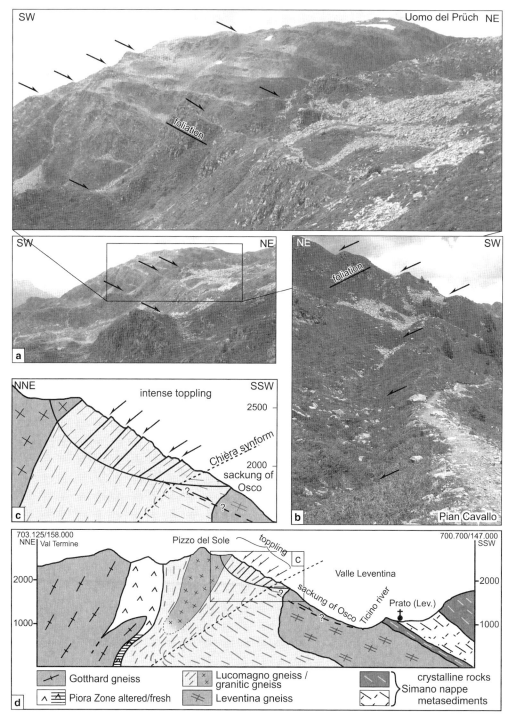

Fig. 6. Gravitational movements at Cari (location indicated in Fig. 3). (a) and (b) show pronounced step-like morphology at Uomo del Prüch and Pian Cavallo. (c) Extensive toppling of gneiss packages. (d) Geological cross-section (depth of Piora zone, after Schaad & Pfiffner 1992; sackung of Osco, after Etter 1992).

drillings for the Alptransit Gotthard Tunnel, it is known that these evaporites are strongly altered and show signs of pervasive dissolution (Herwegh & Pfiffner 1999; Pfiffner 2000). The altered and therefore weak zone reaches more than 1 km down to 800 m.a.s.l. (Schaad & Pfiffner 1992).

A $c.$ 25 km^2 large part of the southern slope of Pizzo del Sole around the villages of Cari and Osco is affected by gravitational movements (Etter 1992). A staircase-like morphology characterizes the summit and crest areas of the mountain ridge and the upper part of the slope (Fig. 6a, b). These steps parallel the N-dipping foliation, occur between 1900 m.a.s.l. and the summit areas and extend over several hundreds of metres along the slope. The step heights can reach a few tens of metres. They are formed gravitationally by large-scale-toppling of packages of Leventina gneiss, decreasing the dip angle of the regional schistosity from about 60 to 15–40° (Fig. 6c). Foliation planes act as slip planes. Movements along these faults are at least partly post-glacial in age, as they dissect and displace moraine screes and talus aprons. Young aprons form where scree is withheld by the scarps (Fig. 6a, b). Additionally, a deep-reaching sackung encompasses the middle and lower parts of the slope (sackung of Osco, after Etter 1992), but does not produce a distinctive morphology.

Tectonic faults. Tectonic faults display linear outcrop traces and do not show a particular correlation with the thalweg of a valley. They often can be traced continuously across several valleys. Tectonic faults either correspond to ductile or brittle shear zones, or reactivated joints or tension cracks. As shown by Laws (2001) and Zangerl (2003), ductile shear zones frequently exhibit a brittle reactivation. In most cases, it proved difficult to precisely constrain the age of activity or reactivation of a tectonic fault.

Unequivocal indications of a young (re-)activation of faults observed in the field are given by the displacement of glacial and postglacial sediments and/or glacial landforms. Such observations are very difficult to make in the Swiss Alps owing to the scarcity of glacial sediments, the weak preservation potential of subtle surface displacements and slow deformation rates. Coupled with high erosion rates in the moderate and humid climate and with anthropogenic overprint in the past hundreds of years, these facts hinder the build-up of recordable offsets of landforms. Nevertheless, two doubtlessly post-glacial faults were detected and are described in the following.

The post-glacially active tectonic fault near Oberwald: this prominent lineament is about 1 km long, strikes NE–SW and perpendicularly crosses the NW–SE-trending mountain ridge between the Geren and Gonerli valleys near Oberwald in the Upper Rhône valley (Fig. 7a, b, for location see Fig. 3). The fault strikes parallel to the NE–SW trend of the regional schistosity of the gneisses of the Gotthard Massif (Fig. 7a). As the fault runs across the mountain ridge, a gravitational reactivation of the fault can be ruled out. Where the fault crosses the top of the mountain ridge, an up to 10 m deep, vegetation-covered and asymmetric incision has formed (Fig. 7e). This may be due to tectonic displacement along the fault coupled with enhanced erosion along the fault zone. The asymmetry of the incision with a higher northwestern block and a lower southeastern block may be the result of tectonic displacement. From the resulting 7 m crest displacement, tectonic slip rates along the fault since the LGM can be calculated to be about 0.4 mm/a (Fig. 7e).

Two parallel moraine ridges of the Daun stage (Aubert 1979; $c.$ 12–14 000 years BP after Preusser 2004) run along the northeastern valley flank of the Gonerli valley at about 2000 m altitude. The lower moraine makes a sharp bend to the southwest a few metres before reaching the fault. The upper moraine ridge is displaced sinistrally by about 3 m (Fig. 7c, d) where it crosses the fault. Sagging features in the valley flank NW of the fault suggest that this displacement is gravitationally induced. Nevertheless, it is thought that a post-glacial reactivation of the fault induced a discrete zone of weakness within the moraine ridge, thus enabling a sliding of the valley flank including the moraine NW of the fault.

The post-glacially active tectonic fault at the Gemmi Pass: a detailed study was performed to unravel the Alpine to post-glacial history of an active tectonic fault in the Gemmi Pass area. An extensive description of this study is given in Ustaszewski *et al.* (2007). The fault forms a prominent, >2.5 km long, NW–SE striking lineament, runs perpendicular to the regional structures, and cuts through the Helvetic nappe stack (Fig. 8a, b). The fault traverses the thalweg at high angle and offsets the bottom of a high-lying valley—two arguments that clearly speak against gravitational reactivation (Fig. 8a). A close examination of the fault rocks reveals a multiphase evolution of this fault starting as a fold-perpendicular joint in a late stage of Alpine nappe emplacement and related deformation (Ustaszewski *et al.* 2007). Tensile deformation and the presence of a fluid led to calcite precipitation in the newly formed cavities. Several cycles of veining and brittle deformation can be observed. Changes in cathodofacies suggest repeated variations in fluid chemistry that speak for episodic fluid pulses. U–Th dating of the calcite cataclasite yielded ages younger than

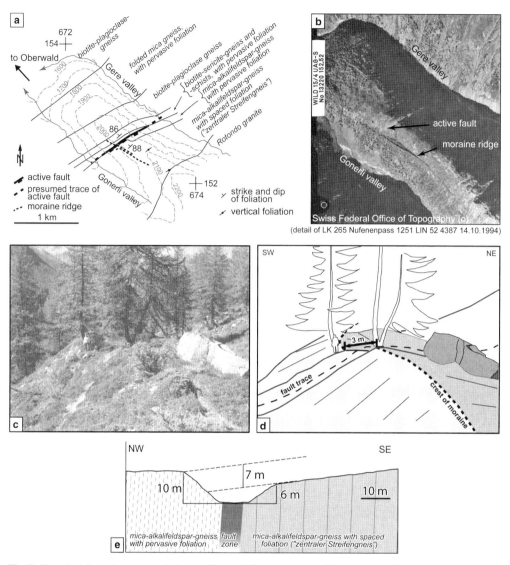

Fig. 7. Post-glacially active tectonic fault near Oberwald (location indicated in Fig. 3). (a) Sketch map of topographic and geological situation. Coordinate numbers correspond to the Swiss national km grid. Geology is drawn after Hafner *et al.* (1975). (b) Non-rectified aerial photograph of area portrayed in (a) showing the active fault and the moraine ridge of the Daun stage (Aubert 1979). (c) Detail photography of intersection of fault with moraine ridge. The crest of the moraine is sinistrally displaced by about 3 m. (d) Line drawing of (c). (e) Profile along the mountain ridge showing vertical displacement of the fault.

2.5 million years (Ustaszewski *et al.* 2007). The fault crosses a small (*c.* 60 m × 30 m) post-glacial, sediment-filled depression (Fig. 8e). A 3D-GPR survey was conducted in order to: (a) image the sediment fill; and (b) determine the position of a trench. Trenching excavation (15.4 m long, 2 m wide and up to 2.2 m deep, Fig. 8c) across the fault allowed verifying the latter's post-glacial reactivation.

The trench bottom reached limestone bedrock along almost the entire trench. The base of the sediment-fill of the depression is made of an up to 1.5 m thick dark brown moraine layer. A very constant, 20 to 30 cm thick, fine-grained (silt to fine sand fraction), yellow loess-like layer was deposited on top of the moraine. The basal contact to the moraine material is sometimes unclear, whereas the upper contact is sharp. This loess

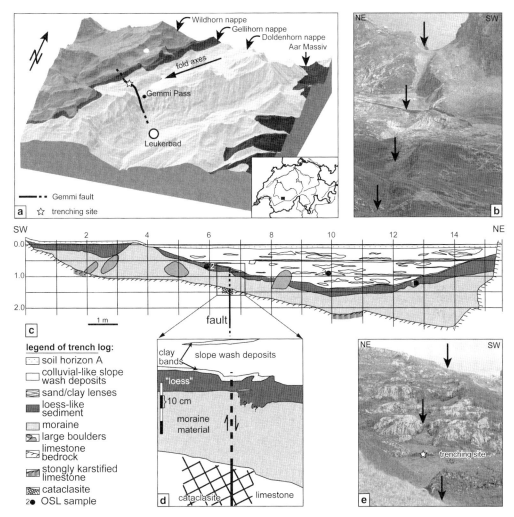

Fig. 8. The post-glacially active tectonic fault at the Gemmi Pass. (**a**) Block diagram from AdS2—Atlas der Schweiz 2 (2004) illustrating geographic position and tectonic setting of the Gemmi fault (coordinates of SW corner: 608.000/133.000; of NE corner: 622.000/145.000). (**b**) Overview photograph of the Gemmi fault (indicated by black arrows), showing its pronounced morphological expression. (**c**) Trench-log across the Gemmi fault. (**d**) Detail of trench log showing offset in loess layer. Fault arrows show apparent offset. (Figure modified after Ustaszewski *et al.* 2007) (**e**) Overview photograph of the Gemmi fault (indicated by black arrows), showing its pronounced morphological expression and trenching location.

layer delineates the basin form. An up to 1.5 m thick grey-brown sequence of colluvial-like slope wash deposits overlies the yellow loess layer. This horizon shows onlap-structures onto the loess on both sides of the basin. The uppermost 5 to 15 cm of the trench wall consists of a soil layer.

A cataclastic fault zone disrupts the partly karstified limestone bedrock in one location on the trench floor (Fig. 8c). This 40 cm wide zone contains an open fissure. No vertical displacement of the bedrock surface is visible. The moraine layer above the bedrock does not show any disturbances in the vicinity of the fault zone. However, the yellow loess layer is heavily disrupted, incorporating moraine material from below, and displays vertical displacements of up to 5 cm at its upper boundary (Fig. 8d). These structures cannot be explained by sedimentary or erosional processes and are not found elsewhere in the trench. We thus conclude that fault motion is strike-slip. The sense of movement could not be determined. The overlying colluvial-like slope-wash deposits are deformed only at the boundary to the loess layer. Some centimetres above the contact, they are not disrupted, thus sealing the movement (Fig. 8d). Samples were taken for optically stimulated luminescence (OSL) dating of the loess layer and the colluvial-like sediments in order to constrain the age of the youngest fault movement. They yielded Late Glacial ages for the loess samples whereas the overlying colluvial sediments are sub-recent (latest Holocene) (Ustaszewski et al. 2007). This clearly indicates a post-glacial activity along the fault. Focal mechanisms of recent instrumentally recorded earthquakes are consistent with our findings and show that the fault at the Gemmi Pass, together with other faults in this area, may be reactivated as a sinistral strike-slip fault in today's stress field (Kastrup et al. 2004).

Composite faults. The identification of post-glacially active tectonic faults in the Alps is exacerbated by the fact that numerous active faults are reactivated and concealed by gravitational movements. As shown by the examples of the faults at Lac de Fully and Cari, gravitational movements (especially DSGSD) often use pre-existing discontinuities such as tectonic faults and foliation planes that may act as slip planes or decoupling horizons (see also Agliardi et al. 2001; Bistacchi et al. 2001; Massironi et al. 2003). Furthermore, a pre-existing foliation also facilitates tectonic reactivation. Additionally, it was discovered that post-glacial differential uplift due to the melting of the LGM ice domes and associated isostatic rebound may reactivate pre-existing faults to a considerable amount (Ustaszewski 2007; Ustaszewski et al. in press). In numerous cases, gravitational, tectonic and isostatic rebound processes act sequentially on one single fault. This complex interplay results in surface displacements which at first are morphologically difficult to distinguish from purely tectonic faults and inhibit the determination of the purely tectonic amount of displacement along such faults. Thus, the estimation of recent crustal deformation is not possible. However, distinguishing between purely tectonic faults, purely gravitational faults and faults, the formation of which involves several processes, is of paramount importance when deciphering the pattern of recent crustal deformation.

Throughout this work, it has emerged that the currently used classification schemes of fault types (e.g. Persaud 2002) are inadequate for a proper description and genetic classification of faults observed in the central and western Swiss Alps. A new type of fault, termed *composite fault*, is therefore proposed by the authors in Ustaszewski 2007 and Ustaszewski et al. (in press). The formation of composite faults involves the complex sequential interplay of several processes (i.e. tectonics, gravitation and differential uplift after deglaciation). This type of fault is often referred to as 'neotectonic fault' in the literature (e.g. Eckardt et al. 1983). Field examples of composite faults are abundant on the northern and southern slopes of the Upper Rhône, the Urseren and Bedretto valley. In all these cases, the bedrock contains a subvertical penetrative foliation. Between c. 1900 m and 2200 m altitude, numerous valley-parallel faults display uphill-facing scarps. These scarps vary in height along strike from less than a metre to up to 10 m over a distance of only a few hundred metres. The young age of these faults is indicated by displaced moraine ridges.

Fault orientations. Figure 3 shows the faults mapped in the study area. The area was separated into different units according to the main lithologies and the regional trend of schistosity. Fault orientations for each region were then plotted as rose diagrams shown in Figure 9. Gravitational faults were not considered. As the majority of observed faults are steeply dipping or even vertical, we chose to present the fault orientation in rose diagrams instead of great circle plots. Thus, the angle of dip only plays a very subordinate role. The regional strike of the main structural features like foliation and fold axes is indicated in the respective rose diagrams.

We first discuss the fault distribution and density in map view and secondly the fault orientations. It has to be kept in mind, however, that the fault map may be incomplete due to reasons mentioned earlier.

(1) Correlation of fault distribution with uplift and seismicity: Comparing the fault distribution with the current surface uplift and seismicity pattern reveals no distinct correlation. The area of the steepest uplift gradient (Rawil Pass) is characterized by a dense fault occurrence, but so is the area of the Grimsel Pass and Gotthard Pass, which reveals only a moderate uplift gradient. The pronounced earthquake alignment directly north of the Rhône river in the western part of the Valais has no correspondence to the fault pattern.

(2) Correlation of fault distribution with valley orientations: The orientation of large valleys is not parallel to pronounced tectonic fault orientations in the whole study area. However, large valleys do follow the trend of major lithological contacts and foliation. This fact is reflected in the clear correlation between valley and fault orientations of the composite lineaments of the Upper Rhône (Goms), Urseren and Bedretto valleys.

(3) Correlation of fault distribution with lithologies: The different lithologies (granites, gneisses, carbonates, sandstone-shale sequences) show very

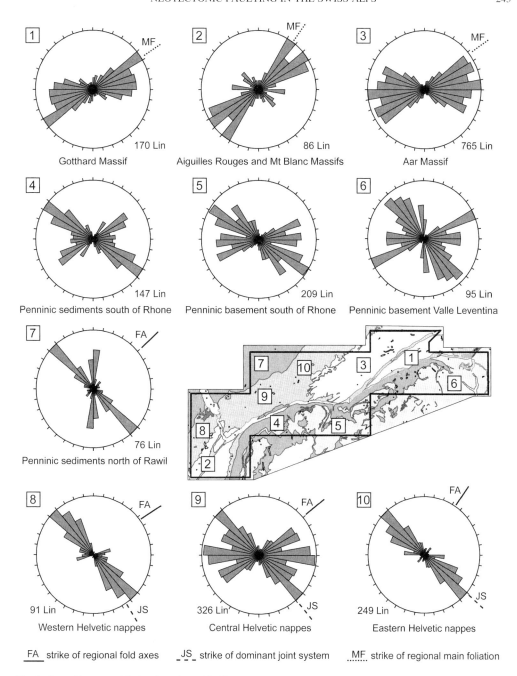

Fig. 9. Rose diagrams of fault orientations. The lineaments were separated regionally and according to the main tectonic units. Regional trends of fold axes, joints and foliation are indicated, where they are uniform and representative for the area (taken from a variety of geological maps).

distinct fault patterns. This presumably reflects the orientation of pre-existing structures and differing preservation potentials of faults in these lithologies. Lithology-dependent fault morphologies and correlations with the different erodibility of lithotypes are described by Persaud & Pfiffner (2004). These differences are summarized in Figure 4. The Penninic nappes, especially in the western part of the study area, are characterized by a rather low fault density (Figs 3 and 9, rose diagrams 4–7). However, basement units consisting of gneisses and schists show a higher density than metasedimentary units. In the Helvetic zone, abundant faults are present in units made up of carbonate rocks (Figs 3 and 9, rose diagrams 8–10). The external basement massifs consisting of gneisses and granites exhibit a dense fault pattern (Figs 3 and 9, rose diagrams 1–3).

(4) Correlation of fault distribution with orientation of main foliation: In a previous study in eastern Switzerland, Persaud & Pfiffner (2004) described vertical faults in lithotypes dominated by schists and sandstones having a gently dipping main foliation. In our study area, gneisses with a gently dipping foliation in the Penninic nappes south of the Rhône river and in the Ticino show long, vertical faults in parts (e.g. in the Ticino), whereas in the Valais (northern Bernhard nappe complex), vertical faults are lacking. Comparable lithologies with a steeply dipping foliation in the Aar and Gotthard massifs show abundant vertical faults (Keller & Schneider 1982; Laws 2001; Zangerl 2003). In the case of the gneisses, faulting is pronounced if the main foliation is favourable, i.e. subvertically to vertically oriented.

(5) Correlation of fault orientation with preexisting structural orientations: A comparison of the fault orientations with the regional trends of schistosity (Fig. 9) shows a strong correlation in the Aar and Gotthard massifs, where the SW–NE trending faults follow the existing foliation. Nevertheless, abundant faults in E–W orientation cross-cut this structure. In the Aiguilles-Rouges/Mont Blanc massif and the Gotthard massif, the main orientations of the faults strike NNE–SSW and NE–SW, respectively, and parallel the foliation orientations. The same fault orientations are also found in the Penninic sediments and basement nappes alongside a second NW–SE orientation. However, a correlation between fault orientations and pre-existing structures is not feasible, because the regional trend of foliation is not uniform in these areas. The same applies to the Leventina valley, where NW–SE and NE–SW fault orientations prevail in the gently dipping gneisses. In the Helvetic nappes, the major fault orientation parallels the dense joint network, especially the NW–SE trending a-c joints, which are perpendicular to the regional trend of fold axes. In the Penninic nappes north of the Rawil depression, NW–SE striking lineaments are perpendicular to the fold axes, similar to the ones in the Helvetic nappes.

(6) Correlation of fault orientation with fault plane solutions: Nodal plane orientations of focal mechanisms of earthquakes that occurred between 1985 and 2004 (Kastrup 2002; Baer et al. 2003, 2005; Deichmann et al. 2000, 2002, 2004), mirror partly the observed fault pattern (Fig. 3 and Table 1). In the western part of the Helvetic nappes studied here (rose diagram 10 in Fig. 9), the most prominent fault orientation strikes NW–SE. This orientation is not observed in the focal plane solutions (FPS). In the central part of the Helvetic nappes (rose diagram 9 in Fig. 9), the first maximum (NW–SE) of fault orientation is not represented in the seismic data, but the subordinate maxima of E–W and NE–SW are (FPS 6, 23, 25 and 26 in Fig. 3). In the western part of the Helvetic nappes, FPS 19 coincides with the NW–SE striking fault orientation maximum (rose diagram 8 in Fig. 9). Strike-slip focal mechanisms prevail in the entire Helvetic zone.

South of the Rhône valley, the correlation between the focal mechanisms and the fault orientations (rose diagrams 4 and 5 in Fig. 9) is less pronounced due to the fact that the faults show a scatter in orientation, unlike in the Helvetic zone. Nevertheless, both major fault orientations (NW–SE and NE–SW) are represented in the focal mechanisms that indicate predominantly normal fault movements.

Other regions, especially the Aar massif, the Gotthard massif and the Penninic nappes in the eastern part of the study area, lack significant seismicity, and no focal mechanisms are available for a comparison with the fault pattern. The seismicity around the culmination zone of the Aar massif and the areas adjacent to the south were described first by Pavoni (1977) to display near-quiescence. Deichmann (1990) suggests possible general reasons for this lack of earthquakes such as high rock temperature, predominance of granitic rocks and low pore pressure and/or few pore fluids.

Discussion and conclusion

The original aim of this study was the identification of post-glacially active faults in the central and western Swiss Alps. Earthquake and surface uplift patterns indicate that this area is characterized by pronounced recent tectonic activity. The uplift pattern suggests that the external basement massifs are still undergoing active uplift, possibly due to compression. It turned out to be difficult to locate post-glacially active tectonic faults even in the area of maximum uplift.

A study of lineaments in this area by means of aerial photograph analysis and fieldwork generally reveals three different types of fault: gravitational faults, tectonic faults and composite faults that follow pre-existing tectonic structures and are

Table 1. *Earthquakes with calculated focal mechanisms occurring in the study area between 1985 and 2004*

No.	Location	Date	hr:min	X-Coord	Y-Coord	z (km)	Mw	1st nodal plane strike	dip	rake	2nd nodal plane strike	dip	rake	P-axis p-az	plunge	T-axis p-az	plunge	data source
1	Zeuzier	9-Oct-1986	10:08	603185.2	130067.7	4	3.6	079	61	167	175	79	030	304	12	041	28	a
2	Montana	7-Jan-1989	2:29	607341.1	132628.1	4	3.4	57	68	170	151	81	022	282	09	016	22	a
3	Anzere	9-Sep-1989	4:41	596100.0	130735.2	6	3.5	*110	90	140	200	50	000	163	27	057	27	a
4	Montana	28-Apr-1990	22:24	605956.2	131737.2	3	2.2	266	46	−145	150	65	−050	108	52	212	11	a
5	Anzere	7-May-1990	16:06	597300.0	130200.0	7	1.6	175	45	−031	288	69	−131	153	49	046	14	a
6	Sanetsch	3-Jun-1990	19:23	587930.8	127411.1	3	2.2	100	60	−151	354	65	−034	315	41	048	03	a
7	Anzere	26-Jul-1990	12:30	596638.8	130401.4	7	2.4	*285	80	−140	187	51	−013	154	35	050	19	a
8	St. Leonard	31-Aug-1990	10:57	601500.0	124400.0	7	2.0	181	53	025	075	70	140	132	11	032	42	a
9	Leukerbad	21-Feb-1996	18:57	610800.8	135190.0	5	3.3	*242	87	−178	152	88	−003	107	4	197	01	a
10	Lötschental	28-Nov-1997	8:30	635072.5	143063.7	12	2.9	250	60	−150	144	64	−034	105	41	198	03	a
11	Vissoie	19-Jan-1986	6:54	616007.0	116081.0	6	3.0	110	40	−080	277	51	−098	143	82	013	05	a
12	Stalden	22-Mar-1987	1:36	633500.0	115700.0	4	2.1	311	51	−047	075	55	−130	286	58	192	02	a
13	St. Niklaus	11-May-1990	8:16	625492.4	118644.4	1	2.0	263	40	−116	115	55	−070	076	72	191	08	a
14	Vissoie	25-Sep-1990	5:19	617105.2	109858.7	5	3.6	070	50	−130	303	54	−052	273	60	007	02	a
15	Vissoie	17-Dec-1990	23:34	615400.0	118600.0	5	1.7	319	42	−049	090	60	−120	310	62	201	10	a
16	Stalden	7-Sep-1990	18:09	638454.7	118738.6	8	2.4	135	55	−019	236	74	−144	101	37	002	12	a
17	Val d'Heremence	7-May-1998	17:16	596471.7	108279.5	6	3.3	092	55	−090	272	35	−090	002	80	182	10	a
18	Grimetz	9-Dec-1998	22:08	608751.3	115510.5	4	3.4	256	28	−080	065	62	−095	323	72	159	17	a
19	Lac de Salanfe	29-Dec-1999	9:29	560100.0	108700.0	4	3.3	-	-	-	-	-	-	249	76	032	11	b
20	Martigny	23-Feb-2001	22:20	568500.0	109600.0	6	3.6	129	52	−009	224	83	−142	093	31	350	20	c
21	Martigny	25-Feb-2001	1:22	568500.0	109600.0	6	3.5	138	56	−011	234	81	−146	101	30	002	16	c
22	Ayer	9-Jul-2001	22:50	614400.0	113300.0	6	3.2	276	42	−072	072	50	−106	283	77	173	04	c
23	Anzere	31-May-2002	16:50	593800.0	129900.0	5	3.5	046	24	−094	231	66	−088	145	69	319	21	d
24	Salgesch	29-Apr-2003	4:55	610000.0	132000.0	10	3.9	067	48	138	188	60	050	305	07	045	55	e
25	Glarey	22-Aug-2003	9:22	590500.0	130000.0	6	3.9	151	54	−027	258	68	−141	120	43	022	08	e
26	Derborence	30-May-2004	9:46	581800.0	126900.0	9	2.9	347	90	000	077	90	−180	302	00	212	00	f

* - Active fault plane; p-az – plunge azimuth; data source: a – Katrup 2002; b – Deichmann *et al.* 2000; c – Deichmann *et al.* 2002; d – Baer *et al.* 2003; e – Deichmann *et al.* 2004; f – Baer *et al.* 2005; - – unknown.

reactivated by gravitational movements and post-glacial rebound processes. Our analysis reveals that the great majority of post-glacial activity along faults is the result of gravitational movements. In this sense, the influence of gravity in the Swiss Alps is more important than hitherto thought. A limitation of this study is the possibly incomplete record of faults due to partly weak preservation of faults, forest and glacier coverage, as well as anthropogenic modifications of the landscape. Nevertheless, two unequivocally post-glacially active tectonic faults displacing Quaternary deposits could be identified.

A comparison of fault distribution with both surface uplift pattern and seismicity does not show clear correlations. Valley orientations and lithologies, on the other hand, correlate quite well. The fault pattern in metamorphic rocks usually is in concordance with the regional trend of schistosity. In carbonate rocks, faults generally follow the joint networks.

Comparing fault orientations with nodal plane orientations from focal mechanisms shows only weak agreement between the strike of lineaments and possibly seismically active fault planes. In contrast to this study, a good agreement between lineament orientation and fault plane solutions was found in eastern Switzerland (Persaud & Pfiffner 2004). In any case, it must be remembered that this correlation is based on a low number of available focal mechanisms and the non-uniform distribution of earthquake occurrence.

We thus conclude that the Swiss Alps may host a large number of faults prone for reactivation in today's stress field. However, for most of these faults, no indications of their last phase of activity exist. Two unambiguously active tectonic faults identified in this study do not allow a general statement about the formation of recent faults in the entire Swiss Alps, but are rather examples of local evolutions of recent structures.

The alignment of recent earthquakes north of the Rhône valley points to the question if a major fault zone could be responsible. In order to roughly estimate a potential post-glacial offset, we delineated a fault area (Swiss coordinates of corners: 584.4/129.3; 610.7/138.0; 586.4/123.5; 612.5/132.4) based on the scatter pattern of instrumentally recorded earthquakes (see Fig. 1) and added historical earthquakes to it. The number and magnitude of the latter is crucial for such an estimate. The included earthquakes have a magnitude greater than 2.0 and occurred between the years 1524 and 2007. Three earthquakes with a magnitude of 6 or larger are recorded in the historic earthquake catalogue in the specified area (April 1524 Ardon/VS: Mw = 6.4; January 1946 Ayent/VS: Mw = 6.1; May 1946 Ayent/VS: Mw = 6.0). The instrumentally recorded seismicity covers the period from 1975 until present. The seismic moments of these earthquakes were converted to cumulative slip following a procedure described by Hinsch & Decker (2003). The resulting total displacement of 1.22 m since 1524 can be converted to slip rates of 0.252 mm/a along the delineated fault zone. This accumulates to approximately 25 m in the Holocene. This displacement could be taken up by many faults found in this area (Fig. 3), thus individual faults may have experienced only 1 m or less of dislocation that may go unnoticed by the geologist's eye. This statement is in concert with our field investigation that has come up with only few unequivocally post-glacially active tectonic faults. It thus seems that the current strain is either predominantly aseismic or, alternatively, cumulated seismic slip is too low for producing surface rupture. This is in agreement with conclusions drawn from the inner western Alps (Sue et al. 2000) and from the Vienna Basin (Hinsch & Decker 2003), where in both cases geodetic deformations exceed seismic deformations by about 90%. Additionally, steep slopes can lead to the camouflage of active fault scarps.

Nevertheless, we do see that the distribution of seismicity in Switzerland reflects gradients of surface uplift rates. In the western part of the study area, pronounced seismicity is accompanied by steep uplift gradients. The eastern part of the study area lacks an enhanced seismicity and is additionally characterized by a uniform surface uplift. In eastern Switzerland, high seismicity coincides with steep uplift gradients again. As suggested by the cross-section in Figure 2b, seismicity and surface uplift could be related to ongoing collision and uplift of the external basement massifs (i.e. Aar, Aiguilles Rouges and Mont Blanc massifs) with the latest phase of collision corresponding to the updoming of the Aar massif (Burkhard 1988; Pfiffner et al. 2002).

This work was financially supported by the Canton of Bern and the Swiss Geophysical Commission (project: Swiss Seismotectonic Atlas), for which we are much obliged. We thank Susan Buiter for the earthquake projection. Our special thanks go to Andreas Ebert for having helped dig the trench at Lac de Fully. We thank Kurt Decker and Andrea Bistacchi for their critical and constructive reviews, which helped improve the manuscript.

References

ADS2-ATLAS DER SCHWEIZ 2. 2004. Version 2, Swiss Federal Office of Topography.
AGLIARDI, F., CROSTA, G. & ZANCHI, A. 2001. Structural constraints on deep-seated slope deformation kinematics. Engineering Geology, 59, 83–102.

AUBERT, D. 1979. *Les Stades de Retrait des Glaciers du Haut-Valais*. PhD thesis, Université de Lausanne.

BADOUX, H., BURRI, M., GABUS, J. H., KRUMMENACHER, D., LOUP, G. & SUBLET, P. 1971. *Geologischer Atlas der Schweiz, Atlasblatt 58: Dent de Morcles*. Schweizerische Geologische Kommission, 1:25 000.

BAER, M., DEICHMANN, N., BRAUNMILLER, J. ET AL. 2003. Earthquakes in Switzerland and surrounding regions during 2002. *Eclogae Geologicae Helvetiae*, **96**, 313–324.

BAER, M., DEICHMANN, N., BRAUNMILLER, J. ET AL. 2005. Earthquakes in Switzerland and surrounding regions during 2004. *Eclogae Geologicae Helvetiae*, **98**, 407–418.

BISTACCHI, A. & MASSIRONI, M. 2000. Post-nappe brittle tectonics and kinematic evolution of the north-western Alps: an integrated approach. *Tectonophysics*, **327**, 267–292.

BISTACCHI, A., EVA, E., MASSIRONI, M. & SOLARINO, S. 2000. Miocene to Present kinematics of the NW-Alps: evidences from remote sensing, structural analysis, seismotectonics and thermochronology. *Journal of Geodynamics*, **30**, 205–228.

BISTACCHI, A., DAL PIAZ, G. V., MASSIRONI, M., ZATTIN, M. & BALESTRIERI, M. 2001. The Aosta-Ranzola extensional fault system and Oligocene-Present evolution of the Austroalpine-Penninic wedge in the northwestern Alps. *International Journal of Earth Sciences*, **90**, 654–667.

BWG (ed.) 2005. *Tektonische Karte der Schweiz 1:500 000*. Bundesamt für Wasser und Geologie, Bern. ISBN 3-906723-56-9.

BURKHARD, M. 1988. L'Helvétique de la bordure occidentale du massif de l'Aar (evolution tectonique et métamorphique). *Eclogae Geologicae Helvetiae*, **81**, 63–114.

CALAIS, E., GALISSON, L., STÉPHAN, J.-F. ET AL. 2000. Crustal strain in the Southern Alps, France, 1948–1998. *Tectonophysics*, **319**, 1–17.

CALAIS, E., NOCQUET, J.-M., JOUANNE, F. & TARDY, M. 2002. Current strain regime in the Western Alps from continuous Global Positioning System measurements, 1996–2001. *Geology*, **30**, 651–654.

CHAMPAGNAC, J.-D., SUE, C., DELACOU, B. & BURKHARD, M. 2003. Brittle orogen-parallel extension in the internal zones of the Swiss Alps (South Valais). *Eclogae Geologicae Helvetiae*, **96**, 325–338.

CHAMPAGNAC, J.-D., SUE, C., DELACOU, B. & BURKHARD, M. 2004. Brittle deformation in the inner NW Alps: from early orogen-parallel extrusion to late orogen-perpendicular collapse. *Terra Nova*, **16**, 232–242.

DEICHMANN, N. 1990. Seismizität der Nordschweiz 1987–1989, und Auswertung der Erdbebenserie von Günsberg, Läufelfingen und Zeglingen. *NAGRA Technischer Bericht*, **90–46**.

DEICHMANN, N., BAER, M., BRAUNMILLER, J. ET AL. 2000. Earthquakes in Switzerland and surrounding regions during 1999. *Eclogae Geologicae Helvetiae*, **93**, 395–406.

DEICHMANN, N., BAER, M., BRAUNMILLER, J. ET AL. 2002. Earthquakes in Switzerland and surrounding regions during 2001. *Eclogae Geologicae Helvetiae*, **95**, 249–261.

DEICHMANN, N., BAER, M., BRAUNMILLER, J. ET AL. 2004. Earthquakes in Switzerland and surrounding regions during 2003. *Eclogae Geologicae Helvetiae*, **97**, 447–458.

DEICHMANN, N., BAER, M., BRAUNMILLER, J. ET AL. 2006. Earthquakes in Switzerland and surrounding regions during 2005. *Eclogae Geologicae Helvetiae*, **99**, 443–452.

DELACOU, B., SUE, C., CHAMPAGNAC, J. & BURKHARD, M. 2004. Present-day geodynamics in the bend of the western and central Alps as constrained by earthquake analysis. *Geophysical Journal International*, **158**, 753–774.

DEMETS, C., GORDON, R. G., ARGUS, D. F. & STEIN, S. 1990. Current plate motions. *Geophysical Journal International*, **101**, 425–478.

DEMETS, C., GORDON, R. G., ARGUS, D. F. & STEIN, S. 1994. Effect of recent revisions to the geomagnetic reversal time scale on estimates of current plate motions. *Geophysical Research Letters*, **21**, 2191–2194.

DRAMIS, F., FARABOLLINI, P., GENTILI, B. & PAMBIANCHI, G. 2002. Neotectonics and large-scale gravitational phenomena in the Umbria-Marche Apennines, Italy. *In*: COMERCI, V. E. (ed.) *Seismically induced ground ruptures and large scale mass movements—field excursion and meeting 21–27 Sept. 2001*. APAT—Italian Agency for Environment Protection and Technical Services, Rome, 17–30.

DRAMIS, F. & SORRISO-VALVO, M. 1994. Deep-seated gravitational slope deformations, related landslides and tectonics. *Engineering Geology*, **38**, 231–243.

ECKARDT, P., FUNK, H. & LABHART, T. 1983. Postglaziale Krustenbewegungen an der Rhein-Rhône-Linie. *Mensuration, Photogrammétrie, Génie rural*, **2**, 43–56.

ETTER, U. 1992. Die Chièra-Synform—Kartierung einer Grossfalte in einer Sackungsmasse. *Bulletin der Vereinigung Schweizerischer Petroleum-Geologen und Ingenieuren*, **59**, 93–99.

EVA, E., PASTORE, S. & DEICHMANN, N. 1998. Evidence for ongoing extensional deformation in the western Swiss Alps and thrust-faulting in the southwestern Alpine foreland. *Journal of Geodynamics*, **26**, 27–43.

EVA, E. & SOLARINO, S. 1998. Variations of stress directions in the western Alpine arc. *Geophysical Journal International*, **135**, 438–448.

FERHAT, G., FEIGL, K. L., RITZ, J.-F. & SOURIAU, A. 1998. Geodetic measurement of tectonic deformation in the southern Alps and Provence, France, 1947–1994. *Earth and Planetary Science Letters*, **159**, 35–46.

GRENERCZY, G., SELLA, G., STEIN, S. & KENYERES, A. 2005. Tectonic implications of the GPS velocity field in the northern Adriatic region. *Geophysical Research Letters*, **32**, L16311.

GUBLER, E. 1991. UELN and the Swiss National Levelling Net. *In: Report on the Geodetic Activities in the years 1987 to 1991*. Presented to the XX General Assembly of the International Union of Geodesy and Geophysics in Vienna, August 1991. Zürich, 1991.

GUTIÉRREZ-SANTOLALLA, F., ACOSTA, E., RIOS, S., GUERRERO, J. & LUCHA, P. 2005. Geomorphology and geochronology of sacking features (uphill-facing scarps) in the Central Spanish Pyrenees. *Geomorphology*, **69**, 298–314.

HAFNER, S., GÜNTHERT, A., BURCKHARDT, C. E., STEIGER, R. H., HANSEN, J. W. & NIGGLI, C. R. 1975. *Geologischer Atlas der Schweiz, Atlasblatt 68: Val Bedretto*. Schweizerische Geologische Kommission (ed.), 1:25 000.

HERWEGH, M. & PFIFFNER, O. A. 1999. Die Gesteine der Piora-Zone (Gotthard-Basistunnel). *In*: LÖW, S. & WYSS, R. (eds) *Vorerkundung und Prognose der Basistunnels*. Balkema, Rotterdam.

HINSCH, R. & DECKER, K. 2003. Do seismic slip deficits indicate an underestimated earthquake potential along the Vienna Basin Transfer Fault System? *Terra Nova*, **15**, 343–349.

HIPPOLYTE, J.-C., BROCARD, G., TARDY, M. ET AL. 2006. the recent fault scarps of the Western Alps (France): tectonic surface ruptures or gravitational sackung scarps? A combined mapping, geomorphic, levelling, and ^{10}Be dating approach. *Tectonophysics*, **418**, 255–276.

HURFORD, A. J., HUNZIKER, J. C. & STÖCKHERT, B. 1991. Constraints on the late thermotectonic evolution of the Western Alps: evidence for episodic rapid uplift. *Tectonics*, **10**, 758–769.

KAHLE, H.-G., GEIGER, B. ET AL. 1997. Recent crustal movements, geoid and density distribution: contribution from integrated satellite and terrestrial measurements. *In*: PFIFFNER, O. A., LEHNER, P., HEITZMANN, P., MÜLLER, S. & STECK, A. (eds) *Deep Structure of the Swiss Alps: Results of NRP 20*. Birkhäuser, Basel.

KASTRUP, U. 2002. *Seismotectonics and stress field variations in Switzerland*. PhD thesis, ETH Zürich.

KASTRUP, U., ZOBACK, M. L., DEICHMANN, N., EVANS, K. F., GIARDINI, D. & MICHAEL, A. J. 2004. Stress field variations in the Swiss Alps and the northern Alpine foreland derived from inversion of fault plane solutions. *Journal of Geophysical Research*, **109**, B01402.

KELLER, F. & SCHNEIDER, T. R. 1982. Geologie und Geotechnik. *In*: Der Furka-Basistunnel: Zur Eröffnung am 25. Juni 1982, *Schweizer Ingenieur und Architekt*, **24/82**.

LAWS, S. 2001. *Structural, geomechanical and Petrophysical Properties of Shear Zones in the Eastern Aar Massif, Switzerland*. PhD thesis, ETH Zürich.

MAHR, T. 1977. Deep-reaching gravitational deformations of high mountain slopes. *Bulletin of the International Association of Engineering Geology*, **16**, 121–127.

MAHR, T. & NEMCOK, A. 1977. Deep-seated creep deformations in the crystalline cores of the Tatry Mts. *Bulletin of the International Association of Engineering Geology*, **16**, 104–106.

MASSIRONI, M., BISTACCHI, A., DAL PIAZ, G. V., MONOPOLI, B. & SCHIAVO, A. 2003. Structural control on mass-movement evolution: A case study from the Vizze Valley, Italian Eastern Alps. *Eclogae Geologica Helvetiae*, **96**, 85–98.

MAURER, H. R., BURKHARD, M., DEICHMANN, N. & GREEN, A. G. 1997. Active tectonism in the central Alps: contrasting stress regimes north and south of the Rhône Valley. *Terra Nova*, **9**, 91–94.

NOCQUET, J.-M. & CALAIS, E. 2003. Crustal velocity field of western Europe from permanent GPS array solutions, 1996–2001. *Geophysical Journal International*, **154**, 72–88.

NOCQUET, J.-M. & CALAIS, E. 2004. Geodetic Measurements of Crustal Deformation in the Western Mediterranean and Europe. *Pure and Applied Geophysics*, **161**, 661–681.

PAVONI, N. 1977. Erdbeben im Gebiet der Schweiz. *Eclogae Geologicae Helvetiae*, **70**, 351–370.

PERSAUD, M. 2002. *Active Tectonics in the Eastern Swiss Alps*. PhD thesis, University of Basel.

PERSAUD, M. & PFIFFNER, O. A. 2004. Active deformation in the eastern Swiss Alps: post-glacial faults, seismicity and surface uplift. *Tectonophysics*, **385**, 59–84.

PFIFFNER, O. A. 2000. Alpine Geotraversen: der Beitrag von Basistunnels und seismischen Profilen zum Verständnis der Alpengeologie am Beispiel der Piora-Zone. *In*: PEDUZZI, R. (ed.) *8. Pubikation der ASSN/SANW*, 67–72.

PFIFFNER, O. A., ELLIS, S. & BEAUMONT, C. 2000. Collision tectonics in the Swiss Alps: Insight from geodynamic modeling. *Tectonics*, **19**, 1065–1094.

PFIFFNER, O. A., SCHLUNEGGER, F. & BUITER, S. 2002. The Swiss Alps and their peripheral foreland basin: stratigraphic response to deep crustal processes. *Tectonics*, **21**, 3.1–3.16.

PREUSSER, F. 2004. Towards a chronology of the Late Pleistocene in the northern Alpine Foreland. *Boreas*, **33**, 195–210.

RADBRUCH-HALL, D. H. 1978. Gravitational creep of rock masses on slopes. *In*: VOIGHT, B. (ed.) *Rockslides and Avalanches, 1—Natural Phenomena. Developments in Geotechnical Engineering*. Elsevier, Amsterdam, Oxford, New York, 607–657 (Chapter 617).

RADBRUCH-HALL, D. H., VARNES, D. J. & SAVAGE, W. Z. 1976. Gravitational spreading of steep-sided ridges ('Sackung') in western United States. *Bulletin of the International Association of Engineering Geology*, **14**, 23–35.

SAVAGE, W. Z. & VARNES, D. J. 1987. Mechanics of gravitational spreading of steep-sided ridges ('sackung'). *Bulletin of the International Association of Engineering Geology*, **35**, 31–36.

SCHAAD, W. & PFIFFNER, O. A. 1992. *Genese des Żuckerkörnigen Dolomits*. Bundesamt für Verkehr, Schweizerische Bundesbahnen, PL Alp Transit, Alpentransit Gotthard, Abschnitt Basistunnel, Bern.

SCHLATTER, A., SCHNEIDER, D., GEIGER, A. & KAHLE, H.-G. 2005. Recent vertical movements from precise levelling in the vicinity of the city of Basel, Switzerland. *International Journal of Earth Sciences*, **94**, 507–514.

SELZER, C. 2006. *Tectonic Accretion Styles at Convergent Margins: A Numerical Modelling study*. PhD thesis, University of Bern.

SERPELLONI, E., ANZIDEI, M., BALDI, P., CASULA, G. & GALVANI, A. 2005. Crustal velocity and strain-rate fields in Italy and surrounding regions: new results from the analysis of permanent and non-permanent GPS networks. *Geophysical Journal International*, **161**, 861–880.

SUE, C., MARTINOD, J., TRICART, P. ET AL. 2000. Active deformation in the inner western Alps inferred from comparison between 1972-classical and 1996-GPS geodetic surveys. *Tectonophysics*, **320**, 17–29.

TESAURO, M., HOLLENSTEIN, C., EGLI, R., GEIGER, A. & KAHLE, H.-G. 2005. Continuous GPS and broad-scale deformation across the Rhine Graben and the Alps. *International Journal of Earth Sciences*, **94**, 525–537.

USTASZEWSKI, M. 2007. *Active Tectonics in the Central and Western Swiss Alps*. PhD thesis, University of Bern.

USTASZEWSKI, M., MCCLYMONT, A., HERWEGH, M., PREUSSER, F. & PFIFFNER, O. A. 2007. Unravelling the evolution of an Alpine to post-glacially active fault in the Swiss Alps. *Journal of Structural Geology*, **29**, 1943–1959.

USTASZEWSKI, M., HAMPEL, A. & PFIFFNER, O. A. in press. Composite fault scarps in the Swiss Alps formed by the interplay of tectonics, gravitation and post-glacial rebound: an integrated field and modelling study. *Swiss Journal of Geology*.

VIGNY, C., CHÉRY, J., DUQUESNOY, T. ET AL. 2002. GPS network monitors the Western Alps' deformation over a five-year period: 1993–1998. *Journal of Geodesy*, **76**, 63–76.

ZANGERL, C. J. 2003. *Analysis of Surface Subsidence in Crystalline Rocks above the Gotthard Highway Tunnel, Switzerland*. PhD thesis, ETH Zürich.

ZISCHINSKY, U. 1967. Bewegungsbilder instabiler Talflanken. *Mitteilung der Gesellschaft der Geologie und Bergbaustudenten*, **17**, 127–168.

ZISCHINSKY, U. 1969. Über Sackungen. *Rock Mechanics*, **1**, 30–52.

On the role and importance of orogen-parallel and -perpendicular extension, transcurrent shearing, and backthrusting in the Monte Rosa nappe and the Southern Steep Belt of the Alps (Penninic zone, Switzerland and Italy)

JAN PLEUGER[1], THORSTEN J. NAGEL[1], JENS M. WALTER[2], EKKEHARD JANSEN[3] & NIKOLAUS FROITZHEIM[1]

[1]*Steinmann-Institut für Geologie, Mineralogie und Paläontologie, Rheinische Friedrich-Wilhelms-Universität Bonn, Nußallee 8, D-53115 Bonn (e-mail: jan.pleuger@uni-bonn.de)*

[2]*Geowissenschaftliches Zentrum der Universität Göttingen, Goldschmidtstraße 3, D-37077 Göttingen*

[3]*Steinmann-Institut für Geologie, Mineralogie und Paläontologie, Rheinische Friedrich-Wilhelms-Universität Bonn, Poppelsdorfer Schloß, D-53115 Bonn*

Abstract: During Europe–Adria collision in Tertiary times, the Monte Rosa nappe was penetratively deformed in several stages after an eclogite-facies pressure peak: (1) top-to-the-NW thrust shearing (Mattmark phase, after 40 Ma); (2) orogen-parallel, top-to-the-SW extensional shearing and folding (Malfatta phase); (3) orogen-perpendicular, top-to-the-SE extensional shearing and folding (Mischabel phase, before 30 Ma); and (4) large-scale, upright, SE-vergent folding (Vanzone phase, c. 29–28 Ma). Structural analysis and neutron texture goniometry of quartz mylonites show that the Stellihorn shear zone in the Monte Rosa nappe accommodated a complex and multidirectional sequence of shearing movements during the Mattmark, Malfatta and Mischabel phases, and was folded in the Vanzone phase. In the tail-shaped eastward prolongation of the Monte Rosa nappe in the Southern Steep Belt of the Alps, both dextral and sinistral mylonites (Olino phase) were formed during and after the formation of the Vanzone fold, reflecting renewed orogen-parallel (SW–NE) extension contemporaneous with NW–SE shortening from c. 29 Ma onward. A similar sequence of deformation stages was identified in the Adula nappe at the eastern border of the Lepontine metamorphic dome. Important consequences arise for the Insubric fault at the southern border of the Lepontine dome: (1) the NW- to N-dipping orientation of the Insubric fault is not a primary feature but resulted from rotation of an originally SE-dipping shear zone after c. 30 Ma; and (2), the strong contrast in metamorphic grade across this fault (upper amphibolite facies to the north versus anchizone to the south) results from north-side-up faulting coupled with orogen-parallel extension of the northern block (Lepontine dome), while no such extension occurred in the southern block (Southern Alps). Extension in the northern block started in the Malfatta phase and continued in the Mischabel phase when the foliation in the area which later became the Southern Steep Belt still dipped towards south. During Vanzone/Olino deformation, further unroofing and uplift of the Lepontine dome relative to the South Alpine block took place while the Southern Steep Belt was progressively rotated into its present, overturned position, changing its character from a normal fault into a backthrust. Complex deformation paths in the Southern Steep Belt resulted from the combination of extension of the northern block with strike-slip motion along the Insubric fault.

Introduction

The Insubric fault is one of the most important Tertiary faults of the Alps (Fig. 1) and an essential part of many geodynamic models of the Alps (Schmid *et al.* 1996, 1997; Beaumont *et al.* 1996; Escher & Beaumont 1997; Pfiffner *et al.* 2000; Handy *et al.* 2005). The term 'Insubric fault' denotes a part of the Periadriatic fault system including the northeastern segment of the Canavese fault (west of Lago Maggiore, Fig. 2) and the western segment of the Tonale fault (east of Lago Maggiore, see Spitz 1919; Gansser 1968). It comprises a steeply dipping greenschist-facies mylonite belt and a brittle fault which is situated partly within, and partly at, the southern boundary of the mylonite belt. Towards the west, the brittle Tonale fault probably continues into the Centovalli line. The Insubric line coincides with a significant break in Alpine metamorphic conditions,

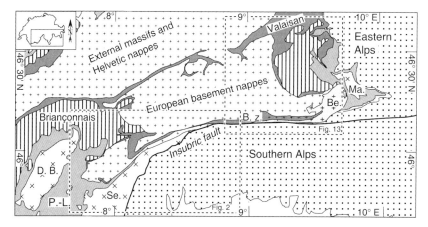

Fig. 1. Overview map of the Central Alps. Map areas of Figures 2 and 13 are marked by the dashed rectangle. D. B. = Dent Blanche nappe; P.–L. = units derived from the Piemont–Ligurian ocean; B. z. = Bellinzona zone; Be. = Bergell pluton; Ma. = Margna nappe; Se. = Sesia zone.

separating, in the Lepontine Alps, the weakly metamorphic to unmetamorphic South Alpine block from the amphibolite-facies Lepontine dome to the north. The Insubric fault is paralleled to the north by the Southern Steep Belt of the Alps (Milnes 1974; Schmid et al. 1987). This structure contains mostly amphibolite-facies mylonites of various Penninic units in steep to overturned orientations.

Movement along the Insubric fault south of the Lepontine Alps is generally interpreted as a combination of south-directed backthrusting with dextral strike-slip (Schmid et al. 1987, 1989). The kinematic interpretation of the Insubric fault, however, still poses some problems. The occurrence of mylonites with steep lineations adjacent to mylonites with shallow lineations (Schmid et al. 1987) indicates either a complex displacement

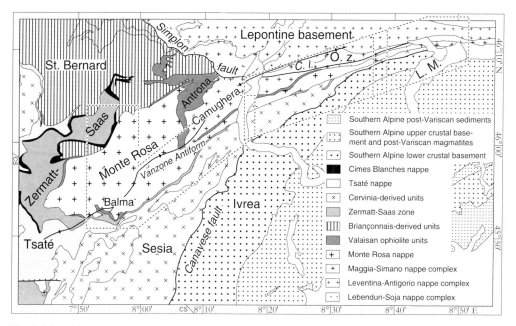

Fig. 2. Map of the Monte Rosa nappe and surrounding units, modified after Steck et al. (1999). The map areas of Figures 9 and 10 are outlined by the dashed lines. O. z. = Orselina zone; C. l. = Centovalli line; L. M. = Lago Maggiore.

path or strain partitioning. Mylonites with a south-side-up vertical displacement component occur locally (Schmid *et al.* 1987; Mancktelow 2003); these do not fit into the scheme of combined backthrusting and dextral strike-slip at all. Reconstruction of deformation in the southern part of the Adula nappe (Nagel *et al.* 2002*a*) suggests that between 35 and 30 Ma, before backthrusting at the Insubric fault, an originally south-dipping normal fault was active in the present-day Southern Steep Belt. In view of the prominent role that the Insubric fault plays in geodynamic reconstructions of the Alps, it is of great importance to better understand the evolution of this fault zone and the Southern Steep Belt.

In this paper, we will present results of structural geological work in the Monte Rosa nappe. The main body of this tectonic unit lies at some distance NW of the Insubric fault. From the main body, however, an elongate 'tail' extends towards ENE into the Southern Steep Belt and progressively approaches the Insubric fault (Fig. 2). We studied Alpine deformation in the Stellihorn shear zone, a shallow-dipping shear zone cross-cutting the main body of the nappe, and in the tail. This allowed us to distinguish between regional deformation of the Penninic nappe stack and Insubric-fault-related deformation. We will show that the tail underwent not only dextral (as expected) but also sinistral shearing related to the Insubric fault. In a next step, we compare the structural evolution of the Monte Rosa nappe with that of the Adula nappe farther east. The results suggest that an important and so far often neglected factor in the kinematics of the Insubric fault is the orogen-parallel extension of the Lepontine dome. The Insubric fault is not only a strike-slip fault and a backthrust, but it is the boundary between a northern block (the Lepontine) which suffered strong and mostly orogen-parallel extension and vertical thinning during the Tertiary, and a southern block (the Southern Alps) where such extension did not take place. This part of the character of the Insubric fault may explain some of its complexities.

Regional geology

The Penninic nappe stack

The Penninic nappe stack (Fig. 3) was imbricated SE-over-NW and was derived from oceanic and continental domains formerly situated north(west) of the Adriatic continent. It formed during south-eastward subduction of this Penninic palaeogeographical domain beneath the Adriatic margin. Consequently, the succession from top to bottom is usually restored in terms of the former palaeogeographical arrangement: From south(east) to north(west), the Piemont–Ligurian (South Penninic) ocean, the Briançonnais microcontinent, the Valais (North Penninic) ocean, and the European margin were progressively subducted (e.g. Schmid *et al.* 1996).

The lowermost part of the Penninic nappe stack is exposed in the core of the Lepontine dome structure in the Central Alps (Fig. 1). Its geometry is determined by rock units dipping moderately away from the core in the eastern and western flanks, the Southern Steep Belt along the Insubric

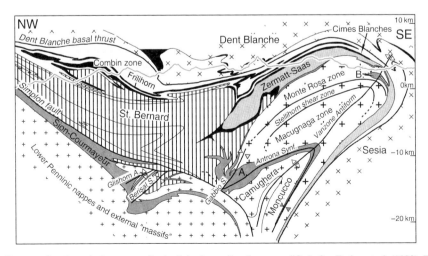

Fig. 3. Cross-section through the western flank of the Lepontine dome, modified after Escher *et al.* (1993). The trace of the cross-section is shown in Figure 2 by lines in the frame labelled 'CS'. For legend, see Figure 2. A. = Antrona zone; B. = Balma unit.

fault, and the Northern Steep Belt parallel to the External massifs (Fig. 1; Milnes 1974). The volumetrically largest part of the Lepontine dome is made up of units derived from subducted European continental crust. These units are fold nappes of metamorphic, pre-Mesozoic basement which are separated by mainly sedimentary units, partly derived from the cover of this basement and partly from the basin fill of the Valais ocean. Lithological contacts were complexly folded during and after formation of the nappes (Milnes 1974; see also Maxelon & Mancktelow 2005 for a recent review). In the eastern flank of the Lepontine dome, the Valaisan units are rooted in the Misox zone and separate the Adula nappe (former most distal European margin: Schmid et al. 1996) below from the Schams and Tambo nappes (Briançonnais sediments and basement) above.

The Simplon fault is commonly taken as the western boundary of the Lepontine dome (Figs 2 and 3). The Simplon fault cuts discordantly through the folded nappe stack. In the north, the Simplon fault is paralleled over some distance by the Valaisan suture marked by flysch-dominated sediments and ophiolitic mélanges of the Sion–Courmayeur zone. Towards the south, this suture continues below the Simplon fault around the Berisal synform in the core of which Briançonnais-derived rocks are contained (Berisal unit). After some distance where it is omitted, the Valaisan suture continues further south in the Antrona unit above the Simplon fault. Finally, it is represented by the Balma unit south and above the Monte Rosa nappe (Figs 2 and 3; Milnes et al. 1981; Froitzheim 2001; Pleuger et al. 2005). In this interpretation, the Monte Rosa nappe is the formerly most distal part of the European margin now exposed in the western flank of the Lepontine dome. Around the Antrona synform, the Monte Rosa nappe is connected with the next lower unit of European origin, the Camughera–Moncucco unit (Milnes et al. 1981; Pleuger et al. 2007). The Monte Rosa nappe is bounded by Valaisan rock units to the NE (Antrona unit) and over a short distance in the south (Balma unit). Note, however, that the palaeogeographical origins of the Monte Rosa and Antrona units are a matter of discussion (for different interpretations, see e.g. Klein 1978; Escher et al. 1993, 1997; Steck et al. 1999; Dal Piaz 1999; Keller & Schmid 2001). To the N and SSW, the Monte Rosa nappe is in direct contact with undisputedly Briançonnais units of the St. Bernard nappe system and Piemont–Ligurian ophiolites of the Zermatt–Saas zone, respectively. Above the Zermatt–Saas zone follows another mostly ophiolitic unit, the Combin zone, which differs lithologically from the Zermatt–Saas zone in containing much larger portions of metasedimentary rocks and in considerably lower Alpine peak pressures (13–18 kbar in the Combin zone, Bousquet et al. 2004; 25–30 kbar in the Zermatt–Saas zone, Bucher et al. 2005). Its palaeogeographical origin was further to the south than that of the Zermatt–Saas nappe, probably even south of the presently uppermost Penninic unit of continental origin—the Sesia–Dent Blanche nappe system (Tsaté basin, Froitzheim et al. 2006; Pleuger et al. 2007). The southern boundary of the Sesia zone largely follows the Canavese fault. Small ophiolitic remnants, which we interpret to represent the Tsaté basin (Froitzheim et al. 2006; Pleuger et al. 2007), occur between the Sesia zone and the Southern Alpine units along the Canavese fault (Elter et al. 1966; Ferrando et al. 2004). However, there is also a large number of different reconstructions concerning the Combin and Sesia zones (e.g. Sartori 1987; Steck 1987; Marthaler & Stampfli 1989; Dal Piaz 1999; Babist et al. 2006).

The Simplon fault does not meet the Insubric fault but curves into parallelism with the latter towards the east (Fig. 2). Keller et al. (2005, 2006) concluded that no discrete continuation of the Simplon fault is present east of Valle d'Ossola. Between the Insubric and Simplon faults, strongly thinned appendices of the Monte Rosa nappe, Zermatt–Saas zone, and Sesia zone can be traced eastward roughly until Locarno, whereas the other units of the Western Alpine nappe stack (Balma unit, Combin zone) wedge out farther to the west. The Camughera–Moncucco unit passes along strike eastwards into the Orselina zone (Fig. 2), and this in turn into the Bellinzona zone (Fig. 1; Composite Lepontine series of Maxelon & Mancktelow 2005). In the Southern Steep Belt, the Sesia zone, Zermatt–Saas zone and Monte Rosa nappe are positioned between the Composite Lepontine series to the north and the brittle Insubric fault to the south. Near Locarno, they disappear below Lago Maggiore and the Quaternary sediments of the Maggia and Magadino valleys. Between Locarno and Bellinzona, the Bellinzona zone is the most internal Penninic unit directly resting against the Insubric brittle fault, whereas east of Bellinzona the Bergell intrusion appears between the Bellinzona zone and the Insubric brittle fault. Geometric modelling by Maxelon & Mancktelow (2005) has shown that the tectonostratigraphic position of the Adula nappe is equivalent or more internal with respect to that of the Bellinzona zone and, hence, comparable to that of the Monte Rosa nappe.

The Monte Rosa nappe

The Monte Rosa nappe consists of continental crustal basement. Paragneisses show in some localities parageneses which testify to a temperature-dominated Variscan metamorphic imprint (Bearth 1952; Dal Piaz 1993) but are largely overprinted by Alpine metamorphism. Granitoid bodies intruded into the metasedimentary rocks during the Late Carboniferous (Hunziker 1970) and/or the Early Permian (Lange et al. 2000) and probably experienced Permian metamorphic equilibration under relatively high temperatures (c. 530 °C, Frey et al. 1976). They have been deformed variably during the Tertiary Alpine orogeny. Metabasic rocks occur as small lenses of eclogite or amphibolite and altogether only make up a small proportion of the nappe. They are almost exclusively contained within the metasedimentary rocks and had probably already been incorporated into these during Variscan orogeny or even in pre-Variscan times.

We follow the concept of Bearth (1952) in subdividing the Monte Rosa nappe into three sub-units. Paragneisses dominate in the western and structurally higher part of the nappe, whereas orthogneisses prevail in the eastern and lower part, including the tail. Bearth (1952) used the terms 'Zone des Monte Rosa' (Monte Rosa zone) and 'Augengneise von Macugnaga' (Macugnaga augengneisses) to distinguish the granitoid rocks of these two subunits. However, by implication Bearth also used these terms to address the subunits in a tectonic sense when he defined a shear zone separating them as the third subunit which comprises both ortho- and paragneisses. This shear zone was named 'Stelli-Zone' by Bearth (1952), but hereafter we will refer to it as Stellihorn shear zone, because it was explicitly named after the Stellihorn mountain. The upper and lower parts of the nappe will be called Monte Rosa zone and Macugnaga zone, respectively (Fig. 3). Whereas Alpine deformation was moderate and unevenly distributed in the Monte Rosa and Macugnaga zones, polyphase shearing was localized in the Stellihorn shear zone. This zone has gradational boundaries with the Monte Rosa and Macugnaga zones and is characterized by thin layers of ortho- and paragneisses extending over large distances. It runs in a north–south direction from the northernmost part of Valsesia through Valle Anzasca (Piemonte, Italy) into the upper Saas valley (Valais, Switzerland). From the upper Saas valley, the Stellihorn shear zone continues towards the NE parallel to the upper Monte Rosa nappe boundary. There, it is less discrete than in Valle Anzasca but seems to form a broader zone of variably intense Alpine deformation. It reaches the contact between the Monte Rosa nappe and the Antrona zone in the Antrona valley at the northeastern tip of the Monte Rosa nappe, where rocks of the Monte Rosa nappe and the Antrona zone are deeply interfingered. In Valle Anzasca, the Stellihorn shear zone is most distinct due to a marked contrast between its mylonites and Variscan paragneisses above and hardly deformed portions of the Macugnaga augengneiss below. South of Macugnaga, the Stellihorn shear zone is exposed in the slopes of the western side of Valle Quarazza. It crosses the hinge of the Vanzone antiform near Colle Turlo and, approximating the southern nappe boundary, grades to a broader zone of Alpine deformation in Valsesia north of Alagna (Fig. 4), Rima and Carcoforo.

The geometry of the Monte Rosa nappe is dominated by the south-vergent Vanzone antiform (Fig. 3). In the main portion of the nappe, i.e. north of the trace of the Vanzone antiform, compositional layering and the main foliation dip moderately to the NW. The southern boundary of the

 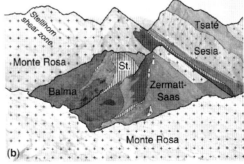

Fig. 4. (a) View over the upper Valsesia towards east. (b) Geological situation of the same view. In the area around Corno d'Olen, the Zermatt–Saas, Stolemberg (St.), Balma, and Monte Rosa units are complexly folded by the Malfatta and Mischabel phases. In the background, the sutures of the Piemont–Ligurian and Valais oceans, interlayered with the Stolemberg unit, are exposed in the western slope of Tagliaferro and trend through the pass between Tagliaferro and Corno Mud. The Stellihorn shear zone, exposed below the summit of Corno Piglimo, approaches the upper boundary of the Monte Rosa nappe. The southward dip of the units is due to their position in the southern limb of the Vanzone antiform, the axial plane trace of which is just outside the view to the left.

main nappe body and the tail-shaped continuation along the Insubric fault (hereafter tail) are within the southern limb of the Vanzone antiform. From west to east, the Vanzone fold axial plane and the southern boundary of the Monte Rosa nappe progressively approach the Insubric fault. The orientation of layering and foliation along the southern nappe boundary changes towards the NE from moderately south-dipping into steeply north-dipping.

Structural analysis of the Monte Rosa nappe

General remarks

Compared to the Monte Rosa and Macugnaga zones, the Stellihorn shear zone and the southern boundary of the Monte Rosa nappe, including its tail, were deformed by relatively strong polyphase shearing and only minor folding. The mylonitic foliation is most often a composite foliation of two or three deformation phases. It is generally parallel with the lithological layering. Mylonitic stretching lineations are marked by preferred orientations of elongate minerals, elongate mineral aggregates or, in orthogneisses, elongate feldspar porphyroclasts with recrystallized tails. In this study, shear senses have been determined from asymmetric porphyroclasts, C'-type shear bands, shape-preferred orientations of dynamically recrystallized quartz, and crystallographic-preferred orientations (CPOs) of quartz.

Since the deformations did not affect much folding in large parts of the shear zones, overprinting relations are often difficult to recognize. In many outcrops, two or three differently oriented stretching lineations on the same flat mylonitic foliation testify to polyphase deformation but cannot be distinguished chronologically. The best opportunities to identify the relative timing of deformation phases are provided by rare occurrences of refolded folds or folded stretching lineations. Within the studied parts of the Monte Rosa nappe, fold axes of the observed deformation phases are almost always (sub)parallel to the stretching lineation of the corresponding deformation phase.

Deformation phases

In the upper Valsesia, the Stellihorn shear zone approaches the southern nappe boundary and its kinematic development can be correlated with polyphase folding and shearing at the contact between the Monte Rosa nappe and overlying units. The following order of deformation phases can be conceived (see also Pleuger et al. 2005): NW-vergent shearing (D_1) was followed by SW-vergent shearing (D_2) and large-scale folding at the nappe boundaries. D_2 in turn was post-dated by another generation of large-scale folds and contemporaneous SE-vergent shearing (D_3). In the following, we will name these deformation phases after 'type localities' where corresponding structures can be observed particularly well and refer to them as Mattmark phase (after Mattmark area in the upper Saas valley; D_1), Malfatta phase (after a mountain in the upper Valsesia; D_2), and Mischabel phase (following Milnes et al. 1981; after a group of mountains between the Saas valley and the valley of Zermatt; D_3). Mattmark-, Malfatta- and Mischabel-phase structures are refolded by the Vanzone antiform (D_4; Vanzone phase; following Milnes et al. 1981). The orientations of the main mylonitic foliation and the stretching lineations of these deformation phases and of the Olino phase (which affected only the tail; see below) are shown in Figure 5.

The relative ages of deformation can be determined also in the Mattmark area in the upper Saas valley. From there towards the NE, the Stellihorn shear zone begins to form a branched pattern of variably strong shear deformation. In some places, shear zones with different displacement direction truncate each other and allow one to establish a similar kinematic sequence as in Valsesia.

The Mattmark phase produced NW- to N-plunging stretching lineation associated with north (west)-vergent shear senses. The next younger structures are related to west-vergent shearing (Malfatta phase; Figs 6b and 7). Generally, the main foliation of mylonites in the Mattmark area is a composite foliation of the Mattmark and Malfatta phases and often carries both stretching lineations. Mischabel-phase shear zones are variably thick but in most cases thin compared to Mattmark- and Malfatta-phase shear zones. They locally generated a new mylonitic foliation at small angles with that of the Mattmark and/or the Malfatta phase. Mischabel-phase stretching lineations trend E–W to NE–SW and yield (north)east-vergent shear senses. From Mattmark towards the NE, the Mischabel phase affected an increasingly larger rock volume and led to folding of the nappe's front (Fig. 3; Gabbio synform; see also Keller & Schmid 2001, their 'Antrona synform'). At Mattmark, there is also some evidence for south-vergent shearing (see also Lacassin 1987) which is often accommodated within shear zones that are only few cm thick. These shear zones are at small angles or parallel to Mattmark- and Malfatta-phase shear zones and clearly younger than the Mattmark and Malfatta phases. Their age relation with the Mischabel phase could not be established. Since this deformation is only weak and locally developed, we will not consider it any further.

Although especially the middle part of the Stellihorn shear zone in the Valle Anzasca and the tail

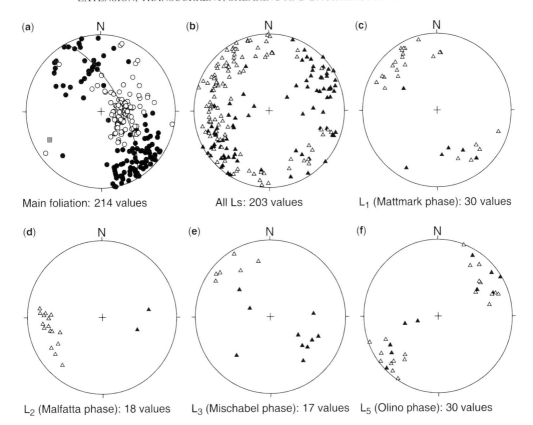

Fig. 5. Structural data from the Stellihorn shear zone and the tail in lower-hemisphere Schmidt projection plots. (**a**) Poles to the main foliation north (white dots) and south (black dots) of the Vanzone fold axial plane. The pole to the great circle through the foliation poles (grey square) is at 242/23 and corresponds to the Vanzone fold axis. (**b**) All stretching lineations from north (white triangles) and south (black triangles) of the Vanzone fold axial plane. (**c**) Mattmark-phase (D_1) stretching lineations from north (white triangles) and south (black triangles) of the Vanzone fold axial plane. (**d**) Malfatta-phase (D_2) stretching lineations from north (white triangles) and south (black triangles) of the Vanzone fold axial plane. (**e**) Mischabel-phase (D_3) stretching lineations from north (white triangles) and south (black triangles) of the Vanzone fold axial plane. (**f**) Stretching lineations of the Olino phase (D_5) related to sinistral (black triangles) and dextral (white triangles) shear senses.

appear homogeneously foliated, the variably oriented stretching lineations are heterogeneously developed. Mischabel-phase structures are increasingly overprinted by younger structures, i.e. folds and mylonites with an intense stretching lineation (Fig. 8), from about Vigino (Fig. 9) in the lower Valle Anzasca to the NE. Keller *et al.* (2005) showed that dextral mylonites in the southern limb of the Vanzone antiform are genetically related to parasitic folds of the Vanzone fold. Therefore, early dextral shearing was contemporaneous with the Vanzone fold. A younger generation of folds exists which post-dates the Vanzone fold. These folds have subvertical axial planes and NNW-dipping enveloping surfaces. They have the wrong vergence for being parasitic folds of the Vanzone fold. Such folds are exposed, e.g., in outcrops in Val di Moneo, west of Rasa (Fig. 10),

and NE above Rio delle Rovine, near Beùra (Fig. 10). There, the main mylonitic foliation carries a gently NE-plunging stretching lineation parallel to the fold axes and associated with dextral shearing. Folds with similar geometries also occur in dextrally sheared mylonites along the northern margin of the Ivrea zone in the area of Corona dei Pinci and Arcegno (Fig. 10). Moreover, they have also been reported by Kruhl & Voll (1976) from a section comprising rocks of Monte Rosa nappe, Sesia zone and the Ivrea zone west of Finero. Shear senses are partly sinistral and partly dextral (Fig. 6c, d). Dextral and sinistral stretching directions are often colinear so that no overprinting relations can be discerned. Since both dextral and sinistral shear senses are related partly to amphibolite-facies and partly to greenschist-facies mineral parageneses, we assume that both

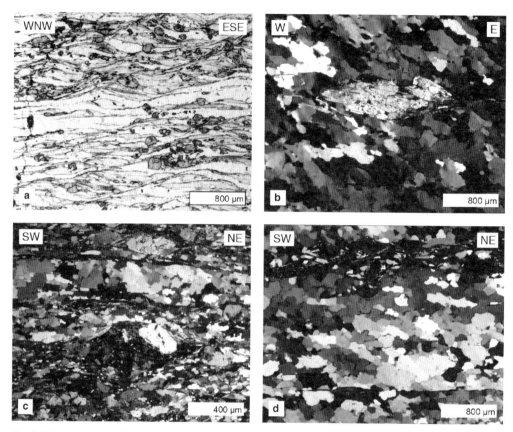

Fig. 6. Photographs of X–Z thin sections of mylonites from the Monte Rosa nappe. Photographs shown in (b), (c) and (d) were taken under crossed polarizers. Sample locations are given in coordinates of the Swiss national coordinate system. (**a**) Detail of MR212 (garnet-bearing gneiss) from the upper boundary of the Monte Rosa nappe in Valsesia (638680/79330). Mattmark-phase C′-type shear bands indicate sinistral shear sense which is WNW-vergent. (**b**) Detail of MR167 (mylonitized quartz vein in whiteschist) from the Stellihorn shear zone at Mattmark (640690/99830). Malfatta-phase sinistral shear sense (west-vergent) is indicated by shape-preferred orientation of quartz grains, an asymmetric kyanite porphyroclast and the CPO shown in Figure 11c. (**c**) Detail of MR78 (alkali feldspar augengneiss) from the tail near Olino/Valle Anzasca (657710/96070). Dextral sense of Olino-phase shearing is indicated by asymmetric feldspar porphyroclasts. (**d**) Detail of MR80 (mylonitized quartz vein in paragneiss) from the tail near Olino/Valle Anzasca (658790/96870). Sinistral sense of Olino-phase shearing is indicated by shape-preferred orientation of quartz grains and the CPO shown in Figure 11j.

shear deformations largely coincided in time or repeatedly replaced each other. We therefore collectively refer to them as Olino-phase structures (after a village in Valle Anzasca; Fig. 9). Early stages of the Olino phase are coeval with the Vanzone phase (dextral shear zones linked to parasitic folds of the Vanzone fold), but late stages post-date the Vanzone phase (younger generation of folds). The Olino-phase stretching lineations (Fig. 8) are generally subhorizontal or plunge gently (Fig. 5f) with azimuths changing between NE and SW along the tail. Steepening, overturning and thinning of the tail are related to an increasing effect of the Olino phase. The Olino phase is therefore partly identical with the Insubric phase of Schmid *et al.* (1987).

However, the Insubric phase according to these authors comprises the entire post-Eocene (i.e. post-34 Ma) shearing along the Insubric fault. This encompasses also pre-Vanzone deformation, i.e. extensional shearing of the Mischabel phase (*c.* 35–30 Ma, see later section on 'Timing of deformation'), subsequently reoriented into a backthrust orientation. In contrast, the term 'Olino phase' is restricted to post-Mischabel deformation.

Representation of structural data

The present study is concerned with three sets of stretching lineations lying within the same main mylonitic foliation in the Stellihorn shear zone.

Fig. 7. Outcrop of Malfatta-phase Stellihorn mylonites at Mattmark (Swiss coordinates 640720/99560). The mylonitic foliation is the axial planar foliation of Malfatta-phase folds as displayed by a deformed granitoid vein in paragneiss. West is to the right. Note hammer for scale.

We assume that in a restored pre-Vanzone-phase situation, the main mylonitic foliation was almost planar (except for some smaller Mattmark-, Malfatta- and Mischabel-phase folds) and that L_1, L_2 and L_3 were more or less straight (see Pleuger et al. 2007). After the Mischabel phase,

Fig. 8. Stretching lineation of the Olino phase defined by tourmaline needles near Olino/Valle Anzasca (659330/97500).

these structural elements were folded around the Vanzone antiform. In Figures 9 and 10, the orientations of the main foliation are mostly averages of about two to ten measured values. They are plotted as surfaces delimited by the strike and the trace of the plane in lower-hemisphere Schmidt projection. Due to folding around the Vanzone antiform and steepening along the Insubric fault, the main foliation dips steeply towards south along the southern nappe boundary and is even overturned along the tail. Plotting stretching lineations in horizontal projection, as a line between the middle of the strike line and the point on the foliation great circle which represents the lineation, would result in a strong distortion of the angle between the lineations wherever the foliation dips steeply (see Pleßmann & Wunderlich 1961). We therefore chose to plot stretching lineations (mostly averages of about two to five measured values) after rotating them into the horizontal about the intersection line of the according foliation with the horizontal. The rotation is performed about the angle $>90°$ or $<90°$ for overturned and normal-lying foliations, respectively. This procedure of a plane projection has earlier been applied by Pleßmann & Wunderlich (1961) and Steck (1990). The greatest advantage of this projection technique is that it conserves the angles between different lines lying in the same plane, e.g. between L_1, L_2 and L_3. A quantitative comparison of the two projection techniques is provided in the Appendix.

In Figures 9 and 10, the orientations of lineation and shear sense from the Stellihorn shear zone and the southern boundary and tail of the Monte Rosa nappe are shown as explained above together with the orientations of the main foliation. The horizontal projections of the stretching lineations show that L_1, L_2 and L_3 with according shear senses also occur in the southern limb of the Vanzone antiform along the southern boundary of the Monte Rosa nappe. Furthermore, it can be seen that lineations related to dextral and sinistral shear of the Olino phase are not everywhere colinear, suggesting that the Olino phase was heterogeneous with the principal axes of strain changing temporally (e.g. west of Rasa and SE of Bèura; Fig. 10).

Neutron texture goniometry of mylonitic quartz veins

Measurement procedure

CPOs of mylonites from the Stellihorn shear zone and the tail were analysed at the neutron texture goniometer SV7-b at Forschungszentrum Jülich (Jansen et al. 2000). The samples were taken from up to a few cm thick foliation-parallel quartz

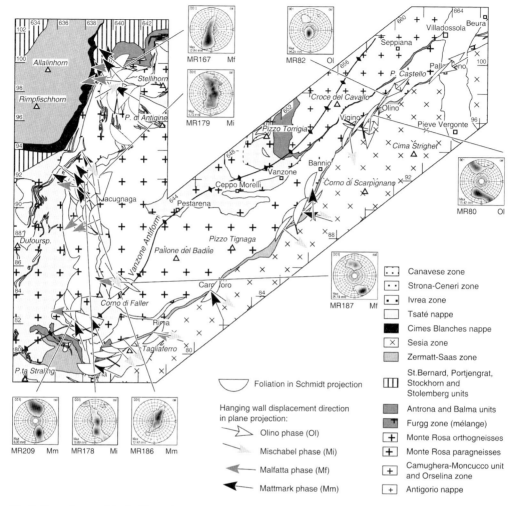

Fig. 9. Tectonic map of the Stellihorn shear zone and the tail of the Monte Rosa nappe west of Valle d'Ossola based on maps of Bearth (1952, 1953, 1954a, b, 1957, 1964), Blumenthal (1952), Mattirolo et al. (1927), Steck et al. (1999), and own observations. The orientation of the main mylonitic foliation is plotted as the surface between the foliation's great circle in lower-hemisphere Schmidt projection and its strike. The hanging wall displacement directions of the Mattmark, Malfatta, Mischabel and Olino phases are plotted after rotation into the horizontal by the procedure described in the sub-section 'Representation of structural data' and the Appendix. Also plotted are the c-axis distributions of mylonitized quartzites in lower-hemisphere Schmidt projection (compare Fig. 11).

veins and were prepared as 1–8 cm³ cubes. The pole figures of {100} (hereafter {m}) and {110} ({a}) are the directly observed ones, whereas the {001} ({c}) pole figures were calculated from the correspondent orientation distribution functions, because {c} is indeterminable in neutron diffraction for quartz.

In the pole figures shown in Figure 11, the traces of foliation trend horizontally, i.e. perpendicular to Z, and the stretching directions are parallel to X. X is generally indicated on the hand specimen by a lineation formed by elongate minerals (e.g. mica) or mineral aggregates (e.g. feldspar). For sample locations, see also Figures 9 and 10.

CPOs from the Stellihorn shear zone. CPOs from the Stellihorn shear zone developed in the Mattmark-, Malfatta- and Mischabel phase (Fig. 11) show with few exceptions incomplete {c} single girdles or point maxima which are elongate along the great circle perpendicular to the dominant {a} maximum. A {c} point maximum close to the edge of the pole figure of sample MR187 (Fig. 11d) indicates basal <a>

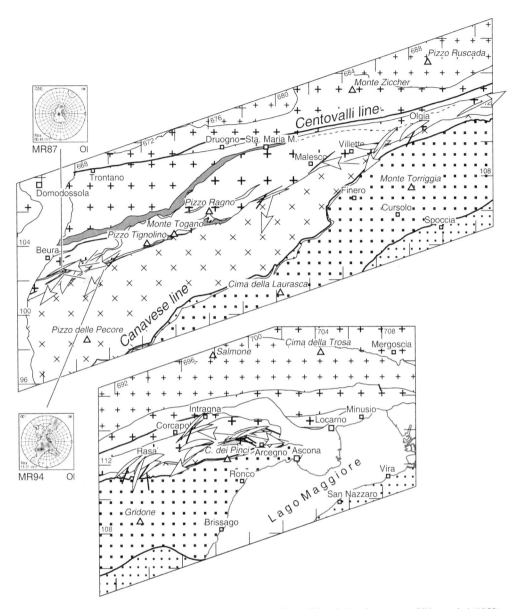

Fig. 10. Tectonic map of the Monte Rosa nappe's tail east of Valle d'Ossola based on maps of Blumenthal (1952), Reinhardt (1966), Schmid et al. (1987), Spicher (1980), Steck et al. (1999), and own observations. The orientation of the main mylonitic foliation is plotted as the surface between the foliation's great circle in lower-hemisphere Schmidt projection and its strike. The hanging wall displacement directions of the Mischabel and Olino phases are plotted after rotation into the horizontal by the procedure described in the sub-section 'Representation of structural data' and the Appendix. Also plotted are the c-axis distributions of mylonitized quartzites in lower-hemisphere Schmidt projection (compare Fig. 11).

slip, while rhomb <a> slip and prism <a> slip lead to {c} maxima between the centre and the edge and in the centre (MR186, MR167; Fig. 11a, c) of the pole figures, respectively. Basal <a> and rhomb <a> slip are the most common slip systems under greenschist-facies metamorphic conditions, whereas prism <a> slip becomes dominant under upper greenschist- to amphibolite-facies conditions (Wilson 1975; Schmid & Casey 1986; Stipp et al. 2002). Petrological data (8 kbar/475 °C, Le Bayon et al. 2006; see also later section on 'Relations between deformation

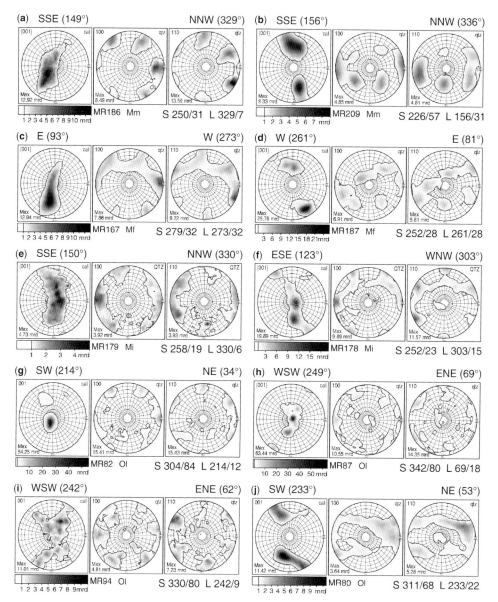

Fig. 11. CPOs of mylonitized quartz veins from the Stellihorn shear zone in lower-hemisphere Schmidt projection. The {100} ({m}) and {110} ({a}) pole figures were directly measured by neutron diffraction; the {001} ({c}) pole figures were calculated from the orientation distribution functions. Below the {a} pole figures, the orientation of the mylonitic foliation (S) and the stretching lineation (L) are given. The background grid of the pole figures shows great circles through Y and small circles around Y in 10° steps. Coordinates of the sample locations refer to the Swiss national coordinate system. (**a**) and (**b**) are interpreted to have formed during the Mattmark phase (labelled 'Ma' behind the sample number), (**c**) and (**d**) during the Malfatta phase ('Mf'), (**e**) and (**f**) during the Mischabel phase ('Mi'), and (**g**)–(**j**) during the Olino phase ('Ol'). (a) Sample MR186 from Valle Anzasca (638730/89510). (b) Sample MR209 from Valsesia (636980/82380). (c) Sample MR167 from Mattmark (640690/99830). The microstructure of this sample is shown in Figure 6b. (d) Sample MR187 from Valle Anzasca (639160/89220). (e) Sample MR179 from Valle Anzasca (639020/92090). (f) Sample MR178 from Valle Anzasca (638080/91800). (g) Sample MR82 from Valle Anzasca (659330/97500). The microstructure of this sample is shown in Figure 12. (h) Sample MR87 from Valle d'Ossola, SE of Beùra (667180/102680). (i) Sample MR94 from the Sesia zone, Valle d'Ossola, E of Beùra (669970/102810). (j) Sample MR80 from Valle Anzasca (658790/96870). The microstructure of this sample is shown in Figure 6d.

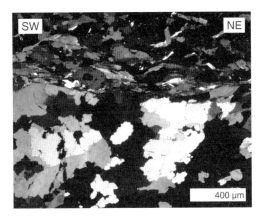

Fig. 12. Photograph of an X–Z thin section of mylonitized quartz vein MR82 from near Olino/Valle Anzasca (659330/97500) under crossed polarizers. Interlobate irregular grain boundaries were formed by grain boundary migration recrystallization under amphibolite-facies conditions during the Olino phase. Compare the CPO shown in Figure 11g whose {c} maximum parallel to Y also suggests temperatures above 500 °C. A weak shape-preferred orientation of the quartz grains indicates dextral sense of shear.

and metamorphism') are only available for MR167, which was sampled from a whiteschist east of the Mattmark reservoir. Its CPO (Fig. 11c) is characterized by a partial {c} girdle close to Y, probably because it formed by prism <a> slip and perhaps additional rhomb <a> slip. For the Mattmark and Malfatta phases, there are examples among our CPOs which show {c} maxima mainly formed by rhomb or basal <a> slip, and others which show {c} maxima formed by dominant prism <a> slip. The CPOs of the Mischabel phase yield incomplete {c} axis girdles resulting from a combination of these three slip systems. This suggests that (1) the three deformation phases occurred under similar thermal conditions, and (2) the activity of slip systems was additionally controlled by other factors, e.g. strain rate.

Almost all our CPOs from the Stellihorn shear zone (Fig. 11) indicate a strong rotational strain component in that the {c} maxima are distributed on great circles inclined to X and Z. Accordingly, most of the {m} and {a} pole figures reveal three more or less sharp single maxima on a great circle perpendicular to the absolute {c} maximum. These asymmetries indicate considerable components of rotational strain for the Mattmark, Malfatta and Mischabel phases. This implies that the Stellihorn shear zone accommodated relative displacement between the Monte Rosa and Macugnaga zones.

CPOs from the Monte Rosa tail. The CPOs from the southern boundary of the Monte Rosa nappe and its tail are very variable. Among them are two with relatively sharp {c} point maxima close to Y (MR82, MR87; Fig. 11g, h) indicating dominant prism <a> slip and therefore amphibolite-facies conditions with temperatures in excess of 500 °C (Stipp *et al.* 2002).

For the samples MR82 (Fig. 12) and MR87, dextral shear senses, in these cases developed during the Olino phase, are identifiable in the thin sections by shape-preferred orientation of quartz. In the case of MR87 (Fig. 11h), a small {c} maximum on the margin of the pole figure *c*. 30° away from Z in a clockwise direction additionally supports a dextral shear sense.

Another CPO (MR94, Fig. 11i) indicates a dextral shear sense by an oblique incomplete {c} cross girdle and the position of the strongest {a} maximum. The sample was taken in the Sesia zone, close to its boundary with the Monte Rosa nappe.

The CPO of a sinistrally sheared mylonite was also obtained. MR80 (Fig. 11j) shows a {c} maximum distributed over half a small circle around Z which we interpret to result from combined basal and rhomb <a> slip.

For all Olino-phase CPOs from the tail, the metamorphic conditions during formation can only roughly be estimated. Amphibolite-facies conditions must be inferred from thin-section observations for all samples discussed here, because they show that biotite, oligoclase and/or staurolite were stable during deformation. This is in line with findings of Stipp *et al.* (2002) that grain boundary migration recrystallization, as observed in the thin sections (Fig. 12), is the main deformation mechanism under amphibolite-facies conditions. An exception is MR80 (Fig. 6d), for which predominant subgrain rotation recrystallization and the stability of chlorite indicate greenschist-facies conditions.

Relations between deformation and metamorphism

Stellihorn shear zone

During the climax of subduction-related burial, the Monte Rosa nappe experienced eclogite-facies metamorphism. Maximum pressure–temperature conditions were determined on various rock types to be *c*. 12–16 kbar/500–550 °C (Chopin & Monié 1984; Dal Piaz & Lombardo 1986; Borghi *et al.* 1996; Engi *et al.* 2001*a*). 24 kbar at 505 ± 30 °C were determined for metapelitic whiteschists from the Monte Rosa zone by Le Bayon *et al.* (2006). However, rocks exhibiting high-pressure mineral assemblages are only sparsely preserved due to an extensive overprint by later stages of metamorphism

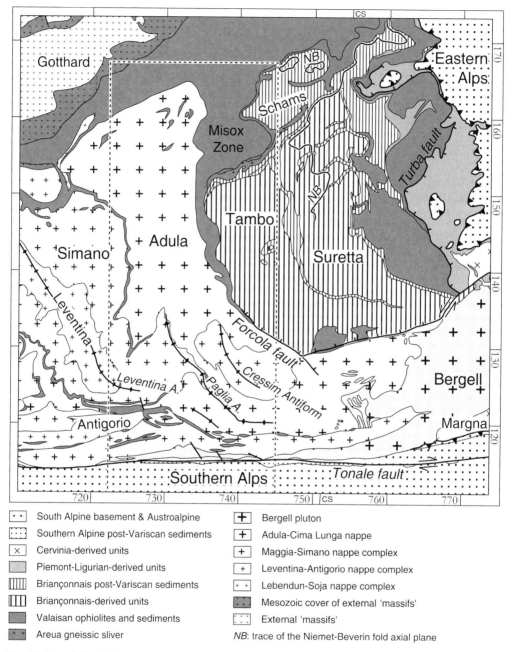

Fig. 13. Map of the Adula nappe and surrounding units in the eastern flank of the Lepontine dome, modified after Spicher (1980). Coordinates in the frame refer to the Swiss national coordinate system. The map area of Figure 15 is marked by the dashed rectangle.

or because not all of the rocks were equilibrated under eclogite-facies conditions. In the Monte Rosa zone, the first part of decompression was related to moderate cooling below 450 °C, while the second part was accompanied by reheating to c. 500 °C (Borghi et al. 1996). We observed inversely zoned plagioclase, i.e. with albitic cores and oligoclase rims (see also Bearth 1958), which confirms a temperature rise during late stages of the metamorphic history.

The maximum temperatures related to this reheating increase slightly with structural depth, i.e. roughly from west to east (Bearth 1958; Wetzel 1972). Within the Monte Rosa nappe, the albite/oligoclase isograd was mapped by Bearth (1958). This isograd, approximately representing the greenschist/amphibolite-facies boundary, roughly coincides with the Stellihorn shear zone in the south. Further north, it acquires a northeastern trend and therefore attains a structural position just below the Stellihorn shear zone. This is in line with our thin-section observations showing that in most cases albite was stable during deformation in the Stellihorn shear zone. Some samples, mostly from the southern part of the shear zone, exhibit inversely zoned plagioclase. Oligoclase growth is not specifically coupled with the Mattmark, Malfatta or Mischabel phases, but it may or may not be present in the samples deformed dominantly by one of these deformation phases. We therefore infer that oligoclase probably grew during a slight late-stage reheating (Bearth 1958; Borghi et al. 1996) that post-dated these deformation phases. Indication that temperatures during the Mattmark to Mischabel phases cannot have been much colder than during oligoclase formation comes from the observation that, in deformed quartz veins, grain boundary migration recrystallization was dominant over subgrain rotation recrystallization. Stipp et al. (2002) have shown that the latter takes over from the former below c. 500 °C. On the other hand, syntectonically grown chlorite-replacing biotite is widely present in the Stellihorn shear zone in gneisses deformed during the Mattmark, Malfatta and Mischabel phase, and suggests that metamorphism during these deformation phases did not exceed the greenschist facies. Petrological pressure–temperature estimations that can be directly connected to a specific part of the retrograde Mattmark-to Mischabel-phase deformation history are sparse. In a kyanite–talc-bearing whiteschist from the east shore of the Mattmark reservoir (see sample location of MR167 in Fig. 9), growth of chlorite and white mica at the expense of the peak-pressure paragenesis took place during retrogression down to c. 8 kbar/475 °C (Le Bayon et al. 2006) and was synkinematic with the Malfatta phase (compare Fig. 6b). Bousquet et al. (2004) related the formation of the above-mentioned albite/oligoclase isograd to a Barrovian metamorphic stage that affected the Penninic nappe stack to variable degree (e.g. Engi et al. 2004).

Monte Rosa tail

Temperatures derived by thermobarometry and reaction equilibria (see Engi et al. 2004, and references therein) reveal that maximum temperatures in the southern Lepontine dome gradually decrease towards the west, north and east, but are truncated by the Insubric fault in the south. Along the tail of the Monte Rosa nappe, temperatures increase from c. 500 °C at the southern end of the Stellihorn shear zone to c. 650 °C near Locarno. The strong overprint of the Monte Rosa tail by Olino-phase mylonites under amphibolite-facies conditions fits well with results of Burri et al. (2005) who showed that, while the above-mentioned thermal peak led to *in-situ* migmatization in eastern parts of the Monte Rosa tail, the Sesia zone is completely devoid of such migmatites. Therefore, these authors postulated a shear zone north of the Sesia zone which effected considerable uplift of the Lepontine units north of the Sesia zone and which we assume to belong to the Olino phase. Many of the Olino-phase mylonites from the Monte Rosa tail continued to be sheared after the temperature peak under greenschist-facies conditions, often with chlorite replacing biotite during deformation. Pervasive greenschist-facies Insubric mylonitization is restricted to a hundreds of metres-thick shear zone comprising rocks of the Sesia zone, Canavese metasediments and Ivrea zone (Schmid et al. 1987). During these late stages of shearing, a N–S metamorphic gradient existed in the Sesia zone east of Valle d'Ossola (Reinhardt 1966; Altenberger et al. 1987). Kruhl & Voll (1976) found that successive generations of upright south-vergent folds in the Sesia zone west of Finero initially formed under rising temperatures within the lower amphibolite facies, as evidenced by inversely zoned plagioclase. Therefore, it appears that the Olino phase initially took place under rising temperatures. However, final steps of Olino-phase deformation took place in the greenschist facies. In the Southern Steep Belt outside the greenschist-facies Insubric mylonite zone, the retrograde overprint does not follow any regular pattern but seems to be the result of heterogeneous deformation that leaves unaffected large rock volumes that were previously sheared under amphibolite-facies conditions.

Timing of deformation

Stellihorn shear zone

Radiometric age data for the above-described shearing and folding stages in the Monte Rosa nappe are scarce. Therefore, the timing of the Mattmark to Mischabel phases and later movements within the tail can only be inferred from interpolation of the few existing data including those of neighbouring units. U/Pb-SHRIMP ages of 40.4 ± 0.7 Ma (Liati & Froitzheim 2006) and 38.5 ± 0.7 Ma (Liati et al. 2005) on metamorphic rims of zircon from eclogites and amphibolites of the Balma and Antrona units have been interpreted as to date the pressure climax within units derived from the Valais ocean. Lapen et al. (2006) dated eclogite-facies metamorphism of the Monte Rosa nappe by U/Pb isochrons on rutile yielding 42.6 ± 0.6 Ma. These data provide an upper limit of c. 40 Ma for the onset of the retrograde history of the Monte Rosa nappe, i.e. the Mattmark phase.

The Vanzone antiform post-dates the Mattmark to Mischabel phases and therefore its formation sets the minimum age for the Mischabel phase. On the one hand, the Vanzone antiform deforms the albite/oligoclase mineral isograd within the Monte Rosa nappe which probably developed during the temperature climax within the Southern Steep Belt c. 28 Ma ago (Engi et al. 2004). On the other hand, its southern limb is deformed by high-grade stages of the Olino phase. Dextral shearing of the Olino phase deformed a 29 Ma-year-old pegmatitic dyke within the tail at Malesco under amphibolite-facies conditions (Schärer et al. 1996). We therefore conclude that the Vanzone antiform also formed c. 28–29 Ma ago and that the Mattmark, Malfatta and Mischabel phases took place before that. This conclusion is broadly in line with the results of $^{40}Ar/^{39}Ar$ dating on hydrothermal muscovite from gold-bearing quartz veins which post-date the ductile deformation (Pettke et al. 1999). Southwest of the Simplon fault, these ages range between c. 24.5 and c. 32.4 Ma for veins which are situated along the fold axial plane of the Vanzone antiform.

Monte Rosa tail, Insubric fault and Simplon fault

Accordingly, the 29.2 ± 0.2 Ma age of Schärer et al. (1996) is also the maximum age for the Olino phase within the tail. This age bracket does not exclude that analogous movements may have commenced earlier in higher structural levels, i.e. closer to the Insubric fault, or at the Insubric fault itself. Approximately 26 Ma-year-old pegmatitic dykes were postkinematically emplaced into rocks of the Monte Rosa nappe tail (at Palagnedra) and the Bosco–Isorno–Orselina zone (at Corcapolo; Fig. 10) that had already cooled below 500 °C (Schärer et al. 1996). However, these authors also state that ductile deformation under lower metamorphic conditions still continued after emplacement of these dykes. Greenschist-facies mylonitization was more localized along the Insubric fault itself and led to differential uplift and cooling of the Penninic units as well as nearly contemporaneous dextral strike-slip shearing (Schmid et al. 1987). In a combined study of apatite and zircon fission track and K/Ar and Rb/Sr ages on biotite and muscovite of samples from the Valle Maggia, Hurford (1986) constrained especially rapid cooling of the Central Alps to have started 23 Ma ago. He assumes that rapid cooling was caused by accelerated uplift from c. 27 Ma onward. Schmid et al. (1989) and Steck & Hunziker (1994) showed by comparison of data that cooling ages for each of these radiometric systems decrease progressively from the eastern to the western Central Alps, implying that uplift was diachronous along the Insubric fault. These movements are probably directly dated by K/Ar formation ages of 26 Ma to 19 Ma obtained from synkinematic white mica by Zingg & Hunziker (1990) on greenschist-facies mylonites of the Insubric fault immediately south of the Insubric brittle fault between Locarno and the Valle d'Ossola. SW-directed extensional shearing in footwall mylonites of the Simplon fault (Mancktelow 1985, 1992) accompanied north-side-up shearing at the Insubric fault from c. 18–19 Ma onwards, as reconstructed from cooling curves across the Simplon fault by Mancktelow (1992) and Grasemann & Mancktelow (1993). Widespread brittle deformation in the hanging wall of the Simplon fault started in the Late Oligocene (Bistacchi et al. 2001). Units in the hanging wall, including the main body of the Monte Rosa nappe, had already been exhumed to c. 300 °C, i.e. the transition from ductile to brittle behaviour of quartz, at c. 25 Ma–30 Ma (see compilation of cooling ages in Mancktelow 1992, his Fig. 14). In the Simplon footwall mylonites, ductile shearing continued until c. 15 Ma and accommodated a total vertical displacement of the footwall of c. 12.5 km (Grasemann & Mancktelow 1993). Contemporaneously with WSW-directed normal faulting, the SSE-vergent Berisal synform and Glishorn antiform pair (Fig. 3) deformed older portions of the footwall mylonites, indicating that orogen-parallel extension was concurrent with orogen-perpendicular shortening (Mancktelow 1992). From 15 Ma on, brittle normal faulting along the Simplon fault effected c. 2.5 km more vertical displacement while the vertical displacement rate was strongly reduced from c. 4.2 mm/a to 0.2 mm/a (Grasemann & Mancktelow 1993). We therefore assume that brittle faulting at the Canavese, Tonale and Centovalli lines did not occur before 15 Ma ago either.

Architecture and evolution in the eastern Lepontine area

We now briefly review the structural architecture of the eastern flank of the Lepontine dome (Fig. 13) in order to compare it with the western flank. We focus on the Adula nappe, which is similar in some ways to

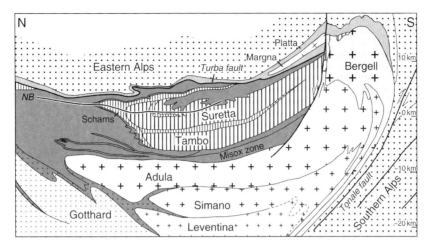

Fig. 14. Cross-section through the eastern flank of the Lepontine dome, modified after Schmid *et al.* (1996). The trace of the cross-section is shown in Figure 13 by lines in the frame labelled 'CS'. For legend see Figure 13.

the Monte Rose nappe. There is a broad consensus that the Adula nappe represents the distal part of the European margin towards the Valais ocean and was subducted southwards beneath the Briançonnias microcontinent in the Eocene (e.g. Schmid *et al.* 1996, 1997). In previous studies, we proposed that the Monte Rosa nappe occupied the same palaeogeographical position as the Adula nappe, i.e. north of the Valais ocean (Froitzheim 2001; Pleuger *et al.* 2005). In terms of lithology and Alpine metamorphic evolution, there are strong similarities between the Adula nappe and the Monte Rosa nappe, as was noted by Bearth (1952).

In contrast to the Monte Rosa nappe, the Adula nappe is not a coherent basement unit but a highly deformed assemblage of various pre-Mesozoic ortho- and paragenic gneisses, partly eclogitic amphibolites, scarce lenses of ultramafics and layers of potentially Mesozoic marbles. It is often described as a 'lithospheric melange' (Trommsdorff 1990) welded together in a subduction channel (e.g. Engi *et al.* 2001*b*). The Adula nappe is sandwiched between the Misox zone above and the Simano nappe below (Fig. 14). The Simano nappe is a classic basement-cover nappe and generally attributed to the European continental margin. The Misox zone mostly consists of Mesozoic sediments and MORB-type metabasalts derived from the Valais basin (Steinmann & Stille 1999). The main body of the Adula nappe dips gently towards the east. North of the Insubric fault, the entire nappe stack is deformed by a series of kilometre-scale backfolds into a steeply north-dipping orientation in the Southern Steep Belt (Fig. 14).

The Adula nappe experienced Eocene high-pressure metamorphism with peak pressures increasing from north to south (Heinrich 1986; Dale & Holland 2003). Maximum pressure–temperature conditions determined on eclogites increased from *c.* 12 kbar/500 °C in the north (Löw 1987) to 25 kbar/750 °C in the south (Dale & Holland 2003). High-pressure parageneses are generally only preserved in the cores of mafic and ultramafic boudins which are typically embedded in garnet–mica schists. The country rocks and the rims of the boudins record a subsequent Barrovian-type metamorphism that reaches upper greenschist-facies conditions in the north and upper amphibolite-facies conditions in the south (Niggli & Niggli 1965; Trommsdorff 1966; Wenk 1970; Frey *et al.* 1974). This Barrovian stage was reached during near-isothermal decompression from peak-pressure conditions (Nagel *et al.* 2002*b*). Very likely, the mafic boudins recording peak pressures are not relicts of a Mesozoic oceanic crust but were part of the continental pre-Mesozoic basement (Santini 1992). For the eclogite-facies metamorphic stage, ages between *c.* 43.9 Ma and *c.* 37.5 Ma were obtained by Sm/Nd dating on peridotites and an eclogite by Becker (1993). In the southern Adula nappe, conditions around 9 kbar/700 °C and 5 kbar/700 °C were reached by 32 Ma and 25 Ma, respectively, based on structural observations (Nagel *et al.* 2002*a*).

In the northern and central Adula nappe, two main deformation phases post-dated peak-pressure conditions (Löw 1987; Meyre 1998; Partzsch 1998; Pleuger *et al.* 2003): The first, very pronounced main foliation S_1, related to D_1 (Zapport phase), is parallel to the lithological layering and associated with north-vergent shearing (Fig. 15). At the end of the Zapport phase, metamorphic conditions reached *c.* 8 kbar/500 °C in the north (Löw 1987) and *c.* 12 kbar/600–650 °C in the

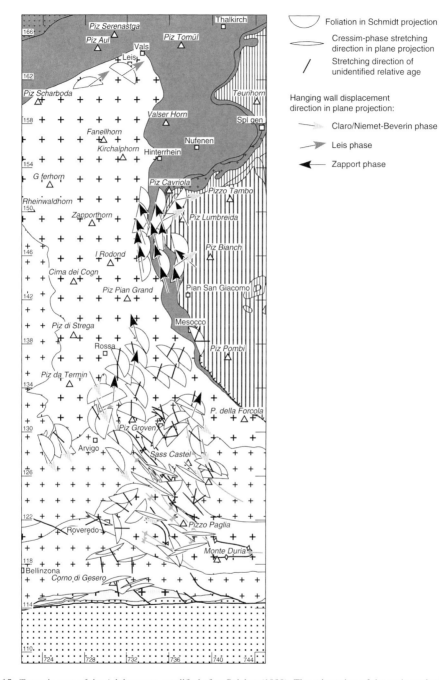

Fig. 15. Tectonic map of the Adula nappe, modified after Spicher (1980). The orientation of the main mylonitic foliation is plotted as the surface between the foliation's great circle in lower-hemisphere Schmidt projection and its strike. The hanging wall displacement directions of the Zapport, Leis and Niemet–Beverin phases and the stretching direction of the Cressim phase are plotted after rotation into the horizontal by the procedure described in the sub-section 'Representation of structural data' and the Appendix. Structural data are partly taken from Babinka (2001), Hundenborn (2001) and Kremer (2001).

south (Nagel et al. 2002a). S$_1$ is overprinted by N- to NNW-vergent folds of D$_2$ (Leis phase). These folds develop a new, ENE–WSW-striking stretching lineation parallel to fold axes in the northeastern part of the nappe (Löw 1987) with top-to-the-ENE shear sense (Fig. 15). The intensity of deformation related to the Leis phase decreases southward (Meyre et al. 1998; Pleuger et al. 2003).

Above the Adula nappe, the Misox zone (Valaisan) and the Tambo and Suretta nappes (Briançonnais) were also strongly affected by NNW-vergent shearing during initial stages of exhumation. These movements (Ferrera phase, Milnes & Schmutz 1978) probably started earlier than the Zapport phase in the Adula nappe but ceased at the same time. The subsequent Niemet–Beverin phase (Milnes & Schmutz 1978; Schreurs 1993, 1995) is characterized by top-to-the-ENE to top-to-the-SE shearing (Fig. 15). Simultaneously with the shearing, the large-scale north-closing Niemet–Beverin fold developed in front of the Tambo and Suretta nappes (Fig. 14). The lower limb contains abundant meso- and small-scale south-vergent parasitic folds of that structure with fold axes parallel to the stretching lineation. The upper boundary of Niemet–Beverin-phase deformation is the Turba mylonite zone (Nievergelt et al. 1996; Marquer et al. 1996; Weh & Froitzheim 2001; Figs 13 and 14) forming the base of the Piemont–Ligurian Platta, Lizun and Forno ophiolitic units. The extensional top-to-the-east Turba mylonite zone is capped by a brittle extensional fault which truncates the hanging wall (Nievergelt et al. 1996).

The Niemet–Beverin phase pre-dates the 30 Ma-old granodiorite of the Bergell pluton (Gulson 1973; von Blanckenburg 1992). It is not clear how the Leis phase is temporally related to the Niemet–Beverin phase which also post-dates the Zapport phase (Nagel et al. 2002a; Pleuger et al. 2002, 2003). This uncertainty is mostly due to the fact that structures of the Leis phase and the Niemet–Beverin phase have almost no spatial overlap, at least within the Adula nappe. Schmid et al. (1996) assumed that the Leis-phase structures in the north formed contemporaneously with Cressim-phase folding at the southern end of the nappe, but this assumption is arbitrary and cannot be confirmed by overprinting relations. In middle parts of the Adula nappe, Leis-phase folds are genetically related to localized NNW-vergent shearing (Pleuger et al. 2003). Leis-phase NW-vergent shearing was also proven by Partzsch (1998) for the lower part of the Adula nappe. In frontal parts of the Adula nappe, shortening perpendicular to the upright Leis fold axial planes and fold-axis-parallel WSW-ENE extension may have accommodated the moderate Leis-phase NNW-vergent shearing further south. Therefore, it is likely that the Leis phase comprises the switch from orogen-perpendicular transport of the Zapport phase to orogen-parallel displacement. Towards the south, the main foliation S$_1$ is progressively overprinted by two subsequent deformation phases, local D$_2$ (Claro phase) and local D$_3$ (Cressim phase). The D$_2$ phase is generally correlated with the Niemet–Beverin phase (Rütti 2001; Nagel et al. 2002a; Nagel 2008) and associated with narrow, SW-vergent folds, the formation of a new axial plane foliation S$_2$ and a shear sense top-to-the-SE (Grond et al. 1995; Nagel et al. 2002a). D$_2$ folds clearly overprint the established upper and lower contacts of the Adula nappe. This deformation was associated with isothermal decompression but pressures remained above 8 kbar throughout D$_2$ in the southern Adula nappe (Nagel et al. 2002a).

Open to tight folds of the Cressim phase intervene between gently dipping structures in the main body of the Adula nappe and the Southern Steep Belt. Immediately north of the Southern Steep Belt, Cressim-phase folds strike E–W but individual large-scale folds turn away from the steep belt in a clockwise sense and continue northwestward (Fig. 13). Thus, in map view Cressim-phase folds form an en echelon sequence of NW–SE-striking folds fanning away from the Southern Steep Belt. This observation suggests bending of the Cressim-phase folds by sinistral movements in the Southern Steep Belt. Where Cressim-phase backfolds are tight, they display a pronounced stretching lineation parallel to fold axes which developed in the upper amphibolite facies, locally under migmatic conditions (Nagel et al. 2002a). Cressim-phase folds affect the 30 Ma-old Bergell granodiorite (Davidson et al. 1996) but are cut by 25 Ma-old dykes (Gebauer 1996) as well as by the smaller Novate intrusion of approximately the same age (Liati et al. 2000). Maxelon & Mancktelow (2005) observed a further generation of backfolds (D$_4$, corresponding to the Vanzone phase) which they place in the same time span as the Cressim phase, i.e. 30 to 25 Ma. These folds are relatively open, large-scale folds with NE–SW-trending axial planes. In the south, the upper contact of the Adula nappe is defined by the Forcola normal fault that cuts out the Misox zone towards south (Fig. 13). The Forcola fault was active c. 25 Ma ago (Ciancaleoni & Marquer 2006) and has a vertical offset of c. 3 km (Meyre et al. 1998).

The Southern Steep Belt is the southern limb of Cressim-phase backfolds north of the Insubric fault and represents a ductile high strain zone along which the northern block has been uplifted tens of kilometres with respect to the Southern Alps. South of the Adula nappe, the mylonitic foliation dips steeply (Fig. 15). The well-developed stretching lineation has an almost down-dip orientation but locally displays a slight pitch towards E or SE. The shear sense in the mylonites is consistently north-side-up in present-day geometry. Some authors (e.g. Milnes 1974; Nagel et al. 2002a) have suggested that the main foliation on the northern limb of the Cressim-phase backfolds, a composite S_1/S_2 foliation, is simply folded around Cressim-phase folds into the subvertical orientation in the Southern Steep Belt. This would imply that north-side-up-directed shearing developed in a south-dipping shear zone (see Fig. 11 in Nagel 2008) and is of the same age as the Niemet–Beverin phase. However, the mylonites show synkinematic metamorphic conditions similar to those recorded during Cressim-phase folding and a considerable portion of the shearing probably occurred during Cressim-phase backfolding in an already steep orientation (Berger et al. 1996). We assume that the orientation of the shear zone was steep but not yet north-dipping. Dextral shearing in the Southern Steep Belt at lower, i.e. greenschist-facies, conditions (Heitzmann 1975, 1987) post-dates the formation of the steep amphibolite-facies mylonitic lineation.

Comparison of the kinematic histories of the Monte Rosa and Adula regions

In both the Monte Rosa and Adula nappes, the oldest preserved, ubiquitous structures developed during exhumation after the subduction-related pressure peak. In both cases, NNW-vergent shearing accommodated the transition from eclogite-facies to greenschist- and amphibolite-facies metamorphism. Mattmark- and Zapport-phase NNW-vergent shearing affected not only the Monte Rosa and Adula nappes but also the entire Penninic part of the overlying nappe stack. The upper age brackets for Mattmark- and Zapport-phase deformations are c. 40 Ma (Liati & Froitzheim 2006; Lapen et al. 2007; Becker 1993).

Given that the Zapport phase in the Adula nappe started at c. 40 Ma, the Niemet–Beverin phase may have commenced c. 35 Ma ago. It ceased before emplacement of the Bergell granodiorite 30 Ma ago (Fig. 16). Since the Niemet–Beverin phase in the Adula nappe and the Mischabel phase in the Monte Rosa nappe are kinematically consistent (Figs. 9, 10 and 15), they may also widely overlap in time. This interpretation is in line with the requirement that the Mischabel phase directly preceded the formation of the Vanzone fold c. 28–29 Ma ago. The Mattmark and Malfatta phases would then have taken place between the pressure peak of the Monte Rosa nappe (40 Ma or slightly younger) and c. 35 Ma. There is no kinematic equivalent of the Malfatta phase in the Adula region. However, for reasons explained in the previous section, we assume that the Leis phase directly succeeded the Zapport

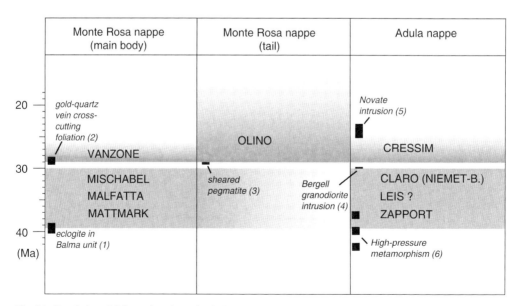

Fig. 16. Correlation of deformation phases in the Monte Rosa and Adula nappes. Sources of time constraints: (1) Liati & Froitzheim (2006); (2) Pettke et al. (1999); (3) Schärer et al. (1996); (4) von Blanckenburg (1992); (5) Liati et al. (2000); (6) Becker (1993). The relative timing of the Leis phase in the Adula nappe is uncertain; it may be older (as assumed here) or younger than the Niemet–Beverin phase.

phase and was broadly coeval with the Malfatta phase. The formation of the Vanzone antiform at c. 28–29 Ma (see discussion in section 'Timing of deformation') post-dates the Mischabel phase and is coeval with or slightly younger than formation of the above-mentioned Cressim-phase folds (see also Maxelon & Mancktelow 2005). It is roughly contemporaneous with the onset of mostly dextral, but also sinistral Olino-phase shearing in its southern limb under amphibolite-facies conditions (Fig. 16).

Kinematic synthesis for the margins of the Lepontine dome

Pre-Vanzone deformation and steepening of the Southern Steep Belt

We have demonstrated that structures related to late stages of nappe stacking (Mattmark/Zapport phase), and exhumation by early orogen-parallel extension (Malfatta phase and Leis phase; Fig. 17b) and SE-directed normal faulting (Mischabel/Niemet–Beverin phase; Fig. 17c) developed prior to formation of the Vanzone antiform and the Southern Steep Belt (Fig. 17d). On the west side of the Lepontine dome, orogen-parallel extension started around 35 Ma ago with the pre-Simplon SW-vergent shearing (Malfatta phase) (Fig. 17b). Ongoing NW–SE convergence during the Malfatta phase is testified by SE-vergent folding with axes parallel to the stretching directions. On the east side of the Lepontine dome, orogen-parallel extension (with opposite displacement direction) and NNW–SSE convergence were accommodated by the Leis phase which we assume to be coeval with the Malfatta phase (see above). At the onset of the Mischabel and Niemet–Beverin phases, the kinematic regime changed to NW–SE extension, perpendicular to oblique to the orogen, by SE-vergent normal faulting which resulted in further exhumation of the Lepontine units.

In the Stellihorn shear zone, structures of the Mattmark to Mischabel phases can be traced around the hinge of the Vanzone antiform into the southern limb of this fold and finally into the Southern Steep Belt (Fig. 9). In a similar way, structures of the Zapport and Niemet–Beverin phases in the Adula nappe are bent around the Cressim antiform into the Southern Steep Belt (Fig. 15). The plane projection technique applied in Figures 9, 10 and 15 is well suited to visualize the approximate original orientations of Mattmark- to Mischabel-phase stretching directions and shear senses. These results are well in line with findings from west of Valle d'Ossola by Steck (1984), who reports structures related to nappe stacking from the contact between the Monte Rosa and Sesia nappes, and Schmid et al. (1989), who found thrust-related structures at the Insubric fault.

The Mattmark to Mischabel and Zapport to Niemet–Beverin deformation phases were related to shearing (sub)parallel to the nappe contacts and generally affected the entire N–S extent of the Penninic nappe pile. When they took place, the Southern Steep Belt did not yet exist but the Penninic nappes dipped more or less evenly to the south. Progressive deformation during retrograde metamorphism led to subvertical shortening of the involved units as well as finally their overturning and localization of deformation in the Southern Steep Belt. A fundamental difference between earlier deformation phases and the Olino phase is that the latter was restricted to southern parts of the Penninic nappe pile in a shear zone that cut the nappe boundaries at a high angle. Therefore, as already pointed out by Milnes (1974), any genetical model assuming temporal coincidence of nappe stacking and formation of the Southern Steep Belt has to be rejected. While the view of the Southern Steep Belt as a root zone from which the Penninic nappes were squeezed out has been abandoned for a long time, models that assume backthrusting at the Insubric fault simultaneously with Penninic nappe formation in front of a South Alpine rigid backstop have recently been published (e.g. Pfiffner et al. 2000). Considering the structural relations of the Olino phase and earlier deformation, however, backthrusting only started when all but the lowermost part of the Penninic nappe stack had already been formed. In terms of dynamics, north-side-up movements along the Insubric fault are probably best explained as a response to blockage of the subduction zone due to incorporation of more and more proximal parts of European continental crust (see e.g. Escher & Beaumont 1997). The transition from Mischabel-phase extensional shearing to steepening in the southern limb of the Vanzone antiform and finally its overturning (Fig. 17e) suggests that north-side-up movements were not primarily accommodated by backthrusting. The small-scale upright folds of the Olino phase with north-dipping enveloping surfaces indicate a synform to the north; however, there is no such synform post-dating the Vanzone antiform. Therefore, in the kinematic context of movements with a north-side-up component and assuming that the southern limb of the Vanzone antiform was originally south-dipping, the X–Y plane of amphibolite-facies deformation parallel to the Insubric fault must have dipped less steeply

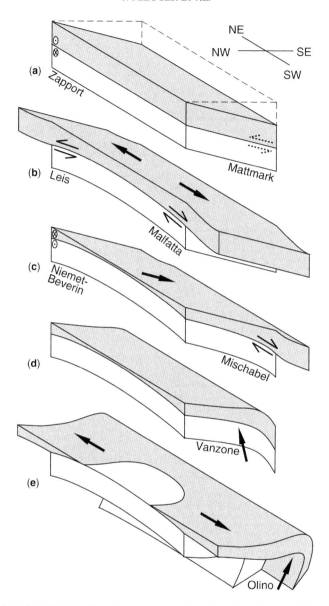

Fig. 17. Strongly schematized diagrams illustrating the exhumation of the Central Alps. (**a**) Situation after the Mattmark/Zapport phase. (**b**) Orogen-parallel stretching of the Malfatta and Leis phases in the western and eastern flank of the Lepontine dome, respectively, accommodate a first stage of unroofing. (**c**) Mischabel-/Niemet–Beverin-phase SE-vergent shearing effects further exhumation by normal faulting. (**d**) Uplift of the nappe stack in the core of the Vanzone antiform. (**e**) Final orogen-parallel stretching and overturning of the Southern Steep Belt are accommodated by Olino-phase shearing in the Southern Steep Belt while extensional shearing, followed by brittle normal-faulting, in the flat-lying part of the nappe stack accommodates further unroofing.

towards the SE than the lithological layering. Looking towards east, this relation between shear zone boundary and layering implies a dextral rotation of the lithological layering (or the fold envelopes) which will eventually be overturned. Thereby, backthrusting rather evolved from steepening and finally overturning of an originally south-dipping extensional shear zone (Fig. 18). The overprint by the Olino phase also led to tightening of the Vanzone antiform.

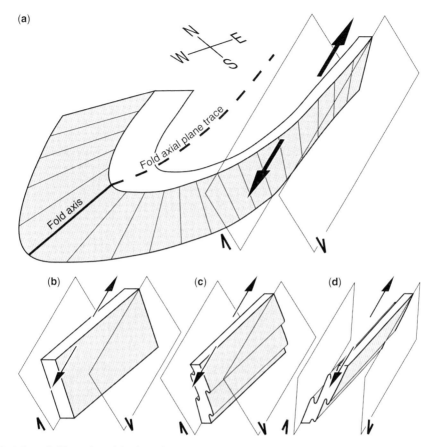

Fig. 18. Schematic illustration of the formation and overturning of Olino-phase folds in the southern limb of the Vanzone antiform. (**a**) Vanzone antiform with the Olino-phase shear zone in the southern limb. (**b**) Steeply south-dipping layering in the southern limb of the Vanzone antiform before onset of Olino-phase folding. (**c**) Olino-phase folding started when the layering in the Southern Steep Belt was south-dipping. (**d**) At the end of the Olino phase, the Southern Steep Belt was overturned into a north-dipping orientation.

Exhumation of the Lepontine dome and lateral movements between the Lepontine and the Southern Alps

Post-Mattmark/Zapport-phase deformation in the Central Alps not only accommodated exhumation of the Penninic nappes but had also a strong lateral component. Estimates of the amount of lateral displacement are controversial. For post-Eocene times, they range from c. 150 km, reconstructed from the offset of large-scale tectonic features by Laubscher (1991), to 30–40 km based on kinematic analyses of the cogenetic Giudicarie fault by Viola et al. (2001) and Müller et al. (2001).

While the Southern Alps did not experience significant orogen-parallel stretching at that time (Handy et al. 2005), ductile deformation in the Penninic nappe stack led to considerable stretching of the present-day Central Alps. With time, extensional shearing of the Malfatta and Niemet–Beverin phases was replaced by shearing in structurally deeper levels, i.e. at the Simplon fault (Mancktelow 1985) and at the Forcola fault (Meyre et al. 1998) which acted with opposite, outward transport directions in the Late Oligocene and Miocene, respectively. Movements at the Simplon and Forcola faults took place during the Olino phase and after formation of the Southern Steep Belt (Fig. 17e). Handy et al. (2005) related orogen-parallel extension in the Central Alps to backthrusting along the frontal surface of the rigid Southern Alps ('Alpine indenter'). However, the kinematics of the Mischabel and Niemet–Beverin phases cannot be reconciled with such an indentation process. The earliest possible onset of backthrusting is when the Southern Steep Belt was overturned during the Oline phase, i.e. not earlier than c. 28 Ma ago. We therefore conclude that pre-Vanzone phases of orogen-parallel extension (Malfatta and Leis phases) are not related to such an indentation.

The Forcola and Simplon faults are ductile mylonite belts capped by brittle faults that do not merge with the Insubric brittle fault. Therefore, we assume that during formation of the Southern Steep Belt, orogen-parallel stretching was ductilely transferred from outward-dipping fault zones in the flat-lying parts of the Lepontine dome to lateral shearing within the evolving Southern Steep Belt. At the same time, N–S shortening went on and interacted with E–W stretching. Such interaction caused the constrictional strain with E–W stretching axis of the famous 'Stengelgneis' at Beùra (Reinhardt 1966) in the tail of the Monte Rosa nappe. Therefore, the boundary zone between Central and Southern Alps can be considered as a complex half-stretching fault in the sense of Means (1989), i.e. a fault accommodating lateral slip while one of the fault blocks (the Penninic nappe stack) changes its length parallel to the slip direction. We assume that overall the field of E–W extension in the Penninic nappe stack was approximately symmetric with respect to a reference point in the middle of the Lepontine dome (Fig. 19). To the north, the Lepontine block interacted with the foreland (External massifs, not shown in the diagram) which was probably underthrust at the same time. To the south, it interacted with the South Alpine block which is here taken to be rigid and moving towards west relative to the centre of the Lepontine block. The shear sense in gneiss of the Southern Steep Belt, with steep foliations and mostly shallow stretching lineations, results from the combination of E–W stretching in the Lepontine block and rigid westward displacement of the South Alpine block (see also Handy et al. 2005). This simple model predicts that in the east part of the Lepontine dome, the Lepontine stretching and the displacement of the South Alpine block both result in dextral transcurrent shearing. On the western side, however, the two processes are in competition, the stretching of the Lepontine dome leading to sinistral and the displacement of the Southern Alps to dextral shearing. We propose that this competition is responsible

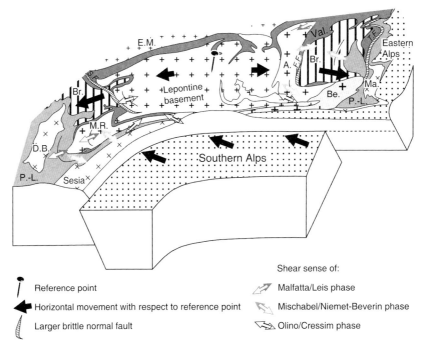

Fig. 19. Schematic block diagram of the Central Alps. The geometry of the Periadriatic line is drawn after Ahrendt (1980). Assumed horizontal relative movement directions of the Olino phase are shown with respect to a reference point in the core of the Lepontine dome. Deformation results in relative stretching of the Central Alps with respect to the Southern Alps. P.-L. = units derived from the Piemont–Ligurian ocean; D.B. = Dent Blanche nappe; M.R. = Monte Rosa nappe; Br. = Briançonnais units; E.M. = External 'massifs'; A. = Adula nappe; Val. = units derived from the Valais ocean; Be. = Bergell pluton; Ma. = Margna nappe; S.F. = Simplon fault; F.F. = Forcola fault; T.F. = Turba fault.

for the alternating dextral and sinistral shearing of the Monte Rosa tail during the Olino phase. Episodes of particularly strong extension in the Lepontine dome led to sinistral shearing, namely during the Olino phase, whereas during slower extension the shear sense was dextral, e.g. the c. 1 km-thick greenschist-facies mylonite belt of the Insubric fault west of Locarno (Schmid et al. 1987) is clearly dextral.

Conclusions

1. The main body of the Monte Rosa nappe records three deformation phases related to nappe stacking, orogen-parallel extension and orogen-perpendicular extension before 30 Ma. This kinematic history shows strong similarities with that of the Adula nappe on the other side of the Lepontine dome. All these deformation phases took place after eclogite-facies metamorphism. In the Monte Rosa nappe, they occurred before the thermal peak of amphibolite-facies Lepontine metamorphism.

2. During these deformation processes, the precursors of the present-day Southern Steep Belt and Insubric fault had a gently SE-dipping orientation. They were subsequently steepened by large-scale folding (Vanzone antiform) and finally overturned to become a south-directed backthrust fault in a relatively late stage of its evolution (Fig. 17).

3. After 30 Ma, the tail of the Monte Rosa nappe in the Southern Steep Belt was overprinted by dextral and sinistral transcurrent shearing (Olino phase) related to movements along the Insubric fault. The alternation of dextral and sinistral shearing may be explained by concurrent westward displacement of the Southern Alps and E–W stretching of the Lepontine dome (Fig. 19).

We thank Alfons Berger and Mark Handy for their reviews and Deutsche Forschungsgemeinschaft for financial support.

Appendix: Projection of lineations in a folded foliation

The angle a between horizontal trends of the same lineation lying in opposite limbs of a fold can be calculated for the plane projection chosen here and the horizontal projection, respectively, by the equations (see Fig. 20 for sign conventions):

$$a = \arcsin\left(\frac{\sin \beta}{\sin \varphi_1}\right) - \arcsin\left(\frac{\sin \beta}{\sin \varphi_2}\right)$$
$$- \arcsin\left(\frac{\tan \beta}{\tan \varphi_1}\right) + \arcsin\left(\frac{\tan \beta}{\tan \varphi_2}\right)$$

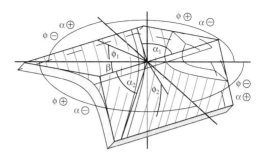

Fig. 20. Variable definitions and sign conventions for the equations given in the Appendix.

and for the horizontal projection:

$$a = \arctan\left(\cos \varphi \times \tan\left(\alpha_1 + \left(\frac{\sin \beta}{\sin \varphi_1}\right)\right)\right) - \arctan$$
$$\left(\cos \varphi \times \tan\left(\alpha_2 + \left(\frac{\sin \beta}{\sin \varphi_2}\right)\right)\right)$$
$$- \arcsin\left(\frac{\tan \beta}{\tan \varphi_1}\right) + \arcsin\left(\frac{\tan \beta}{\tan \varphi_2}\right)$$

For the plane projection, the angle between lineation trends depends only on the plunge of the fold axis and the dip of the fold limbs, whereas in the horizontal projection it also depends on the angles between the fold axis and folded lines. The different effects of the plane and horizontal projections shall be demonstrated for the Vanzone antiform. We simplified the fold geometry to two planes whose orientations are the mean orientations of the foliation in the northern and southern limbs of the fold, i.e. 269/26 and 325/79, respectively. The intersection line of these planes is 240/23. Lines lying within the planes are kinked around the intersection line in such a manner that for each line α_1 equals α_2. Figure 21 shows the

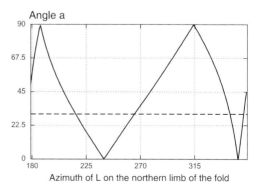

Fig. 21. Angle a for plane projection (dashed line) and horizontal projection (solid line) plotted versus the azimuths of lines on the plane oriented 269/26. The lines are bent around an axis oriented 240/23 onto a plane oriented 325/79. See Appendix for further explanation.

angle (a) between trends of one line projected from both limbs into the horizontal by horizontal and plane projection against the azimuth of the line on the northern plane. Since the plot of stretching lineations L_1, L_2 and L_3 shall visualize kinematic directions of pre-Vanzone deformation, an 'ideal' projection procedure should result in a = 0° for our simplified two-plane model. Although such projections might be carried out on our example (e.g. by rotation of one plane into the orientation of the other around their intersection line), these obviously cannot be applied to geological situations with a greater spread of foliation plane orientations. The example of Figure 21 suggests that for the Vanzone antiform our projection would result in angles a between cogenetic lineations close to 30° if the angles between fold axis and lineation, α_1 and α_2 (Fig. 20), would be equal and constant. This would only be the case if the Vanzone antiform was a pure flexural slip fold without any additional 'within-foliation' deflection of the lineation (e.g. Sander 1948; Ramsay 1960), which it was certainly not. Moreover, L_1, L_2 and L_3 need not have been rectilinear in the pre-Vanzone configuration and the angle a will vary due to variable foliation orientations. For these reasons, it is not possible to correct our lineation trend plots by a calculable angle. Nevertheless, the example of Figure 21 reveals that the plane projection, compared to the horizontal projection, will straighten most lineation orientations and therefore more adequately display L_1, L_2 and L_3.

References

AHRENDT, H. 1980. Die Bedeutung der Insubrischen Linie für den tektonischen Bau der Alpen. *Neues Jahrbuch für Geologie und Paläontologie Abhandlungen*, **160**, 336–362.

ALTENBERGER, U., HAMM, N. & KRUHL, J. H. 1987. Movements and Metamorphism North of the Insubric Line between Val Loana and Val d'Ossola (Italy). *Jahrbuch der Geologischen Bundesanstalt*, **130**, 365–374.

BABINKA, S. 2001. *Geologische und strukturelle Kartierung des Gebietes südlich von San Bernardino*. Unpublished diploma mapping report, Bonn.

BABIST, J., HANDY, M. R., KONRAD-SCHMOLKE, M. & HAMMERSCHMIDT, K. 2006. Precollisional, multistage exhumation of subducted continental crust: The Sesia Zone, western Alps. *Tectonics*, **25**, TC6008, doi: 10.1029/2005TC001927.

BEARTH, P. 1952. Geologie und Petrographie des Monte Rosa. *Beiträge zur geologischen Karte der Schweiz (Neue Folge)*, **96**, 1–94.

BEARTH, P. 1953. *Geologischer Atlas der Schweiz, 1:25000, Blatt Zermatt (Nr. 29)*. Schweizerische Geologische Kommission, Bern.

BEARTH, P. 1954a. *Geologischer Atlas der Schweiz, 1:25000, Blatt Monte Moro (Nr. 30)*. Schweizerische Geologische Kommission, Bern.

BEARTH, P. 1954b. *Geologischer Atlas der Schweiz, 1:25000, Blatt Saas (Nr. 31)*. Schweizerische Geologische Kommission, Bern.

BEARTH, P. 1957. Die Umbiegung von Vanzone (Valle Anzasca). *Eclogae Geologicae Helvetiae*, **50**, 161–170.

BEARTH, P. 1958. Über einen Wechsel der Mineralfazies in der Wurzelzone des Penninikums. *Schweizerische Mineralogische und Petrographische Mitteilungen*, **38**, 363–373.

BEARTH, P. 1964. *Geologischer Atlas der Schweiz, 1:25000, Blatt Randa (Nr. 43)*. Schweizerische Geologische Kommission, Bern.

BEAUMONT, C., ELLIS, S., HAMILTON, J. & FULLSACK, P. 1996. Mechanical model for subduction–collision tectonics of Alpine-type compressional orogens. *Geology*, **24**, 675–678.

BECKER, H. 1993. Garnet peridotite and eclogite Sm–Nd mineral ages from the Lepontine dome (Swiss Alps): New evidence for an Eocene high-pressure metamorphism in the Central Alps. *Geology*, **21**, 599–602.

BERGER, A., ROSENBERG, C. & SCHMID, S. M. 1996. Ascent, emplacement and exhumation of the Bergell pluton within the Southern Steep Belt of the Central Alps. *Schweizerische Mineralogische und Petrographische Mitteilungen*, **76**, 357–382.

BISTACCHI, A., DAL PIAZ, G. V., MASSIRONI, M., ZATTIN, M. & BALESTRIERI, M. L. 2001. The Aosta–Ranzola extensional fault system and Oligocene–Present evolution of the Austroalpine–Penninic wedge in the northwestern Alps. *International Journal of Earth Sciences*, **90**, 654–667.

BLUMENTHAL, M. M. 1952. Beobachtungen über Bau und Verlauf der Muldenzone von Antrona zwischen der Walliser Grenze und dem Locarnese. *Eclogae geologicae Helvetiae*, **45**, 219–263.

BORGHI, A., COMPAGNONI, R. & SANDRONE, R. 1996. Composite P–T paths in the Internal Pennidic Massifs of the Western Alps: Petrological constraints to their thermo-mechanical evolution. *Eclogae Geologicae Helvetiae*, **89**, 345–367.

BOUCHEZ, J. L. 1978. Preferred orientations of quartz a-axes in some tectonites: kinematic inferences. *Tectonophysics*, **49**, T25–T30.

BOUSQUET, R., ENGI, M. & GOSSO, G. ET AL. 2004. Explanatory notes to the map: Metamorphic structure of the Alps Transition from the Western to the Central Alps. *Mitteilungen der Österreichischen Mineralogischen Gesellschaft*, **149**, 145–156.

BUCHER, K., FAZIS, Y., DE CAPITANI, C. & GRAPES, R. 2005. Blueschists, eclogites, and decompression assemblages of the Zermatt–Saas ophiolite: High-pressure metamorphism of subducted Tethys lithosphere. *American Mineralogist*, **90**, 821–835.

BURRI, T., BERGER, A. & ENGI, M. 2005. Tertiary migmatites in the Central Alps: Regional distribution, field relations, conditions of formation, and tectonic implications. *Schweizerische Mineralogische und Petrographische Mitteilungen*, **85**, 215–232.

CHOPIN, C. & MONIÉ, P. 1984. A unique magnesio-chloritoide-bearing, high-pressure assemblage from the Monte Rosa, Western Alps; petrologic and ^{40}Ar–^{39}Ar radiometric study. *Contributions to Mineralogy and Petrology*, **87**, 388–398.

CIANCALEONI, L. & MARQUER, D. 2006. Syn-extension leucogranite deformation during convergence in the

Eastern Central Alps: example of the Novate intrusion. *Terra Nova*, **18**, 170–180.

DALE, J. & HOLLAND, T. J. B. 2003. Geothermobarometry, P–T paths and metamorphic field gradients of high-pressure rocks from the Adula nappe, Central Alps. *Journal of Metamorphic Geology*, **21**, 813–829.

DAL PIAZ, G. V. 1993. Evolution of Austro-Alpine and Upper Penninic Basement in the northwestern Alps from Variscan Convergence to Post-Variscan Extension. *In*: VON RAUMER, J. F. & NEUBAUER, F. (eds) *Pre-Mesozoic Geology in the Alps*. Springer, Berlin, 327–344.

DAL PIAZ, G. V. 1999. The Austroalpine–Piedmont nappe stack and the puzzle of Alpine Tethys. *Memorie di Scienze Geologiche*, **51**, 155–176.

DAL PIAZ, G. V. & LOMBARDO, B. 1986. Early Alpine eclogite metamorphism in the Penninic Monte Rosa–Gran Paradiso basement nappes of the northwestern Alps. *Geological Society of America Memoir*, **164**, 249–265.

DAVIDSON, C., ROSENBERG, C. & SCHMID, S. M. 1996. Synmagmatic folding of the base of the Bergell pluton, Central Alps. *Tectonophysics*, **265**, 213–238.

ELTER, G., ELTER, P., STURANI, C. & WEIDMANN, M. 1966. Sur la prolongation du domaine ligure de l'Apennin dans le Montferrat et les Alpes et sur l'origine de la Nappe de Simme s.l. des Préalpes romandes et chablaisiennes. *Archives des Sciences*, **19**, 279–377.

ENGI, M., SCHERRER, N. C. & BURRI, T. 2001*a*. Metamorphic evolution of pelitic rocks of the Monte Rosa nappe: Constraints from petrology and single grain monazite age data. *Schweizerische Mineralogische und Petrographische Mitteilungen*, **81**, 305–328.

ENGI, M., BERGER, A. & ROSELLE, G. T. 2001*b*. Role of the tectonic accretion channel in collisional orogeny. *Geology*, **29**, 1143–1146.

ENGI, M., BOUSQUET, R. & BERGER, A. 2004. Explanatory notes to the map Metamorphic Structure of the Alps Central Alps. *Mitteilungen der Österreichischen Mineralogischen Gesellschaft*, **149**, 157–173.

ESCHER, A. & BEAUMONT, C. 1997. Formation, burial and exhumation of basement nappes at crustal scale: a geometric model based on the Western Swiss–Italian Alps. *Journal of Structural Geology*, **19**, 955–974.

ESCHER, A., MASSON, H. & STECK, A. 1993. Nappe geometry in the Western Swiss Alps. *Journal of Structural Geology*, **15**, 501–509.

ESCHER, A., HUNZIKER, J. C., MARTHALER, M., MASSON, H., SARTORI, M. & STECK, A. 1997. Geologic framework and structural evolution of the western Swiss–Italian Alps. *In*: PFIFFNER, O. A., LEHNER, P., HEITZMANN, P., MUELLER, S. & STECK, A. (eds) *Deep Structure of the Swiss Alps*. Birkhäuser, Basel, CH, 205–221.

FERRANDO, S., BERNOULLI, D. & COMPAGNONI, R. 2004, The Canavese zone (internal Western Alps): a distal margin of Adria. *Schweizerische Mineralogische und Petrographische Mitteilungen*, **84**, 237–256.

FREY, M., HUNZIKER, J. C., FRANK, W., BOCQUET, J., DAL PIAZ, G. V., JÄGER, E. & NIGGLI, E. 1974. Alpine metamorphism of the Alps. A review. *Schweizerische Mineralogische und Petrographische Mitteilungen*, **54**, 247–290.

FREY, M., HUNZIKER, J. C., O'NEIL, J. R. & SCHWANDER, H. W. 1976. Equilibrium–disequilibrium relation in the Monte Rosa granite, Western Alps: Petrological, Rb–Sr and stable isotope data. *Contributions to Mineralogy and Petrology*, **55**, 147–179.

FROITZHEIM, N. 2001. Origin of the Monte Rosa nappe in the Pennine Alps—A new working hypothesis. *Geological Society of America Bulletin*, **113**, 604–614.

FROITZHEIM, N., PLEUGER, J. & NAGEL, T. J. 2006. Extraction faults. *Journal of Structural Geology*, **28**, 1388–1395.

GANSSER, A. 1968. The Insubric Line, a major geotectonic problem. *Schweizerische Mineralogische und Petrographische Mitteilungen*, **48**, 123–143.

GEBAUER, D. 1996. A P–T–t path for (ultra?-) high-pressure ultramafic/mafic rock-association and their felsic country-rocks based on SHRIMP-dating of magmatic and metamorphic zircon domains. Example: Alpe Arami (Central Swiss Alps). *In*: BASU, A. & HART, S. R. (eds) *Earth Processes: Reading the Isotopic Code*. Geophysical Monograph Series, **95**, 307–329.

GRASEMANN, B. & MANCKTELOW, N. S. 1993. Two-dimensional thermal modelling of normal faulting; the Simplon fault zone, Central Alps, Switzerland. *Tectonophysics*, **225**, 155–165.

GROND, R., WAHL, F. & PFIFFNER, M. 1995. Mehrphasige alpine Deformation und Metamorphose in der nördlichen Cima–Lunga–Einheit, Zentralalpen (Schweiz). *Schweizerische Mineralogische und Petrographische Mitteilungen*, **75**, 371–386.

GULSON, B. L. 1973. Age relations in the Bergell region of the southeast Swiss Alps: with some geochemical comparisons. *Eclogae Geologicae Helvetiae*, **66**, 293–313.

HANDY, M. R., BABIST, J., WAGNER, R., ROSENBERG, C. L. & KONRAD, M. 2005. Decoupling and its relation to strain partitioning in continental lithosphere—insight from the Periadriatic Fault system (European Alps). *In*: GAPAIS, D., BRUN, J. P. & COBBOLD, P. R. (eds) *Deformation mechanisms, rheology and tectonics: From minerals to the lithosphere*. Geological Society, London, Special Publications, **243**, 249–276.

HEINRICH, C. A. 1986. Eclogite facies regional metamorphism of hydrous mafic rocks in the Central Alpine Adula nappe. *Journal of Petrology*, **27**, 123–154.

HEITZMANN, P. 1975. Zur Metamorphose und Tektonik im südöstlichen Teil der Lepontinischen Alpen. *Schweizerische Mineralogische und Petrographische Mitteilungen*, **55**, 467–522.

HEITZMANN, P. 1987. Calcite mylonites in the Central Alpine 'root zone'. *Tectonophysics*, **135**, 207–215.

HUNDENBORN, R. 2001. *Entstehung, Metamorphose und Deformation der basischen und ultrabasischen Gesteine der Adula–Decke und der Misoxer Zone, unter besonderer Berücksichtigung der Adula–Eklogite*. Unpublished diploma thesis, Bonn.

HUNZIKER, J. C. 1970. Polymetamorphism in the Monte Rosa, Western Alps. *Eclogae Geologicae Helvetiae*, **63**, 151–161.

HURFORD, A. J. 1986. Cooling and uplift patterns in the Lepontine Alps South Central Switzerland and an age of vertical movement on the Insubric fault line. *Contributions to Mineralogy and Petrology*, **92**, 413–427.

JANSEN, E., SCHÄFER, W. & KIRFEL, A. 2000. The Jülich neutron diffractometer and data processing in rock texture investigations. *Journal of Structural Geology*, **22**, 1559–1564.

KELLER, L. M. & SCHMID, S. M. 2001. On the kinematics of shearing near the top of the Monte Rosa nappe and the nature of the Furgg zone in Val Loranco (Antrona valley, N. Italy): tectonometamorphic and paleogeographical consequences. *Schweizerische Mineralogische und Petrographische Mitteilungen*, **81**, 347–367.

KELLER, L. M., HESS, M., FÜGENSCHUH, B. & SCHMID, S. M. 2005. Structural and metamorphic evolution SW of the Simplon line. *Eclogae Geologicae Helvetiae*, **98**, 19–49.

KELLER, L. M., FÜGENSCHUH, B., HESS, M., SCHNEIDER, B. & SCHMID, S. M. 2006. Simplon fault zone in the western and central Alps: Mechanism of Neogene folding revisited. *Geology*, **34**, 317–320.

KLEIN, J. 1978. Post-nappe folding southeast of the Mischabelrückfalte (Pennine Alps) and some aspects of the associated metamorphism. *Leidse Geologische Mededelingen*, **51**, 233–312.

KREMER, K. 2001. *Geologische und strukturelle Kartierung des Gebietes um San Bernardino (Graubünden/Schweiz)*. Unpublished diploma mapping report, Bonn.

KRUHL, J. H. & VOLL, G. 1976. Fabrics and Metamorphism from the Monte Rosa Root Zone into the Ivrea Zone near Finero, Southern margin of the Alps. *Schweizerische Mineralogische und Petrographische Mitteilungen*, **56**, 627–633.

LACASSIN, R. 1987. Kinematics of ductile shearing from outcrop to crustal scale in the Monte Rosa nappe, Western Alps. *Tectonics*, **6**, 69–88.

LANGE, S., NASDALA, L., POLLER, U., BAUMGARTNER, L. & TODT, W. 2000. Crystallization age and metamorphism of the Monte Rosa granite, Western Alps. *Abstract volume of 17th Swiss Tectonic Studies Group meeting*, Zürich.

LAPEN, T. J., JOHNSON, C. M., BAUMGARTNER, L. P., DAL PIAZ, G. V., SKORA, S. & BEARD, B. L. 2007. Coupling of oceanic and continental crust during Eocene eclogite-facies metamorphism: evidence from the Monte Rosa nappe, western Alps. *Contributions to Mineralogy and Petrology*, **153**, 139–157.

LAUBSCHER, H. 1991. The Arc of the Western Alps today. *Eclogae Geologicae Helvetiae*, **84**, 631–659.

LE BAYON, R., DE CAPITANI, C. & FREY, M. 2006. Modelling phase-assemblage diagrams for magnesian metapelites in the system $K_2O-FeO-MgO-Al_2O_3-SiO_2-H_2O$: geodynamic consequences for the Monte Rosa nappe, Western Alps. *Contributions to Mineralogy and Petrology*, **151**, 395–412.

LIATI, A. & FROITZHEIM, N. 2006. Assessing the Valais ocean, Western Alps: U–Pb SHRIMP zircon geochronology of eclogite in the Balma unit, on top of the Monte Rosa nappe. *European Journal of Mineralogy*, **18**, 299–308.

LIATI, A., GEBAUER, D. & FANNING, C. M. 2000. U–Pb SHRIMP dating of zircon from the Novate granite (Bergell, Central Alps): evidence for Oligocene–Miocene magmatism, Jurassic–Cretaceous continental rifting and opening of the Valais trough. *Schweizerische Mineralogische und Petrographische Mitteilungen*, **80**, 305–316.

LIATI, A., FROITZHEIM, N. & FANNING, C. M. 2005. Jurassic ophiolites within the Valais domain of the Western and Central Alps: geochronological evidence for re-rifting of oceanic crust. *Contributions to Mineralogy and Petrology*, **149**, 446–461.

LÖW, S. 1987. Die tektono-metamorphe Entwicklung der nördlichen Adula–Decke (Zentralalpen, Schweiz). *Beiträge zur Geologischen Karte der Schweiz (Neue Folge)*, **161**.

MACCREADY, T. 1996. Misalignment of quartz c-axis fabrics and lineations due to oblique final strain increments in the Ruby Mountains core complex, Nevada. *Journal of Structural Geology*, **18**, 765–776.

MANCKTELOW, N. S. 1985. The Simplon Line: a major displacement zone in the western Lepontine Alps. *Eclogae Geologicae Helvetiae*, **78**, 73–96.

MANCKTELOW, N. S. 1992. Neogene lateral extension during convergence in the Central Alps: Evidence from interrelated faulting and backfolding around the Simplon Pass (Switzerland). *Tectonophysics*, **215**, 295–317.

MANCKTELOW, N. S. 2003. The Val d'Ossola to Rimella section of the Periadriatic Fault: a major anomaly in the regional kinematic model. *Abstract volume of 1st Swiss Geosciences meeting*, Basel, 69.

MARQUER, D., CHALLANDES, N. & BAUDIN, T. 1996. Shear zone patterns and strain distribution at the scale of a Penninic nappe: the Suretta nappe (Eastern Swiss Alps). *Journal of Structural Geology*, **18**, 753–764.

MARTHALER, M. & STAMPFLI, G. M. 1989. Les schistes lustrés à ophiolites de la nappe du Tsaté: un ancien prisme d'accrétion issu de la marge active apulienne? *Schweizerische Mineralogische und Petrographische Mitteilungen*, **69**, 211–216.

MATTIROLO, E., NOVARESE, V., FRANCHI, S. & STELLA, A. 1927. *Carta geologica d'Italia, 1:100000, foglio 30 (Varallo)*. Servizio Geologico d'Italia.

MAXELON, M. & MANCKTELOW, N. S. 2005. Three-dimensional geometry and tectonostratigraphy of the Pennine zone, Central Alps, Switzerland and Northern Italy. *Earth-Science Reviews*, **71**, 171–227.

MEANS, W. D. 1989. Stretching faults. *Geology*, **17**, 893–896.

MEYRE, C. 1998. *High-pressure metamorphism and deformation of the middle Adula nappe*. Unpublished PhD thesis, Basel.

MEYRE, C., MARQUER, D., SCHMID, S. M. & CIANCALEONI, L. 1998. Syn-orogenic extension along the Forcola fault: Correlation of Alpine deformations in the Tambo and Adula nappes (Eastern

Penninic Alps). *Eclogae geologicae Helvetiae*, **91**, 409–420.

MILNES, A. G. 1974. Structure of the Pennine Zone (Central Alps): A New Working Hypothesis. *Geological Society of America Bulletin*, **85**, 1727–1732.

MILNES, A. G. & SCHMUTZ, H. U. 1978. Structure and history of the Suretta nappe (Pennine Zone, Central Alps)—a field study. *Eclogae Geologicae Helvetiae*, **71**, 19–23.

MILNES, A. G., GRELLER, M. & MÜLLER, R. 1981. Sequence and style of major post-nappe structures, Simplon–Pennine Alps. *Journal of Structural Geology*, **3**, 411–420.

MÜLLER, R. 1983. Die Struktur der Mischabelfalte (Penninische Alpen). *Eclogae Geologicae Helvetiae*, **76**, 391–416.

MÜLLER, W., PROSSER, G., MANCKTELOW, N. S., VILLA, I. M., KELLEY, S. P., VIOLA, G. & OBERLI, F. 2001. Geochronological constraints on the evolution of the Periadriatic Fault System (Alps). *International Journal of Earth Sciences.*, **90**, 623–653.

NAGEL, T. 2008. Tertiary subduction, collision and exhumation recorded in the Adula nappe, central Alps. *In*: SIEGESMUND, S., FÜGENSCHUH, B. & FROITZHEIM, N. (eds) *Tectonic Aspects of the Alpine-Dinaride-Carpathian System*. Geological Society, London, Special Publications, **298**, 365–392.

NAGEL, T., DE CAPITANI, C., FREY, M., FROITZHEIM, N., STÜNITZ, H. & SCHMID, S. M. 2002a. Structural and metamorphic evolution during rapid exhumation in the Lepontine dome (southern Simano and Adula nappes, Central Alps, Switzerland). *Eclogae Geologicae Helvetiae*, **95**, 301–321.

NAGEL, T., DE CAPITANI, C. & FREY, M. 2002b. Isograds and P–T evolution in the eastern Lepontine Alps (Graubünden, Switzerland). *Journal of Metamorphic Geology*, **20**, 309–324.

NIEVERGELT, P., LINIGER, M., FROITZHEIM, N. & FERREIRO MÄHLMANN, R. 1996. Early to mid Tertiary crustal extension in the Central Alps: The Turba Mylonite Zone (Eastern Switzerland). *Tectonics*, **15**, 329–340.

NIGGLI, E. & NIGGLI, C. R. 1965. Karten der Verbreitung einiger Mineralien der alpidischen Metamorphose in den Schweizer Alpen (Stilpnomelan, Alkali- Amphibol, Chloritoid, Staurolith, Disthen, Sillimanit). *Eclogae Geologicae Helvetiae*, **58**, 335–368.

PARTZSCH, J. H. 1998. *The tectono-metamorphic evolution of the middle Adula nappe, Central Alps, Switzerland*. Unpublished PhD thesis, Basel.

PETTKE, T., DIAMOND, L. W. & VILLA, I. M. 1999. Mesothermal gold veins and metamorphic devolatilization in the northwestern Alps: The temporal link. *Geology*, **27**, 641–644.

PFIFFNER, O. A., ELLIS, S. & BEAUMONT, C. 2000. Collision tectonics in the Swiss Alps: Insight from geodynamic modeling. *Tectonics*, **19**, 1065–1094.

PLEßMANN, W. & WUNDERLICH, H. G. 1961. Eine Achsenkarte des inneren Westalpenbogens. *Neues Jahrbuch für Geologie und Paläontologie Monatshefte*, 199–210.

PLEUGER, J., JANSEN, E., SCHÄFER, W., OESTERLING, N. & FROITZHEIM, N. 2002. Neutron texture study of a natural gneiss mylonite affected by two phases of deformation. *Applied Physics A*, **74**, S1058–S1060.

PLEUGER, J., HUNDENBORN, R., KREMER, K., BABINKA, S., KURZ, W., JANSEN, E. & FROITZHEIM, N. 2003. Structural evolution of Adula nappe, Misox zone, and Tambo nappe in the San Bernardino area: Constraints for the exhumation of the Adula eclogites. *Mitteilungen der Österreichischen Geologischen Gesellschaft*, **94**, 99–122.

PLEUGER, J., FROITZHEIM, N. & JANSEN, E. 2005. Folded continental and oceanic nappes on the southern side of Monte Rosa (western Alps, Italy): Anatomy of a double collision suture. *Tectonics*, **24**, TC4013; doi: 10.1029/2004TC001737.

PLEUGER, J., ROLLER, S., WALTER, J. M., JANSEN, E. & FROITZHEIM, N., 2007. Structural evolution of the contact between two Penninic nappes (Zermatt–Saas zone and Combin zone, Western Alps) and implications for exhumation mechanism and palaeogeography. *International Journal of Earth Sciences*, **96**, 229–252.

RAMSAY, J. G. 1960. The deformation of early lineation structure in areas of repeated folding. *Journal of Geology*, **68**, 75–93.

REINHARDT, B. 1966. Geologie und Petrographie der Monte Rosa-Zone, der Sesia-Zone und des Canavese im Gebiet zwischen Valle d'Ossola und Valle Loana (Prov. di Novara, Italien). *Schweizerische Mineralogische und Petrographische Mitteilungen*, **46**, 553–678.

RÜTTI, R. 2001. Tectono-metamorphic evolution of the Simano–Adula nappe boundary, Central Alps, Switzerland. *Schweizerische Mineralogische und Petrographische Mitteilungen*, **81**, 115–129.

SANDER, B. 1948. *Einführung in die Gefügekunde geologischer Körper, Band 1: Allgemeine Gefügekunde und Arbeiten im Bereich Handstück bis Profil*. Springer, Vienna and Innsbruck.

SANTINI, L. 1992. *Geochemistry and geochronology of the basic rocks of the Penninic Nappes of East-Central Alps (Switzerland)*. Unpublished PhD thesis, Lausanne.

SARTORI, M. 1987. Structure de la zone du Combin entre les Diablons et Zermatt (Valais). *Eclogae Geologicae Helvetiae*, **80**, 789–814.

SCHÄRER, U., COSCA, M., STECK, A. & HUNZIKER, J. 1996. Termination of major ductile strike-slip shear and differential cooling along the Insubric line (Central Alps): U–Pb, Rb–Sr and $^{40}Ar/^{39}Ar$ ages of cross-cutting pegmatites. *Earth and Planetary Science Letters*, **142**, 331–351.

SCHMID, S. M. & CASEY, M. 1986. Complete fabric analysis of some commonly observed quartz c-axis patterns. *In*: HOBBS, B. E. & HEARD, H. C. (eds) *Mineral and rock deformation: Laboratory Studies—the Paterson Volume*. Geophysical Monograph Series, **36**, 263–286.

SCHMID, S. M., ZINGG, A. & HANDY, M. 1987. The kinematics of movements along the Insubric Line and the emplacement of the Ivrea Zone. *Tectonophysics*, **135**, 47–66.

SCHMID, S. M., AEBLI, H. R., HELLER, F. & ZINGG, A. 1989. The role of the Periadriatic Line in the tectonic evolution of the Alps. *In*: COWARD, M. P., DIETRICH,

D. & PARK, R. G. (eds) *Alpine Tectonics*. Geological Society, London, Special Publications, **45**, 153–171.

SCHMID, S. M., PFIFFNER, O. A., FROITZHEIM, N., SCHÖNBORN, G. & KISSLING, E. 1996. Geophysical–geological transect and tectonic evolution of the Swiss–Italian Alps. *Tectonics*, **15**, 1036–1064.

SCHMID, S. M., PFIFFNER, O. A. & SCHREURS, G. 1997. Rifting and collision in the Penninic zone of eastern Switzerland. *In*: PFIFFNER, O. A., LEHNER, P., HEITZMANN, P., MUELLER, S. & STECK, A. (eds) *Deep structure of the Swiss Alps*. Birkhäuser, Basel, 160–185.

SCHREURS, G. 1993. Structural analysis of the Schams nappes and adjacent tectonic units: Implications for the orogenic evolution of the Penninic zone in Eastern Switzerland. *Bulletin de la Société géologique de France*, **164**, 415–435.

SCHREURS, G. 1995. Geometry and kinematics of the Schams nappes and adjacent tectonic units in the Penninic zone. *In*: Die Schamser Decken, part 2. *Beiträge zur geologischen Karte der Schweiz (Neue Folge)*, **167**. (pt II), 1–111.

SPICHER, A. 1980. *Tektonische Karte der Schweiz, 1:500 000*. Schweizerische Geologische Kommission, Bern.

SPITZ, A. 1919. Fragmente zur Tektonik der Westalpen und des Engadins. *Verhandlungen der Geologischen Reichsanstalt*, **4**, 104–122.

STECK, A. 1984. Structures de deformation tertiaires dans les Alpes centrales. *Eclogae Geologicae Helvetiae*, **77**, 55–100.

STECK, A. 1987. Le massif du Simplon—Réflexions sur la cinématique des nappes de gneiss. *Schweizerische Mineralogische und Petrographische Mitteilungen*, **67**, 27–45.

STECK, A. 1990. Une carte des zones de cisaillement ductile des Alpes Centrales. *Eclogae Geologicae Helvetiae*, **83**, 603–627.

STECK, A. & HUNZIKER, J. 1994. The Tertiary structural and thermal evolution of the Central Alps—compressional and extensional structures in an orogenic belt. *Tectonophysics*, **238**, 229–254.

STECK, A., BIGIOGGERO, B., DAL PIAZ, G. V., ESCHER, A., MARTINOTTI, G. & MASSON, H. 1999. *Carte tectonique des Alpes de Suisse occidentale et des régions avoisinantes, 1:100 000, Carte spéc. n. 123* (4 maps). Service hydrologique et géologique national, Bern.

STEINMANN, M. & STILLE, P. 1999. Geochemical evidence for the nature of the crust beneath the eastern North Penninic basin of the Mesozoic Tethys ocean. *Geologische Rundschau*, **87**, 633–643.

STIPP, M., STÜNITZ, H., HEILBRONNER, R. & SCHMID, S. M. 2002. The eastern Tonale fault zone: a 'natural laboratory' for crystal plastic deformation of quartz over a temperature range from 250 to 700 °C. *Journal of Structural Geology*, **24**, 1861–1884.

TROMMSDORFF, V. 1966. Progressive Metamorphose kieseliger Karbonatgesteine in den Zentralalpen zwischen Bernina und Simplon. *Schweizerische Mineralogische und Petrographische Mitteilungen*, **46**, 431–460.

TROMMSDORFF, V. 1990. Metamorphism and tectonics in the Central Alps: The Alpine lithospheric mélange of Cima Lunga and Adula. *Memorie della Società Geologica Italiana*, **45**, 39–49.

VIOLA, G., MANCKTELOW, N. S. & SEWARD, D. 2001. Late Oligocene–Neogene evolution of Europe–Adria collision: New structural and geochronological evidence from the Giudicarie fault system (Italian Eastern Alps). *Tectonics*, **20**, 999–1020.

VON BLANCKENBURG, F. 1992. Combined high-precision chronometry and geochemical tracing using accessory minerals: applied to the Central-Alpine Bergell intrusion (central Europe). *Chemical Geology*, **100**, 19–40.

WEH, M. & FROITZHEIM, N. 2001. Penninic cover nappes in the Prättigau half-window (Eastern Switzerland): Structure and tectonic evolution. *Eclogae Geologicae Helvetiae*, **94**, 237–252.

WENK, E. 1970. Zur Regionalmetamorphose und Ultrametamorphose der Zentralalpen. *Fortschritte der Mineralogie*, **47**, 555–565.

WETZEL, R. 1972. Zur Petrographie und Mineralogie der Furgg-Zone (Monte Rosa-Decke). *Schweizerische Mineralogische und Petrographische Mitteilungen*, **52**, 161–236.

WILSON, C. J. L. 1975. Preferred orientation in quartz ribbon mylonites. *The Geological Society of America Bulletin*, **86**, 968–974.

ZINGG, A. & HUNZIKER, J. C. 1990. The age of movements along the Insubric Line West of Locarno (northern Italy and southern Switzerland). *Eclogae Geologicae Helvetiae*, **83**, 629–644.

Metamorphic evolution of a very low- to low-grade metamorphic core complex (Danubian window) in the South Carpathians

MAGDA CIULAVU[1,2,*], RAFAEL FERREIRO MÄHLMANN[1,3,§], STEFAN M. SCHMID[4], HEIKO HOFMANN[3], ANTONETA SEGHEDI[2] & MARTIN FREY[1,†]

[1]*Mineralogisch-Petrographisches Institut, Universität Basel, Bernoullistrasse 30, CH-4056 Basel, Switzerland*

[2]*Institutul Geologic al Romaniei, 1 Caransebes Street, RO 78 344 Bucharest 32, Romania*

[3]*Institut für Angewandte Geowissenschaften, TU Darmstadt, Schnittspahnstraße 9, D 64287 Darmstadt, Germany*

[4]*Geologisch-Paläontologisches Institut, Universität Basel, Bernoullistrasse 32, CH-4056 Basel, Switzerland*

[*]*Now at Canadian Natural Resources Ltd., Calgary, Alberta, Canada*

[†]*Martin Frey died in a mountain accident during fieldwork in the Alps in September 2000. We thank him for stimulating our research on low-grade metamorphism.*

[§]*Current address: Darmstadt University of Technology, Faculty of Material und Geosciences, Institute of Applied Geosciences, Technical Petrology, Schnittspahnstraße 9, D-64287 Darmstadt, Germany (e-mail: ferreiro@geo.tu-darmstadt.de)*

Abstract: The Danubian window, characterized by diagenetic to low greenschist facies conditions at a high thermal gradient, is evidently of great interest for methodological studies, because high metamorphic thermal gradient conditions during low grade metamorphism have received little attention so far. The general increase in metamorphic grade from SW to NE in the Danubian window is indicated by mineral Parageneses studies, as well as by illite Kübler index (KI) measurements and organic matter reflectance (OMR). For the first time, this study distinguishes between metamorphic conditions related to Jurassic ocean floor, Cretaceous nappe stacking, post-collisional accommodation and syn-kinematic Getic detachment metamorphism and cooling after Oligocene exhumation.

The occurrence of the prehnite–pumpellyite facies in the Severin–Cosustea units in the southeastern area is the result of Cretaceous metamorphism. Remnants of ocean floor metamorphism prevailed. The highest pressure is constrained by the upper stability limit of prehnite to be at around 4.0 kbar. The Danubian units situated within the diagenetic zone were not below 200 °C, due to epidote formation. The KI, OMR and mineral data, indicate diagenetic conditions. Assuming temperatures between >200 and <250 °C, pressures between 1.8 and 2.6 kbar were calculated using kinetic and numerical maturity models.

Orogenic collisional Cretaceous peak pressure conditions of 4.0 ± 1.0 kbar are found in the Danubian nappes not altered by a subsequent syn-detachment metamorphic overprint. Highest temperatures in chloritoid schists and epidote–hornblende-bearing mylonites have been inferred for samples from the northern border of the Danubian window (between >300 and <400 °C). Along a syn- to post-detachment retrograde pressure path, post-dating the chloritoid formation, the occurrence of clinozoisite + chlorite + quartz suggests temperatures >300 °C in the northwest, while the association andalusite + quartz and biotite + muscovite indicates temperatures between 370 and 400 °C at <3.5 kbar in the northeast.

It is demonstrated that the slope of the regression lines between KI and OMR data gives valuable qualitative information about the relative magnitudes of P and T: the slope of the regression line for the Danubian window samples indicates normal heat flow conditions during nappe stacking and hyperthermal conditions during the formation of the Getic detachment.

High thermal gradient conditions can easily be explained by partly isothermal decompression during the Getic detachment event, the elevation of the geotherm being caused by crustal thinning and rapid exhumation of the Danubian units. Probably, also a higher heat-flux prevailed at the end of the Getic detachment, at a time when the retrograde chloritoid decomposition reactions took place, documenting late-stage HT greenschist facies metamorphism.

From: SIEGESMUND, S., FÜGENSCHUH, B. & FROITZHEIM, N. (eds) *Tectonic Aspects of the Alpine-Dinaride-Carpathian System.* Geological Society, London, Special Publications, **298**, 281–315.
DOI: 10.1144/SP298.14   0305-8719/08/$15.00 © The Geological Society of London 2008.

Location of the study area and geological Setting

The South Carpathians represent a roughly east–west-striking segment of the Carpathian loop, situated between the foredeep of the Getic depression in the south and the Transylvanian basin in the north (Fig. 1). Eastward, the mountain chain prolongates in the East Carpathian range in the bending area NE of Bucharest. Southwestward, they find their continuation in eastern Serbia and ultimately in the Balkanides of Bulgaria.

The South Carpathians consist of a pile of basement and cover nappes formed during Cretaceous orogeny (e.g. Săndulescu 1984, 1994). From bottom to top, this nappe stack comprises, the following units used as a reference to locate the metamorphic data of Figures 3 to 6: (1) a stack of Danubian thrust sheets (Upper and Lower Danubian nappes), consisting of pre-Mesozoic metamorphic and magmatic rocks (basement) and their Mesozoic cover; (2) due to a differing lithofacies compared to the other Danubian nappes, the Arjana cover nappe is shown, which occupies the structurally highest position within the Danubian nappe stack; (3) the Cosustea mélange, mostly together with locally preserved tectonic slices of an ophiolitic-sedimentary sequence, representing remnants of the oceanic Severin nappe lithosphere (in Figs 2 and 3, both are shaded undifferentiated); and (4) the Getic and Supragetic nappes which mostly consist of Precambrian and Variscan basement, overlain by a Palaeozoic to Mesozoic cover.

The Severin Ocean opened during the Jurassic between the Danubian and Getic–Supragetic units, which were parts of the European Plate (Sandulescu 1975, 1994). Convergence in the Cretaceous led to the suturing (subduction and obduction) of the Severin oceanic lithosphere (Stånoiu 1972), followed by collision of the Severin and Getic–Supragetic units with the Danubian continental block during an orogenic stage in the Late Senonian. The final nappe stack was subsequently affected by orogen-parallel extension, which, according to fission-track data, led to final exhumation of the Danubian window during Eocene times (Schmid et al. 1998; Fügenschuh & Schmid 2005). The post-orogenic (i.e. post-Eocene) movements did not substantially alter the tectonic edifice as observed today, except for sinistral strike-slip motions in the order of 35 km along the Cerna–Jiu strike-slip fault (Berza & Drăgănescu 1988). The curved Cerna–Jiu fault (Fig. 2) can be

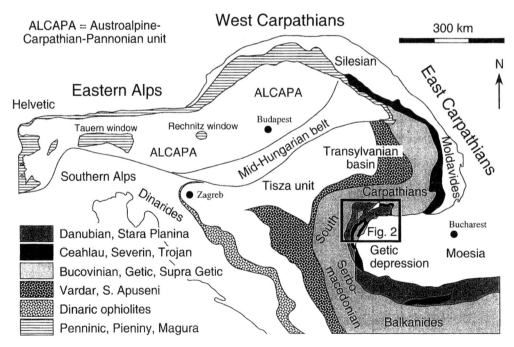

Fig. 1. Sketch map of the Alpine–Carpathian mountain chain and the main tectonic units (modified after Săndulescu 1994).

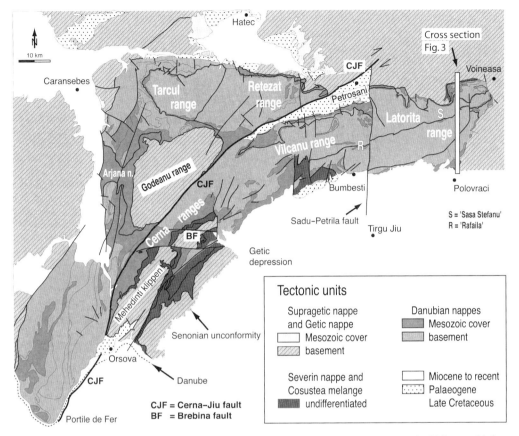

Fig. 2. Tectonic map of the Danubian window between Portile de Fer (Iron Gate) and the eastern end at Voineasa with the mountain ranges, localities and tectonic structures referred in the text (modified after Fügenschuh & Schmid 2005).

followed across the entire Danubian window, changing strike direction from NNE to SSW towards NE to SW following sub-parallel to the Carpathian bend. This strike-slip fault is directly related to the opening of the intramontane Petrosani basin, which formed during the Chattian to Badenian (Berza & Drăgănescu 1988; Ratschbacher et al. 1993). A south-directed thrusting onto the Getic depression (Fig. 2) occurred in the Miocene (Matenco et al. 1997).

Zircon fission-track data show that pervasive Cretaceous low-grade metamorphism (Alpine metamorphism) is restricted to the northern and northeastern part of the Danubian units and the Severin nappe. They are found below an extensional detachment (the Getic detachment), with the Getic and Supragetic units forming the non-metamorphic hanging wall (Schmid et al. 1998; Matenco & Schmid 1999; Fügenschuh & Schmid 2005). The Getic detachment is identical with the tectonic boundary between the Danubian–Severin nappes and the Getic–Supragetic units. Consequently, the studied area is restricted to the Danubian and Severin units.

The pioneering work of Mrazec (1898) and Murgoci (1905, 1912) indicated Alpine (Cretaceous) low-grade metamorphism in the Danubian units, based on the occurrence of chloritoid in Liassic sediments. This metamorphism was assumed to take place under anchimetamorphic to lowermost greenschist facies conditions (Savu 1970; Stănoiu et al. 1982), with a metamorphic grade increasing from southwest to northeast (Stănoiu et al. 1988). Part of the information available on this low-grade metamorphism in the northern and northeastern part of the window is also based on the occurrence of pyrophyllite (Paliuc 1972; Ianovici et al. 1981), kaolinite (thought to be Alpine) and chloritoid (Iancu et al. 1984; Iancu 1986). Alpine metamorphism was related to nappe thrusting by many authors (e.g. Manolescu 1937; Pop 1973; Berza et al. 1983).

Ocean-floor metamorphism was suggested in literature due to the studies of Savu et al. (1985) and Maruntiu (1987) for the mafic rocks of the Severin nappe, based on prehnite + pumpellyite ± epidote ± actinolite assemblages and K–Ar ages

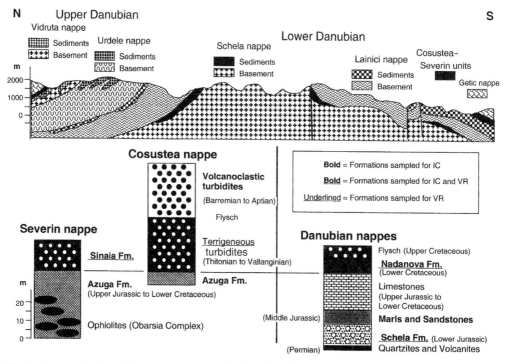

Fig. 3. Cross-section through the easternmost part of the Danubian window (for location see Fig. 2, strongly simplified after Schmid *et al.* 1998). Lithostratigraphic sections are compiled after Seghedi & Berza 1994). Flysch rocks in stratiform contact in Danubian nappes was not recognized during fieldwork, but often described in literature.

of the Jurassic/Cretaceous boundary (127 to 143 Ma according to Lemne *et al.* 1983). In the southwest, the regional occurrence of pumpellyite in successions derived from an accretionary complex and the local transition to pumpellyite–actinolite facies in the northeast were interpreted as evidence for subduction-related Alpine metamorphism of the Severin and Cosustea units (Seghedi *et al.* 1996; Ciulavu & Seghedi 1997). Petrological mineral data from literature need a more critical discussion corresponding to our present knowledge.

Recently, a series of papers used a combined structural and fission-track (FT) data approach in order to discuss the tectono-metamorphic evolution of the South Carpathians (e.g. Schmid *et al.* 1998; Bojar *et al.* 1998; Willingshofer 2000; Fügenschuh & Schmid 2005). Zircon and apatite fission-track ages from the Getic nappes indicate that the Getic nappes were not metamorphosed. Locally, the Getic basis and the Danubian nappes are generally characterized by Cretaceous ages. Cooling to near-surface temperatures of these units immediately followed Late Cretaceous orogeny. Eocene and Oligocene zircon and apatite ages from the part of the Danubian nappes situated SE of the Cerna–Jiu fault monitor Late Palaeogene tectonic exhumation in the footwall of the Getic detachment, while zircon fission-track data from northwest of this fault indicate that slow cooling started during the Latest Cretaceous. The Danubian window is therefore a metamorphic core complex. The change from extension (Getic detachment) to strike-slip-dominated tectonics along the curved Cerna–Jiu fault allowed for further exhumation on the concave side of this strike-slip fault, while exhumation ceased on the convex side. The available FT data consistently indicate that the change to fast cooling associated with tectonic denudation by core complex formation did not occur before Late Eocene times, i.e. long after the cessation of Late Senonian thrusting. Core complex formation in the Danubian window is related to a larger-scale scenario that is characterized by the NNW-directed translation, followed by a 90° clockwise rotation of the Tisza–Dacia block due to rollback of the Carpathian embayment (Fügenschuh & Schmid 2005).

Based on a previous work (Ciulavu 2001), this paper represents a first systematic study attempting to map the grade of metamorphism in the Danubian window on Romanian territory.

Samples from Mesozoic units of the Danubian and Severin units have been analysed by optical microscopy, X-ray diffraction and electron-microprobe. Special attention has been paid to illite Kübler index, bituminite reflectance and vitrinite reflectance data because these methods proved to be efficient for the quantification of metamorphic grade under sub-greenschist facies conditions.

Studied lithologies and sampling

The Danubian unit, subdivided due to lithofacies changes into Upper and Lower Danubian nappes (Berza et al. 1983, 1994, fig. 4), partly comprises Variscan basement nappes. The Variscan cycle ends with locally preserved Late Carboniferous molasse-like sediments and Permian red beds (Fig. 3) followed by Jurassic unconformity and syn-rift sediments. To study Alpine (Cretaceous) metamorphism, the work is focused on the post-Variscan cover.

Coal-bearing Liassic terrigenous deposits in Gresten facies (Schela Formation) directly overlie the basement or Permian rocks. In the western part of the Danubian window, this Schela Formation is overlain by Liassic black belemnite-bearing shales (Ohaba Beds), deposited under anoxic conditions (Nåståseanu 1979). These sediments and meta-sediments were sampled for all analytical methods used (Fig. 3).

The Middle Jurassic is represented by shallow marine quartzites or carbonate sandstones (Stånoiu 1973; Nåståseanu et al. 1985) in the central part of the Danubian window and by Bajocian limestones, Bathonian marls and Upper Jurassic nodular limestones 'ammonitico rosso' in the western part of the window (Codarcea & Raileanu 1960). Late Jurassic to Early Cretaceous carbonate platform sediments overlie these formations (Pop 1973; Berza et al. 1988a, b). Organic matter is very dispersed and mostly of inertinite and alginite composition not suitable for organic matter reflectance studies. Samples for clay mineral studies were collected.

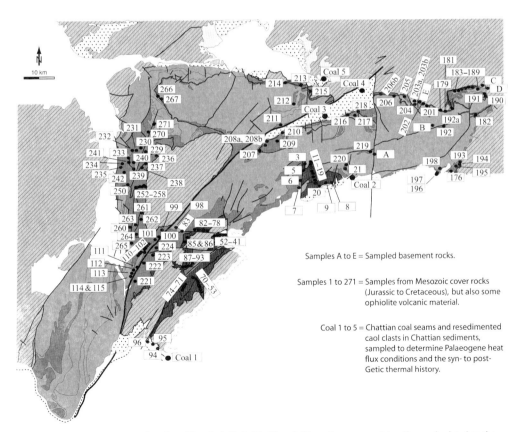

Fig. 4. Sample location map: data from Figs 5, 6, 7, 9, 11, 13 and 17 can be compared together and related to the corresponding location.

In the Arjana nappe west of the Godeanu range (Fig. 2), however, the Middle and Upper Jurassic succession is represented by volcano-sedimentary deposits with basic and alkaline lava flows (Russo-Sǎndulescu, pers. comm.), pyroclastic and epiclastic deposits, stromatactic limestones and red shales (Codarcea 1940). The pelagic limestones and marls of the Nadanova Beds (Albian to Turonian age) and locally flysch sediments (Fig. 4) represent the youngest formations of the Danubian carbonate platform (e.g. Codarcea 1940; Pop 1973). In these lithological units, organic matter was locally observed but mostly altered by oxidation, as also the pelitic rocks and no authigenic mineral could be found mesoscopically. Sampling was very difficult.

Senonian 'wildflysch', represented by terrigenous and volcanoclastic turbidites (Fig. 4) with pelitic background sedimentation, forms the main compound of a tectonic mélange complex. The Cosustea mélange is shown undifferentiated with the Severin nappe in Figures 2, 3, 5, 7, 9 and 13. This mélange complex is characterized by a block-in-sheared matrix structure formed during early stages of tectonic accretion of these sediments to the overlying Severin and Getic units. Its contact to the underlying Nadanova Beds is always tectonic (Seghedi et al. 1996). Organic matter is very dispersed but sufficient and samples for clay mineral studies were collected. Sample locations are shown in Figure 4; the results shown on map Figures 5, 7, 9 and 13 can be related to these sample numbers and the numbers referred to in the text.

The Severin nappe overlies the accretionary wedge in some areas. It includes Late Jurassic radiolarian sediments (Azuga Beds) followed by Early Cretaceous terrigenous turbidites (Sinaia and Comarnic Flysch; see Codarcea 1940). The base of these sediments is formed by a strongly dismembered ophiolite mélange (Savu et al. 1985; Maruntiu 1987), consisting of ocean floor tholeiite basalts (Cioflica et al. 1981), pillow basalts, gabbros and harzburgitic ultramafites.

The Getic–Supragetic nappes found above this accretionary wedge consist of a high-grade metamorphic Proterozoic basement (Iancu & Maruntiu 1994), locally covered by Late Carboniferous to Turonian sediments (a few samples were collected). They were already foliated during the Cretaceous but not overprinted by Alpine orogenic metamorphism. The contact at their base is deformed by the Getic detachment, which produced an extremely discrete jump in metamorphic grade between the non-metamorphic and brittle-deformed Getic–Supragetic nappes and the mylonitized greenschist to sub-greenschist facies units of the Danubian window (Schmid et al. 1998; Matenco & Schmid 1999).

Analytical methods

Optical microscopy and electron-microprobe analysis

Index mineral associations and facies-critical mineral parageneses in basalts, volcanics and volcaniclastics of the Severin and Cosustea nappes were investigated. In order to determine relationships between mineral growth and deformation, optical microscopy was combined with electron-microprobe analysis. For a better qualitative determination and quantification of metamorphic grade, the chemistry of the index minerals has been determined by electron-microprobe analysis (EMPA). Chlorite and white mica chemical compositions were used for approximate temperature and pressure estimates based on the methods of Cathelineau (1988) and Massonne & Schreyer (1987), respectively. Special interest has also been paid to a combined optical and electron-microprobe analysis of the pelitic rocks from the Schela formation in order to determine the lower stability limit of the chloritoid. The determination of chloritoid-formation reactions led to the mapping of a chloritoid-in isograd.

For chemical analyses, a Jeol JXA-8600 Superprobe with an accelerating voltage of 15 kV and a beam current of 10 nA was used. The data were reduced using the PROZA correction. The standards used are: olivine for Si and Mg, rutile for Ti, graphtonite for Fe and Mn, orthoclase for K, albite for Na, wollastonite for Ca, gehlenite for Al and F-topaz for F.

X-ray powder diffraction (XRD)

XRD analyses on pelitic lithologies were used to determine clay-mineral associations and the illite Kübler index (KI) on 170 samples from the Schela and Ohaba Formations, the Nadanova Beds, the Cosustea mélange, the Azuga Bed and Sinaia Flysch. The samples were crushed, decarbonated and prepared using the procedure described in detail by Schmidt et al. (1997).

All X-ray diffraction analyses were performed on a Bruker-AXS (Siemens) D-5000 diffractometer with the following settings: CuKα, 40 kV, 30 mA, V20-V20-2.0 mm slits, step size 0.02, counting time 2 seconds and rotation of the sample during measurement of run interval 2 to 50°2θ. For samples with superposition of peaks in the 10 Å region, the interval 2 to 21°2θ was remeasured with a 0.05° step size and 10 seconds counting time to obtain better diffractogram peak resolution. The clay mineral determination was assisted by Siemens software, namely 'Evaluation' and 'Profile fitting'. Mineral identification and profile analyses were performed using the Siemens software 'Diffrac plus'.

The KI is defined as the full width at half maximum intensity (FWHM) of the first illite basal reflection. FWHM data obtained were transformed into KI values using the correlation from Frey (1986) in order to relate them to literature data with anchizone limits KI at 0.25 and 0.42 $\Delta°2\theta$. We used the KI calibration from the Frey standards. The conservative reason is that most papers on illite 'crystallinity' (the Kübler index was called illite 'cristallinity' (IC) in earlier publications) in the Alpine–Carpathian system refer to Frey (1986, 1987; for details see also Ferreiro Mählmann 1994).

Organic matter reflectance

Organic matter reflectance has been determined on some samples used for the XRD study, for 31 samples from the Schela Formation, 5 from the Ohaba Formation, 4 from the Nadanova Beds and 8 from the Cosustea mélange, but also 3 from Chattian coal seams, and 2 from Chattian coaly clays. Mean random reflectance ($R_r\%$) under non-polarized light of 546 nm wavelength was used for the Tertiary samples of the lignite to sub-bituminous stage A. Maximum reflectance ($\%R_{max}$) and minimum reflectance ($\%R_{min}$) have been measured on Jurassic and Cretaceous samples of the semi-anthracite to meta-anthracite stage (ASTM classification) using monochromatic polarized light (546 nm). Bireflectance ($\%R_{max} - \%R_{min}$) has been used to distinguish the maceral group vitrinite and secondary maceral group bituminite (e.g. Stach *et al.* 1982; Robert 1988; Ferreiro Mählmann 2001).

The analyses were performed on a Leitz Orthoplan-photometer microscope with an ocular ($\times 10$), an oil-immersion objective ($\times 125$), and a photomultiplier with an aperture of 2.5 μm^2 using resin-mounted polished sections and applying standard methods (Robert 1988; Ferreiro Mählmann 1994, 2001).

Results

Detrital mineral associations

Samples from the pelitic background sedimentation of the Sinaia Flysch from the Azuga Bed, Cosustea mélange and Nadanova rocks only occasionally contain detrital mineral assemblages. Thin sections of the Schela and Ohaba Formations from the southern and central area of the Danubian window revealed clasts of white mica (phengite, muscovite–Ms), chlorite (Chl), zoned tourmaline (Tur), red allanite (All), rutile (Rt) and zircon (Zr).

In the central part there was also graphite (Gr) as mineral inclusion or determined by the polymodal highest reflecting organic matter population, light red garnet (Grt) and titano-hematite (Ti–Hem) or hematite (Hem), sometimes together with very minor amounts of K-feldspar (Kfs), plagioclase (Plg), kyanite (Ky) and staurolite (St). All minerals are strongly rounded, and micas and graphite show kinking, disintegration and oxidation due to reworking. The detrital mineral assemblages are related to the basement rocks (also samples A and B) of the Variscan MP-MT metamorphic Dragsan Group (Berza & Seghedi 1983) and will be partly of interest regarding the educts of Cretaceous metamorphic mineral reactions.

Diagenetic and metamorphic mineral associations from the Danubian nappes

EMPA data were used to verify optical studies of very small mineral phases and therefore chemical raw data are not presented in this study. A thermodynamic re-examination and geochemical investigation has to follow our petrogenetic determinations. Because also chemical chrotite-thermometry and phengite-barometry were only of informative character in the present paper, a chemical data presentation was thought to be of minor interest in the study.

Samples of Nadanova Beds consist of illite, chlorite, quartz and rare albite. In contrast, the Schela and Ohaba Formations often have a more complex modal composition and have additionally pyrophyllite, paragonite and chloritoid (Fig. 5).

Kaolinite (Kln) has been identified in some samples of the Schela Formation by XRD. Kaolinite was treated with dimethyl-sulphoxide (causing shift of the kaolinite (001) peak and, consequently, discriminating it from the chlorite (002) peak) or heated at 550 °C (causing the breakdown of the kaolinite structure) and confirmed in a HRTEM study (D. Schmidt 'pers. comm.'). In the northeastern part of the Danubian window, *pyrophyllite* is frequently found in the Schela Formation (Al-rich schists).

Illite-smectite mixed-layer clay minerals, identified after a treatment with ethylene-glycol and heated to 30 °C during 72 h, together with kaolinite and *corrensite* (in one sample), point to diagenetic conditions in the Severin nappe and Cosustea mélange south of the Cerna range and the Brebina fault (Fig. 5). The smectite content in illite was determined using air dry and glycolated specimens. The smectite content is temperature-dependent and the conversion of smectite into illite and the extent of reaction is used as an indicator of diagenetic to metamorphic grade (Frey 1987). Illite of the

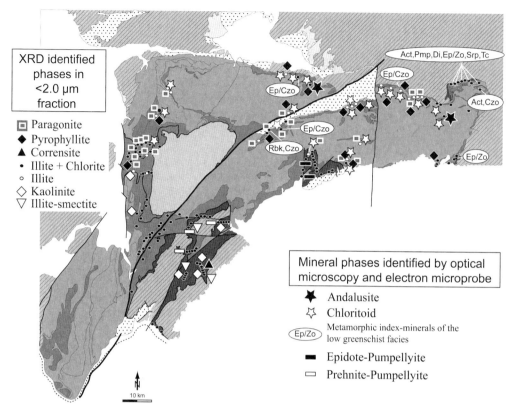

Fig. 5. Distribution of diagenetic and metamorphic mineral phases in Mesozoic sedimentary rocks of the Danubian window. The legend order reflects the increase in metamorphic grade from bottom to top. For mineral abbreviations (after Kretz 1983), see Bucher & Frey (1994) and the text.

diagenetic zone and low anchizone commonly show a marked asymmetry of the 10 Å diffraction peak, with a low-angle tail, indicating the presence of expandable layers (Kübler 1967). In the high anchizone, the tail disappears. After glycol treatment, the smectite content in mixed-layer illite/smectite is recognized by a peak shift (due to swelling) to 16.8 ± 0.1 Å. Swelling of the smectite component leaves the 10 Å reflection free of interference. Surprisingly, most samples also in the diagenetic zone were smectite-free.

North of the Godeanu outlayer and at Schela, north of Bumbesti (Fig. 2), *chloritoid* (Cld) was found (Fig. 5) in the Schela Formation. At these locations, and in the central part of the window, chloritoid rosettes and sheafs are preserved without deformation in massive mesoscopically unfoliated rocks. Mesoscopically also well-crystallized idioblastic chloritoid was found at Rafaila and the southern Tarcul range (Fig. 6b). In both areas, chloritoid is cross-cutting the slaty cleavage with lenticular aggregates of mineral grains (flaser structure) or a poorly developed schistosity with mineral grains, which display a weak preferred orientation. The chloritoid shows a greenish to bluish pleochroism, a lamellar twinning and partly a zonation with a fainter pleochroic rim.

Along the southwestern border of the Petrosani basin (Fig. 2) and in the northeastern part of the Danubian window, colourless chloritoid crystals (Cld1) have also been observed, strongly deformed and partly syn-kinematically recrystallized (Cld2) within a second foliation (Fig. 6e).

Andalusite (And) was found in two samples of the Schela Formation rich in pyrophyllite and Al-white mica at the northern border of the Danubinan window. North of the Petrosani basin (Fig. 5), microprobe observations revealed an aluminosilicate, presumed to be andalusite, intergrown between tabular chloritoid crystals in contact with quartz and pyrophyllite. In the backscatter image, this aluminosilicate is slightly darker in the grey scale than quartz, has euhedral crystals and a poikiloblastic texture. A second occurrence is situated in the Latorita range at Sasa Stefanu (Fig. 2) where poikiloblastic andalusite (Fig. 6c, d) was

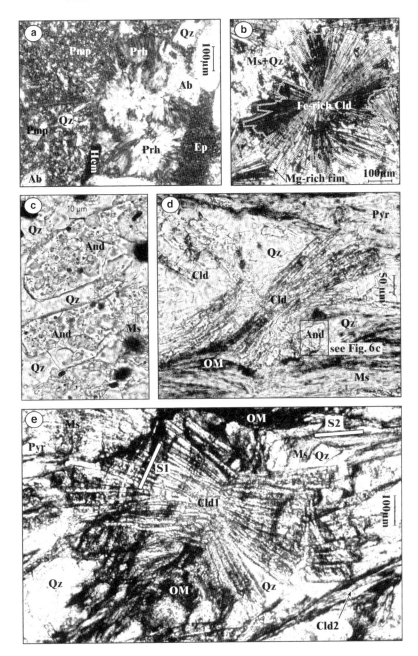

Fig. 6. (a) Prehnite (Prh) in an amygdule replacing plagioclase grains (Albite, Ab) occurring together with pumpellyite (Pmp) and epidote (Ep) in pillow basalt rocks of the prehnite–pumpellyite facies (sample 52 (Fig. 4) close to the Brebina fault). (b) Chloritoid (Cld) rosettes at Rafaila (Fig. 2, sample 219) well preserved without deformation in massive unfoliated rocks. In a combined optical and EMPA study, chloritoid shows a fainter pleochroic Mg-rich rim. (c) Poikiloblastic andalusite (And) at Sasa Stefanu (Fig. 2, sample 192) well identified by the high relief in quartz (see Fig. 6d). (d) Andalusite was identified in a quartz (Qtz) pressure shadow of chloritoid together with Al-rich muscovite (Ms). Pyrophyllite (Pyr) shows rotation and kinking and is part of the chloritoid mineral assemblage, but not in paragenesis with andalusite. A reaction involving andalusite is speculative. (e) Sample from the south of Petrosani (Fig. 2, sample 218) at Hotel Gambrinus, located close to the footwall of the Getic detachment. In the thin section, a rotated, deformed and broken chloritoid (Cld1) is shown. Organic matter (OM), sheet silicate and opaque mineral inclusions oriented in a parallel planar order reflect an old foliation (s_1). The deformation of the Cld1 sheaf was caused by a second foliation (s_2). The formation of quartz, muscovite and a second chloritoid generation (Cld2) in s_2 is shown.

identified by optical microscopy in the pressure shadow of chloritoid (Fig. 6d), formed during deformation related to the Getic detachment. These represent the first occurrences of andalusite found in Al-rich Schela schists of the Southern Carpathians. No metamorphic kyanite or sillimanite was identified in the Mesozoic cover rocks. A blue amphibole was identified as *riebeckite* (Rbk) by EMPA (Fig. 5).

Metamorphic mineral associations from the Cosustea mélange and the Severin nappe

The mineral association identified by XRD in the clay fraction from the Cosustea mélange, from the Azuga and Sinaia Beds of the Severin nappe consist of illite, chlorite, quartz, mixed white mica–paragonite minerals and rare albite. Intermediate peaks between paragonite and muscovite reflections in powder diffraction data are consistent with intimate mixtures of Na- and K-rich micas (Livi *et al.* 1997). These authors have shown in an HRTEM-work that the majority of these complex minerals are best described as domains and not as intercalation (*mixed layer K-mica/paragonite, sensu strictu* Frey 1969). Powder XRD spectra alone do not allow the determination of the exact nature of this metastable phase. Therefore, the name *mixed Na-K white mica* is suggested. Hematite is frequent in samples from the Azuga beds. Some other minerals were detected by transmitted light microscopy:

- *Prehnite* (Prh) is usually present as irregular vein fillings and amygdules (Fig. 6a), or it may replace plagioclase grains. Microscopic folds and/or faults overprinted some prehnite-filled veins. In the southwestern part of the window, prehnite occurs together with pumpellyite (prehnite–pumpellyite facies) in volcanic and volcanoclastic rocks. Prehnite1 is overgrown by fibro- to diablastic partly nematoblastic prehnite2 + pumpellyite + epidote2;
- *Pumpellyite* (Pmp) replaces plagioclase and clinopyroxene. Under prehnite–pumpellyite facies, also pumpellyite has been observed in amygdules (Fig. 6a), or as a relict phase, being broken and elongated along the schistosity and having a lenticular shape. Usually, pumpellyite is preserved only in the core of the crystals, while the rim is transformed to chlorite. In the northeastern part of the Danubian window, unaltered pumpellyite forms patches and is associated with newly formed actinolite (pumpellyite–actinolite facies) and sheet silicates (Fig. 5). In some volcanic rocks from the northeastern part of the Danubian window, granular aggregates of pumpellyite are overgrown by fibrous crystals of the same mineral. It is possible that granular pumpellyite is cogenetic with yellow granular epidote, which is overgrown by colourless epidote–zoisite in the same samples;
- *Actinolite* (Act), together with chlorite, *epidote* (Ep) and pumpellyite, replaces clinopyroxene (augite–aegerinaugite) in gabbros from the northeastern border of the window. Actinolite may also be intergrown with clinozoisite. In gabbroic rocks and basalts, olivine and clinopyroxene are *serpentinized*. The colour and the needle-like habit suggest *antigorite* as the serpentine mineral (Srp, Fig. 5);
- *Zoisite/clinozoisite* (Zo/Czo) has been observed in the assemblage Act + Pmp + Chl + Ab + Ms + Kfs. In the northeast, Zo/Czo is found in pressure shadows related to the deformation of the Getic detachment, mostly in connection with relict feldspars. Euhedral plagioclase (also plagioclase in flysch rocks) and sometimes hornblende in the ophitic texture of the gabbros is partly saussuritized to epidote/clinozoisite + chlorite + albite. Granular epidote is partly overgrown by a colourless epidote–zoisite;
- *Diopside* (Di) occurs as a rim of relict clinopyroxene (Severin nappe, northern border of the Danubian window). Diopside is associated with Srp + Chl + Ep + Pmp + Act (Fig. 5); and
- *Chlorite* (Chl) occurs within volcanic and volcanoclastic rocks, both within the prehnite–pumpellyite and the pumpellyite–actinolite facies areas. However, chlorite is absent in many samples from the two Lower Jurassic formations, specifically at the northern border of the Danubian window (Fig. 5). The ultramafites are strongly serpentinized, and *talc* (Tc) is frequent. Epidote and clinozoisite are found in nearly all rocks of the northeasternmost part of the Severin nappe.

Mineral chemistry

Microprobe analyses indicate that *pumpellyite* commonly found in Severin and Cosustea units is low in Fe. There is a decrease in FeO content, comparing samples from the prehnite–pumpellyite (FeO = 4.52 wt%) and the pumpellyite–actinolite association (FeO = 3.59 wt%).

The chemistry of *actinolite* in mafic rocks of the Severin nappe overprinted by pumpellyite-actinolite facies metamorphism varies between $Mg/(Mg + Fe^{2+})$ ratios 0.69 and 0.81. A trend in chemical variation related to degree of metamorphism (e.g. Schmidt *et al.* 1997) cannot be detected since actinolite was only found at two localities.

Undeformed *chloritoid* rosettes from the central part of the window have an Mg-rich rim, MgO content increasing from 0.72 wt% in the core to 1.69 wt% on the rim (Fig. 6b). The paragenesis is Ms ($\pm$ paragonite) + Al–Chl + Qtz + Hem (Ti–Hem) + Pyr $\pm$ Rt $\pm$ Ep/Zo, and in rare cases paragonite and Fe-chlorite are formed at the expense of chloritoid. The same mineral assemblage without hematite is found at the northern margin of the window, where it is strongly deformed by a second schistosity (main foliation) as shown in Figure 6d and 6e. In many of these samples, chlorite is also missing. Chloritoid (Cld2) formed in the main foliation (Fig. 6e) is Fe-rich or has an iron-rich rim in many samples. Chloritoid in the main foliation may also be replaced by Ms (partly ferri-muscovite) + Al–Fe–Chl + Qtz $\pm$ Ep ($\pm$ Bt).

Paragonite (Pg) composition ranges from that of the Na-mica end-member to a low Na content in cases where interlayer positions are occupied by K-cations. Sometimes, interlayer positions are filled by similar amounts of Na and K. This probably represents a mica/paragonite interstratification (Frey 1969). Towards the muscovite end-member, the interlayer sheet composition is K-rich, still with a relatively high content in Na. No compositional trend as a function of metamorphic grade was found. Mostly paragonite occurs parallel to muscovite and often also parallel in the schistosity plane s_1. Here paragonite grows fibroblastic on muscovite in the shear sense and describes K-mica dissolution and Na-mica cleavage-parallel precipitation as shown by Livi et al. (1997).

Chlorite analyses, when plotted as non-interlayer cation versus Al, scatter along a line from close to clinochlore toward sudoite. In a chlorite $Si-Fe^{2+}-Al$ chemography, the analyses are situated between clinochlore and chamosite. No trend of chemical variation related to degree of metamorphism is evident. Also, the chemical composition gives no indication about a smectite contamination. The chlorite chemistry (Al–Si substitution) was used to determine temperatures with the 'geothermometer' proposed by Cathelineau (1988). Microprobe analyses of authigenic chlorites were normalized to 28 oxygens and fulfill the criterion $\Sigma(Na + Ca + K) < 0.20$ constraining the contamination by other phases. Furthermore, all iron was assigned to Fe^{2+}, because the electron microprobe does not distinguish between Fe^{2+} and Fe^{3+} (see also Cathelineau 1988). In recent years, chlorite 'geothermometry' has been used to determine palaeo-temperatures in sub-greenschist facies metamorphic rocks. According to Cathelineau (1988), the increasing tetrahedral Al content of trioctahedral chlorite is a function of increasing temperature. Schmidt et al. (1997) discussed the 'geothermometer' by a critical revision on the reliability. The decrease of tetrahedral Al in authigenic chlorites as a function of decreasing temperature can be explained by interstratification of smectite with chlorite. TEM studies support this view (Schmidt et al. 1997). Interstratifications in chlorite were not detected by XRD analysis in the samples used. Nevertheless, the chlorite 'geothermometry' is interpreted with caution in this study.

White mica is ubiquitous in all samples analysed. White micas were normalized to 20 oxygens. In samples from the northeastern border of the study area, characterized by the high anchizone to epizone (see below), the interlayer cation content reaches values of 8.6 to 11 wt.% (1.45 to 1.9 atoms per formula unit (pfu)). The Si/2 ratio is 3.3 atoms pfu, and illites from the central part of the Danubian window show a ratio of 2.8 to 3.2 pfu. It is assumed that Si substitutes Al in the tetrahedral position as a function of pressure. Si content of white mica has been used to estimate pressure by applying the Massonne & Schreyer (1987) method only for those samples where the chemical composition indicated a phengitic component with a Si/2 ratio >3.2 pfu. However, because the assemblage phengite + K-feldspar + quartz + phlogopite is not present in our samples, the celadonite content has to be interpreted with caution. Detrital phengite in some samples show higher Si pfu values (>6.8 pfu).

Illite Kübler index

Cosustea mélange, Nadanova Beds, Azuga Beds and Sinaia Flysch (Severin–Cosustea unit) yielded KI values which indicate a trend (Fig. 7) ranging between the high diagenetic zone in the southwest at the Danube, the low anchizone at Bumbesti, the high anchizone at Polovraci and the epizone found in the northern and northeastern part of the Danubian window (in the Retezat range and at Voineasa). From a previous study, the trend is known (Ciulavu & Seghedi 1997). Intra- and intersample variations and petrovariances (dependency on the rock chemistry, Ferreiro Mählmann (1994)) decrease from the diagenetic zone to the anchi- and epizone from KI $\pm$ 0.11 to $\pm$ 0.06 and $\pm$ 0.012 $\Delta°2\theta$.

Since samples from the Schela Formation in the Danubian nappes often contain paragonite and pyrophyllite, the (001) reflection of illite is frequently broadened (the presence of mixed-layer phases was not detected). In anchi- and epizonal illite-muscovites, the $2M_1$ polytype predominates. In these cases, we used two methods to determine KI: (1) deconvolution of the reflection in the 10Å region; or (2) measurement of the FWHM of the (00,10) illite reflection. This latter reflection is not influenced by the presence of paragonite and pyrophyllite. The transformation of the values obtained for the (00,10) illite reflection into values corresponding to the (002) reflection was

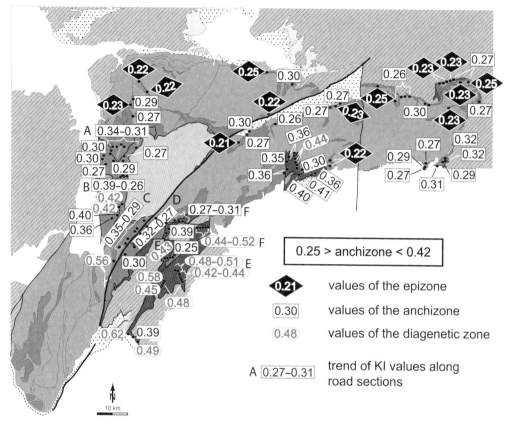

Fig. 7. Illite Kübler index (KI) map of the Danubian window. Values are indicated as Kübler index data $\Delta°2\theta$. The anchizone limits are given by KI values based on a calibration with standards provided by Bernard Kübler and Martin Frey.

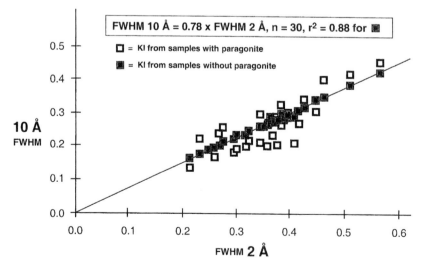

Fig. 8. Correction of illite Kübler index data (FWHM in $\Delta°2\theta$) using the transformation of the values obtained for the (00,10) reflection (2 Å) into values corresponding to the (002) reflection (10 Å). The (002) and (00,10) regression indicated is based only on those samples that lack an illite interference with paragonite and pyrophyllite.

achieved with an empirical regression equation, which correlates (002) and (00,10) reflections (Fig. 8). The results given by both methods are similar, although the values obtained by deconvolution are usually slightly lower (in the second decimal order).

In the area of the northern border of the Danubian window (Tarcul-, Retezat- and northern Latorita range, Fig. 2), the KI values obtained from samples with high organic matter content (determined from samples used also for VR) indicate both high anchizone and epizone conditions. However, the high anchizone values can be attributed to the retardation of illite aggradations. The organic matter forms a hydrophobic mantle around illite crystals, preventing them from aggradations (Frey 1987a). The organic coat is well recognized by reflected light microscopy around phyllosilicates. These samples were not used for the further study.

In the western area (Arjana nappe), diagenetic KI values in samples from the Ohaba Formation were found to be associated with the presence of paragonite and pyrophyllite. Both minerals are typical anchizone metamorphic index minerals (Frey 1987a). In general, in these samples the illite (002) reflection is very asymmetric, with a very long tail towards low 2 theta angles, indicating the presence of an illite/smectite component. In the Ohaba Formation, gypsum is abundant and was probably produced by reaction between sulphuric acid with calcium. Sulphuric acid may result from pyrite weathering. An organic petrography study revealed that pyrite is transformed to an hydroxide. Organic matter is also strongly oxidized. This intense alteration is not visible in the hand-specimen due to the high organic content, which preserves the black colour of the sample. Low diagenetic values from these samples were not used for further correlation and therefore not shown in Figure 7.

Clays from the same area, intercalated in sandstones of the Ohaba Formation, show KI values of the epizone. Mesoscopically, a coarse detrital mica fraction is observed in these samples from the Arjana nappe (Fig. 2). To avoid an over-interpretation due to a possible detrital contamination, these samples were not used for further correlation and therefore not shown in Figure 7.

Using only pelites for a detailed study, most values show anchizonal conditions. Along three road sections (10 samples each) at the east of the Godeanu outlayer, a KI decrease to the east is evident (A, B and C in Fig. 7). At the eastern margin of the window, the KI variation is 0.34 ± 0.06 and at the outlayer it is 0.29 ± 0.03 $\Delta°2\theta$.

In spite of the difficulties regarding non-metamorphic influences on KI data from the Lower Jurassic formations, the trend from the diagenetic/low anchizone in the southwestern part of the Danubian window towards epizonal values for the northeastern parts is shown by the KI study in the Danubian nappes (Fig. 7). This metamorphic gradient indicated by the KI distribution pattern is disturbed across three tectonic structures:

(1) In Figure 7 at the south of the Godeanu outlayer (Fig. 2), diagenetic zone–low anchizone values west of the Cerna–Jiu fault show a SW to NE trend from KI 0.56 to 0.29 $\Delta°2\theta$ (road section C) and are juxtaposed with a KI trend from 0.33 to 0.27 $\Delta°2\theta$ at the east of this fault (road section D, Fig. 7). If the data uncertainty is taken into account, the discontinuity is not pronounced;

(2) A road section (section E) crosses from west to east the Getic outlayer at the south of the Brebina fault (Mehedinti klippen area). In the west, in the Danubian nappes, KI from 0.33 to 0.27 (section D) differs from KI values of 0.39 ± 0.03 $\Delta°2\theta$ (5 samples) in the Cosustea mélange and 0.43 ± 0.02 $\Delta°2\theta$ (5 samples) in the Severin nappe. In the Danubian nappes in the eastern footwall of the Getic outlayer, KI values of 0.25 ± 0.04 $\Delta°2\theta$ (3 samples) differ again from KI values of 0.48 to 0.51 $\Delta°2\theta$ in the Cosustea mélange and 0.42 to 0.48 $\Delta°2\theta$ in the Severin nappe. The diagenetic zone values observed in Severin nappe and Cosustea mélange in the southernmost area contrast with the predominantly anchizonal KI data obtained from the Schela Formation of the Danubian nappes. With a much smaller and not well-defined hiatus, a similar contrast is found between Cosustea mélange (KI = 0.44 to 0.40 ± 0.06 $\Delta°2\theta$, n = 10) and the Schela Formation (KI = 0.30 to 0.36 $\Delta°2\theta$, n = 2) in an area close to Bumbesti; and

(3) The Brebina fault rather abruptly delimits the diagenetic zone (KI = 0.44 to 0.52 $\Delta°2\theta$) found in the Severin–Cosustea units to the south versus the anchizone to the north (KI = 0.27 to 0.31 $\Delta°2\theta$, road section F, Fig. 7).

Organic matter reflectance (OMR)

Organic matter reflectance measured on samples from the Schela and Ohaba Formations, the Cosustea mélange, and Palaeogene coals are shown on the map of Figure 9. The organic matter in pelitic siltstones of the Schela Formation consists of the following percentages of macerals:

- *vitrinite*: 0–20% in the west and south, 70 up to 80% in the east;
- *liptinite*: recognized by its characteristic shape: 2%;

- *inertinite*: (mostly semifuzinite + fuzinite ± macrinite ± micrinite): 40–50% in the west and south, 10% in the east;
- *bituminite*: 30–40%, in pelitic samples and coals from the western area (Ohaba Beds). In the southern areas, bituminite can yield much higher percentages. In the east, the bituminite content is between 10–20%. At high coalification stages, some bituminite varieties with a homogeneous anisotropic reflection can easily be mistaken for vitrinite (Koch 1997; Ferreiro Mählmann 2001). The trend towards more anoxic conditions towards the west and within the Ohaba Beds observed by Nåståseanu (1979) is confirmed in this paper.

Due to the low vitrinite content in the west and south of the Danubian window, bituminite was also considered for maturity determination. Solid homogeneous bituminite occurs as large reworked fragments, or was observed *in situ* in layers parallel to bedding (cata-bituminite). A homogeneous anisotropic variety can be used for maturity studies and not only in diagenetic studies (Jacob & Hiltmann 1985; Jacob 1989) but also under very low-grade metamorphic conditions (Ferreiro Mählmann et al. 2001; Belmar et al. 2002). Solid bituminite is also found to fill pores or late veins (migra-bituminite), which cut through the host rock. In cata-bituminite-rich siltstones of the Schela Formation, vitrinite is sometimes strongly impregnated by migra-bituminite, as described by Radke et al. (1980) and Ernst & Ferreiro Mählmann (2004). This leads to bireflectance values, which are characteristic for bituminite rather than vitrinite (Wolf & Wolff-Fischer 1984; Ferreiro Mählmann 1994). This is well recognizable in coal seams at the mining locality of Schela (samples 21, 220) and in some samples west of the Godeanu outlayer. To avoid an over-interpretation due to a possible bituminite contamination, these samples were not used for further correlation and therefore not shown on Figure 9.

In the eastern part of the window, organic maturation was determined by VR measurements. Many samples, especially in the western part of the

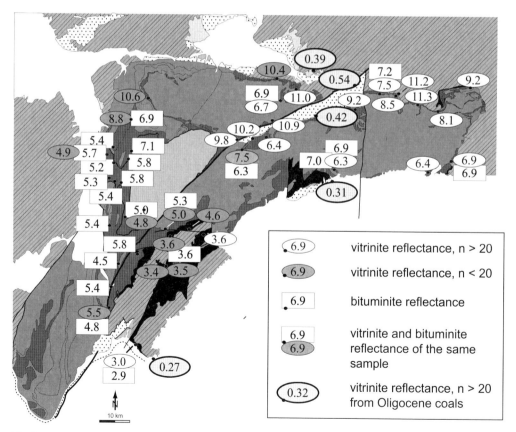

Fig. 9. Coalification map of the Danubian window based on vitrinite and bituminite reflectance %R_{max} data (organic matter reflectance). The Oligocene coals are measured as R_r%.

Danubian window, contained little or no vitrinite but abundant cata-bituminite. In the central and southern part also, only a few VR single values were obtained in one sample, but mostly the mean value is not statistically significant (n < 20 measuring points, $s_2 > 10\%$, Fig. 9). Therefore, a correlation between vitrinite reflectance and cata-bituminite reflectance has been calculated on the basis of Schela samples, which contain both constituents (Fig. 10). In Figure 10, the values are superposed with the correlation given by Ferreiro Mählmann (1994, 2001). Cata-bituminite reflectance correlates well with vitrinite reflectance. Consequently, cata-bituminite reflectance (BR) is used for the evaluation of metamorphic grade. Due to the well-constrained correlation between VR and BR, the combination of the reflectance data will be denominated as organic matter reflectance (OMR $\%R_{max}$).

In strongly sheared samples from the epizone, the organic matter yields much higher reflectance values compared to unfoliated rocks, and bireflectance is anormal. In shear zones, R_{max} reaches 9.0 to 12% for bituminite and 7.0 to 11.3% for vitrinite. These high reflectance values are due to the presence of pre-graphitic vitrinite and bituminite. Pre-graphitization, as described by Ramdohr (1928), Teichmüller (1987) and many others, was optically found (anisotropic or undulatory extinction and a high bireflectance, occurrence of rounded nearly submicroscopic spheres with poorly developed or no Brewster cross and helicitic extinction in cross-polarized light and graphitoid spheroliths). The highest VR were obtained in samples very close to the northern margin of the window deformed by the Getic detachment (Fig. 10). The detachment footwall is formed by mylonites, suggesting a strong influence of strain on the graphitization of organic matter (Bustin 1983; Teichmüller 1987; Suchy et al. 1997; Ferreiro Mählmann et al. 2002). At the east of Petrosani, a sample from a mylonite composed of Schela rocks (sample 205-1) shows a value of 11.2% R_{max}, a nearby sample in Schela schists (sample 205-2), only 100 m away from the mylonite, shows a VR of 8.5% R_{max}.

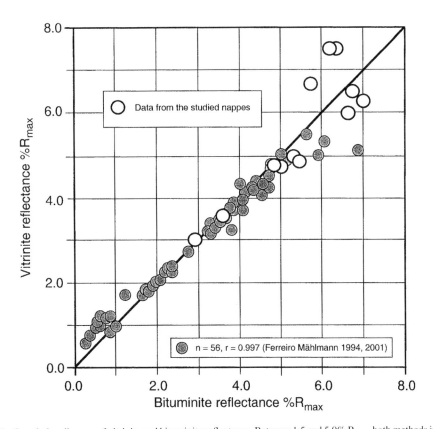

Fig. 10. Correlation diagram of vitrinite and bituminite reflectance. Between 1.5 and 5.0% R_{max}, both methods indicate the same rank of maturation.

All VR reflectance values of 9.2 to 11.3% R_{max} are taken from ductile deformed and penetratively foliated detachment rocks and a strong maturity enhancement to >11% R_{max} is found in mylonites (samples 203b, 205-1, 211, 215). The lowest peak in these samples with a bi- or polymodal distribution (Fig. 11c, d, sample 218 and 203b) is mostly around 7.5 to 8.5% R_{max}. In less-foliated epizonal rocks from the same areas, values of 6.4 to 8.5% R_{max} are found for bituminite and vitrinite. Therefore, the mean values of >8.5% R_{max} close to the detachment do not reflect the grade of metamorphism.

In all parts of the Danubian window, a population with vitrinite reflectance of the meta-anthracite to graphite stage, found in several samples without oxidation, is distinguished as detrital organic matter (e.g. samples 206 and 207, Fig. 11a, b). This is best explained from sample 83 at the north of the Cerna range. A clay-rich foliated siltstone shows silt layers with abundant organic matter, giving a bimodal VR of 8.5 and 10.7% R_{max}. In the clay-rich part, a unimodal peak is determined with a mean value of VR 5.0 and BR 5.3% R_{max}. Few organoclasts show also values of 7.0 to 11.5% R_{max}. Such detrital particles are rotated and have an inhomogeneous extinction; also maximum extinction is not oriented parallel to the foliation or bedding. Usually, they are boudinaged and kinked.

In the epizonal part of the Danubian window, the difference between detrital and pre-graphitic particles (semi-graphite) is not visible due to a complete reorientation of organic matter. Also, the variance of the mean reflectance values of both vitrinite/graphite populations becomes more similar (Fig. 11d) making it impossible to discriminate between detrital and pre-graphitic populations in the reflectance histogram. From the diagenetic zone to the epizone (Fig. 11a, b) in a polymodal reflectance distribution, the lowest rank population is diagnostic for the determination of maturity and rock metamorphism. These populations were shown in the coalification map (Fig. 9). In the epizone, the histograms have a more unimodal shape.

In summary, an increasing metamorphic grade from southwest to northeast is indicated by the OMR data. However, again the distribution pattern is disturbed across three tectonic structures:
(1) the metamorphic discontinuity across the Cerna–Jiu fault is not well recognizable due to scarce OMR data. South of the Godeanu outlayer, an increase in BR from 4.5% R_{max} in the west to 5.4 to 5.8% R_{max} in the east is poorly constrained.

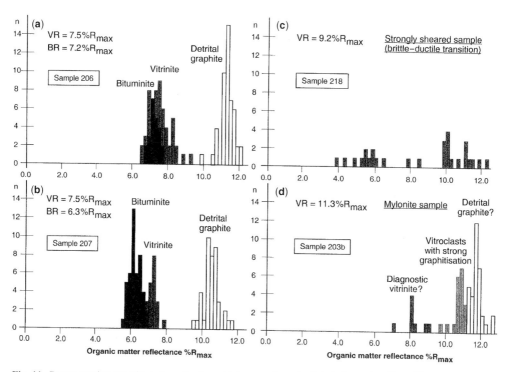

Fig. 11. Representative organic matter reflectance histograms from the Schela formation. From Figure 11a–d, an increase in deformation is indicated (from brittle to ductile). For discussion, see the text. n = absolute number of measurements.

Nevertheless, the change in rank of organic maturation is significant;

(2) The anthracite stage values found in Severin nappe and Cosustea mélange from the Mehedinti klippen (VR 3.0, BR 2.9% R_{max}, southern area and VR 3.4, 3.5, 3.6, BR 3.6% R_{max}, northern area) contrast with mostly high meta-anthracite data from the Schela Formation of the Danubian nappes (VR 4.8, BR 5.5% R_{max}, southern area and VR 5.0, BR 5.3, 5.8% R_{max}, northern area). A sudden increase in rank of organic maturation going across the tectonic contact between Severin nappe/Cosustea mélange and the Danubian nappes in the footwall is evident; and

(3) The zone of anthracite maturation in the south is abruptly limited by the Brebina fault towards the north. OMR values immediately north of this fault show a much higher reflectance (VR 4.6, 5.0, BR 5.3% R_{max}) than values from the same tectonic unit at the south close to the fault (VR 3.6% R_{max}) and further south (VR 3.6, 3.5, 3.4, BR 3.6% R_{max}). The OMR data provide a significant argument to postulate a metamorphic hiatus coinciding with this tectonic structure (Fig. 9).

Discussion

Correlation of metamorphic field trends and establishment of a metamorphic map

Illite Kübler index versus organic matter reflectance. All data of this study, as well as published data on mineral assemblages (Stånoiu et al. 1988; Ciulavu & Seghedi 1997) and FT data (Schmid et al. 1998; Bojar et al. 1998; Willinghofer 2000; Fügenschuh & Schmid 2005) indicate an overall increase in metamorphic grade from SW–NE.

The KI values obtained for Severin nappe and Cosustea mélange range from the diagenetic zone or lower anchizone in the southwest, up to the high anchizone in the east and finally to the epizone north and northeast of the Danubian window. Near the southern border of the Danubian window, KI values always remain anchizonal. Within the Danubian window, the Schela and Ohaba Formations yield high anchizonal to epizonal values along the northern rim of this window, but only anchizonal values along the western border and in the southern part. South of the Godeanu outlayer, some values indicating the diagenetic zone also occur. Overall, the trend given by the KI values is gradual but discontinuous across the Cerna–Jiu fault, across the Brebina fault and in the southern part also across the boundary between the Severin–Cosustea and Danubian nappe systems.

This same trend and discontinuities were confirmed by the measurements of OMR. High values were obtained in the northern and eastern parts of the Danubian window, the values decreasing towards the SW. The correlation between KI and OMR data (correlation coefficient, $r^2 = 0.72$) is presented in Figure 12. The comparison with other IC/KI-VR/OMR, mostly linear regressions is not dependent from different VR measuring techniques,

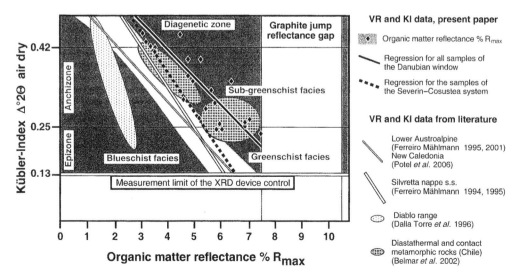

Fig. 12. Correlation diagram of organic matter reflectance and illite Kübler index values. The regressions for different sample groups from the Danubian window are compared with literature data (see text).

illite-calibrations and standardization procedures, because all studies cited were performed by the low-grade working group in Basel or by collaborating groups.

The KI-OMR regression line obtained for the Danubian window has a different slope and position compared to correlations obtained for samples from different tectonic units of the Alps and other areas (Fig. 12). In the white area of Fig. 12, all data of KI-VR correlations are integrated from studies in which the metamorphic imprint was generated by pressure and temperature conditions intimately related to LT–LP orogenesis (Frey et al. 1980; Krumm et al. 1988; Ferreiro Mählmann 1994, 1995, 1996; 2001; Rahn et al. 1994; Schmidt et al. 1997; Árkai et al. 2002). The area of orogenetic diagenesis/metamorphism is limited by the 90% confidence boundary.

Correlation trends situated to the right side of this white area are derived from areas where high thermal gradient sub-greenschist to greenschist facies events were recognized (Krumm et al. 1988; Ferreiro Mählmann 1994, 1995; Belmar et al. 2002). Metamorphism is triggered by hyperthermal conditions as diastathermal-, contact-, hydrothermal- or ocean floor metamorphism. Correlation trends on the left side, however, comprise areas where hypothermal orogenic metamorphism is described ('high pressure sub-greenschist facies' conditions, Ferreiro Mählmann 1994; Petrova et al. 2002) and data from the Diablo range are characteristic for a blueschist facies event (Dalla Torre et al. 1996).

Steady-state equilibrium is documented in the Lower Austroalpine (Fig. 12) as shown by Ferreiro Mählmann (2001). With an identical regression, the one from New Caledonia (Potel et al. 2006) is given by a complete re-equilibration under normal-thermal conditions.

Compared with the correlation from the Lower Austroalpine and New Caledonia, that derived for the Danubian window points to enhanced VR all the way from the diagenetic zone to epizone (Fig. 12). Hence, on the basis of the previous discussion, this indicates 'high temperature gradient sub-greenschist facies' conditions. Furthermore, the data from the Danubian window are similar to those derived for samples from contact metamorphic rocks (Belmar et al. 2002). Note that samples from the Severin nappe and Cosustea melange give a regression line (Fig. 12) noticeably different from that composed by all samples of the Danubian window. The regression falls in the diagram on the right border of the white area related to orogenic (LT–LP) conditions (temperature-dominated). The slope is sub-parallel to that of the Silvretta nappe. A diastathermal regime describes Ferreiro Mählmann (1995) in the latter Alpine tectonic unit.

Sub-greenschist facies metamorphism of the meta-basalts

The prehnite–pumpellyite facies (Fig. 14), restricted between 200 to 270 °C and 2.0 to 3.8 kbar (Frey et al. 1991), is characteristic for the volcanic and volcanoclastic rocks of the Severin–Cosustea nappe system adjacent to the southern part of the Danubian window (Mehedinti klippen) and is correlated with the zone of diagenesis (Fig. 13). In some samples, epidote is found together with prehnite and/or pumpellyite (Fig. 6a), post-dating an older Chl–Prh, Ep–Prh and Ep–Chl mineralization. The feldspars show no sericitization and are not partly saussuritized as is the case under epizonal conditions prevailing northeast of the Danubian window. The lower stability limit of epidote at low pressures <4.0 kbar is given at 200 to 220 °C (Seki 1972) by the following reactions, both possible in the study area (Fig. 6a):

$$2Prh + Hem \rightarrow 2Ep + H_2O \text{ (Seki 1972)} \quad (1)$$

$$6Pmp + 4Qtz + 3Hem \rightarrow 12Ep + Chl + 11H_2O \quad (2)$$

Reactions (1) and (2) are found in the high diagenetic zone to lower anchizone, and OMR data indicate semi-anthracite to low anthracite coal rank.

In the northeast, in the basalts of the Severin nappe, the mineral association $Ep + Zo/Czo + Pmp + Act + Chl + Qtz$ points to the pumpellyite–actinolite facies, which indicates temperatures in excess of 300 °C. Diopside was also identified in this area. The lower limit of the diopside stability field (Bucher & Frey 1994) also points to a minimum temperature of 300 °C. The maximum temperature is roughly estimated by the upper limit of the pumpellyite stability field at 250 to 350 °C (at 2 to 9 kbar, Frey et al. 1991).

The kaolinite-pyrophyllite reaction isograd

The clay mineral fraction from pelitic rocks of the Schela Formation from the southern Danubian window consists of illite + chlorite ± kaolinite in the diagenetic zone. Additionally, mica/paragonite, pyrophyllite and Al-chlorite appear in the anchizone. In the northern and eastern epizonal parts of the Danubian window, pyrophyllite, paragonite, mixed-layer K-mica/paragonite or mixed Na–K white mica and chloritoid are frequently found (Fig. 5) as a typical assemblage of Al-rich pelites.

With increasing diagenetic to incipient anchizonal grade, pyrophyllite appears as the first index mineral. The pyrophyllite (Pyr)-in isograd in the Danubian units is located west of the Godeanu outlayer. The reaction:

$$Kln + 2Qtz \rightarrow Pyr + H_2O \text{ (Thompson 1970)} \quad (3)$$

postulated for the western border of the window (Fig. 5) is pressure-independent (Hemley et al. 1980; Frey 1987b) but strongly dependent on water activity (Frey 1978). In Al-rich sediments, this reaction is reported at water in excess of temperatures ≥ 260 °C (Fig. 14). Water in excess is indicated, as in the Danubian window, by the large amount of different phyllosilicates (Theye 1988) formed during metamorphism.

In contrast, based on a different kinetic approach, Iancu et al. (1984) and Iancu (1986) suggested metamorphic conditions of 390 to 420 °C at PH$_2$O = 2.0 kbar for the Schela Formation, based on the same reaction Kln + 2Qtz $\rightarrow$ Pyr + H$_2$O. These estimates of metamorphic conditions, also referred to in some recent papers, are too high for the following reasons:

(1) at 370 to 400 °C, the association Ky + Cld would be expected in pyrophyllite-rich rocks. Instead, we observed And + Qtz, indicating that the temperature was lower than 370 °C at pressures lower than 4 kbar (Fig. 14). Note that andalusite is found far away to the north of the pyrophyllite reaction isograde (Fig. 5), being part of a syn- to post-kinematic Getic detachment mineral association (Fig. 6c, d);

(2) the high content in organic matter suggests a water activity less than 1.0 a(H$_2$O), which leads to a lowering of temperature for all dehydration reactions. Therefore, Frey (1987b) and Livi et al. (1997) give temperatures of 260 to 280 °C for the kaolinite-pyrophyllite association in the presence of CH$_4$ or CO$_2$ at a(H$_2$O) of 0.6 to 0.9;

(3) pyrophyllite has been reported at the transition from the high diagenetic zone to the anchizone (Frey 1987b; Merriman & Frey 1999), but also at temperatures as low as 200 °C (Frey 1987a; Ferreiro Mählmann 1994);

(4) based on new thermodynamic calculations, temperatures of more than 300 °C postulated by Thompson (1970) and Iancu (1986) with a(H$_2$O) $\leq$ 1 are unrealistic (Theye 1988). The assemblage Ka + Chl + Ms + Qtz is stable between a(H$_2$O) 0.5 and 1.0 below 3.5 and 5.5 kbar, but also temperature is strongly a(H$_2$O) dependent. At an a(H$_2$O) of 0.6, temperature is close to 240 °C (Potel et al. 2006); and

(5) kaolinite in the anchi- and epizone is related to a post-metamorphic P-T retrograde event. Some thin sections from the Schela Formation taken from localities close to the Petrosani basin and along the Cerna–Jiu fault show fibrolitic, wool fibre-like or cloudy kaolinite aggregates between pyrophyllite and chloritoid. This is confirmed by microprobe studies and high magnification images obtained by HRTEM studies (David Schmidt, pers. comm.). The HRTEM study indicates that the kaolinite is newly formed between pyrophyllite layers and more abundant in screw dislocations, and in twin re-entrant angles of chloritoid were the potential of screw dislocations is amplified, providing suitable growth sites for kaolinite and some smectite. Feldspars were also altered to kaolinite.

The pyrophyllite-chloritoid reaction isograd

In the Schela Formation from the northern epizonal part of the area, chlorite is present only in 40% of the samples analysed. Chlorite is absent in those samples from the high anchizone to epizone, where both pyrophyllite and chloritoid are present in the clay fraction (Fig. 5). In samples 202, 203, 206b, 217, 219, 266 and 267 from the epizone, phyrophyllite is also missing. A breakdown of chlorite and phyrophyllite is well known for the epizone (Frey & Wieland 1975). This indicates that chlorite and pyrophyllite have probably been consumed by the reaction:

$$Pyr + Chl \rightarrow Cld + Qtz + H_2O \text{ (Zen 1960)} \quad (4)$$

The choritoid–chlorite thermometer (Vidal et al. 1999) could not be used due to the fact that chlorite is either entirely consumed in the above reaction or represents a post-chloritoid phase (see below). It is suggested that the initial sediments were Al-rich and poor in iron and magnesium. Based on thermodynamic calculations, a temperature of 300 °C (independent of pressure) is indicated for this reaction (Theye et al. 1996). Sometimes, hematite is still present in the same samples. At slightly higher grade (increasing VR up to values typical for the anthracite/meta-anthracite stage), the following reaction was observed by optical microscopy (Fig. 14):

$$Chl + Pyr + Hem \text{ (Ti-Hem)} \rightarrow Cld + Qtz \\ + TiO_2\text{-phase} + H_2O \text{ (Frey 1969)} \quad (5)$$

Due to this reaction, rutile is formed in the Schela Formation. Theye & Seidel (1991) showed that chlorite reacts with hematite at about 310 to 320 °C and produces chloritoid, quartz and some magnetite. In the northern Latorita range, in the epizone (meta-anthracite/semi-graphite stage), bipyramidal opaque minerals are also found, the habitus being typical for magnetite. Therefore, the hematite–magnetite reaction (Fig. 14) is also present in the samples 203, 205 and 206b:

$$Chl + Pyr + Hem \rightarrow Cld + Qz + Mag + H_2O \text{ (Frey 1969)} \quad (6a)$$

An additional reaction is proposed for some samples by combining reactions (5) and (6a):

$$Chl + Pyr + Ti\text{-}Hem \rightarrow Cld + Qz + Mag + Rt + H_2O \quad (6b)$$

An iron oxide was observed under the microprobe and in thin sections, but could not be found mesoscopically. Magnetite is a typical product of chloritoid formation as hematite is consumed at the same time (Bucher & Frey 1994). The chloritoid is Fe-rich, hematite being an important source for its formation (Frey 1969). From reactions (1) and (2) to (6a) and (6b), an increase of metamorphic grade is evidenced (Fig. 14), correlating well with KI and OMR results.

In two samples, Fe-chlorite, together with biotite and mica, was found to form the product of a retrograde reaction decomposing chloritoid. In two other samples, andalusite was also formed due to retromorphism.

Metamorphic map

Similar trends indicated by the results obtained from different methods, as well as the good correlation between these trends, allow the construction of the metamorphic map presented in Figure 13. Due to the lack of data in the central part of the Danubian window and to both sides of the Cerna–Jiu fault, the known dextral movement of about 35 km along the Cerna–Jiu fault (Berza & Drăgănescu 1988; Ratschbacher et al. 1993), which must have been post-metamorphic (Schmid et al. 1998), was additionally taken into account

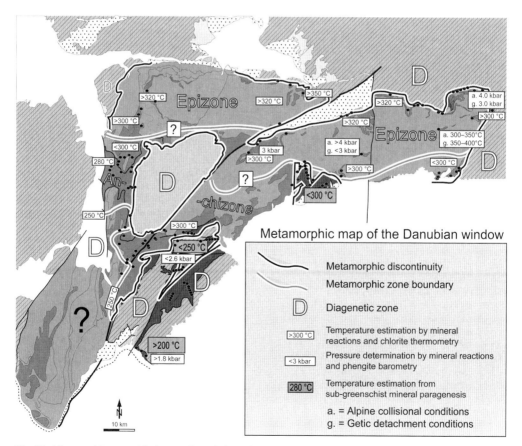

Fig. 13. Metamorphic map with the zone boundaries defined by illite crystallinity (diagenetic zone, anchizone and epizone) and the correlation with vitrinite reflectance, also including some temperature and pressure estimation.

when extrapolating the anchizone–epizone limit in Figure 13. Here, all metamorphic minerals are shown, independent of the chronology of their appearance. Hence the map only reflects maximum P–T conditions.

Relations between metamorphism and deformation

The base of the *Getic units* generally exhibits brittle deformation. In the northeast, syn-detachment deformation near the quartz brittle–ductile transition is restricted to the first ten metres. Within shear zones and specifically within the basement rocks near the Severin–Danubian boundary, a strong mineral banding, oriented parallel to the detachment foliation, is observed. Bands of (a) Ep/Czo + Chl, (b) Act − hornblende + Chl ± Qtz, and (c) Qtz + Ab/Plag ± Bt indicate syn-detachment greenschist facies conditions.

The metamorphic minerals of the prehnite–pumpellyite–epidote assemblage (Fig. 6a) from the *Severin-Cosustea unit* (Mehedinti klippen) were overprinted by the foliation related to the Getic detachment, or they were brittle deformed. The prehnite–pumpellyite facies assemblage is therefore pre-kinematic, and pre-dates the Getic extensional detachment.

In the Severin nappe of the northeastern part of the window, most minerals of the Act–Pmp–Di–Ep/Zo–Srp, Ep–Czo and Act–Czo assemblage + Cc–Qtz–Ap–Tc (Fig. 5) were affected by the detachment foliation and therefore also pre-kinematic to normal faulting. In samples exhibiting a strong detachment foliation, the mineral assemblage shows syn-kinematic deformation, as indicated by some pressure shadows, kinking of phyllosilicates, fragmented crystals and displacement-controlled fibres. A second generation of actinolite and epidote/clinozoisite is syn-kinematically formed. The feldspars were probably saussuritized synchronously. In a very few cases, both minerals, together with Fe-chlorite, cross-cut the detachment foliation and are associated with a late crenulation cleavage.

In the *Danubian units*, it is very difficult to evaluate metamorphic grades during Cretaceous nappe stacking and Palaeogene detachment formation separately. The mineral paragenesis Pg + Ms + Qtz + Cld found in the Tarcul range and at 'Rafaila' (Fig. 2) roughly indicates greenschist facies conditions. The minerals are well preserved with an idioblastic shape. Such undeformed assemblages are found in lower structural units, i.e. further away from the Getic detachment (samples 219, 266, 267). It can be thought that they were unaffected by deformation related to the Getic detachment or that deformation diminishes with increasing depth (Fig. 17). It is also possible that higher temperatures prevailed during the detachment so that the paragenesis continued to recrystallize. Chlorite, hematite and pyrophyllite are missing in these samples (Fig. 6b), therefore also the following reaction, not involving pyrophyllite, can be assumed:

$$Ms/Pg + Chl + Hem \rightarrow Cld + Qtz + H_2O \text{ (Ghent } et\ al.\ 1989) \quad (7)$$

For the first time, blue riebeckite was found in one sample (Fig. 5), together with clinozoisite. Both minerals are affected by solutions related to the syn-detachment foliation. In some other samples, old rotated epidote/zoisite, chlorite and white mica were found. Chemical EMPA composition and thermo-barometric interpretation roughly indicate that temperatures of 300 to 350 °C and pressures around 3 to 5 kbar (Ms Si/2 (pfu) = 3.25 ± 0.05) prevailed prior to the detachment event. This is in good agreement with the chloritoid reactions (5) and (7) observed in some nearby samples (220, 218, 271). In summary, the mineral assemblage, the calculated temperatures, and the chloritoid (Cld1) formation reaction with and/or without pyrophyllite are probably related to a first greenschist facies and deformation event (Fig. 6e) that pre-dates the Getic detachment.

In the northern and northeastern Danubian window, greenschist facies conditions prevailed during the activity of the Getic detachment, since indicative mineral assemblages (samples 179 to 191) are stable within Getic detachment mylonites (Schmid et al. 1998). In syn-detachment micas, the Si content is commonly lower compared to that of the older and rotated micas, pointing to lower pressure conditions. In most cases, the mica is ferri-muscovite.

The thermal evolution during the Getic detachment can be only reconstructed in the Latorita range. In the other areas of the Danubian nappes, the results concerning P–T data related to pre-, syn- and post-Getic detachment stages are only scanty. Regarding the retrograde path (syn- to post-Getic), the occurrence of clinozoisite suggests temperatures >300 °C, while biotite + muscovite + chlorite indicate temperatures around 400 °C (Fig. 14). Hence, temperatures after chloritoid (Cld1) formation are slightly higher and evidenced by the syn-detachment Ep/Czo-Act/hornblende-Plag-Bt mylonites. Syn-detachment pressures of around 3 kbar, based on the phengite barometry (Ms Si/2 (pfu) = 3.20 ± 0.05) are indicated with caution for the northeastern border of the Danubian window. Andalusite formed along a pressure retrograde path and limits the maximum pressure to 4.5 kbar (Bucher & Frey 1994) at the aluminium–silicate triple point, but to 3.5 kbar at 400 °C (Fig. 14).

After formation of the Getic detachment and during rapid cooling, post-Eocene fluid circulation and a low temperature hydrothermal activity is evidenced by the presence of small veins cutting the detachment-related deformation with Qz + Chl + Sm + Kln. Zones with an intense but very local kaolinitization are revealed along the main Oligocene to Miocene faults and related to a relative high maturity of Oligocene coals (Fig. 9), specifically in the area around the Petrosani basin.

Interpretation of metamorphic data

In the diagenetic zone (Fig. 13), between the Cerna range and the Danube (Fig. 2), mineral associations, including clay minerals, represent mostly a primary sedimentary or volcanic origin. Because the entire stratigraphic section (Permian to Cretaceous) occurs under metamorphic conditions all over the Danubian window, the lithologies of this area are used as reference for reaction educts.

Metamorphic evolution of the Severin-Cosustea nappe system

Ocean-floor and Alpine metamorphism in the Mehedinti klippen area. In basites and ultramafites of the Mehedinti klippen area, index minerals epidote + prehnite, chlorite + prehnite and chlorite + epidote are related to an old fabric. Epidote veins and amigdule fillings together with prehnite, chlorite, albite and also smectite, are remnants of a typical ocean-floor metamorphism (Robinson & Bevins 1999). The assemblages have been described in several earlier papers (e.g. Cioflica et al. 1981; Maruntiu 1987; Savu et al. 1985). However, diagenesis from the sedimentary rocks was high enough to overprint a burial KI and OMR trend. Rocks of different age at the same structural level show identical grade of diagenesis. In the slices of the ophiolitic-sedimentary sequence in the accretionary wedge, a top-down increase of diagenesis–anchimetamorphism is not recognized. The increase of diagenesis and maturity to the northeast is a post-wedge tectonic trend within the Severin–Cosustea units.

Attempting a P–T determination, diagenesis cannot have been below 200 °C in the entire diagenetic zone studied since discrete smectite was not identified and liptinite and bituminite do not show a lower reflectance than vitrinite phytoclasts (Ferreiro Mählmann 1996, 2001). In the southern Mehedinti klippen area, pumpellyite was not found; therefore we can conclude that minimum conditions for orogenic diagenesis are established at >200 °C, in agreement with reaction (1) and the formation of prehnite, not exceeding 2.5 kbar (point S-SC in Fig. 14, samples 94 to 96), in agreement with the epidote–phrenite stability field.

The prehnite–pumpellyite facies identified in sandstones from the Cosustea mélange and in basalts from the Severin nappe (point N-SC in Fig. 14, samples 52, 85, 87), is characteristic for the northern part of the Mehedinti klippen area (Fig. 5). Orogenic (Cretaceous) diagenesis (e.g. Manolescu 1937; Stånoiu 1972) has to be distinguished versus ocean-floor metamorphism (e.g. Maruntiu 1987; Savu et al. 1994). Ocean-floor relicts are again present in epidote veins and amigdules. Prehnite is overgrown by prehnite2 + pumpellyite + epidote2. The prehnite2–pumpellyite facies assemblage reflects re-equilibration to higher grade.

Pumpellyite is extremely rare from oceanic crust and related hydrothermal systems. Pressure is also generally too low for the stability of pumpellyite in mid-ocean ridge systems. This commonly results in the formation of prehnite + epidote (Frey et al. 1991; Alt 1999). Prehnite–pumpellyite facies is indicative for a LT-HP trend in the sub-greenschist facies (Frey et al. 1991) and typical in the LP zone (<4 kbar) for convergent settings (Alt 1999; Robinson & Bevins 1999).

KI data from the prehnite–pumpellyite facies in the northern part of the Mehedinti klippen area correspond to the diagenetic-/anchizone limit. Temperatures around 250 °C, based on the high diagenesis zone-anchizone temperature limit (Kisch 1987), are not very reliable due to the data uncertainty, but also kaolinite is present in nearby Al-rich samples (Fig. 5). Also the use of the chlorite thermometry is controversial, giving unrealistic high and inconsistent results, with a mean of 300 to 330 °C and variations in the range of ±50 °C. Bulk rock chemistry, especially the Al/(Fe–Mg) ratios of rocks, strongly control the chemistry of chlorite (Árkai et al. 2002b). The original thermometer (Cathelineau & Nieva 1985) was elaborated for meta-andesites. Chlorite geothermometry values from pelites in the diagenetic zone often appear too high when compared with those indicated by the other methods (Schmidt et al. 1997) and should not be discussed, or with caution, at higher grade (De Caritat 1993).

In Figure 14, the southern (S-SC) and northern (N-SC) part of the Severin–Cosustea unit of the Mehedinti klippen are presented in a P–T diagram. The T °C estimation discussed were related with the Si (pfu) mica values as a first approach and plotted on the MP gradient of the

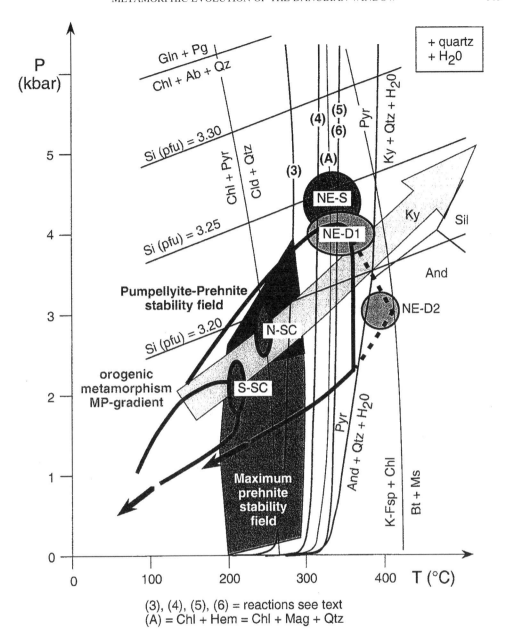

Fig. 14. Petrogenetic grid of the reactions found in the Danubian window. S-SC = southern part of the Severin–Cosustea nappe system; N-SC = northern part of the Severin–Cosustea nappe system (both at the south of the Godeanu outlayer); NE-S = northeastern Severin nappe; NE-D1 = Alpine conditions during the first deformation (nappe stacking) in the northestern Danubian nappes; NE-D2 = conditions during and after the second deformation (Getic detachment) in the northeastern Danubian nappes. Shown also is the typical LT part of the orogenic MP path from Bucher & Frey (1994).

orogenic metamorphism path (Bucher & Frey 1994). The phengitic mica points to low pressures of <3.0 kbar, but mostly the mica is a ferri-muscovite. Muscovites with very small celadonite component cannot be used for pressure estimation.

With the available petrological and clay mineralogical data, a precise pressure determination is not possible.

On the basis that temperatures are around >200 °C (S-SC) and <250 °C (N-CS), VR/OMR is 3.0 (S-SC) and 3.6% R_{max} (N-SC), and pressures probably below 2.5 (S-SC) and 3.0 kbar (N-SC) a kinetic maturity and numerical model calculation for a better P–T approximation is used. The thermo-barometry proposed by Dalla Torre et al. (1997) was applied (see also Ernst & Ferreiro Mählmann 2004), combined for forward coalification simulation with the EASY%R_o model from PDI-PC 2.2-1D (IES GmbH Jülich). From discussed field evidence, the diagenetic field gradient between S-SC and N-SC is synchronous in the klippen area. Therefore, the model is calibrated between the lowest temperature from S-SC and the highest from N-SC. In the OMR/VR-T °C plot (Fig. 15), the maturity range must fit with the temperature range from Figure 14. Pressure was calculated for a diagenetic interval of 1.0, 5.0, 10 and 15 million years. Below 5 million years, modelled temperatures are consistently too high to simulate OMR and above 15 million years pressures do exceed the stability field of prehnite and much above the Si/2 (pfu) 3.20 line (Fig. 14). The best fit results use 10 million years for modelling pressures of 2.1 ± 0.2 kbar (S-SC) at 215 to 245 °C for 3.0% R_{max} and 2.6 ± 0.2 kbar (N-SC) at 245 to 265 °C for 3.6% R_{max} (Fig. 15). The plotted ovals for S-SC and N-SC represent the range of error from measuring and modelling (Fig. 15).

In Figure 16, a very conservative approach is done using the EASY%R_o model from PDI-PC 2.2-1D. Calibration as for the kinetic maturity model is applied. Additionally, burial during sedimentation is modelled assuming a heat flow value of 1.5 HFU = 63 mWm^{-2}. The sedimentary burial modelling step was very important to synthesize maturity in some Alpine nappes (Ferreiro Mählmann 2001; Árkai et al. 2002a). In the presented model, changes of ± 5.0 mWm^{-2} do not significantly influence the result, but during tectonic burial variations of ± 1.0 are much more important. During nappe stacking, a heat flow value of 60 mWm^{-2} was chosen, giving credit to a collisional scenario on the kyanite path (Bucher & Frey 1994). Time in the collisional part of the model is the main factor. At maximum depth, heating time during 10 million years is again proposed, reflecting the age determinations of 100 Ma (mica ages) and 80 Ma (zircon FT ages, see discussion for the Danubian nappes) and the results from the kinetic model. The EASY%R_o PDI-PC 2.2-1D result is very similar to that of the kinetic model (Fig. 15) and the petrogenetic grid (Fig. 14). At 3.0% R_{max}, temperature is 220 °C and eroded tectonic overburden is determined around 6 km (Fig. 16), therefore pressure should be 1.75 kbar for S-SC and 2.55 kbar for N-NC (8 to 9 km overburden).

The estimated metamorphic path shown in Figure 14 connecting S-SC with N-SC leads to a heat-flow path through the P–T field of 40 to 60 mWm^{-2} (Bucher & Frey 1994) and is therefore normal- to hypothermal.

The results obtained in this area show a coherent thermal pattern related to regional diagenesis postdating ocean-floor metamorphism. The comparison

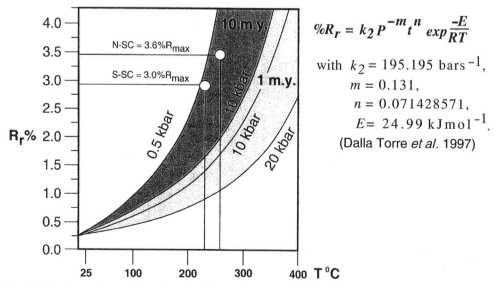

Fig. 15. Kinetic maturity model to substantiate the petrological and clay mineralogical study on Cretaceous metamorphism. For explanations, see text.

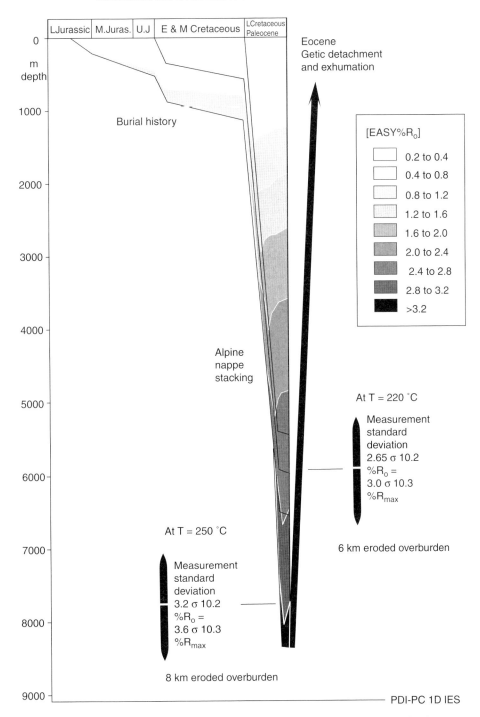

Fig. 16. EASY%R$_o$ PDI-PC 2.2-1D (IES GmbH Jülich) basin evolution to nappe stacking model for the Mehedinti klippen area.

between grades of diagenesis, together with FT data on zircon (220 to 173 Ma) and apatite (55 to 50 Ma) from the very detailed study by Bojar et al. (1988) in the Mehedinti klippen area, suggest that an important Palaeogene orogenic thermal imprint did not affect the area south of the Brebina fault. Therefore, the indication by pumpellyite, together with the result that the values of OMR correlate with KI (Fig. 12), showing a typical trend for orogenic metamorphic conditions at LP on the kyanite path (Fig. 14), post-ocean floor metamorphism is related to Cretaceous Alpine orogeny. While the slope of the P–T path, defined by S-SC and N-SC, is typical for a LT–HP trend in the sub-greenschist facies between the kyanite and glaucophane P–T path.

Subduction-related Cretaceous metamorphism in the eastern Severin–Cosustea unit. While KI and OMR show an increase of metamorphism, the epidote + pumpellyite assemblage at Bumbesti (Fig. 5, samples 3, 5, 11 to 20) is not facies indicative and strongly dependent on fO_2, $a(H_2O)$, Fe_2O_3 and whole rock composition (Potel et al. 2002). The lack of pyrophyllite and paragonite in clastic sediments, and also the lack of actinolite in volcanic rocks of the anchizone, indicate that metamorphism did not exceed 300 °C (Merriman & Frey 1999; Robinson & Bevins 1999) in this area.

At the northernmost corner of the Danubian window, the metamorphic mineral assemblages in the Severin nappe (NE–S) are partly deformed by the fabrics related to the Getic detachment, as is the case for the older granular population of pumpellyite. In case of the replacement of clinopyroxene by actinolite and diopside in gabbros, the relationship to deformation is not known.

For the ophiolite and flysch rocks, temperatures between 300 and 350 °C are roughly estimated from the Alpine mineral paragenesis as shown above. Chlorite in the s-c fabric of s_1 in foliated volcanics (basalts–andesites to prasinites) give 310 to 345 °C using the chlorite thermometry. Chlorite thermometry from volcanic rocks of the epizone is in agreement with other temperature estimations as was found for the Severin nappe. Although the use of the chlorite 'geothermometer' proposed by Cathelineau (1988) for volcanics is uncertain when applied to meta-sediments, the temperatures obtained are often in good agreement with temperatures derived by other methods (Schmidt et al. 1997; Ferreiro Mählmann 2001). In case of the pelite samples (mostly silty marls) from the Cosustea mélange and the Severin nappe, temperatures increase from 300 ± 20 °C at Bumbesti to 340 ± 20 °C in the eastern Latorita range. Chlorite found in the Nadanova beds of the same area indicates temperatures around 340 ± 20 °C. Chlorite2 in the syn-Getic detachment s_2 fabric is not very frequent; the fabric is dominated by mica. A few temperature determinations show no significant difference to chlorite1.

For the older mica population, the phengitic mica (Si/2 pfu > 3.23) points to pressures of ≥ 4.0 kbar (point NE-S in Fig. 14, see also Fig. 13). In general, the Si/2 pfu differences of white mica indicate a relative pressure increase from southwest to northeast, but the celadonite content differs strongly in different samples. Pressure conditions during s_1 based only on phengite barometry are not well established. Using the P–T estimate, the pre-Getic orogenic metamorphism was probably pressure dominated and is located in the P–T diagram (Fig. 14) at the transition of the HP greenschist to blueschist facies. Because of many uncertainties, the P–T area for NE-S is much larger than shown for S-SC and N-SC, but also reflects the same thermal path in the P–T diagram.

Comparing all areas studied in the Severin-Cosustea units, it is demonstrated that metamorphism post-dates ocean-floor metamorphism and sedimentary burial, but is pre-kinematic in respect to the Getic detachment. Hence it is a subduction-related sub-greenschist to greenschist facies Cretaceous metamorphic event (Seghedi et al. 1996; Ciulavu & Seghedi 1997) close to the glaucophane path, related to the 'Austrian' phase of the older literature (e.g. Stǎnoiu 1972). Based on radiometric ages of amphibole K/Ar-ages, biotite K/Ar- and Rb/Sr- ages, the metamorphic peak is dated at around 100 ± 10 and 110 ± 10 Ma (Ratschbacher et al. 1993; Grunenfelder et al. 1993; Dallmeyer et al. 1996).

The increase of pressure with increasing metamorphic grade from south to north results in a subduction geometry to the north and an accretionary wedge thrusting top-south. Therefore, a Late Cretaceous nappe stacking top-south/southeast, as postulated by Schmid et al. (1998) and not top-east/northeast as reported by Ratschbacher et al. (1993) and Willingshofer (2000) is substantiated by the data of this paper.

Regarding the syn-detachment conditions in the northeastern part of the Danubian window, basalts from the Severin nappe show the following mineralogical association characteristic for the pumpellyite–actinolite facies: Chl + Ep/Czo + Ab + Act + Pmp + Qtz. The Zo/Czo + Pmp + Act assemblage, together with paragonite (determined by XRD), is typical for a HP transition (higher pressure bathozone) from sub-greenschist to greenschist facies (Robinson & Bevins 1999). The conditions did not change considerably, but the celadonite content in mica is much lower, indicating a pressure decrease from pre- to the syn-detachment metamorphism.

The replacement of hornblende by actinolite with epidote and rutile in syn-Getic mylonites (samples C, D and E) of the same area probably points towards higher pressures (Maruyama et al. 1986) than those expected for common greenschist conditions, or were controlled by oxygen fugacity and/or the bulk composition of the host rock. The transition from green hornblende to actinolite in the basement rocks may also indicate pressures decreasing to conditions <3.0 kbar during the detachment, together with an increase in temperature (Black 1977; Sperlich 1988). A systematic geochemical study of the basement rocks has to verify this hypothesis.

Metamorphic evolution of the Danubian nappes

Accretion and nappe stacking. For the transition from diagenetic zone to anchizone, temperatures of 250 °C are inferred. The kaolinite–pyrophyllite reaction isograd (3) indicates temperatures >260 °C, probably around 280 °C (Fig. 13, between samples 230 and 250) due to the high a (H_2O) and the close vicinity of reaction (4).

The maximum metamorphic temperatures, such as indicated by reactions (4) and (5) in the Schela Formation of the Tarcul and Vilcanu range (Fig. 2), are between 300 and ≥ 320 °C (Figs 13 and 14). Cooling in the west of the Cerna–Jiu fault is dated at 81 in the west to 51 Ma in the east (Willinghofer 2000; Fügenschuh & Schmid 2005). The reaction products (4) and (5) show deformation by s_2 close to the Getic detachment but cross-cut the flaser structure or a poorly developed schistosity with also mineral grains displaying a weak preferred orientation s_1. Metamorphism occurred after last sedimentation in the Early Cretaceous and prior to cooling at 81 Ma. Cretaceous deformation s_1 is related to nappe thrusting and the mineral paragenesis accommodated due to tectonic burial. Metamorphism can be related syn- to post-kinematic to the deformation during plate convergence posterior to the subduction of the Severin 'ocean' and collision of the Danubian continental block with the Getic–Supragetic units (Berza et al. 1994).

In the northeastern corner of the Danubian window, lower greenschist facies conditions (Stănoiu et al. 1982; Iancu 1986) in the semi-graphite stage (Popescu et al. 1982) at maximum temperatures of 400 °C were also inferred by Ratschbacher et al. (1993) for the northeastern corner of the Danubian window. This assumption is not evidenced by mineral reaction in the s_1 foliation. Relicts of the Cretaceous metamorphism (Pg1–Ms1–Chl1–Cld1–Czo1–Rbk) are typical for Barrovian orogenic low greenschist facies metamorphism but an increase in metamorphic grade is not demonstrated. Relicts are found, as in the locality of Rafaila far away from the detachment (Fig. 17).

Only Ms1, Chl1 and Czo1 grow in many samples syn-kinematic in s_1. Therefore, the EMPA study was focused on the regional variations in the chemistry of Chl1 and Ms1. 300 to 320 ± 50 °C were obtained in Schela Formation samples from the north of the Petrosan basin (Retezat range, samples 212, 213) and in the northern Cerna range, increasing to 350 ± 30 °C in the Vilancu range (samples 3, 210) and in the northern Latorita range (samples 181, 189, 192) and 350 ± 50 °C at Schela (sample 21). In a Nadanova marl at Sasa Stefanu (sample 192b) from a homogenous composition analysis, a temperature of 325 ± 25 °C was calculated. The wide range of inferred temperatures is probably due to a detrital contamination, which is not recognized in the backscatter image. Big crystals of probably detrital origin gave calculated values around 400 to 450 °C. Compared with the other methods, the temperatures calculated are much too high in the Al-rich Schela Formation. The results confirm that the use of the chlorite 'geothermometer' proposed by Cathelineau (1988) for volcanites is unreliable when applied to Al-rich sedimentary rocks (Jiang et al. 1994). In case of Al-rich metapelites, a re-calibration of the temperature values is needed (Árkai et al. 2002b).

The result from the Nadanova marl may be close to realistic temperatures (see discussion related to samples from the Cosustea–Severin units). From phengite barometry, an increase of pressure from the Retezat range (>3.0 kbar) to Schela (>4.0 kbar) is evident on the same samples used for the Chl1 study. The Ms1 NW to SE trend may reflect the accretionary wedge geometry explaining increasing pressure with structural depth pointing to Late Cretaceous nappe stacking top-south/southeast, as postulated by Schmid et al. (1998).

Getic detachment and syn-kinematic metamorphism. Illite Kübler index, vitrinite reflectance, mineral associations and mineral chemistry, but also zircon fission-track (FT) data (Fügenschuh & Schmid 2005), all present the same trend, which indicates, different to that of s_1, an increase in metamorphic grade from southwest to northeast within the Danubian units. Shown by zircon FT ages, cooling started in the west in the Cretaceous and in the east during Palaeogene combined with a trend from partial to complete annealing (Fügenschuh & Schmid 2005). The clay mineral and maturity trends in the Danubian nappes, as

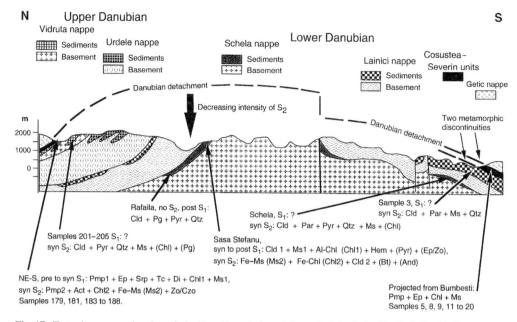

Fig. 17. Tectonic cross-section through the Danubian window at the east of the Sadu–Petrila fault with the metamorphic results from the Latorita range. Data from the Severin–Cosutea unit west of Bumbesti were projected into the section showing two progressive metamorphic discontinuities at nappe boundaries. Also the reason for preservation of Cretaceous Alpine conditions (s_1) at a structural deep position at Rafaila is indicated. Getic deformation during s_2 could not blur the Alpine fingerprint in Mesozoic cover rocks.

also the s_2 mineral paragenesis, is distinguished by a different distribution pattern and not controlled by the structural level of the Cretaceous nappe system or topographic elevation.

The metamorphic pattern shown on Figures 13 and 17 cannot be related to a Cretaceous orogenic scenario, as shown in the last chapter, but was generally postulated (e.g. Mrazec 1898; Murgoci 1912; Manolescu 1937; Pop 1973; Berza *et al.* 1983), with a thermal peak during the Cenomanian to Turonian (Iancu *et al.* 1984; Berza *et al.* 1988a), and more recently dated at 100 Ma (Grünenfelder *et al.* 1983; Ratschbacher *et al.* 1993; Dallmeyer *et al.* 1996; Bojar *et al.* 1998). The main metamorphic pattern needs re-interpretation.

The lower temperature boundary of the anchizone at $\pm 250\,°C$ at the north of the Cerna–Jui fault (not found in the south of the fault) in the Danubian nappes is extraordinarily high (Fig. 13). In a collisional orogenic setting without magmatism, temperatures limit the anchizone in the range between $220 \pm 20\,°C$ and $300 \pm 20\,°C$ (Frey *et al.* 1980; Kisch 1987; Merriman & Frey 1999; Frey & Ferreiro Mählmann 1999; Ferreiro Mählmann 2001). The correlation between OMR and KI points to hyperthermal heat flow conditions and to a high thermal gradient sub-greenschist facies.

Data from Chl2 chlorite and Ms1 muscovite–phengite thermometry and barometry have to be discussed with the same caution as for the s_1 study. The temperature calculations from Chl2, applied on the same samples, are in the identical range of error showing higher temperatures, but a higher standard deviation. In the marl sample 192a from Sasa Stefanu, a homogeneous Chl2 chemistry resulted in temperatures of 350 to 400 °C. Regarding Ms2, the Si/2 (pfu) values scatter around 3.2 (most being lower and identified as ferri-muscovite). Only three samples (213, 220, 189) fulfil the criteria for the estimation, indicating 3.0 kbar in the Retezat range (identical to s_1 conditions), <3.0 at Schela and 3.0 at Voineasa. It can be inferred that the metamorphic temperatures during the Cretaceous evolution and during the Eocene Getic event were very similar while pressure decreased.

Cld1 formation is post-kinematic to Alpine nappe stacking (samples 219 at Rafaila, 266 and 267 in the Tarcul range) and pre- to syn-kinematic to the Getic detachment (northeastern Danubian nappes). In the s_2 foliation, Cld2 is present from south of the Hatec basin to Voineasa, all along the northern margin of the Danubian window and in the epizone east of the Petrosani basin. In the Retezat Mountain and in the east and south of

Petrosani, reactions (A), (5) and (6) are defined from thin and EMPA sections. A temperature of >320 °C is defined by these reactions in agreement with the minimum values of Chl2 thermometry (by accident?).

As described above, the neoformation of And + Bt + Chl + Ms (Fe-Ms) + Qtz in the pressure shadow of chloritoid (Cld1) was syn-kinematic to Cld2 within the schistosity related to the Getic detachment (Fig. 6c, d). K-Fsp in grain contact is also involved in the reaction (Fig. 14). It is possible that peak temperatures of 380 to 410 °C measured by chlorite thermometry and derived from Figure 14 were reached close to the end of the Getic detachment movement. This points to a temperature increase after Alpine orogenesis. Most data are in accordance with a P–T loop typical for a setting between orogenic collisional and HT-LP metamorphic conditions (Fig. 14) on the andalusite path. Maximum pressure (Cretaceous) and temperature (Eocene) were not reached simultaneously, but this is typical for all clockwise loops without isothermal decompression.

Under low-pressure conditions, OMR/VR is particularly dependent on temperature and time because the maturation of thermodynamically metastable organic matter is a mere expression of reaction rate (e.g. Hood et al. 1975; Waples 1980; Hunt et al. 1991; Ernst & Ferreiro Mählmann 2004). Therefore, the OMR can provide very important information about the thermal regime. The thermobarometry proposed by Dalla Torre et al. (1997) and Ernst and Ferreiro Mählmann (2004) could not be applied because for OMR >5.0% R_{max} kinetic modelling is not calibrated.

OMR is very high compared to KI and mineral paragenesis data. As deduced from Figures 12 and 14, the metamorphic imprint was dominated by a thermal event under low-pressure conditions and high heat flow. Comparing our data (Figs 12 and 13) with published data (e.g. Frey et al. 1980; Rahn et al. 1994; Ferreiro Mählmann 1994, 1996, 2001; Belmar et al. 2002), the high VR values (3.5 to 4.5% R_{max}) at the upper limit of the diagenesis zone are associated with either hyperthermal (>70 m Wm^{-2}; Robert 1988) heat flow conditions or high geotherms (>35 °C km^{-1}). From the general correlation (Fig. 12), a VR value of 4.4% R_{max} is calculated, possibly indicating enhanced heat flow during the detachment evolution.

A short heating time is indicated by the fact that OMR suggests a higher metamorphism than that deduced from the corresponding KI values and the mineral assemblages. The retardation of mineral reactions and illite aggradation versus OMR, specifically under short thermal events or during short time with elevated heat-flux, is well demonstrated (see Wolf 1975; Teichmüller 1987; Barker 1989). A short heating time is also explained by the local detection of HT reactions and a poor mineral re-accommodation to the new P–T conditions. During crustal thinning in the South Carpathians and exhumation of the Danubian nappes, the chemical composition of minerals probably did not change and nearly no mineral neo-formation occurred. However, during the temperature increase, OMR was enhanced due to the high reaction rate of the organic maturation processes. The observed disequilibrium between KI, mineral paragenesis and OMR, similar to conditions known from contact metamorphic settings (e.g. Bostick 1973; Barker & Pawlewicz 1994; Elliot et al. 1999; Belmar et al. 2002), can only be preserved during a short time of temperature and/or heat flow increase.

The relatively elevated OMR in Danubian window can be well explained by isothermal decompression during the detachment related to a slight elevation of the geotherm due to crustal thinning and rapid exhumation of Danubian units. High heat flow can be enhanced by the rapid uplift of hot basement rocks with a high thermal conductivity. An elevated higher heat flux probably followed during maximum exhumation at the end of the Getic detachment, a time when the retrograde chloritoid decomposition reactions took place. During exhumation, heat flow increases to >150 mWm^{-2} during uplift of the Tauern dome (Fig. 1), as is well demonstrated by maturity studies (Sachsenhofer 2001). Thermal effects of exhumation of a metamorphic core complex on hanging wall syn-rift sediments were also well studied from the Rechnitz window (Fig. 1, Dunkl et al. 1998). The scenario demands a thermal influence to be considered in the Getic-Supragetic units. In this paper, it is evidenced: (1) by a ductile deformation of the basis of the Getic–Supragetic units in the northeast of the Danubian window (deformation is found in the first 10 m on top of the detachment Ep–Chl–Act–mylonites); and (2) by sediments on the Getic basement showing a high maturity. Coal seams and resedimented coal particles of Oligocene age were coalified to the sub-bituminous to high volatile bituminous stage at the southeast of the Hatec basin and in the Petrosani basin. In the area of Bumbesti and close to the Danube, the coal remains in the lignite stage (Fig. 9).

A slight increase in temperature in the Danubian nappes is supported in Figure 12 by the occurrence of andalusite and biotite (NE-D2). Fresh biotite, actinolite, blue-green amphibole and chlorite cross-cutting all microstructures are known from the basement rocks (Tudor Berza, pers. comm.). An initial increase of temperature is typical during the thermal evolution of a metamorphic core complex footwall (e.g. Genser et al. 1996; Dunkl et al. 1998). An extremely rapid shortening combined with an advective heat transport causes an elevation of isotherms and thus the increase of near-surface geotherms (Koons 1987; Mancktelow & Grasemann 1997; Fügenschuh et al. 1997). Therefore, the main metamorphic pattern in Figure 13 is caused by a syn-Getic, Eocene metamorphism, and the Danubian window can be regarded as a very low- to low-grade high thermal gradient metamorphic core complex. Cooling below zircon annealing is determined for the Retezat and Vilcanu range at 64 to 54 and 67 to 46 Ma (Bojar et al. 1998; Schmid et al. 1998; Willingshofer 2000; Fügenschuh & Schmid 2005). East of the Sadu–Petrila fault, thermal metamorphism did not decrease below zircon annealing at 31 to 21 Ma (Fügenschuh & Schmid 2005). A much longer preservation of metamorphic conditions in this area may be the reason for establishment of equilibrium conditions in the microscopic scale and for identification of poikiloblastic andalusite (Fig. 6c, d) at Sasa Stefanu (Fig. 2) by optical microscopy together with biotite and chloritoid2.

Getic detachment and cooling history of the Danubian window. Fission-track data from the literature (Fügenschuh & Schmid 2005) show three main trends: age decrease toward the north (east), towards lower tectonostratigraphic levels and towards the Tarcul–Retezat dome. The temperatures obtained in the northeastern part of the Danubian window confirm the temperatures in excess of 320 °C indicated by FT analysis (Schmid et al. 1998; Willingshofer 2000). These data emphasize a north(east)-ward-directed unroofing of the dome that was coupled with an increase in strain intensity at its northern margin (Bojar et al. 1998; Schmid et al. 1998; Willingshofer 2000; Fügenschuh & Schmid 2005).

In the Tarcul–Retezat dome, zircon FT data indicate cooling below 250 °C during the Campanian to Oligocene (Schmid et al. 1998). Therefore, the reactions (3) and (4) were of pre-Campanian age but post-nappe tectonic (Cenomanian to Turonian, Ratschbacher et al. 1993; Grünenfelder et al. 1983; Dallmeyer et al. 1996). In the east of the Cerna–Jui fault, cooling is much younger as determined by zircon data (Maastrichtian to Oligocene, Fügenschuh & Schmid 2005) and reactions (4), (5), (6) and (A) may have occurred later or reaction progress continued. Therefore, the zone boundaries on both sides of the fault do not need to be synchronous.

The trends shown by Figure 13 can be well explained by asymmetric extension, starting in the south(western) part of the Danubian window and propagating towards north(east). Asymmetric extension led to the rise of an asymmetric thermal dome, which results in higher temperatures for a longer time in north(eastern) parts of the Danubian window (Fig. 14). In Figure 14, this is shown by thermal re-equilibration to NE-D2 instead of an isothermal decompression.

In the Mehedinti klippen area and at Bumbesti, a diagenetic–metamorphic discontinuity separates diagenetic to low anchizone rocks of the Severin–Cosustea system from hot high anchizonal (Cerna range) to epizonal rocks (Vilcanu range) of the Danubian nappes. The disturbed metamorphic pattern between Severin–Cosustea nappes and Danubian nappes can be explained by a progressive metamorphic hiatus (Fig. 17). A sudden P–T increase towards the footwall is typical for the displacement of a cool upper structural unit on a hotter lower unit along a normal fault. Probably normal faulting between the Severin–Cosustea system and the Danubian nappes is cogenetic with the Getic detachment.

In the Mehedinti klippen area and at Bumbesti, the ocean-floor metamorphism and the diagenetic to anchizonal metamorphic pattern related to Cretaceous nappe stacking in the Severin–Cosustea system was preserved and remained unaffected by the hyperthermal Getic low-pressure event documented for the central and northern part of the metamorphic core complex. The normal faults caused a displacement of the higher nappes towards the southeast, to the extremity of the dome. Consequently, this area escaped the high heat flux in the central part of the dome and the stage of the Alpine KI-VR equilibrium (note the different trend in Fig. 12) was not altered.

Conclusions (geochronology)

(1) Jurassic ocean-floor (prehnite–epidote facies) and Cretaceous collisional orogenic metamorphism (prehnite–pumpellyite facies in the southwest to pumpellyite–actinolite facies/incipient HP greenschist facies in the northeast) is constrained in the Severin–Cosustea nappe system. The trend in increasing metamorphic grade is related to a subduction geometry top-north/northwest. Due to the metamorphic discontinuity at the base of the Cosustea wildflysch and Severin nappe in the Mehedinti klippen area and the displacement to southwest by a normal fault (probably a syn-Getic detachment fault), the Jurassic and Cretaceous metamorphic high diagenetic conditions were also not blurred in sedimentary rocks by the Maastrichtian to Oligocene thermal event.

(2) Two steps in the Cretaceous metamorphic path can be distinguished. The peak metamorphic conditions in the syn-nappe stacking epizone of the subducted Severin unit are close to those of the blueschist–greenschist transition. Also the syn

s_1 (first deformation) metamorphism in the Danubian nappes appears characteristic of LT-HP collisional orogenic conditions between the kyanite and glaucophane path. Peak pressure conditions ≥ 4 kbar in the structural lowest units under sub-greenschist to greenschist facies were reached in the Mesozoic cover during top-southeast Cretaceous nappe stacking. The accretionary wedge was formed during Cenomanian to Turonian time.

(3) In the west of the Cerna–Jiu fault, reaction isograds and the diagenetic to metamorphic zone-boundary pattern are established due to metamorphic accommodation after nappe thrusting (diagensis to greenschist facies, diagenetic zone to epizone). Metamorphic conditions indicate an increase in heat flow and temperature, related to an early stage of Getic detachment movements in the Senonian. Cooling started at Maastrichtian time. The metamorphic pattern is re-equilibrated between the Cenomanian and Maastrichtian.

(4) In the east of the Cerna–Jiu fault peak temperature conditions were reached during or slightly after formation of the Eocene Getic detachment. Decompression during the detachment event caused a high thermal gradient greenschist facies metamorphism with the characteristics of hyperthermal conditions between the kyanite and andalusite path, but far away from a contact metamorphic path. This is best explained with crustal thinning and exhumation, because no magmatism is known. Cooling followed in the west during Maastrichtian and Palaeocene time and in the east during the Eocene and Oligocene. In this area, reaction progress may have continued until the Oligocene.

(5) Crustal thinning is located between the Getic-Supragetic nappes and the Severin–Cosustea nappes. However, part of it must be located between the Severin-Cosustea nappes and the Danubian nappes. The amount of missing crust increases from the southwest to the north and northeast indicated by the metamorphic progressive discontinuity between the foot- and hanging wall of the detachment and at the normal fault below the Mehedinti klippen area.

(6) Oligocene to Miocene faults (Cerna–Jiu, Brebina) disturb the metamorphic pattern in the Danubian window. A very low-grade hydrothermal event is related to Oligocene–Miocene faulting.

We thank Susanne Th. Schmidt and Willem B. Stern for their introduction to microprobe and X-ray diffraction work, respectively. B. Fügenschuh is thanked for his cooperation regarding comparisons with fission-track data. David Schmidt is thanked for HRTEM analyses. Our work benefited from discussions with Tudor Berza who generously shared with us his comprehensive knowledge regarding the Danubian window. M. C. thanks Sebastién Potel and Mauricio Belmar for their help during the last years. R. F. M. thanks Patrik Mathys and Ulrich Szagun for their kind help in terms of computational support. We also thank Ronan Le Bayon for trying several thermodynamic calculations on chlorite and chloritoid. Willy Tschudin, our patient expert, prepared our thin and polished sections. The Swiss National Science Foundation is thanked for financing the project through Research Grant for the Cooperation in Science and Research with Central and Eastern European Countries: CCES/NIS joint research project No. 7RUPJ048623 and for support through SCOPES 2000–2003 project 7RUPJ062307. RFM wants to note that the project benefited very much from the extraordinary good relation between structural geologists and metamorphic petrologists in Basel; therefore, the political decision to close geosciences is very tragic. The authors are grateful to two anonymous reviewers of the first manuscript, but deeply indebted to Prof. Peter Arkai for critical reviewing and constructive corrections and suggestions. Thanks are extended to PD Dr. Ulrich Glasmacher for very helpful comments.

References

ALT, J. C. 1999. Very low grade hydrothermal metamorphism of basic igneous rocks. *In*: FREY, M. & ROBINSON, D. (eds) *Low-Grade Metamorphism*, Blackwell Sciences, Oxford, 169–201.

ÁRKAI, P., FENNINGER, A. & NAGY, G. 2002a. Effects of lithology and bulk chemistry on phyllosilicate reaction progress in the low-T metamorphic Graz Paleozoic, Eastern Alps, Austria. *European Journal of Mineralogy*, **14**, 673–686.

ÁRKAI, P., FERREIRO MÄHLMANN, R., SUCHY, V., BALOGH, K., SYKOROVA, I. & FREY, M. 2002b. Possible effects of tectonic shear strain on phyllosilicates: a case study from the Kandersteg area, Helvetic domain, Central Alps, Switzerland. *Schweizerische Mineralogische und Petrographische Mitteilungen*, **82**, 273–290.

BARKER, C. E. 1989. Temperature and time in the thermal maturation of sedimentary organic matter. *In*: NAESER, N. D. & MCCULLOH, T. H. (eds) *Thermal History of Sedimentary Basins. Methods and Case Histories*. Springer, Heidelberg, 73–98.

BARKER, C. E. & PAWLEWICZ, M. H. 1994. Calculation of Vitrinite reflectance from thermal histories and peak temperature: A Comparison of methods. *American Chemical Society Symposium*, 216–229.

BELMAR, M., SCHMIDT, S. T., FERREIRO MÄHLMANN, R., MULLIS, J., STERN, W. B. & FREY, M. 2002. Diagenesis, low-grade and contact metamorphism in the Triassic-Jurassic of the Vichuquen-Tilicura and Hualane-Gualleco Basins, Coastal Range of Chile. *Schweizerische Mineralogische und Petrographische Mitteilungen*, **82**, 375–392.

BERZA, T. & DRĂGĂNESCU, A. 1988. The Cerna–Jin fault system (South Carpathians, Roumania), a major Tertiary transcurrent lineament. *Dari de Seama ale Sedintelor Institutul Geologie si Geofizica*, **72**, 43–57.

BERZA, T., IANCU, V., BALINTONI, I. ET AL. 1994. Excursions to South Carpathians, Apuseni Mountains and Transylvanian Basin. Description of stops, *Field Guidebook ALCAPA II, Romanian Journal of Tectonics and Regional Geology*, 105–146.

BERZA, T., KRAEUTNER, H. G., DIMITRESCU, R. & ANON. 1983*a*. Nappe structure in the Danubian Window of the central South Carpathians; Travaux du douzieme congres de l'Association geologique Carpatho-Balkanique. *Anuarul Institutului de Geologie si Geofizica (Annuaire de l'Institut de Geologie et de Geophysique)*, **60**, 31–39.

BERZA, T., SEGHEDI, A. & ANON. 1983*b*. The crystalline basement of the Danubian units in the central South Carpathians; constitution and metamorphic history; Travaux du XIIeme congres de l'association geologique Carpatho-Balkanique; Metamorphisme-magmatisme-geologie isotopique. *Anuarul Institutului de Geologie si Geofizica (Annuaire de l'Institut de Geologie et de Geophysique)*, **61**, 15–22.

BERZA, T., SEGHEDI, A. & DRĂGĂNESCU, A. 1988*a*. Unitatite danubienne din versantul nordic al muntilor Vilcan (Carpati Meridionali). *Dari de Seama ale Sedintelor Institutul Geologie si Geofizica*, **72**, 23–41.

BERZA, T., SEGHEDI, A. & STANOIU, I. 1988*b*. Unitatite danubiene din partea estica a muntilor Retezat (Carpatri Meridionali). *Dari de Seama ale Sedintelor Institutul Geologie si Geofizica*, **72**, 5–22.

BLACK, P. M. 1977. Regional high-pressure metamorphism in New Caledonia; phase equilibria in the Ouegoa District. *Tectonophysics*, **43**, 89–107.

BOJAR, A. V., NEUBAUER, F. & FRITZ, H. 1998. Cretaceous to Cenozoic thermal evolution of the southwestern South Carpathians: evidence from fission-track thermochronology. *Tectonophysics*, **297**, 229–249.

BOSTICK, N. H. 1973. Time as a factor in thermal metamorphism of phytoclasts (coaly particles); Septieme Congres International de Stratigraphie et de Geologie du Carbonifere. *Compte Rendu Congres International de Stratigraphie et de Geologie du Carbonifere (International Congress on Carboniferous Stratigraphy and Geology)*, **7**, 183–193.

BUCHER, K. & FREY, M. 1994. *Petrogenesis of Metamorphic Rocks.* 6th edn. Heidelberg, Springer Verlag.

BUSTIN, R. M. 1983. Heating during thrust faulting in the Rocky Mountains; friction or fiction? *Tectonophysics*, **95**, 309–328.

CATHELINEAU, M. 1988. Cation site occupancy in chlorites and iilites as a function of temperature. *Clay Minerals*, **23**, 471–485.

CATHELINEAU, M. & NIEVA, D. 1985. A chlorite solid solution geothermeter. The los Azufres (Mexico) geothermal system. *Contributions to Mineralogy and Petrology*, **91**, 235–244.

CIOFLICA, G., SAVU, H., NICOLAE, I., LUPU, M. & VLAD, S. 1981. Alpine ophiolitic complexes in South Carpathians and South Apuseni Mountains. *Carpato-Balcanic Geological Association, XII Congress, Guidebook 18, Excursion A. 3, Bucharest.*

CIULAVU, M. 2001. *Alpine metamorphism of danubian and severin nappe system and related ore mineral deposits.* Unpublished Ph.D thesis. Bucharest University.

CIULAVU, M. & SEGHEDI, A. 1997. Very low-grade Alpine metamorphism in the Danubian Window (South Carpathians, Romania). *In*: GRUBIC, A. & BERZA, T. (eds) *Geology of the Djerdap Area.* Geoinstitut, Belgrade, Special edition, 291–295.

CODARCEA, A. 1940. Vues nouvelles sur la tectonique du Banat meridional et du plateau de Mehedinti. *Anuarul Geologic Institutul Al României*, **XX**, 1–74.

CODARCEA, A. & RAILEANU, G. 1960. Mezozoicul din Carpatii Meridionali. *Studii si cercetari de Geologie, Geolizică, Geografie, seria Geologie*, **4**.

DALLA TORRE, M., DE CAPITANI, C., FREY, M., UNDERWOOD, M. B., MULLIS, J. & COX, R. 1996. Very low-temperature metamorphism of shales from the Diablo Range, Franciscan Complex, California; new constraints on the exhumation path. *Geological Society of America Bulletin*, **108**, 578–601.

DALLA TORRE, M., FERREIRO MAEHLMANN, R. & ERNST, W. G. 1997. Experimental study on the pressure dependence of vitrinite maturation. *Geochimica et Cosmochimica Acta*, **61**, 2921–2928.

DALLMEYER, R. D., NEUBAUER, F., HANDLER, R., FRITZ, H., PANA, D., PUTIS, M. & ANON. 1996. Tectonothermal evolution of the internal Alps and Carpathians; evidence from $^{40}Ar/^{39}Ar$ mineral and whole-rock data; *International Geological Congress, Abstracts—Congres Geologique International, Resumes*, **1**, 180.

DE CARITAT, P., HUTCHEON, I. & WALSHE, J. L. 1993. Chlorite geothermometry; a review. *Clays and Clay Minerals*, **41**, 219–239.

DUNKL, I., GRASEMANN, B. & FRISCH, W. 1998. Thermal effects of exhumation of a metamorphic core complex on hanging wall syn-rift sediments: an example from the Rechnitz window, eastern Alps. *Tectonophysics*, **297**, 31–50.

ELLIOT, W. C., EDENFIELD, N. M., WAMPLER, J. M., MATISOFF, G. & LONG, P. E. 1999. The kinetics of the smectite to illite transformation in Cretaceous bentonites, Cerro Negro, New Mexico. *Clays and Clay Minerals*, **47**, 286–296.

ERNST, W. G. & FERREIRO MÄHLMANN, R. 2004. Vitrinite alteration rate as a function of temperature, time, starting material, aqueous fluid pressure, and oxygen fugacity–Laboratory corroboration of prior work. *In*: HILL, R. J., LEVENTHAL, J., AIZENSHTAT, M. J. ET AL. (eds) *Geochemical Investigations in Earth and Space Sciences: A Tribute to Isaac R. Kaplan.* The Geochemical Society, 341–357.

FERREIRO MÄHLMANN, R. 1994. *Zur Bestimmung von Diagenesehöhe und beginnender Metamorphose–Temperaturgeschichte und Tektogenese des Austroalpins und Südpenninikums in Voralberg und Mittelbünden* [Published Ph.D. thesis]. *Frankfurter Geowissenschaftliche Arbeiten, Serie C*, **14**, 1–498.

FERREIRO MÄHLMANN, R. 1995. The pattern of diagenesis and metamorphism by vitrinite reflectance and illite-crystallinity in Mittelbunden and in the Oberhalbstein. 1. The relationship to stockwerk tectonics. *Schweizerische Mineralogische und Petrographische Mitteilungen*, **75**, 85–122.

FERREIRO MÄHLMANN, R. 1996. The pattern of diagenesis and metamorphism by vitrinite reflectance and illite 'crystallinity' in Mittelbünden and in the Oberhalbstein. 2. Correlation of coal petrography and of mineralogical parameters. *Schweizerische Mineralogische und Petrographische Mitteilungen*, **76**, 23–46.

FERREIRO MÄHLMANN, R. 2001. Correlation of very low grade data to calibrate a thermal maturity model in a nappe tectonic setting, a case study from the Alps. *Tectonophysics*, **334**, 1–33.

FERREIRO MÄHLMANN, R., BELMAR, M. & CIULAVU, M. 2001. Bituminite reflectance in very low grade metamorphic studies. Key Note Lecture, Terra Abstracts. *Journal of Conference Abstracts*, EUG 11, **6/1**, 230.

FERREIRO MÄHLMANN, R., PETROVA, T. V., PIRONON, J., STERN, W. B., GHANBAJA, J., DUBESSY, J. & FREY, M. 2002. Transmission electron microscopy study of carbonaceous material in a metamorphic profile from diagenesis to amphibolite facies (Bündnerschiefer, Eastern Switzerland). *Schweizerische Mineralogische und Petrographische Mitteilungen*, **82**, 253–272.

FREY, M. 1968. Zur Metamorphose des Keupers vom Tafeljura bis zum Lukmanier-Gebiet. *Schweizerische Mineralogische und Petrographische Mitteilungen*, **48**, 829–831.

FREY, M. 1969. A mixed layer paragonite/phengite of low-grade metamorphic origin. *Contributions in Mineralogy and Petrology*, **24**, 63–65.

FREY, M. 1978. Progressive low-grade metamorphism of a black shale formation, central Swiss Alps, with special reference to pyrophyllite and margarite bearing assemblages. *Journal of Petrology*, **19**, 95–135.

FREY, M. 1986. Very low-grade metamorphism of the Alps: an introduction. *Schweizerische, Mineralogische und Petrographische Mitteilungen*, **66**, 13–27.

FREY, M. 1987a. The reaction-isograd kaolinite + quartz = pyrophyllite + H_2O, Helvetic Alps, Switzerland. *Schweizerische Mineralogische und Petrographische Mitteilungen*, **67**, 1–11.

FREY, M. 1987b. Very low-grade metamorphism of clastic sedimentary rocks. *In*: FREY, M. (ed.) *Low Temperature Metamorphism*. Blackie and Son Ltd., Glagow and London, 9–58.

FREY, M., DE CAPITANI, C. & LIOU, J. G. 1991. A new petrogenetic grid for low-grade metabasites. *Journal of Metamorphic Geology*, **9**, 497–509.

FREY, M., DESMONS, J. & NEUBAUER, F. 1999. Alpine Metamorphic Map 1:50 0000; Pre-Alpine Metamorphic Map 1:1000 000. *In*: FREY, M., DESMONS, J. & NEUBAUER, F. (eds) The new metamorphic map of the Alps. *Schweizerische Mineralogische und Petrographische Mitteilungen*, **79**, 1–4.

FREY, M. & FERREIRO MÄHLMANN, R. 1999. Alpine metamorphism of the Central Alps. *Schweizerische Mineralogische und Petrographische Mitteilungen*, **79**, 135–154.

FREY, M. & ROBINSON, D. 1999. *Low-Grade Metamorphism*. Oxford, Blackwell Sciences.

FREY, M., TEICHMUELLER, M., TEICHMUELLER, R. ET AL. 1980. Very low-grade metamorphism in external parts of the Central Alps; illite crystallinity, coal rank and fluid inclusion data. *Eclogae Geologicae Helvetiae*, **73**, 173–203.

FREY, M. & WIELAND, B. 1975. Chloritoid in autochthon-parautochthonen Sedimenten des Aarmassivs. *Schweizerische Mineralogische und Petrographische Mitteilungen*, **55**, 407–418.

FÜGENSCHUH, B. & SCHMID, S. M. 2005. Age and significance of core complex formation in a very curved orogen: Evidence from fission track studies in the South Carpathians (Romania). *Tectonophysics*, **404**, 33–53.

FÜGENSCHUH, B., SEWARD, D. & MANCKTELOW, N. 1997. Exhumation in a convergent orogen: the western Tauern window. *Terra Nova*, **9**, 213–217.

GENSER, J., VAN WEES, J. D., CLOETINGH, S. & NEUBAUER, F. 1996. Eastern Alpine tectonometamorphic evolution: Constraints from two-dimensional P–T–t modeling. *Tectonics*, **15**, 584–604.

GHENT, E. D., STOUT, M. Z. & FERRI, F. 1989. Chloritoid, paragonite, pyrophyllite and stilpnomelane-bearing rocks near Blackwater Mountain, Western Rocky Mountains, British Columbia. *Canadian Mineralogist*, **27**, 59–66.

GRUNENFELDER, M., POPESCU, G., SOROIU, M., ARSENESCU, V. & BERZA, T. 1983. K–Ar and U–Pb dating of the metamorphic formations and the associated igneous bodies of the Central South Carpathians. *Anuarul Geologic Institutul Al României*, **61**, 37–46.

HEMLEY, J. J., MONTOYA, J. W., MARINENKO, J. W. & LUCE, R. W. 1980. Equilibria in the system $Al_2O_3 - SiO_2 - H_2O$ and some general implications for alteration/mineralization processes. *Economic Geology and the Bulletin of the Society of Economic Geologists*, **75**, 210–228.

HOOD, A., GUTJAHR, C. C. M. & HEACOCK, R. L. 1975. Organic metamorphism and the generation of petroleum. *American Association of Petroleum Geologists Bulletin*, **59**, 986–996.

HUNT, J. M., LEWAN, M. D. & HENNET, R. J. C. 1991. Modeling oil generation with Time-Temperature Index graphs based on the Arrhenius Equation. *American Association of Petroleum Geologists Bulletin*, **75**, 795–807.

IANCU, V. 1986. Mineral assemblages in low grade metamorphic rocks in the South Carpathians; Mineral parageneses. *Theophrastus Publications S. A.*, Athens, 503–519.

IANCU, V. & MARUNTIU, M. 1994. Pre-Alpine lithotectonic units and related shear zones in the basement of the Getic-Supragetic nappes (South Carpathians). *Romanian Journal of Tectonics and Regional Geology (Suppl. 2: ALCAPA II Field Guidebook)*, **75**, 87–92.

IANCU, V., UDUBASA, G., RADAN, S., VISARION, A. & ANON. 1984. Complex criteria of separating weakly metamorphosed formations; an example, the South Carpathians; Volume special; Congres geologic international. *Anuarul Institutului de Geologie si Geofizica*, **64**, 51–60.

IANOVICI, V., NEACSU, G., NEACSU, V. ET AL. 1981. Pyrophyllite occurrences and their genetic relations with the kaolin minerals in Romania. *26eme Congres geologique international: Symposium interactions fluides, mineraux, roches. Bulletin de Mineralogie*, **104**, 768–775.

JACOB, H. 1989. Classification, structure, genesis and practical importance of natural solid oil bitumen (Migrabitumen). *International Journal of Coal Geology*, **11**, 65–79.

JACOB, H. & HILTMANN, W. 1985. Disperse, feste Erdölbitumina als Migrations- und Maturitätsindikatoren im Rahmen der Erdöl- und Erdgas-Prospektion. Eine Modellstudie in Nordwestdeutschland. *Deutsche Wissenschaftliche Gesellschaft für Erdöl, Erdgas und Kohle, Forschungsberichte*, **267**, 1–54.

JIANG, W. T., PEACOR, D. R. & BUSECK, P. R. 1994. Chlorite Geothermometry—Contamination and Apparent Octahedral Vacancies. *Clays and Clay Minerals*, **42**, 593–605.

KISCH, H. J. 1987. Correlation between indicators of very low-grade metamorphism. *In*: FREY, M. (ed.) *Low Temperature Metamorphism*. Blackie and Son Ltd., Glagow and London, 227–300.

KOCH, J. 1997. Organic petrographic investigations of the Kupferschiefer in northern Germany. *International Journal of Coal Geology*, **33**, 301–316.

KOONS, P. O. 1987. Some thermal and mechanical consequences of rapid uplift; an example from the Southern Alps, New Zealand. *Earth and Planetary Science Letters*, **86**, 307–319.

KRUMM, H., PETSCHICK, R. & WOLF, M. 1988. From diagenesis to anchimetamorphism, Upper Austroalpine sedimentary cover in Bavaria and Tyrol. *Geodinamica Acta*, **2**, 33–47.

KRETZ, R. 1983. Symbol for rock-forming minerals. *American Mineralogist*, **68**, 277–279.

KÜBLER, B. 1967. La cristallinité de l'illite et les zones tout à fait supérieures du métamorphisme. Etage tectoniques. Colloquium Neuchâtel, 18–21 avril 1967, 105–122.

LEMNE, M., SAVU, H., STEFAN, A., BORCOS, M., SĂNDULESCU, D., UDUBASA, G., VAJDEA, E., ROMANESQUE, O., TANASESCU, A. & IOSIPENCO, N. 1983. Date geocronologie privind formatiuni metamatice si magmatice din Romania. Report Arhiva Geologic Institutul al României, Bucuresti.

LIVI, K. J. T., VEBLEN, D. R., FERRY, J. M. & FREY, M. 1997. Evolution of 2:1 layered silicates in low-grade metamorphosed Liassic shales of Central Switzerland. *Journal of Metamorphic Geology*, **15**, 323–344.

MANCKTELOW, N. S. & GRASEMANN, B. 1997. Time-dependent effects of heat advection and topography on cooling histories during erosion. *Tectonophysics*, **270**, 167–195.

MANOLESCU, G. 1937. Etude geologique et petrographique dans les Muntii Vulcan (Carpates meridionales, Rumanie). *Anuarul Geologic Institutul Al României*, **XVIII**, 79–172.

MARUNTIU, M. 1987. *Studiul geologic complex al rocilor ultrabazice din Carpatii Meridionali*. [unpublished Ph.D. thesis], Bucharest University.

MARUYAMA, S., CHO, M. & LIOU, J. G. 1986. Experimental investigations of blueschist-greenschist transition equilibria; pressure dependence of Al_2O_3 contents in sodic amphiboles; a new geobarometer; Blueschists and eclogites. *Memoir Geological Society of America*, **164**, 1–16.

MASSONNE, H. J. & SCHREYER, W. 1987. Phengite geobarometry based on the limiting assemblage with K-feldspar, phlogopite, and quartz. *Contributions to Mineralogy and Petrology*, **96**, 212–224.

MATENCO, L., BERTOTTI, G., DINU, C. & CLOETINGH, S. 1997. Tertiary tectonic evolution of the external South Carpathians and the adjacent Moesian platform (Romania). *Tectonics*, **16**, 896–911.

MATENCO, L. & SCHMID, S. 1999. Exhumation of the Danubian nappes system (South Carpathians) during the Early Tertiary: inferences from kinematic and paleostress analysis at the Getic/Danubian nappes contact. *Tectonophysics*, **314**, 401–422.

MERRIMAN, R. J. & FREY, M. 1999. Patterns of very low-grade metamorphism in metapelitic rocks. *In*: FREY, M. & ROBINSON, D. (eds) *Low-Grade Metamorphism*. Blackwell Sciences, Oxford, 61–107.

MRAZEC, L. 1898. Dare de sema asupra cercetarilor geologice din vara 1897. I: Partea de E a muntilor Vulcan. *Buletinul Ministerului Agriculturii si Domeniilor*, 39p.

MURGOCI, G. M. 1905. *Sur l'existence d'une grande nappe de recouvrement dans les Carpathes Méridionales*. Unpublished thesis, Comptes Rendus de l'Académie des Sciences, Paris.

MURGOCI, G. M. 1912. The geological synthesis of the South Carpathians. *Compte Rendu du XI-eme Congres Geologique International*, 871–881.

NĂSTĂSEANU, S. V. 1979. Geologie des Monts Cerna. *Anuarul Institutului de Geologie si Geofizica*, **54**, 153–280.

NĂSTĂSEANU, S. V., RUSSO-SANDULESCU, D., BERCIA, I., SAVU, H., MARUNTEANU, M., RUSU, A. & BERCIA, E. 1985. *Geological Map of Romania, scale 1:50 000, sheet Cornereva*. Geologic Institutul al României, Bucuresti.

PALIUC, G. 1972. Pirofilitul din formatiunea de Schela. *Dari de Seama Ale Sedintelor, Comitetul de Stat el Geologiei*, **58**, 45–65.

PETROVA, T. V., FERREIRO MÄHLMANN, R., STERN, W. B. & FREY, M. 2002. Application of combustion and DTA–TGA analysis to the study of metamorphic organic matter. *Schweizerische Mineralogische und Petrographische Mitteilungen*, **82**, 33–53.

POP, G. 1973. *Depozitele mezozoice din Muntii Valcan*. Editura Academiei, Bucuresti.

POPESCU, G. C., TATU, M. & DAMIAN, G. 1982. La reflectivite et les caracteristiques oroeentgenostructurale de l'Anthracite de la formation de Schela. *Analele Universitatii Bucuresti, Geologie*, **31**, 13–20.

POTEL, S., FERREIRO MÄHLMANN, R., STERN, W. B., MULLIS, J. & FREY, M. 2006. Very low-grade metamorphic evolution of pelitic rocks under high-pressure/low-temperature conditions, NW New Caledonia (SW Pacific). *Journal of Petrology*, **47**, 991–1015.

POTEL, S., SCHMIDT, S. T. & DE CAPITANI, C. 2002. Composition of pumpellyite, epidote and chlorite from New Caledonia — How important are metamorphic grade and whole-rock composition? *Schweizerische Mineralogische und Petrographische Mitteilungen*, **82**, 229–252.

RADTKE, M., SCHAEFFER, R. G., LEITHAUSER, D. & TEICHMÜLLER, M. 1980. Composition of soluble organic matter in coals: relation to rank and liptinite fluorescence. *Geochimica Cosmochimica Acta*, **44**, 1787–1800.

RAHN, M., MULLIS, J., ERDELBROCK, K. & FREY, M. 1994. Very low-grade metamorphism of the Taveyanne Greywacke, Glarus Alps, Switzerland. *Journal of Metamorphic Geology*, **12**, 625–641.

RAMDOHR, P. 1928. Mikroskopische Beobachtungen an Graphiten und Koksen. *Archiv für Eisenhüttenwesen*, **1**, 669–672.

RATSCHBACHER, L., LINZER, H. G., MOSER, F., STRUSIEVICZ, R. O., BEDELEAN, H., HAR, N. &

MOGOS, P. A. 1993. Cretaceous to Miocene thrusting and wrenching along the Central South Carpathians due to a corner effect during collision and orocline formation. *Tectonics*, **12**, 855–873.

ROBERT, P. 1988. Organic metamorphism and geothermal history—Microscopic study of organic matter and thermal evolution of sedimentary basins. *Elf-Aqitaine and D. Reidel Publishing Company*, Nordrecht.

ROBINSON, D. & BEVINS, R. E. 1999. Patterns of regional low-grade metamorphism in metabasites. *In*: FREY, M. & ROBINSON, D. (eds) *Low-Grade Metamorphism*. Blackwell Sciences, 143–168.

SACHSENHOFER, R. F. 2001. Syn- and post-collisional heat flow in the Cenozoic Eastern Alps. *International Journal of Earth Sciences*, **90**, 579–592.

SĂNDULESCU, M. 1975. Essai de synthese structurale des Carpathes. *Bulletin de la Societe Geologique de France*, **17**, 299–358.

SĂNDULESCU, M. 1984. Geotectonica Romaniei. *Editura Tehnica*, Bucharest, 334p.

SĂNDULESCU, M. 1994. Oveview of Romanian geology. Abstracts volume ALCAPA II Symposium, Covasna 1994. *Romanian Journal of Tectonics and Regional Geology*, **75**, supplement number 2, 57.

SAVU, H. 1970. Structura plutonului Susita si relatiile sale cu formatiunile autohtonului danubian (Carpatii meridionali). *Dari de Seama ale Institutului Geologic*, **LVI/5**, 123–153.

SAVU, H., UDRESCU, C., NEACSU, V., BRATOSIN, I. & STOIAN, M. 1985. Origin, geochemistry and tectonic position of the Alpine ophiolites in the Severin Nappe (Mahedinti Plateau, Romania); Ophiolites through time; proceedings. *Ofioliti*, **10**, 423–440.

SCHMID, S. M., BERZA, T., DIACONESCU, V., FROITZHEIM, N. & FÜGENSCHUH, B. 1998. Orogen-parallel extension in the Southern Carpathians. *Tectonophysics*, **297**, 209–228.

SCHMIDT, D., SCHMIDT, S. T., MULLIS, J., FERREIRO MÄHLMANN, R. & FREY, M. 1997. Very low grade metamorphism of the Taveyanne formation of western Switzerland. *Contributions to Mineralogy and Petrology*, **129**, 385–403.

SEGHEDI, A. & BERZA, T. 1994. Duplex interpretation for the structure of the Danubian Thrust Sheets. Abstracts volume ALCAPA II Symposium, Covasna 1994. *Romanian Journal of Tectonics and Regional Geology*, **75**(1), 57.

SEKI, Y. 1972. Lower-grade stability limit of epidote in the light of natural occurrences. *Journal of the Geological Society of Japan*, **78/8**, 405–413.

SPERLICH, R. 1988. The transition from crossite to actinolite in metabasites of the Combin unit in Vallee St. Barthelemy (Aosta, Italy). *Schweizerische Mineralogische und Petrographische Mitteilungen—Bulletin Suisse de Mineralogie et Petrographie*, **68**, 215–224.

STACH, F., MACKOWSKY, M. T., TEICHMÜLLER, M., TAYLOR, G. H., CHANDRA, D. & TEICHMÜLLER, R. 1982. *Textbook of Coal Petrology*, Borntrager, Stuttgart.

STĂNOIU, I. 1972. Incercare de reconstituire a succesiunii Paleozoicului din partea de est a autohtonului danubian, cu privire speciala asupra regiunii de la obirsia vaii Motru (Carpatii Meridionali). *Dari de Seama ale Institutului Geologic*, **LVIII/4**, 57–71.

STĂNOIU, I. 1973. Zona Mehedinti-Retezat, o unitate paleogeografica si tectonica distincta a Carpatilor Meridionali. *Dari de Seama ale Institutului Geologic*, **59/5**, 127–171.

STĂNOIU, I., UDUBASA, G., RADAN, S. & VANGHELIE, I. 1982. Cercetarea complexa a formatiunilor purtatoare de carbuni si sisturi combustibile din Carpatii Meridionali—Versantul nordic al masivului Vilcan. Conturarea domeniului de extindere a formatiunii liasice purtatoare de antracit din versantul nordic al masivului Vilcan. *Arhiva Geologic Institutul Al României*, Bucuresti.

STĂNOIU, I., CONOVICI, M., MARINESCU, F., RUSSO-SĂNDULESCU, D., AERBĂNESCU, A. & VANGHELIE, I. 1988. Harta geologice a României, scara 1 : 50000, foaia Balta: Arh. IGG.

SUCHY, V., FREY, M. & WOLF, M. 1997. Vitrinite reflectance and shear-induced graphitization in orogenic belts: a case study from the Kandersteg area, Helvetic Alps, Switzerland. *International Journal of Coal Geology*, **34**, 1–20.

TEICHMÜLLER, M. 1987. Organic material and very low-grade metamorphism. *In*: FREY, M. (ed.) *Low Temperature Metamorphism*. Blackie and Son Ltd., Glasgow and London, 114–160.

THEYE, T. 1988. *Aufsteigende Hochdruckmetamorphose in Sedimenten der Phyllit-Quarzit-Einheit Kretas und des Peloponnes*. Unpublished Ph.D. thesis, Technische Universität, Braunschweig.

THEYE, T., SCHREYER, W. & FRANSOLET, A. M. 1996. Low-temperature, low-pressure metamorphism of Mn-rich rocks in the Lienne syncline, Venn-Stavelot Massif (Belgian Ardennes), and the role of carpholite. *Journal of Petrology*, **37**, 767–783.

THEYE, T. & SEIDEL, E. 1991. Petrology of low-grade high-pressure metapelites from the external Hellenides (Crete, Peloponnese)—a case-study with attention to sodic minerals. *European Journal of Mineralogy*, **3**, 343–366.

THOMPSON, A. B. 1970. A note on the kaolinite-pyrophyllite equilibrium. *American Journal of Science*, **268**, 454–458.

VIDAL, O., GOFFÉ, B., BOUSQUET, R. & PARRA, T. 1999. Calibration and testing of an empirical chloritoid-chlorite Mg–Fe exchange thermometer and thermodynamic data for daphnite. *Journal of Metamorphic Geology*, **17**, 25–39.

WAPLES, D. W. 1980. Time and temperature in petroleum formation; application of Lopatin's method to petroleum exploration. *American Association of Petroleum Geologists Bulletin*, **64**, 916–926.

WILLINGSHOFER, E. 2000. Extension in collisional orogenic belts: the Late Cretaceous evolution of the Alps and Carpathians. Unpublished Ph.D. thesis, Vrije Universiteit Amsterdam, Amsterdam.

WOLF, M. 1975. Über die Beziehungen zwischen Illit-Kristallinität und Inkohlung. *Neues Jahrbuch für Geologie und Paleontologie*, 437–447.

WOLF, M. & WOLFF-FISCHER, E. 1984. Alginit in Humuskohlen karbonischen Alters und sein Einfluss auf die optischen Eigenschaften des begleitenden Vitrinits. *Glückauf Forschungshefte*, **45**, 243–246.

ZEN, E. A. 1960. Metamorphism of lower Paleozoic rocks in the vicinity of the Taconic Range in west-central Vermont. *American Mineralogist*, **45**, 129–175.

Tectonics and sedimentation during convergence of the ALCAPA and Tisza–Dacia continental blocks: the Pienide nappe emplacement and its foredeep (N. Romania)

M. TISCHLER[1,4], L. MATENCO[2], S. FILIPESCU[3], H. R. GRÖGER[1,4],
A. WETZEL[1] & B. FÜGENSCHUH[5]

[1]*Geologisch-Paläontologisches Institut, Universität Basel, Bernoullistr. 35, 4056 Basel, Switzerland (e-mail: matthias.tischler@gmail.com)*

[2]*Netherlands Centre for Integrated Solid Earth Sciences, Vrije Universiteit, Faculty of Earth and Life Sciences, De Boelelaan 1085, 1081 HV Amsterdam, The Netherlands*

[3]*Department of Geology, Babeş-Bolyai University, Str. Kogălniceanu 1, 400084 Cluj-Napoca, Romania*

[4]*Present address: Statoil ASA, Statoil Head Office, Forusbeen 50, 4035 Stavanger, Norway*

[5]*Geology and Paleontology, Innsbruck University, Innrain 52f, A-6020 Innsbruck, Austria*

Abstract: The juxtaposition of the ALCAPA and Tisza–Dacia continental blocks, although one of the key issues in the evolution of the Carpathians, is not well known in terms of associated effects on the sedimentary systems during frontal foreland development. Most of the contact between ALCAPA and Tisza–Dacia being covered by post-tectonic deposits, these effects can best be observed in northern Romania. Sedimentological data on facies, palaeocurrents and modal composition of sandstones combined with micropalaeontological data and 2D well-calibrated seismic lines constrain the tectonic history of the contact zone between ALCAPA and Tisza–Dacia. Pervasive deposition of sand-dominated siliciclastics beginning in late Early Oligocene (Late Rupelian) times is interpreted to reflect the onset of convergence between ALCAPA and Tisza–Dacia in the study area. The depocentre of coarse siliciclastic material migrates southward, finally forming a southeastward-thinning clastic wedge in the Transylvanian Basin. This Burdigalian-age clastic wedge is interpreted as fill of a flexural foreland basin that formed in response to the coeval thrusting of parts of ALCAPA (Pienides) over Tisza–Dacia. A shift from an E–W to SE–NW striking basin axis during Oligocene times towards a WSW–ENE oriented basin axis during Burdigalian times is interpreted as a result of clockwise rotation of Tisza–Dacia during basin formation.

The Miocene geological history of the Carpathians is characterized by the emplacement of continental blocks, ALCAPA and Tisza–Dacia, into the so-called Carpathian embayment, a large-scale bight in the European margin between the Moesian and Bohemian promontories (Fig. 1). Emplacement of these continental blocks and their subsequent extensional collapse was coeval to thin-skinned nappe stacking at the exterior of the thrusted chain (e.g. Săndulescu 1988; Horvath 1993). Lithospheric slab retreat of the distal parts of the Carpathians foreland (Royden 1993) is considered to represent the principal driving force for the invasion of the two continental blocks (Fig. 1). They were separated by a broad zone of deformation, the Mid-Hungarian fault zone (e.g. Fodor et al. 2005). Along this fault zone, substantial strike-slip movements and block rotations are documented for Palaeogene to Early Miocene times (e.g. Csontos & Vörös 2004; Horváth et al. 2006 and references therein). Convergence between the invading continental blocks ultimately led to thrusting of ALCAPA onto Tisza–Dacia (Csontos & Nagymarosy 1998). This process culminated during the Early Miocene (Pienide nappes, Săndulescu et al. 1981; Fig. 1). However, the onset of thrusting remains rather poorly constrained to Late Oligocene times in the Pannonian Basin (Csontos & Nagymarosy 1998; Fodor et al. 1999) and Transylvanian Basin (Krezsek & Bally 2006), respectively.

A very suitable place to study the tectonics and sedimentary processes linked to the convergence of ALCAPA and Tisza–Dacia along the Mid-Hungarian fault zone is located at its ENE-most tip, in northern Romania. This is the only place where the thin-skinned thrust units and associated foredeep sediments are exposed (Fig. 1). An

From: SIEGESMUND, S., FÜGENSCHUH, B. & FROITZHEIM, N. (eds) *Tectonic Aspects of the Alpine-Dinaride-Carpathian System.* Geological Society, London, Special Publications, **298**, 317–334.
DOI: 10.1144/SP298.15   0305-8719/08/$15.00 © The Geological Society of London 2008.

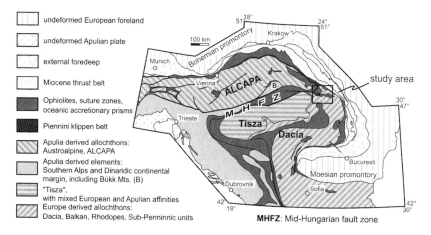

Fig. 1. Major tectonic units of the Alps, Carpathians and Dinarides (simplified after Schmid *et al.* 2006).

additional advantage is the availability of detailed subsurface data obtained during exploration of the Transylvanian Basin (e.g. De Broucker *et al.* 1998). The goal of this study is to provide new constraints on the thrusting of ALCAPA onto Tisza–Dacia at the northeasternmost tip of the Mid-Hungarian fault zone (i.e. the Pienide nappe contact) by analysing syntectonic sediments that accumulated on Tisza–Dacia.

Geological setting

The arcuate belt of the Carpathians acquired its present shape during the Cenozoic emplacement of continental blocks into the Carpathian embayment. These continental blocks are referred to as ALCAPA (e.g. Csontos 1995) and Tisza–Dacia (e.g. Balla 1986; Internal and Median Dacides of Săndulescu 1980, 1988), respectively (Fig. 1). They differ in their contrasting Triassic and Jurassic sedimentary facies and fossil assemblages (e.g. Csontos & Vörös 2004 and references therein).

Evolution of the continental blocks ALCAPA and Tisza–Dacia

Between the name-giving constituents of the Tisza–Dacia block, Tisza and Dacia, remainders of an oceanic domain (Transylvanides, Sandulescu 1984) occur. After the closure of this oceanic domain starting in Albian times (e.g. Săndulescu & Visarion 1977; Csontos & Vörös 2004), Tisza–Dacia can be considered as a single block.

The Dacia block (or Rhodopian fragment, Burchfiel & Bleahu 1976) was separated from the European platform by a partially oceanic basin, whose remnants are presently found in the highly deformed Ceahlău and Severin nappes (Săndulescu 1984). Following an initial Albian shortening, this oceanic domain has been completely subducted in latest Cretaceous times (Săndulescu 1988). Shortening continued eastwards and ended after Late Miocene continental collision (*c.* 11 Ma, Matenco & Bertotti 2000).

Cenozoic slab-retreat (Royden 1988) in the Carpathians, favoured by deep lithospheric processes related to the subducted oceanic slab of the Ceahlău–Severin domain, is acting as the principal driving force for the emplacement of the two continental blocks (ALCAPA and Tisza–Dacia) into the Carpathian embayment. A number of recent studies addressed these deep lithospheric processes (e.g. Wortel & Spakman 2000; Cloetingh *et al.* 2004; Knapp *et al.* 2005; Weidle *et al.* 2005).

The emplacement of ALCAPA was additionally affected by 'lateral extrusion', eastward escape along conjugate strike-slip faults coupled to indentation processes in the eastern Alps (Ratschbacher *et al.* 1991*a, b*). The northeastward movement of ALCAPA was guided by strike-slip zones oriented subparallel to the collision suture (Nemčok 1993). Corner effects at the Bohemian promontory led to counter-clockwise rotation of the ALCAPA block.

The Tisza–Dacia block started its emplacement into the Carpathian embayment during the Eocene (Fügenschuh & Schmid 2005). Corner effects at the Moesian promontory (e.g. Ratschbacher *et al.* 1993; Schmid *et al.* 1998) led to large clockwise rotations during the emplacement of Tisza–Dacia (e.g. Panaiotu 1998). The convergence of ALCAPA with the European margin (Krzywiec 2001) is terminated by 'soft-collision' (*sensu* Royden 1988) simultaneously with the last

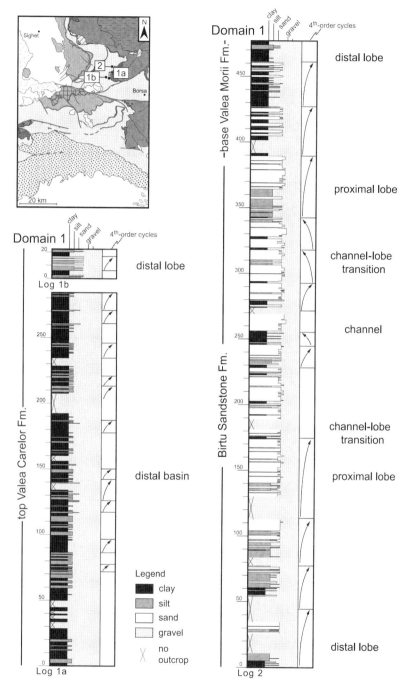

Fig. 4. Simplified overview of logs 1 and 2. Following the gradual progradation from basin plain deposits towards a distal lobe setting (log 1a, 1b), a complete 2nd-order depositional sequence (progradation and retreat) is documented in log 2 (Lower Oligocene Bîrțu Sandstone).

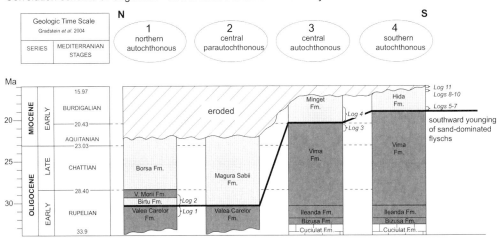

Fig. 3. Correlation scheme of Oligocene to Lower Miocene deposits in the study area, illustrating progressive younging of sand-dominated flysch towards the south. Approximate locations of sedimentary logs are indicated; width of the brackets corresponds to the approximate percentage of the interval covered. The number above the columns refers to Domain number given in Figure 2. The ages of the respective units are compiled from official 1:50 000 map sheets (Geological Survey of Romania; Domains 1 and 3), Săndulescu pers. comm. (2002, Domain 2). Ages for Domain 4 are compiled from Rusu (1989), Popescu (1984) and Moiescu (1981) in Györfi et al. (1999) as well as De Broucker et al. (1998). Unfortunately, the age of the Minget Fm. is rather poorly constrained to the Early Miocene. However, the base of the Minget Fm. is considered to be older (around the Aquitanian–Burdigalian boundary) than the base of the Hida Fm. (mid-Burdigalian), since all available geological maps show the Hida Fm. to overlie the Minget Fm.

might have prevailed in the Paratethys (i.e. including the study area) from mid-Oligocene to Burdigalian times (Rögl 1999). A significant short-term sea-level drop occurred during the Late Oligocene (Haq et al. 1987). It probably coincided with the deposition of the Borşa Sandstone Fm. (Fig. 3). However, this apparent compatibility is put into perspective by the continuing progradation during the Burdigalian (Hida Fm.), a time of short-term sea-level rise (Haq et al. 1987).

Data

In order to constrain the depositional setting of the Oligocene to Burdigalian siliciclastic units and evaluate their relationship to the Pienide nappe emplacement, sedimentological studies have been combined with seismic data. Facies analysis based on detailed lithostratigraphic logs, assessment of the detrital modes of the siliciclastic units and palaeocurrent analysis are combined with new biostratigraphical data. These data will be related to the insights gained by the interpretation of well-calibrated 2D seismic records from the Transylvanian Basin.

Facies analysis

Based on the lithofacies classification of Pickering et al. (1989, 1995), the depositional trends have been defined by using the hierarchical scheme of Mutti & Normark (1987). First order features are at the scale of basin fills (turbidite complex), while second order features (turbidite system, c. 400 m thickness) include depositional sequences, commonly bounded by highstand mud facies. Third order features (turbidite stage, c. 250 m thickness) are facies associations reflecting different development phases of the system. Fourth order features comprise bed packages (turbidite sub-stage, c. 15 m) and characterize sub-stages of system development. The smallest, fifth division, is defined by individual lithofacies.

Continuous, long sections, even when sparse, reflect the development of the depositional setting (Figs 4 and 5). Correlation in between the outcrops is based on the corresponding units shown on geological maps. As an example, the Birţu Sandstone Fm. overlies the Valea Carelor Fm. and the sections are correlated accordingly. In the case of the Hida Fm., the relationship of the three sections covering the basal part has been established in the field, while

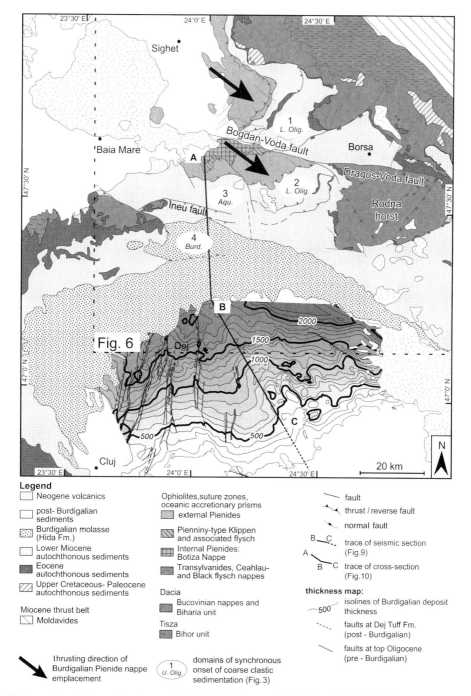

Fig. 2. Map of the study area, based on published geological maps (1 : 50 000 and 1 : 200 000) of the Geological Survey of Romania (Dicea *et al.* 1980; Săndulescu 1980; Săndulescu *et al.* 1981; Aroldi 2001). In the southern part of the study area, isopachs of the Burdigalian molasse deposits, based on well-calibrated 2D seismic lines, are shown.

moments of the emplacement of Tisza–Dacia over the European and Moesian foreland during Middle–Late Miocene (Matenco et al. 2003).

Palaeogene to Lower Miocene sedimentary architecture near the ENE contact between ALCAPA and Tisza–Dacia

The study area lies internally in the main Carpathian mountain chain at the northern border of the Transylvanian Basin, at the contact of Tisza–Dacia and units assigned to ALCAPA (Fig. 2, Pienides). The large-scale, thick-skinned nappe emplacement of the basement units outlined above was followed by the deposition of significant post-tectonic covers (e.g. Săndulescu 1988). Sedimentation started with uppermost Cretaceous to Paleocene continental to shallow marine deposits (Jibou Fm., e.g. Popescu 1984), well exposed on the NW margin of the Transylvanian Basin.

The Pienide nappe stack. Top to SE emplacement of the Pienide nappe stack (Fig. 2) commenced during Burdigalian times (Săndulescu et al. 1981; Tischler et al. 2006). The Pienides mainly consist of Eocene to Oligocene turbiditic siliciclastic units and can be divided into internal (Botiza nappe and Pienniny Klippen type of units, Săndulescu et al. 1993) and external nappes (Petrova, Leordina and Wildflysch nappes). While the internal nappes of the Pienides are correlated to the Inacovce–Krichevo units, the external Pienides correspond to the Magura flysch of the Western Carpathians (Săndulescu et al. 1981; Săndulescu 1994).

The Pieninny Klippen belt originally represented the outermost rim of ALCAPA (e.g. Csontos & Vörös 2004). More external units such as the external Pienides have been accreted during the invasion of ALCAPA into the Carpathian embayment (e.g. Fodor et al. 1999). Hence, the Pienides represent the easternmost tip of ALCAPA in Burdigalian times. It is worthwhile to note that during Oligocene times sand-dominated siliciclastics of similar facies and petrography as the ones in the foreland have been deposited in the external thrust sheets of the Pienides (Aroldi 2001). This suggests a genetic connection, i.e. limited post-dating thrusting (Săndulescu & Micu 1989).

Eocene. During the Eocene, epicontinental deposits formed in the NE part of the study area (Săndulescu et al. 1981). Carbonate platforms in the area of the Rodna Horst indicate relatively shallow marine environments (Lutetian to Priabonian Iza Limestone, 1982). Conglomerates and sandstones prevail in the NE (Prislop Fm., Kräutner et al. 1982, 1983; Săndulescu et al. 1991). They grade westwards into marly deposits (Vaser Fm., Săndulescu et al. 1991) developed in a littoral–neritic facies (Dicea et al. 1980). A further deepening towards the west is indicated by basal-slope turbidites ('hieroglyphic flysch'), outcropping north of the Bogdan Voda fault (Pîrîul Mocilnei Fm., Săndulescu et al. 1991).

In the southern part of the study area, two sedimentary cycles formed during the Eocene (Popescu 1984). Lower Eocene red-bed continental deposits were followed by two Middle to Upper Eocene shallow-marine highstand deposits that are separated by a continental red-bed and lacustrine lowstand interval (De Broucker et al. 1998). During the second, Late Eocene highstand, a carbonate platform developed during Priabonian times (Culmea Cozlei Fm., Cluj limestone, De Broucker et al. 1998; Rusu et al. 1983). Several local cycles of subsidence and deposition in the N–NW part of the Transylvanian Basin are related to rather-reduced Eocene tectonic events such as the Puini thrust (e.g. De Broucker et al. 1998; Krezsek & Bally 2006).

Oligocene to Lower Miocene. A phase of regional subsidence is recorded near the Eocene–Oligocene boundary. Following successive drowning of the carbonate platform from north (earliest Oligocene, Kräutner et al. 1982) to south (Early Oligocene, Györfi et al. 1999), muds became the prevailing deposits. In the north (Fig. 3), these Lower Oligocene strata are associated with slumps and olistoliths and have large variations in thicknesses (25 to 1200 m, Dicea et al. 1980). In the south black shales dominate.

The shales are progressively overlain by sandy turbidites, deposited within a basin having an initially W–E to NW–SE trending basin axis. The base of these sandy turbidites becomes increasingly younger towards the south (Fig. 3). This coarse sedimentation started during the Late Rupelian (Săndulescu et al. 1991) in the north (Fig. 3, Domains 1, 2), Early Miocene in the central autochthonous domain (Fig. 3, Domain 3) and Burdigalian in the south (Fig. 3, Domain 4). In the southern part of the study area, the siliciclastic deposits are grading into molasse-type deposits (Hida Fm., Ciupagea et al. 1970) with a southwestwards thinning, wedge-shaped geometry (Fig. 2, see also Ciulavu et al. 2002). The axis of the Burdigalian Basin is oriented WSW–ENE. Within the Hida Fm., a general shallowing-up trend can be observed, starting with deep marine outer fan turbidites and ending with coarse sands and conglomerates deposited by fandeltas (Koch 1900; Popescu 1975; Krézsek & Bally 2006).

The eustatic sea-level variations are bound to affect the study area only in limited time intervals due to intermittent changes between open marine and restricted conditions. Open marine conditions

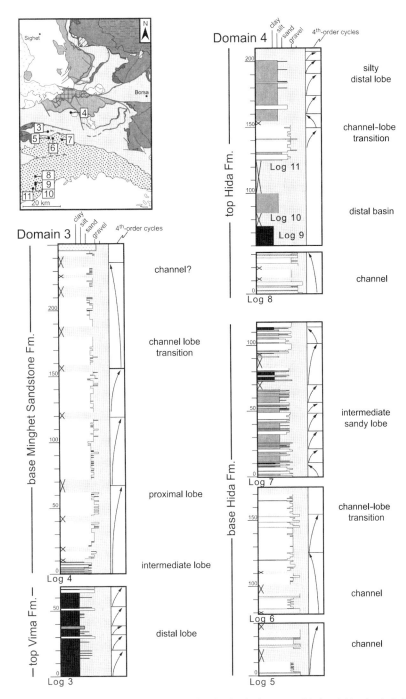

Fig. 5. Simplified overview of logs 3–11. A progradational pulse is documented in log 4 (Aquitanian). Further to the south (logs 5–7), coarse clastic input is documented later, in Burdigalian times. Logs 8–11 are situated at the top of the Hida Fm., featuring coarse intercalations of reduced thickness in a silty matrix.

the relationship of the four logs constituting the top part has been established by projection of the outcrop locations using the dip of the recorded strata.

The characterization of the used facies classes and the detailed lithostratigraphic logs and their description are published separately and these are available online at http://www.geolsoc.org.uk/SUP18312. A hard copy can be obtained from the Society Library.

Facies development. The earliest coarse clastic input is Late Rupelian in age; it occurs in the northern part of the study area (Fig. 4). The lithofacies associations documented in Log 1 suggest a gradual change of the depositional setting from a basin plain towards a distal lobe (Fig. 4). The trend towards more and coarser sediment input continues with the deposition of the overlying Birţu Sandstone Fm. Log 2, covering the Birţu Sandstone Fm., shows a gradual progradation from distal, intermediate and proximal lobe towards a channel-lobe transitional setting (Fig. 4). After an interval possibly indicating channel facies, the depositional setting steps towards a distal lobe setting.

Although this retreat occurs on a larger scale (it corresponds to the strata mapped as Valea Morii Fm.), it is not documented in the central parautochthonous realm (Domain 2, Fig. 3). Due to insufficient temporal resolution, it remains unclear if this retreat reflects a basinwide decrease in clastic input due to a moment of decreased tectonic activity (Mutti *et al.* 1999) or represents a localized event and is not related to regional tectonics.

Approximately 30 km further to the south (Domain 3), coarse clastics arrive around the Aquitanian–Burdigalian boundary, documented by prograding lobe fringes (Fig. 5, log 3). Further progradation of the system is recorded by deposits suggesting a channel-lobe transitional setting followed by channelized deposits typical for a middle-fan setting (Fig. 5, log 4). When comparing the northern and central area, the lobe-dominated sedimentation shifted more than 30 km towards the south between Late Rupelian and Late Aquitanian to Early Burdigalian times.

About 10 km still further to the south (Domain 4), Burdigalian channel-fill deposits (Fig. 5, logs 5/6) and intermediate, sand-rich lobes (Fig. 5, log 7) confirm the southward migration of proximal depositional settings.

In the upper part of the coarse-grained siliciclastics of Domain 4 (Fig. 5), the transition from channel-fill deposits (Fig. 5, log 8) to silt-rich basin plain deposits (Fig. 5, logs 9 and 10) suggests a decrease in clastic input. A renewed progradational pulse is documented by the development from an intermediate lobe to a channel-lobe transitional setting (Fig. 5). Log 11 covering the top of the Burdigalian strata is composed of silty distal lobe deposits. Note, however, that the Hida Fm. in general is characterized by a shallowing-up trend, which results in very coarse fan-delta deposits being found at the top of the Hida Fm. in other places.

Interpretation of the facies analysis. Since submarine fans may have a very complex architecture, local observations derived from discontinuous outcrop data may not be characteristic for the entire system. Since the lithostratigraphic logs only cover a part of the complete Oligocene to Lower Miocene succession, these rather local observations have to be interpreted as merely indicative for the regional development, and their extension to a regional scale has to be done very carefully.

The depositional environments match those known in foreland basins. Significant amounts of sand, a generally high sand:mud ratio and the development of well-defined 4th and 3rd order coarsening-up trends typical for lobes, fit to an active margin setting (Shanmugam & Moiola 1988). Hence, the general development of the depositional environment is interpreted to be dominated by tectonic activity.

Palaeocurrents

Palaeocurrent indicators (such as flute casts and scour marks, subordinate tool marks, groove casts and parting lineations) have been measured focusing on Oligocene to Burdigalian strata (Fig. 6). In the case of conglomeratic layers, long axes of imbricated pebbles have been measured, and are displayed in the form of a rose diagram (Fig. 6). Palaeocurrent indicators derived from beds with a dip of more than 30° have been backtilted.

A pronounced change in the palaeocurrent directions occurs between W-directed transport in Eocene times and to NE- to ESE-directed transport in Oligocene to Burdigalian times (Fig. 6). This latter distribution is evident in individual outcrops as well as on the regional scale (Fig. 6, locations 1–3). The observed directions are in good accordance with existing data from Oligocene deposits (Jipa 1962; Mihailescu & Panin 1962). A log in the Oligocene sandy flysch of the most external Pienide nappe (Wildflysch nappe) yielded compatible palaeocurrent directions.

The persistence of the observed palaeocurrent directions suggests a similar basin geometry and depositional setting for Oligocene through Burdigalian strata. The two dominant directions are longitudinal, respectively transversal, to the maximum visible thickness within the clastic wedge (Fig. 2).

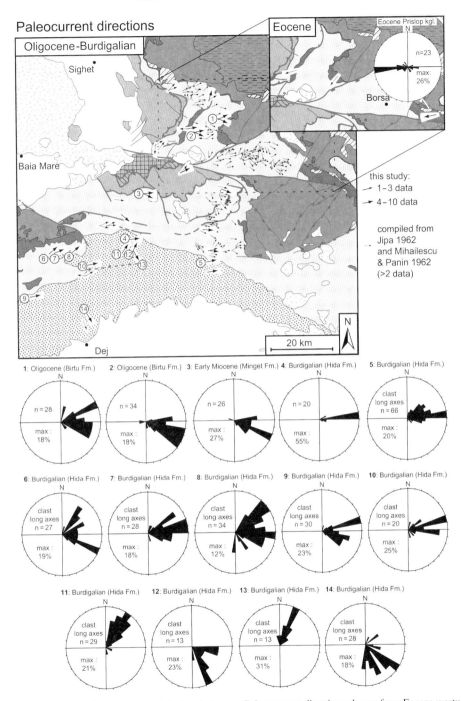

Fig. 6. Map showing palaeocurrent trends in the study area. Palaeocurrent directions change from Eocene westward (inset) to Oligocene southeastward transport. During Oligocene to Burdigalian times, northeastward and east–southeastward directions are dominant. Locations 1–4 comprise indicators from logs covering 100–400 m of sediment thickness. For legend and location of the map, compare Figure 2.

Petrography of the studied units

Thirty-two thin sections from Lower Oligocene to Burdigalian sandstones have been analysed and reveal only minor changes in sandstone composition. The sandstones are moderately to well-sorted litharenites with a dominantly pseudosparitic matrix (classification of Pettijohn et al. 1987).

Grains are subangular to subrounded and have a low sphericity. Most samples only show weak signs of compaction, and point contacts between grains dominate. Compaction features like bent micas and solution of carbonate grains are common in the Lower Oligocene to Lower Miocene strata, while rarer in Burdigalian samples. The lithic clasts are mainly limestone (10–20%) and chert (c. 5%), subordinately volcanic clasts. Micaschists/gneiss and mud-/siltstones occur rarely.

A rough estimate of the modal composition of the samples has been obtained by using visual estimation charts (Flügel 1978; Tucker 1981). An increase in the overall percentage of lithics implies decreasing compositional maturity for the Burdigalian samples. While the average composition of the Oligocene and Early Miocene sands is roughly $Qt_{50}F_{20}L_{30}$ (Total Quartz–Feldspar–Lithics), the average composition of the Burdigalian samples is estimated to about $Qt_{50}F_{15}L_{35}$.

The modal compositions of the individual samples are compared to the compositional fields indicative for different provenance types using standard triangular diagrams (Dickinson 1985; Fig. 7). The analysed samples plot in the QtFL and QmFLt diagrams in the 'recycled orogenic' and 'continental block' provenance category fields. A trend towards the 'recycled orogenic' field with decreasing rock age can be inferred, reflecting the decrease in compositional maturity.

The abundance of sedimentary lithic grains (Ls, QpLvLs–plot) reflects a relatively short transport suggestive of a fold-and-thrust-belt setting, where

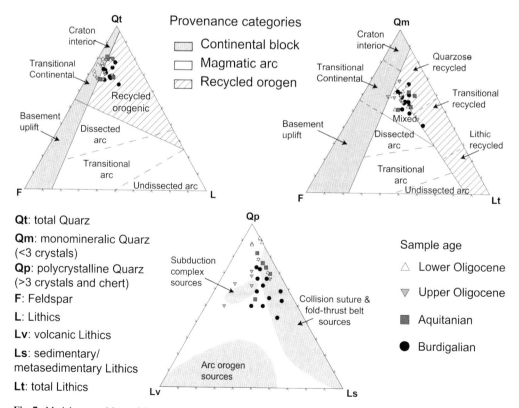

Fig. 7. Modal compositions of the analysed samples. The compositions and low maturity are suggestive for a foreland basin setting. The Burdigalian samples show a decrease in compositional maturity and an affinity towards fold-and-thrust belt sources, well comparable to their interpretation as molasse deposits.

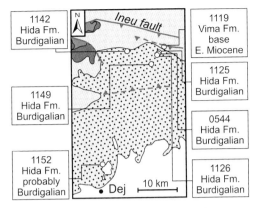

Fig. 8. Sample locations and ages of micropalaeontological dating samples.

detritus from a recycled orogenic source is deposited (Dickinson 1985). The relative proximity to the source area is also reflected by the low degree of rounding of the constituent grains. The modal compositions of the analysed sandstones show an affinity to recycled orogenic source areas and a decreasing compositional maturity with decreasing rock age. These modal compositions as well as the abundance of calcareous lithoclasts (not included into the Ls pole, 10–25%) are well compatible with a foreland basin setting (Dickinson 1985).

Micropalaeontological data

Only few data are available regarding the age of the Hida Formation, therefore samples have been collected for microfossil dating. The preparation of samples followed standard procedures (Wissing & Herrig 1999). Sample locations are indicated in Figure 8, the results are shown in Table 1.

All analysed samples yielded Burdigalian (in a few cases Early Miocene in a broad sense only) ages, based on foraminifera assemblages consisting mainly of benthic agglutinated and planktonic taxa. In two cases, a more exact age could be inferred. Sample 1119, collected below the base of the Hida Fm. (Fig. 8), yielded an age closer to the beginning of the Early Miocene, while sample 1125, collected from the basal Hida Fm., yielded a probable mid-Burdigalian age.

The probable mid-Burdigalian age of the basal parts of the Hida Fm. (sample 1125) confirms the generally accepted Burdigalian age of the Hida Fm. The micropalaeontological assemblage found in this sample, together with the lithofacies associations at this locality (basal part log 7, Fig. 5), suggest a deep marine turbiditic environment. Sample 1149 yielded a Burdigalian age. The identified lagenids are typical for the Chechiş Formation (Early Burdigalian).

The Burdigalian ages of the Hida Fm. clearly show that the onset of coarse clastic input in the southern part of the study area commences later than in the north. The microfossil assemblages indicate a general shallowing-up trend for the Hida Fm., while showing a transition from bathyal to offshore and marginal assemblages.

Subsurface data

In the southern part of the study area, the Upper Oligocene to Lower Miocene foredeep sediments are buried below the Middle to Upper Miocene sediments (Fig. 9) related to the evolution of Transylvania as a backarc basin (e.g. Ciulavu et al. 2002; Krezsek & Bally 2006). Here, these sediments rest on the Jurassic to Eocene deposits linked to the continental break-up, passive margin phase, subduction/obduction and continental collision of the Transylvanides ocean. In the north of the Transylvanian Basin, small patches of obducted ophiolites and/or Jurassic limestones are embedded in Cretaceous sediments deposited as pre-, syn- or post-tectonic sediments in relationship to the Albian continental collision (see also Săndulescu & Visarion 1977). The last pulses related to the active margin evolution were recorded during the Senonian and at the end of the Eocene, when the Puini Basin was inverted and finally thrusted (Fig. 9, De Broucker et al. 1998).

Depth interpretations (well-calibrated 2D seismic data) in north Transylvania indicate an overall unconformity between the syn-Puini Eocene thrusting and subsequent (Upper) Oligocene to Lower Miocene strata. The most apparent feature is the Lower Miocene wedge onlapping over older strata, subsequently tilted during the Middle to Late Miocene events (Krezsek & Bally 2006).

Although the overall wedge appears unitary, small internal reflector terminations (onlaps) suggest at least two pulses of loading in the foredeep (Fig. 9) prior to the deposition of the Middle Miocene Dej Fm. and Ocna Dejului Fm (salt). In map view, the Lower Miocene wedge strikes WSW–ENE. The deposits are up to 2200 m thick in the NE (Fig. 2).

The orientation of onlaps inside the wedge indicates a sediment transport towards the SE, a direction compatible with the SE thinning geometry of the clastic deposits. The SE-directed sediment transport within the wedge, as well as the WSW–ENE oriented basin axis (Fig. 2), are in good agreement with the SE-directed emplacement direction of the Pienides (Tischler et al. 2006). The internal unconformities in the wedge are interpreted to reflect successive tilting of the basin floor.

Table 1. Microfossil assemblages of the analysed samples and their probable environment. Compare Figs 8 and 10 for location of samples. Coordinates are in lat/long WGS 84

Sample No.	X	Y	Micropalaeontological assemblage	Environment	Age
0554 Hida Fm.	24.055	47.4374	Agglutinated foraminifera (*Nothia robusta*—very frequent, *Bathisyphon filiformis*, *Rhabdammina discreta*, *Reticulophragmium rotundidorsatum*, *Reticulophragmium acutidorsatum*).	bathyal	Burdigalian (lower Hida Fm.)
1119 Vima Fm.	24.0749	47.4571	Flattened planktoni and other poorly preserved foraminifera (*Catapsydrax unicavus*, *Globigerina eurapertura*, *Globigerina officinalis*, *Globigerina praebulloides*, *Globigerina wagner*).	pelagic	Probably base of the Early Miocene (upper Vima Fm.)
1125 Hida Fm.	24.0582	47.4408	Agglutinated foraminifera (*Retigulophragmium acutidorsatum*, *Cyclammina bradyi*, *Rhizammina algaeformis*, *Karrerulina horrida*) and rare, poorly preserved, planktonic foraminifera (*Globigerinoides trilobus*, and other indeterminable species).	bathyal	Burdigalian (lower Hida Fm.)
1126 Hida Fm.	24.0558	47.4362	Flattened and poorly preserved planktonic foraminifera (*Globigerina* cf. *anguliofficinalis*, *Globigerina* cf. *falcomensis*, *Globigerinoides* cf. *trilobus*, and other indeterminable species).	pelagic	Probably Burdigalian (?base of Hida Fm.)
1142 Hida Fm.	23.9497	47.4434	Agglutinated foraminifera (*Nothia excelsa*, *Reophax scorpiurus*, *Haplophragmoides* cf. *fragilis*) together with fish bones and teeth.	bathyal	Burdigalian (lower Hida Fm.)
1149 Hida Fm.	24.0303	47.4145	Agglutinated foraminifera (*Rhizammina algaeformis*, *Ammodiscus miocenicus*, *Glomospira charoides*) and calcareous foraminifera (*Pyramidulina latejugata*, *Lenticulina arcuatostriata*—probably reworked).	upper bathyal	Burdigalian (lower Hida Fm.)
1152 Hida Fm.	23.8648	47.197	Very rare calcareous foraminifera tolerant to salinity fluctuations (*Ammonia* div. sp.), reworkings from Cretaceous, fish bones and teeth.	deltaic	Probably Burdigalian (mid to upper Hida Fm.)

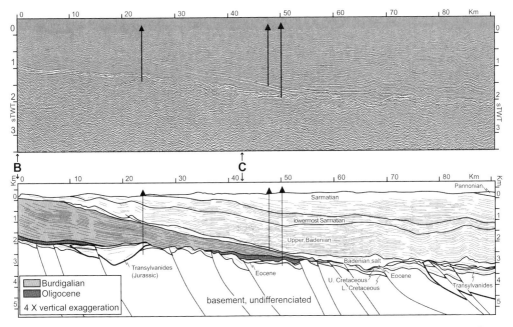

Fig. 9. Interpreted 2D seismic line (B–C), showing the wedge-shaped Burdigalian deposits. The shaded area correlates to the Burdigalian wedge. The two major unconformities within the wedge are interpreted as result of progressive loading and resultant flexure of the northern part of the Transylvanian Basin, acting as a foredeep. See Figure 2 for the trace of the section.

Comparable to the Lower Miocene wedge, the Oligocene deposits also show an increase in thickness towards the north, although with a WNW–ESE striking basin axis (De Broucker et al. 1998; Krezsek & Bally 2006). Hence, the following interpretation is put forward: Subsidence and wedging of Oligocene to Lower Miocene sediments coeval with the emplacement of the Pienides nappes implies that the clastic wedge represents the fill of the frontal foredeep of the Pienides (Fig. 10). Thrusting of the Pienide nappes led to progressive loading and flexure of the northern sector of the Transylvanian Basin causing the formation of a foredeep. This took place in successive tilting stages due to the flexural response of the loaded plate during southeastward emplacement of the Pienides (Fig. 9).

Summary

It is likely that eustatic sea-level variations accentuated sedimentation during Late Oligocene to Burdigalian times. On the other hand, the study area only intermittently showed open marine conditions, and the progradation phases of sand-dominated siliciclastics do not match the eustatic sea-level curve of Haq et al. (1987). Hence, eustatic sea-level variations cannot exclusively account for the overall southeastward progradation of the sand-dominated siliciclastics documented in the study area.

An overall southward migration of subsidence followed by the deposition of coarse terrigenous clastics is documented in the study area. The onset of this deposition is of late Early Oligocene (Late Rupelian) age in the northern part of the study area, continuously younging to Burdigalian age at the northern rim of the Transylvanian Basin. The facies associations and their depositional trends are suggestive for an active margin setting as confirmed by compatible sandstone compositions. Bad exposure conditions and insufficient temporal resolution prevented the clear recognition of depositional sequences related to enhanced tectonic activity as apparent for later (Burdigalian) stages from the analysis of seismic lines (Fig. 9).

Within the Burdigalian Hida Fm. a shallowing-up trend can be recognized, evidenced by changing micropalaeontological assemblages as well as by the transition from predominantly deep-water turbidites towards a succession locally showing fluvial–deltaic character. The southeastward thinning clastic wedge, as seen in seismic records from the Transylvanian Basin, shows an

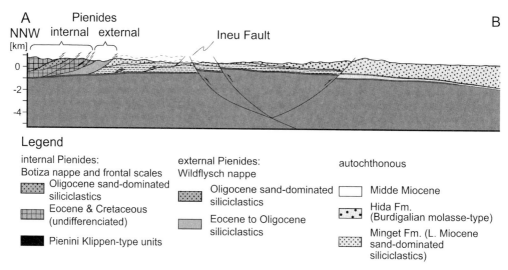

Fig. 10. Schematic geological cross-section A–B–C. The thrust front of the Pienide nappe pile, interpreted to cause the formation of the flexural basin, can be seen in the northernmost part of the section. The reverse faults in the middle of Section A–B (e.g. Ineu fault) are attributed to a later stage of deformation. The cross-section has been compiled by integrating information from available map sheets (Geological Survey of Romania), seismic data (Fig. 9) and unpublished data (C. Krézsek pers. comm. 2004).

internal geometry suggestive for a foreland basin fill (Fig. 9). Two important intra-Burdigalian unconformities indicate phases of tectonic activity, forcing the fan system to further prograde towards the SE. The geometry of the clastic deposits is suggestive for an elongate basin (WSW–ENE striking axis) with major clastic input derived from provenances from the NW.

This geometry is consistent with the bimodal palaeocurrent distribution measured within the Burdigalian strata, showing NE directed (longitudinal) transport as well as SE directed current directions. The predominance of NE directed transport is interpreted as an effect of deposition within a WSW–ENE striking, elongated basin. Regarding source areas, there are two likely candidates. The first possibility is to derive the siliciclastic material from the SW, i.e. the shallower parts of the Transylvanian Basin. The second possibility is to derive the material from the approaching accretionary wedge of the Pienides situated in the NW. An argument against a derivation of the Oligocene siliciclastic rocks (Domains 1 and 2) from the SW is that Domain 3, situated in the SW, features mud-dominated deposits during this time. Therefore a source area in the NW is more likely. Regarding Burdigalian times, the SE thinning of the Burdigalian wedge is a powerful argument for a source area located in the NW.

The most likely cause for the formation of the flexural basin is the emplacement of the Pienides,

as already suggested by Ciulavu (1998). Their SE-directed emplacement direction is in good agreement with the basin axis of the suspected flexural foreland basin, which is oriented roughly perpendicular. It seems highly likely that the Oligocene to Burdigalian clastic deposits at the northern border of the Transylvanian Basin are related to a continuously south to southeastward migrating flexural foreland basin. The initiation of convergence causing flexural foreland basin formation is indirectly dated by the onset of sand-dominated clastic input to Late Rupelian times.

The Pienide nappe emplacement and its foredeep in the regional context

Convergence between the continental blocks ALCAPA and Tisza–Dacia during their emplacement into the Carpathian embayment finally resulted in the thrusting of parts of ALCAPA onto Tisza–Dacia (Csontos & Nagymarosy 1998). While the Pieninny Klippen belt originally represents the outermost rim of ALCAPA (e.g. Csontos & Vörös 2004), more external units such as the external Pienides have been accreted during the invasion of ALCAPA into the Carpathian embayment (e.g. Fodor et al. 1999). In the final stages of the juxtaposition of these continental blocks, the easternmost tip of ALCAPA (i.e. the

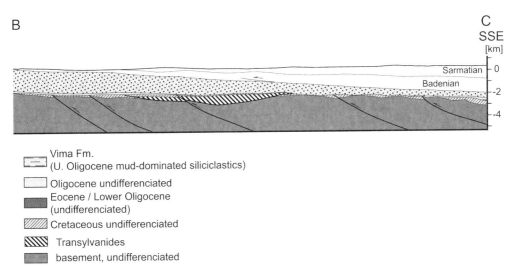

Fig 10. *Continued.*

Pienides) has been thrust onto Tisza–Dacia (e.g. Csontos & Nagymarosy 1998).

The formation of the Burdigalian flexural foredeep and its wedge-shaped clastic infill are highly likely to be related to the coeval last stages of Pienide nappe emplacement (e.g. Ciulavu et al. 2002). At the northern rim of the Transylvanian Basin, the Oligocene-age onset of turbidite deposition at a high rate suggests an earlier stage of thrusting of ALCAPA onto Tisza–Dacia as already proposed by Györfi et al. (1999). Hence, the interpretation is put forward, that the Late Rupelian to Burdigalian flexural foredeep development and subsequent deposition of coarse-grained siliciclastics reflects a continuous process resultant of the convergence of ALCAPA and Tisza–Dacia. The Late Rupelian onset of extensive sand-dominated sedimentation in the study area suggests a coeval timing for the convergence of ALCAPA and Tisza–Dacia across the northeasternmost segment of the Oligocene Mid-Hungarian fault zone (i.e. the Pienide nappe contact).

The change in strike of the respective basin axes (Oligocene: E–W to SE–NW, De Broucker et al. 1998; Burdigalian: WSW–ENE, Fig. 2) is likewise interpreted as a continuous development. Palaeomagnetic studies indicate (e.g. Panaiotu 1998; Márton et al. 2007) that Tisza–Dacia underwent significant clockwise rotation after the Oligocene, i.e. during emplacement into the Carpathian embayment. Such a clockwise rotation of Tisza–Dacia during ongoing flexural foreland basin formation could result in a passive turning of older basin axes (Fig. 11), while the younger basin axis remains in its original orientation.

Figure 11 summarizes the proposed model for Oligocene to Early Miocene times. Initial convergence between ALCAPA and Tisza–Dacia in Late Rupelian times results in the formation of a flexural foredeep on Tisza–Dacia being in the lower plate position. The overriding plate, ALCAPA, constitutes the source area for sand-dominated siliciclastic deposits, which are shed into an initially W–E to NW–SE (present-day coordinates) trending foredeep. During further convergence, Tisza–Dacia rotates in a clockwise sense, which results in a progressive migration of deformation southwards (present-day coordinates). By this migration of deformation, the loci of topography and foredeep formation also move towards the south, promoting longitudinal transport in the foredeep basin. The rotation of the lower plate results in a passive rotation of the foredeep sediments. As convergence continues, intervening units are accreted to ALCAPA. The last increment of convergence between ALCAPA and Tisza–Dacia results in the formation of the Burdigalian Basin striking WSW–ENE in present-day coordinates. Successive filling of this basin is demonstrated by the general shallowing-up trend of the sedimentary record.

The authors are indebted to M. Săndulescu for his support and fruitful discussions during the project. M. T. is particularly grateful to S. M. Schmid and L. Csontos for the careful and constructive review of a first version of the manuscript. C. Krézsek is thanked for very stimulating discussions and his contagious enthusiasm. C. Krézsek and D. Ciulavu are thanked for their constructive and careful reviews.

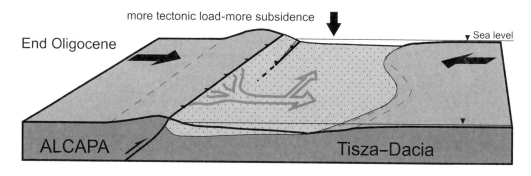

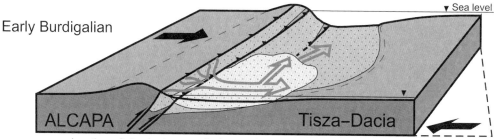

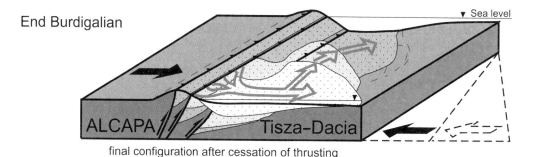

Fig. 11. Schematic block diagram of the proposed tectonic setting causing the development from an E–W to NW–SE trending Oligocene basin axis towards a WSW–ENE trending Burdigalian Basin axis. Rotation of Tisza–Dacia during convergence with ALCAPA results in a migration of shortening as well as a passive rotation of the older basin axes.

Financial support by the Swiss National Science Foundation (NF-project Nr. 21-64979.01, granted to B.F.) is gratefully acknowledged.

References

AROLDI, C. 2001. *The Pienides in Maramures—Sedimentation, Tectonics and Paleogeography*. PhD thesis, Cluj.

BALLA, Z. 1986. Paleotectonic reconstruction of the central Alpine-Mediterranean belt for the Neogene. *Tectonophysics*, **127**, 213–243.

BURCHFIEL, B. C. & BLEAHU, M. 1976. *Geology of Romania*. Geological Society of America, Special Paper, **158**, Boulder Colorado.

CIULAVU, D. 1998. *Tertiary Tectonics of the Transylvanian Basin*. PhD thesis, Vrije Universiteit Amsterdam.

CIULAVU, D., DINU, C. & CLOETINGH, S. A. P. L. 2002. Late Cenozoic tectonic evolution of the Transylvanian basin and northeastern part of the Pannonian basin (Romania): constraints from seismic profiling and numerical modelling. *EGU Stephan Mueller Special Publication Series*, **3**, 105–120.

CIUPAGEA, D., PAUCĂ, M. & ICHIM, T. 1970. *Geologia Depresiunii Transilvaniei*. Editura Academiei Romane, Bucuresti.

CLOETINGH, S. A. P. L., BUROV, E., MATENCO, L. ET AL. 2004. Thermo-mechanical controls on the mode of continental collision in the SE Carpathians

(Romania). *Earth and Planetary Science Letters*, **218**, 57–76.

CSONTOS, L. 1995. Cenozoic tectonic evolution of the Intra-Carpathian area: a review. *Acta Vulcanologica*, **7**, 1–13.

CSONTOS, L. & NAGYMAROSY, A. 1998. The Mid-Hungarian line: a zone of repeated tectonic inversions. *Tectonophysics*, **297**, 51–71.

CSONTOS, L. & VÖRÖS, A. 2004. Mesozoic plate tectonic reconstruction of the Carpathian region. *Paleogeography, Paleoclimatology, Paleoecology*, **210**, 1–56.

DE BROUCKER, G., MELLIN, A. & DUINDAM, P. 1998. Tectonostratigraphic evolution of the Transylvanian Basin, Pre-Salt sequence, Romania. *In*: DINU, C. (ed.) *Bucharest Geoscience Forum*. Special volume, **1**, 36–70.

DICEA, O., DUŢESCU, P., ANTONESCU, F. *ET AL*. 1980. Contributii la cunoasterea stratigrafiei zonei transcarpatice din Maramures. *Dări de seamă ale şedinţelor Institutul de Geologie şi Geofizică*, **LXV (4)**, 21–85.

DICKINSON, W. R. 1985. Interpreting provenance relations from detrital modes of sandstones. *In*: ZUFFA, G. G. (ed.) *Provenance of Arenites*. NATO ASI Series. Series C: Mathematical and Physical Sciences, **148**, 333–361.

FLÜGEL, E. 1978. *Mikrofazielle Untersuchungsmethoden von Kalken*. Springer-Verlag, Berlin Heidelberg New York.

FODOR, L., CSONTOS, L., BADA, G., GYÖRFI, I. & BENKOVICS, L. 1999. Cenozoic tectonic evolution of the Pannonian basin system and neighbouring orogens: a new synthesis of paleostress data. *In*: DURAND, B., JOLIVET, L., HORVÁTH, F. & SÉRANNE, M. (eds) *The Mediterranean Basins: Cenozoic Extension within the Alpine Orogen*. Geological Society, London, Special Publication, **156**, 295–334.

FODOR, L., BADA, G., CSILLAG, G. *ET AL*. 2005. An outline of neotectonic structures and morphotectonics of the western and central Pannonian Basin. *Tectonophyiscs*, **410**, 15–41.

FÜGENSCHUH, B. & SCHMID, S. M. 2005. Age and significance of core complex formation in a highly bent orogen: evidence from fission track studies in the South Carpathians (Romania). *Tectonophysics*, **404**, 33–53.

GRADSTEIN, F., OGG, J. & SMITH, A. 2004. *A Geologic Time Scale*. Cambridge University Press, Cambridge.

GYÖRFI, I., CSONTOS, L. & NAGYMAROSY, A. 1999. Early Cenozoic structural evolution of the border zone between the Pannonian and Transylvanian basins. *In*: DURAND, B., JOLIVET, L., HORVÁTH, F. & SÉRANNE, M. (eds) *The Mediterranean Basins: Cenozoic Extension within the Alpine Orogen*. Geological Society, London, Special Publication, **156**, 251–267.

HAQ, B. U., HARDENBOL, J. & VAIL, P. R. 1987. The chronology of fluctuating sea level since the Triassic. *Science*, **235**, 1156–1167.

HORVÁTH, F. 1993. Towards a mechanical model for the formation of the Pannonian basin. *Tectonophysics*, **226**, 333–357.

HORVÁTH, F., BADA, G., SZAFIÁN, P., TARI, G., ÁDÁM, A. & CLOETINGH, S. 2006. Formation and deformation of the Pannonian basin: constraints from observational data. *In*: GEE, D. G. & STEPHENSON, R. (eds) *European Lithosphere Dynamics*. Geological Society, London, Memoirs, 1–672.

JIPA, D. 1962. Directii de aport în gresia de Borsa (Maramures). *Comunicarile Academiei Replublicii Populare Române*, **12**, 1363–1368.

KNAPP, J. H., KNAPP, C. C., RAILEANU, V., MATENCO, L., MOCANU, V. & DINU, C. 2005. Crustal constraints on the origin of mantle seismicity in the Vrancea Zone, Romania: the case for active continental lithospheric delamination. *Tectonophysics*, **410**, 311–323.

KOCH, A. 1900. *Die Tertiärbildungen des Beckens der Siebenbürgische Landestheile. II Neogene Abtheilung*, Budapest.

KRÄUTNER, H. G., KRÄUTNER, F. & SZASZ, L. 1982. *Geological Map 1:50.000 No 20a Pietrosul Rodnei*. Institutul de Geologie si Geofizica, Bucharest.

KRÄUTNER, H. G., KRÄUTNER, F. & SZASZ, L. 1983. *Geological Map 1:50.000 No 20b Ineu*. Institutul de Geologie si Geofizica, Bucharest.

KRÉZSEK, C. & BALLY, A. 2006. The Transylvanian Basin (Romania) and its relation to the Carpathian fold and thrust belt: insights in gravitational salt tectonics. *Marine and Petroleum Geology*, **23**, 405–442.

KRZYWIEC, P. 2001. Contrasting tectonic and sedimentary history of the central and eastern parts of the Polish Carpathian foredeep basin—results of seismic data interpretation. *Marine and Petroleum Geology*, **18**, 13–38.

MÁRTON, E., TISCHLER, M., CSONTOS, L., FÜGENSCHUH, B. & SCHMID, S. M. 2007. The contact zone between the ALCAPA and Tisza-Dacia mega-tectonic units of Northern Romania in the light of new paleomagnetic data. *Swiss Journal of Geosciences*, **100**, 109–124.

MATENCO, L. & BERTOTTI, G. 2000. Tertiary tectonic evolution of the external East Carpathians (Romania). *Tectonophysics*, **316**, 255–286.

MATENCO, L., BERTOTTI, G., CLOETINGH, S. & DINU, C. 2003. Subsidence analysis and tectonic evolution of the external Carpathian-Moesian Platform during Neogene times. *Sedimentary Geology*, **156**, 71–94.

MIHAILESCU, N. & PANIN, N. 1962. Directii de curent în depozitele Eocen-Oligocene din Regiunea Telciu-Romuli (Maramures). *Comunicarile Academiei Replublicii Populare Române*, **12**, 1357–1362.

MOIESCU, V. 1981. Oligocene deposits of Transylvania and their correlation in Parathethys. *Révues Roumanien de Géologie, Géophysique et Géographie, ser Géologique*, **25**, 161–169.

MUTTI, E., TINTERRI, R., REMACHA, E., MAVILLA, N., ANGELLA, S. & FAVA, L. 1999. *An Introduction to the Analysis of Ancient Turbidite Basins from an Outcrop Perspective*. AAPG Continuing Education Course Note Series, **39**, Tulsa.

MUTTI, E. & NORMARK, W. R. 1987. Comparing examples of modern and ancient turbidite systems; problems and concepts. *In*: LEGGET, J. K. & ZUFFA, G. G. (eds) *Marine Clastic Sedimentology; Concepts and Case Studies*. Graham and Trotman, London, 1–38.

NEMČOK, M. 1993. Transition from convergence to escape: field evidence from the West Carpathians. *Tectonophysics*, **217**, 117–142.

PANAIOTU, C. 1998. Paleomagnetic constraints on the geodynamic history of Romania. *In*: IOANE, D. (ed.)

Monograph of Southern Carpathians. Reports on Geodesy, **7**, 205–216.
PETTIJOHN, F. J., POTTER, P. E. & SIEVER, R. 1987. *Sand and Sandstone*. Springer-Verlag, New York.
PICKERING, K. T., HISCOTT, R. N. & HEIN, F. J. 1989. *Deep Marine Environments*. Unwin Hyman. London, United Kingdom.
PICKERING, K. T., CLARK, J. D., SMITH, R. D. A., HISCOTT, R. N., RICCHI LUCCI, F. & KENYON, N. H. 1995. Architectural element analysis of turbidite systems, and selected topical problems for sand-prone deep-water systems. *In*: PICKERING, K. T., HISCOTT, R. N., KENYON, N. H., RICCHI LUCCI, F. & SMITH, R. D. A. (eds) *Atlas of Deep Water Environments—Architectural Style in Turbidite Systems*. Chapman and Hall, London, 1–10.
POPESCU, Gh. 1975. Etudes des foraminiferes du Miocene inferieur et moyen du nord-ouest de la Transylvanie. *Memorii Institutul de Geologie si Geofizica*, **23**, 5–121.
POPESCU, B. M. 1984. Lithostratigraphy of cyclic continental to marine Eocene deposits in NW Transylvania, Romania. *In*: POPESCU, B. M. (ed.) *The Transylvanian Paleogene Basin*. Geneva, 37–73.
RATSCHBACHER, L., MERLE, O., DAVY, P. & COBBOLD, P. 1991*a*. Lateral extrusion in the Eastern Alps; Part 1, Boundary conditions and experiments scaled for gravity. *Tectonics*, **10**, 245–256.
RATSCHBACHER, L., FRISCH, W., LINZER, H. G. & MERLE, O. 1991*b*. Lateral extrusion in the Eastern Alps; Part 2, Structural analysis. *Tectonics*, **10**, 257–271.
RATSCHBACHER, L., LINZER, H. G. & MOSER, F. 1993. Cretaceous to Miocene thrusting and wrenching along the Central South Carpathians due to a corner effect during collision and orocline formation. *Tectonics*, **12**, 855–873.
RÖGL, F. 1999. Mediterranean and Paratethys. Facts and Hypotheses of an Oligocene to Miocene Paleogeography (Short Overview). *Geologica Carpathica*, **50**, 339–349.
ROYDEN, L. H. 1988. Late Cenozoic Tectonics of the Pannonian Basin System. *In*: ROYDEN, L. H. & HORVÁTH, F. (eds) *The Pannonian Basin; A Study in Basin Evolution*. AAPG Memoir, **45**, 27–48.
ROYDEN, L. H. 1993. The tectonic expression of slab pull at continental convergent boundaries. *Tectonics*, **12**, 303–325.
RUSU, A. 1989. Problems of correlation and nomenclature concerning the Oligocene formations in NW Transylvania. *In*: PETRESCU, I. (ed.) *The Oligocene from the Transylvanian Basin*. Cluj-Napoca, 67–78.
RUSU, A., BALINTONI, I., BOMBITA, G. & POPESCU, G. 1983. *Geological Map 1 : 50.000 No 18c Preluca*, Institutul de Geologie si Geofizica, Bucharest.
SĂNDULESCU, M. 1980. Sur certain problèmes de la corrélation des Carpathes orientales Roumaines avec les Carpathes Ucrainiennes. *Dări de seamă ale şedinţelor Institutul de Geologie şi Geofizică*, **LXV**, 163–180.
SĂNDULESCU, M. 1984. *Geotectonica Romaniei*. Editura Tehnica, Bucharest, 1–450.
SĂNDULESCU, M. 1988. Cenozoic Tectonic History of the Carpathians. *In*: ROYDEN, L. H. & HORVÁTH, F. (eds) *The Pannonian Basin; A Study in Basin Evolution*. AAPG Memoir, 17–25.
SĂNDULESCU, M. & VISARION, M. 1977. Considerations sur la structure tectonique du soubassement de la depression de Transylvanie. *Dari de seama ale sedintelor Institutul de Geologie si Geofizica*, **LXIV**, 153–173.
SĂNDULESCU, M. & MICU, M. 1989. Oligocene Paleography of the east Carpathians. *In*: PETRESCU, I. (ed.) *The Oligocene from the Transylvanian Basin*. Cluj-Napoca, 79–86.
SĂNDULESCU, M., KRÄUTNER, H. G., BALINTONI, I., RUSSO-SĂNDULESCU, D. & MICU, M. 1981. *The Structure of the East Carpathians. (Guide Book B1)*, Carpatho-Balkan Geological Association 12th Congress, Bucharest.
SĂNDULESCU, M., SZASZ, L., BALINTONI, I., RUSSO-SĂNDULESCU, D. & BADESCU, D. 1991. *Geological Map 1 : 50.000 No 8d Viseu*. Institutul de Geologie si Geofizica, Bucharest.
SĂNDULESCU, M., VISARION, M., STANICA, D., STANICA, M. & ATANASIU, L. 1993. Deep structure of the inner Carpathians in the Maramures-Tisa zone (East Carpathians). *Romanian Journal of Geophysics*, **16**, 67–76.
SCHMID, S. M., BERZA, T., DIACONESCU, V., FROITZHEIM, N. & FÜGENSCHUH, B. 1998. Orogen-parallel extension in the South Carpathians. *Tectonophysics*, **297**, 209–228.
SCHMID, S. M., FÜGENSCHUH, B., MATENCO, L., SCHUSTER, R., TISCHLER, M. & USTASZEWSKI, K. 2006. The Alps-Carpathians-Dinarides-connection: a compilation of tectonic units. *18th Congress of the Carpathian-Balkan Geological Association*. Belgrade, Serbia, Sept. 3–6, 2006, Conference Proceedings, 535–538.
SHANMUGAN, G. & MOIOLA, R. J. 1988. Submarine fans; characteristics, models, classification, and reservoir potential. *Earth Science Reviews*, **24**, 383–428.
SPERNER, B., RATSCHBACHER, L. & NEMČOK, M. 2002. Interplay between subduction retreat and lateral extrusion: tectonics of the Western Carpathians. *Tectonics*, **21**, 1051.
TISCHLER, M., GRÖGER, H. R., FÜGENSCHUH, B. & SCHMID, S. M. 2006. Miocene tectonics of the Maramures area (Northern Romania)—implications for the Mid-Hungarian fault zone. *International Journal of Earth Sciences*, Online first: doi: 10.1007/s00531-006-0110-x.
TUCKER, M. E. 1981. *Sedimentary Petrology: An Introduction to the Origin of Sedimentary Rocks*. Blackwell Science Publications, Oxford.
WEIDLE, C. & WIDIYANTORO, S. 2005. CALIXTO Working Group. Improving depth resolution of teleseismic tomography by simultaneous inversion of teleseismic and global P-wave traveltime data—application to the Vrancea region in Southeastern Europe. *Geophysical Journal International*, **162**, 811–823.
WISSING, F. N. & HERRIG, E. 1999. *Arbeitsmethoden der Mikropaläontolgie: Eine Einführung*. Enke-Verlag Stuttgart.
WORTEL, M. J. R. & SPAKMAN, W. 2000. Subduction and slab detachment in the Mediterranean-Carpathian region. *Science*, **290**, 1910–1917.

Calcareous nannofossil age constraints on Miocene flysch sedimentation in the Outer Dinarides (Slovenia, Croatia, Bosnia-Herzegovina and Montenegro)

TAMÁS MIKES[1], MÁRIA BÁLDI-BEKE[2], MIKLÓS KÁZMÉR[3], ISTVÁN DUNKL[1] & HILMAR VON EYNATTEN[1]

[1]*Sedimentologie/Umweltgeologie, Geowissenschaftliches Zentrum der Universität Göttingen, Goldschmidtstrasse 3, D-37077 Göttingen, Germany (e-mail: tamas.mikes@geo.uni-goettingen.de)*

[2]*Rákóczi utca 42, H-2096 Üröm, Hungary*

[3]*Department of Palaeontology, Eötvös University, Pázmány Péter sétány 1/c, H-1117 Budapest, Hungary*

Abstract: Flysch deposits are associated with the Outer Dinaride nappe front. They overlie Eocene platform carbonate to bathyal marl successions that subsequently cover Cretaceous platform carbonates of Apulia and the Dinaride nappes. Planktonic foraminifer biostratigraphy indicates Eocene age of flysch sedimentation. New calcareous nannofossil data reveal that several assemblages are present; besides the dominant Mid-Eocene species, Cretaceous, Paleocene, Oligocene and Miocene taxa were also identified throughout the entire flysch belt. Widespread occurrence of nannofossil species of zone NN4-6 indicates that flysch deposition lasted up to at least the Mid-Miocene. Ubiquitous occurrence of various pre-Miocene taxa demonstrates that extensive, possibly submarine, sediment recycling has occurred in the Cenozoic. As flysch remnants are typically sandwiched between thrust sheets, these new stratigraphic ages give a lower bracket on deformation age of the coastal range. The data provide a link between Cretaceous compression in the Bosnian Flysch and recent deformation in the Adriatic offshore area.

Cenozoic synorogenic clastic rocks overlie an upward-deepening Eocene carbonate platform to bathyal marl succession of the Apulian foreland and of the outermost parts of the SW-vergent Outer Dinaride thrust belt (Fig. 1). Established mostly on the basis of planktonic foraminifera, and locally by calcareous nannofossils, the stratigraphic age of the deposits has been traditionally placed into the Mid- or Late Eocene (Table 1). A SE-directed orogen-parallel younging of Palaeogene sedimentation has been inferred by Piccoli & Proto Decima (1969).

Recent calcareous nannofossil studies indicate, however, that at several locations in the central and SE part of the basin system clastic deposition lasted up to the Middle Miocene (review in de Capoa & Radoičić 2002). In addition, tectonic slices of older flysch series dated or inferred to be of Late Cretaceous and Palaeogene age are found in the inner part of the Dinaride imbricate thrust belt occupying structurally higher positions.

These contradictory data pose a series of important questions that need to be addressed in detail:

(1) Can ages younger than Eocene be demonstrated in the NW parts of the coastal flysch zone, too?

(2) Why do the age data from planktonic foraminifera and calcareous nannofossil studies seem to mismatch?

(3) Where and when did flysch deposition start? Are the deposits in the main belt and those preserved in narrow thrust slices related?

(4) Does revised biostratigraphy support the idea of diachronic onset of deposition along the orogenic front?

(5) What is the bearing of such young stratigraphic ages on the understanding of the Outer Dinaride geodynamic evolution?

In the present paper, new results of a calcareous nannofossil study are reported, obtained from several flysch localities throughout the Outer Dinaride region. Areas of large-scale regional sampling included the Trieste–Koper and Pazin Basins on the Istrian Peninsula, the Northern Kvarner Islands (Krk and Rab), Pag Island, Šopot section near Benkovac in Northern Dalmatia, Central Dalmatia, Southern Dalmatia, Montenegro coast as well as the 'Dalmatian-Herzegovinian Zone' of southern Bosnia-Herzegovina and Montenegro inland (Fig. 1 and Appendix B).

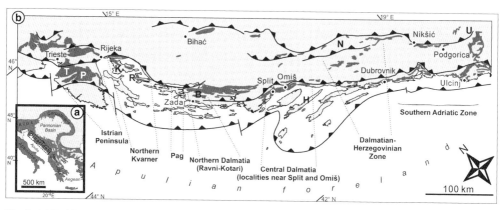

Fig. 1. (a) Position of the Dinarides within the European Alpine chain. Dark grey: Alpine orogen; light grey: post-orogenic basins. (b) Schematic geological setting of the Outer Dinarides with locations of the major sampling areas. Light grey: substrate of the flysch—mainly Palaeozoic to Mesozoic formations, dominantly platform carbonates. Dark grey: Cenozoic flysch and associated shallow marine sediments in the Outer Dinaride foreland basin system. T, Trieste-Koper Basin; P, Pazin Basin; K, Krk Island; R, Rab Island; B, Benkovac town; H, Hvar Island; U, Kuči Thrust; N, Nevesinjsko Polje. Position of thrusts after the Geological Map of Yugoslavia (F.G.I. 1970), Tari (2002) and Schmid *et al.* (2006).

Geological setting and sedimentology

The Outer Dinarides and their foreland are dominated by the thick deposits of the Adriatic Carbonate Platform (AdCP – Vlahović *et al.* 2005) that existed throughout the entire Mesozoic until its final drowning in the Mid-Eocene. Structurally, it consists of two parts: the lower plate corresponds to autochthonous Apulia, while the upper plate forms a broad, c. 100 km wide zone made up of the Dinaride nappes and imbricate thrust sheets. These units are built up mainly of Cretaceous rudistid limestones and Eocene foraminiferal limestones and marls, and are covered by Tertiary flysch that interfingers with and underlies shallow marine to continental clastic sediments. For extensive reviews of the Cenozoic stratigraphy, the reader is referred to Drobne (1977), Marjanac & Ćosović (2000) and Ćosović *et al.* (in press). The carbonate nappes are thrust by the folded, Mesozoic units of the Bosnian Flysch and its low-grade metamorphic, Palaeozoic basement, the Bosnian Schist Mountains (BSM). They underwent Early Cretaceous metamorphism and the BSM was exhumed during the Eocene–Oligocene (Pamić *et al.* 2004; Petri 2007).

The Outer Dinaride flysch is part of the fill of a large foreland basin system at the front of the thrust wedges (Fig. 1). Sub-basins stretch from the southern Alps along the Adriatic Sea coastline as far to the SE as the Hellenides. Stratigraphic position of the flysch is described in terms of a Paleogene marine sequence overstepping the Cretaceous platform. The underlying foraminiferal ramp covers a regionally widespread unconformity and started to develop in the Paleocene with paralic deposits. Facies development indicates transgression throughout the Early to Middle Eocene but the sequence is punctuated with a number of short-lived subaerial exposure events probably due to the interplay of eustatic oscillations in a shallow marine environment and the effects of ongoing compressional tectonics affecting the Adriatic area already since the Cretaceous (Channell *et al.* 1979; Mindszenty *et al.* 1995; Pamić *et al.* 1998). The Eocene carbonate unit is rarely thicker than 200 m and is overlain by the 'Transitional Beds'—a deepening-upward shelf to shallow bathyal sequence several tens of metres thick, characterized by increasing amount of pelagic biota, glauconite and silt upsection. Its thin lower part is referred to as 'Marl with crabs', passing upwards into the thick '*Globigerina* Marl' (e.g. Juračić 1979; Ćosović *et al.* 2004).

The flysch rests upon the '*Globigerina* Marl'. Their contact is conformable at places but angular or erosional unconformities have often been reported (e.g. Marinčić 1981; Marjanac *et al.* 1998; Marjanac 2000). Due to repetitive thrusting, flysch profiles are usually truncated and less than 100–400 m thick. The offshore succession resting on the Apulian Plate is gently folded, reaches up to the Neogene, and may exceed 1000 m in thickness (e.g. Tari-Kovačić 1998).

Turbidite beds are dominantly composed of siliciclastics with variable (0–50%) amounts of carbonate admixture. In the N part of the basin, palaeocurrent data were interpreted as resulting from largely SW-directed primary, and SE-directed

Table 1. Overview of published age data from the Outer Dinaride flysch. Offshore and sporadic mainland data are not included. Whenever available, standard planktonic foraminifer or calcareous nannoplankton biozones are also indicated. Results yielding Neogene age are marked bold for clarity. Abbreviations: np, nannofossils; pf, planktonic foraminifera; lf, larger foraminifera; po, palynomorphs; mo, mollusc macrofauna

Area	Age	Locality	Source	Age based on	Biozone	Notes
Istria	Upper Oligocene	Pićan	Šparica et al. (2005)	lf, np, po		
	Priabonian	Oprtalj	Benić (1991)	np	NP18	
	Upper Eocene/Lower Oligocene	Pazin, Motovun	Marinčić (1981)	pf		1
	Upper Eocene	Pićan	Drobne et al. (1979)	pf, np		
	Middle Lutetian to Lower Priabonian	several localities, review	Ćosović et al. (in press)	pf, np	P11–P15	
	Bartonian	Pričejak (Učka)	Benić (1991)	np	NP17	
	Bartonian	Pićan	Benić (1991)	np	NP17	
	Upper Lutatian-Bartonian	Pićan	Hagn et al. (1979)	np, lf, pf	NP16	
	Upper Middle Eocene	several localities	Kraševnikov et al. (1968)	pf		2
	Middle Eocene	Vranja, Velanov brijeg	Stradner (1962)	np		3
	Middle Lutetian	Pićan	Pavlovec et al. (1991)	pf, np		4
	Middle Lutetian	Buzet, Paz, Vranje, Kotle, Draguć	Muldini-Mamužić (1965)	pf		
	Upper Lutatian-Bartonian	Izola, Piran	Pavšič (1981)	np	NP16	
	Upper Lutatian-Bartonian	Piran	Pavšič & Peckmann (1996)	np	NP16–NP17	
	Middle Eocene	several localities	Piccoli & Proto Decima (1969)	pf		
	Middle Lutetian to Lower Bartonian	Pićan–Gračišće	Živković & Babić (2003)	pf	P11–P13	
	Cuisian to Middle/Late Lutetian	several localities	Živković (2004)	pf		
Northern Kvarner: Krk	Upper Mid-Eocene	several localities	Šikić (1963)	mo		
	Upper Mid-Eocene	Murvenica	Schubert (1905)	mo		
	Lower Eocene	Omišalj, Baška, Dobrinj	Piccoli & Proto Decima (1969)	pf		
Northern Kvarner: Rab	U. Mid-Eocene to U. Eocene	Lopar	Muldini-Mamužić (1962)	pf		
Pag	Middle Lutetian-Bartonian	Dinjiška	Benić (1975)	np	NP15–17	5
	Middle Eocene	Gorica, Vrčići	Piccoli & Proto Decima (1969)	pf		

(Continued)

Table 1. Continued

Area	Age	Locality	Source	Age based on	Biozone	Notes
Northern Dalmatia (Ravni-Kotari)	Upper Eocene	Šopot	Kraševinnikov et al. (1968)	pf	P13-P16/17	
	Bartonian to Priabonian	Šopot	Drobne et al. (1991)	pf	NP16-17	
	Uppermost Lutetian to Bartonian	Šopot	Benić (1983, fide Marjanac et al. 1998)	np		
	Middle Eocene	Zadar, Zemunik	Piccoli & Proto Decima (1969)	pf		
Central Dalmatia	**Lower Tortonian**	'Split E'	de Capoa et al. (1995)	np	NN9	
	Middle Miocene	Mravince	de Capoa et al. (1995)	np		
	Lower Aquitanian	Hvar Island	Puškarić (1987)	np	NN1	
	Upper Oligocene to Lower Miocene	Jadro quarry	de Capoa et al. (1995)	np	NP25-NN2	
	Upper Rupelian	Gornje Sitno	de Capoa et al. (1995)	np	NP23	
	Bartonian	Hvar Island	Puškarić (1987)	np	NP17	
	Upper Eocene/Lower Oligocene	Split-Omiš	Grubić & Komatina (1963)	pf		
	Upper Eocene/Lower Oligocene	Hvar Island	Marinčić (1981)	pf		
	Upper Eocene	Marjan Peninsula in Split	Piccoli & Proto Decima (1969)	pf		
	Priabonian	Hvar Island	Kraševinnikov et al. (1968)	pf		6
	Upper Eocene	Hvar Island	Herak et al. (1976)	pf		
	Upper Priabonian	Split	Jerković & Martini (1976)	np	NP19/20	
	Bartonian to Upper Priabonian	Hvar Island	Marjanac et al. (1998)	pf	NP17-NP19	
	Bartonian	Orebić (Pelješac Peninsula)	Benić (1983, fide Marjanac et al. 1998)	np	NP17	7
Dalm.-Herz. Zone	**Lower Tortonian**	Vukov Klanac	Radoičić et al. (1991)	np	NN9	
	Lower Tortonian	Bađula	Radoičić et al. (1991)	np	NN9	
	Lower Tortonian	Moševići	Radoičić et al. (1991)	np	NN9	
	Lower Serravallian	Žitomislići	Radoičić et al. (1991)	np	NN5 top	
	Lower Serravallian	Gradnići	Radoičić et al. (1991)	np		
	Lower Serravallian (?Upper Serravallian)	Dabarsko-Fatničko Polje	Radoičić et al. (1991)	np		
	Upper Burdigalian	Glavatovići	de Capoa & Radoičić (1994b)	np	NN5 (?NN7)	
	Middle Eocene	Ljubuški	Kraševinnikov et al. (1968)	pf	NN4	
	Middle Eocene	Gornji Studenci	Kraševinnikov et al. (1968)	pf		
	Middle Eocene	Lukavačko polje	Kraševinnikov et al. (1968)	pf		

S-Adriatic Zone	Age	Locality	Reference		Nanno zone
	Lower Tortonian	Možura North	de Capoa et al. (1995)	np	NN9b
	Upper Serravallian (?Lower Tortonian)	Možura-Saško Brdo	Radoičić et al. (1989)	np	
	Serravallian	W of Grbalj	de Capoa et al. (1995)	np	
	Lower Serravallian (?Upper Serravallian)	Konavle	de Capoa et al. (1995)	np	
	Lower Serravallian	Kotor-Trojica	de Capoa & Radoičić (1994a)	np	NN5 top
	Lower Serravallian	Trojica-Grbalj	de Capoa & Radoičić (1994a)	np	NN5
	Lower Serravallian	Tivat	de Capoa & Radoičić (1994a)	np	NN5
	Lower Serravallian	Petrovac	de Capoa & Radoičić (1994a)	np	NN5
	Lower Serravallian	Kruševica	de Capoa & Radoičić (1994a)	np	NN5
	Langhian	Grbalj	de Capoa et al. (1995)	np	
	Miocene	Kotor-Vrmac	de Capoa & Radoičić (1994a)	pf	
	Oligocene	Ulcinj	Čanović & Džodžo-Tomić (1958)		
	Upper Bartonian to Lower Oligocene	Ulcinj	Luković & Petković (1952)	lf	
	Upper Eocene to Oligocene	Cavtat	Krašeninnikov et al. (1968)	pf	
	Middle to Upper Eocene	several localities	Pavić (1970)	pf, lf	
	Bartonian	Izvor Česma	de Capoa & Radoičić (1994a)	np	NP17

(1) Drobne et al. (1979) report Lower, Middle and Upper Lutetian as well as Upper Eocene planktonic foraminifera and Middle Lutetian nannoplankton from the Pićan flysch. They conclude that age of the strata is Middle Lutetian.
(2) Age younger than the LO of *Discoaster lodoensis* and *D. kuepperi*, but older than the FO of *Isthmolithus recurvus* and may correspond to the zones NP12-15 as re-interpreted by Jerković & Martini (1976).
(3) Pelagic foraminifera with different biozonal ranges within Middle Lutetian.
(4) Flysch contains arenaceous species of Foraminifera, which markedly contrast the *Globigerina* in the 'Transitional Beds'. The change appears just above the 'Nummulite breccia' and was observed at several localities in Istria (Buzet, Kotle, Draguć).
(5) Summarizing description of 20 samples. Reworked Paleocene and Lower Eocene forms are also reported.
(6) Reworked Middle Eocene forms are also reported.
(7) 'Similarly as in Split region' (Marjanac et al. 1998).

deflected (longitudinal) flows (e.g. Magdalenić 1972; Babić & Zupanič 1983; Orehek 1991). Radial current directions are commonly found in Central Dalmatia, resulting from complex basin floor topography and multiple flow reflection (Marjanac 1990).

A clear NE-directed flow direction can be observed in the N part of the basin on carbonate debrites and calciturbidites that intercalate into the siliciclastic succession (Engel 1974; Babić & Zupanič 1996). The coarser-grained debrites range in composition from breccia consisting exclusively of well-cemented Eocene foraminiferal limestone lithoclasts, through mixed ones having much isolated larger foraminifer tests and rhodoliths beside the lithoclasts, to pure grain- or matrix-supported debrites made up of *Nummulites* tests. Marl and Upper Cretaceous limestone clasts are subordinate (Skaberne 1987; Magdalenić 1972; Hagn et al. 1979; Marjanac & Marjanac 1991; Radoičić et al. 1991; Tunis & Venturini 1992; Babić et al. 1995; Marjanac 1996; Tomljenović 2000; Bergant et al. 2003; Pavlovec 2003).

Based on its narrow appearance in map view and the—largely scattered—uniform, longitudinal, SE-directed palaeoflow indicators, the flysch basin has been interpreted as a single major elongated trough (e.g. Marinčić 1981). However, flysch deposits at places rapidly grade upsection into thick sandstone beds deposited in shallow shelf environments, pointing to a complex, dissected basin floor topography with different subsidence histories in the individual domains (Zupanič & Babić 1991; Babić et al. 1993; Babić & Zupanič 1998). Rapid upward decrease of water depth in the upper part of the thin flysch sequences has also been observed in other localities at the Island of Pag and in Northern Dalmatia (Lj. Babić, pers. comm. 2005).

Present status of flysch biostratigraphy

Traditionally, Cenozoic clastic strata stretching along the Adriatic coast have been regarded as Middle to Late Eocene in age. Ages based on planktonic foraminifera and partly on nannofossils range from Early Eocene to Early Oligocene, mostly Bartonian to Priabonian, as summarized in Table 1. Piccoli & Proto Decima (1969) recognized that the ages become progressively younger towards the SE. Since then, deposition of the flysch in the coastal zone has been commonly explained in terms of a SE-directed diachroneity (Marjanac & Ćosović 2000; Ćosović et al. in press, and references therein).

Upper Eocene to Lower Oligocene planktonic foraminifera described from Pazin and Motovun localities in Istria and from Hvar Island by Marinčić (1981), and from sites near Split by Grubić & Komatina (1963), received little attention in subsequent works. Recently, Šparica et al. (2005) reported Upper Oligocene larger foraminiferal, calcareous nannofossil and pollen assemblages from the Pićan profile in Istria.

The first notion of onshore Neogene is from Puškarić (1987) who proved distinct biozones in two nearby profiles on Hvar Island by means of calcareous nannofossils: NP17 (Bartonian) and NN1 (Uppermost Chattian to Lower Aquitanian). Nannofossil studies of de Capoa revealed Early to Middle Miocene ages up to Serravallian from a considerable number of localities in the central and SE part of the flysch basin (Radoičić et al. 1989, 1991; de Capoa & Radoičić 1994a, b; de Capoa et al. 1995; de Capoa & Radoičić 2002). Their results are summarized in Table 1. Quantitative test data from nannofossil counting by Radoičić et al. (1989) suggest that Miocene forms constitute only a few percent of the dominantly reworked nannofossil assemblage at any locality.

Methods

Pelitic rocks were collected for nannofossil analysis throughout the flysch belt (Appendix B). This study does not replace detailed sectionwise biostratigraphic work, yet it represents an exemplary sampling of the most suitable outcrops in the entire basin, performed as such for the first time. Rocks were sampled in five various facies: (1) laminated hemipelagic pelite, (2) pelite rip-up clasts found within sandstone turbidite beds, as well as (3) plastically deformed pelite fragments included in clast-supported carbonate breccia, or in (4) *Nummulites* debrite, and finally (5) the pelite matrix of matrix-supported debrites made up of limestone clasts and *Nummulites* tests. The small sampled volume of the clasts (a few mm^3) required extremely clean conditions during preparation to avoid contamination.

Standard smear slides were prepared from a total of 69 crushed samples using no chemical treatment or centrifugation. Slides were examined under the microscope in normal and cross-polarized lights at ×1250 magnification.

Stratigraphic evaluation was performed for each sample individually, since correlated or thick continuous profiles were not sampled. Evaluation was based on stratigraphic ranges of the taxa alone (from the first (FO) to the last (LO) occurrences, see Appendix A), without using any additional geological information. Species older than the youngest assemblage were also determined and registered, so as to gain information on recycling. In flysch deposits where recycled forms are typically the most abundant, species LO-s are only relevant to the age of the 'original' assemblage in the sediment they were eroded from. Sedimentation ages were always established by forms having the youngest FO. In cases where a narrow biozone was proven (e.g. the most frequent NP16), it does not necessarily follow that a 'peak sedimentation event' occurred within that zone.

As many long-lived Cenozoic taxa reach into the Neogene, the identified specimens could be either autochthonous or allochthonous, but both types may also occur together in the sample and cannot be distinguished from each other. This is crucial insofar as abundance of the youngest zonal markers was often found to be extremely low.

Calcareous nannofossil classification in this paper follows Bown & Young (1997) for the Mesozoic and Young & Bown (1997) for the Cenozoic. Ranges of Cretaceous species are from Burnett (1998) and Perch-Nielsen (1985a), while Palaeogene species ranges are from Perch-Nielsen (1985b) and Báldi-Beke (1977, 1984). With respect to the Neogene, the latest summary of Young (1998) was used, a work that also took results from the Mediterranean into consideration (Fornaciari et al. 1996; Fornaciari & Rio 1996). The applied nannoplankton zonation is from Martini (1971).

The nannofossil assemblages examined are mostly of poor preservation and allowed the estimation of abundances only, without exact counting. Our experience has shown that this procedure is good enough for stratigraphic evaluation if flysch samples are dealt with (e.g. Nagymarosy & Báldi-Beke 1993). Species abundances were variable but generally low, which may depend on their preservation upon long-lasting depositional, diagenetic and weathering processes. Special care was taken to search and identify forms smaller than 10 μm, too, as most Neogene taxa occur in this size range.

Results of nannofossil analyses

A total of 69 samples were analysed along the Outer Dinaride coastal range from various tectonostratigraphic units. Four of them were barren of calcareous nannofossils. Estimated taxon abundances are summarized in Table 2. Established stratigraphic ranges for each sample are shown in Fig. 2.

The youngest nannofossil assemblages correspond to the zones NN4-6, placing most of the flysch into the Lower to Middle Miocene, most probably the upper part of this interval, i.e. Langhian to Early Serravallian. In addition, there are many reworked specimens from the Upper Cretaceous, and from the Middle and Upper Eocene—many of them having non-overlapping stratigraphic ranges. The obtained Miocene ages of deposition are rather uniform throughout the flysch zone.

Istrian Peninsula: Trieste–Koper and Pazin Basins

Fourteen samples were analysed from the Istrian Peninsula; seven each from the Trieste–Koper Basin (localities Izola, Dekani, Babiči and two nearby sites each at Korte and Momjan) and from the Pazin Basin (Baredine, Kašćerga, Žlepčari, Pićan and Lukačići). Most of them are dominated by Bartonian nannoflora.

At Dekani village (sample TD11), *Pemma* sp. ind., *Chiasmolithus* cf. *modestus* and *Sphenolithus spiniger* indicate Middle Eocene, whereas at the coastal cliffs of Izola (TD13) the co-occurrence of *Chiasmolithus grandis* and *Reticulofenestra placomorpha* corresponds to zones NP16-17, Uppermost Lutetian to Bartonian. The same age was proven at Lukačići (TD171), together with a variable Middle Eocene assemblage. At Pićan (PIC-2), a Bartonian to Priabonian age can be established. Few older Eocene species also occur whose ranges do not reach into zone NP16 (*Tribrachiatus orthostylus*: NP11-12, middle part of the Ypresian; *Discoaster septemradiatus*: NP12-14, Upper Ypresian to Lowermost Lutetian). Scarce but ubiquitous reworked Cretaceous forms include Lower and Upper Cretaceous markers as well.

In subordinate quantity, Miocene forms were also discovered throughout Istria, in all but one sample. The assemblage *Calcidiscus leptoporus, C. premacintyrei, Coccolithus miopelagicus, Reticulofenestra pseudoumbilicus* (partly >7 μm) and, possibly also six-rayed *Discoaster* spp., hint at a Miocene age. In the Pićan (PIC-2), Korte (TD15), Babiči (TD16) and Baštini (TD178) localities, specimens of *Helicosphaera carteri* were also identified. *Helicosphaera carteri* has its FO at the base of the Miocene worldwide and is a very characteristic form (Photos 13–14 in Fig. 4). Overall, the identified Neogene species define the zones NN4-6 which correspond to Late Burdigalian to Serravallian age. However, at Izola (TD13), Korte (TD15, TD18), Momjan (TD22) and Babiči (TD16), *Reticulofenestra pseudoumbilicus* is dominated by specimens larger than 7 μm, such as occur first in the Langhian, close to base NN5. Neogene nannofossils found near Dekani (TD11) are somewhat poorer than at Pićan, indicating either similar or slightly older Early Miocene age. At Lukačići (TD171), the Neogene is only represented by few and small specimens of *Reticulofenestra*

Table 2. *Distribution of calcareous nannofossils in the Outer Dinaride flysch. Abbreviations: N-Kv: Northern Kvarner area, DHZ: Dalmatian-Herzegovinian zone. See table notes*

* forms with six narrow arms, probably Neogene species
** Neogene species, see text for discussion

Observed nannofossil abundance per 1000 fields of view

□	1-2 specimens observed	1-2
■	very rare	<10 specimen
◆	rare	10-100
▣	frequent	>100
◀	common	>>100

?: questionable occurrence. In case of entries cf and ? 1-2 specimens were observed

Sample type

l	laminated pelite
r	rip-up pelite clast at turbidite base
b	pelite clast in limestone breccia
n	pelite clast in *Nummulites* debrite
x	pelite matrix of matrix-supported debrite of limestone clasts + *Nummulites* tests

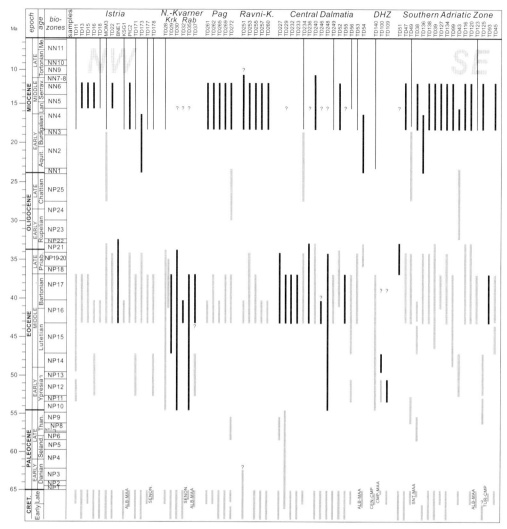

Fig. 2. Stratigraphic position of zonal marker nannoplankton species in the Outer Dinarides. Samples are arranged in columns; bars in each column represent the ranges of zonal markers found in that single sample. *N.B.*: only species with short zonal ranges are displayed; persistent species living through several epochs are omitted. Black bars mark the youngest assemblages indicating the most probable age of sedimentation. Narrow black lines: Neogene assemblages only comprising species that range beyond the Miocene. Grey bars represent reworked nannoflora. Abbreviations: Ravni-K.: Northern Dalmatia (Ravni–Kotari area), DHZ: Dalmatian-Herzegovinian Zone.

pseudoumbilicus. Of all, the most diverse Miocene assemblage is identified in the Pićan sample (PIC-2), within a rather fresh pelite rip-up clast from the base of a sandstone turbidite bed. *Sphenolithus conicus*, identified in Istria only in sample MOM-3, ranges from NP25 to NN3 (Upper Chattian to Middle Burdigalian) and is probably reworked due to the presence of *Coccolithus miopelagicus*, *Reticulofenestra haqii* and *R. pseudoumbilicus* in MOM-3.

Northern Kvarner

Sampling sites are located on Krk and Rab Islands and on the mainland near Crikvenica. Three out of nine samples are barren of nannofossils, the rest being rather poor. They always contain reworked Cretaceous nannofloral elements.

On Krk, near Draga Baščanska (TD28), only a poor assemblage was found with *Arkhangelskiella* sp. and abundant *Watznaueria barnesae* suggesting a Late Cretaceous age, together with rare Eocene forms. In the slide, only two specimens of *Reticulofenestra pseudoumbilicus* >7 μm were found, pointing to zone NN4 or younger, i.e. not older than Burdigalian. A poorly preserved but rather rich Middle Eocene assemblage is found near Bribir (TD29). Although in this sample most species range from Early to Middle Eocene and *Sphenolithus radians* is rather rare from the NP17 upwards, the FO of the larger forms of *Reticulofenestra placomorpha* is in NP16. The nannoflora of Bribir likely indicates the zones NP15-17 of Mid-Lutetian to Bartonian age. Another sample from Štale near Bribir (TD30) also contains a very poor assemblage, with the identified Upper Cretaceous, Eocene and Miocene taxa being represented with only one specimen each.

Three samples are from the clastic sediments of Rab Island. Near the town of Rab (TD32), a scarce but diverse Palaeogene nannoflora was established. The characteristic forms of *Reticulofenestra placomorpha* (FO in NP16), together with *Sphenolithus furcatolithoides* (LO in NP16), indicate the zone NP16, uppermost Lutetian to Bartonian. A single specimen of Miocene *Reticulofenestra pseudoumbilicus* was also found here. Similarly, in a poor nannoflora at the port of Lopar (TD35) which contains mostly Senonian and Lower to Middle Eocene nannofloral elements, only one specimen is encountered which resembles *Reticulofenestra pseudoumbilicus*. A likewise poor assemblage was obtained from sample TD37 from Dumići, 1 km NW of Supetarska Draga. Among its Eocene species, *Discoaster lodoensis* has the shortest range, and indicates Late Ypresian to Early Lutetian age (NP12-14). Alternatively, based on *Reticulofenestra* cf. *placomorpha*, the Dumići (TD37) assemblage can be evaluated as a younger one (Bartonian), with *Discoaster lodoensis* then being in reworked position.

Pag Island

Two profiles of the undisturbed, SW-dipping subvertical flysch succession were sampled on Pag Island. At Stara Vas, 10 km SE of Pag town, three samples were taken 9, 45 and 92 m above the foraminiferal limestone (TD261, TD262 and TD265, respectively). Two samples stem from the coastal profile at Vlašići, collected 17 and 34 m above the limestone top (TD269 and TD272, respectively).

In all samples, the nannoflora is very diverse, and a particular abundance was observed in the middle portion of the Stara Vas profile (TD262). Overall, the observed Cretaceous elements are rare, but indicate reworking of Upper Cretaceous assemblages (*Arkhangelskiella* sp., *Microrhabdulus* sp.). The majority of the nannofossils are Middle Eocene and most probably belong to zones NP16-17, i.e. Uppermost Lutetian to Bartonian. In the upper sample from Vlašići (TD272), the dominant Middle Eocene assemblage and the isolated Lower and Upper Cretaceous forms are accompanied by single specimens of *Heliolithus kleinpelli* from the Late Paleocene (NP6-9) and *Cyclicargolithus abisectus* most probably from the Chattian (NP24-NN1).

The Neogene part of the assemblages consists of three to five species which are identical in all samples throughout the basin. Of these, *Calcidiscus premacintyrei* is characterized by the shortest range, i.e. Late Burdigalian and Early Serravallian (zones NN4-6). The remaining Neogene species have their FO mainly in the Early Miocene but FO of *Reticulofenestra pseudoumbilicus* >7 μm is close to the base of zone NN5, in the Langhian. Although Neogene taxa make up only a very small part of the whole of the assemblages, in sample TD262 they appear to be more abundant than elsewhere in Pag together with a similarly higher abundance of Palaeogene forms.

Northern Dalmatia (Ravni–Kotari)

The nearly complete profile exposed by the Šopot railway cut was sampled at four sites (28, 116, 130 and 383 m above the foraminiferal limestone) and an additional sample is from a road cut of the recently constructed Zagreb–Split motorway near Islam Latinski, beneath the Suhovare bridge. Except the section top, all Šopot samples were extracted from fresh rip-up pelite clasts at the base of graded turbidite beds.

The observed nannofossil assemblages bear close resemblance to each other. Reworked Cretaceous coccoliths are rare compared to other parts of the basin system, being represented by *Watznaueria barnesae*. The Palaeogene nannoflora is rather diverse, yet uniform between samples. It is not older than latest Lutetian (zone NP16), as proven by samples bracketing the Šopot section at its base and top (TD251, TD257) and by the Suhovare outcrop (TD260). Although in none of the two intermediate samples in the Šopot section can the Eocene nannoflora be narrowed to NP16, a younger Palaeogene age for these assemblages is unlikely.

Neogene species identified in this unit are similar to those appearing in the Istrian Peninsula and Pag Island. *Calcidiscus premacintyrei, Coccolithus miopelagicus* and *Reticulofenestra pseudoumbilicus* occur in each sample in low yet meaningful amounts. Six-rayed *Discoaster* spp. occur in both profiles, while *Calcidiscus leptoporus* and *Helicosphaera carteri* were identified in the Šopot section only. The Neogene assemblage is assigned to the NN4-6 zones, indicating Late Burdigalian to Early Serravallian age.

Central Dalmatia

In Central Dalmatia, near the cities of Split and Omiš, 15 samples were analysed. The very scarce reworked Cretaceous nannofossils are represented mostly by *Watznaueria barnesae*, and Eocene nannofossils constitute the vast majority of the diverse and mostly rich assemblages.

Four sites near Mravince (TD227, TD232, TD233 and TD234) reveal a rather similar Palaeogene nannoflora, composed of very abundant but poorly preserved uppermost Lutetian to Priabonian forms. In addition, older index species also appear in Mravince: *Heliolithus kleinpelli* (NP6-9, Upper Paleocene) and *Cyclagelosphaera reinhardtii*, Albian to Paleocene. In the marl quarry at Mravince (TD229), recrystallized coccoliths dominate the sample and indicate Middle to Late Eocene age (NP16-21). The three most frequent species (*Reticulofenestra bisecta*, *Coccolithus pelagicus* and *Cyclococcolithus formosus*) share a common size and shape, strongly suggesting hydrodynamic control on species composition. Only a few specimens of *Calcidiscus premacintyrei* were found in the sample from the Mravince quarry (TD229), which may place the marl to the zones NN4-6.

We established Miocene ages with more confidence only in a part of the localities. In Jadro Valley, at Vrilo Jadro (TD240), besides Eocene species which define only a broad range of zones NP16-21 (Bartonian to Priabonian), five Neogene taxa also occur. Of these, *Calcidiscus premacintyrei* and *Reticulofenestra pseudoumbilicus* define the zones NN4-6, placing the age of these beds between Late Burdigalian and Early Serravallian.

Of the samples taken at Mravince, a pelite lithoclast extracted from a grain-supported *Nummulites* debrite (TD234) yielded comparatively abundant Neogene nannoflora: *Sphenolithus conicus* (Upper Chattian to Middle Burdigalian; probably reworked), as well as *Coccolithus miopelagicus*, *Reticulofenestra haqii* and *R. pseudoumbilicus*, corresponding to an age not older than Late Burdigalian.

Another pelite lithoclast (TD56) from a limestone breccia exposed by the large abandoned quarry of Omiš also yielded rather diverse Neogene nannofossils; *Helicosphaera carteri*, *Coccolithus miopelagicus*, *Reticulofenestra haqii*, *R. pseudoumbilicus* and *Umbilicosphaera rotula*, which suggest biozone NN5 or younger, i.e. at least Langhian age.

In other Central Dalmatian localities examined, either an abundant but monotonous and poorly preserved nannoflora occurs displaying Middle to Late Eocene age (the larger pit of the Jadro quarry, TD236), or solely a poor Eocene assemblage is encountered. Such extremely scarce nannoflora was found at several localities over larger along-strike distances (80 km): in the town of Solin in the Voljak Street section (TD248), a laminated pelite sample in the large abandoned quarry of Omiš (TD55), road cut at Porat near Živogošće (TD52) and in a metre-sized, grey, angular marl block included in thick limestone debrite at Gizdići near Klis (TD246).

Dalmatian-Herzegovinian Zone

In the densely imbricated thrust belt of the Dinaride carbonate platform in Southern Herzegovina, a single sample was analysed from Crnići village near the Neretva valley (TD140). The diverse, reworked Cretaceous nannofossil association hints at an Upper Cretaceous source older than Maastrichtian. A scarce Lower to Middle Eocene nannoflora is also present. Indication for the Neogene age of the rocks is provided by several specimens of *Calcidiscus leptoporus* ranging from the Early Miocene to recent (NN2-21).

Further to the SE, a narrow flysch zone is exposed in front of the Kuči Thrust that streches NW–SE from the Nevesinjsko Polje to Podgorica. Two samples were taken close to Podgorica; from a fault-bounded block standing out from the Zeta Valley 3 km NW of Spuž village (TD100), and in Medun village, 2 km ENE of Podgorica (TD109).

Both profiles contain reworked Upper Cretaceous taxa; age ranges can be probably narrowed to Campanian–Maastrichtian at Spuž. Here the relatively rich nannoflora is Lower to Middle Eocene and consists of several taxa with overlapping stratigraphic ranges (Table 2). *Discoaster lodoensis* and *D. septemradiatus* are markers of zones NP12-14 indicating Late Ypresian to Early Lutetian age. Important is a single specimen of *Discoaster* cf. *sublodoensis*, which marks the base NP14, the base of the Lutetian. The Ypresian *Tribrachiatus orthostylus* (NP11-12) is also part of the assemblage. Such a complexity is best explained by a multiple reworking history and will be discussed later.

In Medun profile (TD109), the Palaeogene assemblage is scarce but *Tribrachiatus orthostylus* occurs here as well, the Ypresian index species of the zones NP11-12.

Southern Adriatic Zone

A common feature of the 17 flysch samples taken in the Southern Adriatic Zone is their paucity of Cretaceous forms. In two samples, they are entirely absent (Ulcinj, TD99; Stari Bar, TD116). Nevertheless, reworked Upper Cretaceous index forms are identified in some cases and they preferentially occur in the southern part of the zone. *Eiffellithus eximius* and *Uniplanarius gothicus* are characterized by the shortest range of all, and indicate reworking from Turonian–Campanian and Santonian–Maastrichtian strata, respectively. Two samples bear *Nannoconus steinmanni*, hinting at reworking from Lower Cretaceous sediments (TD125, TD136).

In all samples, Eocene forms are predominant, albeit different in origin. Often the zones NP16-17 were registered (Uppermost Lutetian and Bartonian) but in other cases merely a longer interval could be given (NP16-20 or NP16-21).

Three, partly overlapping biozones can be proven at the base of the Klezna profile (TD125) from the Upper Paleocene to the Lutetian (range of *Discoaster multiradiatus* is NP9-11, that of *Tribrachiatus orthostylus* is NP11-12 and that of *Discoaster lodoensis* and *Discoaster septemradiatus* is NP12-14). This peculiar overlap offers a wide range of interpretations with respect to the age and recycling history of the sediment and will be discussed in the subsequent part.

A variety of reworked Palaeogene zone markers have been encountered throughout the Southern Adriatic Zone. Specimens of *Discoaster multiradiatus* (NP9-11; Paleocene) occur in sample TD49 (Dubravka). The NP12-14 zones are proven from the Zaljevo profile (TD43) corresponding to Upper Ypresian and earliest Lutetian. In addition, sample TD45 yielded *Nannotetrina* sp., a marker for NP15 (Mid-Lutetian). In TD51, an Upper Eocene zonal marker was encountered, *Chiasmolithus oamaruensis* (NP18-22). Oligocene marker taxa are extremely rare in the entire flysch belt, thus the finding of *Reticulofenestra lockeri* (NP23-NN1; Middle Rupelian to Lower Aquitanian) in sample TD43 and *Sphenolithus conicus* (NP25-NN3; Chattian to Middle Burdigalian) in sample TD49 is of particular importance.

Thirteen out of 17 samples of Miocene forms are sufficiently represented, whereas in TD93 they are missing, in TD51 there is one specimen and in TD45 and TD136 there are two specimens. Miocene species characteristic of zones NN4-6 indicate depositional ages not older than Late Burdigalian. In Zaljevo (TD43), occurrence of two specimens of *Helicosphaera ampliaperta* (range NN2-4) is of particular importance as together with other Neogene species (*Calcidiscus leptoporus, C. premacintyrei* and possibly also six-rayed *Discoaster* spp.) the stratigraphic position can be ascertained to the zone NN4 (Upper Burdigalian to Lower Langhian).

Nannofossil preservation

Our results reveal that the Neogene calcareous nannofossil assemblages are surprisingly low, both in abundance and diversity. Accurate dating of the Outer Dinaride Neogene successions awaits further bio- and chronostratigraphic control.

Physical and chemical processes operating during flysch formation and diagenesis influence the composition of the nannoflora (disintegration, dissolution, recrystallization) and probably account for the observed overall scarcity of the nannoflora (e.g. Thierstein 1980; de Kaenel & Villa 1996). High degree of reworking of older species, together with evidence for diagenetic processes as indicated by carbonate-cemented turbidite beds, carbonate veinlets dissecting the turbidites, bent mica plates and severely etched surfaces of susceptible heavy mineral grains (amphibole, staurolite, garnet), are in accordance with the poor preservation of the nannoflora. Indeed, samples from pelite clasts, embedded in breccias or well-cemented sandstones and presumably preserved from corrosive solutions, proved to yield more diverse Neogene assemblages than those found in laminated pelites (e.g. samples MOM-3, KSG-1, PIC-2, TD253, TD255, TD234, TD56; see Table 2).

Cretaceous to Palaeogene nannofloral elements

Most nannofossil assemblages are mixed, with evidence for recycling of specimens from pre-existing sediments. Cretaceous forms commonly occur together with the Cenozoic ones. Common long-lived Cretaceous species, e.g. *Watznaueria barnesae*, are ubiquitous. Among the shorter-range taxa, *Nannoconus steinmanni* indicates Lower Cretaceous, and several Upper Cretaceous markers occur as well (*Arkhangelskiella* sp., *Microrhabdulus* sp., *Eiffellithus turriseiffelii*). A limited number of characteristic taxa prove the availability of Lower Palaeogene sediments to erosion: *Heliolithus kleinpelli, Discoaster multiradiatus, Discoaster lenticularis* (Paleocene), *Tribrachiatus orthostylus* (Lower Eocene) and *Discoaster lodoensis* (Ypresian to lowermost Lutetian). All these Lower Palaeogene species are large in size and, according to our experience, fairly resistant against dissolution.

In spite of reworking, an attempt was made to identify characteristic Palaeogene assemblages preserved in the entire sample material. The most frequent Middle Eocene forms can be placed into the NP16 zone in several localities. Often, NP16 zone was recognized using the FO of *Reticulofenestra placomorpha* and the LO-s of *Sphenolithus furcatolithoides* and *Chiasmolithus solitus*. Whenever the latter species were not registered, the possible age of that assemblage could only be bracketed with lower precision, which might then extend from the NP16 upward into NP17 or even longer to the Late Eocene or Early Oligocene. We used the LO of *Chiasmolithus grandis* for the end of NP17, that of *Discoaster saipanensis* and/or *Discoaster barbadiensis* for the NP20/NP21 boundary (close to the Eocene/Oligocene boundary), and that of *Cyclococcolithus formosus* and *Reticulofenestra placomorpha* for the end of NP 21 and NP 22 zones, respectively, in the Early Oligocene. As for Late Eocene, a single, poorly preserved specimen of *Chiasmolithus oamaruensis* occurred in the entire studied material, as previously reported from Oprtalj, Istria (Benić 1991), Split (Jerković & Martini 1976) and Hvar (Marinčić 1981) in Central Dalmatia, and the North Mozura section in the Southern Adriatic Zone as well (de Capoa *et al.* 1995). *Isthmolithus recurvus* (NP20-22), which indicates Late Priabonian to Middle Rupelian, was found at Omiš (sample TD54).

Species of even longer ranges spanning from the Middle Eocene to various levels in the Miocene include the ubiquitous and highly abundant *Cyclicargolithus floridanus* and *Coccolithus pelagicus*. Also *Sphenolithus moriformis*, *Helicosphaera euphratis*, *H. intermedia*, *Discolithina* div. sp., and *Transversopontis* sp. appear in a number of samples. It is evident that among them there may exist specimens indistinguishably that lived in a given part of the Eocene, Oligocene or Miocene.

In the Oligocene nannofossil zonation, there are only a few markers to appear, due to the global cooling trend following the terminal Eocene events (e.g. Pomerol 1985). These few FOs are typically represented by the low-latitude forms of *Sphenolithus*. Of them, *Sphenolithus conicus*, ranging from NP25 to NN3, has been found in Istria, Central Dalmatia and the Southern Adriatic Zone. Further short-range index species include *Cyclicargolithus abisectus* and *Reticulofenestra lockeri*, and in this study both species have been encountered on Pag Island and in the Southern Adriatic Zone.

The scarcity of Paleocene nannofloral elements is probably due to the unavailability of Paleocene sediments for the reworking processes. Normally, both the diversity of Paleocene zonal markers and their resistance would permit them to be well-represented in Cenozoic flysch sediments.

Distribution of nannofossil age ranges and sediment recycling

Apart from significant Upper Cretaceous and Middle Eocene nannofloral components, and the presence of Lower to Middle Miocene nannoflora, there are index species of Paleocene, Lower and Upper Eocene and Oligocene present, too (Fig. 3). Unusual composition of some assemblages depicts characteristic age distribution patterns, whether or not Neogene forms are recognized in that sample (Fig. 2). We can discern two types.

(1) Joint occurrence of Palaeogene species with no overlapping age ranges. We observe this in Istria (TD11, TD171), Northern Kvarner area (TD37), Pag Island (TD272) at Mravince in Central Dalmatia (TD227), and in the Southern Adriatic Zone (TD38, TD43, TD45).

(2) A blurred, consecutive overlap pattern of several index species: *Discoaster multiradiatus* (NP9-11), *Tribrachiatus orthostylus* (NP11-12), *Discoaster lodoensis* and *D. septemradiatus* (NP12-14) and *Discoaster* cf. *sublodoensis* (NP14-15) occur together, in addition to Lower and Upper Cretaceous markers. Such mixed assemblages were encountered both in the Zeta Valley in the Dalmatian–Herzegovinian Zone (TD100) and in the Southern Adriatic Zone (TD125). Evidently, the partly overlapping ranges of these species can be combined to a number of species coexistence patterns, but recycling remains necessary to explain overall species composition. The resulting maximum time span for the Early Palaeogene 'cascade' alone may cover *c.* 9 Ma, but presence of further non-overlapping taxa suggests a more or less continuous and by all means cannibalistic sedimentation. Both types of age distribution provide an insight into the progressive Cretaceous to Cenozoic sediment reworking history. Actual sedimentation ages of the flysch probably become younger towards the foreland, along with the increasing number of reworked 'fossil' biozones. During thrusting, the deposited flysch was off-scraped and reworked, its material acting as a source for the yet younger flysch deposits.

We need to stress that the few, comparatively thick flysch profiles available (e.g. Brkini, Šopot, Skradin, Split, Petrovac) all display a limited thickness (*c.* 300–500 m), which is not enough to accommodate the time represented by the full nannofossil age spectrum assuming typical flysch sedimentation rates (50–2000 m/Ma; Einsele 1992, p. 389). De Capoa *et al.* (1995) documented the nannofossils of the 140 m thick Petrovac profile in the Southern Adriatic. With nearly equidistant sampling, they obtained successively younger assemblages upsection from the Upper Thanetian to the Langhian, which spans *c.* 40 Ma implying a sedimentation rate of only *c.* 3.5 m/Ma, which is not realistic for submarine fan depositional setting. Judging from their data, at Petrovac there are either undiscovered, short-lived unconformities or, more likely, most of the older ages come from reworked nannofossils due to extensive dilution.

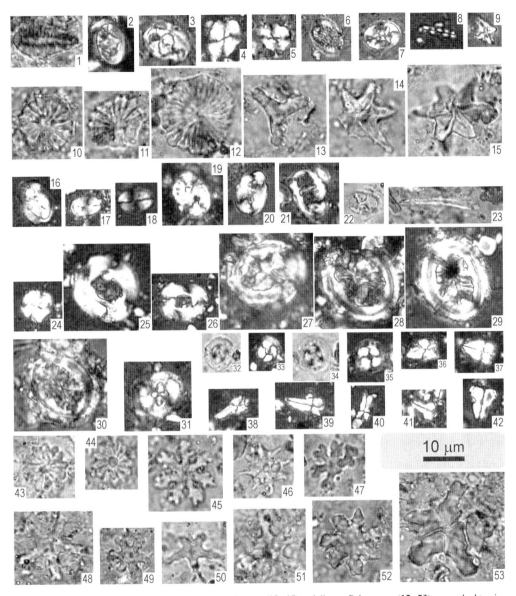

Fig. 3. Cretaceous (**1–9**), Paleocene (**10–12**), Lower Eocene (**13–15**) and diverse Palaeogene (**15–53**) nannoplankton in the Outer Dinaride flysch. Abbreviations refer to light types used: PP, plane-polarized; SX, slightly cross-polarized; XP: cross-polarized. **1**, *Nannoconus steinmanni* (Sample TD43, SX); **2–3**, *Zeugrhabdotus embergeri* (2: TD16, XP; 3: TD123, XP); **4–5**, *Watznaueria barnesae* (4: TD16, XP; 5: TD262, XP); **6**, *Cribrosphaerella ehrenbergii* (TD123, SX); **7**, *Eiffellithus turriseiffelii* (TD43, XP); **8**, *Microrhabdulus* sp. (TD18, XP); **9**, *Micula* sp. (TD123, XP); **10–11**, *Discoaster multiradiatus* (10: TD262, PP; 11: TD123, PP); **12**, *D. lenticularis* (TD43, PP); **13**, *Tribrachiatus orthostylus* (TD262, PP); **14–15**, *Discoaster lodoensis* (14: TD16, PP; 15: TD43, PP); **16**, *Helicosphaera compacta* (TD43, XP); **17–18**, *Discolithina plana* (17: TD261, XP; 18: TD16, XP); **19**, *Transversopontis pulcher* (TD262, XP); **20**, *Transversopontis* sp. (TD15, XP); **21**, *Lophodolithus nascens* (TD262, XP); **22**, *Neococcolithes dubius* (TD253, PP); **23**, *Blackites* sp. (TD123, PP); **24**, *Reticulofenestra bisecta* (PIC-2, XP); **25–26**, *R. placomorpha* (25: TD16, XP; 26: TD18, XP); **27–30**, *Chiasmolithus grandis* (27: TD16, XP; 28: TD16, XP; 29: TD15, XP; 30: TD262, SX); **31**, *Chiasmolithus* sp. (TD43, XP); **32–35**, *Cyclococcolithus formosus* (32–33: TD43, PP and XP resp.; 34–35, TD16, PP and XP resp.); **36–39**, *Sphenolithus radians* (36: TD123, XP; 37: TD123, XP; 38: TD16, XP; 39: TD16, XP); **40**, *S. furcatolithoides* (TD16, XP); **41–42**, *Zygrhablithus bijugatus* (41: TD261, XP; 42: TD16, XP); **43–44**, *Discoaster barbadiensis* (43: TD262, PP; 44: TD262, PP); **45**, *D. mirus* (TD262, PP); **46**, *D. saipanensis* (TD262, PP); **47** and **49**, *D. deflandrei* (both TD262, PP); **48** and **50–51**, *D. tanii* (48: TD123, PP; 50: TD123, PP; 51: TD123, PP); **52–53**, *D. nodifer* (52: TD262, PP; 53: TD18, PP).

Miocene nannofossils

The majority of samples contains nannofossils, suggesting Miocene age. In the NW portion of the Dinaride foreland basin, this paper reports such fossils for the first time from onshore outcrops. Unfortunately, these crucial forms are small, very rare, often poorly preserved and tend to comprise morphologically variable species.

In our experience however, most standard flysch nannofossil stratigraphic studies typically ignore or overlook such 'inconvenient'-forms. As they have outstanding importance in Dinaride flysch stratigraphy, they have to be discussed in more detail.

The most common placolith species encountered in the Outer Dinarides are species of the *Calcidiscus* group, *Reticulofenestra pseudoumbilicus*, *R. haqii* and *Coccolithus miopelagicus*. As demonstrated by Young (1998), however, these often exhibit a wide variety in shape and size, hampering exact taxonomic identification. In fact, morphologically related forms do occur in the Palaeogene, albeit a very rare phenomenon, e.g. *Reticulofenestra dictyoda* (Deflandre *in* Deflandre & Fert) Stradner *in* Stradner & Edwards (see Varol 1998). The comparatively high frequency of this type of placolith met in the Dinaride samples strongly argues for their Miocene rather than Eocene age. Among these species, *Reticulofenestra pseudoumbilicus* >7 μm is the most common one. Regarding the *Calcidiscus* species, *C. leptoporus* and *C. macintyrei* are of very low frequency, and only a few specimens were found in the entire material. On the contrary, *C. premacintyrei* has often been registered, even though it occurs in its less typical varieties. The occurrence of *Coccolithus miopelagicus* is well demonstrated (Fig. 4).

The genus *Helicosphaera* is represented by several species, and two of them have a Miocene FO—*Helicosphaera carteri* and *H. ampliaperta*. Unfortunately, *Helicosphaera* is just moderately resistant to dissolution and overcalcification, resulting in poor preservation. This makes *H. carteri* difficult to identify, but the flange and the central area with the two pores are visible (Fig. 4). A third species of *Helicosphaera* is characterized by prominent central openings and a conjunct central bar which suggests Neogene age despite the poorly preserved flange at the rim of the coccolith (Photo 12 in Fig. 4).

Among the *Discoaster* species, a rather frequent type occurs with five or six narrow rays. Their preservation is always very poor, with heavily overcalcified specimens, and ends of their arms mostly broken off. Although this type exists also in the Eocene, *Discoaster tanii* and similar forms are generally very rare. In fact, in the Miocene the most frequent *Discoaster* spp. are 5- or 6-rayed with bifurcations at the end of their arms, represented by a wide variety of species. Among them the *Discoaster exilis* group is the most probable candidate (e.g. with *D. aulakos* Gartner 1967 *in* Young 1998, p. 257), but they bear close resemblance to *D. variabilis* Martini & Bramlette 1963 as well. If bifurcation is not visible, it is either broken off or it is a primary characteristic (e.g. such as that of *D. bellus* Bukry & Percival 1971). With respect to decisive specific characteristics, the central area (with or without knob), the inter-ray area (rounded or rather V-shaped), the arms (with the sides tapering, parallel or curving) and the ending of the arms (bifurcating or not) were considered. In spite of poor preservation of the *Discoaster* spp. under discussion, their characteristics suggest Miocene age.

Overall, the identified Miocene nannofossil species correspond to the zones NN4-6 (based on *Calcidiscus premacintyrei*), but they are probably not older than NN5 as *Reticulofenestra pseudoumbilicus* >7 μm occurs first in this zone. Consequently, whenever Neogene taxa are present, the age of clastic sedimentation in the Adriatic onshore area is probably not older than Langhian.

As a result of the surprisingly rare occurrence and poor preservation of the nannofossils and occasionally wide specific morphological variability, the Neogene age of deposition cannot be unambiguously confirmed. Noticeably, the observed problematic forms are unknown in the Bakony Eocene Basin in Hungary although it exhibits rather comparable Middle Eocene nannofossil assemblages to those in the Outer Dinarides (Báldi-Beke 1984; Báldi-Beke & Báldi 1991).

Research is in progress to address this stratigraphic problem further: to look at whether the roles of hydraulic, diagenetic and weathering processes exert a fundamental control on the assemblage compositions.

Implications for palaeoenvironment

Pelagic environment with neritic influence

Despite sediment mixing and reworking, it is evident that the most frequent forms pertain to a Middle Eocene assemblage. The species composition is rather complex and exhibits common placoliths. Some species possess a specific habitat and are thus of ecological importance, such as the nearshore taxa *Neococcolithes dubius*, *Pemma* sp., *Discolithina*, *Transversopontis* and holococcoliths, e.g. *Zygrhablithus bijugatus* (e.g. Perch-Nielsen 1985*b*). Although abundance of these nearshore taxa is low in all samples, as many of them are at the same time particularly susceptible to dissolution, it can be assumed that their initial abundance

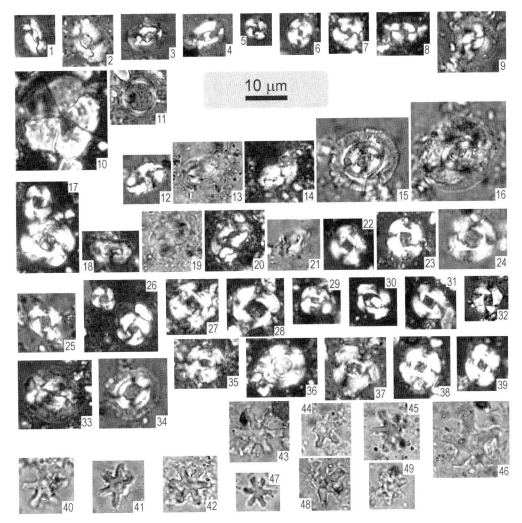

Fig. 4. Palaeogene to Neogene (1–11) and Neogene (12–46) nannoplankton of the Outer Dinaride flysch. Designations as in Fig. 3. 1–2, *Helicosphaera euphratis* (1: Sample TD43, XP; 2: Sample TD262, XP); 3–4, *H. intermedia* (3: TD127, XP; 4: TD43, XP); 5–7, *Cyclicargolithus floridanus* (5–6: PIC-2, XP; 7: TD261, XP); 8–9, *Coccolithus pelagicus* (8: TD261, XP; 9: TD262, SX); 10, *Braarudosphaera bigelowi* (TD127, XP); 11, *Coronocyclus nitescens* (TD262, SX); 12, *Helicosphaera* sp. (flange poorly preserved but note prominent central openings and the conjunct central bar, TD127, XP); 13–14, *H. carteri* (13: TD15, PP; 14: TD15, XP); 15–16, *Coccolithus miopelagicus* (15: TD262, PP; 16: TD16, XP); 17, *Reticulofenestra pseudoumbilicus* >7 μm (above; together with the Palaeogene *R. placomorpha*, below, TD15, XP); 18–20, *Helicosphaera carteri* (18: TD16, XP; 19: TD262, PP; 20: TD262, XP); 21, *H. obliqua* (TD262, SX); 22–26, *Reticulofenestra haqii* (22: TD16, XP; 23: TD18, XP; 24: TD262, XP; 25: TD123, XP; 26: TD43, XP, 2 specimens); 27–32, *R. pseudoumbilicus* >7 μm (27: TD261, XP; 28: TD261, XP; 29: TD261, XP; 30: TD251, XP; 31: TD16, XP; 32: TD15, XP); 33–34, *Coccolithus miopelagicus* (33: TD16, SX; 34: TD16, SX); 35–37, *Calcidiscus premacintyrei* (35: TD261, XP; 36: TD262, XP; 37: TD261, XP); 38–39, *Calcidiscus* sp. (38: TD15, XP; 39: TD261, XP); 40–44 and 48–49, *Discoaster* spp. 6-ray (40: PIC-2, PP; 41: TD15, PP; 42: TD18, PP; 43: TD123, PP; 44: TD253, PP; 48 and 49: TD18, PP); 45–46, *Discoaster* spp. 5-ray (45: TD15, PP; 46: TD123, PP); 47, *Discoaster* sp. 7 μm, 6-ray (TD261, PP).

was higher in the sediment, which in turn calls for a remarkable neritic influence on sedimentation in the Eocene. The scarce Neogene assemblage does not provide useful environmental information.

Neogene: absence of planktonic foraminifera, presence of nannofossils

Preliminary examination of our samples containing Miocene nannofossils did not yield any planktonic foraminifera younger than Middle Eocene in Istria and Kvarner or younger than Early Oligocene in the Southern Adriatic (V. Ćosović, pers. comm. 2006; F. Rögl, pers. comm. 2007). To our knowledge, no such data have been published from onshore localities, either (Table 1).

The characteristic depth habitat of several planktonic foraminifer species can preclude their survival in relatively shallow waters. Most species require oceanic salinities near 35–36‰, and a few of them can tolerate salinities down to 30.5‰, and can thus only penetrate coastal waters if they are sufficiently clear and lacking in turbidity (Haynes 1981, p. 330). Nannoplankton, in contrast, tolerate reduced salinity although diversity can be reduced in brackish water under elevated freshwater input (e.g. Olszewska & Garecka 1996; Schulz et al. 2005).

In places, the Outer Dinaride flysch deposits rapidly grade into shallow marine deposits. Sand-rich shallow shelf environments influenced by nearby river mouths have been described from the Islands of Rab and Pag and from Northern Dalmatia (Zupanič & Babić 1991; Babić et al. 1993; Babić & Zupanič 1998). Facies architecture of clastic deposits in Northern and Central Dalmatia is also well documented (e.g. Postma et al. 1988; Mrinjek 1993; Marjanac 1996), and show fan deltas prograding on the flysch succession. The increasing proximity of the alluvial environments implies significant freshwater input which could have a profound effect on salinity and thus adversely influenced foraminifer distribution.

Heterochronous redeposition

As a prominent feature, debrites of several metres of thickness that intercalate the turbiditic flysch successions of the Outer Dinarides often contain isolated tests of Eocene larger foraminifera. In spite of attempts to date the depositional age by means of such fossils (Pavić 1970), they are evidently allochtonous in a submarine fan setting. Pavlovec (2003) noted that Nummulites assemblages found in flysch deposits Dinarides-wide were in fact no true biocoenoses but rather mixed in composition with unusual species proportion as compared to those in the limestones. Indeed, derived fossils may occur in a state of preservation as good as or better than that of the original rock, and the reworked fauna may be more diverse and contain a much higher proportion of planktonics than that of the source rock (Curry 1982).

Ample examples from the Outer Dinarides and from comparable settings in the surrounding areas show that larger, smaller and also pelagic foraminifer tests can survive diagenesis in unconsolidated sediment and be reworked into younger strata: (i) Lower Eocene (Upper 'Cuisian') flysch of Trnovo, SW Slovenia contains Lower to Middle Cuisian larger foraminifera: Nummulites subdistans (Pavlovec 2006); (ii) planktonic foraminifera in the flysch of Pićan are uniformly Middle Lutetian in age but if examined in detail, they belong to various biozones (Pavlovec et al. 1991); (iii) Lower Oligocene (NP21-22) calcareous turbidites in Budapest, Hungary, contain Upper Eocene (Priabonian) larger foraminifera: Chapmanina gassinensis and Nummulites fabianii (Varga 1982, 1985; Nagymarosy 1987); (iv) Lower Oligocene flysch in the Carpathians yielded Eocene Nummulites (Kulka 1985); (v) the Frazzanò Flysch of the Calabria–Peloritani arc previously dated by Upper Eocene foraminifera yielded Upper Oligocene calcareous nannofossils (de Capoa et al. 1997); (vi) similarly, flysch of the Sicilian Maghrebids containing Lower Oligocene planktonic foraminifera were dated by means of calcareous nannofossils to be at least Aquitanian (de Capoa et al. 2000); (vii) microfossils of Lower Miocene strata of the Zawada Formation in the Carpathians are dominated by reworked Middle Eocene planktonic foraminifera and calcareous nannoplankton (Oszczypko et al. 1999); and (viii) flysch of the Ionian Zone in the Hellenides, to the SE of the Outer Dinaride flysch basins, is Early Miocene in age and contains abundant reworked Cretaceous and Eocene nannofossils (Piper et al. 1978; Bellas 1997). Evidence for a considerable time gap between foraminifer and nannofossil ages arises also from our new results. For instance, isolated Lower to Middle Eocene Nummulites tests from various biozones occur together with older, massive Palaeogene and Cretaceous carbonate lithoclasts (Hagn et al. 1979) as well as with Middle to Upper Eocene planktonic foraminifera and Cretaceous to Palaeogene nannofossils from a number of biozones (Drobne et al. 1979) and with Upper Oligocene palynomorphs, nannofossils and larger foraminifera (Šparica et al. 2005) in the Pićan flysch profile on Istria. These strata have been dated herein to be not older than Late Burdigalian. Table 1 and Fig. 2 illustrate the contradiction of microfossil age data from the entire Outer Dinaride flysch.

Weak early-stage diagenesis

Redeposition of micro- and macrofossils such as larger, smaller and planktonic foraminifera without intense signal of wearing or abrasion can be attributed to subaqueous mud volcanoes linked to dewatering, in what might be a dynamic, accretionary wedge-type environment (e.g. Kohl & Roberts 1994). Probably, high water content of sediment prevented diagenesis initially, and allowed easy removal of carbonate particles from the siliciclastic matrix downslope of the submarine fans, over a longer time span. Such a mechanism for a continuous redeposition and accumulation of Eocene foraminifera until at least the Middle Miocene is very likely to have taken place in the imbricated frontal thrust belt of the Dinarides, interpreted as an accretionary wedge (Tari-Kovačić 1998).

Implications for palaeogeography

We have demonstrated that our new calcareous nannofossil data from the Outer Dinaride Cenozoic prove the presence of a wide range of Lower Cretaceous, Upper Cretaceous, Paleocene, several non-overlapping Eocene, Oligocene and Miocene species in these strata.

The significance of our results is twofold. On the one hand, they indicate that the Cretaceous platform carbonate nappes have been covered by pelitic sediments, connected with progressive sediment reworking during the Cenozoic. It is essential to assume that before, during, and after the deposition of the Eocene foraminiferal limestones in the present-day Outer Dinaride coastal range, there existed widespread marine environments, covering the Cretaceous platform carbonates in the inner imbricate belt, where the nannofossils can be derived from.

On the other hand, the data strongly suggest a considerably younger, at least Burdigalian sedimentation age throughout the onshore flysch deposits, with clear implications on Outer Dinaride tectonics. Overall, this picture is best explained by a series of wedge-top basins (see DeCelles & Giles 1996) progressively migrating towards the Apulian foreland.

The flysch develops from the underlying '*Globigerina* Marl' (Marjanac & Ćosović 2000). These shelf to shallow bathyal deposits are dated by means of micro- and macrofauna and by calcareous nannofossils (Muldini-Mamužić 1965; Benić 1991; Drobne & Pavlovec 1991; Pavšič & Premec-Fuček 2000; Schweitzer *et al.* 2005) ranging in age from Paleocene in the NW to Upper Eocene in the SE.

Biostratigraphic results obtained in our study for Central Dalmatia and the Southern Adriatic Zone agree well with the findings of Puškarić (1987), Radoičić *et al.* (1989, 1991), de Capoa *et al.* (1995),

and de Capoa & Radoičić (1994a, b, 2002) who first demonstrated Neogene nannofossil age in a number of flysch localities of the SE part of the basin system. Further to the SE, flysch in the Ionian Zone in the Hellenides is likewise dated to be Lower Miocene (Piper *et al.* 1978; Bellas 1997). Our new data, however, lead us to extend the Miocene sedimentation ages to other portions of the coastal onshore localities as well, from Northern Dalmatia through Pag Island untill the Istrian Peninsula. As a consequence, proposing SE-directed diachroneity along the Outer Dinaride front (Piccoli & Proto Decima 1969) is no longer reasonable. Furthermore, superposition of Neogene flysch onto the Eocene 'Marl with crabs' or *Globigerina* marl implies a widespread, basin-scale regional unconformity, but hitherto this issue has not received much attention. In fact, the thickness of the *Globigerina* marl varies extremely from 10 to 150 m (Marjanac & Ćosović 2000), and contacts with angular (Marjanac 2000) or erosional (Marjanac *et al.* 1998) unconformity to the overlying flysch both exist. Šikić (1963, 1968) also argued that deposition of both the units was interrupted by a deformational event and suggested the existence of a basin-wide unconformity. An abrupt change in the composition of the planktonic foraminifer fauna above a carbonate debrite horizon separating the *Globigerina* marl and the flysch was noted by Muldini-Mamužić (1965). De Capoa *et al.* (1995) noted that a hardground is developed on top of the 'Marl with crabs', directly overlain by flysch. All these data suggest a break in sedimentation and possibly slight submarine erosion as well, in spite of the classical view dealing with the progressive transition of the *Globigerina* marl to the flysch (see review in Marjanac & Ćosović 2000). Although remnants of Palaeogene flysch can exist in the coastal range, our data clearly imply that, at most localities in the basin, the flysch is of Neogene age and separated from the *Globigerina* marl or from older flysch strata by unconformity (Fig. 5).

The heavy mineral composition of the flysch (Magdalenić 1972; Mikes *et al.* 2004, 2005) and especially the nannofossil 'age spectra' both exhibit a remarkable degree of basinwide homogenization. In the course of wedge-top deposition, a series of small, relatively shallow basins could have been developed on and in front of the advancing thrust sheets. The 'smoothing' is interpreted as a result of multiple reworking from the precursor flysch slices and sediment dispersal that occurred within the westward-propagating, complex thrust wedge (Fig. 6).

Implications for Cenozoic deformation history

Flysch remnants are typically sandwiched between thrust sheets, and occur in different structural

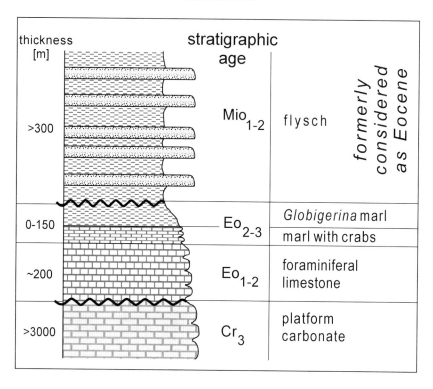

Fig. 5. Schematic stratigraphic setting of the Outer Dinaride flysch in the coastal range according to the new nannofossil age data. As it directly overlies well-dated Eocene *Globigerina* marls, their relationship requires the presence of a major, basin-wide unconformity separating them. At places however, small erosional remnants of older, Palaeogene flysch may also still exist in the same orogenic strike.

position in the Outer Dinaride nappe pile. The Bosnian Flysch is in uppermost position, and ranges in age up to Turonian to Paleocene (Dimitrijević 1997 p. 38, Hrvatović 1999; Christ 2007). Turbiditic sequences of comparable age are found in the same structural position along orogenic strike in isolated outcrops near Zagreb, Bosanski Novi and Bihać (Jelaska et al. 1969; Babić 1974; Babić & Zupanič 1976; Crnjaković 1981) and in the Slovenian Trough (e.g. Buser 1987). These units are thrust on the AdCP, an imbricated pile of Mesozoic carbonates, the detachment surfaces being marked by a series of extremely narrow slices of Cenozoic flysch. Here, limited evidence suggests Early Eocene age of deposition (samples TD100 and TD109; planktonic foraminifer data of Kraseninnikov et al. 1968). In the thrust slices below, available biostratigraphic and sedimentological data suggest a foreland-directed migration of clastic facies zones in the Palaeogene (e.g. Bignot 1972; Engel 1974; Chorowicz 1977; Drobne 1977; Cadet 1978; Marinčić 1981; Košir 1997).

Our new biostratigraphic data imply post-mid-Miocene deformation in the Outer Dinaride coastal range. Older thrusting events to the NE are recorded by the Lower and Upper Cretaceous units of the folded Bosnian Flysch (Dimitrijević 1997; Hrvatović 1999; Christ 2007; Petri 2007) and by Lower Eocene flysch slices thrust by Mesozoic carbonates (Kraseninnikov et al. 1968). On the other hand, to the SW of our sample sites, deformed Pliocene sediments and GPS measurements indicate ongoing shortening along the coastal range (Tari-Kovačić 1998; Pribičević et al. 2002; Picha 2002; Tari 2002; Prelogović et al. 2003; Altiner et al. 2006; Mantovani et al. 2006; Vrabec & Fodor 2006). Therefore, the new Miocene nannofossil depositional age of most flysch units and their structural position indicate that both sedimentation and deformation have been a long-lived continuous process in the Dinarides. The formation, subsidence and inversion of the sub-basins and thus sediment cannibalization are well-documented by the high proportion of recycled nannofossils.

Conclusions

Sixty-nine samples taken along the 700 km onshore sector of the Outer Dinaride flysch belt were analysed for calcareous nannofossils. Most of them yielded

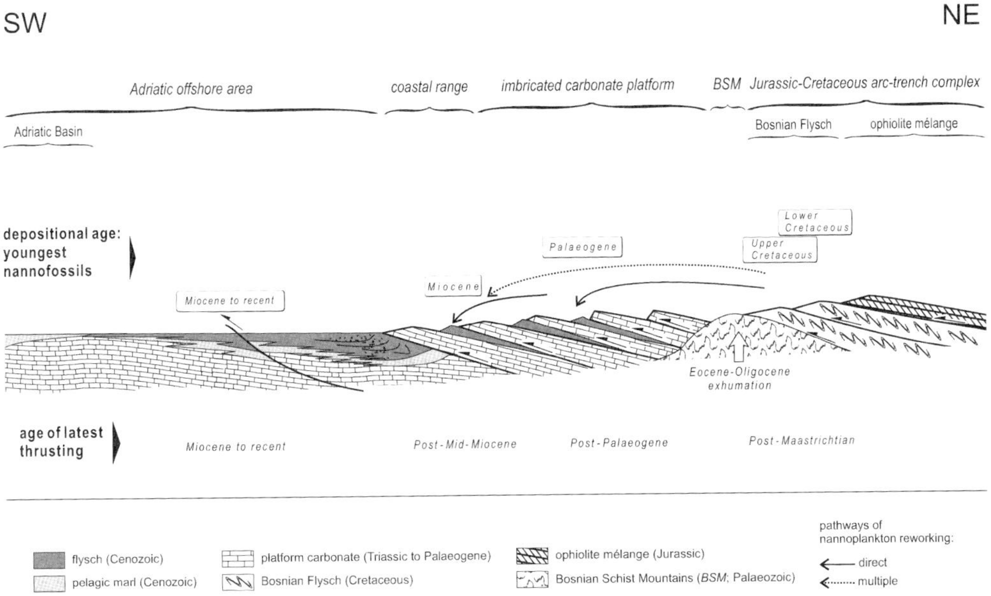

Fig. 6. Schematic sketch showing the implications of biostratigraphic data. With progressive nappe propagation towards Apulia, shallow foredeeps develop at the actual nappe front. Continuous thrusting and associated sediment off-scraping result in consecutive reworking of planktonic fossils into younger strata. Flysch remnants preserved at the base of thrust sheets become progressively younger towards Apulia. This process lasted until at least the late Early Miocene in the present-day Outer Dinaride coastal range, as indicated by deformed flysch sediments yielding Neogene calcareous nannoplankton. In the offshore Adriatic basin, clastic sedimentation has lasted until recent time. Age of the Bosnian Flysch taken from Hrvatović (1999) and Christ (2007). In the upper nappe unit of the Bosnian Flysch, an Albian deformation phase is also supposed (Petri 2007). Thermal history of the Bosnian Schist Mountains after Pamić et al. (2004).

mixed nannoflora with Cretaceous, Palaeogene and Neogene species. In the light of the new nannofossil data presented herein, we suggest that the flysch sedimentation lasted at least up to the Mid-Miocene all along the Dinaride coastal range. In spite of disagreement with existing planktonic foraminifer biostratigraphic data, our results are well in line with recently published nannofossil data from a number of localities within the flysch basin.

A majority of the nannoflora consists of Middle Eocene taxa, together with less abundant Cretaceous, Paleocene and Oligocene nannofossils. Wherever Neogene species have been discovered, all these older floras are implied to be recycled from older flysch units of the Outer Dinaride accretionary wedge. Cretaceous platform carbonate nappes of the Outer Dinarides were extensively covered by Cenozoic marine sediments. Remnants are preserved at the base of thrust sheets while the reworked microfossils also testify to the existence of Cenozoic basins on top of the accretionary wedge.

The stratigraphic relation of well-dated Eocene *Globigerina* marls lacking evidence for reworked fossil content and the Neogene flysch allows us to propose a working hypothesis on a major, widespread, hitherto largely ignored unconformity along the entire Dinaride coastal range, which requires further field evidence.

The Miocene onshore Outer Dinaride flysch suffered severe compressional deformation and is typically preserved below the imbricate thrust sheets of the Dinaride carbonate platform. Along the coastal zone, the deformation post-dates Middle Miocene. This compressional phase provides evidence for the continuity of deformational events between the more internal, older Late Cretaceous to Palaeogene compressional phases and the ongoing thrusting at the Adriatic front.

Our most sincere gratitude is due to a number of colleagues in the University of Zagreb for their invaluable and friendly help, especially to Lj. Babić, V. Ćosović and T. Marjanac. Essential local literature, maps and field discussions all formed a solid basis to perform this study. Substantial aid with field work in Montenegro was received from D. Čađenović (Podgorica). F. Rögl (Vienna) and V. Ćosović kindly offered their expertise on foraminifer biostratigraphy. Responsibility for the conclusions presented are borne by the authors alone. Constructive criticism by M. Wagreich (Vienna) and an anonymous reviewer significantly improved the manuscript. The work was supported by the Deutsche Forschungsgemeinschaft (DFG Ey 23/4).

Appendix A. List of all taxa cited in the text and figures

Arkhangelskiella cymbiformis Vekshina 1959	Campanian–Maastrichtian
Arkhangelskiella sp.	Upper Cretaceous
Blackites creber (Deflandre 1954) Roth 1970	Eocene
Blackites sp.	
Braarudosphaera bigelowi (Gran & Braarud 1935) Deflandre 1947	Cretaceous–recent
Braarudosphaera sp.	
Broinsonia sp.	Albian–Maastrichtian
Calcidiscus leptoporus (Murray & Blackman 1898) Loeblich & Tappan *1978*	NN2-21
Calcidiscus macintyrei (Bukry & Bramlette 1969) Loeblich & Tappan 1978	NN7-19
Calcidiscus premacintyrei Theodoridis 1984	NN4-6
Calcidiscus tropicus Kamptner 1956 *sensu* Gartner 1992	NN4-10
Chiasmolithus grandis (Bramlette & Riedel 1954) Radomski 1968	NP11-17
Chiasmolithus modestus Perch-Nielsen 1971	NP16
Chiasmolithus oamaruensis (Deflandre 1954) Hay, Mohler & Wade 1966	NP18-22
Chiasmolithus solitus (Bramlette & Sullivan 1961) Locker 1968	NP10-16
Chiasmolithus sp.	
Chiastozygus sp.	
Clausiococcus fenestratus (Deflandre & Fert 1954) Prins 1979	Palaeogene–NN1
Coccolithus eopelagicus (Bramlette & Riedel, 1954) Bramlette & Sullivan 1961	Eocene
Coccolithus miopelagicus Bukry 1971	?NN5-8
Coccolithus pelagicus (Wallich 1871) Schiller 1930	Eocene–Recent
Coronocyclus nitescens (Kamptner 1963) Bramlette & Wicoxon 1967	Eocene–Miocene
Cribrocentrum reticulatum (Gartner & Smith 1967) Perch-Nielsen 1971	NP16-20
Cribrosphaerella ehrenbergii (Arkhangelsky 1912) Deflandre *in* Piveteau 1952	Albian–Maastrichtian
Cyclagelosphaera reinhardtii (Perch-Nielsen 1968) Romein 1977	Cretaceous–Paleocene
Cyclicargolithus abisectus (Müller 1970) Wise 1973	NP24-NN1
Cyclicargolithus floridanus (Roth & Hay *in* Hay *et al.* 1967) Bukry 1971	Middle Eocene–NN7
Cyclicargolithus luminis (Sullivan 1965) Bukry 1971	Middle Eocene–Oligocene
Cyclicargolithus sp.	
Cyclococcolithus formosus Kamptner 1963	Eocene–NP21
Discoaster aster Bramlette & Riedel 1954	Miocene
Discoaster barbadiensis Tan 1927	NP10-20
Discoaster binodosus Martini 1958	Lower to Middle Eocene
Discoaster deflandrei Bramlette & Riedel 1954	Eocene–Oligocene
Discoaster distinctus Martini 1958	NP12-14
Discoaster exilis Martini & Bramlette 1963	NN4-9
Discoaster lenticularis Bramlette & Sullivan 1961	NP9-10
Discoaster lodoensis Bramlette & Riedel 1954	NP12-14
Discoaster mirus Deflandre in Deflandre & Fert 1954	?NP13-14
Discoaster multiradiatus Bramlette & Riedel 1954	NP9-11
Discoaster nodifer (Bramlette & Riedel, 1954) Bukry 1973	NP15-22
Discoaster saipanensis Bramlette & Riedel 1954	NP15-20
Discoaster septemradiatus (Klumpp 1953) Martini 1958	NP12-14
Discoaster sublodoensis Bramlette & Sullivan 1961	NP14-15
Discoaster tanii Bramlette & Riedel 1954	NP16-22
Discoaster sp.	
Discolithina multipora (Kamptner 1948) Martini, 1965	Eocene–Miocene
Discolithina plana (Bramlette & Sullivan, 1961) Perch-Nielsen 1971	?Palaeogene
Discolithina sp.	
Eiffellithus eximius (Sover 1966) Perch-Nielsen 1968	Turonian–Campanian

(Continued)

Appendix A. *Continued*

Eiffellithus sp.	
Eiffellithus turriseiffelii (Deflandre *in* Deflandre & Fert 1954) Reinhardt 1965	Upper Albian–Maastrichtian
Helicosphaera ampliaperta Bramlette & Wilcoxon 1967	NN2-4
Helicosphaera carteri (Wallich 1877) Kamptner 1954	NN1–recent
Helicosphaera compacta Bramlette & Wicoxon 1967	Middle Eocene–NP24
Helicosphaera euphratis Haq 1966	Middle Eocene–Miocene
Helicosphaera intermedia Martini 1965	Middle Eocene–Miocene
Helicosphaera obliqua Bramlette & Wilcoxon 1967	NP24–NN6
Helicosphaera seminulum Bramlette & Sullivan 1961	Lower–Middle Eocene
Heliolithus kleinpelli Sullivan 1964	NP6-9
Isthmolithus recurvus (Deflandre *in* Deflandre & Fert 1954)	NP 20-22 (marker of base NP20; Martini 1971)
Lanternithus minutus Stradner 1962	NP16-22
Lophodolithus nascens Bramlette & Sullivan 1961	NP9-15
Markalius inversus (Deflandre *in* Deflandre & Fert 1954) Bramlette & Martini 1964	Cretaceous–Eocene
Marthasterites sp.	Upper Cretaceous
Micrantholithus vesper Deflandre 1954	Eocene–Miocene
Microrhabdulus sp.	Cenomanian–Maastrichtian
Micula sp.	Coniacian–Maastrichtian
Nannoconus steinmanni Kamptner 1931	Uppermost Jurassic–Lower Cretaceous
Nannotetrina sp.	NP15
Neococcolithes dubius (Deflandre *in* Deflandre & Fert 1954) Black 1967	NP13-16 (?17-18)
Pemma papillatum Martini 1959	Middle Eocene
Pemma rotundum Klumpp 1953	Middle Eocene
Pemma sp.	Middle Eocene
Reticulofenestra bisecta (Hay, Mohler & Wade 1966) Roth 1970	NP16-25 (or NN1)
Reticulofenestra haqii Backman 1978	NN2-15
Reticulofenestra lockeri Müller 1970	NP23 rare, NP24–?NN1
Reticulofenestra placomorpha (Kamptner 1948) Stradner *in* Stradner & Edwards 1968 [actual valid synonym: *R. umbilica* (Levin 1965) Martini & Ritzkowski 1968)]	NP16-22
Reticulofenestra pseudoumbilicus (Gartner 1967) Gartner 1969	NN4-15 (<7 µm occurs from NN5 in the Mediterranean)
Rhabdolithus sp.	
Sphenolithus conicus Bukry 1971	NP25–NN3
Sphenolithus furcatolithoides Locker 1967	NP15-16
Sphenolithus moriformis (Brönnimamm & Stradner 1960) Bramlette & Wilcoxon 1967	Lower Eocene–Miocene
Sphenolithus radians Deflandre *in* Deflandre & Fert 1954	Lower to Middle Eocene (rare in Upper Eocene)
Sphenolithus sp.	
Sphenolithus sp. (?*calyculus*: Bukry 1985)	*S. calyculus*: Palaeogene–NN1
Sphenolithus spiniger Bukry 1971	Middle Eocene
Transversopontis pulcher (Deflandre *in* Deflandre & Fert 1954) Hay, Mohler & Wade 1966	Eocene–Oligocene
Transversopontis sp.	
Tribrachiatus orthostylus Shamrai 1963	NP11-12 (?13-14)
Umbilicosphaera rotula (Kamptner 1956) Varol 1982	NN2-16
Uniplanarius gothicus (Deflandre 1959) Hattner & Wise, 1980	Santonian–Maastrichtian
Watznaueria barnesae (Black 1959) Perch-Nielsen 1968	Bajocian–Masstrichtian
Zeugrhabdotus embergeri (Noël 1958) Perch-Nielsen 1984	Tithonian–Maastrichtian
Zeugrhabdotus sp.	
Zygrhablithus bijugatus (Deflandre *in* Deflandre & Fert 1954) Deflandre 1959	Eocene–NP25

Note: Stratigraphic ranges of species taken from the following sources. Cretaceous: Burnett (1998) and Perch-Nielsen (1985*a*); Palaeogene: Perch-Nielsen (1985*b*) and Báldi-Beke (1977, 1984); Neogene: Fornaciari *et al.* (1996), Fornaciari & Rio (1996) and Young (1998).

Appendix B. Geographic position of sampling localities

Area	Sample	Locality	Latitude (N)	Longitude (E)
Istria	TD11	Dekani	45° 33′ 4.3″	13° 48′ 25.0″
	TD13	Izola	45° 31′ 58.5″	13° 38′ 24.2″
	TD15	Korte	45° 29′ 14.7″	13° 40′ 14.7″
	TD16	Babiči	45° 30′ 55.0″	13° 46′ 55.0″
	TD18	Korte	45° 29′ 35.3″	13° 40′ 35.3″
	MOM-3	Momjan	45° 26′ 3.2″	13° 42′ 18.5″
	TD22	Momjan	45° 25′ 8.0″	13° 42′ 8.0″
	BNE-1	Zrenj-Baredine	45° 25′ 14.6″	13° 53′ 32.3″
	KSG-1	Kaščerga	45° 18′ 38.9″	13° 54′ 46.2″
	PIC-2	Pićan	45° 12′ 16.2″	14° 2′ 46.2″
	TD171	Lukačići	45° 10′ 53.5″	14° 0′ 25.9″
	TD173	Škrbani	45° 10′ 17.8″	14° 0′ 54.7″
	TD177	Baštini near Draguć	45° 20′ 16.2″	14° 0′ 18.2″
	TD178	Baštini near Draguć	45° 20′ 16.2″	14° 0′ 18.2″
Northern Kvarner: Krk Island + nearby mainland areas	TD28	Draga Bašćanska	44° 59′ 34.0″	14° 43′ 0.7″
	TD29	Bribir (road crossing Grižane/Selce)	45° 13′ 8.8″	14° 41′ 31.1″
	TD30	Bribir (Štale)	45° 9′ 59.9″	14° 45′ 11.9″
Northern Kvarner: Rab Island	TD32	Rab	44° 46′ 11.9″	14° 45′ 21.1″
	TD35	Lopar	44° 50′ 23.8″	14° 43′ 10.9″
	TD37	Dumići	44° 48′ 12.9″	14° 42′ 29.9″
Pag Island	TD261	Stara Vas, 10 km SE of Pag town	44° 22′ 53.5″	15° 9′ 12.1″
	TD262	Stara Vas, 10 km SE of Pag town	44° 22′ 53.5″	15° 9′ 12.1″
	TD265	Stara Vas, 10 km SE of Pag town	44° 22′ 51.2″	15° 9′ 8.0″
	TD269	coastal cliffs at the SE tip of Pag Island, near Vlašići	44° 19′ 5.1″	15° 13′ 39.7″
	TD272	coastal cliffs at the SE tip of Pag Island, near Vlašići	44° 19′ 4.8″	15° 13′ 38.8″
Northern Dalmatia (Ravni-Kotari)	TD251	Šopot railway cut near Benkovac	44° 1′ 24.1″	15° 35′ 36.5″
	TD253	Šopot railway cut near Benkovac	44° 1′ 25.5″	15° 35′ 42.5″
	TD255	Šopot railway cut near Benkovac	44° 1′ 25.5″	15° 35′ 43.2″
	TD257	Šopot railway cut near Benkovac	44° 1′ 27.1″	15° 35′ 50.6″
	TD260	Zagreb–Split motorway, 3 km SSW of Islam Latinski exit	44° 10′ 21.2″	15° 25′ 33.5″
Central Dalmatia	TD227	Mravince, 200 m E of the limestone olistolith	43° 32′ 9.3″	16° 30′ 41.5″
	TD229	Mravince, marl quarry	43° 32′ 16.0″	16° 31′ 3.9″
	TD232	Mravince abandoned quarry near police	43° 32′ 7.3″	16° 31′ 26.8″
	TD233	Mravince abandoned quarry near police	43° 32′ 3.6″	16° 31′ 23.0″
	TD234	Mravince abandoned quarry near police	43° 32′ 3.6″	16° 31′ 23.0″
	TD236	Jadro creek right side, larger one out of two quarries	43° 32′ 34.9″	16° 31′ 17.0″
	TD240	Jadro valley, Vrilo Jadro	43° 32′ 39.3″	16° 31′ 31.6″
	TD246	Gizdići near Klis	43° 33′ 27.1″	16° 30′ 14.5″
	TD248	Solin, Voljak Street	43° 33′ 14.9″	16° 29′ 26.6″
	TD249	Solin, Voljak Street	43° 33′ 14.9″	16° 29′ 26.6″
	TD52	Živogošće	43° 11′ 15.6″	17° 9′ 43.8″
	TD55	Omiš quarry, E wall	43° 25′ 33.8″	16° 42′ 53.0″
	TD56	Omiš quarry, E wall	43° 25′ 33.8″	16° 42′ 53.0″
	TD53	Medići near Omiš	43° 24′ 19.9″	16° 48′ 20.4″
	TD54	Mala Luka near Omiš	43° 25′ 10.0″	16° 42′ 59.6″

(Continued)

Appendix B. *Continued*

Dalm.-Herz. Zone	TD100	Spuž	42° 31′ 31.3″	19° 11′ 2.0″
	TD109	Medun	42° 28′ 16.1″	19° 21′ 51.7″
	TD140	Crnići	43° 7′ 34.1″	17° 51′ 32.3″
S-Adriatic Zone	TD51	Konavle hills	42° 33′ 44.9″	18° 18′ 28.5″
	TD47	Dubravka	42° 31′ 1.4″	18° 25′ 17.1″
	TD49	Dubravka	42° 31′ 57.8″	18° 24′ 56.9″
	TD38	Sutorina	42° 28′ 34.1″	18° 28′ 45.2″
	TD136	Sutorina	42° 28′ 20.0″	18° 25′ 36.9″
	TD138	Sutorina	42° 28′ 20.0″	18° 25′ 36.9″
	TD39	Tivat	42° 24′ 54.6″	18° 43′ 6.6″
	TD127	Radanovići	42° 20′ 45.3″	18° 30′ 37.9″
	TD134	Radanovići	42° 20′ 50.1″	18° 30′ 43.7″
	TD99	Stari Bar	42° 5′ 19.5″	19° 8′ 34.3″
	TD43	Zaljevo	42° 4′ 23.0″	19° 7′ 58.7″
	TD116	Ulcinj	41° 40′ 51.4″	19° 0′ 52.0″
	TD120	Donja Klezna - Gornja Klezna	41° 40′ 46.3″	19° 0′ 19.6″
	TD123	Donja Klezna - Gornja Klezna	41° 40′ 46.3″	19° 0′ 19.6″
	TD125	Donja Klezna - Gornja Klezna	41° 40′ 42.2″	19° 0′ 18.8″
	TD93	Kravari	42° 3′ 51.9″	19° 21′ 34.4″
	TD45	Vladimir	42° 0′ 39.1″	19° 17′ 15.8″

References

ALTINER, Y., MARJANOVIĆ, M., MEDVED, M. & RASIĆ, L. 2006. Active deformation of the Northern Adriatic region: Results from the CRODYN geodynamical experiment. *In*: PINTER, N., GRENERCZY, G., WEBER, J., STEIN, S. & MEDAK, D. (eds) *The Adria Microplate: GPS Geodesy, Tectonics and Hazards*. NATO Science Series IV, **61**, Springer, 257–269.

BABIĆ, LJ. 1974. Hauterivian to Cenomanian time in the region of Žumberak, Northwestern Croatia: stratigraphy, sediments, paleogeographic and paleotectonic evolution. *Geološki vjsenik*, **27**, 11–33.

BABIĆ, LJ. & ZUPANIČ, J. 1976. Sediments and paleogeography of the *Globotruncana calcarata* Zone (Upper Cretaceous) in Banija and Kordun, central Croatia. *Geološki vjsenik*, **29**, 49–74.

BABIĆ, LJ. & ZUPANIČ, J. 1983. Palaeogene clastic formations in Northern Dalmatia: Excursion A2. *In*: BABIĆ, LJ. & JELASKA, V. (eds) *Contributions to sedimentology of some carbonate and clastic units of the coastal Dinarides*. Excursion Guidebook, 4th IAS Regional Meeting, Split 1983. Zagreb.

BABIĆ, LJ. & ZUPANIČ, J. 1996. Coastal Dinaric flysch belt: paleotransport model for the Pazin Basin, and the role of a foreland uplift (Istria, Croatia). *Natura Croatica*, **5**, 317–327.

BABIĆ, LJ. & ZUPANIČ, J. 1998. Nearshore deposits in the Middle Eocene clastic succession in Northern Dalmatia (Dinarides, Croatia). *Geologia Croatica*, **51**, 175–193.

BABIĆ, LJ., ZUPANIČ, J. & CRNJAKOVIĆ, M. 1993. An association of marine tractive and gravity flow sandy deposits in the Eocene of the Island of Pag (Outer Dinarides, Croatia). *Geologia Croatica*, **46**, 107–123.

BABIĆ, LJ., ZUPANIČ, J. & JURAČIĆ, M. 1995. Supply from an Outer Carbonate Platform to the Foreland Basin of the Coastal Dinarides: the Pazin Flysch Basin (Eocene, Croatia). *1. Hrvatski geološki kongres Opatija*, **1**, 43–45.

BÁLDI-BEKE, M. 1977. Stratigraphical and faciological subdivisions of the Oligocene as based on nannoplankton. *Földtani Közlöny*, **107**, 59–89.

BÁLDI-BEKE, M. 1984. *The nannoplankton of the Transdanubian Palaeogene formations*. Geologica Hungarica, Series Palaeontologica, **43**.

BÁLDI-BEKE, M. & BÁLDI, T. 1991. Palaeobathymetry and palaeogeography of the Bakony Eocene Basin in western Hungary. *Palaeogeography, Palaeoclimatology, Palaeoecology*, **88**, 25–52.

BELLAS, S. M. 1997. Calcareous nannofossils of the Tertiary Flysch (Post Eocene to Early Miocene) of the Ionian Zone in Epirus, NW-Greece: Taxonomy and biostratigraphical correlations. *Berliner Geowissenschaftliche Abhandlungen*, **E(22)**, 1–173.

BENIĆ, J. 1975. Calcareous nannoplankton from the Eocene flysch on Pag Island. *Geološki vjsenik*, **28**, 19–23.

BENIĆ, J. 1991. The age of the Istria flysch deposits based on calcareous nannofossils. *In: Introduction to the Paleogene SW Slovenia and Istria Field-Trip Guidebook IGCP Project 286 'Early Paleogene Benthos'*, 2nd Meeting, Postojna.

BERGANT, S., TIŠLJAR, J. & ŠPARICA, M. 2003. Istrian flysch and its relationship with flysch from NE Italy, SW Slovenia and the Adriatic coastal region in Croatia (Ravni Kotari area and Dalmatia). *In: Evolution of Depositional Environments from the Palaeozoic to the Quaternary in the Karst Dinarides and the Pannonian Basin. 22nd IAS Meeting of Sedimentology, Opatija—Sept. 17–19, 2003, Field Trip Guidebook.*

BIGNOT, G. 1972. Recherches stratigraphiques sur les calcaires du Crétacé supérieur et de lÉocéne d'Istrie et des régions voisines. Essai de révision du Liburnien. *Travaux du Laboratoire de Micropaléontologie*, **2**, 1–353.

BOWN, P. R. & YOUNG, J. R. 1997. Mesozoic calcareous nannoplankton classification. *Journal of Nannoplankton Research*, **19**, 21–36.

BOWN, P. R. & YOUNG, J. R. 1998. Techniques. *In*: BOWN, P. R. (ed.) *Calcareous Nannofossil Biostratigraphy*. Chapman & Hall, London.

BURNETT, J. A. 1998. Upper Cretaceous. *In*: BOWN, P. R. (ed.) *Calcareous Nannofossil Biostratigraphy*. Chapman & Hall, London, 132–199.

BUSER, S. 1987. Development of the Dinaric and the Julian carbonate platforms and of the intermediate Slovenian basin (NW Yugoslavia). *Memorie della Società Geologica Italiana*, **40**, 313–320.

CADET, J. P. 1978. Essai sur l'évolution alpine d'une paléomarge continentale. *Mémoires de la Société géologique de France*, **57**, 1–83.

ČANOVIĆ, M. & DŽODŽO-TOMIĆ, R. 1958. Vorläufige Mitteilung Über die Oligozäne Mikrofauna aus der Bohrung Us-6 bei Ulcinj (Montenegro). *Geološki Glasnik (Titograd)*, **2**, 203–213.

CHANNELL, J. E. T., D'ARGENIO, B. & HORVÁTH, F. 1979. Adria, the African Promontory, in Mesozoic Mediterranean Palaeogeography. *Earth Science Reviews*, **15**, 213–292.

CHOROWICZ, J. 1977. Étude géologique des Dinarides le long da la structure transversale Split-Karlovac (Yougoslavie). *Publications de la Société Géologique du Nord*, **1**, 1–331.

CHRIST, M. D. 2007. Sedimentologie und Liefergebietsanalyse an ausgewählten Profilen des Bosnischen Flysches. *M.Sc. Thesis*, University of Göttingen, 89pp. + Appendix.

CRNJAKOVIĆ, M. 1981. Maastrichtian flysch sediments in the south-west part of Mt. Medvednica. *Geološki vjsenik*, **34**, 47–61.

ĆOSOVIĆ, V., DROBNE, K. & MORO, A. 2004. Paleoenvironmental model for Eocene foraminiferal limestones of the Adriatic carbonate platform (Istrian Peninsula). *Facies*, **50**, 61–75.

ĆOSOVIĆ, V., MARJANAC, T., DROBNE, K. & MORO, A. In press. Eastern Adriatic Coast – External Dinarids. *In*: MCCANN, T. (ed.) *Geology of Central Europe*. Geological Society London, Special Publications.

CURRY, D. 1982. Differential preservation of foraminiferids in the English Upper Cretaceous – consequental observations. *In*: BANNER, F. T. & LORD, A. R. (eds) *Aspects of Micropalaeontology. Papers presented to Professor Tom Barnard*. George Allen & Unwin, London.

DE CAPOA, P., GUERRERA, F., PERRONE, V. & SERRANO, F. 1997. New biostratigraphic data on the Frazzanò Formation (Longi-Taormina Unit): consequences on defining the deformation age of the Calabria-Peloritani arc southern sector. *Rivista Italiana di Paleontologia e Stratigrafia*, **103**, 343–356.

DE CAPOA, P. & RADOIČIĆ, R. 1994a. Calcareous nannoplankton biostratigraphy of Tertiary sequences of the Cukali-Budva Basin (Montenegro, External Dinarides, Yugoslavia). *Rivista Española de Micropaleontologia*, **26**, 101–116.

DE CAPOA, P. & RADOIČIĆ, R. 1994b. Tertiary nannoplankton biostratigraphy of the Zeta Intraplatform Furrow (Montenegro). *Palaeopelagos*, **4**, 289–294.

DE CAPOA, P. & RADOIČIĆ, R. 2002. Geological implications of biostratigraphic studies in the external and internal domains of the Central-Southern Dinarides. *Memorie della Società Geologica Italiana*, **57**, 185–191.

DE CAPOA, P., RADOIČIĆ, R. & D'ARGENIO, B. 1995. Late Miocene deformation of the External Dinarides (Montenegro and Dalmatia): New biostratigraphic evidence. *Memorie di Scienze Geologiche*, **47**, 157–172.

DE CAPOA, P., GUERRERA, F., PERRONE, V., SERRANO, F. & TRAMONTANA, M. 2000. The onset of the syn-orogenic sedimentation in the flysch basin of the Sicilian Maghrebids; state of the art and new biostratigraphic constraints. *Eclogae Geologicae Helvetiae*, **93**, 65–79.

DECELLES, P. G. & GILES, K. A. 1996. Foreland basin systems. *Basin Research*, **8**, 105–123.

DE KAENEL, E. & VILLA, G. 1996. Oligocene-Miocene calcareous nannofossil biostratigraphy and paleoecology from the Iberia abyssal plain. *Proceedings of the Ocean Drilling Program, Scientific Results*, **149**, 79–145.

DIMITRIJEVIĆ, M. D. 1997. *Geology of Yugoslavia*. Geological Institute GEMINI, Belgrade, 187pp.

DROBNE, K. 1977. Alvéolines paléogènes de la Slovénie et de l'Istrie. Schweizerische Paläontologische Abhandlungen, **99**.

DROBNE, K. & PAVLOVEC, R. 1991. Paleocene and Eocene beds in Slovenia and Istria. *In: Introduction to the Paleogene SW Slovenia and Istria Field-Trip Guidebook IGCP Project 286 'Early Paleogene Benthos', 2nd Meeting*, Postojna.

DROBNE, K., PAVLOVEC, R., ŠIKIĆ, L. & BENIĆ, J. 1979. Pićan, Istria – Cuisian, Lutetian. *In*: DROBNE, K. (ed.) *16th European Micropaleontological Colloquium, Guidebook*. Ljubljana, F177–F182.

EINSELE, G. 1992. *Sedimentary Basins*. Springer, Berlin.

ENGEL, W. 1974. Sedimentologische Untersuchungen im Flysch des Beckens von Ajdovščina (Slowenien). *Göttinger Arbeiten zur Geologie und Paläontologie*, **16**, 1–65.

Federal Geological Institute (F.G.I.) 1970. *SFR Yugoslavia Geological Map 1:500 000*. Belgrade.

FORNACIARI, E. & RIO, D. 1996. Latest Oligocene to early middle Miocene quantitative calcareous nannofossil biostratigraphy in the Mediterranean region. *Micropaleontology*, **42**, 1–36.

FORNACIARI, E., DI STEFANO, A., RIO, D. & NEGRI, A. 1996. Middle Miocene quantitative calcareous nannofossil biostratigraphy in the Mediterranean region. *Micropaleontology*, **42**, 37–63.

GRUBIĆ, A. & KOMATINA, M. 1963. Properties of the Eocene-Oligocene Flysch between Split and Makarska. *Sedimentologija*, **2–3**, 21–38.

HAGN, R., PAVLOVEC, R. & PAVŠIČ, J. 1979. Gračišće near Pićan, Istria – Eocene. *In*: DROBNE, K. (ed.)

16th European Micropaleontological Colloquium, Guidebook. Ljubljana.
HAYNES, J. R. 1981. *Foraminifera*. Macmillan, London.
HERAK, M., MARINČIĆ, S. & POLŠAK, A. 1976. Geology of the Island of Hvar. *Prirodoslovna Istraživanja*, **42**, 5–14.
HRVATOVIĆ, H. 1999. *Geološki vodić kroz Bosnu i Hercegovinu*. Zavod za Geologiju, Sarajevo.
JELASKA, V., AMŠEL, V., KAPOVIĆ, B. & VUKSANOVIĆ, B. 1969. Sedimentological characteristics of the clastic Upper Cretaceous of the western part of Bosanska Krajina. *Nafta*, **20**, 487–495.
JERKOVIĆ, L. & MARTINI, E. 1976. Upper Eocene calcareous nannoplankton from Split and Dugi Rat (Yugoslavia). *Nafta*, **27**, 67–70.
JURAČIĆ, M. 1979. Sedimentation depth of the 'Marls with crabs' based on planktonic to benthic foraminifer ratio. *Geološki vjesnik*, **31**, 61–67.
KOHL, B. & ROBERTS, H. H. 1994. Fossil Foraminifera from four active mud volcanoes in the Gulf of Mexico. *Geo-Marine Letters*, **14**, 126–134.
KOŠIR, A. 1997. Eocene platform-to-basin depositional sequence, southwestern Slovenia. *Gaea heidelbergensis*, **3**, 205.
KRAŠENINNIKOV, V., MULDINI-MAMUŽIĆ, S. & DŽODŽO-TOMIĆ, R. 1968. Signification des foraminifères planctoniques pour la division du paléogène de la Yougoslavie et comparaison avec les autres régions examinées. *Geološki vjesnik*, **21**, 117–145.
KULKA, A. 1985. *Nummulites* from Poręba near Myślenice (Polish Western Carpathians). *Kwartalnik Geologiczny*, **29**, 199–236.
LUKOVIĆ, M. & PETKOVIĆ, K. 1952. Geology and tectonics of the area of Ulcinj (littoral of Montenegro); an analysis of previous works and some new contributions. *SANU Posebna izdanja, knjiga 197*, **4**, 1–62.
MAGDALENIĆ, Z. 1972. Sedimentology of Central Istria flysch deposits. *Prirodoslovna Istraživanja*, **39**, 1–34.
MANTOVANI, E., BABBUCCI, D., VITI, M., ALBARELLO, D., MUGNAIOLI, E., CENNI, N. & CASULA, G. 2006. Post-Late Miocene kinematics of the Adria microplate: Inferences from geological, geophysical and geodetic data. *In*: PINTER, N., GRENERCZY, G., WEBER, J., STEIN, S. & MEDAK, D. (eds) *The Adria Microplate: GPS Geodesy, Tectonics and Hazards*. NATO Science Series IV, **61**, Springer, 51–69.
MARINČIĆ, S. 1981. Eocene flysch of the Adriatic area. *Geološki vjesnik*, **34**, 27–38.
MARJANAC, T. 1990. Reflected sediment gravity flows and their deposits in flysch of middle Dalmatia, Yugoslavia. *Sedimentology*, **37**, 921–929.
MARJANAC, T. 1996. Deposition of megabeds (megaturbidites) and sea-level change in a proximal part of the Eocene-Miocene Flysch of central Dalmatia (Croatia). *Geology*, **24**, 543–546.
MARJANAC, T. 2000. Kaštela-Split flysch region. *Vijesti Hrvatskoga geološkog društva*, **37**, 109–116.
MARJANAC, T. & ĆOSOVIĆ, V. 2000. Tertiary depositional history of Eastern Adriatic realm. *Vijesti Hrvatskoga geološkog društva*, **37**, 93–103.
MARJANAC, T. & MARJANAC, LJ. 1991. Shallow-marine clastic Paleogene on the Island of Rab (Northern Adriatic). *In*: BOSELLINI, A., BRANDNER, R., FLÜGEL, E., PURSER, B., SCHLAGER, W., TUCKER, M. & ZENGER, D. (eds) Dolomieu Conference on Carbonate Platforms and Dolomitization, Ortisei, 159–160.
MARJANAC, T., BABAC, D., BENIĆ, J. *ET AL*. 1998. Eocene carbonate sediments and sea-level changes on the NE part of Adriatic Carbonate Platform (Island of Hvar and Pelješac Peninsula, Croatia). *Dela-Opera SAZU 4. razreda*, **34**, 243–254.
MARTINI, E. 1971. Standard Tertiary and Quaternary calcareous nannoplankton zonation. *In*: FARINACCI, A. (ed.) *Proceedings of the Second Planktonic Conference, Roma 1970*, 739–785.
MIKES, T., DUNKL, I. & FRISCH, W. 2004. Provenance mixing in a foreland basin as revealed by sandstone geochemistry and detrital zircon fission track analysis: the Eocene flysch of the NW Dinarides. *In*: PENA DOS REIS, R., CALLAPEZ, P. & DINIS, P. (eds) 23rd IAS Meeting of Sedimentology, Coimbra—Sept. 15–17, 2004, Abstracts Book, Coimbra.
MIKES, T., DUNKL, I. & VON EYNATTEN, H. 2005. Significance of ophiolitic detritus in the Tertiary External Dinaride flysch belt. *Schriftenreihe der Deutschen Gesellschaft für Geowissenschaften*, **38**, 110–111.
MINDSZENTY, A., D'ARGENIO, B. & AIELLO, G. 1995. Lithospheric bulges recorded by regional unconformities. The case of Mesozoic-Tertiary Apulia. *Tectonophysics*, **252**, 137–161.
MRINJEK, E. 1993. Sedimentology and depositional setting of alluvial Promina Beds in northern Dalmatia, Croatia. *Geologia Croatica*, **46**, 243–261.
MULDINI-MAMUŽIĆ, S. 1962. Mikrofaunistische untersuchungen des Eozän-Flysches der Insel Rab. *Geološki vjesnik*, **15**, 149–159.
MULDINI-MAMUŽIĆ, S. 1965. The microfauna of limestones and of the clastic development in the Paleogene of central Istria. *Geološki vjesnik*, **18**, 281–289.
NAGYMAROSY, A. 1987. *Geological key profile of Hungary. Buda Hills, Budapest, Róka-hegy, Ibolya Street, quarry. Tard Clay Formation*. Hungarian Geological Institute, Budapest.
NAGYMAROSY, A. & BÁLDI-BEKE, M. 1993. The Szolnok Unit and its probable paleogeographic position. *Tectonophysics*, **226**, 457–470.
OLSZEWSKA, B. & GARECKA, M. 1996. Foraminifer and nannoplankton biostratigraphy of the lower Miocene of the Carpathian Foredeep. *Przegląd Geologiczny*, **44**, 1049–1053.
OREHEK, S. 1991. Palaeotransport of SW Slovenian flysch. *In*: *Introduction to the Paleogene SW Slovenia and Istria Field-Trip Guidebook IGCP Project 286 'Early Paleogene Benthos', 2nd Meeting*, Postojna, 27–31.
OSZCZYPKO, N., ANDREYEVA-GRIGOROVICH, A. S., MALATA, E. & OSZCZYPKO-CLOWES, M. A. 1999. The lower Miocene deposits of the Rača Subunit near Nowy Sącz (Maguma Nappe, Polish Outer Carpathians. *Geological Carpathica*, **50**, 419–433.
PAMIĆ, J., BALOGH, K., HRVATOVIĆ, H., BALEN, D., JURKOVIĆ, I. & PALINKAŠ, L. 2004. K-Ar and Ar-Ar dating of the Palaeozoic metamorphic complex from the Mid-Bosnian Schist Mts., Central Dinarides, Bosnia and Hercegovina. *Mineralogy and Petrology*, **82**, 65–79.
PAMIĆ, J., GUŠIĆ, I. & JELASKA, V. 1998. Geodynamic evolution of the Central Dinarides. *Tectonophysics*, **297**, 251–268.

PAVIĆ, A. 1970. *Paléogène marin du Montenegro*. PhD. Thesis. Zavod za geološka Istraživanja Crne Gore. Titograd, 208pp.

PAVLOVEC, R. 2003. The type of nummulitins localities in the Dinarides. *Materials and Geoenvironment*, **50**, 777–788.

PAVLOVEC, R. 2006. Lower Eocene Nummulits from Trnovo in surroundings of Ilirska Bistrica (SW Slovenia). *Geologija*, **49**, 45–52.

PAVLOVEC, R., KNEZ, M., DROBNE, K. & PAVŠIČ, J. 1991. Profiles: Košana, Sv. Trojica and Leskovec; the disintegration of the carbonate platform. *In: Introduction to the Paleogene SW Slovenia and Istria Field-Trip Guidebook IGCP Project 286 'Early Paleogene Benthos'*, 2nd Meeting, 69–72.

PAVŠIČ, J. 1981. Nannoplankton of the flysch in the Slovenian coast. *Radovi Znanstvenog Savjeta za Naftu, Sekcija za Primjenu Geologije, Geofizike i Geokemije, JAZU, Serija A.*, **8**, 257–266.

PAVŠIČ, J. & PECKMANN, J. 1996. Stratigraphy and sedimentology of the Piran Flysch Area (Slovenia). *Annales (Annals for Istrian and Mediterranean Studies)*, **9**, 123–138.

PAVŠIČ, J. & PREMEC-FUČEK, V. 2000. Calcareous nannoplankton and planktonic foraminiferal Zones during the Middle and Upper Eocene of the Transitional Beds on the Adriatic platform. *Annali del Museo Civico di Storia Naturale di Ferrara*, **3**, 22–23.

PERCH-NIELSEN, K. 1985a. Mesozoic calcareous nannofossils. *In*: BOLLI, H. M., SAUNDERS, I. B. & PERCH-NIELSEN, K. (eds) *Plankton stratigraphy*. Cambridge University Press, Cambridge, 329–426.

PERCH-NIELSEN, K. 1985b. Cenozoic calcareous nannofossils. *In*: BOLLI, H. M., SAUNDERS, I. B. & PERCH-NIELSEN, K. (eds) *Plankton stratigraphy*. Cambridge University Press, Cambridge, 427–554.

PETRI, R. 2007. Grad und Alter der schwachmetamorphen Überprägung an ausgewählten Profilen des Bosnischen Flysches. MSc. Thesis, University of Göttingen, 111pp. + Appendix.

PICCOLI, G. & PROTO DECIMA, F. 1969. Ricerche biostratigrafiche sui depositi flyschoidi della regione Adriatica settentrionale e orientale. *Memorie degli Instituti di Geologia e Mineralogia dell' Università di Padova*, **27**, 3–21.

PICHA, F. J. 2002. Late orogenic strike-slip faulting and escape tectonics in frontal Dinarides-Hellenides, Croatia, Yugoslavia, Albania and Greece. *AAPG Bulletin*, **86**, 1659–1671.

PIPER, D. J. W., PANAGOS, A. G. & PE, G. G. 1978. Conglomeratic Miocene flysch of Western Greece. *Journal of Sedimentary Petrology*, **48**, 117–126.

POMEROL, C. 1985. La transition Eocène-Oligocène: est-elle un phénomène progressif ou brutal? *Bulletin de la Société géologique de France, série 8*, **1**, 263–267.

POSTMA, G., BABIĆ, LJ., ZUPANIČ, J. & RØE, S.-L. 1988. Delta-front failure and associated bottomset deformation in a marine, gravelly Gilbert-type fan delta. *In*: NEMEC, W. & STEEL, R. J. (eds) *Fan deltas; sedimentology and tectonic settings*. Blackie and Son, Glasgow, 91–102.

PRELOGOVIĆ, E., PRIBIČEVIĆ, B., IVKOVIĆ, Ž., DRAGIČEVIĆ, I., BULJAN, R. & TOMLJENOVIĆ, B. 2003. Recent structural fabric of the Dinarides and tectonically active zones important for petroleum-geological exploration in Croatia. *Nafta*, **55**, 155–161.

PRIBIČEVIĆ, B., MEDAK, D. & PRELOGOVIĆ, E. 2002. Determination of the recent structural fabric in the Alps-Dinarides area by combination of geodetic and geologic methods. *Raziskave s področja geodezije in geofizike, Zbornik predavanja*, **2002**, 57–64.

PUŠKARIĆ, S. 1987. Calcareous nannoplankton from clastic sediments of the Island of Hvar. *RAD Jugoslavenske akademije znanosti i umjetnosti*, **431**, 7–16.

RADOIČIĆ, R., DE CAPOA, P. & D'ARGENIO, B. 1989. Late rather than early Tertiary deformation of External Dinarides. Stratigraphic evidence from Montenegro. *Rendiconto dell' Accademia delle Scienze Fisiche e Matematiche Napoli*, **56**, 41–59.

RADOIČIĆ, R., DE CAPOA, P. & D'ARGENIO, B. 1991. Middle-Late Miocenic age of the preorogenic sedimentation in the Dinaric carbonate platform domain of Herzegovina. *Annales Géologiques de la Péninsule Balkanique*, **55**, 1–21.

SCHMID, S. M., FÜGENSCHUH, B., MATENCO, L., SCHUSTER, R., TISCHLER, M. & USTASZEWSKI, K. 2006. The Alps-Carpathians-Dinarides-connection: a compilation of tectonic units. *In*: SUDAR, M., ERCEGOVAC, M. & GRUBIĆ, A. (eds) *XVIII Congress of the Carpathian-Balkan Geological Association*, Belgrade.

SCHUBERT, R. J. 1905. Zur Stratigraphie des istrisch-norddalmatinischen Mitteleocäns. *Jahrbuch der Kaiserlich-Königlichen Geologischen Reichsanstalt*, **55**, 153–188.

SCHULZ, H.-M., BECHTEL, A. & SACHSENHOFER, R. F. 2005. The birth of the Parathetys during the Early Oligocene: From Tethys to an ancient Black Sea analogue? *Global and Planetary Change*, **49**, 163–176.

SCHWEITZER, C. E., ĆOSOVIĆ, V. & FELDMANN, R. M. 2005. *Harpactocarcinus* from the Eocene of Istria, Croatia, and the paleoecology of the Zanthopsidae Via, 1959 (Crustacea: Decapoda: Brachyura). *Journal of Paleontology*, **79**, 663–669.

SKABERNE, D. 1987. Megaturbidites in the Paleogene flysch in the region of Anhovo (W. Slovenia, Yugoslavia). *Memorie della Società Geologica Italiana*, **40**, 231–239.

ŠIKIĆ, D. 1963. Eine vergleichende Darstellung der Entwicklung des jüngeren klastischen Paläogens in Istrien, dem kroatischen Küstenland und Dalmatien. *Geološki vjesnik*, **15**, 329–336.

ŠIKIĆ, D. 1968. The development of the Paleogene and the Lutetian movements in northern Dalmatia. *Geološki vjesnik*, **22**, 309–331.

ŠPARICA, M., KOCH, G., IBRAHIMPAŠIĆ, H., GALOVIĆ, I. & BERGANT, S. 2005. New data to the Palaeogene stratigraphy of the clastic-carbonate beds in SE Istria, Croatia. *In*: VELIĆ, I., VLAHOVIĆ, I. & BIONDIĆ, R. (eds) *Third Croatian Geological Congress. Abstracts Book*, Opatija, 29.09–01.10.2005, Zagreb.

STRADNER, H. 1962. Über das fossile Nannoplankton des Eozän-Flysch von Istrien. *Verhandlungen der Geologischen Bundesanstalt*, **Jg. 1962**, 176–186.

TARI, V. 2002. Evolution of the northern and western Dinarides: a tectonostratigraphic approach. *EGU Stephan Mueller Special Publication Series*, **1**, 1–21.

TARI-KOVAČIĆ, V. 1998. Geodynamics of the Middle Adriatic offshore area, Croatia, based on stratigraphic and seismic analysis of Paleogene beds. *Acta Geologica Hungarica*, **41**, 313–326.

THIERSTEIN, H. R. 1980. Selective dissolution of Late Cretaceous and Earliest Tertiary calcareous nannofossils: Experimental evidence. *Cretaceous Research*, **2**, 165–176.

TOMLJENOVIĆ, B. 2000. Map-scale folding and thrust faulting in Eocene flysch sediments (The Bay of Omiš). *Vijesti Hrvatskoga geološkog društva*, **37**, 106–108.

TUNIS, G. & VENTURINI, S. 1992. Evolution of the southern margin of the Julian Basin with emphasis on the megabeds and turbidites sequence of the southern Julian Prealps (NE Italy). *Geologia Croatica*, **45**, 127–150.

VARGA, P. 1982. The lower marine member of the Tard Clay: Its age on the foraminiferal evidence of allodapic limestone beds. *Földtani Közlöny*, **112**, 177–184.

VARGA, P. 1985. Turbiditic limestone intercalations of the Buda Marl and Tard Clay. *Őslénytani Viták*, **31**, 93–99.

VAROL, O. 1998. Palaeogene. *In*: BOWN, P. R. (ed.) *Calcareous nannofossil biostratigraphy*. Chapman and Hall, London, 200–224.

VLAHOVIĆ, I., TIŠLJAR, J., VELIĆ, I. & MATIČEC, D. 2005. Evolution of the Adriatic carbonate platform: Palaeogeography, main events and depositional dynamics. *Palaeogeography, Palaeoclimatology, Palaeoecology*, **220**, 333–360.

VRABEC, M. & FODOR, L. 2006. Late Cenozoic tectonics of Slovenia: Structural styles at the northwestern corner of the Adriatic microplate. *In*: PINTER, N., GRENERCZY, G., WEBER, J., STEIN, S. & MEDAK, D. (eds) *The Adria Microplate: GPS Geodesy, Tectonics and Hazards*. NATO Science Series IV, **61**, Springer, 151–168.

YOUNG, J. R. 1998. Neogene. *In*: BOWN, P. R. (ed.) *Calcareous nannofossil biostratigraphy*. Chapman and Hall, London, 225–265.

YOUNG, J. R. & BOWN, P. R. 1997. Cenozoic calcareous nannoplankton classification. *Journal of Nannoplankton Research*, **19**, 36–47.

ŽIVKOVIĆ, S. 2004. *Smaller benthic foraminifers from the Eocene clastics from western Croatia: Paleoecology of sedimentary basin*. PhD. Thesis, University of Zagreb, Zagreb.

ŽIVKOVIĆ, S. & BABIĆ, LJ. 2003. Paleoceanographic implications of smaller benthic and planktonic foraminifera from the Eocene Pazin Basin (coastal Dinarides, Croatia). *Facies*, **49**, 49–60.

ZUPANIČ, J. & BABIĆ, LJ. 1991. Cross-bedded sandstones deposited by tidal currents in the Eocene of the Outer Dinarides (Island of Rab, Croatia). *Geološki vjesnik*, **44**, 235–245.

Tertiary subduction, collision and exhumation recorded in the Adula nappe, central Alps

THORSTEN J. NAGEL

Geologisches Institut der Universität Bonn, Nussallee 8, D-53115 Bonn, Germany
(e-mail: tnagel@uni-bonn.de)

Abstract: The Adula nappe in the Central Alps represents a lithospheric mélange assembled in a south-dipping subduction zone during the Tertiary orogenic cycle. It consists of several heterogeneous lobes which are stacked in a forward-dipping duplex geometry. Eclogites, garnet peridotites and garnet-white-mica schists record southward-increasing peak pressure conditions which culminate at 12–17 kbar/500–600 °C in the north and 30 kbar/800–850 °C in the south. Some studies infer even higher peak pressures for the garnet peridotite body of Alpe Arami. The present-day metamorphic field gradient for peak pressures exceeds the lithostatic pressure gradient. So far, only eclogites and garnet peridotites from the Cima Lunga complex in the south and the adjacent Southern Steep Belt have yielded Tertiary metamorphic ages for the peak-pressure stage. Some recent studies propose that the Adula nappe got assembled after the formation of high-pressure assemblages in eclogites and garnet peridotites and reject regional high-pressure conditions in Tertiary times. This scenario, however, is in conflict with the observed continuity of metamorphic field gradients and post-peak-pressure structures. Amphibolite facies conditions post-date formation of the Central Alpine nappe stack. In this paper, the associated field gradient is explained through southward-increasing temperatures during near-isothermal decompression.

The main mylonitic foliation in the Adula nappe post-dates peak-pressure conditions. It is associated with top-to-the-north shearing and southward-increasing amounts of decompression from eclogite facies to amphibolite facies conditions. Also, the present-day supra-lithostatic field gradient for peak pressures probably results from this deformation phase and is here related to substantial vertical flattening during northward shearing. All subsequent structures affect established nappe boundaries. Pervasive Oligocene deformation events in the Adula nappe are coeval with intense shearing along the so-called Insubric mylonites and occur during ongoing isothermal decompression to around 5 kbar. They are associated with orogen-oblique to orogen-parallel stretching of unspecified amount which may considerably contribute to the exhumation of the Lepontine dome already before the onset of the well-known Miocene extension.

One of the best-referenced studies on Alpine geology is Schmid *et al.* (1996a). It summarizes the results of the NFP20 seismic cross-section through the central Alps and of several studies performed in the wake of this campaign. Schmid *et al.* (1996a) present a unifying theory of the central Alps as resulting from Eocene, southward subduction of the European margin beneath the Apulian continental margin and subsequent nappe stacking and backthrusting along the Insubric mylonites. The famous cross-section from Lake Constance to Milano (Fig. 1) and the proposed Tertiary evolution of the central Alps (Fig. 2) have served as a reference model not only for Alpine studies but for collision orogens in general (e.g. Willett *et al.* 1993; Chemenda *et al.* 1995; Ellis *et al.* 1999). Data sets from the Adula nappe in SE Switzerland and northern Italy have an important role in the reconstruction shown in Figure 2. The Adula nappe represents the only continental unit along the NFP20 transect which got subducted to high-pressure and probably even ultra-high-pressure metamorphic conditions during the Alpine cycle (e.g. Evans & Trommsdorff 1978; Heinrich 1986; Nimis & Trommsdorff 2001) and the associated metamorphic rocks have drawn the attention of petrologists throughout the world. The recognition of high-pressure conditions as being Tertiary in age (Becker 1993; Gebauer *et al.* 1992; Gebauer 1996) led to the conclusion that the entire evolution of the central Alpine nappe stack might be much younger than previously assumed.

Due to the great variety of rocks, the Adula nappe contains not only the relicts of the high-pressure stage but also an excellent record of the decompression history and the subsequent amphibolite facies stage of metamorphism, which affected the entire Lepontine nappe stack (Niggli & Niggli 1965; Trommsdorff 1966; Wenk 1970; Frey *et al.* 1974; Frey *et al.* 1980; Todd & Engi 1997; Frey & Ferreiro-Mählmann 1999). The Adula nappe also provides spectacular outcrop conditions with more than 2 km of local relief which allows study of its internal architecture and the relation to the surrounding units in three dimensions. North of the Insubric line, the central

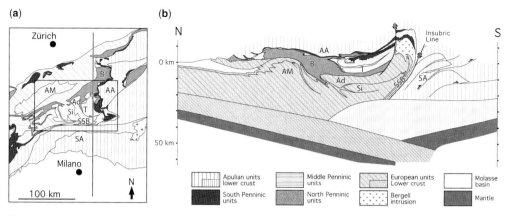

Fig. 1. (a) Tectonic sketch map of the central Alps (after Froitzheim et al. 1996). Line indicates trace of main NFP20 seismic line and geological cross-section in Fig. 1(b). Frame outlines area displayed in Fig. 3. (b) Geological cross-section through the eastern central Alps based on surface geology and seismic reflection data of the NFP20 project (Schmid et al. 1996a). AA, Austroalpine nappes; Ad, Adula nappe; AM, Aare massif; B, Bündnerschiefer; SA, southern Alps; SSB, Southern Steep Belt; Si, Simano nappe; T, Tambo nappe.

Alpine nappe stack is folded into the Southern Steep Belt (Fig. 1), a several km wide mylonite zone separating the central from the southern Alps. The Insubric mylonites are considered as the main backthrust of the European wedge on top of the Adriatic backstop (Milnes 1974, 1978; Schmid et al. 1987, 1989; Pfiffner et al. 2000). Again, the topographic and petrological conditions in the Adula nappe are among the best to study the formation of this structure. Stefan Schmid has supervised several PhD and Diploma theses in and around the Adula nappe and the present-day picture of this unit is to a large extent based on his and his students' work. Here, I summarize the metamorphic and structural record of the Adula nappe and its significance for tectonic models. Some aspects of Schmid et al. (1996a) have been challenged in the meantime and I will present and discuss alternative views. The following paragraph briefly introduces the edifice of the central Alps (Fig. 2). After that, four paragraphs address the overall architecture of the Adula nappe, the metamorphic record, the structural record, and the regional context of events. In the final paragraph, I describe what, in my view, are the most important open questions in that area.

The edifice of the central Alps

The Adula nappe is located on the eastern flank of the so-called Lepontine dome, a structural half-window and metamorphic high in southern Switzerland and northern Italy (Fig. 3). Main foliation and nappe-dividing surfaces dip gently towards northeast to east, so that progressively higher structural levels are exposed in an eastward direction. The units in the Lepontine dome and the subsequently higher nappes in the east are derived from at least three continental domains separated by two Mesozoic oceanic basins, the North and the South Penninic ocean (e.g. Frisch 1979; Schmid et al. 1996a; Froitzheim et al. 1996; Stampfli et al. 1998) (Fig. 2). In Tertiary times, all these units were accreted in one or two south-dipping subduction zones leading to structurally higher units representing more southerly located palaeogeographic domains. The Adula nappe is the highest of the so-called lower Penninic nappes (Milnes 1974). It is derived from the distal European margin towards the North Penninic ocean, which started opening in the late Jurassic. The nappe constitutes a lithospheric mélange assembled in a subduction channel (Trommsdorff 1990; Engi et al. 2001). Beneath and west of the Adula nappe are the Simano and the next lower Leventina nappe. These nappes consist of pre-Mesozoic, European upper crust with a strong alpine metamorphic overprint (Preiswerk 1921; Preiswerk et al. 1934; Niggli & Niggli 1965). Above the Adula nappe is the so-called Misox zone (Ganser 1937), a mixed unit of dominantly North-Penninic provenance. It consists of metamorphic clastic sediments (Bündnerschiefer) interlayered with MORB-derived amphibolites and a few slivers of continental basement (Ganser 1937; Steinmann 1994; Steinmann & Stille 1999). The sediments are wrapped around the front of the Adula nappe and also found in synclines separating the deeper lower Penninic nappes (e.g. Preiswerk et al. 1934; Maxelon & Mancktelow 2005). Also the Chiavenna ophiolite complex further south is considered as a sliver of North Penninic lithosphere

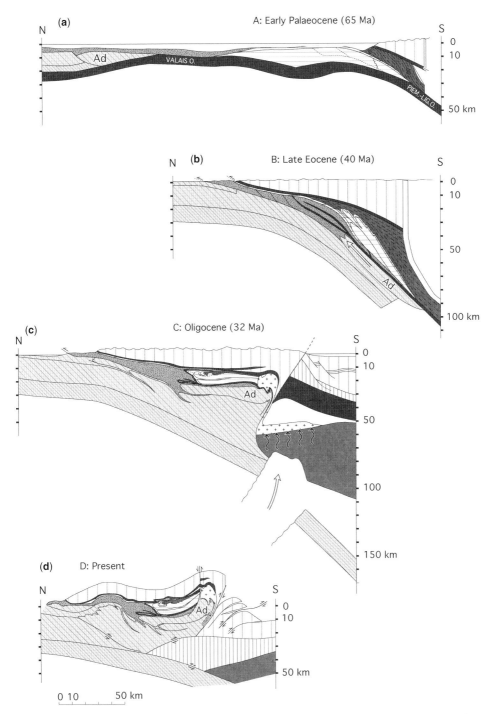

Fig. 2. Cross-sections from Schmid *et al.* (1996*a*) picturing the formation of the central Alps as proposed by the authors. Key is the same as in Fig. 1. The present edifice results from Palaeocene and Eocene southward subduction of the palaeogeographic domains between the European and the Apulian margin followed by Oligocene backthrusting along the Insubric mylonites and associated indentation of Adriatic lower crust into Europe. Rocks now assembled in the Adula nappe are mainly derived from the distal European margin. They extruded within the subduction channel in Late Eocene times.

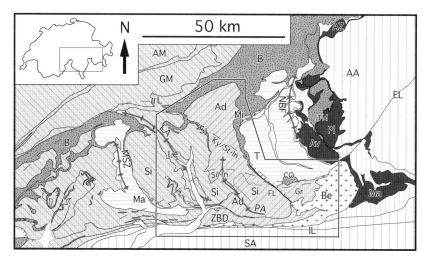

Fig. 3. Tectonic sketch map of the central Alps, after Spicher (1980) and Froitzheim et al. (1996). Mineral zone boundaries after Niggli & Niggli (1965). Key same as in Fig. 1. Frame outlines area of Fig. 4. AA, Austroalpine nappes; Ad, Adula nappe; AM, Aare massif; Av, Avers unit; B, Bündnerschiefer; Be, Bergell pluton; CO, Chiavenna ophiolite; EL, Engadin line; FL, Forcola line; Gr, Gruf complex; GM, Gotthard massif; IL, Insubric line; Ma, Maggia nappe; LA, Leventina nappe; Ma, Malenco unit; Mi, Misox zone; Pl, Platta nappe; SA, Southern Alps; Si, Simano nappe; Sj, Soja syncline; T, Tambo nappe; TM, Turba mylonite; ZBD, zone of Bellinzona–Dascio. Large post-nappe folds: MSZ, Maggia steep zone; La, Leventina antiform; PA, Paglia antiform; NBF, Niemet-Beverin fold.

(Schmutz 1976; Huber & Marquer 1998; Liati et al. 2004). The next higher units above the Misox zone are the Tambo, the Suretta, and the Schamps nappes (Ganser 1937; Milnes & Schmutz 1978; Baudin et al. 1993; Schreurs 1995). They are continental basement and/or cover nappes derived from the Briançonnais domain, a ribbon continent and a promontory of the Iberian microplate that got separated from Europe during the opening of the North Penninic ocean (Frisch 1979; Stampfli et al. 1998).

Units of the former Briançonnais domain are usually referred to as Middle Penninic units. Even further in the east and structurally above the middle Penninic units are the remnants of the South Penninic ocean which opened in early Jurassic times, namely the Arosa zone, the Avers unit, the Platta nappe and the Malenco unit. The highest units in the stack are the Austroalpine nappes representing the margin of the Apulian Plate. According to Schmid et al. (1996a), this nappe stack was assembled in a single, south-dipping subduction zone beneath the Apulian Plate in the Early Tertiary (Fig. 2a, b). North of Lake Verbano, the entire nappe stack is wrapped around a series of large-scale antiforms into an upright and even overturned, i.e. north-dipping orientation. The southern limb of these folds is called Southern Steep Belt. It represents a complex ductile mylonite zone—also called the Insubric mylonite belt or Insubric mylonites—with at least 25 km vertical and several tens of km strike-slip offset that was active from high amphibolite to sub-greenschist facies conditions (Heitzmann 1975; Milnes 1980; Vogler & Voll 1981; Hurford 1986; Heitzmann 1987; Schmid et al. 1987, 1989; Viola et al. 2001; Handy et al. 2005). The southern, brittle termination of the Insubric mylonites is called the Tonale fault and represents a segment of the Periadriatic line. Today, the Southern Steep Belt separates unmetamorphic rocks derived from the Apulian continent to the south from high amphibolite facies units of the European margin to the north. Most studies interpret the Insubric mylonites as an Oligocene, mid-crustal backthrust on top of a north-dipping Adriatic backstop in the course of oblique plate convergence (Schmid et al. 1996a; Pfiffner et al. 2000) (Fig. 2c). However, tectonic setting, kinematics, amount of offset, and age of mylonitization are subjects of ongoing discussion (e.g. Bradbury & Nolen-Hoeksema 1985; Viola et al. 2001; Nagel et al. 2002a; Maxelon & Mancktelow 2005; Pleuger et al. 2008). In the southeastern Lepontine dome, the nappe stack is penetrated by the Bergell pluton which consists of a large, 30 Ma old granodiorite body enveloped by a 32 Ma old Tonalite (von Blanckenburg 1992).

In the eastern Lepontine Alps, the Adula nappe is the only continental unit containing rocks that

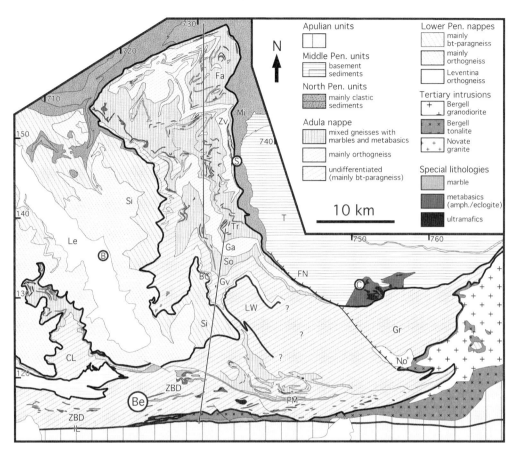

Fig. 4. Tectonic map of the Adula nappe and the adjacent units after Jenny *et al.* (1923), Löw (1987), Davidson (1996), Huber & Marquer (1998), Nagel *et al.* (2002*a*), and Berger *et al.* (2005). B, Blenio; Be, Bellinzona; BG, basal gneiss; C, Chiavenna; CL, Cima Lunga complex; Fa, Fanella lobe; FN, Forcola normal fault; Ga, Ganan lobe; Gr, Gruf complex; Gv, Groveno lobe; IL, Insubric line; Le, Leventina nappe; LW, Lostallo window; Mi, Misox zone; No, Novate intrusion; PM, Paina marble; S, San Bernardino; Si, Simano nappe; So, Soazza lobe; T, Tambo nappe; Tr, Trescolmen lobe; ZBD, zone of Bellinzona–Dascio; Zv, Zervreila lobe. Coordinates are Swiss coordinates.

experienced high-pressure and ultra-high-pressure conditions during Tertiary subduction. Peak pressures in the higher Tambo nappe as well as in the deeper lower Penninic nappes are at least 10 kbar less (Marquer *et al.* 1994; Rütti *et al.* 2003; Galli *et al.* 2007). Sediments derived from the North Penninic basin in the northeastern continuation of the Misox zone underwent a high-pressure–low-temperature metamorphic stage under blueschist facies conditions which is indicated by the frequent observation of Mg-carpholite (Oberhänsli *et al.* 1995; Goffe & Oberhänsli 1992; Bousquet *et al.* 2002). No unambiguous traces of eclogite facies metamorphism have been found in Misox zone *sensu stricto* immediately on top of the Adula nappe. However, it may well be that such a record has been erased by intense subsequent overprint at amphibolite facies conditions. The entire Lepontine nappe stack was affected by strong Barrovian metamorphism post-dating high-pressure assemblages in the Adula nappe. Isograds and mineral zone boundaries of this stage cross-cut nappe boundaries and describe a metamorphic dome which largely coincides with the structural dome. Metamorphic grade reaches high amphibolite facies conditions in the Southern Adula nappe immediately north of and within the Insubric mylonites.

The architecture of the Adula nappe

Circumference

Despite research over one-and-a-half centuries, the boundaries of the Adula nappe are still uncertain in some areas, particularly the lower contact to the

Simano nappe (Figs 3 and 4). The frontal termination of the Adula nappe is well defined by mixed gneisses of the Adula nappe bordering the particular calcareous schists of the Bündnerschiefer. The upper contact towards the Misox zone can be traced southwards to the Swiss–Italian border. There, the schists of the Misox zone are cut off by an east-dipping normal fault of Miocene age, the Forcola normal fault (Meyre et al. 1999a; Ciancaleoni & Marquer 2006). The Forcola normal fault separates the Adula and the Tambo nappe over a short distance and continues southeastward between the Adula nappe sensu stricto and the Gruf complex, which is generally considered as a part of the Adula nappe (e.g. Davidson et al. 1996). The Forcola normal fault finally runs into the 25–24 Ma old Novate intrusion where fault-associated deformation occurred under synmagmatic conditions (Ciancaleoni & Marquer 2006), thus giving a robust age constraint on the fault. The boundary between the Gruf complex and the Tambo nappe is readily traceable because of distinct rocks in the Tambo nappe and the Chiavenna ophiolite complex, a large sliver of oceanic basement derived from the North Penninic lithosphere (e.g. Schmutz 1976) (Fig. 3). Towards the southeast, the Suretta and Tambo nappes are cut out by the Bergell intrusion (Fig. 3). The contact between the Gruf complex and the magmatic rocks can be traced around the backfolds into the Insubric mylonites. Mesozoic sediments between the northern Adula nappe and the underlying Simano nappe soon disappear in a southward direction and the contact becomes all but clear since the lower Adula nappe and the upper Simano nappe both consist of monotonous biotite–plagioclase gneisses occasionally interlayered with amphibolites (Jenny et al. 1923; Partzsch 1998).

Sporadic layers of marbles have been interpreted as defining the nappe contact; however, marbles are also present within the lihologically diverse Adula nappe. Tracing the contact is further complicated by tight SE-facing folds affecting the entire nappe stack in the south. In Val Calanca, a distinct K-feldspar Augengneiss, the so-called Basalgneiss, has been proposed as to represent the lowest portion of the Adula nappe (Kündig 1926) but this interpretation is not indisputable. The Basalgneiss can be followed for some 30 km and defines the so-called Lostallo window in the lower Mesolcina valley (Kündig 1926; Nagel et al. 2002a) (Fig. 4). The area further southeast in Italy is lithologically less varied and without obvious marker horizons. The eastern border of the window has to be considered unknown. The Cima Lunga unit west of Valle Leventina is generally recognized as a part of the Adula nappe. There, the contact to the Simano nappe is relatively well defined by thick, massive orthogneisses in the Simano nappe underlying a heterogeneous sequence in the Adula nappe. This sequence, the Cima Lunga unit, contains some of the best studied of all high-pressure rocks. Within the Southern Steep Belt, the so-called Paina marble has been viewed as representing an early nappe-dividing syncline and thus to mark the upper contact of the Adula nappe (Kündig 1926; Fumasoli 1974; Heitzmann 1975). Accordingly, Schmid et al. (1996a) viewed the sequence south of the Paina marble, the zone of Bellinzona–Dascio, as a structurally higher unit and as the continuation of the North Penninic Misox zone. However, the diverse rock assemblage south of the Paina marble is identical to the one north of it and both are typical for the Adula nappe. The zone of Bellinzona–Dascio contains abundant eclogites, which recently yielded Tertiary metamorphic ages (Forster 1947; Brouwer et al. 2005). Figure 4 follows the suggestion that the zone of Bellinzona–Dascio and the Adula nappe are one and the same unit (Engi et al. 2001, 2004; Berger et al. 2005). This paper focuses on the Swiss side of the Adula nappe, which is much better investigated than the Italian side and petrologically more interesting. Critical information from further east, however, is included in the discussions.

Internal architecture

The internal architecture of the Adula nappe is very different from the deeper lower Penninic nappes (Figs 4 and 5). The underlying Simano and Leventina nappes consist largely of prealpine granitoid rocks affected only by little alpine deformation (Rütti et al. 2003). Only towards nappe contacts are these units mylonitized. In contrast, the entire Adula nappe displays a strong mylonitic foliation and is of extremely heterogeneous composition. It is built of various ortho- and paragneisses, frequently interlayered with amphibolites, metapelites, and marbles. Metapelites often contain boudins of more or less retrogressed eclogites typically between 1 and 100 m in size. Small and large bodies of ultramafics (up to a couple of km in the case of the Arami body) are scattered throughout the nappe. In the central part of the nappe, the tight foliation is almost unaffected by subsequent folding, creating a particular sliced appearance in the field (Fig. 6a). In early studies, some of the rocks have been viewed as to be of Mesozoic origin (e.g. Jenny et al. 1923) but so far all isotopic formation ages of zircon obtained in eclogites, ortho- and paragneisses have yielded pre-Mesozoic and even Proterozoic ages (e.g. Hänny et al. 1975; Santini 1992; Liati et al. 2005). The Adula nappe is generally considered as a lithospheric mélange assembled in a subduction channel (e.g. Trommsdorff 1990;

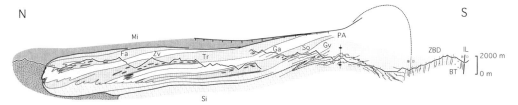

Fig. 5. Tectonic cross-sections through the Adula nappe, modified from Jenny *et al.* (1923) and Nagel *et al.* (2002*a*). BT, Bergell tonalite; Fa, Fanella lobe; Ga, Ganan lobe; Gv, Groveno lobe; IL, Insubric line; Mi, Misox zone; PA, Paglia antiform; Si, Simano nappe; So, Soazza lobe; Tr, Trescolmen lobe; ZBD, Zone of Bellinzona Dascio; Zv, Zervreila lobe.

Engi *et al.* 2001). However, especially the upper portion in Switzerland displays a well mappable architecture at a large scale. The Adula nappe is made up of several lobes that can be traced for tens of km (Figs 4 and 5). Although individual lobes have a heterogeneous interior, they are well separable (Jenny *et al.* 1923; Kündig 1926; Nagel *et al.* 2002*a*; Berger *et al.* 2005). The Zervreila and Ganan lobes consist dominantly of thick orthogneisses, whereas the Fanella, Trescolmen and Soazza lobes are rich in paragneisses, amphibolites and eclogite-bearing metapelites. These lobes occupy successively higher structural positions towards the north and thus describe a forward-dipping-duplex geometry (Figs 4 and 5).

Petrological record

The metamorphic record of the Adula nappe is certainly one of the best studied on Earth. The strong metamorphic field gradient and the collection of spectacular rocks including metamafics, metaultramafics and metapelites (Fig. 7) have attracted petrologists over decades. Dozens of high-pressure locations and several hundreds of mostly amphibolite facies metapelites are recorded in the literature with many of them investigated geochemically (e.g. Niggli & Niggli 1965; Klein 1976; Thompson 1976; Heinrich 1986; Löw 1987; Meyre & Puschnig 1993; Parztsch 1998; Engi *et al.* 1995; Todd *et al.* 1997; Nagel *et al.* 2002*b*; Dale & Holland 2003). As a result, mineral zones and metamorphic isograds are now mapped with unparalleled accuracy (Fig. 8).

Many studies come to the conclusion that the Adula nappe experienced a single pressure–temperature loop during alpine orogeny (Heinrich 1986; Löw 1987; Meyre *et al.* 1997; Nagel *et al.* 2002*a*; Dale & Holland 2003). Peak-pressure conditions increase from about 12 kbar/500–600 °C in the north to 25–30 kbar/750–850 °C in the south (e.g. Heinrich 1986; Dale & Holland 2003) (Fig. 9). For the garnet peridotite from Alpe Arami, peak pressure conditions are highly controversial and range between 30 kbar/850 °C (Nimis & Trommsdorff 2001) and >100 kbar (Dobrzhinetskaya *et al.* 1996). Mineral-zone boundaries and metamorphic isograds of the subsequent amphibolite facies stage also show southward-increasing temperatures but cross-cut the entire Lepontine nappe edifice (Niggli & Niggli 1965; Trommsdorff 1966; Thompson 1976). Thus, the Adula nappe got welded together with the underlying Simano nappe during decompression and the metamorphic field gradient during the Barrovian stage can be viewed as a result of rapid isothermal exhumation from southward-increasing high-pressure conditions. However, several recent studies associate the amphibolite facies metamorphism with a reheating event after exhumation and cooling (e.g. Toth *et al.* 2000; Brouwer *et al.* 2005; Brouwer & Engi 2005).

High-pressure conditions

High-pressure conditions are preserved in eclogites, garnet amphibolites, garnet peridotites and metapelites. The mentioned peak pressure conditions between 12 kbar/500–600 °C in the north and 25–30 kbar/750–850 °C in the south are mainly derived from eclogites (e.g. Heinrich 1986; Löw 1987; Meyre *et al.* 1997; Dale & Holland 2003) (Figs 8 and 9) but also from metapelites, i.e. garnet-mica-schists (Heinrich 1982; Meyre *et al.* 1999*b*; Nagel *et al.* 2002*a*). Several ultramafic bodies are distributed over the entire Adula nappe but garnet peridotites have been found only in the southern part of the nappe and in the Cima Lunga unit, namely at the localities Cima di Gagnone, Alpe Arami and Monte Duria (Fig. 8). There is robust microstructural evidence that at least the garnet peridotites at Cima di Gagnone evolved from a serpentinized body during prograde metamorphism (Evans & Trommsdorff 1978; Evans *et al.* 1979; Pfiffner & Trommsdorff 1998). Also at Monte Duria and at the famous Alpe Arami,

Fig. 6. Photographs from the Adula nappe. (**a**) View from Passo San Bernardino towards southwest. This photo shows a landscape in the upper lobes of the Adula nappe with the typical sliced appearance resulting from strong regional mylonitization in diverse rocks. (provided by Niko Froitzheim). (**b**) K-feldspar Augengneiss from the central Adula nappe (San Bernardino) with Zapport-phase foliation. Sigma clast indicates left-lateral, top-to-the-north shearing. (**c**) Stretching lineation in orthogneisses associated with the Leis phase (from the frontal anticline of the Adula nappe near Vals). (**d**) Typical Claro-phase fold in metapelites from the southern Adula nappe (Pizzo Claro area, view is towards SE). (**e**) Superposition of Claro- and Zapport-phase-related folds in the southern Adula nappe (Auriglia gorge/Val Calanca) (from Nagel *et al.* 2002*a*). (**f**) Cressim-phase-related stretching lineation in garnet-mica schists from the southern Adula nappe (upper Val Grono/Mesolcina). Pressure shadows of garnets consist of quartz and sillimanite. (From Nagel *et al.* 2002*a*.)

body subduction from near-surface conditions was inferred (e.g. Trommsdorff *et al.* 2000; Nimis & Trommsdorff 2001) but the observations are less striking. Garnet peridotites yielded the highest peak-pressure conditions in the Adula nappe— around 30 kbar/850 °C (Nimis & Trommsdorff 2001) and, in the case of the Alpe Arami peridotite, even 50 kbar/1120 °C (Brenker & Brey 1997) and 59 kbar/1180 °C (Paquin & Altherr 2001) based on aluminium exchange between orthopyroxene and garnet and chromium exchange between clinopyroxene and garnet in association with various thermometers. Pressure estimations in excess of 100 kbar have been inferred from exsolved ilmentite rods in olivine (Dobrzhinetskaya *et al.* 1996) and clinoenstatite in diopside (Bozhilov *et al.* 1999). However, data and interpretation of these studies have been severely challenged (e.g. Arlt *et al.* 2000; Hacker *et al.* 1997; Trommsdorff *et al.* 2000; Risold *et al.* 2001; Nimis & Trommsdorff 2001).

The most common rocks preserving high-pressure conditions are eclogites with various degrees of retrogressive overprint. They are present throughout the higher lobes of the Adula nappe and in the vicinity of the Southern Steep Belt (Figs 4 and 8). Fresh eclogites are preserved in cores of isolated mafic boudins, metres to tens of metres in size and often embedded in metapelites. Towards the boudin rims, eclogite is typically transformed into garnet amphibolite. Preserved peak-pressure assemblages display garnet + omphacite ± amphibole ± phengite ± paragonite ± zoisite ± kyanite ± quartz and a few more accessory minerals such as rutile. They often show some degree of retrogression, usually small domains of intergrowing barroisitic to pargasitic hornblende and plagioclase (Fig. 7d). Pristine, coarse-grained eclogite can be found at several locations, mainly in the central portion of the nappe (Heinrich 1986; Löw 1987; Meyre *et al.* 1997; Dale & Holland 2003). The mineralogical and chemical variation of mafic high-pressure assemblages throughout the Adula nappe is limited but significant. In the north, eclogites contain glaucophane in peak-pressure assemblages (van der Plas 1959; Heinrich 1986, Löw 1987; Dale & Holland 2003). Kyanite + omphacite-bearing assemblages in the central and southern portion of the nappe (Heinrich 1986) indicate that peak pressures exceed the upper stability limit of paragonite around 20 kbar. Metamorphic conditions for specific samples have been identified using thermobarometry and petrogenetic grids (e.g. Heinrich 1986; Meyre *et al.* 1997; Toth *et al.* 2000; Dale & Holland 2003; Brouwer & Engi 2005). Overall southward-increasing peak-pressure conditions have been inferred in two nappe-scale studies (Heinrich 1986; Dale & Holland 2003) and follow from compiling local findings. The results of several studies agree reasonably well (Fig. 9). Noteworthy is the difference in obtained peak-pressure conditions for the northernmost Adula nappe, 12 kbar/500 °C according to Löw (1987) or Heinrich (1986) and 17 kbar/580 °C (with subsequent temperatures of 620 °C at 15 kbar) according to Dale & Holland (2003).

The conditions proposed by Löw and Heinrich do not differ significantly from peak-pressure conditions found in the surrounding nappes (Marquer *et al.* 1994; Rütti *et al.* 2003; Galli *et al.* 2007) whereas the conditions of Dale & Holland (2003) do. Eclogite facies conditions are occasionally preserved in metapelites that show the assemblage quartz + silica-rich phengite + garnet + paragonite ± kyanite and derived peak-pressure conditions accord with the findings in eclogites (Heinrich 1982; Meyre *et al.* 1999b; Nagel *et al.* 2002b). Orthogneisses display silica-rich phengite in biotite-free assemblages, especially in the north (Löw 1987; Meyre *et al.* 1997; Nagel *et al.* 2002). They do not, however, show the high-pressure assemblages predicted by equilibrium thermodynamics for the conditions recorded in the eclogites, notably no formation of sodium-bearing pyroxene or coesite (so far, coesite has not been found at all in the Adula nappe). It is still debated whether the acidic gneisses experienced the same peak pressures as the above-described high-pressure rocks or not. If not, the Adula nappe would constitute a mélange that was assembled after peak-pressure conditions in individual samples were reached and several petrologists actually favour this scenario (Toth *et al.* 2000; Engi *et al.* 2001; Brouwer *et al.* 2005). Recently, this hypothesis got additional support from isotopic data (Liati 2005; Hermann *et al.* 2006). I will return to this problem after reviewing structures and isotopic ages.

Barrovian conditions

Metamorphic grade at intermediate to lower pressures increases from upper-greenschist facies conditions in the northern Adula nappe to high-amphibolite facies conditions associated with migmatization in the south (Wenk 1970; Niggli & Niggli 1965) (Fig. 8). In the Gruf unit in the south-easternmost Adula nappe, even granulite facies conditions are recorded in sapphirine-bearing assemblages (Droop & Bucher-Nurminen 1984). The most complete record of the Barrovian history is preserved in metapelites, which are very frequent throughout the Fanella, Trescolmen and Soazza lobes in the middle and northern Adula nappe (Fig. 4) and is identifiable in the deeper portions and in the Southern Steep Belt. In these rocks, the metamorphic field gradient is expressed through distinct mineral zone boundaries, i.e. the progressive appearance of aluminium-rich index minerals (staurolite, kyanite and sillimanite) and the disappearance of paragonite. Early studies have mapped a single kyanite–staurolite mineral-zone boundary and a sillimanite mineral-zone boundary (e.g. Niggli & Niggli 1965). However, kyanite and rarely also staurolite occur also north of the classic mineral zone boundary in extremely aluminium-rich rocks (Klein 1976; Löw 1987; Nagel 2002b). The proposed boundary is defined by a sudden appearance of cm-sized staurolite intergrown with plagioclase in sodium-rich samples (Figs 7b and 8).

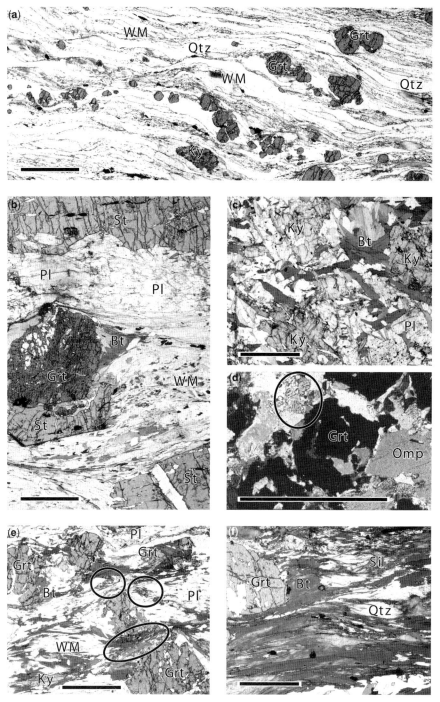

Fig. 7. Photomicrographs of samples from the Adula nappe. Scale bar is 1 mm in all photos. (**a**) Aluminium-rich metapelite with Zapport-phase foliation. Shear bands indicate left-lateral, top-to-the-north shear sense. The synkinematic assemblage is quartz + garnet + white mica + kyanite with small amounts of biotite in the form of tiny crystals replacing phengite. (**b**) Sodium–aluminium-rich metapelite from the central Adula nappe south of the staurolite/kyanite mineral zone boundary (Calvarese, Val Calanca). Sample contains quartz, garnet, some biotite,

Further south, already in the northern sillimanite zone, kyanite becomes the most abundant aluminium-rich phase (Figs 7c and 8). Only in the southern sillimanite zone does the amount of sillimanite exceed that of kyanite (Fig. 7f). Amphibolite facies mineral growth in the Lepontine has traditionally been viewed as to record a late orogenic thermal event (Wenk 1970; Thompson 1976; Todd & Engi 1997; Roselle et al. 2002; Brouwer et al. 2004). Nagel et al. (2002b) have mapped four zones of different staurolite appearance in the Adula nappe (Fig. 8) and inferred that southward-increasing temperatures during isothermal decompression would account for the observed variations. They attributed the growths of all aluminium-rich index minerals to the continuous decomposition of garnet and white mica with the classic kyanite–staurolite mineral zone boundary being related to the breakdown of paragonite at about 8 kbar/650 °C (Fig. 7b). From north to south, there is a gradual but manifest increase of biotite and plagioclase at the expense of garnet and white mica in metapelites (Fig. 7a–c, e, f). Hence, in addition to increasing amphibolite facies conditions towards the south, there is also an increasing degree of post-peak-pressure overprint. North of the classic staurolite/kyanite mineral zone boundary (Fig. 8), biotite–plagioclase-free high-pressure assemblages are often preserved and phengites usually show high silica contents (Heinrich 1982; Löw 1987; Meyre et al. 1999b). Further south, garnet and white mica become progressively corroded and in the southernmost Adula nappe metapelites sometimes display the assemblage quartz + biotite + feldspar ± kyanite/sillimanite with phengite and garnet being preserved only as inclusions in plagioclase and kyanite (Fig. 7c). The overall better record of high-pressure conditions in metamafics is probably due to the fact that reactions leading to the transition from eclogite-facies to amphibolite-facies assemblages in metapelites release water. In mafic rocks, these reactions consume water. The most important water-releasing reaction in metapelites is paragonite breakdown which is also documented by abundant alumosilicate-bearing quartz segregations south of the staurolite–kyanite mineral-zone boundary (Klein 1976).

Engi et al. (1995) and Todd & Engi (1997) have performed a regional study of amphibolite facies metapelites in the Lepontine area that included many samples from the Adula nappe. They used the results of multi-equilibrium thermobarometry in metapelites to calculate contour lines of pressure and temperature during an assumed Barrovian equilibration event (Fig. 8). In the Adula nappe, these maps show a continuous temperature rise from 500 °C in the north to >650 °C in the south. Isobars display a slight hump in the central part of the nappe but altogether equilibration pressures are around 5.5 kbar. The authors interpret the pressure decrease in the southeasternmost part as to reflect younger equilibration. Metamorphic conditions obtained by thermobarometry and displayed in the pressure–temperature isograds in Figure 8 are reasonably well located on pressure–temperature paths proposed for the respective area (Fig. 9). However, they do not appear to reflect peak temperatures, nor do they represent conditions of staurolite/kyanite growth in the central portion of the nappe which took place at higher pressures. Engi et al. (1995) and Todd & Engi (1997) have attributed formation of isotherms and isobars to reheating after exhumation, a view supported by phase relations in garnet amphibolites from the Southern Steep Belt (Toth et al. 2000; Brouwer et al. 2005; Brouwer & Engi 2005) and thermomechanical models of subduction and exhumation (Roselle et al. 2002; Brouwer et al. 2004; Brouwer & Engi 2005). However, in the Adula nappe, biotite and plagioclase in metapelites clearly formed at the expense of garnet which represents a corroded relict from the peak-pressure assemblage (Fig. 7b, c, e). Garnets are of paramount importance for thermobarometry and in view of their typically corroded appearance, the consistency of thermobarometric results obtained for the amphibolite facies stage is surprising.

All records of amphibolite facies conditions, mineral zone boundaries as well as thermobarometric isograds, are continuous across Lepontine nappe boundaries (Fig. 8). Staurolite and kyanite in the upper Simano nappe including the famous occurrences at Alpe Sponda (Irouschek 1983) result from the same paragonite–garnet-consuming reactions as

Fig. 7. (Continued). and cm-sized crystals of staurolite and plagioclase overgrowing white mica. (**c**) Metapelite from the southern Simano nappe (Pian di Renten/Val Calanca). Sample contains biotite, kyanite, plagioclase and small amounts of quartz. Garnet and white mica are preserved as inclusions in kyanite and plagioclase. (**d**) Eclogite from the southern Adula nappe (Upper Val Grono/Mesolcina). Main assemblage is garnet and omphazite. Circles denote domains of symplectitic plagioclase and pargasitic amphibole replacing omphazite. (**e**) Metapelite with Claro-phase foliation. Shear bands indicate left-lateral, top-to-the-southeast sense of shear. Synkinematic assemblage is garnet, white mica, plagioclase, biotite, and kyanite. Circles denote intergrown biotite and kyanite replacing white mica and garnet in shear bands. (**f**) Metapelite cut parallel to Cressim-phase stretching lineation from the southern Adula nappe (Upper Val Grono/Mesolcina). Left side is to the east. Synkinematic assemblage is garnet + biotite + plagioclase + sillimanite with some relicts of white mica and kyanite. Garnets in the section are irregularly shaped, small and large fragments.

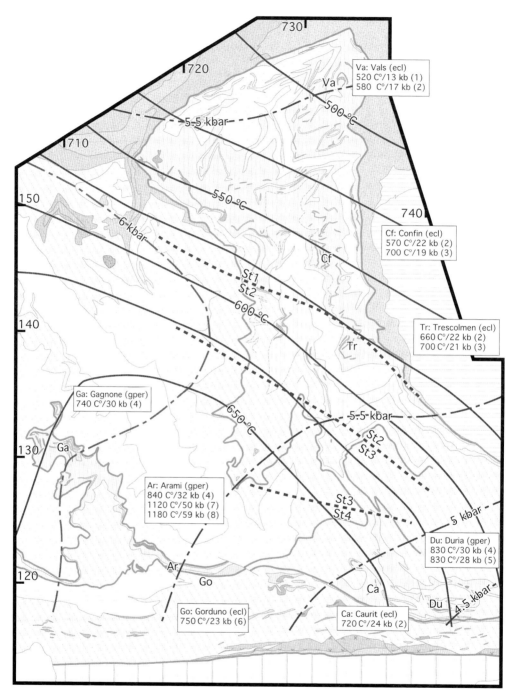

Fig. 8. Tectonic map shown in Fig. 4 with additional petrological data. Contour lines of temperature (solid lines) and pressure (widely dashed) are from Todd & Engi (1997) and derived from many thermobarometric results in amphibolite facies metapelites. Narrow-dashed lines indicate boundaries of four staurolite zones (st1-st4) of Nagel *et al.* (2002*b*). The classic staurolite/kyanite mineral zone boundary (e.g. Niggli & Niggli 1965) corresponds to the st1/st2 boundary. Frames denote peak pressure conditions inferred from eclogites (ecl) or garnet peridotites (gper) at well-investigated locations (Va, Vals; Cf, Alp de Confin; Tr, Alp de Trescolmen; Du, Piz Duria; Ca, Alp de Caurit; Go, Gorduno; Ar, Alpe Arami; Ga, Cima di Gagnone). Numbers indicate authors: (1) Löw 1987; (2) Dale & Holland 2003; (3) Meyre *et al.* 1997; (4) Nimis & Trommsdorff 2001; (5) Hermann *et al.* 2006; (6) Toth *et al.* 2000; (7) Brenker & Brey 1997; (8) Paquin & Altherr 2001.

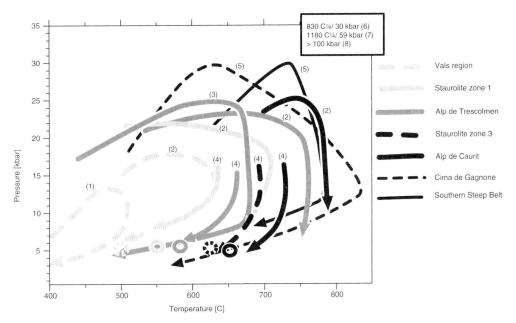

Fig. 9. Pressure–temperature loops for different localities and from different studies. Line styles indicate sample locations. Numbers denote authors: (1) Löw 1987; (2) Dale & Holland 2003; (3) Meyre et al. 1997; (4) Nagel et al. 2002b; (5) Brouwer et al. 2005. Frame displays peak-pressure conditions obtained for the garnet peridotite at Alpe Arami: (6), Nimis & Trommsdorff 2001; (7) Paquin & Altherr 2001; (8) Dobrzhinetskaya et al. 1996. Circles indicate thermobarometrically derived metamorphic conditions for the location defined by the line style (Todd & Engi 1997). Locations and staurolite zones are indicated in Fig. 8.

in the Adula nappe (Nagel et al. 2002b). Hence, the two nappes were welded together at pressures in excess of 8 kbar, which is close to peak pressures in the Simano nappe around 10–12 kbar (Rütti et al. 2003). This proposition is confirmed by structural observations, further discussed in the next paragraph. Isoclinal folds formed at conditions between 7–9 kbar/650–750 °C in the southern Adula nappe (Nagel et al. 2002a; Rüti et al. 2003) overprint the established nappe contact.

Structural record

The oldest pervasive and altogether dominant deformation event in the Adula nappe is usually referred to as the Zapport phase (Löw 1987; Meyre & Puschnig 1993; Partzsch 1998; Nagel et al. 2002a; Pleuger et al. 2003). Structures related to the Zapport phase affect all of the Adula nappe and cause the typical regional foliation. The Zapport phase is associated with isoclinal to dismembered folds, and a tight, more or less north–south-striking stretching lineation parallel to fold axes (Fig. 10). Shear sense indicators consistently denote a top-to-the-north sense of motion (Figs 6b and 7a). The Zapport foliation displays a uniform NW-dipping orientation in the central portion of the Adula nappe (Fig. 10), as this area is only weakly overprinted by subsequent deformation. Zapport structures are superposed by at least two generations of folds along the frontal antiform in the north and in the vicinity of the Southern Steep Belt. Deformation events in the north are called Leis phase and Carassino phase (Löw 1987; Partzsch 1998). In the south, the main backfolding phase is generally referred to as the Cressim phase and I propose the name Claro phase for another, well-documented generation of folds pre-dating the Cressim phase (Table 1). Correlations between the post-Zapport folding events in the north and in the south are ambiguous and maybe neither phase has a corresponding event on the other side of the nappe.

Pervasive deformation associated with the Leis phase affects and, according to Löw (1987), actually forms the frontal antiform in the Adula nappe and the enveloping middle Penninic sediments. As for the other Lepontine nappes, the present northern termination of the Adula nappe seems to be related to post-nappe folding rather than to represent the original front of a thrust sheet (Milnes 1974; Ayrton & Ramsay 1974; Maxelon & Mancktelow 2005). Along the northern front of the Adula

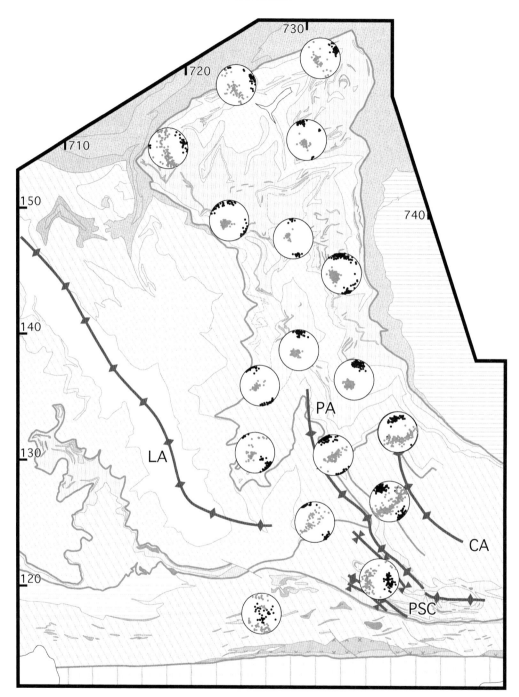

Fig. 10. Tectonic map shown in Fig. 4 with additional structural data. Grey and black dots indicate poles of main foliation and stretching lineation, respectively. Data from Löw (1987), Partzsch (1998), and Nagel *et al.* (2002*a*). Traces of antiforms are for major Cressim-phase folds (LA, Leventina antiform; PA, Paglia antiform; CA, Cressim antiform) according to Maxelon & Mancktelow (2005) and Nagel *et al.* (2002*a*); PSC, Paglia Schlingenkomplex.

Table 1. *Correlation of local deformation phases observed in the Alps. Deformation phases are taken from Pleuger et al. (2008) (Monte Rosa nappe), Maxelon & Mancktelow (2005) (Central Lepontine), Löw (1987) (Northern Adula nappe), Nagel et al. (2002a) (Southern Adula nappe), and Marquer et al. 1996 (Tambo nappe and higher)*

Monte Rosa nappe	Central Lepontine	Northern Adula nappe	Southern Adula nappe	Tambo nappe and higher units	Tectonics	Age
Mattmark	D1	Zapport	Zapport	D1	Top-N shearing	<40 Ma
Malfatta		↑			Orogen-parallel and orogen-oblique stretching	<40 Ma
Mischabel	D2	Leis?	Claro	D2	(Niemet–Beverin)	>30 Ma
Vanzone & Olino	D3	↓	Cressim	D3	Backfolding/backthrusting	30–25 Ma
	D4	Carassino		D4	Open large-scale folding	<25 Ma

nappe, the Leis phase is associated with tight to isoclinal, WSW–ENE-striking folds, a new axial plane foliation and a pronounced new stretching lineation parallel to the fold axes (Löw 1987) (Fig. 6c). The strong overprint during the Leis phase causes the marked change in orientation of the dominant stretching lineation in the northern Adula nappe (Fig. 10). Towards the south, intensity of Leis deformation rapidly fades and in the central portion of the nappe, the Leis phase is represented only by occasional, north-facing folds (Partzsch 1998; Pleuger *et al.* 2003). The Carassino phase is expressed in a couple of large-scale, east–west-striking folds, the Carassino antiform and the Alpettas synform (Löw 1987). It is not associated with penetrative structures except for some local crenulation (Löw 1987).

Two generations of intense deformation (the Claro and the subsequent Cressim phase) are found in the southern Adula nappe and the underlying Simano nappe (Grond *et al.* 1995; Grujic & Mancktelow 1996; Nagel *et al.* 2002*a*). The deformation front of the Claro phase is located some 10 km north of the Southern Steep Belt. Tight to isoclinal, NW–SW-striking and SW-facing folds (Fig. 6d) affect the entire area. Because of the pervasive presence of these folds, the contact to the deeper Simano nappe must be affected by the Claro phase despite the uncertain location. West of Valle Leventina, the better-defined contact between the Cima Lunga unit and the underlying orthogneisses of the Simano nappe is folded in the same style (Grond *et al.* 1995; Pfiffner & Trommsdorff 1998; Grujic and Mancktelow 1996). Claro-phase-related folds display a well developed, gently dipping axial plane foliation, a stretching lineation parallel to fold axes and are associated with top-to-the-SE shearing (Grond *et al.* 1995; Nagel *et al.* 2002*a*). Folds of the younger Cressim phase cause the transition from the gently dipping nappe pile to the north into the Southern Steep Belt. Large-scale examples for Cressim phase backfolds are the Cressim antiform (Hänny 1972) and the Paglia antiform, which can be traced over 25 km (Bruggmann 1965; Nagel *et al.* 2002*a*) (Fig. 10). Immediately north of the Southern Steep Belt, these folds strike east–west. They usually do not show a new axial plane foliation, but a well-defined, east-dipping stretching lineation where Cressim-phase-related deformation is intense (Fig. 6f). Again, the stretching lineation is parallel to the fold axes.

Towards the south, Cressim-phase fold axes and associated stretching lineations become progressively steeper and apparently translate continuously into a subvertical orientation in the Southern Steep Belt (Fig. 10). Towards the north, fold axes turn in an anticlockwise way into a NNW–SSE-striking orientation defining an *en-echelon*-type array of folds (Fig. 10). In the Adula nappe, this change of orientation is associated with a manifest reduction of strain intensity. Folds become open, and finally die out 10–20 km north of the Southern Steep Belt. Further west, in the deeper Lepontine nappes, backfolds of the same generation undergo the same anticlockwise turn and are traceable through the entire Lepontine dome. Prominent examples are the Maggia cross fold or the Leventina antiform (Preiswerk *et al.* 1934; Grujic & Mancktelow 1996; Steck 1998; Berger *et al.* 2005; Maxelon & Mancktelow 2005) (Fig. 3). Superposition of large-scale isoclinal folds of the Claro and Cressim phase cause the so-called Paglia–Schlingenkomplex in the Swiss–Italian border region around Piz Duria (Bruggmann 1965; Fumasoli 1974; Nagel *et al.* 2002*a*) (Figs 10 and 11). Folds of all three deformation phases (Zapport, Claro and Cressim phase) are coaxial, which allows to map a dominant stretching lineation, although displayed lineations belong to

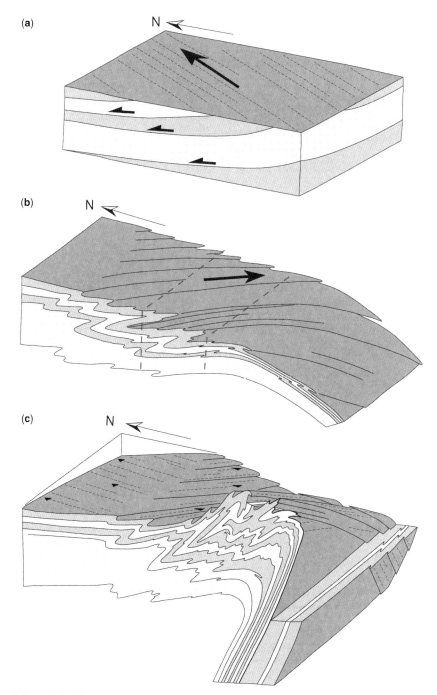

Fig. 11. Schematic sketches illustrating successive deformation in the Southern Adula nappe. (**a**) After Zapport phase. (**b**) After Claro phase. Dashed lines indicate orientation of subsequently forming Cressim-phase antiforms. (**c**) After Cressim phase.

different deformation phases (Wenk 1955; Nagel et al. 2002a; Maxelon & Mancktelow 2005). The existence of three independent deformation phases is, however, clearly documented by type 3 fold-interference patterns of Ramsay & Huber (1987) (Fig 6e), a reversal of the dominant shear-direction towards south and differing synkinematic metamorphic conditions (see below).

Most rocks in the Southern Steep Belt show a tight mylonitization, a more or less vertical stretching lineation, and a north-block-up sense of motion. Many studies acknowledge that these mylonites are related to backthrusting during the Cressim phase and that Cressim-phase backfolds would actually represent drag folds (e.g. Pfiffner et al. 2000). In contrast, the lack of a new axial plane foliation in Cressim-phase backfolds has guided several authors to the conclusion that these folds would refold a pre-existing foliation into the subvertical orientation in the Southern Steep Belt and that consequently the mylonitization would be older than the backfolds (Milnes 1978; Nagel et al. 2002a; Maxelon & Mancktelow 2005). The asymmetry of small-scale folds in the Southern Steep Belt with fold axes parallel to the steep stretching lineation typically indicate a synform in the NE, not consistent with the overall Cressim-phase antiformal structure (Maxelon & Mancktelow 2005). Shear-sense and facing direction are, however, in line with the mylonitization belonging to the Claro phase (Fig. 11). Therefore, the Southern Steep Belt could also represent a Claro-phase-related, south-dipping shear zone with a normal sense of motion that got refolded during the Cressim phase (Nagel et al. 2002a; Maxelon & Mancktelow 2005). The matter is not resolved. Petrological evidence presented in the next paragraph, as well as the continuity of stretching lineations in the Southern Adula nappe, suggest that the mylonites in the Southern Steep Belt and associated exhumation of the northern block are mainly related to the Cressim phase.

Deformation phases and metamorphic conditions

All structures described above post-date peak-pressure conditions and thus all of the deformation events account in some way for the observed decompression. In this paragraph, I summarize pressure–temperature ranges proposed for each deformation event and derive a couple of consequences arising from that data.

Metamorphic conditions during the Zapport phase can best be studied north of the staurolite/kyanite mineral zone boundary. South of it, intense overprint at amphibolite facies conditions obstructs the study of Zapport-related assemblages. Formation of the Zapport foliation is associated with retrogression of unfoliated eclogites to foliated garnet amphibolites along the rims of mafic boudins. These boudins are typically elongated in the Zapport foliation suggesting that boudinage occurred during the Zapport phase. Hence, the Zapport phase is the oldest deformation event post-dating peak-pressure conditions. In paragneisses, the Zapport phase is associated with decompressional growth of biotite and plagioclase at the expense of phengite and garnet. Aluminum-rich metapelites in the north occasionally preserve a synkinematic assemblage quartz + garnet + Si-rich phengite + kyanite (e.g. Löw 1987) (Fig. 7a) indicating deformation at elevated pressures above the stability range of biotite and plagioclase in these rocks. Cores of phengites aligned in the Zapport foliation can be extremely silica-rich (suggesting pressures up to 20 kbar in the area of Trescolmen) while rims and recrystallized fine-grained phengite display intermediate silica contents (Meyre et al. 1999b; Nagel et al. 2002a). The formation of typical amphibolite-facies assemblages including staurolite growths and complete paragonite breakdown clearly post-dates the Zapport phase.

The above-mentioned studies come to the conclusion that Zapport deformation started close to peak-pressure conditions and was associated with considerable exhumation. The amount of exhumation related to the Zapport phase increases in a southward direction. During this phase, pressures dropped from 13–9 kbars in the Northern Adula nappe (Löw 1987) whereas pressures decreased from 20–10 kbars in the central portion of the nappe (Meyre et al. 1999b; Nagel et al. 2002a). Thus, the amount of exhumation increases towards deeper structural levels in the south and the pressure gradient associated with peak-pressure conditions gets largely obliterated during the Zapport phase (Pleuger et al. 2003). According to Löw (1987), Leis deformation in the north occurred in a relatively narrow pressure interval at around 8 kbar/500 °C whereas the Carassino phase took place under greenschist facies conditions. In the south, metamorphic assemblages and quartz microstructures in metapelites were used to infer conditions around 650–720 °C/8–9 kbar during the Claro phase (Nagel et al. 2002a). Synkinematic aluminium release during the decomposition of garnet and white mica in metapelites consistently produced kyanite (Fig. 7e) indicating the relatively high pressures. Staurolite and sillimanite if present are post-kinematic with respect to the Claro phase. In contrast, the alumosilicate formed during Cressim-phase backfolding in the Paglia–Schlingenkomplex is always sillimanite (Figs 6f and 7f) and synkinematic metamorphic conditions are around 6–7 kbar/680–740 °C (Nagel et al.

2002a). Also in the Southern Steep Belt, mylonites contain abundant sillimanite. In addition, frequent synkinematic migmatization and K-feldspar formation as a result of white-mica breakdown can be observed (Burri et al. 2005). Pressures during mylonitization are in the same range as for the Cressim-phase backfolding but certainly lower than for the Claro phase, i.e. 6–7 kbar at temperatures of around 700 °C (Burri et al. 2005). The above-summarized metamorphic conditions thus support the view that mylonitization in the Southern Steep Belt largely post-dates top-to-the-SE shearing during the Claro phase. In the Southern Steep Belt, high-temperature mylonites are consistently associated with subvertical stretching lineations. Structures associated with dextral strike-slip subsequently developed at greenschist facies conditions and show subhorizontal slip vectors (e.g. Heitzmann 1975, 1987; Schmid et al. 1987, 1989).

Regional context of deformation events and isotopic ages

Over the past two decades, the rapid development of techniques to date high-pressure conditions had profound impact on models of the central Alps. Dating garnet (Becker et al. 1993) and metamorphic domains in zircon (Gebauer et al. 1992; Gebauer 1996) led to the recognition that peak pressures in the Cima Lunga unit are Eocene in age and not Upper Cretaceous to Palaeocene as had been assumed before (e.g. Hunziker et al. 1989; Ring 1992; Steck & Hunziker 1994; Stampfli et al. 1998). SHRIMP dating of zircon rims in garnet peridotites and eclogites yielded ages between 43 and 35 Ma (Gebauer et al. 1992; Gebauer 1996) and the authors favour the younger portion of this range as to represent peak-pressure conditions. For the same rock types, Becker (1993) obtained ages around 40 Ma using a garnet–clinopyroxene–whole-rock Sm–Nd isochron. He interprets these as cooling ages close to peak pressure conditions. Lately, Brouwer et al. (2005) found Lu–Hf garnet ages of 36–38 Ma in three of four eclogite samples from the Southern Steep Bank. A fourth sample yielded an age of 47.5–70 Ma and all four ages are interpreted as prograde growth ages close to peak pressure conditions. Brouwer et al. (2005) view the dissenting age as to support their interpretation of the Southern Steep Belt and the Adula nappe as a mélange of rocks with very different Alpine histories. So far, high-pressure rocks from the main body of the Adula nappe have not yielded Eocene ages. In an eclogite sample from Trescolmen, Liati et al. (2005) found two generations of metamorphic rims in zircon yielding ages of 371 ± 8 Ma and 33.2 ± 1.1, respectively. The authors propose eclogite facies conditions for the Palaeozoic generation and amphibolite facies conditions for the Oligocene one. They conclude that the studied rocks probably did not experience Alpine high-pressure conditions at all. Hermann et al. (2006) presented metamorphic zircon ages of 34.2 ± 0.2 and 32.9 ± 0.3 Ma obtained in garnet peridotite from Monte Duria and viewed these ages as indicating exhumation from eclogite facies to amphibolite facies conditions.

Forty years ago, Emilie Jäger performed a series of classic studies. For white micas sampled north of the staurolite/kyanite mineral zone boundary she obtained Rb–Sr ages of 38–35 Ma and interpreted these as formation ages when compared to younger cooling ages further south (Jäger et al. 1967; Jäger 1970, 1973; Hurfort et al. 1989). For decades, this so-called Bern age was attributed to the amphibolite facies stage (although Jäger emphasizes white mica formation). However, in all the Lepontine and especially in the Adula nappe, white mica is most abundant in peak-pressure assemblages. White mica in gneisses of the northern Adula is typically silica-rich.

Amphibolite facies minerals such as biotite, staurolite or kyanite grow during partial decomposition of white mica (Fig. 7b, c, e). White mica in amphibolite-facies assemblages often displays silica-rich cores surrounded by silica-poorer, i.e. re-equilibrated, rims suggesting growth of the crystal at higher pressures. The 38 Ma of Emilie Jäger is perfectly sensible if viewed as representing peak pressures or recrysallization associated with early exhumation during the Zapport phase. Dating further deformation events or any particular point on the decompressive pressure–temperature path is a challenging task. Since temperatures remained essentially constant during exhumation, cooling ages will not help. So far, the best time constraints may be provided by the spatial relationship between particular deformation phases and Alpine intrusives in combination with conventional and robust dating of these magmatic rocks. This approach has been quite successful in the southern Adula nappe, where two generations of intrusive rocks can be put into relation to the observed deformation events. These intrusions are the 32–30 Ma old Bergell pluton (von Blanckenburg 1992) and the 24–25 Ma Novate intrusion (Liati et al. 2000), which is coeval with acidic dykes in the Southern Steep Belt (Gebauer 1999). Most authors correlate the Claro phase in the lower Adula and Simano nappe with the so-called Niemet–Beverin phase in higher structural levels north of the Bergell intrusion further east (Grond et al. 1995; Grujic & Mancktelow 1996; Nagel et al. 2002a; Maxelon & Mancktelow 2005; Pleuger et al. 2008)

(Table 1). The Niemet–Beverin phase displays the same geometry as the Claro phase as it is associated with significant top-to-the-SE extensional shearing, namely along the Turba mylonite at the base of the Austroalpine nappe stack (Nievergelt *et al.* 1996) and in the upper limb of the Niemet–Beverin fold in the Middle-Penninic Suretta nappe (Schreurs 1993; Baudin *et al.* 1993; Marquer *et al.* 1996) (Fig. 3).

Top-to-the-SE shearing south of the Turba mylonite affects the 32 Ma-old tonalite of the Bergell pluton but is cross-cut by the slightly younger granodiorite (Nievergelt *et al.* 1996). Accordingly, deformation associated with the Claro phase would have ceased between 32 and 30 Ma. Subsequent backfolding and backthrusting during the Cressim phase affects the base of the Bergell pluton under synmagmatic conditions (Rosenberg *et al.* 1995; Davidson *et al.* 1996). Also, the tonalitic tail of the intrusion extending into the Southern Steep Belt is affected by the main mylonitization event (Vogler & Voll 1981; Berger *et al.* 1996). The 25–24 Ma magmatic events, however, appear to cross-cut (or are only very little affected by) Cressim-phase-related folding and shearing in the Southern Steep Belt. Thus, the Cressim-phase mainly occurred between 30 Ma and 25 Ma (Berger *et al.* 1996).

In the north, integrating the Leis phase into the regional context is much less conclusive. Löw (1987) connects the Leis-related frontal antiforms of the Adula nappe with the so-called Lunschania antiform in the North Penninic sediments. This fold can be traced into the Prättigau half-window in northern Graubünden, where it post-dates top-to-the-SE shearing of the Niemet–Beverin phase (Weh & Froitzheim 2001). Accordingly, the Leis phase would post-date the Claro phase (which corresponds to the Niemet–Beverin phase). Several studies correlate the Leis phase with the Cressim phase in the south (Schmid *et al.* 1996*a*; Partzsch 1998; Nagel *et al.* 2002*a*). There are two observations putting this view into question. First, WSW–ENE-oriented folds and stretching lineations of the Leis phase strike almost perpendicular to Cressim-related, large-scale folds further west such as the Leventina antiform or the Maggia cross-fold and, secondly, Löw (1987) inferred elevated pressures around 8 kbar to be associated with the Leis phase. Based on geometric similarities, Pleuger *et al.* (2008) correlate the Leis phase in the Adula nappe with the so-called Malfatta phase in the Monte-Rosa nappe (Pleuger *et al.* 2005, 2007), which would make it older than the Niemet–Beverin phase. At present, age and tectonic environment of the Leis phase have to be considered unknown. The Carassino phase is probably related to the formation of the so-called Northern Steep Belt south of the Gotthard massif further west and thus would be younger than 25 Ma and post-date the Cressim phase in the south (Maxelon & Mancktelow 2005).

Discussion and open questions

In this last paragraph, I will present what in my view are the main open problems associated with the above-reviewed information:

1. Are there one or two Tertiary subduction zones recorded in the Central Alps?
2. What is the relative age of mélange assembling—before or after peak-pressure conditions? What is the absolute age of the high-pressure conditions in the main Adula nappe?
3. What is the significance of the Zapport phase? How does its strain field relate to the exhumation from high-pressure conditions?
4. What is the tectonic scenario of top-to-the-SE-shearing during the Claro phase in the central Alps? What is the contribution of orogen-parallel to orogen-oblique stretching to the exhumation of the Lepontine dome? What causes this extension?

One versus two subduction zones

Most studies explain the Tertiary edifice of the Alps through a single subduction zone dipping southeastward beneath the Apulian margin (e.g. Schmid *et al.* 1996*a*, 2004; Handy & Oberhänsli 2004). A second subduction zone north of the Briançonnais continent was first proposed by Frisch (1979, 1981) and again favoured by Froitzheim *et al.* (2003) (Fig. 12) in order to explain exhumation from eclogite facies conditions. In this scenario, the Adula nappe gets partly exhumed from peak pressure conditions by ongoing subduction of the structurally higher slab after that plate (the Briançonnais Plate) is completely consumed at the surface (Fig. 12c, d). This problem does not represent a problem particularly associated with the record in the Adula nappe. It certainly requires treatment beyond the scale of the central Alps alone.

Assemblage of the Adula nappe

Recent petrological and geochemical studies have suggested that the Adula nappe would represent a mélange accreted in a subduction channel after peak pressures in individual components were attained (e.g. Engi *et al.* 2001; Brouwer *et al.* 2005; Hermann *et al.* 2006) and some authors even doubt any Alpine eclogite facies conditions in the main body of the nappe. The main arguments supporting this conception are: (1) varying Alpine ages and pressure–temperature paths inferred for

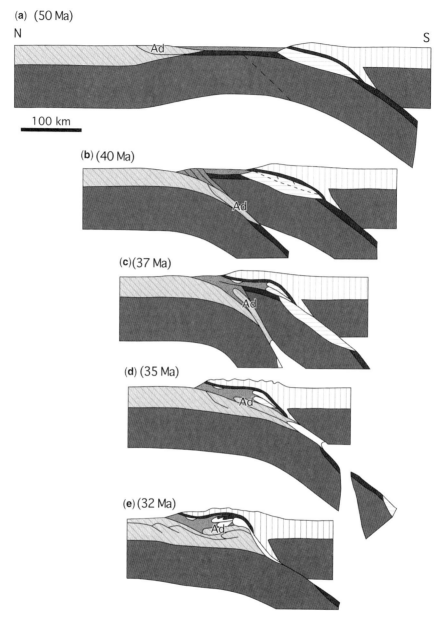

Fig. 12. Cross-sections from Froitzheim *et al.* (2003) picturing the formation of the central Alps as proposed by the authors. Key is the same as in Fig. 1 (lower crust is not specified). Main difference to Fig. 2 is that the central Alps result from accretion in two subduction zones.

individual samples from the Southern Steep Belt (e.g. Toth *et al.* 2000; Brouwer *et al.* 2005); (2) no petrological evidence for eclogite facies conditions in most of the acidic rocks; and (3) lack of isotopic ages representing an Alpine high-pressure stage from the main portion of the nappe (Santini 1992; Liati *et al.* 2005; Hermann *et al.* 2006). However, most structural geologists still believe that the high-pressure evolution recorded in all the eclogites and the pressure–temperature paths depicted in Figure 9 are related to the Alpine orogenic cycle and valid for the entire nappe (e.g. Froitzheim *et al.* 2003; Schmid *et al.* 2004). There are three lines of evidence supporting this view: (1) the continuity of the petrological record; (2) the continuity of the post peak-pressure structural record; and (3) growing evidence of Eocene high-pressure conditions in similar structural levels throughout the Alps. Peak-pressure conditions in the Adula nappe display a well-documented gradient, which fits the overall polarity of the Alpine edifice. Furthermore, amphibolite facies conditions of unquestionable Tertiary age gradually develop from eclogite-facies assemblages under apparently isothermal conditions. The Zapport phase, which appears to have started at eclogite facies conditions, shows an inhomogeneous but continuous strain over the entire nappe and, in my view, cannot possibly assemble a mélange of pieces with very different histories. Finally, high-pressure conditions in other units derived from the distal European margin and the adjacent Penninic domains have been repeatedly dated as being Eocene in age, e.g. in the Tauern window (Zimmerman 1994; Thöni 2006) and in the Monte Rosa area in the western Alps (Liati & Froitzheim 2006; Lapen *et al.* 2007). However, more isotopic studies addressing the eclogite facies conditions in the main Adula nappe are certainly needed.

Significance of the Zapport phase for exhumation from peak pressures

Kinematic models of exhumation from high pressures have to account for the fact that the Adula nappe is sandwiched between units displaying much lower peak pressures. As a consequence, all reconstructions have proposed some kind of extrusion as the exhumation mechanism (e.g. Schmid *et al.* 1996*a*) (Fig. 2b). These models are more or less indifferent about nappe internal deformation and assume opposite directed shear zones at the top and at the bottom of an extruding sheet. This would mean a top-to-the-south shear zone in the upper Adula nappe or Misox zone. Such a shear zone is not observed. Early exhumation appears to be associated with top-to-the north mylonitization in all of the Adula nappe during the Zapport phase. A possible top-to-the-south shear zone either got wiped out by subsequent deformation or cut out by a shear zone. The Zapport deformation is paradoxical for further reasons.

As mentioned above, the pressure difference between the northern and southern Adula nappe gets basically obliterated during the Zapport phase. This has to be accommodated by any combination of bulk-nappe rotation or nappe-internal strain. It seems that Zapport-related nappe internal deformation should amplify the synkinematic pressure difference, since higher structural levels would have been sheared further northward, i.e. further upward in the subduction channel according to an extrusion model. Moreover, Zapport deformation lengthens the Adula nappe in a nappe-parallel direction, thus it appears to reduce the metamorphic field gradient for peak pressures. The observed field gradient, however, is larger than the lithostatic gradient and there is no subsequent structure that can balance the Zapport mylonitization. I believe that the intense nappe-internal deformation during the Zapport phase has an important role in the exhumation process and that the Adula nappe experienced a major shape change during the Zapport phase (Fig. 13). In this scheme, the Zapport foliation primarily indicates vertical flattening. This flattening is partly accommodated by northward shearing of successively higher lobes. The pre-Zapport position of each lobe cannot be located south of the next deeper subunit, as is indicated by the polarity of the preserved pressure gradient. However, it can be located on top of the deeper unit. Accordingly, subunits would experience successively more exhumation towards the south and peak pressures could be telescoped during top-to-the-north shearing, since the bulk shape of the nappe is completely altered. This scheme is certainly not a full exhumation model but a step forward in relating Zapport kinematics to the metamorphic record. So far, nappe-internal strain is neglected in exhumation models.

Tectonic significance of Oligocene orogen-parallel to orogen-oblique stretching

Over the past two decades, estimations of the amount and regional distribution of Early Oligocene top-to-the-southeast shearing in the central Alps during the Claro and Cressim phases and the corresponding events elsewhere (Table 1) have steadily increased. The discussion about its cause is still unsettled. Schmid *et al.* (1996*b*) propose that top-to-the-southeast shearing in the upper limb of the Niemet–Beverin fold north of the Bergell intrusion would result from northward indentation of the Suretta and Tambo nappes. This motion would have been driven by the emplacement of the Bergell pluton south of these two nappes. This hypothesis now seems to have been proven wrong. The amount of SE displacement above the Niemet–Beverin axial plane accumulates to tens of km (Marquer *et al.* 1996; Weh & Froitzheim 2001), which cannot be provided by push from the Bergell intrusion. In addition, top-to-the-southeast shearing is now found in deeper structural levels such as in the Adula and Simano nappes. Rütti *et al.* (2003) proposed gravitational collapse of the thickening European nappe stack as a cause. This, however, seems in conflict

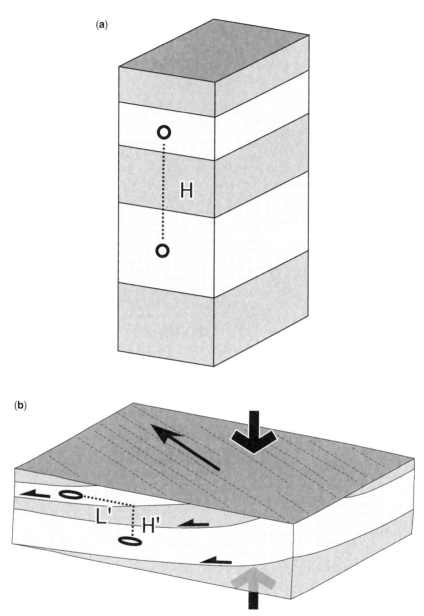

Fig. 13. Schematic sketch illustrating nappe-internal deformation during the Zapport phase. Two reference points are placed vertically above each other at a distance H at the peak-pressure stage. (**a**) During the Zapport phase, the Adula nappe is flattened and sheared with a top-to-the-north sense of motion resulting in the observed forward-dipping-duplex geometry (Figs 4 and 5). (**b**) After the Zapport phase, the two reference points are separated by (1) a vertical distance H′ smaller than H and (2) a horizontal distance L′ which can be smaller than the pre-deformation vertical distance H. This would explain why the Zapport phase reduced the pressure difference between northern and southern Adula nappe and why the present metamorphic field gradient for pressure is larger than the lithostatic pressure gradient.

with relatively high pressures associated with the Claro phase in the Adula nappe indicating a depth of at least 30 km (Nagel *et al.* 2002*a*). If anything, vertical collapse had to be driven by negative buoyancy of detached and extruding upper crust in order to explain such a depth. There is no consensus whether the Southern Steep Belt existed as a south-dipping shear zone in the Early Oligocene or not, and solving this problem would certainly help to understand the dynamics. Such a study would require careful petrological and structural reinvestigation in the Southern Steep Belt and the adjacent areas further north, preferably at places where penetrative deformation associated with Cressim backfolding is relatively weak. Also, the Cressim phase and Leis phase in the north are associated with orogen-parallel to orogen-oblique stretching. Oligocene to Miocene orogen-parallel stretching in all of the Lepontine (Steck 1984, 1990; Mancktelow 1990, 1992) has been attributed to right lateral shearing coeval with north–south shortening and associated backthrusting along the Southern Steep Belt (e.g. Schmid et al. 2004; Keller et al. 2006). Pleuger et al. (2008) propose considerable E–W extension of the Lepontine dome in the Late Oligocene in order to explain reversing horizontal shear senses in the tail of the Monte Rosa nappe. I am unaware of a study focusing on the Cressim-phase-related E–W to NW–SE-directed stretching in the central Alps and how this deformation is accommodated in the Southern Steep Belt. It may be that the Insubric mylonites are driven by a rapidly uplifting and extending northern block rather than by N–S compression.

I thank Jürgen Konzett and Romain Bousquet for detailed and helpful reviews.

References

ARLT, T., KUNZ, M., STOLZ, J., ARMBRUSTER, T. & ANGEL, R. J. 2000. P–T–X data on P2$_1$/c-clinopyroxenes and their displacive phase transitions. *Contributions to Mineralogy and Petrology*, **138**, 35–45.
AYRTON, S. N. & RAMSAY, J. G. 1974. Tectonic and metamorphic events in the Alps. *Schweizerische Mineralogische und Petrographische Mitteilungen*, **54**, 609–639.
BAUDIN, T., MARQUER, D. & PERSOZ, F. 1993. Basement-cover relationships in the Tambo nappe (Central Alps, Switzerland): geometry, structure and kinematics. *Journal of Structural Geology*, **15**, 543–553.
BECKER, H. 1993. Garnet peridotite and eclogite Sm–Nd ages from the Lepontine Dome (Swiss Alps): new evidence for Eocene high pressure in the central Alps. *Geology*, **21**, 599–602.
BERGER, A., ROSENBERG, C. & SCHMID, S. M. 1996. Ascent, emplacement and exhumation of the Bergell pluton within the Southern Steep Belt of the Central Alps. *Schweizerische Mineralogische und Petrographische Mitteilungen*, **76**, 357–382.
BERGER, A., MERCOLLI, I. & ENGI, M. 2005. The central Lepontine Alps: notes accompanying the tectonic-petrographic map sheet Sopra Ceneri (1:100 000).
Schweizerische Mineralogische und Petrographische Mitteilungen, **85**, 109–146.
BOUSQUET, R., GOFFÉ, B., VIDAL, O., OBERHÄNSLI, R. & PATRIAT, M. 2002. The tectono metamorphic history of the Valaisan domain from the Western to the Central Alps: new constraints on the evolution of the Alps. *Geological Society of America Bulletin*, **114**, 207–225.
BOZHILOV, K. N., GREEN, H. W. & DOBRZHINETSKAYA, L. 1999. Clinoenstatite in Alpe Arami peridotite: additional evidence of very high pressure. *Science*, **284**, 128–132.
BRADBURY, H. J. & NOLEN-HOEKSEMA, R. C. 1985. The Lepontine Alps as an evolving metamorphic core complex during A-type subduction—evidence from heat-flow, mineral cooling ages, and tectonic modeling. *Tectonics*, **4**, 187–211.
BRENKER, F. E. & BREY, G. P. 1997. Reconstruction of the exhumation path of Alpe Arami peridotite body from depths exceeding 160 km. *Journal of Metamorphic Geology*, **15**, 581–592.
BROUWER, F. M. & ENGI, M. 2005. Staurolite and other aluminous phases in alpine eclogite from the central Swiss Alps: analysis of domain evolution. *Canadian Mineralogist*, **43**, 105–128.
BROUWER, F. M., VON DE ZEDDE, D. M. A., WORTEL, M. J. R. & VISSERS, R. L. M. 2004. Late-orogenic heating during exhumation: Alpine PTt trajectories and thermomechanical models. *Earth and Planetary Science Letters*, **220**, 185–199.
BROUWER, F. M., BURRI, T., ENGI, M. & BERGER, A. 2005. Eclogite relicts in the Central Alps: PT-evolution, Lu-Hf ages and implications for formation of tectonic mélange zones. *Schweizerische Mineralogische und Petrographische Mittelungen*, **85**, 147–174.
BRUGGMANN, H. O. 1965. Geologie und Petrographie des südlichen Misox. PhD thesis, Universität Zürich.
BURRI, T., BERGER, A. & ENGI, M. 2005. Tertiary migmatites in the Central Alps: regional distribution, field relations, conditions of formation, and tectonic implication. *Schweizerische Mineralogische und Petrographische Mitteilungen*, **85**, 215–232.
CHEMENDA, A. I., MATTAUER, M., MALAVIEILLE, J. & BOKUN, A. N. 1995. A mechanism for syn-collisional rock exhumation and associated normal faulting—results from physical modeling. *Earth and Planetary Science Letters*, **132**, 225–232.
CIANCALEONI, L. & MARQUER, D. 2006. Syn-extension leucogranite deformation during convergence in the Eastern Central Alps: example of the Novate intrusion. *Terra Nova*, **18**, 170–180.
DALE, J. & HOLLAND, T. J. B. 2003. Geothermobarometry, P-T paths and metamorphic field gradients of high-pressure rocks from the Adula nappe, Central Alps. *Journal of Metamorphic Geology*, **21**, 813–829.
DAVIDSON, C., ROSENBERG, C. & SCHMID, S. M. 1996. Synmagmatic folding of the base of the Bergell pluton, Central Alps. *Tectonophysics*, **265**, 213–238.
DOBRZHINETSKAYA, L., GREEN, H. W. & WANG, S. 1996. Alpe Arami: a peridotite massif from depths of more than 300 kilometers. *Science*, **271**, 1841–1845.
DROOP, G. T. R. & BUCHER-NURMINEN, K. 1984. Reaction textures and metamorphic evolution of

sapphirine-bearing granulites from the Gruf complex, Italian Central Alps. *Journal of Petrology*, **25**, 766–803.

ELLIS, S., BEAUMONT, C. & PFIFFNER, O. A. 1999. Geodynamic models of custal-scale episodic tectonic accretion and underplating in subduction zones. *Journal of Geophysical Research*, **104**, 15169–15190.

ENGI, M., TODD, S. C. & SCHMATZ, D. R. 1995. Tertiary metamorphic conditions in the eastern Lepontine Alps. *Schweizerische Mineralogische und Petrographische Mitteilungen*, **75**, 347–396.

ENGI, M., BERGER, A. & ROSELLE, G. T. 2001. Role of the tectonic accretion channel in collisional orogeny. *Geology*, **29**, 1143–1146.

ENGI, M., BOUSQUET, R. & BERGER, A. 2004. Explanatory notes to the map metamorphic structure of the Alps: Central Alps. *Mitteilungen der Österreichischen Mineralogischen Gesellschaft*, **149**, 157–173.

EVANS, B. W. & TROMMSDORFF, V. 1978. Petrogenesis of garnet lherzolite, Cima di Gagnone, Lepontine Alps. *Earth and Planetary Science Letters*, **40**, 333–348.

EVANS, B. W., TROMMSDORFF, V. & RICHTER, W. 1979. Petrology of an eclogite-metarodingite suite at Cima di Gagnone, Ticino, Switzerland. *American Mineralogist*, **64**, 15–31.

FORSTER, R. 1947. Geologisch-petrographische Untersuchungem im Gebiete nördlich Locarno. *Schweizerische Mineralogische und Petrographische Mitteilungen*, **27**, 1–249.

FREY, M. & FERREIRO-MÄHLMANN, R. 1999. Alpine metamorphism in the Central Alps. *Schweizerische Mineralogische und Petrographische Mitteilungen*, **79**, 135–154.

FREY, M., HUNZIKER, J. C., FRANK, W., BOCQUET, J., DAL PIAZ, G. V., JÄGER, E. & NIGGLI, E. 1974. Alpine metamorphism of the Alps. A review. *Schweizerische Mineralogische und Petrographische Mitteilungen*, **54**, 247–290.

FREY, M., BUCHER, K., FRANK, E. & MULLIS, J. 1980. Alpine metamorphism along the geotraverse Basel-Chiasso–a review. *Eclogae Geologicae Helvetiae*, **73**, 527–546.

FRISCH, W. 1979. Tectonic progradation and plate tectonic evolution of the Alps. *Tectonophysics*, **60**, 121–139.

FRISCH, W. 1981. Plate motions in the Alpine region and their correlation to the opening of the Atlantic ocean. *Geologische Rundschau*, **70**, 402–411.

FROITZHEIM, N., SCHMID, S. M. & FREY, M. 1996. Mesozoic paleogeography and the timing of eclogite-facies metamorphism in the Alps: a working hypothesis. *Eclogae Geologicae Helvetiae*, **89**, 81–110.

FROITZHEIM, N., PLEUGER, J., ROLLER, S. & NAGEL, T. 2003. Exhumation of high- and ultrahigh-pressure metamorphic rocks by slab extraction. *Geology*, **31**, 925–928.

FUMASOLI, M. W. 1974. Geologie des Gebietes nördlich und südlich der Jorio-Tonale-Linie im Westen von Gravedone (Como, Italia). PhD thesis, Universität Zürich.

GALLI, A., MANCKTELOW, N., REUSSER, E. & CADDICK, M. 2007. Structural geology and petrology of the Naret region (northern Valle Maggia, N. Ticino,

Switzerland). *Swiss Journal of Geoscience*, **100**, 53–70.

GANSSER, A. 1937. Der Nordrand der Tambodecke. *Schweizerische Mineralogische und Petrographische Mitteilungen*, **17**, 291–522.

GEBAUER, D. 1996. A P-T-T-t path for an (ultra?-) high-pressure ultramafic/mafic rock association and their felsic country-rocks based on SHRIMP-dating of magmatic and metamorphic zircon domains. Example: Alpe Arami (Central Swiss Alps). *In*: BASU, A. & HART, S. (eds) *Earth Processes: Reading the Isotopic Code*. Geophysical Monograph Series Special AGU, **95**, 307–329.

GEBAUER, D. 1999. Alpine geochronology of the Central and Western Alps: new constraints for a complex geodynamic evolution. *Schweizerische Mineralogische und Petrographische Mitteilungen*, **79**, 191–208.

GEBAUER, D., GRÜNENFELDER, M., TILTON, G., TROMMSDORFF, V. & SCHMID, S. M. 1992. The geodynamic evolution of garnet-peridotites and eclogites of Alpe Arami and Cima di Gagnone (Central Alps) from Early Proterozoic to Oligocene. *Schweizerische Mineralogische und Petrographische Mitteilungen*, **72**, 107–111.

GOFFÉ, B. & OBERHÄNSLI, R. 1992. Ferro- and magnesiocarpholite in the 'Bündnerschiefer' of the eastern Central Alps (Grisons and Engadine window). *European Journal of Mineralogy*, **4**, 835–838.

GROND, F., WAHL, F. & PFIFFNER, M. 1995. Mehrphasige alpine Deformation und Metamorphose in der nördlichen Cima-Lunga-Einheit, Zentralalpen (Schweiz). *Schweizerische Mineralogische und Petrographische Mitteilungen*, **75**, 371–386.

GRUJIC, D. & MANCKTELOW, N. 1996. Structure of the northern Maggia and Lebendun Nappes, Central Alps, Switzerland. *Eclogae Geologicae Helvetiae*, **89**, 461–504.

HACKER, B. R., SHARP, T., ZHANG, R. Y., LIOU, J. G. & HERVIG, R. L. 1997. Determining the origin of ultrahigh-pressure lherzolites. *Science*, **278**, 702–704.

HANDY, M. & OBERHÄNSLI, R. 2004. Age map of the metamorphic structure of the Alps-Tectonic interpretation and outstanding problems. *In*: OBERHÄNSLI, R. (ed.) *Explanatory notes to the map: metamorphic structure of the Alps*. Mitteilungen der Österreichischen Mineralogischen Gesellschaft, **149**, 97–121.

HANDY, M. R., BABIST, J., WAGNER, C., ROSENBERG, C. & KONRAD, M. 2005. (eds) *Decoupling and Its Relation to Strain Partitioning in Continental Lithosphere: Insight from the Periadriatic Fault System (European Alps)*. Geological Society, London, Special Publications, **243**, 249–276.

HÄNNY, R. 1972. Das Migmatitgebiet in der Valle Bodengo (östliches Lepontin). *Beiträge zur geologischen Karte der Schweiz (NF)*, **145**, 1–109.

HÄNNY, R., GRAUERT, B. & SOPTRAJANOVA, G. 1975. Paleozoic migmatites affected by high grade Tertiary metamorphism in the central Alps (Valle Bodengo, Italy). *Contributions to Mineralogy and Petrology*, **51**, 173–196.

HEINRICH, C. A. 1982. Kyanite-eclogite to amphibolite facies evolution of hydrous mafic and pelitic rocks,

Adula nappe, Central Alps. *Contributions to Mineralogy and Petrology*, **81**, 30–38.

HEINRICH, C. A. 1986. Eclogite facies regional metamorphism of hydrous mafic rocks in the Central Alpine Adula nappe. *Journal of Petrology*, **27**, 123–154.

HEITZMANN, P. 1975. Zur Metamorphose und Tektonik im südöstlichen Teil der Lepontinischen Alpen. *Schweizerische Mineralogische und Petrographische Mitteilungen*, **55**, 467–522.

HEITZMANN, P. 1987. Calcite mylonites in the Central Alpine 'root zone'. *Tectonophysics*, **135**, 207–215.

HERMANN, J., RUBATTO, D. & TROMMSDORFF, V. 2006. Sub-solidus Oligocene zircon formation in garnet peridotite during fast decompression and fluid infiltration (Duria, Central Alps). *Mineralogy and Petrology*, **88**, 181–206.

HUBER, R. K. & MARQUER, D. 1998. The tectonometamorphic history of the peridotitic Chiavenna unit from Mesozoic to Tertiary tectonics: a restoration controlled by melt polarity indicators (Eastern Swiss Alps). *Tectonophysics*, **269**, 205–223.

HUNZIKER, J. C., DESMONS, J. & MARTINOTTI, G. 1989. Alpine thermal evolution in the central and western Alps. *In*: COWARD, M. P., DIETRICH, D. & PARK, R. G. (eds) *Alpine Tectonics*. Geological Society, London, Special Publication, **45**, 353–367.

HURFORD, A. J. 1986. Cooling and uplift patterns in the Lepontine Alps, South Central Switzerland and an age of vertical movement on the Insubric fault line. *Contributions to Mineralogy and Petrology*, **92**, 413–427.

HURFORT, A. J., FLISCH, M. & JÄGER, E. 1989. Unravelling the thermo-tectonic evolution of the Alps: a contribution from fission track analysis and mica dating. *In*: COWARD, M. P., DIETRICH, D. & PARK, R. G. (eds) *Alpine Tectonics*. Geological Society, London, Special Publication, **45**, 369–398.

IROUSCHEK, A. 1983. *Mineralogie und Petrologie von Metapeliten der Simano-Decke mit besonderer Berücksichtigung cordieritführender Gesteine zwischen Alps Sponda und Biasca*. PhD. thesis, Basel.

JÄGER, E. 1970. Rb/Sr systems in different degrees of metamorphism. *Eclogae Geologicae Helvetiae*, **63**, 163–172.

JÄGER, E. 1973. Die alpine Orogenese im Lichte der radiometrischen Altersbestimmung. *Eclogae Geologicae Helvetiae*, **66**, 11–21.

JÄGER, E., NIGGLI, E. & WENK, E. 1967. Rb–Sr Altersbestimmungen an Glimmern der Zentralalpen. *Beiträge zur geologischen Karte der Schweiz*, **134**.

JENNY, H, FRISCHKNECHT, G. & KOPP, J. 1923. Geologie der Adula. *Beiträge zur geologischen Karte der Schweiz (Neue Folge)*, **51**.

KELLER, L. M., FUGENSCHUH, B., HESS, M., SCHNEIDER, B. & SCHMID, S. M. 2006. Simplon fault zone in the western and central Alps: Mechanism of Neogene faulting and folding revisited. *Geology*, **34**, 317–320.

KLEIN, H. H. 1976. Metamorphose von Peliten zwischen Rheinwaldhorn und Pizzo Paglia (Adula- und Simano-Decke). *Schweizerische Mineralogische und Petrographische Mitteilungen*, **56**, 457–479.

KÜNDIG, E. 1926. Beiträge zur Geologie und Petrographie der Gebirgskette zwischen Val Calanca und Misox. *Schweizerische Mineralogische und Petrographische Mitteilungen*, **4**, 1–99.

LAPEN, T. J., JOHNSON, C. M., BAUMGARTNER, L. P., DAL PIAZ, G. V., SKORA, S. & BEARD, B. L. 2007. Coupling of oceanic and continental crust during Eocene eclogite-facies metamorphism: evidence from the Monte Rosa nappe, western Alps. *Contributions to Mineralogy and Petrology*, **153**, 139–157.

LIATI, A. & FROITZHEIM, N. 2006. Assessing the Valais ocean, Western Alps: U–Pb SHRIMP zircon geochronology of eclogite in the Balma unit, on top of the Monte Rosa nappe. *European Journal of Mineralogy*, **18**, 299–308.

LIATI, A., GEBAUER, D. & FANNIMG, M. 2000. U-Pb SHRIMP-dating of zircon from the Novate granite (Bergell, Central Alps): evidence for Oligocene-Miocene magmatism, Jurassic-Cretaceous continental rifting and opening of the Valais trough. *Schweizerische Mineralogische und Petrographische Mitteilungen*, **80**, 305–316.

LIATI, A., GEBAUER, D. & FANNING, C. M. 2004. The youngest basic oceanic magmatism in the Alps (Late Cretaceous; Chiavenna unit, Central Alps): geochronological constraints and geodynamic significance. *Contributions to Mineralogy and Petrology*, **146**, 144–158.

LIATI, A., GEBAUER, D. & FANNING, C. M. 2005. Pre-Alpine and Alpine metamorphism in the Adula nappe, Central Alps: Constraints by SHRIMP-dating and REE of zircon. *Mitteilungen der Österreichischen Mineralogischen Gesellschaft*, **150** (Abstract).

LÖW, S. 1987. Die tektono-metamorphe Entwicklung der Adula-Decke (Zentralalpen, Schweiz). *Beiträge zur Geologischen Karte der Schweiz (Neue Folge)*, **161**, 1–84.

MANCKTELOW, N. S. 1990. The Simplon fault zone. *Beiträge zur Geologischen Karte der Schweiz (NF)*, **163**, 1–74.

MANCKTELOW, N. S. 1992. Neogene lateral extension during convergence in the Central Alps. Evidence from interrelated faulting and backfolding around the Simplon Pass (Switzerland). *Tectonophysics*, **215**, 293–317.

MARQUER, D., BAUDIN, T., PEUCAT, J.-J. & PERSOZ, F. 1994. Rb-Sr mica ages in the Alpine shear zones of the Truzzo granite: timing of the Tertiary Alpine P-T-deformations in the Tambo nappe (Central Alps, Switzerland). *Eclogae Geologicae Helvetiae*, **87**, 225–239.

MARQUER, D., CHALLANDES, N. & BAUDIN, T. 1996. Shear zone patterns and strain distribution at the scale of a Penninic nappe: the Suretta nappe (Eastern Swiss Alps). *Journal of Structural Geology*, **18**, 753–764.

MAXELON, M. & MANCKTELOW, N. S. 2005. Three-dimensional geometry and tectonostratigraphy of the Pennine zone, Central Alps, Switzerland and Northern Italy. *Earth-Science Reviews*, **71**, 171–227.

MEYRE, C. & PUSCHNIG, A. R. 1993. High-pressure metamorphism and deformation at Trescolmen, Adula nappe, Central Alps. *Schweizerische Mineralogische und Petrographische Mitteilungen*, **73**, 277–283.

MEYRE, C., DE CAPITANI, C. & PARTSCH, J. H. 1997. A ternary solid solution model for omphacite and its application to geothermobarometry of eclogites from the Middle Adula nappe (Central Alps, Switzerland). *Journal of Metamorphic Geology*, **15**, 687–700.

MEYRE, C., MARQUER, D., SCHMID, S. M. & CIANCALEONI, L. 1999a. Syn-orogenic extension along the Forcola fault. *Eclogae Geologicae Helvetiae*, **91**, 409–420.

MEYRE, C., DE CAPITANI, C., ZACK, T. & FREY, M. 1999b. Petrology of high-pressure metapelites from the Adula nappe (Central Alps, Switzerland). *Journal of Petrology*, **40**, 199–213.

MILNES, A. G. 1974. Structure of the Pennine Zone (Central Alps): a new working hypothesis. *Bulletin of the Geological Society of America*, **85**, 1727–1732.

MILNES, A. G. 1978. Structural Zones and Continental Collision, Central Alps. *Tectonophysics*, **47**, 369–392.

MILNES, A. G. 1980. Tectonic evolution of the Central Alps in the cross section St. Gallen-Como. *Eclogae Geologicae Helvetiae*, **73**, 619–633.

MILNES, A. G. & SCHMUTZ, H. U. 1978. Structure and history of the Suretta nappe (Pennine Zone, Central Alps)—a field study. *Eclogae Geologicae Helvetiae*, **71**, 19–23.

NAGEL, T., DE CAPITANI, C., FREY, M., FROITZHEIM, N., STÜNITZ, H. & SCHMID, S. M. 2002a. Structural and metamorphic evolution during rapid exhumation in the Lepontine dome (southern Simano and Adula nappes, Central Alps, Switzerland). *Eclogae Geologicae Helvetiae*, **95**, 301–321.

NAGEL, T., DE CAPITANI, C. & FREY, M. 2002b. Isograds and P-T evolution in the eastern Lepontine Alps (Graubünden, Switzerland). *Journal of Metamorphic Geology*, **20**, 309–324.

NIEVERGELT, P., LINIGER, M., FROITZHEIM, N. & FERREIRO MÄHLMANN, R. 1996. Early to mid Tertiary crustal extension in the Central Alps: the Turba Mylonite Zone (Eastern Switzerland). *Tectonics*, **15**, 329–340.

NIGGLI, E. & NIGGLI, C. R. 1965. Karten der Verbreitung einiger Mineralien der alpidischen Metamorphose in den Schweizer Alpen (Stilpnomelan, Alkali-Amphibol, Chloritoid, Staurolith, Disthen, Sillimanit). *Eclogae Geologicae Helvetiae*, **58**, 335–368.

NIMIS, P. & TROMMSDORFF, V. 2001. Revised thermobarometry of Alpe Arami and other garnet peridotites from the Central Alps. *Journal of Petrology*, **42**, 103–115.

OBERHÄNSLI, R., GOFFE, B. & BOUSQUET, R. 1995. Record of a HP–LT metamorphic evolution in the Valais zone: Geodynamic implications. *Bollettino del Museo Regionale di Scienze Naturali di Torino*, **13**, 221–240.

PAQUIN, J. & ALTHERR, R. 2001. New constraints on the P-T evolution of the Alpe Arami Garnet Peridotite Body (Central Alps, Switzerland). *Journal of Petrology*, **42**, 1119–1140.

PARTZSCH, J. H. 1998. The tectono-metamorphic evolution of the middle Adula nappe, Central Alps, Switzerland. PhD. thesis, Basel.

PFIFFNER, M. & TROMMSDORFF, V. 1998. The high-pressure ultramafic-mafic-carbonate suite of Cima Lunga-Adula, Central Alps: Excursions to Cima di Gagnone and Alpe Arami. *Schweizerische Mineralogische und Petrographische Mitteilungen*, **78**, 337–354.

PFIFFNER, O. A., ELLIS, S. & BEAUMONT, C. 2000. Collision tectonics in the Swiss Alps: Insight from geodynamic modeling. *Tectonics*, **19**, 1065–1094.

PLEUGER, J., HUNDENBORN, R., KREMER, K., BABINKA, S., KURZ, W., JANSEN, E. & FROITZHEIM, N. 2003. Structural evolution of Adula nappe, Misox zone, and Tambo nappe in the San Bernardino area: Constraints for the exhumation of the Adula eclogites. *Mitteilungen der Österreichischen Geologischen Gesellschaft*, **94**, 99–122.

PLEUGER, J., FROITZHEIM, N. & JANSEN, E. 2005. Folded continental and oceanic nappes on the southern side of Monte Rosa (western Alps, Italy): anatomy of a double collision suture. *Tectonics*, **24**, TC4013; doi: 10.1029/2004TC001737.

PLEUGER, J., ROLLER, S., WALTER, J. M., JANSEN, E. & FROITZHEIM, N. 2007. Structural evolution of the contact between two Penninic nappes (Zermatt-Saas zone and Combin zone, Western Alps) and implications for exhumation mechanism and palaeogeography. *International Journal of Earth Sciences*, **96**, 229–252.

PLEUGER, J., NAGEL, T. J., WALTER, J. M., JANSEN, E. & FROITZHEIM, N. 2008. On the role and importance of orogen-parallel and -perpendicular extension, transcurrent shearing, and backthrusting in the Monte Rosa nappe and the Southern Steep Belt of the Alps (Penninic zone, Switzerland and Italy). *In*: SIEGESMUND, S., FÜGENSCHUH, B. & FROITZHEIM, M. (eds) *Tectonic Aspects of the Alpine-Dinaride-Carpathian System*. Geological Society, London, Special Publication, **298**, 251–280.

PREISWERK, H. 1921. Die zwei Deckenkulminationen Tosa-Tessin und die Tessiner Querfalte. *Eclogae Geologicae Helvetiae*, **16**, 485–496.

PREISWERK, H., BOSSARD, L., GRÜTTER, O., NIGGLI, P., KÜNDIG, E. & AMBÜHL, E. 1934. Geologische Karte der Tessiner Alpen zwischen Maggia- und Bleniotal. *Schweizerische Geologische Kommission*, **116**.

RAMSAY, J. G. & HUBER, M. I. 1987. *The Techniques of Modern Structural Geology, Vol. 2: Folds and Fractures*. Academic Press.

RING, U. 1992. The kinematic history of the Pennine nappes east of the Lepontine dome—implications for the Tectonic evolution of the Central Alps. *Tectonics*, **11**, 1139–1158.

RISOLD, A.-C., TROMMSDORFF, V. & GROBÉTY, B. 2001. Genesis of ilmenite rods and palisades along humite-type defects in olivine from Alpe Arami. *Contributions to Mineralogy and Petrology*, **140**, 619–628.

ROSELLE, G. T., THÜRING, M. & ENGI, M. 2002. Melonpit: A finite element code for simulating tectonic mass movement and heat flow within subduction zones. *American Journal of Sciences*, **302**, 381–409.

ROSENBERG, C. L., BERGER, A. & SCHMID, S. M. 1995. Observations from the floor of a granitoid pluton: Inferences on the driving force of final emplacement. *Geology*, **23**, 443–446.

RÜTTI, R., MAXELON, M. & MANCKTELOW, N. S. 2003. Structure and kinematics of the northern Simano nappe, Central Alps, Switzerland. *Eclogae Geologicae Helvetiae*, **98**, 63–81.

SANTINI, L. 1992. Geochemistry and geochronology of the basic rocks of the Penninic Nappes of East-Central Alps (Switzerland). PhD thesis, Lausanne.

SCHMID, S. M., ZINGG, A. & HANDY, M. 1987. The kinematics of movements along the Insubric Line and the emplacement of the Ivrea Zone. *Tectonophysics*, **135**, 47–66.

SCHMID, S. M., AEBLI, H. R., HELLER, F. & ZINGG, A. 1989. The role of the Periadriatic Line in the tectonic evolution of the Alps. *In*: COWARD, M. P., DIETRICH, D. & PARK, R. G. (eds) Alpine Tectonics. *Geological Society, London, Special Publication*, **45**, 153–171.

SCHMID, S. M., PFIFFNER, O. A., FROITZHEIM, N., SCHÖNBORN, G. & KISSLING, E. 1996a. Geophysical-geological transect and tectonic evolution of the Swiss-Italian Alps. *Tectonics*, **15**, 1036–1064.

SCHMID, S. M., BERGER, A. & DAVIDSON, C. *ET AL*. 1996b. The Bergell pluton (Southern Switzerland, Northern Italy): overview accompanying a geological-tectonic map of the intrusion and surrounding country rocks. *Schweizerische Mineralogische und Petrographische Mitteilungen*, **76**, 329–355.

SCHMID, M. S., FÜGENSCHUH, B., KISSLING, E. & SCHUSTER, R. 2004. Tectonic map and overall architecture of the Alpine orogen. *Eclogae Geologicae Helvetiae*, **97**, 93–117.

SCHMUTZ, H. U. 1976. Der Mafitit – Ultramafitit Komplex zwischen Chiavenna und Bondasca. *Beiträge zur Geologischen Karte der Schweiz*, **149**.

SCHREURS, G. 1993. Structural analysis of the Schams nappes and adjacent tectonic units: Implications for the orogenic evolution of the Penninic zone in Eastern Switzerland. *Bulletin de la Société géologique de France*, **164**, 415–435.

SCHREURS, G. 1995. Geometry and kinematics of the Schams nappes and adjacent tectonic units in the Penninic zone. *In*: Die Schamser Decken, part 2. *Beiträge zur geologischen Karte der Schweiz (Neue Folge)*, **167**.

SPICHER, A. 1980. Tektonische Karte der Schweiz. *Schweizerische Geologische Kommission*.

STAMPFLI, G. M., MOSAR, J., MARQUER, D., MARCHANT, R., BAUDIN, T. & BOREL, G. 1998. Subduction and obduction prosesses in the Swiss Alps. *Tectonophysics*, **296**, 159–204.

STECK, A. 1984. Structures de deformation tertiaires dans les Alpes centrales. *Eclogae Geologicae Helvetiae*, **77**, 55–100.

STECK, A. 1990. Une carte des zones de cisaillement ductile des Alpes Centrales. *Eclogae Geologicae Helvetiae*, **83**, 603–626.

STECK, A. 1998. The Maggia cross-fold: an enigmatic structure of the lower Penninic nappes of the Lepontine Alps. *Eclogae Geologicae Helvetiae*, **91**, 333–343.

STECK, A. & HUNZIKER, J. 1994. The Tertiary structural and thermal evolution of the Central Alps—compressional and extensional structures in an orogenic belt. *Tectonophysics*, **238**, 229–254.

STEINMANN, M. 1994. Ein Beckenmodell für das Nordpenninikum der Ostschweiz. *Jahrbuch der Geologischen Bundesanstalt*, **137**, 675–721.

STEINMANN, M. & STILLE, P. 1999. Geochemical evidence for the nature of the crust beneath the eastern North Penninic basin of the Mesozoic Tethys ocean. *Geologische Rundschau*, **87**, 633–643.

THOMPSON, H. P. 1976. Isograd patterns and pressure-temperature distribution during regional metamorphism. *Contributions to Mineralogy and Petrology*, **57**, 277–295.

THÖNI, M. 2006. Dating eclogite-facies metamorphism in the Eastern Alps—approaches, results, interpretations: a review. *Mineralogy and Petrology*, **88**, 123–148.

TODD, C. S. & ENGI, M. 1997. Metamorphic field gradients in the Central Alps. *Journal of Metamorphic Geology*, **15**, 513–530.

TOTH, T. M., GRANDJEAN, V. & ENGI, M. 2000. Polyphase evolution and reaction sequence of compositional domains in metabasalt: a model based on local chemical equilibrium and metamorphic differentiation. *Geological Journal*, **35**, 163–183.

TROMMSDORFF, V. 1966. Progressive Metamorphose kieseliger Karbonatgesteine in den Zentralalpen zwischen Bernina und Simplon. *Schweizerische Mineralogische und Petrographische Mitteilungen*, **46**, 431–460.

TROMMSDORFF, V. 1990. Metamorphism and tectonics in the Central Alps: the Alpine lithospheric mélange of Cima Lunga and Adula. *Memorie della Società Geologica Italiana*, **45**, 39–49.

TROMMSDORFF, V., HERMANN, J., MUNTENER, O., PFIFFNER, M. & RISOLD, A. C. 2000. Geodynamic cycles of subcontinental lithosphere in the Central Alps and the Arami enigma. *Journal of Geodynamics*, **30**, 77–92.

VAN DER PLAS, L. 1959. Petrology of the northern Adula region, Switzerland (with particular reference to the glaucophane-bearing rocks). *Leidse Geologische Medelingen*, **24**, 413–603.

VIOLA, G., MANCKTELOW, N. S. & SEWARD, D. 2001. Late Oligocene-Neogene evolution of Europe-Adria collision: new structural and geochronological evidence from the Giudicarie fault system (Italian Eastern Alps). *Tectonics*, **20**, 999–1020.

VOGLER, W. S. & VOLL, G. 1981. Deformation and Metamorphism at the South margin of the Alps East of Bellinzona, Switzerland. *Geologische Rundschau*, **70**, 1232–1262.

VON BLANCKENBURG, F. 1992. Combined high-precision chronometry and geochemical tracing using accessory minerals: applied to the Central-Alpine Bergell intrusion (central Europe). *Chemical Geology*, **100**, 19–40.

WEH, M. & FROITZHEIM, N. 2001. Penninic cover nappes in the Prättigau half-window (Eastern Switzerland): Structure and tectonic evolution. *Eclogae Geologicae Helvetiae*, **94**, 237–252.

WENK, E. 1955. Eine Strukturkarte der Tessineralpen. *Schweizerische Mineralogische und Petrographische Mitteilungen*, **35**, 311–319.

WENK, E. 1970. Zur Regionalmetamorphose und Ultrametamorphose der Zentralalpen. *Fortschritte der Mineralogie*, **47**, 555–565.

WILLETT, S., BEAUMONT, C. & FULLSACK, P. 1993. Mechanical model for the tectonics of doubly vergent compressional orogens. *Geology*, **21**, 371–374.

ZIMMERMANN, R., HAMMERSCHMIDT, K. & FRANZ, G. 1994. Eocene high pressure metamorphism in the Pennine units of the Tauern Window (Eastern Alps): evidence from ^{40}Ar-^{39}Ar dating and petrological investigations. *Contributions to Mineralogy and Petrology*, **117**, 175–186.

Metamorphism of metasediments at the scale of an orogen: a key to the Tertiary geodynamic evolution of the Alps*

ROMAIN BOUSQUET[1], ROLAND OBERHÄNSLI[1], BRUNO GOFFÉ[2],
MICHAEL WIEDERKEHR[3], FRIEDRICH KOLLER[4], STEFAN M. SCHMID[3],
RALF SCHUSTER[5], MARTIN ENGI[6], ALFONS BERGER[6] & GIORGIO MARTINOTTI[7]

[1]*Institut für Geowissenschaften, Universität Potsdam, Germany*
(e-mail: romain@geo.uni-potsdam.de)

[2]*Laboratoire de Géologie, ENS Paris-UMR 8538, France*

[3]*Geologisches Institut, Universität Basel, Switzerland*

[4]*Department für Lithoshpärenforschung, Universtät Wien, Austria*

[5]*Geological Survey, Vienna, Austria*

[6]*Institute für Geologie, Universität Bern, Switzerland*

[7]*Dipartimento di Scienze della Terra, Università degli Studi di Torino, Italy*

Abstract: Major discoveries in metamorphic petrology, as well as other geological disciplines, have been made in the Alps. The regional distribution of Late Cretaceous–Tertiary metamorphic conditions, documented in post-Hercynian metasediments across the entire Alpine belt from Corsica–Tuscany in the west to Vienna in the east, is presented in this paper. In view of the uneven distribution of information, we concentrate on type and grade of metamorphism; and we elected to distinguish between metamorphic paths where either pressure and temperature peaked simultaneously, or where the maximum temperature was reached at lower pressures, after a significant temperature increase on the decompression path.

The results show which types of process caused the main metamorphic imprint: a subduction process in the western Alps, a collision process in the central Alps, and complex metamorphic structures in the eastern Alps, owing to a complex geodynamic and metamorphic history involving the succession of the two types of process. The western Alps clearly show a relatively simple picture, with an internal (high-pressure dominated) part thrust over an external greenschist to low-grade domain, although both metamorphic domains are structurally very complex. Such a metamorphic pattern is generally produced by subduction followed by exhumation along a cool decompression path. In contrast, the central Alps document conditions typical of subduction (and partial accretion), followed by an intensely evolved collision process, often resulting in a heating event during the decompression path of the early-subducted units. Subduction-related relics and (collisional/decompressional) heating phenomena in different tectonic edifices characterize the Tertiary evolution of the Eastern Alps. The Tuscan and Corsica terrains show two different kinds of evolution, with Corsica resembling the western Alps, whereas the metamorphic history in the Tuscan domain is complex owing to the late evolution of the Apennines. This study confirms that careful analysis of the metamorphic evolution of metasediments at the scale of an entire orogen may change the geodynamic interpretation of mountain belts.

After more than a century of investigations, the Alps still represent an outstanding natural laboratory for the study of geodynamic processes linked to the evolution of mountain belts in general. The integration of regional geology and metamorphic evolution provides highly needed constraints for increasingly complex quantitative models (e.g. Escher & Beaumont 1996; Henry *et al.* 1997; Pfiffner *et al.* 2000).

Major discoveries in metamorphic petrology, as well as other geological disciplines, have been and are still made in the Alps. For example, eclogites were described for the first time in the eastern Alps (Koralpe, Saualpe massifs) by Haüy (1822). More recently, the discovery of coesite in the Dora Maira unit (Chopin 1984, 1987) proved that continental crust went into subduction, contrary to a still widely held opinion, and returned from

---

*This paper is dedicated to Martin Burkhard who tragically died during work in the Alps.

great depths. Many others, occasionally less spectacular yet important, petrological discoveries, were also made in the Alps. For instance, studies of metapelites in the Alps starting at the beginning of the 1970s, revealed a specific mineralogy reflecting high-pressure conditions. The most emblematic minerals found in such rocks are ferro- and magnesiocarpholites (Goffé et al. 1973). Besides such discoveries, many holistic attempts have been made to assess the dynamics of this orogen. Niggli & Niggli (1965) applied Barrow's concept to the central Alps and presented a mineral distribution map with mineral isograds reflecting a Lepontine high-temperature event. Zwart (1973) and Zwart et al. (1978) compiled mineral distributions at the scale of the orogen, using the facies concept and available age data in the Alps and elsewhere.

Based on such mineralogical work, Ernst (1971) was able to use the plate tectonic concept for proposing a first modern model for the evolution of the Alps. Such ideas developed further on the basis of work such as that of Dal Piaz et al. (1972), Dal Piaz (1974a, b) and Hunziker (1974), just to name a few. Frey (1969), as well as Trommsdorff (1966), started to investigate metamorphism in isochemical systems provided by shales and siliceous carbonates, respectively. This allowed for quantitatively constraining the Cenozoic temperature evolution in the central Alps. Frey et al. (1999) compiled all available information on the peak temperature distribution, and used the occurrence of eclogites to display the dynamics of the Alpine evolution. Previous works of this kind led to the compilation of a new map showing the metamorphic structure of the Alps (Oberhänsli et al. 2004). This new map was also based on: (1) new tectonic concepts and maps (Schmid et al. 2004); (2) a wealth of new radiogenic age data (for references see e.g. Handy & Oberhänsli 2004, and Berger & Bousquet 2008); and (3) an extension of the facies concept based on mafic to metapelitic rock compositions.

In the Alps, many areas are devoid of index minerals classically observed in mafic and quartzo-feldspathic rocks systems, allowing a direct comparison to be made. Petrological investigation on metasediments greatly helps to constrain geodynamic evolution of such areas (see Bousquet 2008). One tool to understand such problems better at the orogen scale is maps (Niggli & Niggli 1965; Niggli 1970; Frey et al. 1999; Oberhänsli et al. 2004). This study combines these different sources of information: presenting metamorphism in maps and combining this with metamorphic evolution data. This provides insights into the geodynamics of metasediments inside the orogen. The metamorphism of metasediments can be subdivided into general geodynamic groups: (1) pressure-dominated metamorphism; (2) temperature-dominated metamorphism at intermediate pressures (Barrovian metamorphism), which is often referenced in the literature as HT metamorphism; and (3) contact metamorphic aureoles which are temperature-dominated metamorphism at low pressures. The latter type will be excluded from this contribution because it is only of minor importance in the Alps.

This paper reviews existing data and presents ongoing work in an attempt to integrate metamorphic studies and Late Cretaceous–Tertiary geodynamic concepts in the Alps. We will illustrate how mineral data obtained from metasediments may constrain the geodynamic evolution of mountain belts in general.

Metamorphic mineralogy of metasediments

In contrast to mafic complexes or meta-igneous rocks, metasediments commonly crop out continuously over very large areas in many mountain belts, such as the Alps (Fig. 1). Since these metasediments cover large areas, this allows to simultaneously observe their structural and metamorphic evolution, and thus to decipher the geodynamic frame. However, since Barrow (1893, 1912) and Eskola (1929), the definition of metamorphic facies, as well as petrographic work on metamorphic rocks, was mainly focused on mafic systems (Evans 1990; Frey et al. 1991; Carswell 1990).

Detailed studies on pelitic systems (Yardley 1989; Koons & Thompson 1985; Spear 1993; McDade & Harley 2001) are only available for medium- to high-temperature metamorphic conditions. Metamorphic studies addressing low-temperature conditions extended methods taken from studies on diagenesis or anchimetamorphism, such as illite crystallinity, vitrinite reflectance or clay mineralogy (Frey & Robinson 1999) which lack good possibilities of pressure and temperature calibration. Spectacular improvements on the knowledge of mountain belt evolution based on the study of metasediments could only be made starting with the discovery of coesite in metasediments (Chopin 1984) and other work on Ultra High Pressure rock systems (UHP) in general (Coleman & Wang 1995; Chopin & Sobolev 1995; Massonne & O'Brien 2003).

Petrogenetic grids

Classical index minerals, such as pumpellyite, glaucophane or jadeite, observed in mafic and quartzo-feldspathic rocks systems, are unfortunately rarely

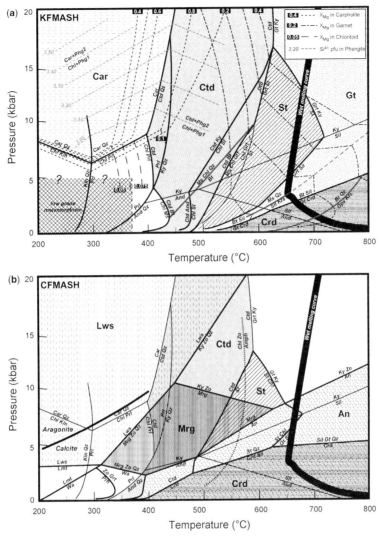

Fig. 2. Petrogenetic grids for metapelites for a temperature range from 200 to 800 °C. (a) In the KFMASH ($K_2O-FeO-MgO-Al_2O_3-SiO_2-H_2O$) system, the grid is strongly temperature-controlled. The appearance of assemblages, from (Fe, Mg)-carpholite assemblage at HP or from chlorite–pyrophyllite assemblage delimits the low-temperature domain from the middle temperature one at around 400 °C. The exact temperature limit depends on rock and mineral chemistry. At higher temperature conditions, the breakdown of chloritoid into garnet or staurolite indicates the transition towards high-temperature domains between 500 and 600 °C depending on pressure conditions as well as on rock and mineral chemistry. (b) In the CFMASH ($CaO-FeO-MgO-Al_2O_3-SiO_2-H_2O$) system, the temperature control is less important. While under LT conditions, lawsonite is the main stable mineral, sometimes coexisting with (Fe, Mg)-carpholite; at middle and HT conditions, margarite and staurolite stability fields are pressure-dependent. We note a large cordierite-stability field in the CFMASH system.

Diagrams drawn from field experience and theoretical studies after Spear & Cheney (1989), Wang & Spear (1991), Vidal et al. (1992), Oberhänsli et al. (1995), Bousquet et al. (2002), Proyer (2003), wei et al. (2004), Wei & Powell (2004), Wei & Holland (2003), Chatterjee (1976), Frey & Niggli (1972), Zeh (2001), Pattison et al. (2002), McDade & Harley (2001), Kohn & Spear (1993), Hébert & Ballèvre (1993) as well as own calculation using the Theriak-Domino software (De Capitani & Brown 1987, De Capitani (1994) using Berman database (1988) completed by recent thermodynamic data: Mg-chloritoid data of B. Patrick (listed in Goffé & Bousquet 1997), Fe-chloritoid data of Vidal et al. (1994), chlorite data of Vidal et al. (2001), and alumino-celadonite data from Massonne & Szpurka (1997). Mineral abbreviations are from Bucher & Frey (1994) except for (Fe, Mg)-carpholite (Car).

observed in Alpine metasediments. Nevertheless, metasediments have highly variable chemical and mineralogical compositions that represent an important geothermobarometric potential. Figure 2 shows petrogenetic grids for the KFMASH and CFMASH subsystems that integrate field observations, experimental data and thermodynamic modelling using an internally consistent database. This new kind of compilation covers a large P–T space, extending from low to high pressure (0–2 GPa) as well as from low to high temperature (200–800 °C).

Mineral assemblage containing ferro-magnesio-carpholite with phengite, chlorite and quartz is one of the most emblematic mineral assemblages of metasediments in the KFMASH system (De Roever 1951; Goffé et al. 1973; Chopin & Schreyer 1983; Goffé & Chopin 1986; Rimmelé et al. 2003). It is encountered in various rock types, such as aluminium-rich metapelites, quartzites, marbles and albite-free pelitic schists in which it is abundant, often in veins. Textural and mineralogical observations in these rocks reveal that at low temperatures the main equilibrium reactions of ferro- and magnesio-carpholite involve quartz, kaolinite, pyrophyllite, kyanite, chlorite, chloritoid and phengite (Fig. 2a). At high temperatures, large P–T fields are dominated by an assemblage containing staurolite, biotite and garnet (Spear & Cheney 1989). Fe–Mg variations in mineral composition, as a function of P and T (Goffé 1982; Spear & Selverstone 1983; Vidal et al. 1992; Theye et al. 1992), as well as Si isopleths in phengite (Massone & Schreyer 1987; Oberhänsli et al. 1995; Massone & Szpurka 1997; Bousquet et al. 2002) allow for relatively precise P–T estimates (Vidal et al. 2001; Parra et al. 2002a, b; Rimmelé et al. 2005) for some metapelitic compositions.

In the CFMASH system, for comparison, there is less resolution at low-temperature conditions. In carbonaceous systems, the stability field of the index mineral lawsonite covers the whole low-grade space, including the stability field of carpholite and the aragonite–calcite transition. While the staurolite field is substantially smaller in the CFMASH system, as compared to the KFMASH system, margarite and zoisite are characteristic of medium P and T conditions.

Oberhänsli et al. (2004) proposed a new type of metamorphic facies grid that better integrated field observations into models of the geodynamic. This facies grid also took into account the importance of metasediments, which is less clear in traditional grids. The proposed grid also involved more subdivisions, which are based on the understanding of the metasediments. Based on these compilations, a revised version of this tool is presented in Figure 3, and it will be used in this paper.

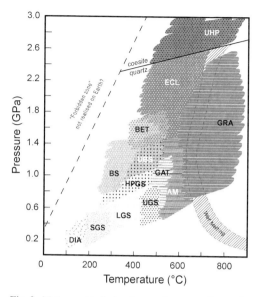

Fig. 3. Metamorphic facies diagram for metapelites and metabasites (modified after Oberhänsli et al. 2004). This diagram has been used for Figures 6 and 7. For abbreviations, see Table 2. This diagram is in good agreement with previous published facies diagrams (e.g. Yardley 1989; Spear 1993; Bousquet et al.1997).

Ambiguity of some mineral assemblages in the Alps

While some parageneses unambiguously grow only within a certain geodynamic context (Table 1a), other mineral assemblages commonly occur over a large field of P–T conditions and may evolve in diverse geodynamic contexts. Examples of unambiguous mineral assemblages containing lawsonite and carpholite only form under low-temperature/high-pressure conditions, typical for subduction processes; mineral assemblages containing staurolite and andalusite are typical high-temperature phases occurring during collision processes (i.e. Barrovian-type metamorphism).

On the other hand, recurrent minerals such as chloritoid, zoisite, kyanite and garnet may form at different P–T conditions. This hampers their use for an interpretation of the geodynamic setting. Chloritoid–phengite assemblage, for example, can be produced in different geodynamic settings. Moreover, it may occur in tectonic units in close spacial juxtaposition although these units formed in different geodynamic scenarios (Oberhänsli et al. 2003). Figure 4 clearly illustrates that two P–T paths may lead to different mineral reactions that produce chloritoid–phengite–chlorite mineral assemblages. The P–T path along a cold geotherm leads to the formation of chloritoid–phengite–chlorite as a result of the breakdown

Table 1. *Mineralogical description of the metamorphic facies presented in Figure 3 for basic rocks as well as for meta-pelitic rocks*

Facies Name	Abb.	Basic rocks	Metapelites
diagenesis/sub anchizone	DIA	zeolite	illite–kaolinite
sub greenschist	SGS	laumontite–prehnite–pumpellyite	illite
lower greenschist	LGS	albite–chlorite	pyrophyllite–chlorite ± chloritoid
upper greenschist	UGS	actinolite–epidote–chlorite	biotite–chlorite–kyanite ± chloritoid
high-pressure greenschist	HPGS	albite–lawsonite–chlorite ± crossite	lawsonite–chlorite ± chloritoid ± kyanite
greenschist–amphibolite transition	GAT	albite–epidote–amphibole	biotite–garnet–chloritoid or phengite–chloritoid–kyanite
amphibolite	AM	plagioclase–hornblende – garnet	biotite–garnet–staurolite–phengite ± kyanite
blueschist	BS	glaucophane–lawsonite	carpholite–phengite ± pyrophyllite
upper blueschist	UBS	glaucophane–epidote–garnet	chloritoid–phengite ± garnet
blueschist–eclogite transition	BET	glaucophane–zoisite–garnet ± clinopyroxene	garnet–Mg–rich chloritoid–phengite
eclogite	ECL	garnet–omphacite–zoisite– quartz ± amphibole ± phengite	garnet–Mg–rich chloritoid–kyanite or garnet–lawsonite
ultrahigh-pressure	UHP	garnet–omphacite–kyanite ± phengite	coesite or Mg–rich chloritoid–talc–phengite
granulite	GRA	plagioclase–clinopyroxene (augite)–orthopyroxene (hypersthene) ± garnet ± olivine ± spinel	clinopyroxene (diopside)–orthopyroxene (enstatite) – K–feldspath ± garnet ± corderite ± sapphirine

of Fe–Mg carpholite (Chopin & Schreyer 1983) while under higher geothermal gradients this assemblage can also form by the breakdown of pyrophyllite (Frey & Wieland 1975).

The high-pressure alumosilicate polymorph, kyanite, is particularly difficult to interpret as an indicator of geodynamic processes. It may occur in UHP associations such as the Dora Maira unit, as well as in temperature-dominated areas such as the Lepontine dome. It indicates a subduction-type low geothermal gradient in the first, but a high collision-related geotherm in the latter case.

Ambiguity of some metamorphic facies

Assignment of a metamorphic rock to a metamorphic facies is based on its mineralogy. In the best case, a rock might undergo a simple metamorphic evolution in a distinct geodynamic setting, leading to a peak metamorphic paragenesis (simultaneous P and T peaks) possibly followed by later retrogression (Fig. 5a). However, a metamorphic pressure peak related to one geodynamical scenario may also have been overprinted by a thermal peak that resulted from a second and different geodynamic scenario (Fig. 5b). Hence, in such a case metamorphic facies is ambiguous in that it is difficult to distinguish between continuous and discontinuous evolutions. It is only the exact shape of the P–T path, details that are often difficult to constrain, which is specific for a complex geodynamic evolution. For example, the significance of the amphibolite facies mineralogy is ambiguous. It may either represent a single path that entirely formed in a collision setting; or alternatively, it may merely represent an exhumation stage that formed during ongoing subduction and before final collision (Fig. 5a). In the case of dual peak paths, details regarding amphibolite facies overprint are crucial for better understanding exhumation processes in general and details of the transition from subduction to collision in particular.

The metamorphic data in a geodynamic context

The above-described importance of metasediments and the presented tool of a facies grid with its characteristic mineral assemblages can be well used in the Alps. The Alps are well suited for such a compilation because they are a relatively small and well-investigated orogen (e.g. Frey *et al.* 1974, 1980; Goffé & Chopin 1986; Roure *et al.* 1990; Dal Piaz 2001; Oberhänsli *et al.* 2004; Schmid *et al.* 2004). The Alps are developed by subduction of two different oceans followed by collision between two main continents (Adria and Europe). The relics of the oceans (Piemont–Liguria, Valais) and the separating microcontinent

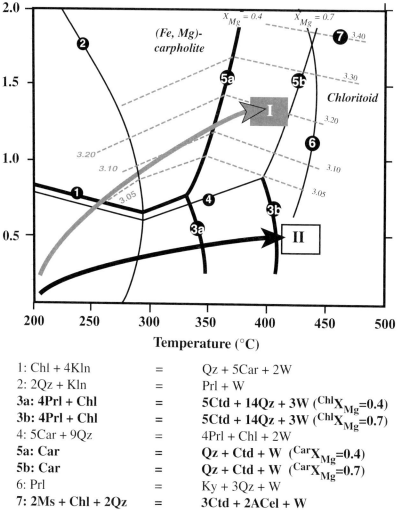

Fig. 4. Ambiguity of chloritoid-quartz as index assemblage. Depending on the P–T path, chloritoid-quartz can be formed either by breakdown of (Fe, Mg)-carpholite at HP conditions (reaction 5) or by breakdown of pyrophyllite at LP conditions (reaction 3). Chloritoid-quartz assemblage alone cannot be used as index for pressure conditions. Its significance depends on the mineral reaction and on the associated minerals.

(Briançonnais) are still existing and are now sandwiched between Adria and Europe. The palaeogeographic (tectonic) overview of the present-day situation is principally inspired from Schmid et al. (2004) and Bigi et al. (1990), and can so be combined with metamorphic information of the different units. This approach will be presented below.

Subduction-related minerals and their distribution

All the way from the Adriatic margin preserved in the lower Austroalpine nappes to Europe-derived nappes, the tectonic units (i.e. Piemont–Liguria, Briançonnais, Valais) contain metasediments that recorded the Late Cretaceous–Cenozoic subduction history. The minerals indicating subduction-related processes are listed in Table 2 and their distribution is shown in Figure 6.

We recognize the HP–LT imprint on the European continental margin (Tauern window, Adula nappe and surrounding covers), on all the metasediments derived from the partly oceanic Valaisan domain, as well as on the Piemont–Ligurian realm (Rechnitz window, Matrei zone, Avers, Zermatt–Saas zone, Entrelor, Cottian Alps, Voltri

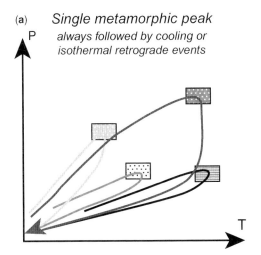 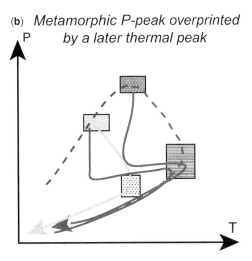

Fig. 5. Ambiguity of some metamorphic facies. (**a**) A rock undergoes a simple metamorphic evolution in a distinct geodynamic setting, leading to only one peak metamorphic paragenesis (simultaneous P and T peaks) followed by later retrogression. (**b**) In some cases, certain metamorphic PT paths show distinct pressure and temperature peaks. The pressure-peak can be related to one geodynamical scenario and it may have been overprinted by a thermal peak that resulted from a second and different geodynamic scenario.

group, Tuscany and Corsica). The situation is more complex for the units derived from the Briançonnais terrane. Going from internal to external, they indicate eclogite and UHP conditions in the 'internal massifs' (Monte Rosa, Gran Paradiso, Dora Maira), blueschist conditions (Suretta cover, Mont Fort, Ruitor, Vanoise, Acceglio, Ligurian Alps, Tenda), and subduction-related greenschist and low-grade conditions (Tasna, Schams, Zone Houillère). Based on structural arguments, some authors dispose of this complexity by changing the palaeogeographic attribution of the 'internal massifs' (Froitzheim 2001; Pleuger et al. 2005). We are agreeing with the fact that the geodynamic evolution is complex and that the 'internal massifs' are part of the Briançonnais (see e.g. Polino et al. 1990; Borghi et al. 1996; Froitzheim et al. 1996; Dal Piaz 1999).

In the following, we will present four examples (Cottian Alps, Ruitor, Entrelor and Valais ocean) that document and demonstrate the use of metamorphic studies on metasediments for unravelling different PT evolutions during the early subduction-related history of the Alps.

Continuous P increase within sediments from one single palaeogeographic unit (Cottian Alps). The Schistes Lustrés complex in the Cottian Alps is formed by intensely folded Upper Jurassic (Malm; De Wever & Caby 1981) to Upper Cretaceous calcschists deposited in the oceanic Piemont–Liguria trough (Coniacian–Santonian; Lemoine & Tricart 1986; Deville et al. 1992),

Table 2. *Main metamorphic minerals or mineral assemblages found in metasediments in the Alps, classified according to their meaning*

HP–LT 'minerals'	'Low-grade'/ greenschists 'minerals'	HT 'minerals'	'Ambiguous' minerals
Fe, Mg-carpholite Car	margarite (Mgr)	andalusite (And)	chloritoid (Ctd)
lawsonite Lws	pyrophyllite (Prl)	sillimanite (Sil)	kyanite (Ky)
aragonite Arg	kaolinite (Kln)–Quartz	cordierite (Crd)	garnet (Gt)
coesite Cs	riebeckite (Rbk)	staurolite (St)	clinozoisite/zoisite (c/Zo)
talc Tlc	glauconite (Glt)	wollastonite (Wo)	cookeite (Cook)
glaucophane Gln	stilpnomelane (Stl)	diopside (Di) *in marble*	sudoite (Sud)
jadeite Jd	albite–quartz–phengite–chlorite	tremolite (Tr) *in marble*	
	quartz–phengite–chlorite		

with a few mantle slivers (mainly serpentinites) representing the floor of this Alpine realm largely devoid of mafic oceanic crust (Lagabrielle & Lemoine 1997). The study of metamorphic sediments shows that carpholite-bearing assemblages are present in the western part (Goffé & Chopin 1986; Agard et al. 2001) while chloritoid-bearing assemblages as well as garnet–lawsonite–glaucophane assemblages in marbles (Ballèvre & Lagabrielle 1994) occur in the eastern part. On the basis of metapelite mineralogy, P–T estimates at maximum pressure increase from west to east across the study area from c. 1.2–1.3 GPa at 300–350 °C (Agard et al. 2001) to 1.4–1.5 GPa at 450–500 °C (Ballèvre & Lagabrielle 1994; Agard et al. 2000).

Bimodal evolution within a single paleogeographic unit (Ruitor area). The metamorphic evolution of the basement units derived from the Briançonnais microcontinent was always a matter of debate (Desmons et al. 1999; Monié 1990). In the southwestern Alps, an HP imprint is well documented by occurrences of Fe-Mg-Carpholite (Goffé 1977, 1984; Goffé et al. 1973, 2004; Goffé & Chopin 1986), and aragonite (Gillet & Goffé 1988) in metasediments and by occurrences of lawsonite and jadeite in metabasites (Lefèvre & Michard 1965; Schwartz et al. 2000). In the northwestern Alps in contrast, only the uppermost unit of the Briançonnais domain (the Mont Fort nappe) displays blueschist facies conditions (Schaer 1959; Bearth 1963).

In the Zone Houillère and in its Permo-Triassic cover as well as in the Briançonnais basement, metamorphic mineral assemblages are mainly composed of white micas with varying chemical composition, chloritoid and garnet. This same assemblage may occur within different lithologies (meta-arkose, meta-pelite, meta-sandstone). The increase in metamorphic grade from greenschist facies conditions in the northwest (Zone Houillère) to the transition between blueschist and eclogite facies conditions in the southeast (Internal Briançonnais) is well documented (Bucher & Bousquet 2007). A major discontinuity in metamorphic grade, as documented by a pressure gap of c. 0.7 kbar, is located at a tectonic contact within the Briançonnais terrane, namely that between the Zone Houillère and Ruitor unit (Caby et al. 1978; Bucher 2003).

Rock associations displaying different metamorphic peak conditions (Entrelor area). Two types of metamorphic rock (blueschist and eclogites) have been described in this area, which is part of the Piemont–Liguria units of the western Alps (Dal Piaz 1999). Recently, the rock assemblage in the Entrelor area has been interpreted as a metamorphic mix, consisting of eclogitic rocks that were embedded into a blueschist facies matrix consisting of metapelites and greenstones (Bousquet 2008). The two kinds of HP metamorphic rock reveal different peak metamorphic conditions (1.2 GPa at 450 °C vs. 2.3 GPa at 550 °C); it is their contemporaneous exhumation within a subduction channel which juxtaposed them at a shallow crustal level. This evolution illustrates that subduction processes cannot be considered as a single-pass process; instead, return flow of a considerable portion of crustal and upper mantle material must be accounted for (Gerya & Stockhert 2002), and the exhumation of the different rock types cannot be considered independently from each other (Engi et al. 2001). The rocks of the Entrelor area can be viewed as an exhumed part of a frozen subduction channel consisting of a mix of metamorphic rocks that have different metamorphic evolutions, and which were accreted at great depths.

Geometry of the subduction (units derived from the Valais ocean). In the eastern and central Alps, blueschist-facies rocks derived from the Valais ocean are exposed structurally below the Austroalpine nappes over an area of 300×20 km^2 (from the Tauern window to the Grisons area) and have a thickness of around 10 km. This large volume of blueschist-facies rocks is in contrast with that of the eclogite-facies rocks of the western Alps that only form a small 2 to 5 km thick slice. The difference in volume and metamorphic conditions from east to west is probably due to a change in style and geometry of subduction.

In the eastern and central Alps, the blueschist metasediments formed within a wide accretionary wedge with a thickness of 40–50 km which underlies the orogenic lid formed by the Austroalpine nappes and they were exhumed before the final collision between the European and Apulian continents (see discussion in Bousquet et al. 2002). Subduction occurred at a high angle to the strike of the orogen. In the western Alps, where only a narrow accretionary wedge formed (Ceriani & Schmid 2004), producing low-grade metamorphic conditions (Ceriani et al. 2003), subduction occurred in a sinistrally transpressive environment, i.e. at a small angle to strike of the orogen (Schmid & Kissling 2000, Ceriani et al. 2001). The blueschist and eclogite facies metasediments of the Versoyen area (Petit St. Bernard and Versoyen units, Goffé & Bousquet 1997) were also subducted and extruded along a N–S direction, i.e. at a small angle to the orogen (Fügenschuh et al. 1999). Moreover, the western Alps were never overlain by an orogenic lid formed by the Cretaceous-age Austroalpine nappe stack, but at best by rather thin basement silvers attributed to the Margna–Sesia fragment (Schmid et al. 2004).

Despite the fact that metamorphism related to Latest Cretaceous to Cenozoic subduction is scattered all over the Alps, information on these processes is unevenly distributed. Areas with wide occurrences of metasediments (Fig. 1) allow for the best insight into the early geodynamic evolution of the Alps. In the western Alps, all stages of a subduction process in PT frame (from UHP to greenschist conditions), as well its evolution in time (from the latest Cretaceous to the Oligocene, i.e. between 70 and 33 Ma), are recorded. Contrarily in the eastern Alps and in the Tauern window (we exclude the Cretaceous-age high-pressure metamorphism from this discussion since it was related to a different orogeny; Froitzheim et al. 1996) the metamorphic record of the metasediments is limited, and the HP rocks only exhibit Eocene ages (45–35 Ma).

Minerals related to collision processes and their distribution

Collision-related minerals apparently do not occur over the whole Alpine edifice. Minerals produced during collision are mainly indicative for temperature-dominated metamorphic conditions (Barrovian-type metamorphism). They occur in the external zones, as well as in the central Alps (Lepontine dome) and in the Tauern window of the eastern Alps (Fig. 7).

In the external zones, the metamorphic evolution reaches maximum lower greenschist conditions, metamorphism resulting from collisional deformation as nappe emplacement, thrusting and folding (Frey & Ferreiro-Mählmann 1999; Burkhard & Goy-Eggenberger 2001; Ferreiro-Mählmann 2001). Burial processes both control metamorphic conditions in the external zones and limit it to low grade. The area of the Lepontine dome and the Tauern window, however, experienced higher metamorphic conditions, at least amphibolite facies.

Three examples will elucidate geodynamic processes that led to high-temperature metamorphic overprinting of metasediments in more internal zones.

Continental underplating (Tauern window). In the post-Variscan metasediments of the Tauern window, a high-temperature event is mainly indicated by the occurrence of Fe–Ca-rich garnet (Droop 1981; Selverstone 1985). Only a few occurrences of staurolite have been reported, indicating maximum amphibolite facies conditions. The mineral distribution pattern indicates two dome-like structures with concentric temperature gradients. This pattern resulted from the underthrusting of Europe-derived continental basement (Kurz et al. 1999) and its accretion to the overlying Austroalpine basement complex (Apulian Plate). This geodynamic scenario led to simple and continuous P–T paths (Fig. 6) that indicate decompressional heating from HP conditions into amphibolite facies conditions.

In Schieferhülle, the earlier HP- stage (blueschist facies conditions, Selverstone & Spear 1985) at 36 Ma was overprinted by HT conditions (amphibolite facies conditions) at around 30–27 Ma ago (Christensen et al. 1994; Zimmermann et al. 1994).

Continental wedging (Lepontine dome). High-temperature metamorphic conditions in the central Alps (the Lepontine area) were based on the study of metasediments in the early works on Alpine geology (e.g. Schmidt & Preiswerk 1908; Preiswerk 1918). Since these pioneering descriptions, several workers have dealt with metasediments from the Lepontine in order to understand progressive metamorphic evolution in isochemical systems (siliceous carbonates system: Trommsdorff 1966; pelitic and marly compositions: Frey 1969). The Lepontine area is characterized by extensive amphibolite facies conditions, reaching migmatization and/or granulite facies conditions. The thermal overprint (Fig. 6) progressively decreases from UHT-conditions in the south to greenschist facies conditions outwards (Streckeisen et al. 1974; Engi et al. 1995; Todd & Engi 1997). The northern margin of the amphibolite grade Lepontine dome is defined by the appearance of staurolite in pelitic systems. However, to the south it is truncated by the Insubric line, along which granulites and migmatites are juxtaposed to rocks of the southern Alps that did not experience substantial Alpine metamorphism.

Thin Mesozoic metasedimentary bands separate large volumes of basement rocks belonging to various nappes stacked below the Austroalpine nappes and in front of the southern Alps (Apulian Plate). The accretion of vast amounts of crustal material derived from the European margin (Adula, Simano, Leventina) and the Briançonnais terrane (i.e. Maggia nappe) allowed for high radiogenic heat production (Verdoya et al. 2001; Roselle & Engi 2002) producing HT assemblages.

The PT paths deduced from the high-grade metasediments often, but not always (see Nagel et al. 2002; Keller et al. 2005a for more simple PT paths), show bimodal trends (Fig. 6): a HP event is followed by a phase of heating (Engi et al. 2001; Berger et al. 2005; Brouwer et al. 2005; Wiederkehr et al. 2007). Shortly after the HP event, which probably occurred at around 45 Ma (47–51 Ma Bucher 2003; 42.6 Ma Lapen et al. 2007 see discussion in Berger & Bousquet 2008), HT metamorphic conditions prevailed over a long period of time until 30 Ma ago in the south and

15 Ma in the north respectively (Hunziker et al. 1992).

Post-orogenic heating (Tuscany). In western Tuscany, Quaternary magmatism is witnessed by volcanic as well as intrusive rocks. This magmatism is associated with crustal thinning and high heat flow values (Scrocca et al. 2003). Consequently, metasediments not only exhibit HP mineral assemblages, but also minerals such as andalusite, staurolite, chloritoid, epidote that document the high-temperature evolution. A bimodal PT path has been reconstructed from Giglio, indicating both an HP and HT event. (Rossetti et al. 1999).

The inferred palaeotemperature distribution pattern resembles an asymmetric thermal high defined by the appearance of kyanite, similar to the present geothermal pattern of the Tuscan crust (Franchescelli et al. 1986), as indicated by a series of geothermal anomalies passing through the northern Apennines (Della Vedoya et al. 2001). The age of the HT event (from 15 to 8 Ma, Kligfield et al. 1986; Brunet et al. 2000; Molli et al. 2000a, b) clearly post-dated the HP stage (31–26.7 Ma, Brunet et al. 2000) in Tuscany.

Summary. Remnants of the high-temperature event are unevenly distributed throughout the Alps. They are localized in the Tauern window, the Lepontine dome, and in Tuscany. In contrast, large areas lack such an HT overprint: the entire western Alps and the Engadine window located between the Lepontine dome and Tauern window. Both the Lepontine and the Tauern domes are made up of continent-derived granitoid upper crustal metamorphic sequences (Lammerer & Weger 1998; Neubauer et al. 1999; Schmid et al. 1996), while the Engadine window (Bousquet et al. 2002) and the southwestern Alps (Agard et al. 2002; Lardeaux et al. 2006) are mainly built up of oceanic-metasedimentary sequences.

Hence we conclude that the high-temperature event in the eastern and central Alps is due to large local accumulations of crustal material during continental collision, while in Tuscany it records a post-orogenic event, associated with thinning of the lithosphere (Rossetti et al. 1999).

Metamorphic structure of the Alps

Sediments occur throughout mountain belts such as the Alps and represent a large variety of palaeo-environments and chemical compositions. Moreover, several of these compositions are very sensitive to temperature and pressure variations. Thus, they have a high potential for registering the different stages of their geodynamic evolution. Mineral distributions in metasediments, combined with previous works on metabasites, allow the deciphering of the complexity of the Late Cretaceous–Tertiary alpine history (Figs 8 and 9).

Map representation (Fig. 8) allows the clear separation of different areas in the Alps. Corsica and the western Alps as well as the far eastern Alps (Rechnitz window) have recorded only the subduction-related evolution, characterized by HP metamorphism. The central Alps and the eastern Alps (Tauern window) are displaying a more complex history. In these areas, the HP phase is overprinted by higher temperature conditions.

Alpine eclogite facies remnants in the central Lepontine area appear to be restricted to a metamorphic mix (Berger et al. 2007). They are isolated occurrences in a belt that includes relics of variegated high-grade metamorphism, from granulite facies to eclogite to amphibolite facies. This structure is interpreted as representing remnants of a tectonic accretion channel (Engi et al. 2001), which had developed along the convergent plate boundary during Alpine subduction.

From the metamorphic map (Fig. 8) and using four major geological transects (Schmid et al. 2004), we propose metamorphic transects across the Alps down to 15–20 km depth (Fig. 9). In the eastern transect (Fig. 9a, along the TRANSALP profile), the main alpine metamorphic features show the thermal overprint. Only scarce relics of the HP history are preserved. In the central Alps (Fig. 9b, NFP 20 east profile), HP and HT metamorphic rocks coexist. The thermal overprinting of different subduction patterns can be observed: eclogites of the Adula complex, rocks undergone into blueschist conditions (European margin—*north of Adula, Simano*—, Valais, Briançonnais —*Tambo, Suretta*—) as well as rocks that have not been subducted (Maggia nappe for the Briançonnais) have been thermally overprinted by the Late Tertiary event. A wedge-type structure built against the Insubric line can be clearly distinguished on the central Alps profile. In the northwestern Alps (Fig. 9c, NFP 20 west profile), while the subduction-related metamorphism is widespread, the thermal overprinting is limited to the European platform (sediments—*Helvetic nappes or Dauphiné*—, basement rocks—e.g. *Mt Blanc*—) or to the structurally lower units (root of the Mte Rosa, Antrona). In the western Alps (Fig. 9d, ECORS-CROP profile), subduction-related metamorphism is the main record. The thermal overprint appears west of the Penninic front and it is limited to the European platform (Belledonne, Pelvoux, Dauphiné). The most internal units are completely lacking an HT event. It can only be assumed for the deepest units (at around 20 km depth). Two subduction zones, indicated by HP metamorphic conditions, can be clearly evidenced. One (in the east) is formed by the Piemont–Liguria rocks, the Gran Paradiso massif and the most internal part of the Briançonnais. The second subduction zone is formed only by the Valais rocks and is rooted at depth. Both zones are separated from each other by the external Briançonnais (Zone Houillère) which lacks evidence for HP metamorphism. The arrangement of the nappe pile in the western Alps clearly shows a subduction-type structure in which most of the tectonic units dip southwestward.

Discussion

Evidence for HP metamorphism recording prolonged subduction processes during at least over 37 Ma (from 70 Ma in Sesia to 33 Ma in Valaisan, see Bousquet et al. 2002 and Berger & Bousquet 2008, for timing constraints). The evidence for these long-lasting processes is also widespread over the entire orogen, from the Rechnitz window in the east to Corsica and Tuscany in the southwest. All palaeogeographic units, of continental and oceanic origin, were involved in these subduction processes. The nappes derived from the Penninic–Austroalpine transition zone (Margna–Sesia fragment), the Piemont–Ligurian ocean, the Briançonnais terrane, the Valais ocean and the European margin all were successively subducted under the Apulian Plate (Berger & Bousquet 2008).

In contrast, high-temperature metamorphism is a relative short-lived process, lasting for some 15 Ma (30 to 15 Ma). The evidence for such HT metamorphism is also localized in specific regions. It is limited to areas where considerable amounts of continental crust were accumulated into accreted nappe stacks. High-temperature conditions (more than 650 °C, up to 800 °C; granulites and migmatites) were reached in the Lepontine dome where a huge amount of continental crust allowed for high radiogenic heat production. In the Tauern, a less important amount of imbricated continental crust led to amphibolite facies conditions (up to 600 °C). In the Engadine window (Hitz 1995, 1996), as well as in the western Alps, the relative scarcity of continental crust involved in the orogenic wedge does not allow for such a high heat production and associated high-temperature overprint.

The transition between the western Alps, lacking such an HT overprint, and the Lepontine dome can be observed in the Monte Rosa area. There, only the lower parts exposed in deep valleys (Domodossola) show an amphibolite facies overprint (650 °C, Keller et al. 2005b).

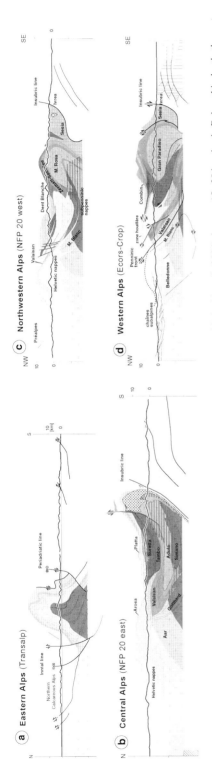

Fig. 9. Metamorphic structure of the Alps. We report metamorphic data along four geological transects of Schmid et al. (2004) down to 15–20 km depth. Below this depth, the past metamorphic history should be overprinted by the present thermal regime. (a) Eastern transect along the TRANSALP seismic profile. (b) Central Alps transect (along the NFP 20 east seismic profile. (c) Northwestern Alps transect along the NFP 20 west seismic profile. (d) Western Alps transect, along the ECORS_CROP seismic profile.

Thermal overprint is primarily related to the amount of crust involved in the subduction and collision processes (Bousquet et al. 1997; Goffé et al. 2003) rather than to processes of shear or viscous heating (Burg & Gerya 2005). The latter mechanism, which suppose high deformation rate, will not allow for the preservation of HP–LT assemblages within high-grade rocks, as is found for example in the southern Adula complex (Nagel et al. 2002).

The relation between the volume of continental crust imbricated and intensity of high-temperature orogenic metamorphism can be generalized over the entire Alpine edifice, except for Tuscany where the late (<8 Ma) thermal overprint is clearly related to lithospheric thinning.

Conclusions

Based on metamorphic studies in metasediments, we evidence substantial differences in the metamorphic and hence the geodynamical evolution along strike of the Alpine orogen.

The western Alps did not reach the mature stage of a head-on colliding belt as is indicated by a continuous metamorphic evolution, representing all the subduction-related processes ranging from lower greenschist to UHP conditions. All the metamorphic rocks behind the Pennine frontal thrust were already exhumed to upper crustal level during ongoing oceanic and continental subduction and before collision with the Dauphinois domain from around 32 Ma onwards (Fügenschuh & Schmid 2003; Leloup et al. 2005). Hence, the western Alps represent a frozen-in subduction zone. Since then, only exhumation by erosional processes affected the inner parts of the orogen.

The rest of the Alpine orogen later underwent a more important collision process due to the ongoing head-on geometry of subduction and collision. It therefore often but not always shows a bimodal metamorphic evolution with two distinct P and T peaks. The intensity of the thermal overprint relates to the amount of crustal material incorporated to the orogenic wedge.

This article benefits from discussions with many colleagues. All cannot be cited, but RB wants to specially thank M. Ballèvre, B. Fügenschuh, S. Bucher, R. Caby, R. Polino, D. Marquer, G. Gosso, M.-I. Spalla, V. Höck, O. Vanderhaeghe, S. Duchêne, R. Ferreiro-Mählmann, J.-M. Lardeaux and P. Rossi. We are indebted to N. Froitzheim for his corrections and comments. He helped us to clarify and improve the manuscript. Despite the cutting tone of his review, we thank K. Stüwe: some of his comments were useful. Without the encouragement, the editorial work and the patience of S. Siegesmund, this paper would not have existed.

Appendix

List of the references used to build the maps on Figures 5–7.

Eastern Alps: Bousquet (this study), Bousquet et al. (1998), Dachs (1986, 1990), Droop (1981, 1985), Franz (1983), Höck (1974, 1980), Leimser & Purtscheller (1980), Matile & Widmer (1993), Miller et al. (2007), Selverstone et al. 1984.

Central Alps: Bousquet (this study), Bousquet et al. (2002), Droop & Bucher-Nurminen (1984), Frey (1969, 1978), Frey & Niggli (1972), Frey & Wiedeland (1975), Frey et al. (1973, 1974, 1980), Frey & Ferreiro-Mählmann (1999), Goffé & Oberhänsli (1992), Irouschek (1983), Oberhänsli et al. (1995, 2003), Niggli & Niggli (1965), Staub (1926), Trommsdorff (1966), Wiederkehr et al. (2007).

Western Alps: Agard et al. (2000, 2001), Ballèvre (1988), Ballèvre & Lagabrielle (1994), Bousquet (2007), Bousquet et al. (2004), Bucher & Bousquet (2007), Caron & Saliot (1969), Caron (1974), Ceriani et al. (2003), Chopin (1981, 1984), Chopin et al. (2003), Cigolini (1995), Desmons et al. (1999), Gillet & Goffé (1988), Goffé (unpub.), Goffé et al. (1973), Goffé (1977, 1982, 1984), Goffé & Chopin (1986), Goffé & Velde (1984), Goffé & Bousquet (1997), Goffé et al. (2004), Henry (1990), Jullien & Goffé (1993), Le Bayon et al. (2006), Leikine et al. (1983), Martinotti (unpub.), Saliot (1979), Venturini (1995).

Corsica and Tuscany: Caron & Péquignot (1986), Caron (1994), Daniel & Jolivet (1995), Daniel et al. (1996), Franceschelli et al. (1986, 1989), Franseschelli & Memmi (1999), Giorgetti et al. (1997), Goffé (unpub.), Goffé (1982), Jolivet et al. (1998), Leoni et al. (1996), Molli et al. (2006), Rossetti et al. (1999, 2001), Theye et al. (1997), Tribuzio & Giacomini (2002).

References

AGARD, P., GOFFÉ, B., TOURET, J. L. R. & VIDAL, O. 2000. Retrograde fluid evolution in blueschist-facies metapelites (Schistes lustrés unit, Western Alps). Contributions to Mineralogy and Petrology, **140**, 296–315.

AGARD, P., JOLIVET, L. & GOFFÉ, B. 2001. Tectonometamorphic evolution of the Schistes Lustrés complex: implications for the exhumation of HP and UHP rocks in the Western Alps. Bulletin de la Société géologique de France, **172**, 617–636.

AGARD, P., MONIÉ, P., JOLIVET, L. & GOFFÉ, B. 2002. Exhumation of the Schistes Lustrés complex: in situ laser probe $^{40}Ar/^{39}Ar$ constraints and implications for the Western Alps. Journal of Metamorphic Geology, **20**, 599–618.

BALLÈVRE, M. 1988. Collision continentale et chemins P–T: l'unité pennique du Grand Paradis, Alpes Occidentales Centre Armoricain d'études structurale des socles, Rennes.

BALLÈVRE, M. & LAGABRIELLE, Y. 1994. Garnet in Blueschist-facies marbles from the Queyras unit (Western Alps) — its occurrence and its significance. *Schweizerische Mineralogische und Petrographische Mitteilungen*, **74**, 203–212.

BARROW, G. 1893. On an intrusion of muscovite-biotite gneiss in the south-eastern Highlands of Scotland, and its accompanying metamorphism. *The Quarterly Journal of the Geological Society*, **49**, 330–358.

BARROW, G. 1912. On the geology of the lower Deeside and the southern Highland border. *Proceedings of the Geologists' Association*, **23**, 268–284.

BEARTH, P. 1963. Contribution à la subdivision tectonique et stratigraphique du cristallin de la nappe du Grand Saint-Bernard dans le Valais (Suisse). *In*: DURAND-DELGA, M. (ed.) *Livre à la mémoire du Professeur Fallot*. Mémoire de la Société géologique de France, Paris, 407–418.

BERGER, A. & BOUSQUET, R. 2008. Subduction related metamorphism in the Alps: Review of isotopic ages based on petrology and their geodynamic consequences. *In*: SIEGESMUND, S., FÜGENSCHUH, B. & FROITZHEIM, N. 2008. *Tectonic Aspects of the Alpine-Dinaride-Carpathian*. Geological Society, London, Special Publication, **298**, 117–144.

BERGER, A., MERCOLLI, I. & ENGI, M. 2005. The central Lepontine Alps: Explanatory notes accompanying the tectonic-geological map sheet Sopra Ceneri (1:100 000). *Schweizerische Mineralogische und Petrographische Mitteilungen*, **85**, 109–146.

BERMAN, R. G. 1988. Internally-Consistent Thermodynamic Data for Minerals in the system Na_2O-K_2O-CaO-MgO-FeO-Fe_2O_3-Al_2O_3-SiO_2-TiO_2-H_2O-CO_2. *Journal of Petrology*, **29**, 445–522.

BIGI, G., CASTELLARIN, A., COLI, M., DAL PIAZ, G. V., SARTORI, R., SCANDONE, P. & VAI, G. B. 1990. Structural model of Italy. *In*: *Progetto Geodinamica*, SELCA, Firenze.

BORGHI, A., COMPAGNONI, R. & SANDRONE, R. 1996. Composite P-T paths in the internal Penninic massifs of the western Alps: petrological constraints to their thermo-mechanical evolution. *Eclogae Geologicae Helvetiae*, **89**, 345–367.

BOUSQUET, R. 2008. Metamorphic heterogeneities within a same HP unit: overprint effect or metamorphic mix? *Lithos*, doi: 10.1016/j.lithos.2007.09.010.

BOUSQUET, R., GOFFÉ, B., HENRY, P., LE PICHON, X. & CHOPIN, C. 1997. Kinematic, Thermal and Petrological Model of the Central Alps: Lepontine Metamorphism in the Upper Crust and Eclogitisation of the Lower Crust. *Tectonophysics*, **273**, 105–127.

BOUSQUET, R., OBERHÄNSLI, R., GOFFÉ, B., JOLIVET, L. & VIDAL, O. 1998. High pressure-low temperature metamorphism and deformation in the Bündnerschiefer of the Engadine window: implications for the regional evolution of the eastern Central Alps. *Journal of Metamorphic Geology*, **16**, 657–674.

BOUSQUET, R., GOFFÉ, B., VIDAL, O., OBERHÄNSLI, R. & PATRIAT, M. 2002. The tectono-metamorphic history of the Valaisan domain from the Western to the Central Alps: New constraints for the evolution of the Alps. *Bulletin of the Geological Society of America*, **114**, 207–225.

BOUSQUET, R., ENGI, M., GOSSO, G. *ET AL*. 2004. Transition from the Western to the Central Alps. *In*: OBERHÄNSLI, R. (ed.) *Explanatory note to the map 'Metamorphic structure of the Alps'*. Mitteilungen Österreichische Mineralogischen Gesellschaft, Wien, 145–156.

BROUWER, F. M., BURRI, T., ENGI, M. & BERGER, A. 2005. Eclogite relics in the Central Alps: PT-evolution, Lu-Hf ages, and implications for formation of tectonic mélange zones. *Schweizerische Mineralogische und Petrographische Mitteilungen*, **85**, 147–174.

BRUNET, C., MONIÉ, P., JOLIVET, L. & CADET, J.-P. 2000. Migration of compression and extension in the Tyrrhenian Sea, insights from 40Ar/39Ar ages on micas along a transect from Corsica to Tuscany. *Tectonophysics*, **321**, 127–155.

BUCHER, S. 2003. *The Briançonnais units along the ECORS-CROP transect (Italian-French Alps): structures, metamorphism and geochronology*. Universität Basel, Basel.

BUCHER, S. & BOUSQUET, R. 2007. Metamorphic evolution of the Briançonnais units along the ECORS-CROP profile (Western Alps): New data on metasedimentary rocks. *Journal of Swiss Geosciences*, **100**, 1–16.

BUCHER, K. & FREY, M. 1994. *Petrogenesis of Metamorphic Rocks*. Springer Verlag, Berlin.

BURG, J.-P. & GERYA, T. V. 2005. The role of viscous heating in Barrovian metamorphism of collisional orogens: thermomechanical models and application to the Lepontine Dome in the Central Alps. *Journal of Metamorphic Geology*, **23**, 75–95.

BURKHARD, M. & GOY-EGGENBERGER, D. 2001. Near vertical iso-illite-crystallinity surfaces cross-cut the recumbent fold structure of the Morcles nappe, Swiss Alps. *Clay Minerals*, **36**, 159–170.

CABY, R. 1996. Low-angle extrusion of high-pressure rocks and the balance between outward and inward displacements of Middle Penninic units in the western Alps. *Eclogae Geologicae Helvetiae*, **89**, 229–267.

CABY, R., KIENAST, J.-R. & SALIOT, P. 1978. Structure, métamorphisme et modèle d'évolution tectonique des Alpes Occidentales. *Revue de Géographie physique et de Géologie dynamique*, **XX**, 307–322.

CARON, J.-M. 1974. Rapports entre diverses 'générations' de lawsonite et les déformations dans les Schistes lustrés des Alpes cottiennes septentrionales (France et Italie). *Bulletin de la Société Géologique de France*, **16**, 255–268.

CARON, J.-M. 1994. Metamorphism and deformation in Alpine Corsica. *Schweizerische Mineralogische und Petrographische Mitteilungen*, **74**, 105–114.

CARON, J.-M. & PEQUIGNOT, G. 1986. The Transition between Blueschists and Lawsonite-Bearing Eclogites Based on Observations from Corsican Metabasalts. *Lithos*, **19**, 205–218.

CARON, J.-M. & SALIOT, P. 1969. Nouveaux gisements de lawsonite et de jadeite dans les alpes franco-italiennes. *Comptes Rendus de l'Académie des Sciences Paris*, **268**, 3153–3156.

CARSWELL, D. A. 1990. *Eclogite Facies Rocks*. Blackie Academic and Professional, London.

CERIANI, S. & SCHMID, S. M. 2004. From N-S collision to WNW-directed post-collisional thrusting and folding: Structural study of the Frontal Penninic Units in Savoie (Western Alps, France). *Eclogae Geologicae Helvetiae*, **97**, 347–369.

CERIANI, S., FÜGENSCHUH, B. & SCHMID, S. M. 2001. Multi-stage thrusting at the 'Penninic Front' in the Western Alps between Mont Blanc and Pelvoux massifs. *International Journal of Earth Sciences*, **90**, 685–702.

CERIANI, S., FUGENSCHUH, B., POTEL, S. & SCHMID, S. M. 2003. Tectono-metamorphic evolution of the Frontal Penninic units of the Western Alps: correlation between low-grade metamorphism and tectonic phases. *Schweizerische Mineralogische und Petrographische Mitteilungen*, **83**, 111–131.

CHATTERJEE, N. D. 1976. Margarite stability and compatibility relations in the system $CaO-Al_2O_3-SiO_2-H_2O$ as a pressure–temperature indicator. *American Mineralogist*, **61**, 699–709.

CHOPIN, C. 1981. Talc-phengite: a Widespread Assemblage in High-Grade Pelitic Blueschists of the Western Alps. *Journal of Petrology*, **22**, 628–650.

CHOPIN, C. 1984. Coesite and pure pyrope in high grade blueschists of the western Alps: a first record and some consequences. *Contributions to Mineralogy and Petrology*, **86**, 107–118.

CHOPIN, C. 1987. Very-High-Pressure Metamorphism in the Western Alps: Implications for Subduction of Continental Crust. *Philosophical Transactions of the Royal Society of London*, **321**, 183–195.

CHOPIN, C. & SCHREYER, W. 1983. Magnesiocarpholite and magnesiochloritoid: two index minerals of pelitic blueschists and their preliminary phase relations in the system $MgO-Al_2O_3-SiO_2-H_2O$. *American Journal of Science*, **283-A**, 72–96.

CHOPIN, C. & SOBOLEV, N. V. 1995. Principal mineralogic indicators of UHP in crustal rocks. In: COLEMAN, R. G. & WANG, X. (eds) *Ultrahigh Pressure Metamorphism*. Cambridge University Press, Cambridge, 96–131.

CHOPIN, C., GOFFÉ, B., UNGARETTI, L. & OBERTI, R. 2003. Magnesiostaurolite and zincostaurolite: mineral description with a petrogenetic and crystal-chemical update. *European Journal of Mineralogy*, **15**, 167–176.

CHRISTENSEN, J. N., SELVERSTONE, J., ROSENFELD, J. L. & DEPAOLO, D. J. 1994. Correlation by Rb-Sr geochronology of garnet growth histories from different structural levels within the Tauern Window, Eastern Alps. *Contributions to Mineralogy and Petrology*, **118**, 1–12.

CIGOLINI, C. 1995. Geology of the Internal Zone of the Grand Saint Bernard Nappe: a metamorphic Late Paleozoic volcano-sedimentary sequence in South-Western Aosta Valley (Western Alps). In: LOMBARDO, B. (ed.) *Studies on metamorphic rocks and minerals of the western Alps. A Volume in Memory of Ugo Pognante*. Bollettino del Museo Regionale di Scienze Naturali (suppl.), Torino, 293–328.

COLEMAN, R. G. & WANG, X. 1995. *Ultrahigh Pressure Metamorphism*. Cambridge University Press, Cambridge.

DACHS, E. 1986. High-pressure mineral assemblages and their breakdown-products in metasediments South of the Grossvenediger, Tauern Window, Austria. *Schweizerische Mineralogische und Petrographische Mitteilungen*, **66**, 145–162.

DACHS, E. 1990. Geothermobarometry in metasediments of the southern Grossvenediger area (Tauern Window, Austria). *Journal of Metamorphic Geology*, **8**, 217–230.

DAL PIAZ, G. V. 1974a. Le métamorphisme alpin de haute pression-basse température dans l'évolution structurale du bassin ophiolitique alpino-appenninique. 1ère partie. *Bolletino della Società Geologica Italiana*, **93**, 437–468.

DAL PIAZ, G. V. 1974b. Le métamorphisme alpin de haute pression-basse température dans l'évolution structurale du bassin ophiolitique alpino-appenninique. 2ème partie. *Schweizerische Mineralogische und Petrographische Mitteilungen*, **54**, 399–424.

DAL PIAZ, G. V. 1999. The Austroalpine-Piedmont nappe stack and the puzzle of Alpine Tethys. In: GOSSO, G., JADOUL, F., SELLA, M. & SPALLA, M. I. (eds) *3rd Workshop on Alpine Geological Studies*. Memorie di Scienze Geologiche, Padova, 155–176.

DAL PIAZ, G. V. 2001. History of tectonic interpretations of the Alps. *Journal of Geodynamics*, **32**, 99–114.

DAL PIAZ, G. V., HUNZIKER, J. C. & MARTINOTTI, G. 1972. La Zona Sesia—Lanzo e l'evoluzione tettonico-metamorfica delle Alpi Nordoccidentali interne. *Memorie della Societa Geologica Italiana*, **11**, 433–460.

DANIEL, J.-M. & JOLIVET, L. 1995. Detachment faults and pluton emplacement—Elba Island (Tyrrhenian Sea). *Bulletin de la Société Géologique de France*, **166**, 341–354.

DANIEL, J. M., JOLIVET, L., GOFFÉ, B. & POINSSOT, C. 1996. Crustal-scale strain partitioning: footwall deformation below the Alpine Oligo-Miocene detachment of Corsica. *Journal of Structural Geology*, **18**, 41–59.

DE CAPITANI, C. 1994. Gleichgewichts-Phasendiagramme: Theorie und Software. *Berichte der Deutschen Mineralogischen Gesellschaft*, **6**, 48.

DE CAPITANI, C. & BROWN, T. H. 1987. The computation of chemical equilibrium in complex systems containing non-ideal solutions. *Geochimica et Cosmochimica Acta*, **51**, 2639–2652.

DE ROEVER, W. P. 1951. Ferrocarpholite, the hitherto unknown ferrous iron analogue of capholite proper. *American Mineralogist*, **36**, 736–745.

DE WEVER, P. & CABY, R. 1981. Datation de la base des schistes lustrés postophiolitiques par des radiolaires (Oxfordien-Kimmeridgien moyen) dans les Alpes Cottiennes (Saint Véran, France). *Comptes Rendus de L'Académie des Sciences*, **292**, 467–472.

DELLA VEDOYA, B., BELLANI, S. G, P. & SQUARCI, P. 2001. Deep temperatures and surface heat flow distribution. In: VAI, G. B. & MARTINI, P. (eds) *Anatomy of an orogen: the Apennines and adjacent Mediterranean basins*. Kluwer Academic Publishers, Dordrecht, The Netherlands, 65–76.

DESMONS, J., APRAHAMIAN, J., COMPAGNONI, R., CORTESOGNO, L. & FREY, M. 1999. Alpine Metamorphism of the Western Alps: I. Middle to highT/P metamorphism. *Schweizerische Mineralogische und Petrographische Mitteilungen*, **79**, 89–110.

DEVILLE, E., FUDRAL, S., LAGABRIELLE, Y., MARTHALER, M. & SARTORI, M. 1992. From oceanic closure to continental collision: A synthesis of the 'Schistes lustrés' metamorphic complex of the Western Alps. *Bulletin of the Geological Society of America*, **104**, 127–139.

DROOP, G. T. R. 1981. Alpine metamorphism of pelitic schists in the Southeast Tauern Window, Austria. *Schweizerische Mineralogische und Petrographische Mitteilungen*, **61**, 237–273.

DROOP, G. T. R. 1985. Alpine metamorphism in the south-east Tauern Window, Austria: 1. P–T variations in space and time. *Journal of Metamorphic Geology*, **3**, 371–402.

DROOP, G. T. R. & BUCHER-NURMINEN, K. 1984. Reaction textures and metamorphic evolution of sapphirine-bearing granulites from Gruf Complex, Italian Central Alps. *Journal of Petrology*, **25**, 766–803.

ENGI, M., TODD, C. S. & SCHMATZ, D. R. 1995. Tertiary metamorphic conditions in the eastern Lepontine Alps. *Schweizerische Mineralogische und Petrographische Mitteilungen*, **75**, 347–369.

ENGI, M., BERGER, A. & ROSELLE, G. T. 2001. Role of the accretion channel in collisional orogeny. *Geology*, **29**, 1143–1146.

ERNST, W. G. 1971. Metamorphic Zonations on Presumably Subducted Lithospheric Plates from Japan, California and the Alps. *Contributions to Mineralogy and Petrology*, **34**, 43–59.

ESCHER, A. & BEAUMONT, C. 1997. Formation, burial and exhumation of basement nappes at crustal scale: a geometric model based on the Western Swiss–Italian Alps. *Journal of Structural Geology*, **19**, 955–974.

ESKOLA, P. 1929. On mineral facies. *Geologiska Föreningens i Stockholm Förhandlingar*, **51**, 157–172.

EVANS, B. W. 1990. Phase relation of epidote blueschists. *Lithos*, **25**, 3–23.

FERREIRO-MÄHLMANN, R. 2001. Correlation of very low grade data to calibrate a thermal maturity model in a nappe tectonic setting, a case study from the Alps. *Tectonophysics*, **334**, 1–33.

FRANCESCHELLI, M. & MEMMI, I. 1999. Zoning of chloritoid from kyanite-facies metapsammites, Alpi Apuane, Italy. *Mineralogical Magazine*, **63**, 105–110.

FRANCESCHELLI, M., LEONI, L., MEMMI, I. & PUXEDDU, M. 1986. Regional distribution of Al-silicates and metamorphic zonation in the low-grade Verrucano metasediments from the Northern Apennines, Italy. *Journal of Metamorphic Geology*, **4**, 309–321.

FRANCESCHELLI, M., MELLINI, M., MEMMI, I. & RICCI, C. A. 1989. Sudoite, a rock-forming mineral in Verrucano of the Northern Apennines (Italy) and the sudoite–chloritoid–pyrophyllite assemblage in prograde metamorphism. *Contributions to Mineralogy and Petrology*, **101**, 274–279.

FRANZ, G. & SPEAR, F. S. 1983. High pressure metamorphism of siliceous dolomites from the central Tauern Window, Austria. *American Journal of Science*, **283-A**, 396–413.

FREY, M. 1969. *Die Metamorphose des Keupers vom Tafeljura bis zum Lukmanier-Gebiet*, Bern.

FREY, M. 1978. Progressive low-grade metamorphism of a black shale formation, Central Swiss Alps, with special reference to pyrophyllite and margarite bearing assemblages. *Journal of Petrology*, **19**, 95–135.

FREY, M. & NIGGLI, E. 1972. Margarite, an important mineral rock-forming mineral in regionally metamorphosed low-grade rocks. *Naturwissenschaften*, **59**, 214–225.

FREY, M. & ROBINSON, D. 1999. *Low-Grade Metamorphism*. Blackwell Science. Wiley-Blackwell, Oxford.

FREY, M. & WIEDELAND, B. 1975. Chloritoid in autochthon-paraautochthonen Sedimenten des Aarmassivs. *Schweizerische Mineralogische und Petrographische Mitteilungen*, **55**, 407–418.

FREY, M., HUNZIKER, J. C., ROGGWILLER, P. & SCHINDLER, C. 1973. Progressive niedriggradige Metamorphose glaukonitführender Horizonte in den helvetischen Alpen der Ostschweiz. *Contributions to Mineralogy and Petrology*, **39**, 185–218.

FREY, M., HUNZIKER, J.-C., FRANK, W., BOQUET, J., DAL PIAZ, G. V., JÄGER, E. & NIGLI, E. 1974. Alpine metamorphic of the Alps: a review. *Schweizerische Mineralogische und Petrographische Mitteilungen*, **54**, 247–291.

FREY, M., BUCHER, K., FRANK, E. & MULLIS, J. 1980. Alpine metamorphism along the GeoTraverse Basel-Chiasso—a review. *Eclogae Geologicae Helvetiae*, **73**, 527–546.

FREY, M., DE CAPITANI, C. & LIOU, J. G. 1991. A new petrogenetic grid for low-grade metabasites. *Journal of Metamorphic Geology*, **9**, 497–509.

FREY, M. & FERREIRO-MÄHLMANN, R. 1999. Alpine metamorphism of the Central Alps. *Schweizerische Mineralogische und Petrographische Mitteilungen*, **79**, 135–154.

FROITZHEIM, N. 2001. Origin of the Monte Rosa nappe in the Pennine Alps – a new working hypothesis. *Bulletin of the Geological Society of America*, **113**, 604–614.

FROITZHEIM, N., SCHMID, S. M. & FREY, M. 1996. Mesozoic paleogeography and timing of eclogitefacies metamorphism in the Alps: a working hypothesis. *Eclogae geologicae Helvetiae*, **89**, 81–110.

FÜGENSCHUH, B. & SCHMID, S. M. 2003. Late stages of deformation and exhumation of an orogen constrained by fission-track: A case study in the Western Alps. *Bulletin of the Geological Society of America*, **115**, 1425–1440.

FÜGENSCHUH, B., LOPRIENO, A., CERIANI, S. & SCHMID, S. M. 1999. Structural analysis of the Subbriançonnais and Valais units in the area of Moûtiers (Savoy, Western Alps): palaeogeographical and tectonic consequences. *International Journal of Earth Sciences*, **88**, 201–218.

GERYA, T. V. & STOCKHERT, B. 2002. Exhumation rates of high pressure metamorphic rocks in subduction channels: The effect of rheology. *Geophysical Research Letters*, **29**, 1261.

GILLET, P. & GOFFÉ, B. 1988. On the significance of aragonite occurrence in the Western Alps. *Contributions to Mineralogy and Petrology*, **99**, 70–81.

GIORGETTI, G., MEMMI, I. & NIETO, F. 1997. Microstructures of intergrown phyllosilicate grains from Verrucano metasediments (northern Apennines, Italy). *Contributions to Mineralogy and Petrology*, **128**, 127–138.

GOFFÉ, B. 1977. Succession de subfacies métamorphiques en Vanoise méridionale (Savoie). *Contributions to Mineralogy and Petrology*, **62**, 23–41.

GOFFÉ, B. 1982. Définition du faciès à Fe, Mg–carpholite–chloritoide, un marqueur du métamorphisme de HP-BT dans les métasédiments alumineux. *Université P. et M. Curie, Paris.*

GOFFÉ, B. 1984. Le facies à carpholite–chloritoide dans la couverture briançonnaise des Alpes Ligures: un témoin de l'histoire tectono-metamorphique régionale. *Memorie della Societa Geologica Italiana*, **28**, 461–479.

GOFFÉ, B. & BOUSQUET, R. 1997. Ferrocarpholite, chloritoïde et lawsonite dans les métapelites des unités du Versoyen et du Petit St Bernard (zone valaisanne, Alpes occidentales). *Schweizerische Mineralogische und Petrographische Mitteilungen*, **77**, 137–147.

GOFFÉ, B. & CHOPIN, C. 1986. High-pressure metamorphism in the Western Alps: zoneography of metapelites, chronology and consequences. *Schweizerische Mineralogische und Petrographische Mitteilungen*, **66**, 41–52.

GOFFÉ, B. & OBERHÄNSLI, R. 1992. Ferro- and magnesiocarpholite in the 'Bündnerschiefer' of the eastern Central Alps (Grisons and Engadine Window). *European Journal of Mineralogy*, **4**, 835–838.

GOFFÉ, B. & VELDE, B. 1984. Contrasted metamorphic evolution in thrusted cover units of the Briançonnais zone (French Alps): a model for the conservation of HP-BT metamorphic mineral assemblages. *Earth and Planetary Science Letters*, **68**, 351–360.

GOFFÉ, B., GOFFÉ-URBANO, G. & SALIOT, P. 1973. Sur la présence d'une variété magnésienne de la ferrocarpholite en Vanoise (Alpes françaises): sa signification probable dans le métamorphisme alpin. *Comptes Rendus de l'Académie des Sciences Paris*, **277**, 1965–1968.

GOFFÉ, B., BOUSQUET, R., HENRY, P. & LE PICHON, X. 2003. Effect of the chemical composition of the crust on the metamorphic evolution of orogenic wedges. *Journal of Metamorphic Geology*, **21**, 123–142.

GOFFÉ, B., SCHWARTZ, S., LARDEAUX, J. M. & BOUSQUET, R. 2004. Explanatory Notes to the Map: Metamorphic structure of the Alps Western and Ligurian Alps. *Mitteilungen Österreichische Mineralogischen Gesellschaft*, **149**, 125–144.

HANDY, M. & OBERHÄNSLI, R. 2004. Age Map of the Metamorphic Structure of the Alps—Tectonic Interpretation and Outstanding Problems. *In:* OBERHÄNSLI, R. (ed.) *Explanatory note to the map 'Metamorphic structure of the Alps'*. Mitteilungen Österreichische Mineralogischen Gesellschaft, Wien, 201–226.

HAUŸ, R.-J. 1822. *Traité de minéralogie. Seconde édition, revue, corrigée et considérablement augmentée par l'auteur.* Bachelier et Huzard, Paris.

HÉBERT, R. & BALLÈVRE, M. 1993. Petrology of staurotide-bearing metapelites from the Cadomian belt, northern Brittany (France): constraints on low-pressure metamorphism. *Bulletin de la Société géologique de France*, **164**, 215–228.

HENRY, C. 1990. L'unité à coesite du massif de Dora maira dans son cadre pétrologique et structural (Alpes occidentales, Italie). *Université Paris VI, Paris.*

HENRY, P., LE PICHON, X. & GOFFÉ, B. 1997. Kinematic, thermal and petrological model of the Himalayas: constraints related to metamorphism within the underthrust Indian crust and topographic elevation. *Tectonophysics*, **273**, 31–56.

HITZ, L. 1995. The 3D crustal structure of the Alps of eastern Switzerland and western Austria interpreted from a network of deep-seismic profiles. *Tectonophysics*, **248**, 71–96.

HITZ, L. 1996. The deep structure of the Engadine Window: Evidence from deep seismic data. *Eclogae Geologicae Helvetiae*, **89**, 657–675.

HÖCK, V. 1974. Coexisting phengite, paragonite and margarite in metasediments of mittlere Hohe-Tauern, Austria. *Contributions to Mineralogy and Petrology*, **43**, 261–273.

HÖCK, V. 1980. Distribution maps of minerals of the Alpine metamorphism in the Penninic Tauern window, Austria. *Mitteilungen österreichische geologische Gesellschaft*, **71/72**, 119–127.

HUNZIKER, J. C. 1974. Rb-Sr and K-Ar age determination and the Alpine history of the Western Alps. *Memoire degli Instituti di Geologia e Mineralogia dell'Università di Padova*, **31**, 1–54.

HUNZIKER, J.-C., DESMONS, J. & HURFORD, A. J. 1992. Thirty-two years of geochronological work in the Central and Western Alps: a review on seven maps. *Mémoires de Géologie*, **13**. Lausanne.

IROUSCHEK, A. 1983. *Mineralogie und Petrographie von Metapeliten der Simano-Decke unter besonderer Berücksichtigung cordieritführender Gesteine zwischen Alpe Sponda and Biasca.* Universität Basel, Basel.

JOLIVET, L., FACCENNA, C., GOFFÉ, B., MATTEI, M., ROSSETTI, F., BRUNET, C., STORTI, F., FUNICIELLO, R., CADET, J.-P., D'AGOSTINO, N. & PARRA, T. 1998. Midcrustal shear zones in postorogenic extension: example from the northern Tyrrhenian Sea. *Journal of Geophysical Research*, **103**, 12123–12161.

JULLIEN, M. & GOFFÉ, B. 1993. Occurrences de cookeite et de pyrophyllite dans les schistes du Dauphinois (Isère, France): Conséquences sur la répartition du métamorphisme dans les zones externes alpines. *Schweizerische Mineralogische und Petrographische Mitteilungen*, **73**, 357–363.

KELLER, L. M., HESS, M., FÜGENSCHUH, B. & SCHMID, S. M. 2005a. Structural and metamorphic evolution of the Camughera, Moncucco, Antrona and Monte Rosa units southwest of the Simplon line, Western Alps. *Eclogae Geologicae Helvetiae*, **98**, 19–49.

KELLER, L. M., ABART, R., SCHMID, S. M. & DE CAPITANI, C. 2005b. Phase Relations and Chemical Composition of Phengite and Paragonite in Pelitic Schists During Decompression: a Case Study from the Monte Rosa Nappe and Camughera-Moncucco Unit, Western Alps. *Journal of Petrology*, **46**, 2145–2166.

KLIGFIELD, R., HUNZIKER, J., DALLMEYER, R. D. & SCHAMEL, S. 1986. Dating of deformation phases using K–Ar and $^{40}Ar/^{39}Ar$ techniques—Results from the Northern Apennines. *Journal of Structural Geology*, **8**, 781–798.

KOHN, M. J. & SPEAR, F. S. 1993. Phase Equilibria of Margarite-Bearing Schists and Chloritoid + Hornblende Rocks from Western New Hampshire, USA. *Journal of Petrology*, **34**, 631–651.

KOLLER, F. & HÖCK, V. 1987. Mesozoic ophiolites in the Eastern Alps. *In*: MALPAS, J., MOORES, E. M., PANAYIOTOU, A. & XENOPHONTOS, C. (eds) *Ophiolites, Oceanic Crustal Analogues.* Proceedings of the Symposium "TROODOS 1987", 253–263.

KOONS, P. O. & THOMPSON, A. B. 1985. Non-mafic rocks in the greenschist, blueschist and eclogite facies. *Chemical Geology*, **50**, 3–30.

KURZ, W., NEUBAUER, F. & UNZOG, W. 1999. Evolution of Alpine Eclogites in the Eastern Alps: Implications for Alpine Geodynamics. *Physics and Chemistry of the Earth, part A: Solid Earth and Geodesy*, **24**, 667–674.

LAGABRIELLE, Y. & LEMOINE, M. 1997. Alpine, Corsican and Apennine ophiolites: the slow-spreading ridge model. *Comptes Rendus de l'Academie des Sciences – Series IIA – Earth and Planetary Science*, **325**, 909–920.

LAMMERER, B. & WEGER, M. 1998. Footwall uplift in an orogenic wedge: the Tauern Window in the Eastern Alps of Europe. *Tectonophysics*, **285**, 213–230.

LAPEN, T. J., JOHNSON, C. M., BAUMGARTNER, L. P., DAL PIAZ, G. V., SKORA, S. & BEARD, B. L. 2007. Coupling of oceanic and continental crust during Eocene eclogite-facies metamorphism: evidence from the Monte Rosa nappe, western Alps. *Contributions To Mineralogy And Petrology*, **153**, 139–157.

LARDEAUX, J. M., SCHWARTZ, S., TRICART, P., PAUL, A., GUILLOT, S., BETHOUX, N. & MASSON, F. 2006. A crustal-scale cross-section of the southwestern Alps combining geophysical and geological imagery. *Terra Nova*, **18**, 412–422.

LE BAYON, B., PITRA, P., BALLEVRE, M. & BOHN, M. 2006. Reconstructing P-T paths during continental collision using multi-stage garnet (Gran Paradiso nappe, Western Alps). *Journal of Metamorphic Geology*, **24**, 477–496.

LEFÈVRE, R. & MICHARD, A. 1965. La jadéite dans le métamorphisme alpin, à propos des gisements de type nouveau, de la bande d'Acceglio (Alpes cottiennes, Italie). *Bulletin de la Société française de Minéralogie et de Cristallographie*, **LXXXVIII**, 664–677.

LEIKINE, M., KIENAST, J.-R., ELTCHANINOFF-LANCELOT, C. & TRIBOULET, S. 1983. Le métamorphisme polyphasé des unités dauphinoises entre Belledonne et Mont-Blanc (Alpes occidentales). Relation avec les épisodes de déformation. *Bulletin de la Société géologique de France*, **XXV**, 575–587.

LEIMSER, W. M. & PURTSCHELLER, F. 1980. Beiträge zur Metamorphose von Metvulkaniten in Pennin des Engadiner Fenster. *Mitteilungen österreichische geologische Gesellschaft*, **71/72**, 129–137.

LELOUP, P. H., ARNAUD, N., SOBEL, E. R. & LACASSIN, R. 2005. Alpine thermal and structural evolution of the highest external crystalline massif: The Mont Blanc. *Tectonics*, **24**, TC4002.

LEMOINE, M. & TRICART, P. 1986. Les schistes lustrés piémontais des Alpes Occidentales: Approches stratigraphique, structurale et sédimentologique. *Eclogae Geologicae Helvetiae*, **79**, 271–294.

LEONI, L., MARRONI, M., SARTORI, F. & TAMPONI, M. 1996. Metamorphic grade in the metapelites of the Internal Liguride Units (Northern Apennines, Italy). *European Journal of Mineralogy*, **8**, 35–50.

MASSONNE, H.-J. & SCHREYER, W. 1987. Phengite geobarometry based on the limiting assemblage with K-feldspar, phlogopite and quartz. *Contributions to Mineralogy and Petrology*, **96**, 212–224.

MASSONNE, H.-J. & SZPURKA, Z. 1997. Thermodynamic properties of white micas on the basis of high-pressure experiments in the systems $K_2O-MgO-Al_2O_3-SiO_2-H_2O$ and $K_2O-FeO-Al_2O_3-SiO_2-H_2O$. *Lithos*, **41**, 229–250.

MASSONNE, H. J. & O'BRIEN, P. J. 2003. The Bohemian Massif and the NW Himalaya. *In*: CARSON, C. J. & COMPAGNONI, R. (eds) *Ultrahigh Pressure Metamorphism.* Eötvös University Press, Budapest, 145–187.

MATILE, L. & WIDMER, T. W. 1993. Contact-metamorphism of siliceous dolomites, marls and pelites in the SE contact aureole of the Bruffione intrusion (SE Adamello, N. Italy). *Schweizerische Mineralogische und Petrographische Mitteilungen*, **73**, 53–67.

MCDADE, P. & HARLEY, S. L. 2001. A petrogenetic grid for aluminous granulite facies metapelites in the KFMASH system. *Journal of Metamorphic Geology*, **19**, 45–60.

MILLER, C., KONZETT, J., TIEPOLO, M., ARMSTRONG, R. A. & THÖNI, M. 2007. Jadeite-gneiss from the Eclogite Zone, Tauern Window, Eastern Alps, Austria: Metamorphic, geochemical and zircon record of a sedimentary protolith. *Lithos*, **93**, 68–88.

MOLLI, G., CONTI, P., GIORGETTI, G., MECCHERI, M. & OESTERLING, N. 2000*a*. Microfabric study on the deformational and thermal history of the Alpi Apuane marbles (Carrara marbles), Italy. *Journal of Structural Geology*, **22**, 1809–1825.

MOLLI, G., GIORGETTI, G. & MECCHERI, M. 2000*b*. Structural and petrological constraints on the tectono-metamorphic evolution of the Massa Unit (Alpi Apuane, NW Tuscany, Italy). *Geological Journal*, **35**, 251–264.

MOLLI, G., TRIBUZIO, R. & MARQUER, D. 2006. Deformation and metamorphism at the eastern border of the Tenda Massif (NE Corsica): a record of subduction and exhumation of continental crust. *Journal of Structural Geology*, **28**, 1748–1766.

MONIÉ, P. 1990. Preservation of Hercynian Ar^{40}/Ar^{39} Ages Through High-Pressure Low-Temperature Alpine Metamorphism in the Western Alps. *European Journal of Mineralogy*, **2**, 343–361.

NAGEL, T., DE CAPITANI, C., FREY, M., FROITZHEIM, N., STÜNITZ, H. & SCHMID, S. M. 2002. Structural and metamorphic evolution in the Lepontine dome. *Eclogae Geologicae Helvetiae*, **95**, 301–322.

NEUBAUER, F., GENSER, J., KURZ, W. & WANG, X. 1999. Exhumation of the Tauern window, Eastern Alps. *Physics and Chemistry of the Earth, part A: Solid Earth and Geodesy*, **24**, 675–680.

NIGGLI, E. C. 1970. Alpine Metamorphose und alpine Gebirgsbildung. *Fortschritte der Mineralogie*, **47**, 16–26.

NIGGLI, E. & NIGGLI, C. 1965. Karten der Verbreitung einiger Mineralien der alpidischen Metamorphose in den Schweizer Alpen (Stilpnomelan, Alkali-Amphibol, Chloritoid, Staurolith, Disthen, Sillimanit). *Eclogae Geologicae Helvetiae*, **58**, 335–368.

OBERHÄNSLI, R., GOFFÉ, B. & BOUSQUET, R. 1995. Record of a HP-LT metamorphic evolution in the Valais zone: Geodynamic implications. *In*: LOMBARDO, B. (ed.) *Studies on metamorphic rocks and minerals of the western Alps. A Volume in Memory of Ugo Pognante*. Bollettino del Museo Regionale di Scienze Naturali (suppl.), Torino, 221–239.

OBERHÄNSLI, R., BOUSQUET, R. & GOFFÉ, B. 2003. Comment to 'Chloritoid composition and formation in the eastern Central Alps: a comparison between Penninic and Helvetic occurrences'. *Schweizerische Mineralogische und Petrographische Mitteilungen*, in press.

OBERHÄNSLI, R., BOUSQUET, R., ENGI, M. ET AL. 2004. Metamorphic structure of the Alps. *In: Explanatory note to the map 'Metamorphic structure of the Alps'*. Commission for the Geological Map of the World, Paris.

PARRA, T., VIDAL, O. & AGARD, P. 2002a. A thermodynamic model for Fe-Mg dioctahedral K white micas using data from phase-equilibrium experiments and natural pelitic assemblages. *Contributions to Mineralogy and Petrology*, **143**, 706–732.

PARRA, T., VIDAL, O. & JOLIVET, L. 2002b. Relation between the intensity of deformation and retrogression in blueschist metapelites of Tinos Island (Greece) evidenced by chlorite-mica local equilibria. *Lithos*, **63**, 41–66.

PATTISON, D. R. M., SPEAR, F. S., DEBUHR, C. L., CHENEY, J. T. & GUIDOTTI, C. V. 2002. Thermodynamic modelling of the reaction muscovite + cordierite Al_2SiO_5 + biotite + quartz + H_2O: constraints from natural assemblages and implications for the metapelitic petrogenetic grid. *Journal of Metamorphic Geology*, **20**, 99–118.

PFIFFNER, O. A., ELLIS, S. & BEAUMONT, C. 2000. Collision tectonics in the Swiss Alps: Insight from geodynamic modeling. *Tectonics*, **19**, 1065–1094.

PLEUGER, J., FROITZHEIM, N. & JANSEN, E. 2005. Folded continental and oceanic nappes on the southern side of Monte Rosa (western Alps, Italy): anatomy of a double collision suture. *Tectonics*, **24**, TC4013.

POLINO, R., DAL PIAZ, G. V. & GOSSO, G. 1990. The alpine cretaceous orogeny: an accretionary wedge model based on integrated statigraphic, petrologic and radiometric data. *In*: F. ROURE, F., HEITZMANN, P. & POLINO, R. (eds) *Deep structure of the Alps*. Mémoire de la Société géologique de France, Paris, 345–367.

PREISWERK, H. 1918. *Geologische Beschreibung der Lepontinischen Alpen Beiträge zur geologischen Karte der Schweiz*, Bern.

PROYER, A. 2003. Metamorphism of pelites in NKFMASH: a new petrogenetic grid with implications for the preservation of high-pressure mineral assemblages during exhumation. *Journal of Metamorphic Geology*, **21**, 493–509.

RIMMELÉ, G., OBERHÄNSLI, R., GOFFÉ, B., JOLIVET, L., CANDAN, O. & CETINKAPLAN, M. 2003. First evidence of high-pressure metamorphism in the 'Cover Series' of the southern Menderes Massif. Tectonic and metamorphic implications for the evolution of SW Turkey. *Lithos*, **71**, 19–46.

RIMMELE, G., PARRA, T., GOFFE, B., OBERHANSLI, R., JOLIVET, L. & CANDAN, O. 2005. Exhumation Paths of High-Pressure-Low-Temperature Metamorphic Rocks from the Lycian Nappes and the Menderes Massif (SW Turkey): a Multi-Equilibrium Approach. *Journal of Petrology*, **46**, 641–669.

ROSELLE, G. T. & ENGI, M. 2002. Ultra high pressure (UHP) terrains: Lessons from thermal modeling. *American Journal of Science*, **312**, 410–441.

ROSSETTI, F., FACCENNA, C., JOLIVET, L., FUNICIELLO, R., TECCE, F. & BRUNET, C. 1999. Syn- versus post-orogenic extension: the case study of Giglio Island (Northern Tyrrhenian Sea, Italy). *Tectonophysics*, **304**, 71–93.

ROSSETTI, F., FACCENNA, C. & JOLIVET, L. ET AL. 2001. Structural signature and exhumation P–T–t path of the Gorgona blueschist sequence (Tuscan archipelago, Italy). *Ofioliti*, **26** (2A), 175–186.

ROURE, F., HEITZMANN, P. & POLINO, R. 1990. *Deep structure of the Alps*. Mémoires de la société géologique de France, Paris.

SALIOT, P. 1979. La jadéite dans les Alpes françaises. *Bulletin de la Société française de Minéralogie et de Cristallographie*, **102**, 391–401.

SCHAER, J.-P. 1959. Géologie de la partie septentrionale de l'éventail de Bagnes (entre le Val d'Hérémence et le Vaal de Bagnes, Valais, Suisse). *Archives des Sciences (Genève)*, **12**, 473–620.

SCHMID, S. M. & KISSLING, E. 2000. The arc of the Western Alps in the light of new data on deep crustal structure. *Tectonics*, **19**, 62–85.

SCHMID, S. M., PFIFFNER, O. A., FROITZHEIM, N., SCHÖNBORN, G. & KISSLING, E. 1996. Geophysical-geological transect and tectonic evolution of the Swiss–Italian Alps. *Tectonics*, **15**, 1036–1064.

SCHMID, S. M., FUGENSCHUH, B., KISSLING, E. & SCHUSTER, R. 2004. Tectonic map and overall architecture of the Alpine orogen. *Eclogae Geologicae Helvetiae*, **97**, 93–117.

SCHMIDT, C. & PREISWERK, H. 1908. Geologische Karte der Simplongruppe. Spezialkarte n°48 mit Erläuterungen.

SCHWARTZ, S., LARDEAUX, J.-M. & TRICART, P. 2000. La zone d'Acceglio (Alpes cottiennes): un nouvel exemple de croûte continentale éclogitisée dans les Alpes occidentales. *Comptes Rendus de l'Académie des Sciences Paris*, **320**, 859–866.

SCROCCA, D., DOGLIONI, C. & INNOCENTI, F. 2003. Constraints for an interpretation of the Italian geodynamics: a review. *Memorie Descrittive della Carta Geologica d'Italia*, **LXII**, 15–46.

SELVERSTONE, J. 1985. Petrological constraints on imbrication metamorphism and uplift in the SW Tauern Window, Eastern Alps. *Tectonics*, **4**, 687–704.

SELVERSTONE, J. & SPEAR, F. S. 1985. Metamorphic P-T paths from pelitic schists and greenstones from the south-west Tauern Window. *Journal of Metamorphic Geology*, **3**, 439–465.

SELVERSTONE, J., SPEAR, F. S., FRANZ, G. & MORTEANI, G. 1984. High-pressure metamorphism in SW Tauern Window, Austria: P–T paths from hornblende-kyanite-staurotide schists. *Journal of Petrology*, **25**, 501–531.

SPEAR, F. S. 1993. *Metamorphic Phase Equilibria and Pressure–Temperature–Time Paths*. Mineralogical Society of America, Washington, D. C.

SPEAR, F. & CHENEY, J. 1989. A petrogenetic grid for pelitic schists in the system $SiO_2-Al_2O_3-FeO-MgO-K_2O-H_2O$. *Contributions to Mineralogy and Petrology*, **101**, 149–164.

SPEAR, F. & SELVERSTONE, J. 1983. Quantitative P–T paths from zoned minerals: Theory and tectonic applications. *Contributions to Mineralogy and Petrology*, **83**, 348–357.

STAUB, R. 1926. Geologische Karte des Avers. In: *Beiträge zur geologischen Karte der Schweiz*. Bern.

STRECKEISEN, A., WENK, E. & FREY, M. 1974. Steep isogradic surface in Simplon area. *Contributions to Mineralogy and Petrology*, **47**, 81–95.

THEYE, T., SEIDEL, E. & VIDAL, O. 1992. Carpholite, sudoite and chloritoid in low-grade high-pressure metapelites from Crete and the Peloponnese, Greece. *European Journal of Mineralogy*, **4**, 487–507.

THEYE, T., REINHARDT, J., GOFFÉ, B., JOLIVET, L. & BRUNET, C. 1997. Ferro- and magnesiocarpholite from the Monte Argentario (Italy): First evidence for high-pressure metamorphism of the metasedimentary Verrucano sequence, and significance for P-T path reconstruction. *European Journal of Mineralogy*, **9**, 859–873.

TODD, C. S. & ENGI, M. 1997. Metamorphic field gradients in the Central Alps. *Journal of Metamorphic Geology*, **15**, 513–530.

TRIBUZIO, R. & GIACOMINI, F. 2002. Blueschist facies metamorphism of peralkaline rhyolites from the Tenda crystalline massif (northern Corsica): evidence for involvement in the Alpine subduction event? *Journal of Metamorphic Geology*, **20**, 513–526.

TROMMSDORFF, V. 1966. Progressive Metamorphose kieseliger Karbonatgesteine in den Zentralalpen zwischen Bernina und Simplon. *Schweizerische Mineralogische und Petrographische Mitteilungen*, **46**, 431–460.

VENTURINI, G. 1995. *Geology, Geochemistry and Geochronology of the inner central Sesia Zone (Western Alps)*. Mémoires de Géologie, Lausanne.

VERDOYA, M., CHIOZZI, P. & PASQUALE, V. 2001. Heat-producing radionucleides in metamorphic rocks of the Briançonnais-Piedmont Zone. *Eclogae Geologicae Helvetiae*, **94**, 213–219.

VIDAL, O., GOFFÉ, B. & THEYE, T. 1992. Experimental study of the stability of sudoite and magnesiocarpholite and calculation of a new petrogenetic grid for the system $FeO-MgO-Al_2O_3-SiO_2-H_2O$. *Journal of Metamorphic Geology*, **10**, 603–614.

VIDAL, O., THEYE, T. & CHOPIN, C. 1994. Experimental study of chloritoid stability at high pressure and various fO_2 conditions. *Contributions to Mineralogy and Petrology*, **118**, 256–270.

VIDAL, O., PARRA, T. & TROTET, F. 2001. A thermodynamic model for Fe–Mg aluminous chlorite using data from phase equilibrium experiments and natural pelitic assemblages in the 100–600 °C, 1–25 kbar range. *American Journal of Science*, **301**, 557–592.

WANG, P. & SPEAR, F. S. 1991. A field and theoretical analysis of garnet + chlorite + chloritoid + biotite assemblages from the tri-state (MA, CT, NY) area, USA. *Contributions to Mineralogy and Petrology*, **106**, 217–235.

WEI, C. J. & HOLLAND, T. J. B. 2003. Phase relations in high-pressure metapelites in the system KFMASH ($K_2O-FeO-MgO-Al_2O_3-SiO_2-H_2O$) with application to natural rocks. *Contributions to Mineralogy and Petrology*, **301**, 301–315.

WEI, C. J. & POWELL, R. 2004. Calculated Phase Relations in High-Pressure Metapelites in the System NKFMASH ($Na_2O-K_2O-FeO-MgO-Al_2O_3-SiO_2-H_2O$). *Journal of Petrology*, **45**, 183–202.

WEI, C. J., POWELL, R. & CLARKE, G. L. 2004. Calculated phase equilibria for low- and medium-pressure metapelites in the KFMASH and KMnFMASH systems. *Journal of Metamorphic Geology*, **22**, 495–508.

WIEDERKEHR, M., BOUSQUET, R., ZIEMANN, M. A., SCHMID, S. M. & BERGER, A. 2007. Thermal structure of the Valaisan and Ultra-Helvetic sedimentary units of the northern Lepontine dome—consequences regarding the tectono-metamorphic evolution. *In*: EGU, Wien.

YARDLEY, B. W. D. 1989. *An Introduction to Metamorphic Petrology*. Longman.

ZEH, A. 2001. Inference of a detailed P–T path from P–T pseudosections using metapelitic rocks of variable composition from a single outcrop, Shackleton Range, Antarctica. *Journal of Metamorphic Geology*, **19**, 329–350.

ZIMMERMANN, R., HAMMERSCHMIDT, K. & FRANZ, G. 1994. Eocene high pressure metamorphism in the Penninic units of the Tauern Window (Eastern Alps): evidence from $^{40}Ar-^{39}Ar$ dating and petrological investigations. *Contributions to Mineralogy and Petrology*, **117**, 175–186.

ZWART, H. J. 1973. *Metamorphic map of Europe, scale 1:2500 000*. Subcommission of Cartography of the Metamorphic Belts of the World, Leiden/ UNESCO, Paris.

ZWART, H. J., SOBOLEV, V. S. & NIGGLI, E. 1978. *Metamorphic map of Europe, scale 1:2 500 000. Explanatory text*. Subcommission of Cartography of the Metamorphic Belts of the World, Leiden/UNESCO, Paris.

Northern Apennine–Corsica orogenic system: an updated overview

GIANCARLO MOLLI

Dipartimento di Scienze della Terra, Università di Pisa, Via S.Maria 53, I - 56126 Pisa, Italy
(e-mail: gmolli@dst.unipi.it)

Abstract: The aim of this paper is to describe some aspects of the Northern Apennine/Corsica orogenic system, a classical and still-debated subject in the geology of the central Mediterranean. After a necessarily short historical outline, a general updated overview of the Northern Apennine and Corsica is presented. The results of recent research on metamorphic units representative of the former continental margins have been used to constrain some key events of the geological history of the Northern Apennine/Corsica orogenic system and its relationship with the western Alps. Many still-controversial topics are tackled in this paper, which calls for further data and studies. In particular, the early stages of the oceanic convergence recorded by the Ligurian units of the Apennine and by Corsican ophiolites need refinement of structural and chronological data sets. Similarly, the deep structure of the inner Northern Apennine is far from being well explored; nevertheless, three crustal-scale cross-sections across the Northern Apennine are presented to show differences in the crustal and near-surface architecture at the orogen scale and to discuss some points of the Apennine history, namely the relationships between shortening in the external and exhumation/extension in the internal domains, from the Oligocene onward.

It is outside the aim of this contribution to give a complete account of the history of research concerning the Northern Apennine and the Northern Apennine/Corsica/western Alps relationships (see the review papers and references in Castellarin 1992, 2001 and Conti & Lazzarotto 2004). Nevertheless, some key points in the more recent evolution of knowledge are highlighted here to give an historical perspective of the problem.

During the 1970s, the contributions of Elter *et al.* (1966), Laubscher (1971), Sturani (1973), Elter & Pertusati (1973), Dal Piaz (1974), Debelmas (1975) and Grandjacquet & Haccard (1977) were the first to analyse in modern terms the relationships between the Northern Apennine and the Alps, laying the basis for present research on the subject. Later on, the papers of Treves (1984) and Principi & Treves (1985), expanding on former propositions of Scholle (1970) and Scandone (1979), presented the interpretation of the Apennine as an accretionary wedge. These authors linked the growth of the Apenninic wedge to a Cretaceous to present west-dipping subduction of Adria below Corsica (Fig. 1). In this interpretation, Western 'Hercynian' Corsica is considered as the backstop of the Apenninic wedge, while Alpine Corsica is part of the Northern Apennines, i.e. the deepest part of the accretionary complex, backthrusted on the 'Hercynian' Corsica by corner flow (Cowan & Silling 1978) in the rear side of the accretionary system. The Northern Apennine units are considered as in-sequence imbricates offscraped from oceanic (until Eocene) and then continental domains (from the Oligocene onward). Principi & Treves (1985) discussed the relationships between the Apennines and the western Alps by suggesting the presence of a wide east–west-striking, distributed zone of deformation north of the Voltri Group, which accommodates the displacement between the opposite-dipping alpine (i.e. eastward) and apenninic (i.e. westward) subduction zones, drawing a direct comparison with the present convergent setting of New Zealand. A similar framework was successively proposed with some variations by Vescovi (1993), Maufrett & Contrucci (1999) and Argnani (2002). The interpretation in Principi & Treves (1985), which originally found its basis in the Oligocene to present evolution of the Apennine and in pre-Oligocene (mainly Late Cretaceous) stratigraphic-sedimentological data and provenaence studies of arenites in the flysch units (Abbate & Sagri 1984; Pandolfi 1998; Argnani *et al.* 2004 and references) was progressively accepted in a large number of papers on the Apennine (Hill & Hayward 1988; Carmignani & Kligfield 1990; Keller & Coward 1996; Decandia *et al.* 1998; Bortolotti *et al.* 2001) and forms the basis of some recent geodynamic contributions (Alvarez 1991; Jolivet *et al.* 1998; Lahondere *et al.* 1999; Rossetti *et al.* 2002; Faccenna *et al.* 2001, 2004). A long-lived (Cretaceous–present) west-dipping subduction for the Northern Apennine/Corsica orogen, however, challenges or does not take into account a large

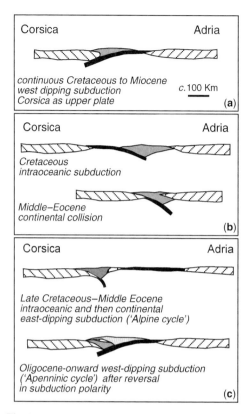

Fig. 1. Tectonic models proposed for the Corsica/Northern Apennine region. (**a**) Continuous Cretaceous to Miocene west-dipping Apenninic subduction (e.g. Principi & Treves 1985; Jolivet *et al.* 1998; Rossetti *et al.* 2002). (**b**) Alpine east-vergent intra-oceanic subduction (Mattauer *et al.* 1981) followed by Middle-Eocene continental collision (e.g. Gibbons & Horak 1984; Warbourton 1986; Finetti *et al.* 2001 and references). According to this model, the Apenninic history is mainly ensialic and did not involve Apenninic subduction. (**c**) Alpine east-dipping intra-oceanic and then continental subduction followed by Late Eocene/Early Oligocene-onward Apenninic subduction (Elter & Pertusati 1973; Doglioni *et al.* 1998; Molli & Tribuzio 2004 and this paper). In black oceanic crust, while in dark grey (a, b) accretionary complex; in (c) dark and light grey indicate Corsican and Apenninic accretionary wedge, respectively.

amount of Corsican regional and structural works published up to the 1980s (e.g. Nardi 1968; Boccaletti *et al.* 1971; Mattauer *et al.* 1981; Cohen *et al.* 1981; Durand-Delga 1984; Harris 1985; Warbourton 1986; Caron 1994; Dallan & Puccinelli 1995), where the early structures of the Alpine Corsica units were interpreted as produced within an east-dipping 'alpine' subduction zone and Corsica itself considered as the southern prolongation of the Alpine orogen (Argand 1924; Staub 1924).

Papers based on Alpine Corsica geology by Jolivet *et al.* (1990, 1998), Lahondere *et al.* (1999) and Rossetti *et al.* (2002) discussed regional, structural and petrological data in a way that was influenced by two points of view:

(1) that the importance of exhumation-related greenschist structures and kinematics (Jolivet *et al.* 1990; Lahondere 1996; Daniel *et al.* 1996; Jolivet *et al.* 1998) made the early contractional history uncertain (Jolivet *et al.* 1998; Rossetti *et al.* 2002); and

(2) that the indications of high pressure/low temperature peak conditions which were classically (Stam 1952; Delcey & Meunier 1966; Gibbons & Horak 1984) attributed to one of the major Corsica-derived continental units, the Tenda Massif, were partially or completely negated.

In the meantime, the pre-1980s classical view (Boccaletti *et al.* 1971; Elter & Pertusati 1973; Alvarez *et al.* 1974; Dal Piaz 1974; Reutter *et al.* 1978) in which Corsica and some Apenninic units (the Ligurian units) are considered as former part of a single pre-Oligocene Alpine orogenic system connected with an east-dipping subduction has been continuously supported by Dutch, French, Swiss and Italian researchers who have worked around the so-called 'Ligurian knot' (Laubscher 1991), i.e. the area comprising the Maritime Alps, the Ligurian and the Emilian Apennines (Cortesogno *et al.* 1979; Vanossi *et al.* 1986; Van Wamel 1987; Hoogerdujin Strating & Van Wamel 1989; Galbiati 1990; Castellarin 1992; Labaume 1992; Hoogerdujin Strating 1994; Vanossi *et al.* 1994; Piana & Polino 1995; Marroni & Pandolfi 1996; Crispini 1996; Elter 1997; Doglioni *et al.* 1998, 1999). Among the most recent papers on the subject—those of Daniele & Plesi (2000); Finetti *et al.* (2001); Capponi & Crispini (2002); Castellarin (2001); Cerrina *et al.* (2004); Carminati *et al.* (2004); Molli & Tribuzio (2004); Del Castello *et al.* (2005)—follow the classical view with some major differences.

The problem of the relationships between the Northern Apennine/Corsica and the western Alps orogenic systems can be connected with the following general aspects:

- presence of tectonic units with similar lithostratigraphic features and comparable structural evolution;
- at least in part coeval age of tectonic events and deformations;
- at least in part common dynamic of orogenic system(s) (in the pre-, syn- and post-collisional evolution) between the two chains; and

- superficial continuities with common and/or polycyclic ('Alpine' and then 'Apennic') evolution of tectonic units across the western Alps/Northern Apennine boundary.

These features were partially derived from a palaeogeographic heritage of the Alpine and Apenninic realms which were laterally continuous and shared the rifting and the drifting stages of the Ligurian Tethys (Elter 1975; Piccardo 1977; Laubscher & Bernoulli 1977; Stampfli et al. 1998; Manatschal & Bernoulli 1999; Rampone & Piccardo 2000; Lemoine et al. 2001), the ocean which during the Late Mesozoic separated palaeoEurope from the southern palaeocontinent Adria/Apulia (the 'Africa promontory' of Argand 1924). The diachronous closure of this ocean (Dewey et al. 1989; Polino et al. 1990; Stampfli et al. 1998; Lemoine et al. 2001; Michard et al. 2002; Schmid et al. 1996, 2004 and references) during the Cretaceous to Eocene and the following Oligocene/Miocene Europe/Adria collision characterized the evolution of the Alps and of the palaeoApennine, while, starting from the Oligocene, the opening of the Provençal and that of the Tyrrhenian Sea (from the Middle Miocene) in the wake of retreating Adria subduction represent the key events of the Apenninic evolution (Elter et al. 1975; Laubscher 1971, 1991; Scandone 1979; Doglioni 1991).

Geophysical signature and present-day tectonic setting of the Northern Apennine

The Northern Apennine is the orogenic belt forming the backbone of the Italian peninsula. In its northwestern part it is in contact with the western Alps and towards the south with the Central and Southern Apennine (Figs 2 and 3). Surface tectonic boundaries are traditionally located along the Sestri/Voltaggio line, a step zone in which high pressure/low temperature ophiolite and continental units (Alpine units of the Voltri group and Sestri-Voltaggio Zone) crop out, whereas toward south the Northern Apennine is separated from the Central Apennine (mainly carbonate–platform sediments) along the Ancona-Anzio (or Olevano-Antrodoco) line and from the Southern Apennine by the Ortona-Roccamorfina line (Patacca & Scandone 1989).

At a regional scale, the Northern Apennine (Elter et al. 1975; Pialli et al. 1998 and references therein) is formed by two crustal domains (Figs 3 and 4): (i) a Ligurian–Tuscan–Tyrrhenian realm forming the inner part of the orogenic belt; and (ii) the foreland (or outer) domain developed from the Adriatic microplate, east of the main orographic divide. The first domain is characterized by positive Bouger anomalies, high heat flow (up to

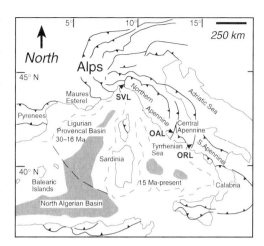

Fig. 2. Tectonic setting of Corsica/Northern Apennine and western Alps within the western Mediterranean. SVL, Sestri–Voltaggio line, traditional surface boundary between Alps and Apennines; OAL, Olevano–Antrodoco line, traditional surface boundary between Northern and Central Apennines; ORL, Ortona–Roccamorfina line, boundary between Northern and Southern Apennines according to Patacca & Scandone (1989).

150 mWm^{-2} in southern Tuscany) and reduced crustal thickness (20–25 km), the second domain is on the contrary characterized by negative Bouger anomalies, low values of heat flow, normal crust thickness (up to 35 km) with a Moho gently dipping west below the main orographic divide. The down-bending and underthrusting of the Adriatic microplate below the Apennine is at the surface expressed by a highly segmented foredeep in which up to 5000 m of Plio-Pleistocene sediments were accumulated (Royden et al. 1987; Patacca & Scandone 1989; Patacca et al. 1992). At deeper structural levels, it is testified by the distribution of subcrustal seismicity down to 90 km and by a high-velocity, W–S dipping zone from 180 down to 250 km (Selvaggi & Amato 1992; Eva & Solarino 1992; Spakman & Wortel 2004 and included references).

The outer domain is seismologically characterized by earthquakes with reverse and strike-slip solutions whereas predominantly normal and strike-slip solutions characterize the internal domain (Elter et al. 1975; Eva & Solarino 1992; Boncio et al. 2000; Meletti et al. 2000; Di Bucci & Mazzoli 2002; Viti et al. 2004; Collettini et al. 2006). This simple scheme gradually changes to become more and more complex in the northern part of the belt toward the so-called 'Ligurian Knot' of Laubscher (1991). Here a crocodile crustal structure has been suggested by interpretations of deep and industrial seismic profiles

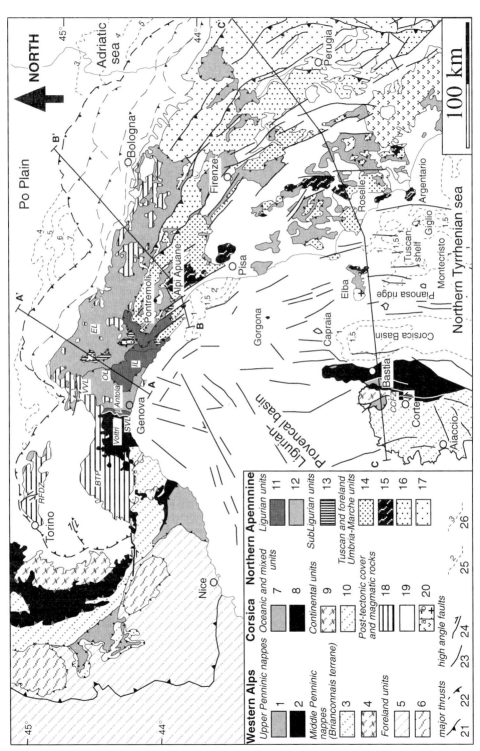

Fig. 3. *Continued.*

(Cassano et al. 1986; Biella et al. 1987, 1988; Laubscher 1991; Laubscher et al. 1992; Piana & Polino 1995; Schumacher & Laubscher 1996; Makris et al. 1999; Piana 2000). Along-strike variation of the deep structure is associated with a complementary change of the superficial structure as shown hereafter.

Geological structure of the Northern Apennine

The inner part of the Apennine is largely below sea level (Ligurian and Northern Tyhrrenian seas), therefore the inner root area of the belt is almost completely hidden. On the base of on-land geology (e.g. Elter 1975) complemented by information of seismic profiles in the Liguria/Northern Po plain area (Cassano et al. 1986; Piana & Polino 1995; Makris et al. 1999) and across the Northern Tyrrhenian Sea (Bartole et al. 1991; Contrucci et al. 2005), the following geometry of the nappe stack can be sketched (Figs 3–5):

High pressure/low temperature ophiolitic and continental units which are observable from north to south (Fig 3) in western Liguria, in some Tuscan islands (especially Gorgona, Giglio) and along or near the Tyrrhenian coast in Tuscany (Argentario promontory and Roselle). Geometrically, these units occupy the uppermost position of the nappe stack in the northwestern part of the belt where they are underthrusted by the Pliocene foredeep of the Apennine. In outcrop exposures, they are juxtaposed along a subvertical tectonic zone, the 'Sestri-Voltaggio line', to the Ligurian units, whereas further south (Gorgona, Argentario, Giglio, Roselle) they directly overlie the Tuscan units or are interleaved with them through low angle contacts.

Ligurian units, classically considered as the uppermost nappes of the Apennine stack, can be subdivided on the basis of stratigraphic and structural features into two main groups (Elter 1975) well defined in the Ligurian–Emilian Apennine (Figs 3 and 4):

(i) *the Internal Ligurian Units* which are characterized by the presence of ophiolites and an Upper Jurassic to Lower Cretaceous sedimentary cover (cherts, Calpionella limestone and Palombini shales) associated with Upper Cretaceous–Paleocene turbiditic sequences (Marroni & Pandolfi 1996). The Internal Ligurian units are considered as remnants of the Liguro-Piemontese ocean or Ligurian Tethys;

(ii) *the External Ligurian Units* which are, on the other hand, distinguishable for the presence of the typical Cretaceous–Paleocene calcareous-dominant sequences (the Helminthoid Flysch) associated with complexes or pre-flysch formations called 'basal complexes'. According to their stratigraphic differences, two main subgroups of units can be recognized (Marroni et al. 1988): those associated with ophiolites and with ophiolite-derived debris, and others without ophiolites and associated with fragments of Mesozoic sedimentary sequences and conglomerates with Adria affinity (Sturani 1973; Zanzucchi 1988; Marroni et al. 1998 and references). Because of their age (Elter et al. 1966; Wildi 1985; Zanzucchi 1988; Gasinski et al. 1997; Daniele & Plesi 2000) and composition, these coarse-grained conglomerates (Salti del Diavolo Conglomerates) have been compared since the early 1970s with those of Préalpes Romandes (Mocausa conglomerates of the Simme Flysch) implying a common palaeotectonic setting on the distal side of the Adria continental margin (e.g. Elter 1997; Stampfli et al. 1998; Lemoine et al. 2001 and references). As a whole, the External Ligurian units can be regarded as relicts of the former ocean–continent transition area and of the distal Adria continental margin in the Apenninic

Fig. 3. (*Continued*) Tectonic map of western Alps/Corsica and Northern Apennine showing the main tectonic and lithostratigraphic units, based on Bigi et al. 1983. **For the western Alps**: (1) Upper Penninic Helminthoid Flysch units of the Ubaye–Embrunais and Maritime Alps; (2) Schistes Lustrés composite nappe system; (3) Middle Penninic Briançonnais nappes; (4) Middle Penninic basement nappes (Dora Maira); (5) Alpine foreland units; (6) external massifs (Argentera and Pelvoux). **Corsica**: (7) upper mixed units (unmetamorphic stack of continental and oceanic units, i.e. Balagne, Nebbio and Macinaggio); (8) Schistes Lustrés composite nappe system; (9) internal continental basement units, i.e. Corsica margin deeply involved in Alpine subduction (only the Tenda massif is shown); (10) external continental units. Northern Apennine; (11) internal Ligurian units, IL (not indicated in southern Tuscany); (12) external Ligurian units (EL) (including Antola Nappe); (13) sub-Ligurian (Canetolo) units; (14) Tuscan nappe; (15) Tuscan metamorphic units; (16) Cervarola unit; (17) Umbria–Marche foreland units; (18) post-tectonic cover of Tertiary Piemontese basin and Epiligurian units; (19) Neogene and Quaternary sediments of Po Plain and inner Tuscany; (20) magmatic rocks of southern Tuscany, (a) volcanic centres, Vulsini and Amiata; (b) Elba and Giglio granites; major thrusts at surface (21) and in subsurface (22); high-angle faults: (23) normal, (24) transcurrent; (25) sediment thickness in seconds TWTt for the Tyrrhenian Sea; (26) Pliocene isobaths (in km) in the Po Plain and Adriatic Sea. In the map are indicated some important fault zones: RFDZ, Rio Freddo Deformation zone (Piana & Polino 1995) separating the Torino Hill with Alpine metamorphic basement from the Apenninic Monferrato; SVL, Sestri–Voltaggio Line; VVL, Villalvernia–Varzi line; OL, Ottone–Levanto Line; CCFZ, Central Corsica Fault Zone.

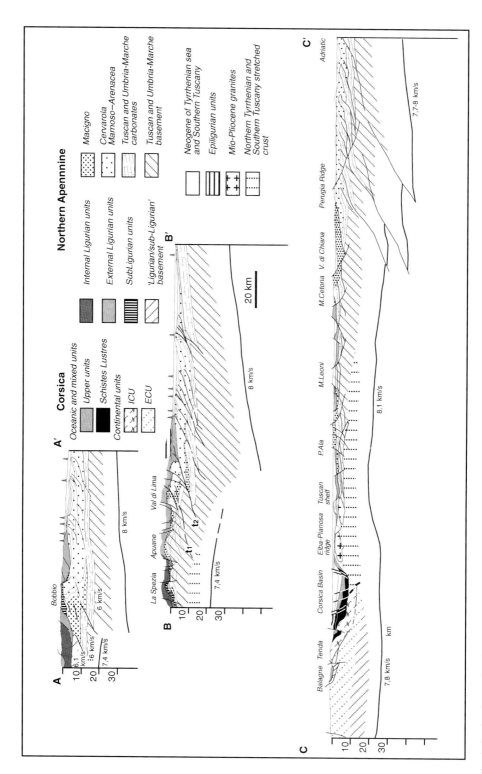

Fig. 4. Regional crustal-scale cross-sections across the Northern Apennine: main patterns indicated. For Corsica continental units, ICU indicates the Internal Continental units (Corsican continental margin strongly involved in deformations), ECU the External Continental units. Subsurface data included in sections are based on: Cassano et al. (1986); Laubscher et al. (1992); Elter et al. (1992); Castellarin (1992); Argnani et al. (2003); Pialli et al. (1998); Bartole (1991); Jolivet et al. (1998); Contrucci et al. (2005) and Labaume (1992). In section C–C', black dashed line between Corsica and Elba represents the possible site of the suture between 'Alpine' (east-vergent) and 'Apennine' (west-vergent) subduction.

transect (Molli 1996; Marroni et al. 1998 and references therein).

The Internal Ligurian units suffered polyphase deformation and metamorphism in sub-greenschist facies conditions (prehnite–pumpellyite in metabasic rocks), whereas the External Ligurian units were deformed at shallow structural levels (diagenesis–anchizone transition in pelites). Among the Ligurian units, the Antola Nappe deserved a special mention. From the lithostratigraphic point of view, this unit can be correlated with the External Ligurian units (Abbate & Sagri 1984 and references; Cerrina et al. 2002; Levi et al. 2006), whereas it occupies a structural position at the top of the Internal Ligurian units, in contrast to the other External Ligurian units which are structurally below the Internal Ligurian units. Moreover, it is classically correlated with the Helminthoid Flysch of the Ligurian and Maritime Alps and therefore played a special role during the pre-Oligocene evolution of the Alps/Apennine orogenic system (Elter & Pertusati 1973; Elter 1997; Corsi et al. 2001).

Sub-Ligurian units can be geometrically observed at the base of the Ligurian units, although with a strong regional thickness variability. The sub-Ligurian units are represented (Plesi 1975; Cerrina Feroni et al. 2002 and references therein) by Cretaceous–Eocene sequences mainly formed by sandstones and shaly-calcareous deposits (Ostia–Scabiazze and Canetolo fms) followed by Oligocene–lower Miocene (Aquitanian) siliciclastic and marly deposits (Aveto–Petrignacola and Coli units). Within the Cretaceous–Paleogene sequence, unconformities and depositional hiatuses (Vescovi 1993, 1998) of Early, Middle and Late Eocene age are documented, whereas volcanoclastic deposits with calc-alkaline affinities (the Aveto–Petrignacola fm.) are dated to the lower Oligocene. For its age and composition, the Aveto–Petrignacola fm., which also includes coarse-grained conglomerates fed by a medium- to high-grade crystalline basement, has been associated with calc-alkaline volcanic centres located on the inner (Adria) side of the Alpine belt (Boccaletti et al. 1971; Ruffini et al. 1995; Cibin et al. 1998). The original substratum of the sub-Ligurian units is unknown, although it can be considered transitional between the Ligurian and Tuscan domains and probably characterized by a thinned continental crust like part of the External Ligurian domain. This interpretation is supported by the observation that the oldest stratigraphic unit includes arenitic turbidites and shales similar to those of the External Ligurian unit (Ostia–Scabiazza fm.) (Ghiselli et al. 1991; Cerrina Feroni et al. 2002). The lower Oligocene–Aquitanian part of the sequence can be connected with the early accretional and thrust top basins of the Apennine wedge and bears similarities with siliciclastic turbidites at the top of the Tuscan units. The sub-Ligurian units were deformed at shallow structural levels (anchimetamorphic conditions in pelites, Cerrina et al. 1985) starting from Rupelian (c. 30 Ma) (Cerrina et al. 2002, 2004 and references).

Tuscan units are representative of the former proximal side of the Adria continental margin (i.e. the Tuscan Domain). These units are formed by continental successions subdivided in different thrust sheets, some of which were deformed at shallow structural levels (e.g. the Tuscan nappe), whereas others were more deeply involved in the collisional stack and metamorphosed in high- and medium-pressure greenschist facies conditions, forming the so-called Tuscan metamorphic units. These units cropout (Fig. 3) in tectonic windows forming an arcuate belt from P. Bianca in the north through the Alpi Apuane, M. Pisani, Montagnola Senese and M Romani in the south, along the so-called Mid-Tuscan metamorphic ridge. The stratigraphic evolution of the Tuscan sequences testifies to the sedimentation on a passive continental margin during Mesozoic rifting and post-rifting stages related with the Ligurian ocean opening (Bernoulli et al. 1979; Bernoulli 2001; Ciarapica & Passeri 2002). Sedimentary response to regional-scale contraction and tectonic inversion is recorded during the Cretaceous and Eocene within the Scaglia fm. where conglomerates and unconformities can be observed (Fazzuoli et al. 1994). The sedimentary history in the Tuscan domain ends during the Oligocene and Early Miocene with siliciclastic turbidites (Pseudomacigno and Macigno) interpreted as clastic wedges of Apennine foredeep basins.

Umbria-Romagna-Marche units are well exposed in the southernmost outer Northern Apennine where they are characterized by Jurassic to Palaeogene carbonates and Mio-Plio-Pleistocene marine clastic sediments deposited in a foredeep and/or in piggyback basins which evolved during thrusting. These units mainly represented in outcrop exposure by a turbiditic clastic wedge (Marnosa–Arenacea fm. and Laga fm.) deformed as a classical foreland fold and thrust belt (Calamita et al. 1994; Tavarnelli 1997; Coward et al. 1999; Barchi et al. 2001).

Late- to post-orogenic sediments (Late Eocene to Pliocene in age) of continental to shallow marine origin can be locally observed. These deposits lie unconformably on the uppermost tectonic units (Western Liguria, Internal and External Ligurian units) in the northern part of the belt where they form the Tertiary Piemonte or Langhe basin (Laubscher et al. 1992; Mutti et al. 1995) and the so-called Epiligurian units (Elter 1975). Further south, they form basins within the Northern

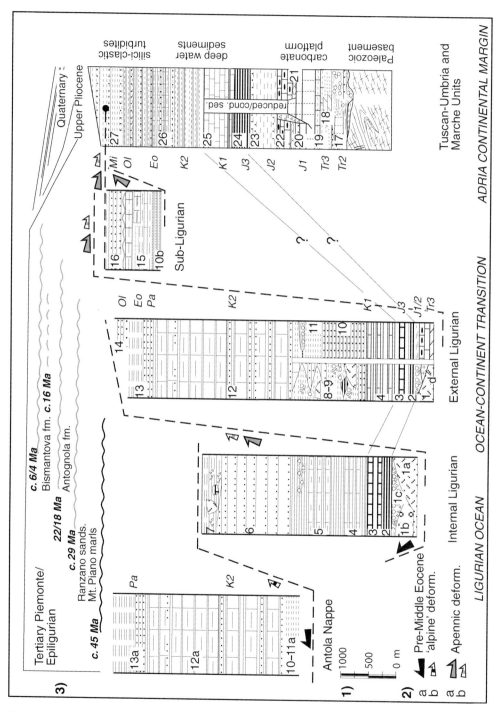

Fig. 5. Schematic sedimentary sequences of the main tectonic systems of the Northern Apennine and related palaeotectonic setting. (1) ophiolitic basement: (1a) mantle serpentinites, (1b) gabbros, (1c) basalts, (1d) continental granites (in the External Ligurian units); (2) cherts; (3) Calpionella limestones; (4) Palombini shales; (5) Val Lavagna schists; (6) Gottero sandstones; (7) Bocco/Colli Tavarone schists; (8–9) Casanova melange and sandstones; (10) Ostia/Scabiazza sandstones; (11) Varicoloured shales with Salti del Diavolo Conglomerates, (11a), Montoggio shales;

Tyrrhenian realm which will be analysed in the following paragraph.

The Northern Tyrrhenian Sea

The Northern Tyrrhenian Sea separates the Northern Apennine from Corsica and is subdivided into two main parts (Figs 3 and 4): a western domain—the Corsica basin—and an eastern one—the Tuscan shelf. The two domains are separated by the north–south elongated Elba–Pianosa Ridge (Bartole et al. 1991; Carmignani et al. 1995; Mauffret & Contrucci 1999; Pascucci et al. 1999; Cornamusini et al. 2002).

The Northern Tyrrhenian Sea developed since early Middle Miocene with the rifting of the half-graben of the Corsica basin and then since the Late Tortonian (Middle–Late Miocene) with the rifting of eastern part of the Tuscan shelf. Offshore deep and industrial seismic profiles and boreholes document the presence of three main stratigraphic and structural units (Bartole et al. 1991; Carmignani et al. 1995; Mauffret & Contrucci 1999; Contrucci et al. 2005):

- a substratum of metamorphic and non-metamorphic accretionary wedge units of oceanic and continental affinities, including both cover and basement rocks;
- a thick mainly clastic succession (Eocene to Oligocene in age) resting on top of the previous units and at least partially comparable with the onland Eocene Flysch of Corsica and for the Oligocene part with the coeval Epiligurian units; and
- a clastic succession (Miocene–Pliocene in age) comprising syn- and post-rift deposits, unconformably resting on the Eocene–Oligocene successions in the Corsica basin and directly on the accretionary wedge units in the Tuscan shelf. Similar sedimentary units can be also observed onland in southern Tuscany and Umbria where they are Pleistocene in age.

The formation of Miocene and younger basins was associated with magmatism which similarly shows an eastward-younging trend between Middle Miocene and Quaternary. Sisco lamproites (alcaline sills) in eastern Corsica represent the oldest magmatic rocks (15–13.5 Ma), intermediate-age magmatism (7.3–2.2 Ma) characterizes the intrusive and volcanic bodies of the Tuscan Archipelago and Central Tuscany (e.g. Capraia, Elba, Giglio, Orciatico, Montecatini Val di Cecina). The most eastern and recent (1.3–0.1 Ma) magmatism can be found in the Tuscan–Latium area at Mt. Cimini, Mt. Vulsini, Mt. Amiata and Larderello (Civetta et al. 1978; Lavecchia & Stoppa 1990; Serri et al. 1993; Musumeci et al. 2002; Rocchi et al. 2002 and references).

Corsica and the Alpine Corsica orogen

At the western side of the Tyrrhenian Sea, Corsica occupies a central position within the western Mediterranean. It is part of the Sardinia–Corsica microblock and is separated by the Ligurian–Provençal basin from the Iberian–European mainland.

The Ligurian–Provençal basin records a complex geological evolution with an early history Late Eocene in age (Late Priabonian), connected with the development of the Pyrenees orogen (Seranne 1999; Lacombe & Jolivet 2005). This early evolution was followed by an Oligocene to Early Miocene (Aquitanian–early Burdigalian) Apennine-related back-arc rifting stage and the Burdigalian (19–16 Ma) oceanic spreading associated with the anticlockwise Corsica block-rotation of $30°$ (Speranza et al. 2002 and references).

Corsica is subdivided into two main structural domains: an 'autochthonous Hercynian' Corsica, constituted by mainly undeformed series of the Corsica microplate (basement rocks intruded by granites and a sedimentary cover up to Eocene) and the so-called 'Alpine' Corsica (Fig. 6).

Fig. 5. (*Continued*) (12) Helminthoid Flysch, (12a) Antola Flysch; (13) Viano/Signano shales, (13a) Pagliaro shales; (14) Val Sporzana fm.; (15) Canetolo fm. (limestone and shale), Groppo del Vescovo and Vico Flysch; (16) Aveto/Petrignacola/Bratica sandstones and Coli/Marra marls; (17) continental 'Verrucano' and marine transgressive deposits on Hercynian and post-Hercynian units; (18) evaporites and dolomites; (19) Rhaetavicula contorta limestone and marls; (20) Massiccio limestone; (21) Rosso Ammonitico; (22) cherty limestone; (23) Posidonia marls; (24) Cherts; (25) Maiolica; (26) Scaglia Toscana, Scisti a Fucoidi, Scaglia Umbra and Bisciaro; (27) siliciclastic turbidites (Macigno, PseudoMacigno, Cervarola, Marnoso–Arenacea, Laga). Also indicated are. (1) approximate scale of thickness; (2) sense of emplacement for (a) thrusting; (b) extensional reactivations; (3) the major units in the Tertiary Piemonte and Epiligurian with bounding unconformity suturing or recording pre-Middle Eocene 'alpine' deformation (in black); Apenninic major events (in grey). This scheme is considered representative only for the northern part of the Northern Apennine. Major differences in southern Tuscany can be found in: (1) the nature of basement of the External Ligurian units (continental crust and cover slices are non-documented); (2) the stratigraphy of some of the Internal Ligurian units where shallow-water sediments and mélange (e.g. Lanciaia fm. of Paleocene age, Cerrina et al. 2002 and ref.) do not found an equivalent in the northern part. Moreover, in southern Tuscany, the Epiligurian units are not documented as well as early pre-Middle Eocene west-vergent deformations.

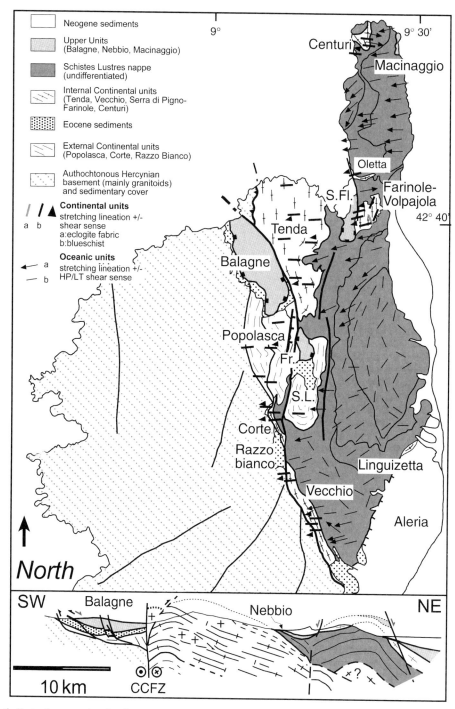

Fig. 6. Tectonic map and regional cross-sections of northern Corsica showing the lithostratigraphic units and the regional trend of stretching/mineral lineations with sense of shear for eclogite (a) and blueschist facies (b) fabrics in continental units (from Mattauer *et al.* 1981; Faure & Malavieille 1981; Malavieille 1983; Bezert & Caby 1988; Bezert 1990; Lahondere 1996 and personal data). HP/LT stretching lineations (and shear sense where available) for oceanic units: (a) from Mattauer *et al.* (1981); Malavieille (1983); Harris (1985); Lahondere (1996); Malavieille *et al.* (1998) and personal data; (b) from Fournier *et al.* 1991; Caron 1994. The dispersed orientation of HP/LT

In simplified terms, Alpine Corsica can be considered as formed by three major composite groups of units (cf. Mattauer et al. 1981; Durand-Delga 1984; Malavieille et al. 1998). From bottom to top, they are:

(1) The Corsica-derived continental units which comprise (see below) the External Continental units (parts of the 'autochthonous' and 'parautochthonous' of the French authors), and the Internal Continent units such as the Tenda Massif and the Centuri/Serra di Pigno/Farinole slices. These units are formed by Hercynian granitoids with minor relicts of host-rock basement and series of metasedimentary rocks of Permian to Eocene ages with Briançonnais affinity (Durand-Delga 1984; Michard & Martinotti 2002). The continental units were affected by metamorphism at different peak metamorphic conditions and are now observable in different positions within the nappe stack (see below);

(2) The 'Schistes Lustrés' composite nappe formed by Ligurian Tethys-derived ophiolitic sequences (mantle ultramafics, gabbros, pillow lavas and associated Jurassic to Cretaceous metasediments) with peak metamorphism of epidote–blueschist and/or eclogite facies variably retrogressed under greenschist facies conditions (Dal Piaz & Zirpoli 1979; Gibbons et al. 1986; Caron 1994; Fournier et al. 1991); and

(3) The Balagne–Nebbio–Macinaggio system represented by ophiolitic and continental units characterized by low grade (prehnite/pumpellyite facies in basic rocks) assemblages and a geometrically high structural position within the nappe stack. While the Balagne nappe is found in an external position directly overriding the External Continental units (Marroni & Pandolfi 2003, and references), the Nebbio and Macinaggio units occupy a more internal position, resting above the Schistes Lustrés and/or the Internal Continental units.

The uppermost nappes and their contacts with the underlying units are locally unconformably sealed by lower Miocene continental or marine sediments, e.g. the Francardo, St. Florent and Aleria basins (Dallan & Puccinelli 1995; Ferrandini et al. 1998).

Structural and metamorphic evolution of the Corsican continental units

The structural and metamorphic evolution of the Corsica-derived continental units can be used to provide first order constraints on the kinematics of the early evolution in the Corsica/Northern Apennine orogenic system.

Based on the tectono-metamorphic evolution and structural position within the nappe stack, two main groups of Corsican-derived continental units are considered here (Fig. 6): (1) the External Continental units (ECU) which include, from north to south, the Popolasca, Corte and Razzo bianco units ('parautochthonous' and part of 'autochthonous' of the French authors); (2) the Internal Continental units (ICU) including the Tenda massif, the Vecchio slices and the westernmost Centuri, Serra di Pigno and Farinole-Volpajola slices (Fig. 6). These units show variable high pressure/low temperature metamorphic imprint which ranges from lower grade, high pressure grenschist to lower blueschist facies in the external slices to eclogite facies in the more internal ones (Fig. 6, Bezert & Caby 1988; Bezert 1990; Caron 1994; Lahondere 1996; Tribuzio & Giacomini 2002; Molli & Tribuzio 2004; Malasoma et al. 2006; Molli et al. 2006).

Pervasive composite mylonitic foliation with relicts of eclogite fabrics can be observed in the internal Serra di Pigno and Farinole–Volpajola units. Stretching lineation and shear sense indicators (Fig. 6) associated with eclogite fabrics provide evidence of scarce top north/northwest-shearing (Lahondere 1996), whereas the blueschist retrograde fabric formed the mappable main foliation associated with an east–west stretching lineation and a dominant top-to-west kinematics (Faure & Malavieille 1981; Mattauer et al. 1981; Malavieille 1983; Harris 1985a, b; Molli, personal data).

In the Tenda and Centuri units, an heterogeneous deformation pattern allowed the preservation of domains without fabric (i.e. isotropic) or with only a magmatic grain-shape fabric locally preserved surrounded by mylonitic orthogneisses and/or mylonites. The dominant fabric in the Tenda unit is represented by greenschist facies exhumation-related tectonites; however, kinematic criteria (stretching/mineral lineations and the associated shear sense, Fig. 6), on well-preserved blueschist relict domains give indications of early deep top-to-west shearing (Molli & Tribuzio 2004; Molli et al. 2006 and references).

The external Corte, Popolasca, Razzo Bianco units, together with the so-called 'parautochthonous' of the French authors are also characterized by high pressure/low temperature metamorphic overprint

Fig. 6. (*Continued*) lineations in oceanic units in central-southern Corsica (e.g. in the Castagniccia antiform) is related, according to Caron 1994, to an originally lower amount of finite strain compared to northern area but also to reorientation effects by late orogenic structures S.Fl., S.Florent; Fr., Francardo; S.L., S. Lucia nappe (here figured as an external continental unit although its origin is still debated, e.g. Durand-Delga 1984 and discussion in the text).

(Bezert & Caby 1988; Malasoma et al. 2006). The high pressure/low temperature parageneses are mainly observable in suitable rock types such as mafic dykes and were formed in granitoids in millimetre-scale shear zones localized by brittle precursors (Molli et al. 2005) and therefore have been overlooked by most authors (except Bezert 1990). These shear zones show evidence of east–west direction of transport associated with top-west kinematics observable all along the boundary between Alpine and Hercynian Corsica (Fig. 6).

As a whole, the present geometries of the Corsican continental units within the nappe stack with their internal structures, deformation styles and metamorphic peak conditions can provide first order constraints on the role of the Corsican continental margin during the early stages of the orogenic history and suggest a progressive underthrusting of the Corsican crust in a framework of east-dipping continental subduction (Mattauer & Proust 1975; Gibbons et al. 1986; Malavieille et al. 1998; Molli et al. 2006 and references).

The age of the early metamorphic events in Corsica is still a matter of debate (Handy & Oberhänsli 2004) and needs further investigation. Nevertheless, available geochronological and geological data sets provide a coherent regional frame. In the Schistes Lustrés, eclogites show a Sm–Nd whole rock Grt–Gla–Cpx isochron age of 84 ± 5 Ma (Lahondere & Guerrot 1997). Phengite from continental-derived eclogitic gneisses of the Farinole unit gives a $^{40}Ar/^{39}Ar$ discordant age spectrum with ages increasing from 55.3 ± 4.3 Ma to 65.3 ± 0.7 Ma, whereas post-eclogitic phengites give ages from 54.3 ± 0.5 Ma to 37.4 ± 0.4 Ma (Brunet et al. 2000). Similar retrogression-related ages were previously obtained by Maluski (1977) and Lahondere (1996).

For the Tenda unit, a crude, two-step discordant $^{40}Ar/^{39}Ar$ age on glaucophane of about 90 Ma was interpreted by Maluski (1977) as the age of the thermal peak. Similarly, a Rb–Sr whole-rock age of 105 ± 8 Ma was considered by Cohen et al. (1981) as the age of the high pressure/low temperature metamorphic event. More recently, a separate of celadonite-rich phengites ($Si = 3.5$ apfu) from a deformed granitoid of the Northern Tenda massif has yielded a discordant $^{39}Ar/^{40}Ar$ spectrum that regularly increases during step-heating, from about 25 Ma to 47 Ma (Brunet et al. 2000). According to Molli & Tribuzio (2004), this suggests that high-pressure metamorphism in the Tenda Massif has a minimum age of 47 Ma.

The underthrusting of the Corsican continental crust is documented to persist during the Eocene. The External Continental Units which are affected by high-pressure greenschist facies to high pressure/low temperature metamorphism (Bezert & Caby 1988; Bezert 1990; Molli et al. 2005; Malasoma et al. 2006), contain a sedimentary cover palaeontologically dated on the basis of the presence of Nummulites biarritzensis; Discociclina sp., to early Mid-Eocene (Bezert & Caby 1988). Therefore, stratigraphic data from the frontal part of the Corsican wedge suggest that blueschist facies associated with underthrusting and subduction of the continental crust continued until at least the Bartonian (40–37 Ma).

A Late Cretaceous to early–Mid Eocene age for the subduction in the Corsican orogenic system is altogether supported by regional data such as: (1) the presence of detritic glaucophane in Maastrichtian sediments of northwest Sardinia (Dieni & Massari 1982; Malavieille et al. 1998); and (2) the observation that in southeast Corsica Upper Eocene sediments lacking high pressure/low temperature assemblages rest unconformably on basement of Western Corsica and Schistes Lustrés units (Egal 1992; Caron 1994).

The subduction- and syn-contractional exhumation-related structures are locally strongly modified and reactivated by a late- to post-orogenic strike-slip deformation whose importance was first highlighted in the early 1970s by Maluski et al. (1973).

A major regional-scale structure is represented by the Central Corsica Fault Zone (CCFZ) of Waters (1990) (see also Jourdan 1988). This structure (recently renamed Ostriconi Fault in the northern sector by Lacombe & Jolivet 2005) can be traced from Ghisoni in the Fiumorbo valley in the south to Ostriconi in the Balagne region to the north. The CCFZ is part of a sinistral-linked faults system formed by anastomosing and overlapping–overstepping faults (Fig. 5) whose tectonic activity can be only roughly constrained between the early Late Eocene and Middle Miocene (Maluski et al. 1973; Jourdan 1988; Waters 1990; Molli & Tribuzio 2004; Lacombe & Jolivet 2005). Post-Middle Miocene to Pliocene and present-day local reactivations characterize segments of this fault zone as testified by deformation of Miocene sediments, levelling measurements and ongoing seismic activity (cf. Lenotre et al. 1996; Fellin et al. 2004).

To sum up, three major tectonic events characterize the evolution of Alpine Corsica:

(1) Late Cretaceous–Middle Eocene east-dipping subduction and progressive underthrusting of the Corsican continental crust. This event ended with the arrival of the thick Corsican crust which blocked the subduction during late Mid-Eocene;

(2) a Mid-Eocene–Early Oligocene(?) sinistral transpressional stage with segmentation of previous nappe structures; and

(3) Oligocene–Miocene (up to present) reactivation of previous systems of thrust and wrench

structures in transtension with development of Miocene basins and their deformation.

Structural and metamorphic evolution of the ophiolitic and continental units in the Northern Apennine

A key role in the interpretation of the Northern Apennine evolution is played by the constraints coming from the uppermost units of the nappe stack (Ligurian units and high-pressure ophiolitic units of Western Liguria and Tuscany) as well as from the deepest exposed levels of the Northern Apennine thrust wedge, i.e. the Tuscan metamorphic units (Figs 4 and 5). The first group of units provides insight into the early stages of the oceanic accretion, its timing and spatial variability from the northern to the southern part of the Northern Apennine, whereas the second allows us to define the stages of continental subduction and exhumation of the distal part of the Adria/Apulia continental margin during collisional and retreating stages.

Western Liguria represents a sector of the Alpine belt (up to Middle Late Eocene) backthrusted toward Adria and included in the Apennine orogen since the Early Oligocene (Elter & Pertusati 1973; Cortesogno et al. 1979; Vanossi et al. 1994; Giglia et al. 1996). Western Liguria includes oceanic slices from the Piemont–Ligurian or Alpine Tethys oceanic and continental basement and cover units (Cortesogno et al. 1979; Vanossi et al. 1986; Schmid et al. 2004). The latter are mainly represented by Briançonnais-derived nappes whereas oceanic units comprise the Helminthoid Flysch of western Liguria (Borghetto, Moglio–Testico and Sanremo Flyschs) and the ophiolitic and metasedimentary units of the Voltri Group (Vanossi et al. 1986; Capponi & Crispini 2002). The Helminthoid Flysch units are generally unmetamorphic (Seno et al. 2005 and references) whereas the ophiolitic and metasedimentary units of western Liguria show a polyphase tectono-metamorphic history with variable peak and retrograde metamorphic conditions from eclogite to prehnite–pumpellyte facies (Vanossi et al. 1986). Early deformation in eclogite to blueschist facies consists of large-scale west-vergent folds and shear zones overprinted by later greenschist-facies exhumation-related structures (Hoogerdujin Strating 1994; Crispini 1996; Capponi & Crispini 2002). Subduction metamorphism in the eclogite- and blueschist-facies field was dated as Early and Mid-Eocene (49–40 Ma, Federico et al. 2004 and references), while later exhumation-related greenschist structures are compatible with a regional transpressional regime in which the dextral Sestri–Voltaggio Zone could play a role of antithetic splay (Capponi & Crispini 2002; Spagnolo 2006). The Internal Ligurian units of the Northern Apennine are characterized by polyphase deformation structures related to a subduction and accretion history. They include west-vergent thrusts and large-scale isoclinal folds formed at peak sub-greenschist facies metamorphic conditions overprinted by retrograde exhumation-related structures (Pertusati & Horremberger 1975; Van Wamel 1987; Hoogerdujin Strating & Van Wamel 1989; Hoogerdujin Strating 1994; Marroni & Pandolfi 1996). The latter include doubly vergent hectometre- to kilometre- scale folds with shallow to moderately dipping axial planar foliations and locally fold structures with vertical axes producing segmentation of early structures and their clockwise and anticlockwise torsion (Pertusati & Horremberger 1975; Meccheri & Antonpaoli 1982; Marroni et al. 1988; Ellero et al. 2001; Molli personal data). These structures, observable for instance at Mt. Verruga, Mt. Ghiffi and Mt. Ramaceto in the Ligurian Apennine, can be interpreted as related to a regional-scale transpression (Elter & Marroni 1991), pre-dating the emplacement of the Ligurian units on top of Tuscan ones, i.e. pre-dating the Late Oligocene–Early Miocene.

The Antola nappe is a part of the Helminthoid Flysch nappe system forming the uppermost unit of the Mid-Eocene western Alps/palaeoNorthern Apennine belt while the Internal Ligurian units can be interpreted as the superficially accreted units on the backside of the intraoceanic Alpine wedge during its broadening stages (Elter & Pertusati 1973; Hoogerdujin Strating 1994; Elter 1997).

The age of early subduction, accretion and intra-wedge deformation, which also includes important backthrusting (see below), is constrained as post-Early Paleocene by the age of the youngest sediments involved in the deformation. Moreover, structures produced during this early history are unconformably overlain by sediments of the Tertiary Piemonte and Epiligurian Basins. These sediments (Monte Piano Marls) are palaeontologically dated as Upper Lutetian–Lower Rupelian (Cerrina Feroni et al. 2002).

These successions of thrust top/rear wedge basins (Mutti et al. 1995; Di Giulio & Galbiati 1995; Marroni et al. 2002; Carrapa et al. 2004) locally sealed the contacts between the exhumed high pressure/low temperature metamorphic units of the Voltri group, the Internal Ligurian units and the Antola nappe, and some regional-scale folds and thrust in the External Ligurian Helminthoid Flysch units (Marroni et al. 2002 and references).

A different history can be reconstructed for the more external part of the External Ligurian domain in the northwest Apennine and for part of the ophiolitic units in southern Tuscany. In the former, deep-water sedimentation is documented as continuous until Late Priabonian–Early Rupelian (Cerrina et al. 2002, 2004), testifying the absence of an early 'Alpine' tectonic evolution. The same setting can be suggested for the southern Tuscany Ligurian units due to the complete absence of structures sealed by Epiligurian sediments and for some ophiolitic units characterized by a high pressure/low temperature metamorphic imprint. These Tuscan ophiolitic units exposed in the Gorgona, Giglio, Argentario promontory and around Roselle (Grosseto inland) show a tectonic evolution and timing different from the ophiolitic units of western Liguria (Voltri Group). In Gorgona, the major structure is represented by the inverted limb of a kilometre-scale east-facing recumbent fold associated with top-to-east high pressure/low temperature shear zones (Molli & Bernardeschi 2005). These subduction-related structures are overprinted by exhumation-related greenschist-facies folds and top-to-east shear zones (Capponi et al. 1990; Jolivet et al. 1998; Orti et al. 2001). The high pressure metamorphism is dated in the Gorgona island at c. 25 Ma by Brunet et al. (2000), implying an 'Apennine' subduction of oceanic crust in agreement with Jolivet et al. (1998), Rossetti et al. (2002) and Faccenna et al. (2001).

The Tuscan metamorphic units represent the exhumed part of the Adria continental crust which was involved in deep underthrusting in the wake of the oceanic subduction of the remaining Ligurian Tethys in southern Tuscany. The Tuscan metamorphic units show high-pressure greenschist-facies peak conditions (Mg-chloritoid and kyanite in metapelites) at 0.6–0.8 GPa and 400–500 °C (Massa unit in the Alpi Apuane region, Molli et al. 2000, 2002 and included references); high pressure/low temperature (Fe–Mg carpholite in metapelites) at 0.6–1 GPa and 350–380 °C (M. Leoni/Monticiano Roccastrada unit in the Montagnola Senese area, Giorgetti et al. 1998) and at 1–1.2 GPa and 350–420 °C (Verrucano of Monte Argentario, Theye et al. 1997; Jolivet et al. 1998). The structural history of these continental units is characterized by a polyphasic deformation with an early generation of structures connected with underthrusting and syn-metamorphic nappe-fold formation and a number of younger overprinting structures (Carmignani et al. 1978, 1995; Molli et al. 2000 and references).

In the Apuane region, early deformation (regional D1) is associated with a main foliation axial planar of isoclinal folds, bearing a regionally southwest/northeast-oriented stretching lineation (Fig. 7) This major lineation trend can also be observed in the other Tuscan metamorphic unit occurrences (Moretti 1986; Elter & Sandrelli 1994; Storti 1995; Carosi et al. 2002; Liotta 2002). Because of the high shear strain values, the lineation is considered parallel to the direction of progressive shearing, and hence parallel to the main transport direction of the inner Northern Apennine orogenic wedge during the early stage of deformation (e.g. Carmignani et al. 1978; Dewey et al. 1989; Molli & Vaselli 2006). In the following deformation history (post-D1), previous structures were reworked with development of different generations of folds and related high strain zones, which accommodate the unroofing and exhumation of the unit toward superficial structural levels where deformation was associated with late D2 brittle structures—kink and open folds, low- and high-angle normal faults. Recent structural work in the Alpi Apuane interpreted the syn-metamorphic D1 structures of the Tuscan Metamorphic Units as related to a progressive deformation of the inner Tuscan continental margin during underthrusting and syn-contractional exhumation while D2 is connected with the vertical movement of the unit within the ductile wedge. Structural data support a constant southwest-to-northeast sense of transport during both D1 and early D2 deformation history (Molli & Vaselli 2006). The recently available data from the Tuscan metamorphic units of the Alpi Apuane region, constraining the timing of exhumation by low-temperature thermochronometry (Balestrieri et al. 2003; Fellin et al. 2004) and available P,T,t data, can be coupled with the timing of sedimentary evolution of the wedge-top and foredeep siliciclastic basins and their eastward migration (Costa et al. 1998; Argnani & Ricci Lucchi 2001; Botti et al. 2004; Cerrina Feroni et al. 2002, 2004 and references therein) to support a model of continuous crustal contraction until the Pliocene. This is well recorded by migration of foredeep depocentres and coeval exhumation of internal metamorphic units (Molli 2005; Molli et al. 2006).

Cross-sections in Fig. 4 show the differences in style and architecture at both crustal and shallow levels in the Northern Apennine orogen. Section A–A' is traced across the Bobbio window, the major tectonic structure of the northwest Apennine. It is mainly based on available geological data contained in Elter et al. (1992); Labaume (1992); Bernini et al. (1997); Cerrina et al. (2002) and references; refraction seismic interpretations of Biella et al. (1987); Cassinis et al. (1990); Laubscher et al. (1992); and borehole-controlled reflection profiles for the external area (Cassano et al. 1986; Toscani et al. 2006). The southwestern part of the section is characterized by two crustal-scale thrusts. The westernmost thrust which brings a

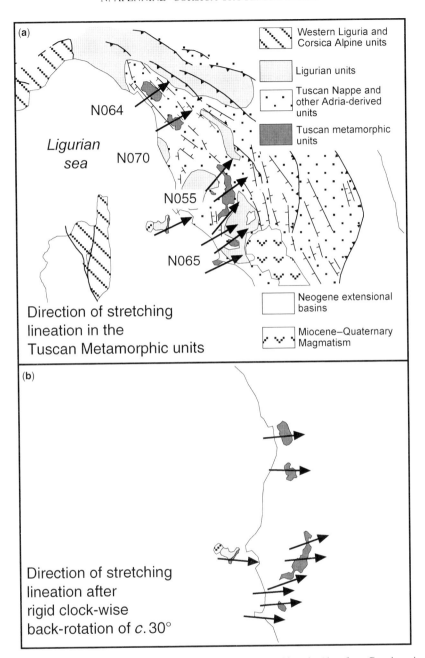

Fig. 7. (a) Regional trend of stretching lineations in the Tuscan metamorphic units (data from Carmignani *et al.* 1978; Moretti 1986; Elter & Sandrelli 1994; Carosi *et al.* 2002; Liotta 2002; Molli & Vaselli 2006 and references). (b) Trend of stretching lineation after *c.* 30° clockwise backrotation (supposed Langhian/pre-late Messinian deformation in the wake of Adria slab-retreat).

6–6.1 km/sec layer to a depth of 5 km is connected at the surface with the sub-Ligurian overthrust surface. The layer, overlain by Ligurian nappes (Antola, Internal and External Ligurian units) can be interpreted as related to sub-Ligurian and External Ligurian basement. The first activation of this thrust which shows out-of-sequence relationships post-dating the synsedimentary emplacement of the Sub-Ligurian unit on top of the Bobbio sandstone (belonging to the Cervarola system) can be constrained as post-Early Burdigalian (Labaume 1992; Elter et al. 1992). The second thrust displaced the top of Mesozoic Adria carbonates (reflector with velocity of 6 km/sec) and bounds in subsurface the Bobbio composite structure. The development of Bobbio crustal antiform produced the final emplacement (Tortonian in age) with eastward sliding away from the crest of the antiform of the Ligurian units on top of Middle Miocene (Serravallian) sandstone reached in the Ponte dell'Olio deep hole (Elter et al. 1992 and references; Toscani et al. 2006). Along the profile, the Moho gently dips west, reaching a depth of c. 40 km southwest of Bobbio where it rises abruptly to a shallower position in the Ligurian–Tyrrhenian basin (Laubscher et al. 1992; Castellarin 1992, 2001).

Section B–B' is traced across the Alpi Apuane window. It includes data of Cassano et al. (1986); Labaume (1992); Castellarin (1992, 2001) for the external part and surface geology of Carmignani et al. (1978); Molli et al. (2000, 2002) and Molli & Vaselli (2006). In the cross-section, the top of basement dips eastward from the Alpi Apuane metamorphic complex and reaches c. 10 km of depth east of the Val di Lima fold (Carmignani & Kligfield 1990; Argnani et al. 2003). The basement below is here subdivided into two parts. The upper one is considered part of the Tuscan metamorphic units exposed in the Alpi Apuane, whereas the lower one is interpreted as an external slice (?Cervarola basement) in analogy with the basement of Pontremoli hole north of the Alpi Apuane (Montanini & Molli 2002). This second basement slice, downward bounded by thrust (t2) whose activity could be Tortonian to Messinian in age, overlies the more external basement and cover thrust sheets. The two internal crustal thrusts (t1 and t2) which correspond to the 'Apuan Alps' and 'Abetone' thrusts of Boccaletti & Sani (1998) are here considered as having a component of motions out of section, as structural data in their surface splays indicate (e.g. Plesi et al. 1998). Along the profile, the Moho is gently west dipping reaching a depth of c. 50 km and jumps to a shallower position in the Tyrrhenian Moho (references in Pialli et al. 1998; Castellarin 2001).

Section C–C' is based on results of the Crop-03 Seismic Profile (Pialli et al. 1998 and included papers) with shallow structures mainly after Patacca et al. (1992); Decandia et al. (1998); Barchi et al. (2001); Carminati et al. (2004); Pertusati et al. (2004) for the Apennine side; data in Bartole et al. (1991); Jolivet et al. (1998); Contrucci et al. (2005) have been added to draw the Tyrrhenian Sea and deep structures in Corsica, where shallow geology is mainly after Jolivet et al. (1990); Rossi et al. (1994, 2003); Malavieille et al. (1998); Molli & Tribuzio (2004); Molli et al. (2006).

Section C–C' can be subdivided into three parts: Corsica, a central domain from the Corsica basin to inner southern Tuscany and an easternmost domain including Umbria and the Adriatic side. In the Corsica domain, some orogenic and late orogenic crustal structures (pre-Late Oligocene in age) are figured notwithstanding the local overprints of Provençal- and Tyrrhenian-related extensional and transtensional structures. The central domain shows a thin crust (22–24 km) and the overall tectonic features are dominated by extension (Patacca et al. 1992; Carmignani et al. 1995; Pialli et al. 1998; Decandia et al. 1998; Carminati et al. 2004). Crustal-scale extensional and transcurrent fault systems are associated with sedimentary basins and with exposed (e.g. Elba, Giglio, Gavorrano, Campiglia) or still at depth granitoids (Larderello). The crustal basement of the central domain includes metamorphic units of the inner side of the Corsican belt as well as those of the pre-Langhian inner Apenninic nappes stack.

The easternmost domain is at the surface characterized by contractional structures active since the Middle Miocene and propagating toward the Adriatic Sea. Its main crustal feature consists of westward–dipping Adriatic crust (c. 35 km in thickness) and thick-skinned crustal thrusts below the Val Chiana–Val Tiberina. At shallow structural levels, this domain includes the Tuscan Nappe thrust front and two major east-dipping normal faults: the western Val di Chiana fault and the eastern Val Tiberina fault which develops above the crustal stack (Barchi et al. 2001; Collettini et al. 2006).

The three cross-sections therefore show different features which can be mainly associated with the different amounts of subduction retreating to the Adriatic Plate and related extension in the internal domains (Doglioni 1991; Patacca et al. 1992). While the northern section is almost unaffected by post-Langhian extension, this is stronger in the southern (Tuscany) section while the central (Alpi Apuane) section represents a transitional domain.

Orogenic history of the Northern Apennine/Corsica system and relationships with the western Alps

Taking into account the major constraints derived from the evolution of units of the former Corsican and Apenninic continental margins, a large-scale scenario of orogenic evolution is here presented and used as a framework to discuss the relationships with the western Alps. The Cretaceous to present palaeogeographic scenario of Gueguen et al. (1997); Stampfli et al. (1998); Neugebauer et al. (2001); Michard et al. (2002); Rosenbaum & Lister (2002, 2004, 2005); and above all the

Alpine-derived palaeotectonic reconstructions of Schmid et al. (1996), Schmid & Kissling (2000), Dezes et al. (2004) form the regional reference frame in which the proposed evolution is inserted.

The older part of the Corsica/Apennine orogenic evolution is recorded at the deepest crustal levels in Corsica, where the tectonic evolution can be framed in a Cretaceous stage of east-dipping intraoceanic subduction (Fig. 8a) related to the east/southeast movement of the Iberian Plate (Lagabrielle & Polino 1988; Schmid et al. 1996; Stampfli et al. 1998; Michard et al. 2002). This stage is recorded in Corsica Schistes Lustres by the development of eclogites in ophiolitic units dated at 84 ± 5 Ma (Lahondere & Guerrot 1997). A low to very low geothermal gradient for these ophiolitic units, related to a very fast and steep subduction by Molli & Tribuzio (2004), is shown by the local occurrence of eclogites with coexisting almost pure jadeite and quartz and by lawsonite-bearing eclogites (Pequignot et al. 1984; Caron 1994; Lahondere 1996). On the other side of the Ligurian ocean, a nearly coeval tectonic event is recorded in the Ligurian domain by the development of Late Cretaceous mélanges (e.g. the Casanova complex; Marroni et al. 2002 and references) characteristic of the stratigraphy of some External Ligurian units (Fig. 5). The development of these sedimentary mélanges has to be related to the slicing and tectonic inversion of the External Ligurian domain in the continent–ocean transition area in connection with transpressive tectonics affecting part of the ocean and the adjacent Adria continental margin during the early formation and evolution of the Late Cretaceous alpine accretionary wedge (Bertotti et al. 1986; Elter 1997; Marroni et al. 2001, 2002). While in the northern transect the accretionary wedge developed close to the Adria continental margin (Elter & Pertusati 1973; Dal Piaz 1974; Pleuger et al. 2005), further south the subduction was closer to the opposite Iberian/Corsica Plate as testified by the 65 Ma age of continental-derived eclogites (Farinole unit, Brunet et al. 2000) at the same time when the

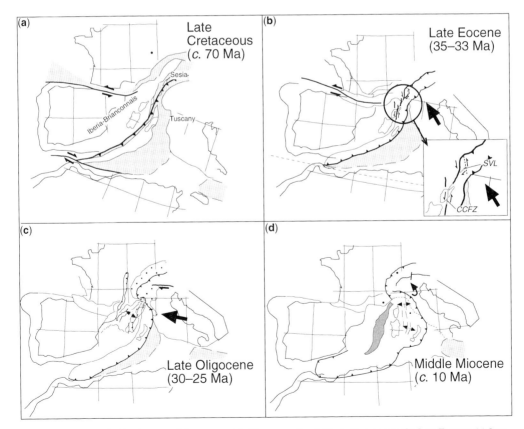

Fig. 8. Palaeotectonic sketch maps of the western Mediterranean for: (**a**) Late Cretaceous; (**b**) Late Eocene; (**c**) Late Oligocene; (**d**) Middle Miocene (mainly based on Gueguen et al. 1997; Stampfli et al. 1998; Séranne 1999; Neugebauer et al. 2001; Michard et al. 2002; Dezes et al. 2004 and references).

Sesia unit in the Alps was deeply accreted (Polino et al. 1990; Dal Piaz et al. 2003; Compagnoni 2003). This can be fitted into a frame of original transversal trace of the subduction (Sturani 1973; Elter & Pertusati 1973; Dal Piaz 1974) which developed an accretionary wedge in the western Alps transect near the Adria margin (metamorphism in the Sesia unit), while southward (east–southeast of Corsica) it was closer to the opposite, Iberia/Corsica continent.

The interpretation in map view of Figure 8a shows the trace of a subduction system in the frame of Late Cretaceous palaeotectonics. Between Iberia and Adria, a continental block is figured bounded by two branches of the Alpine ocean. To this continental block, which corresponds to the Alkapeca domain of Boullin et al. (1986); Guerrera et al. (1993); Michard et al. (2002), can be related tectonic units and terranes in the western Mediterranean orogens from Alboran, Kabylias, Peloritani to Calabria. Triassic–Liassic stratigraphic features of the cover units close to the coeval Adria stratigraphy (e.g. Tuscan domain) and the Hercynian basement evolution and rock types, including granulite-facies rocks, suggest that the Alkapeca microcontinent drifted away from the Apulia lower plate of the Tethyan rift (Michard et al. 2002 and references). In Corsica, remnants of this domain can be found within the Nebbio units where Triassic–Liassic stratigraphy shows Adria-like features (Dallan & Puccinelli 1995 and references) and possibly, for the original westernmost side of the former domain, in the S. Lucia Nappe in central Corsica (Durand-Delga 1984; Zibra 2006 and references), where granulite-facies continental rocks can be observed. It is significant to point out that tectonic interpretations including an intra-oceanic microblock (Nebbio microcontinent in Fig. 9a) are not new for Corsica and have been suggested since the 1980s (Durand-Delga 1984 and references; Dallan & Puccinelli 1995 and references; Lahondere 1996; Malavieille et al. 1998). What is proposed here (Figs 8a and 9a) follows these interpretations and moreover suggests that the Sesia, further northwest, and possibly the Triassic–Liassic cover unit in the Sestri–Voltaggio zone (Crispini 1996 and references) could be interpreted similarly as microblocks and/or extensional allochthonous, derived from the Tethyan passive rifting (e.g. Molli 1996; Marroni et al. 1998; Dal Piaz 1999; Manatschal & Bernoulli 1999; Pleuger et al. 2005).

During the Paleocene–Early Eocene, the ophiolitic and oceanic cover units of western Liguria and the Ligurian units of the Northern Apennine (Fig. 8b) were involved in the 'Alpine' accretionary wedge. Western Ligurian units were involved deeper in the accretionary structure than Internal Ligurian units, while the Helminthoid Flysch nappes of western Liguria, Maritime Alps and the Antola Nappe (plus parts of the External Ligurian units according to Daniele & Plesi 2000) were the leading edge and the uppermost unit of the Eocene Alpine wedge along the Ligurian transect (Grandjacquet & Haccard 1977; Elter 1997; Seno et al. 2005).

The arrival of the Corsican distal continental margin at the subduction zone (Serra di Pigno/Farinole units, Lahondere 1996; Malavieille et al. 1998) progressively reduced the subduction rate of the Iberia Plate (Molli & Tribuzio 2004). The arrival of the thicker continental crust (Tenda Massif) and the more external continental units, then halted the subduction during the Early–Middle Eocene (Molli & Tribuzio 2004). Due to buoyancy forces, the Corsican crust failed and started to be exhumed, whereas break-off and detachment of subducted mantle lithosphere took place (Malavieille et al. 1998; Tribuzio & Giacomini 2002; Molli & Tribuzio 2004). Erosion of Early–Middle Eocene Flysch originally covering a large part of the autochthonous Hercynian basement (Danisik et al. 2007) can be directly connected to a strong surface uplift related to slab break off. This event was enhanced by the changed relative motion of Iberia and Europe, which became north–south convergence according to Séranne (1999); Rosenbaum et al. (2002); Dezes et al. (2004); Lacombe & Jolivet (2005) producing collision and formation of the Pyrenees thrust and fold belt.

As a result of blocked subduction and far-field stress from of the Pyrennean collision, a wrench-dominated tectonics (north–south striking in present coordinates) developed in Corsica while the ongoing regional shortening (Figs 8b and 9b) and the presence of the still-open oceanic domain east of the Alpine accretionary wedge (central Tuscany and southern Italy) allowed subduction flip and the initiation of the west-northwestward Apenninic subduction. In western Liguria and in the Apenninic Ligurian units, this event was associated with a deformation characterized by back-thrusting and back-sliding superimposed on older top-west kinematics. The uppermost units of the western Liguria Alpine wedge moved with west-northwest kinematics toward the foreland areas (Merle & Brun 1984; Seno et al. 2005 and references) as well as backward (east-southeastward) with Apenninic vergences (Elter 1975; Daniele & Plesi 2000; Marroni et al. 2002; Levi et al. 2006). The final exhumation of the high pressure/low temperature Western Liguria units of the Voltri Group was achieved by northeast-vergent thrusting (Mutti et al. 1995; Crispini 1996; Capponi & Crispini 2002; Mosca et al. 2005; Spagnolo 2006)

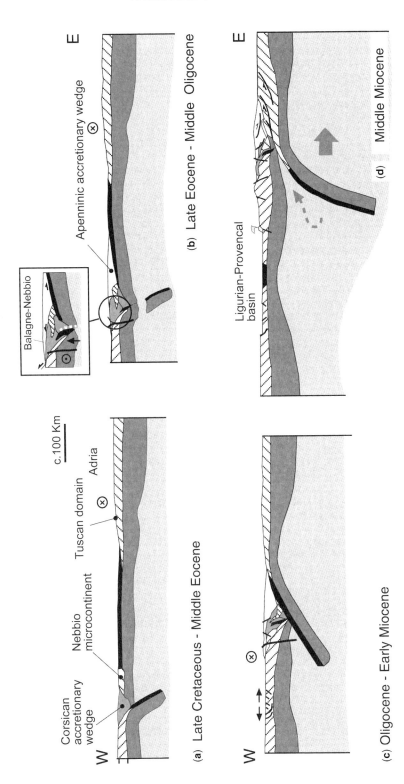

Fig. 9. Tectonic evolution of the Corsica/Northern Apennine orogenic system: (a) subduction of distal Corsica margin and eclogitic metamorphism (Late Cretaceous), followed by the early-Mid Eocene involvement of thick Corsica crust (Tenda and External continental units); (b) blocking of subduction, slab break-off and flip of subduction polarity (Mid–Late Eocene–Early Oligocene). According to Malavieille et al. (1998) during the early development of Apenninic wedge top-to-west back-thrusting of Balagne-Nebbio units above HP/LT Corsican units occurred; (c) Oligocene–Early Miocene Apenninic subduction with regional upper-plate crustal extension west of Sardinia/Corsica; (d) Middle Miocene beginning of retreat of the Apenninic subduction zone with transtension in Corsica and Northern Tyrrhenian sea.

over Ligurian units coupled with erosion during Late Eocene–Early Oligocene, as testified by detritus derived from such units in Upper Eocene–Oligocene sediments of the Tertiary Piemonte and Epiligurian basins (Di Biase et al. 1997; Carrapa et al. 2004).

With the late Priabonian/Rupelian, the complete closure of the oceanic area east of the Alpine accretionary wedge was accomplished (Figs 8b and 9c). This closure was diachronous from north to south due to the V-shaped remnants of Ligurian Tethys. High pressure/low temperature ophiolitic units of the Tuscan Islands (Gorgona and Giglio) and Argentario represent the preserved deeply underplated units of the Apenninic accretionary wedge (Jolivet et al. 1998; Rossetti et al. 2002) according to radiometric data (Ar/Ar whole rock age of 25 Ma in Brunet et al. 2000; Rossetti et al. 2002) and structural work in Gorgona (Molli & Bernardeschi 2005).

The final closure of the remnants of Ligurian Tethys was initially controlled by a north–south Adria movement (Schmid & Kissling 2000; Schmid et al. 2004) which induced the early Apenninic accretion in sinistral transpression while, starting from the Late Oligocene, a change in Adria movement direction to northwestward controlled the following evolution. In the wake of oceanic crust subduction, part of the distal western edge of Adria i.e. the Apenninic continental margin (sub-Ligurian and Tuscan domains) was involved in the underthrusting from Late Oligocene–Early Miocene, in agreement with radiometric data spanning between 27 and 25 Ma (Kligfield et al. 1986). The different P/T peak conditions of the deeply underplated continental-derived Tuscan units give further indication of along-strike variations of the subduction system. The different geothermal gradient associated with peak metamorphism from north (higher) to south (lower) can be connected with a difference in crustal and lithospheric thickness indicating a more advanced stage of continental subduction in the north (Alpi Apuane) versus a more initial stage of continental subduction in the south (metamorphic units of southern Tuscany).

The Apenninic subduction produced Early Oligocene to Miocene calc-alkaline volcanism on the western side of Sardinia–southern Corsica, the associated back-arc rifting (30–21 Ma), the formation of oceanic crust (21–18 Ma) in the Ligurian–Provençal basin, and then (19–16 Ma) the rotation of the Sardinian–Corsica block away from the Iberia/Europe mainland (Scandone 1979; Rehault et al. 1984; Serri et al. 1993; Speranza et al. 2002 and references).

The internal deformation of the inner Apennine wedge is reflected at superficial structural levels in the Epiligurian basins by two major events (Aquitanian and Burdigalian phases, Cerrina et al. 2004 and references) which are roughly coeval with the counterclockwise Corsica–Sardinia block rotation away from Iberia. The rotation is recorded, although with smaller intensities, by the palaeomagnetic studies on clastic sediments of the Epiligurian basin which suggest nearly coeval counterclockwise rotation of Ligurian thrust sheets of $c.$ 24° (Muttoni et al. 2000 and references).

The blocking of rotation around 16 Ma (Speranza et al. 2002) and the change of style in the Adria subduction with beginning of delamination and retreat (Doglioni 1991; Serri et al. 1993) induced an important wedge reshaping (Figs 8d and 9d) with regional-scale thinning of the wedge associated with the exhumation of metamorphic units. This is recorded by widespread low angle normal faulting in the upper units of the Northern Apennine (Carmignani et al. 1995; Molli et al. 2000; Cerrina et al. 2004; Molli 2005; Molli et al. 2006).

The process of retreating subduction and slab bending (Patacca & Scandone 1989; Doglioni 1991; Castellarin 1992; Pialli et al. 1998; Lucente & Speranza 2001) induced contraction in the external part of the Northern Apennine with progressive migration of foredeeps (sensu Costa et al. 1998; Botti et al. 2004) and their involvement in the accretionary front. Two major stages of thrust activity occurred during the Langhian–Serravalian (15–11 Ma) and Tortonian–Messinian (11–5 Ma). In the northern corner of the accretionary wedge (inner Northern Appenine), roto-translational kinematics (Doglioni 1991; Patacca et al. 1992) were induced by the retreat and produced the strain pattern and geometry of late deformation around and north of the Alpi Apuane and surroundings (Plesi et al. 1998; Ottria & Molli 2000; Vescovi 1998, 2005; Carosi et al. 2002; Cerrina et al. 2004; Molli, personal data). High-angle normal faulting dissected the previous structural building during Plio- and Pleistocene when intramontane to marine basins developed (Patacca et al. 1992; Bossio et al. 1998; Decandia et al. 1998; Barchi et al. 2001).

To sum up, five major stages characterized the development of the Corsica/Northern Apennine orogenic system; in each of them, the relationships with the growing Alpine wedge show different characters and structures. Complex reactivation of former thrusts, segmentation and overprinting of structures coeval with development and internal deformation of basins characterized western Liguria and Ligurian units of the Northern Apennine, giving way to the strain pattern of the 'Ligurian Knot'. Moving southward, the effects of the retreating Adriatic slab produced the major crustal and near-surface features and architecture of the chain in Tuscany with Miocene–Pliocene contraction in the external domain associated with extension and magmatism in the internal domains:

STAGE 1. Late Cretaceous–Middle Eocene east-dipping intra-oceanic and then continental subduction with progressive underthrusting of Corsican continental crust. This event ended with the arrival of thick Corsican crust which blocked the subduction during late Mid-Eocene at the same time when Iberia–Europe convergence began and the Pyrenees started to develop;

STAGE 2. Late Mid-Eocene to early-Mid-Oligocene sinistral transpression stage in Corsica coeval with inversion of subduction and beginning of Apenninic history. This was initially controlled by northwestward Adria movement (Schmid et al. 1996; Schmid & Kissling 2002) which induced in the rear of the previously formed western Alps/Corsican orogenic prism, the development of the Apenninic wedge in an oblique convergence setting (Elter & Marroni 1991; Del Castello et al. 2005), while the western Alps were affected by a regional-scale sinistral transpression;

STAGE 3. Between the Late Oligocene and the Burdigalian, the Northern Apennine developed as a stable accretionary wedge, with consumption of relict Ligurian Tethys ocean in southern Tuscany and development of foredeep-thrust top basins longitudinally fed by the uplift of the central Alps (Macigno–Modino–Cervarola). As a result of the early oceanic subduction stage, back-arc opening produced the Liguro-Provençal basin and the Corsica–Sardinia block rotation took place; in the internal part of the wedge, syn-contractional exhumation and tectonic thinning of the nappe systems occurred whereas contraction still existed at depth. This Apenninic evolution is coeval and kinematically related with dextral strike-slip movement along the east–west segment (Tonale line) of the Periadriatic or Insubric fault system (Laubscher 1991; Schmid et al. 1996; Schmid & Kissling 2000);

STAGE 4. Starting from the Langhian, delamination of subducted continental crust and beginning of retreat of subducting slab toward the east occurred. This process promoted extension in the overriding plate associated with basin formation and magmatism (Langhian to Pliocene) and a migration of the contractional front. Complex rotational movement in the northern corner of the accretionary wedge produced the strain pattern and geometry of late deformation in the Alpi Apuane. In particular, a pre-Messinian rigid body rotation of nearly 30° of the Tuscan metamorphic units in the hanging wall of a major thrust can restore the main transport direction in the Tuscan Units to east–west trend, thus fitting the Adria movement during Early–Middle Miocene (stage 3); and

STAGE 5. A change of the former regime (stage 4) recorded by the progressive submarine advance of Ligurian nappes toward the foredeep basins took place during Late Pliocene–Early Pleistocene and continued up to the present. During this time, the outward thrust-front propagation decreased and vertical movement started to be dominant in the inner as well toward the outer domain where the present-day orographic front started to be uplifted (Castellarin 1992, 2001; Bertotti et al. 1997; Argnani et al. 2003). Whether this change in tectonic regime has to be related with a decreasing rate of passive sinking of the Adria Plate (Patacca et al. 1992), Pliocene slab-break off ending with the present-day absence of contraction at the mountain front (Carminati et al. 1998; Di Bucci & Mazzoli 2002), or a continuous retreating process with mountain fronts still active (Brandon 2002; Carminati et al. 2004; Collettini et al. 2006), is still a matter of investigations.

This paper is mainly derived from two invited talks held at the ETH of Zurich in spring 2004 and at the Department of Geology of the University of Basel in January 2005. All the attendees at those Meetings are deeply acknowledged for discussions and suggestions with special thanks to Stefan Schmid, who strongly encouraged me to write the paper. Piero Elter and colleagues of Pisa, Genova, Siena, Napoli, Padova, Parma, Pavia, Perugia and Rome are thanked for the many discussions about the treated topics over the past years as well as M. Brandon and the Retreat Project members. N. Froitzheim and J. Malavieille are acknowledged for their very useful comments and suggestions that improved the paper. This paper is dedicated to the memory of my father.

References

ABBATE, E. & SAGRI, M. 1984. Le unità torbiditiche cretacée dell'appennino settentrionale ed i margini continentali della Tetide. *Memorie della Società Geologica Italiana*, **24**, 359–375.

ALVAREZ, W. 1991. Tectonic evolution of the Corsica-Apennines-Alps region studied by the method of successive approximations. *Tectonics*, **10**, 936–947.

ALVAREZ, W., COCOZZA, T. & WEZEL, F. C. 1974. Fragmentation of the Alpine orogenic belt by microplate dispersal. *Nature*, **248**, 309–314.

ARGAND, E. 1924. La tectonique de l'Asie. *Comptes Rendus Congrés Géologique International, XIII, Belgique 1922*, **1**, 171–372.

ARGNANI, A. 2002. The Northern Apennines and the kinematics of Europe-Africa convergence. *Bollettino della Società Geologica Italiana*, Special Volume, **1**, 47–60.

ARGNANI, A. & RICCI LUCCHI, F. 2001. Tertiary Siliciclastic Turbidite Systems of the Northern Apennines. *In*: VAI, G. B. & MARTINI, I. P. (eds) *Anatomy of an Orogen*. Kluwer Academic Publishers, 327–350.

ARGNANI, A., BARBARACINI, G., BERNINI, M. ET AL. 2003. Gravity tectonics driven by Quaternary uplift in the Northern Apennines: insights from La Spezia-Reggio Emilia geo-transect. *Quaternary Journal*, **101**, 13–26.

ARGNANI, A., FONTANA, D., STEFANI, C. & ZUFFA, G. G. 2004. Late Cretaceous Carbonate Turbidites of the Northern Apennines: shaking Adria at the onset of Alpine Collision. *Journal of Geology*, **112**, 251–259.

BALESTRIERI, M. L., BERNET, M., BRANDON, M. T., PICOTTI, V., REINERS, P. & ZATTIN, M. 2003. Pliocene and Pleistocene exhumation and uplift of two key areas of the Northern Apennines. *Quaternary International*, **101–102**, 67–73.

BARCHI, M., LANDUZZI, A., MINELLI, G. & PIALLI, G. 2001. Outer Northern Apennines. *In*: VAI, G. B. & MARTINI, I. P. (eds) *Anatomy of an Orogen*. Kluwer Academic Publishers, 215–254.

BARTOLE, R., TORELLI, L., MATTEI, G., PEIS, D. & BRANCOLINI, G. 1991. Assetto stratigrafico-strutturale del Tirreno Settentrionale. *Studi Geologici Camerti*, *volume speciale*, **1**, 115–140.

BERNINI, M., VESCOVI, P. & ZANZUCCHI, G. 1997. Schema strutturale dell'Appennino Nord-Occidentale. *Acta Naturalia de 'L'Ateneo Parmense'*, **33**, 43–54.

BERNOULLI, D. 2001. Mesozoic-Tertiary Carbonate Platforms, Slopes and Basins of the external Apeninnes and Sicily. *In*: VAI, G. B. & MARTINI, I. P. (eds) *Anatomy of an Orogen*. Kluwer Academic Publishers, 307–326.

BERNOULLI, D., KÄLIN, O. & PATACCA, E. 1979. A sunken continental margin of the Mesozoic Tethys: the northern and central Apennines. *Association Geologique Français, Publication Special*, **1**, 197–210.

BERTOTTI, G., ELTER, P., MARRONI, M., MECCHERI, M. & SANTI, R. 1986. Le Argilliti a blocchi di M.Veri: considerazioni sulla evoluzione tettonica del bacino Ligure nel Cretaceo Superiore. *Ofioliti*, **11**, 193–220.

BERTOTTI, G., CAPOZZI, R. & PICOTTI, V. 1997. Extensional controls on Quaternary tectonics, geomorphology and sedimentations of the N-Apennines foothills and adjacent Po Plain (Italy). *Tectonophysics*, **282**, 291–301.

BEZERT, P. 1990. Les unites Alpines a la marge du Massif Cristallin Corse: Nouvelles donnes structurales, metamorphiques et contraintes cinematiques. *These de 3ème cycle, Université Montepelier II*.

BEZERT, P. & CABY, R. 1988. Sur l'age post-bartonien des événements tectono-métamorphiques alpines en bourdure orientale de la Corse cristalline. *Bulletin de la Société Géologique de France*, **8**, 965–971.

BIELLA, G. A., GELATI, R., MAISTRELLO, M., MANCUSO, M., MASSIOTTA, P. & SCARASCIA, S. 1987. The structure of the upper crust in the Alps-Apennines boundary region deduced from refraction seismic data. *Tectonophysics*, **142**, 71–85.

BIELLA, G. A., GELATI, R., LOZEJ, A., ROSSI, P. M. & TABACCO, I. 1988. Sezioni Geologiche nella zona limite Alpi Occidentali-Appennino Settentrionale ottenute da dati geofisici. *Rendiconti Società Geologica italiana*, **11**, 287–292.

BIGI, G., COSENTINO, D., PAROTTO, M., SARTORI, R. & SCANDONE, P. 1983. Structural model of Italy, Sheets 1–6. Scale 1:500 000. *CNR, Quaderni de 'La Ricerca Scientifica*, **114**, 3.

BOCCALETTI, M., ELTER, P. & GUAZZONE, G. J. P. 1971. Plate tectonic models for the development of the western Alps and Northern Apennines. *Nature*, **234**, 108–111.

BOCCALETTI, M. & SANI, F. 1998. Cover thrust reactivations related to internal basement involvement during Neogene-Quaternary evolution of the northern Apennines. *Tectonics*, **17**, 112–130.

BONCIO, P., BROZZETTI, F. & LAVECCHIA, G. 2000. Architecture and seismotectonics of a regional Low-Angle Normal Fault zone in Central Italy. *Tectonics*, **19**, 1038–1055.

BORTOLOTTI, V., FAZZUOLI, M., PANDELI, E., PRINCIPI, G., BABBINI, A. & CORTI, S. 2001. Geology of central and eastern Elba island, Italy. *Ofioliti*, **26**, 97–150.

BOSSIO, A., CONSTANTINI, A., FORESI, L. M. ET AL. 1998. Neogene-Quaternary sedimentary evolution in the western side of the Northern Apennines (Italy). *Memorie della Società Geologica Italiana*, **52**, 513–526.

BOTTI, F., ALDEGA, L. & CORRADO, S. 2004. Sedimentary and tectonic burial evolution of the Northern Apennines in the Modena-Bologna area: constraints from combined stratigraphic, structural, organic matter and clay mineral data of Neogene thrust-top basins. *Geodinamica Acta*, **17**, 185–203.

BOULLIN, J. P., DURAND-DELGA, M. & OLIVIER, P. 1986. Betic-Rifain and Tyrrhenian Arcs: distinctive features, genesis and development stages. *In*: WEZEL, F. C. (ed.) *The Origins of Arcs*. Elsevier, 281–304.

BRANDON, M. 2002. Retreat Project, NSF. http://earth.geology.yale.edu/RETREAT/

BRUNET, C., MONIÉ, P., JOLIVET, L. & CADET, J. P. 2000. Migration of compression and extension in the Tyrrhenian Sea, insights from $^{40}Ar/^{39}Ar$ ages on micas along a transect from Corsica to Tuscany. *Tectonophysics*, **321**, 127–155.

CALAMITA, F., CELLO, G., DEIANA, G. & PALTRINIERI, W. 1994. Structural styles, chronology rates of deformation and time-space relationships in the Umbria-Marche thrust system (central Apennines, Italy). *Tectonics*, **13**, 873–881.

CAPPONI, G., GIANMARINO, S. & MAZZANTI, R. 1990. Geologia e morfologia dell'isola di Gorgona. *Quaderni Museo Storia Naturale Livorno*, **11**, 115–137.

CAPPONI, G. & CRISPINI, L. 2002. Structural and metamorphic signature of alpine tectonics in the Voltri Massif (Ligurian Alps, North-Western Italy). *Eclogae Geologique Helvetique*, **95**, 31–42.

CARMIGNANI, L., GIGLIA, G. & KLIGFIELD, R. 1978. Structural evolution of the Apuane Alps: An example of continental margin deformation in the Northern Apennine. *Journal of Geology*, **86**, 487–504.

CARMIGNANI, L. & KLIGFIELD, R. 1990. Crustal extension in the Northern Apennines: the transition from compression to extension in the Alpi Apuane Core Complex. *Tectonics*, **9**, 1275–1303.

CARMIGNANI, L., DECANDIA, F. A., DISPERATI, L., FANTOZZI, P. L., LAZZAROTTO, A., LIOTTA, D. & OGGIANO, G. 1995. Relationships between the Tertiary structural evolution of the Sardinia-Corsica-Provençal Domain and the Northern Apennines. *Terra Nova*, **7**, 128–137.

CARMINATI, E., WORTEL, M. J. R., SPAKMAN, W. & SABADINI, R. 1998. The role of slab detachment process in the opening of the western-central Mediterranean basins: some geological and geophysical evidence. *Earth and Planetary Science Letters*, **160**, 651–665.

CARMINATI, E., DOGLIONI, C. & SCROCCA, D. 2004. Alps vs Apennine. In: CRESCENTI, U., D'OFFIZI, S., MERLINO, S. & SACCHI, L. (eds) *Geology of Italy*. Special volume of the Italian Geological Society for the IGC 32 Florence-2004, 141–152.

CARON, J. M. 1994. Metamorphism and deformation in Alpine Corsica. *Schweizerische Mineralogische und Petrographische Mitteilungen*, **7**, 105–114.

CAROSI, R., MONTOMOLI, C. & PERTUSATI, P. C. 2002. Late orogenic structures and orogen-parallel compression in the Northern Apennines. *Bollettino Società Geologica Italiana Volume Speciale*, **1**, 167–180.

CARRAPA, B., DI GIULIO, A. & WIJBRANS, J. 2004. The early stages of the Alpine collision: an image derived from the upper Eocene-lower Oligocene record in the Alps-Apennines junction area. *Sedimentary Geology*, **171**, 181–203.

CASSANO, E., ANELLI, L., FICHERA, R. & CAPPELLI, V. 1986. Pianura Padana. Interpretazione integrata di dati geofisici e geologici. *Agip S Donato Milanese*.

CASSINIS, R., LOZEJ, A., TABACCO, I., GELATI, R., BIELLA, G., SCARASCIA, S. & MAZZOTTI, A. 1990. Reflection and refraction seismic data in areas of complex geology. An example in the Northern Apennines. *Terra Nova*, **2**, 351–362.

CASTELLARIN, A. 1992. Strutturazione Eo- e Mesoalpina dell'Appennino Settentrionale attorno al 'nodo Ligure'. *Studi Geologici Camerti, volume speciale*, **(1992/1) CROP 1-1A**, 99–108.

CASTELLARIN, A. 2001. Alps-Apennines and Po Plain-Frontal Apennines Relationships. In: VAI, G. B. & MARTINI, I. P. (eds) *Anatomy of an Orogen*. Kluwer Academic Publishers, 177–196.

CERRINA FERONI, A., MARTINELLI, P., PLESI, G., GIANMATTEI, L., FRANCESCHELLI, M. & LEONI, L. 1985. La cristallinità dell'illite nelle argille e calcari (Unità di Canetolo) tra La Spezia e l'alta Val Parma (Appennino Settentrionale). *Bollettino della Società Geologica Italiana*, **104**, 421–427.

CERRINA FERONI, A., MARTELLI, L., MARTINELLI, P., OTTRIA, G. & CATANZARITI, R. 2002. Carta Geologico-Strutturale dell'Appennino Emiliano-Romagnolo-Note Illustrative. *Regione Emilia-Romagna*. Selca Firenze.

CERRINA FERONI, A., OTTRIA, G. & ELLERO, A. 2004. The Northern Apennine, Italy: Geological Structure and Transpressive evolution. In: CRESCENTI, U., D'OFFIZI, S., MERLINO, S. & SACCHI, L. (eds) *Geology of Italy*. Special volume of the Italian Geological Society for the IGC 32 Florence-2004, 14–32.

CIARAPICA, G. & PASSERI, L. 2002. The paleogeographic duplicity of the Apennines. *Bollettino della Società Geologica Italiana, Volume Speciale*, **1**, 67–75.

CIBIN, U., TATEO, F., CATANZARITI, R., MARTELLI, L. & RIO, D. 1998. Composizione, origine ed età del vulcanismo andesitico Oligocenico inferiore dell'Appennino Settentrionale: le intercalazioni vulcano-derivate nella formazione di Ranzano. *Bollettino della Società Geologica Italiana*, **117**, 569–591.

CIVETTA, L., ORSI, G., SCANDONE, P. & PECE, R. 1978. Eastwards migration of Tuscan anatectic magmatism due to anticlockwise rotation of the Apennines. *Nature*, **276**, 604–606.

COHEN, C. R., SCHWEICKERT, R. A. & ODOM, A. L. 1981. Age of emplacement of the Schistes Lustrés nappe, alpine Corsica. *Tectonophysics*, **72**, 276–284.

COLLETTINI, C., DE PAOLA, N., HOLDSWORTH, R. E. & BARCHI, M. R. 2006. The development and behaviour of low-angle normal faults during Cenozoic asymmetric extension in the Northern Apennines, Italy. *Journal of Structural Geology*, **28**, 333–352.

COMPAGNONI, R. 2003. HP metamorphic belt of the western Alps. *Episodes*, **26**, 200–204.

CONTI, P. & LAZZAROTTO, A. 2004. Geology of Tuscany: evolution of the state-of-knowledge presented by geological maps and the new geological map of Tuscany, 1:250 000 scale. In: MORINI, D. & BRUNI, P. (eds) *The 'Regione Toscana' Project of Geological Mapping: Case Histories and Data Acquisition*, 33–50.

CONTRUCCI, R., MAUFFRET, A., BRUNET, C., NERCESSIAN, A., BÉTHOUX, N. & FERRANDINI, J. 2005. Deep structure of the Northern Tyrrhenian Sea from multi-channel seismic profiles and on land wide angle reflection/refraction seismic recording (LISA cruise): Geodynamic implications. *Tectonophysics*, **406**, 141–163.

CORNAMUSINI, G., LAZZAROTTO, A., MERLINI, S. & PASCUCCI, V. 2002. Eocene-Miocene evolution of the north Tyrrhenian Sea. *Bollettino della Società Geologica Italiana, Volume Speciale*, **1**, 769–787.

CORSI, B., ELTER, F. M. & GIANMARINO, S. 2001. Structural fabric of the Antola unit (Riviera di Levante, Italy) and implications for its Alpine versus Apennine origin. *Ofioliti*, **26**, 1–8.

CORTESOGNO, L., GRANDJACQUET, C. & HACCARD, D. 1979. Contribution à l'étude de la liaison Alpes-Appennins. Evolution tectono-métamorphique des principaux ensembles ophiolitiques de Ligurie (Appennins du Nord). *Ofioliti*, **4**, 157–172.

COSTA, E., PIALLI, G. & PLESI, G. 1998. Foreland basins of the Northern Apennines: relationships with passive subduction of the Adriatic lithosphere. *Memorie della Società Geologica Italiana*, **52**, 595–606.

COWAN, D. & SILLING, R. M. 1978. A dynamic scaled model of accretion at trenches and its implications for the tectonic evolution of subduction complexes. *Journal of Geophysical Research*, **83**, 5389–5396.

COWARD, M. P., DE DONATIS, M., MAZZOLI, M., PALTRINIERI, W. & WEZEL, F.-C. 1999. Frontal part of the northern Apennines fold and thrust belt in the Romagna-Marche area (Italy): shallow and deep structural styles. *Tectonics*, **18**, 559–574.

CRISPINI, L. 1996. Evoluzione strutturale dei metasedimenti del Gruppo di Voltri e della zona

Sestri-Voltaggio: implicazioni nell'evoluzione tettonica e geodinamica alpina. *Tesi di Dottorato, E.R.S.U., Università di Genova*.

DAL PIAZ, G. V. 1974. Le métamorphisme alpine de haute pression et basse température dans l'évolution structurale du bassin ophiolitique alpino-apenninique. 1e partie. *Bollettino della Società Geologica Italiana*, **93**, 437–468.

DAL PIAZ, G. V. 1999. The Austroalpine–Piedmont nappe stack and the puzzle of Alpine Tethys. *Memorie di Scienze Geologiche di Padova*, **51/1**, 155–176.

DAL PIAZ, G. V. & ZIRPOLI, G. 1979. Occurrence of eclogites relics in the ophiolite nappe from Marine d'Albo, Northern Corsica. *Neus Jahrbuch fur Mineralogie*, **3**, 118–122.

DAL PIAZ, G. V., BISTACCHI, A. & MASSIRONI, M. 2003. Geological outline of the Alps. *Episodes*, **26**, 175–180.

DALLAN, L. & PUCCINELLI, A. 1995. Geologia della regione tra Bastia e St-Florent (Corsica Settentrionale). *Bollettino della Società Geologica Italiana*, **114**, 23–66.

DANIEL, J. M., JOLIVET, L., GOFFÉ, B. & POINSOTT, C. 1996. Crustal-scale strain partitioning: footwall deformation below the Alpine Oligo-Miocene detachment of Corsica. *Journal of Structural Geology*, **18**, 1841–1859.

DANIELE, G. & PLESI, G. 2000. The Helminthoid flysch units of the Emilian Apennines: stratigraphic and petrographic features, paleogeographic restoration and structural evolution. *Geodinamica Acta*, **13**, 313–333.

DANISIK, M., KUHLEMANN, J., DUNKL, I., SZEKELY, B. & FRISCH, W. 2007. Burial and exhumation of Corsica (France) in the light of fission track data. *Tectonics*, **26**, TC1001, doi:10.1029/2005TC001938.

DEBELMAS, J. 1975. Réflexions et hypotheses sur la paléogéographie crétacée des confines alpino-apenniniques. *Bulletin de la Sociéte Géologique de France*, **17**, 1002–1012.

DECANDIA, A., LAZZAROTTO, A., LIOTTA, D., CERBORI, L. & NICOLICH, R. 1998. The CROP03 traverse: insights on post-collisional evolution of the Northern Apennines. *Memorie della Società Geologica Italiana*, **52**, 427–440.

DEL CASTELLO, M., MCCLAY, K. R. & PINI, G. A. 2005. Role of pre-existing topography and overburden on strain partitioning of oblique doubly vergent convergent wedges. *Tectonics*, **24**, TC6004, doi:10.1029/2005TC001816.

DELCEY, R. & MEUNIER, R. 1966. Le massif du Tenda (Corse) et ses bourdes: la série volcano-sédimentaire, les gneiss et les granites; leurs rapports avec les schistes lustrés. *Bullettin de la carte Géologique de la France*, **278**, 237–251.

DEWEY, J. F., HELMAN, M. L., TURCO, E., HUTTON, D. H. W. & KNOTT, S. D. 1989. Kinematics of the western Mediterranean. *In*: COWARD, M. P., DIETRICH, D. & PARK, R. G. (eds) *Alpine Tectonics*. Geological Society, London, Special Publication, **45**, 265–283.

DEZES, P., SCHMID, S. & ZIGLER, P. A. 2004. Evolution of European Cenozoic Rift System: interaction of the Alpine and Pyrennean orogens with their foreland lithosphere. *Tectonophysics*, **389**, 1–33.

DI BIASE, D., MARRONI, M. & PANDOLFI, L. 1997. Age of the deformation phases in the Internal Liguride Units: evidence from Lower Oligocene Val Borbera Conglomerates of Tertiary Piedmont Basin (Northern Italy). *Ofioliti*, **21**, 172–178.

DI BUCCI, D. & MAZZOLI, S. 2002. Active tectonics of the Northern Apennines and Adria geodynamics: new data and a discussion. *Journal of Geodynamics*, **34**, 687–707.

DI GIULIO, A. & GALBIATI, B. 1995. Interaction between tectonics and deposition into an episutural basin in the Alps-Apennine knot. *In*: POLINO, R. & SACCHI, R. (eds) *Atti del Convegno sul tema 'Rapporti tra Alpi e Appennino'*, 113–128.

DIENI, I. & MASSARI, F. 1982. Présence de glaucophane détritique dans le Maastrichtien inférieur de Sardaigne orientale. Implications géodynamiques. *Compte Rendu Académie Science Paris*, **295**, 679–682.

DOGLIONI, C. 1991. A proposal for the kinematic modelling of W-dipping subduction—possible applications to the Tyrrhenian-Apennines system. *Terra Nova*, **3**, 423–434.

DOGLIONI, C., MONGELLI, F. & PIALLI, G. 1998. Boudinage of the Alpine belt in the Apenninic back-arc. *Memorie della Società Geologica Italiana*, **52**, 457–468.

DOGLIONI, C., GUEGUEN, E., HARABAGLIA, P. & MONGELLI, F. 1999. On the origin of west-directed subduction zones and applications to western Mediterranean. *In*: DURAND, B., JOLIVET, L., HORVÁT, F. & SERANNE, M. (eds) *Mediterranean Basins*. Geological Society, London, Special Publication, **156**, 541–561.

DURAND-DELGA, M. 1984. Principaux traits de la Corse Alpine et correlations avec les Alpes Ligures. *Memorie della Società Geologica Italiana*, **28**, 285–329.

EGAL, E. 1992. Structures and tectonic evolution of the external zone of Alpine Corsica. *Journal of Structural Geology*, **14**, 1215–1228.

ELLERO, A., LEONI, L., MARRONI, M. & SARTORI, F. 2001. Internal Liguride Units from Central Liguria, Italy: new constraints to the tectonic setting from white mica and chlorite studies. *Schwizerische Mineralogische Petrographische Mittelungen*, **81**, 39–53.

ELTER, P. 1975. Introduction à la géologie de l'Apennin septentrional. *Bulletin de la Sociéte Géologique de France*, **7**, 956–962.

ELTER, P. 1997. Detritismo ofioltico e subduzione: riflessioni sui rappporti Alpi e Appennino. *Memorie della Società Geologica Italiana*, **49**, 205–215.

ELTER, P. & PERTUSATI, P. C. 1973. Considerazioni sul limite Alpi-Appennino e sulle sue relazioni con l'arco delle Alpi Occidentali. *Memorie della Società Geologica Italiana*, **12**, 359–375.

ELTER, F. M. & SANDRELLI, F. 1994. La fase post-nappe nella Toscana Meridionale: nuova interpretazione sull'evoluzione dell'Appennno Settentrionale. *Atti Ticinensi di Scienze della Terra*, **37**, 173–193.

ELTER, P. & MARRONI, M. 1991. Le unità Liguri dell'Appenino Settentrionale: sintesi dei dati e nuove interpretazioni. *Memorie Descrittive Carta Geologica d'Italia*, **46**, 121–138.

ELTER, G., ELTER, P., STURANI, C. & WEIDMANN, M. 1966. Sur la prolongation du domaine ligure de l'Apennin less dans le Monferrat et les Alpes et sur l'origine de la Nappe de la Simme e.l. et des Préalps romande et chablasiennes. *Archives Science Genève*, **19**, 279–377.

ELTER, P., GIGLIA, G., TONGIORGI, M. & TREVISAN, L. 1975. Tensional and compressional areas in the recent (Tortonian to present) evolution of the Northern Apennines. *Bollettino Geofisica Teorica e Applicata*, **17**, 3–18.

ELTER, P., GHISELLI, F., MARRONI, M., OTTRIA, G. & PANDOLFI, L. 1992. Il profilo Camogli-Ponte dell'Olio: assetto strutturale e problematiche connesse. *Studi Geologici Camerti, volume speciale*, **(1992/1) CROP 1-1A**, 9–15.

EVA, C. & SOLARINO, S. 1992. Alcune considerazioni sulla sismotettonica dell'Appennino nord-occidentale ricavate dall'analisi dei meccanismi focali. *Studi Geologici Camerti*, **2**, 75–83.

FACCENNA, C., BECKER, T. W., LUCENTE, F. P., JOLIVET, L. & ROSSETTI, F. 2001. History of subduction and back-arc extension in the Central Mediterranean. *Geophysics Journal International*, **145**, 809–820.

FACCENNA, C., PIROMALLO, C., CRESPO-BLANC, L., JOLIVET, L. & ROSSETTI, F. 2004. Lateral slab deformation and the origin of Western Mediterranean arcs. *Tectonics*, **23**, TC1012, doi:10.1029/2002TC001488.

FAURE, M. & MALAVIEILLE, J. 1981. Etude structurale d'un cisallement ductile: le chariagge ophiolitique Corse dans la région de Bastia. *Bulletin de la Sociéte Géologique de France*, **23**, 335–343.

FAZZUOLI, M., PANDELI, E. & SANI, F. 1994. Considerations on the sedimentary and structural evolution of the Tuscan Domain since early Liassic to Tortonian. *Memorie Società Geologica Italiana*, **48**, 31–50.

FEDERICO, L., CAPPONI, G., CRISPINI, L. & SCAMBELLURI, M. 2004. Exhumation of alpine high-pressure rocks: insights from petrology of eclogite clasts in the Tertiary Piedmontese basin (Ligurian Alps, Italy). *Lithos*, **74**, 21–40.

FELLIN, M. G., REINER, P. W., BRANDON, M. T., MOLLI, G., BALESTRIERI, M. L. & ZATTIN, M. 2004. Exhumation of the Northern Apennines core: new thermochronological data from the Alpi Apuane. 32nd International Geological Congress Florence, Italy, August 20–28, Abstract, 297–299.

FERRANDINI, M., FERRANDINI, J., LOYE-PYLOT, M. D., BUTTERLIN, J., CRAVETTE, J. & JANIN, M. C. 1998. Le Miocene du Basin de Saint-Florent (Corse): Modalités de la trasgression du Burdigalien Superior et mise en evidence du Serravalien. *Geobios*, **31**, 125–137.

FINETTI, I. R., BOCCALETTI, M., BONINI, M., DEL BEN, A., GELETTI, R., PIPAN, M. & SANI, F. 2001. Crustal section based on CROP seismic data across the North Tyrrhenian-Northern Apennines-Adriatic Sea. *Tectonophysics*, **343**, 135–163.

FOURNIER, M., JOLIVET, L., GOFFÈ, B. & DUBOIS, R. 1991. The Alpine Corsica metamorphic core complex. *Tectonics*, **10**, 1173–1186.

GALBIATI, B. 1990. Considerazioni sulle fasi iniziali dell'orogenesi nell'Appennino Settentrionale. *Atti Ticinensi Scienze della Terra*, **33**, 255–266.

GASINSKI, M. A., SLACZKA, A. & WINKLER, W. 1997. Tectono-sedimentary evolution of the Upper Prealpine Nappe (Switzerland and France): nappe formation by Late Cretaceous-Paleogene accretion. *Geodinamica Acta*, **10**, 137–157.

GHISELLI, F., OTTRIA, G. & PERILLI, N. 1991. Nuovi dati biostratigrafici sulle Arenarie di Scabiazza in base ai nannofossili calcarei (Val Trebbia, Appennino Settentrionale). *Atti Ticinesi di Scienze della Terra*, **34**, 75–82.

GIBBONS, W. & HORAK, J. 1984. Alpine metamorphism of Hercynian hornblende granodiorite beneath the blueschist facies schistés lustrés nappe of NE Corsica. *Journal of Metamorphic Geology*, **2**, 95–113.

GIBBONS, W., WATERS, C. & WARBOURTON, J. 1986. The blueschist facies schistés lustrés of Alpine Corsica: a review. *Geological Society of America Memoir*, **164**, 301–331.

GIGLIA, G., CAPPONI, G., CRISPINI, L. & PIAZZA, M. 1996. Dynamics and seismotectonics of the western-Alpine arc. *Tectonophysics*, **267**, 143–175.

GIORGETTI, G., GOFFÉ, B., MEMMI, I. & NIETO, F. 1998. Metamorphic evolution of Verrucano metasediments in northern Apennines: new petrological constraints. *European Journal of Mineralogy*, **9**, 859–873.

GRANDJACQUET, C. & HACCARD, D. 1977. Position structural et rôle palèogéographique de l'unité du Bracco au sein du contexte ophiolitique ligure-piémontais (Apennin-Italie). *Bulletin de la Sociéte Géologique de France*, **19**, 901–908.

GUEGUEN, E., DOGLIONI, C. & FERNANDEZ, M. 1997. On the post-25 Ma geodynamic evolution of the western Mediterranean. *Tectonophysics*, **298**, 259–269.

GUERRERA, F., MARTIN-ALGARRA, A. & PERRONE, V. 1993. Late Oligocene-Miocene syn-/late-orogenic successions in Western and Central Mediterranean Chains from Betic Cordillera to the Southern Apennines. *Terra Nova*, **6**, 525–544.

HANDY, M. R. & OBERHANSLI, R. 2004. Explanatory notes to the Map: Metamorphic structure of the Alps age map of metamorphic structure of the Alps—Tectonic interpretation and outstanding problems. *Mitteilungen der Osterrichischen Mineralogischen Gesellschaft*, **149**, 201–225.

HARRIS, L. B. 1985a. Progressive and polyphase deformation of the Schistes Lustrés in Cap Corse, Alpine Corsica. *Journal of Structural Geology*, **7**, 637–650.

HARRIS, L. B. 1985b. Direction changes in thrusting of the Schistes Lustrés in Alpine Corsica. *Tectonophysics*, **120**, 37–56.

HILL, K. C. & HAYWARD, A. B. 1988. Structural constraints on the Tertiary tectonic evolution of Italy. *Marine and Petroleum Geology*, **5**, 2–16.

HOOGERDUJIN STRATING, E.H. 1994. Extensional faulting in an intraoceanic subduction complex-working hypothesis for the Paleogene of the Alps-Apennine system. *Tectonophysics*, **238**, 255–273.

HOOGERDUJIN STRATING, E. H. & VAN WAMEL, W. A. 1989. The structure of the Bracco Ophiolite complex (Ligurian Apennines, Italy): a change from Alpine to

Apennine polarity. *Journal of Geological Sociey of London*, **146**, 933–944.

HOOGERDUJIN STRATING, E. H., VAN WAMEL, W. A. & VISSERS, R. L. M. 1991. Some constraints on the kinematics of the Tertiary Piemonte Basin. *Tectonophysics*, **198**, 47–51.

JOLIVET, L., DUBOIS, R., FOURNIER, M., MICHARD, A. & JOURDAN, C. 1990. Ductile extension in Alpine Corsica. *Geology*, **18**, 1007–1010.

JOLIVET, L., FACCENNA, C., GOFFÉ, B. ET AL. 1998. Midcrustal shear zones in postorogenic extension: Example from the northern Tyrrhenian Sea. *Journal of Geophysical Research*, **103**, 12123–12160.

JOURDAN, C. 1988. Balagne orientale et massif du Tenda (Corse septentrionale): étude structurale, interpretation des accidents et des déformations, reconstitutions géodynamiques. *These de 3ème cycle, Université Paris-Sud, Orsay*.

KELLER, J. V. & COWARD, M. P. 1996. The structure and evolution of the Northern Tyrrhenian Sea. *Geological Magazine*, **133**, 1–16.

KLIGFIELD, R., HUNZIKER, J., DALLMEYER, R. D. & SCHAMEL, S. 1986. Dating of deformational phases using K-Ar and $^{40}Ar/^{39}Ar$ techniques: results from the Northern Apennines. *Journal of Structural Geology*, **8**, 781–798.

LABAUME, P. 1992. Evolution tectonique et sedimentarie des fronts de chaine sous-marins. Exemples des Apennins du Nord, des Alpes Francaises et de Sicile. *These Doctarate d'Etate, Université Montepelier II*.

LACOMBE, O. & JOLIVET, L. 2005. Structural relationships between Corsica and the Pyrenees-Provence domain at the time of Pyrennean orogeny. *Tectonics*, **24**, TC1003, doi:10.1029/2004TC001673.

LAGABRIELLE, Y. & POLINO, R. 1988. Un shéma structural du domaine des Schistes Lustrés ophiolitiféres au nord-ouest du massif du mont Viso (Alpes sud-occidentales) et ses implications. *Comptes Rendu Académie Science, Paris*, **323**, 957–964.

LAHONDÈRE, D. 1996. Les schistes blues et les èclogites à lawsonite des unités continentals et océanique de la Corse alpine: Nouvelles donnée pétrologique et structurales (Corse). *Documents du BRGM*, **240**.

LAHONDÈRE, D. & GUERROT, C. 1997. Datation Nd-Sm du métamorphisme éclogitique en Corse alpine: un argument pour l'existence, au Crétacé supérieur, d'une zone de subduction active localisée le long du block corse-sarde. *Géologie de la France*, **3**, 3–11.

LAHONDÈRE, D., ROSSI, Ph. & LAHONDÈRE, J. C. 1999. Structuration alpine d'une marge continentale externe: le massif du Tenda (Haute-Corse). Implications géodynamiques au niveau de la transversale Corse-Apennines. *Gèologie de la France*, **4**, 27–44.

LAUBSCHER, H. P. 1971. The large scale kinematics of the western Alps and the northern Apennines and its palinspastic implications. *American Journal of Sciences*, **271**, 193–226.

LAUBSCHER, H. P. 1991. The arcs of western Alps today. *Eclogae Geologique Helvetique*, **84**, 613–651.

LAUBSCHER, H. P. & BERNOULLI, D. 1977. Mediterranean and Tethys. In: NAIRN, A. E. M., KANES, W. H. & STEHLI, F. G. (eds) *The Ocean Basins and Margins*. Vol. 4A Plenum Publishing Corporation, New York, 1–28.

LAUBSCHER, H. P., BIELLA, G. C., CASSINIS, R., GELATI, R., LOZEJ, A., SCARASCIA, S. & TABACCO, I. 1992. The collisional knot in Liguria. *Geologische Rundschau*, **81**, 275–289.

LAVECCHIA, G. & STOPPA, F. 1990. The Tyrrhenian zone: a case of lithosphere extension control of intra-continental magmatism. *Earth Planetary Science Letters*, **99**, 336–350.

LEMOINE, M., DE GRACIANSKY, P. C. & TRICART, P. 2001. De l'océan à la chaîne de montagnes. *Société Géologique de France, Collection Géosciences*.

LENÒTRE, N., FERRANDINI, J., DELFAU, M. & PANIGHI, J. 1996. Mouvements verticaux de la Corse (France) par comparaison de nivellements. *Compte Rendu Académie Science Paris*, **323**, 957–964.

LEVI, N., ELLERO, A., PANDOLFI, L. & OTTRIA, G. 2006. Polyorogenic deformation history recognized at shallow structural levels: the case of the Antola unit (Northern Apennine, Italy). *Journal of Structural Geology*, **28**, 1694–1709.

LIOTTA, D. 2002. D2 asymmetric folds and their vergence meaning in the Montagnola Senese metamorphic rocks (inner northern Apennines, central Italy). *Journal of Structural Geology*, **24**, 1479–1490.

LUCENTE, F. P. & SPERANZA, F. 2001. Belt bending driven by lateral bending of subducting lithospheric slab: geophysical evidence from the Northern Apennines (Italy). *Tectonophysics*, **337**, 53–64.

MAKRIS, J., EGLOFF, F., NICOLICH, R. & RIHM, R. 1999. Crustal structure from the Ligurian Sea to the Northern Apennines—a wide angle seismic transect. *Tectonophysics*, **301**, 305–319.

MALASOMA, A., MARRONI, M., MUSUMECI, G. & PANDOLFI, L. 2006. High pressure mineral assemblage in granitic rocks from continental units in Alpine Corsica, France. *Geological Journal*, **41**, 49–59.

MALAVIEILLE, J. 1983. Étude tectonique et microtectonique de la nappe de socle de Centuri (zone des Schistes Lustrés de Corse), Conséquence pour la géométrie de la châine alpine. *Bulletin de la Sociéte Géologique de France*, **25**, 195–204.

MALAVIEILLE, J., CHEMENDA, A. & LARROQUE, C. 1998. Evolutionary model for Alpine Corsica: mechanism for ophiolite emplacement and exhumation of high-pressure rocks. *Terra Nova*, **10**, 317–322.

MALUSKI, H. 1977. Application des methods de datation $^{39}Ar/^{40}Ar$ aux mineraux des roches cristallins perturbes par les évènementes thermiques et tectoniques en Corse. *These de 3ème cycle, Université Montpelier, France*.

MALUSKI, H., MATTAUER, M. & MATTE, Ph. 1973. Sur la présence de decrochement alpins en Corse. *Compte Rendu Académie Science Paris, serie D*, **276**, 709–712.

MANATSCHAL, G. & BERNOULLI, D. 1999. Architecture and tectonic evolution of non-volcanic margins: present-day Galicia and ancient Adria. *Tectonics*, **18**, 1099–1119.

MARRONI, M., DELLA CROCE, G. & MECCHERI, M. 1988. Structural evolution of the M.Gottero Unit in the M.Zatta/M.Ghiffi sector. *Ofioliti*, **13**, 29–42.

MARRONI, M. & PANDOLFI, L. 1996. The deformation history of an accreted ophiolite sequence: the Internal

Liguride units (Northern Apennines, Italy). *Geodinamica Acta*, **9**, 13–29.

MARRONI, M. & PANDOLFI, L. 2003. Deformation history of the ophiolite sequence from the Balagne Nappe, northern Corsica: insights into the tectonic evolution of Alpine Corsica. *Geological Journal*, **38**, 67–83.

MARRONI, M., MOLLI, G., MONTANINI, A. & TRIBUZIO, R. 1998. The association of continental crust rocks with ophiolites in the Northern Apennines (Italy): implications for continent-ocean transition in the Western Tethys. *Tectonophysics*, **292**, 43–66.

MARRONI, M., MOLLI, G., OTTRIA, G. & PANDOLFI, L. 2001. Tectono-sedimentary evolution of the External Liguride units (Northern Apennine, Italy): insights into the pre-collisional history of a fossil ocean-continent transition zone. *Geodinamica Acta*, **14**, 307–320.

MARRONI, M., CERRINA FERONI, A., DI BIASE, D., OTTRIA, G., PANDOLFI, L. & TAINI, A. 2002a. Polyphase folding at upper structural levels in the Borbera valley (Northern Apennines, Italy): implications for the tectonic evolution of the linkage area between Alps and Apennines. *Compte Rendu Geoscience*, **334**, 565–572.

MARRONI, M., MOLLI, G., MONTANINI, A., OTTRIA, G., PANDOLFI, L. & TRIBUZIO, R. 2002b. The External Liguride units (Northern Apennine, Italy) from rifting to convergence of a fossil ocean-continent transition zone. *Ofioliti*, **27**, 119–131.

MATTAUER, M. & PROUST, F. 1975. Sur quelques problèmes géneraux de la chaine alpine en Corse. *Bulletin de la Société Géologique de France*, **18**, 1177–1178.

MATTAUER, M., FAURE, M. & MALAVIEILLE, J. 1981. Transverse lineation and large-scale structures related to Alpine obduction in Corsica. *Journal of Structural Geology*, **3**, 401–409.

MAUFFRET, G. & CONTRUCCI, R. 1999. Crustal structure of the Northern Tyrrhenian Sea: first result of the multichannel seismic LISA cruise. *In*: DURAND, B., JOLIVET, L., HORVÁTH, F. & SÉRANNE, M. (eds) *Mediterranean Basins*. Geological Society of London, Special Publication, **156**, 169–194.

MECCHERI, M. & ANTONPAOLI, M. L. 1982. Analisi strutturale ed evoluzione delle deformazioni della regione di M.Verruga, M.Porcile e Maissana (Appennino Ligure, La Spezia). *Bollettino della Società Geologica Italiana*, **101**, 117–140.

MELETTI, C., PATACCA, E. & SCANDONE, P. 2000. Construction of a seismotectonic model: the case of Italy. *Pure and Applied Geophysics*, **157**, 11–35.

MERLE, O. & BRUN, J. P. 1984. The curved translation path of the Parpaillon nappe (French Alps). *Journal of Structural Geology*, **6**, 711–719.

MICHARD, A. & MARTINOTTI, G. 2002. The Eocene unconformity of the Briançonnais domain in the French-Italian Alps, revisited (Marguareis massif, Cuneo), a hint for Late Cretaceous-Middle Eocene frontal bulge setting. *Geodinamica Acta*, **15**, 289–301.

MICHARD, A., CHALOUAN, A., FEINBERG, H., GOFFÉ, B. & MONTIGNY, R. 2002. How does the Alpine belt end between Spain and Morocco? *Bulletin de la Société Géologique de France*, **173**, 3–15.

MOLLI, G. 1996. Pre-orogenic tectonic framework of the northern Apennine ophiolites. *Eclogae Geologique Helvetique*, **89**, 163–180.

MOLLI, G., GIORGETTI, G. & MECCHERI, M. 2000. Structural and petrological constraints on the tectono-metamorphic evolution of the Massa Unit (Alpi Apuane, NW Tuscany, Italy). *Geological Journal*, **35**, 251–264.

MOLLI, G., GIORGETTI, G. & MECCHERI, M. 2002. Tectono-metamorphic evolution of the Alpi Apuane Metamorphic Complex: new data and constraints for geodynamic models. *Bollettino della Società Geologica Italiana, Volume Speciale*, **1**, 789–800.

MOLLI, G. & TRIBUZIO, R. 2004. Shear zones and metamorphic signature of subducted continental crust as tracers of the evolution of the Corsica/Northern Apennine orogenic system. *In*: ALSOP, I., HOLDSWORTH, R. E., MCCAFFREY, J. W. & HAND, M. (eds) *Flow Process in Faults and Shear Zones*. Geological Society of London, Special Publication, **224**, 321–335.

MOLLI, G. & BERNARDESCHI, A. 2005. Carta geologica Isola di Gorgona. Progetto Cartografia Regione Toscana.

MOLLI, G., MALASOMA, A. & MENEGHINI, F. 2005. Brittle precursors of HP/LT microscale shear zone: a case study from Alpine Corsica. *15th Conference on Deformation Mechanisms, Rheology and Tectonics, ETH Zurich, 2-4 May 2005*, Abstract volume, 153.

MOLLI, G. 2005. Storia Geologico-strutturale ed esumazione delle unità metamorfiche delle Alpi Apuane. *Istituto Italiano di Speleologia, Memorie*, **18**, 13–18.

MOLLI, G. & VASELLI, L. 2006. Structures, interference patterns and strain regime during mid-crustal deformation in the Alpi Apuane (Northern Apennines, Italy). *In*: MAZZOLI, S. & BUTLER, R. (eds) *Styles of Continental Contraction*. Geological Society of America Special Paper, **414**, 79–93.

MOLLI, G., BOTTI, F., FELLIN, G., ZATTIN, M. & BALDACCI, F. 2006a. Evolution of the Northern Apennine orogenic wedge: combining data from structures of deep and shallow units, thrust-top and foredeep deposits. EUG06-A-08457; TS7.1-1MO3P-0576.

MOLLI, G., TRIBUZIO, R. & MARQUER, D. 2006b. Deformation and metamorphism at the eastern border of the Tenda Massif (NE Corsica): a record of subduction and exhumation of continental crust. *Journal of Structural Geology*, **29**, 1748–1766.

MONTANINI, A. & MOLLI, G. 2002. New petrologic and $^{39}Ar/^{40}Ar$ data on the Paleozoic metamorphic basement of the Northern Apennines (NW Italy): evidence from surface and subsurface units. *81a Riunione Società Geologica Italiana, Volume Riassunti*, 247–248.

MORETTI, A. 1986. La virgazione della dorsale Medio-Toscana: nuovi dati strutturali. *Bollettino della Società Geologica Italiana*, **35**, 555–567.

MOSCA, P., POLINO, R. & ROGLEDI, S. ET AL. 2005. Oligocene-Neogene kinematic constraints on the Alps and Apennine interaction in the western Po Plain from subsurface data. *Rendiconti della Società Geologica Italiana*, **1**, 148–149.

MUSUMECI, G., BOCINI, L. & CORSI, R. 2002. Alpine tectonothermal evolution of the Tuscan Metamorphic complex in the Larderello geothermal field. *Journal Geological Society of London*, **159**, 443–456.

MUTTI, E., PAPANI, L., DI BIASE, V., DAVOLI, G., MORA, S., SEGADELLI, S. & TINTERRI, R. 1995. Il bacino terziario epi-mesoalpino e le sue implicazioni sui rapporti Alpi ed Appennino. *Memorie Scienze Geologiche*, **47**, 217–244.

MUTTONI, G., LANCI, L., ARGNANI, A., HIRT, A. M., CIBIN, U., ABRAHAMSEN, N. & LOWRIE, W. 2000. Paleomagnetic evidence for a Neogene two-phase counterclockwise tectonic rotation in the Northern Apennines (Italy). *Tectonophysics*, **326**, 241–253.

NARDI, R. 1968. Le unità alloctone della Corsica e la loro correlazione con le unità delle Alpi e dell'Appennino. *Memorie della Società Geologica Italiana*, **7**, 323–344.

NEUGEBAUER, J., GREINER, B. & APPEL, E. 2001. Kinematics of the Alpine-West Carpathian orogen and paleogeographic implications. *Journal Geological Society of London*, **158**, 97–110.

ORTI, L., MORELLI, M., PANDELI, E. & PRINCIPI, G. 2001. New geological data from Gorgona island (Northern Tyrrhenian Sea). *Ofioliti*, **27**, 133–144.

OTTRIA, G. & MOLLI, G. 2000. Superimposed brittle structures in the late orogenic extension of the Northern Apennine: results from Carrara area (Alpi Apuane, NW Tuscany). *Terra Nova*, **12**, 52–59.

PANDOLFI, L. 1998. Le successioni dell'unità Due Ponti (Unità Liguri Interne, Appennino Settentrionale): evidenze di un'area sorgente carbonatica nelle successioni torbiditiche cretaciche della Tetide Occidentale. *Bollettino della Società Geologica Italiana*, **117**, 593–612.

PASCUCCI, V., MERLINI, S. & MARTINI, I. P. 1999. Seismic stratigraphy of the Miocene-Pleistocene sedimentary basins of the Northern Tyrrhenian Sea and western Tuscany (Italy). *Basin Research*, **11**, 337–356.

PATACCA, E. & SCANDONE, P. 1989. Post-Tortonian mountain building in the Apennines. The role of passive sinking of a relict lithospheric slab. *In*: BORIANI, A., BONAFEDE, M., PICCARDO, G. B. & VAI, G. B. (eds) *The Lithosphere in Italy*. Accademia Nazionale Lincei, Atti Convegni Lincei, **80**, 157–176.

PATACCA, E., SARTORI, R. & SCANDONE, P. 1992. Tyrrhenian basin and Apenninic arcs: kinematic relationships since Late Tortonian times. *Memorie della Società Geologica Italiana*, **45**, 425–451.

PEQUIGNOT, G., LARDEAUX, J. M. & CARON, J. M. 1984. Recrystallization d'éclogites de basse temperature dans les metabaltes corse. *Compte Rendu Académie Science Paris, Serie II*, **299**, 871–874.

PERTUSATI, P. C. & HORREMBERGER, J. C. 1975. Studio strutturale degli Scisti della Val Lavagna (Unità del Gottero, Appennino ligure). *Bollettino della Società Geologica Italiana*, **94**, 1375–1436.

PERTUSATI, P. C., MUSUMECI, G., BONINI, L. & FRANCESCHI, M. 2004. The serie ridotta in southern Tuscany: a cartographic example and consideration on the origin. *In*: MORINI, D. & BRUNI, P. (eds) *The 'Regione Toscana' Project of Geological Mapping: Case Histories and Data Acquisition*, 183–185.

PIALLI, G., BARCHI, M. & MINELLI, G. 1998. Results of the CROP03 Deep Seismic Reflection Profile. *Memorie della Società Geologica Italiana*, **52**, 427–439.

PIANA, F. 2000. Structural setting of western Monferrato (Alps-Apennines Junction Zone, NW Italy). *Tectonics*, **19**, 943–960.

PIANA, F. & POLINO, R. 1995. Tertiary structural relationships between Alps and Apennines: the critical Torino Hill and Monferrato area, northwest Italy. *Terra Nova*, **7**, 138–143.

PICCARDO, G. B. 1977. Le ofioliti dell'areale Ligure: petrologia ed ambiente geodinamico di formazione. *Rendiconti Società Geologica Italiana*, **33**, 221–252.

PLESI, G. 1975. La giacitura del Complesso Bratica-Petrignacola nella serie di Roccaferrara (Val Parma) e dei Flysch arenacei tipo Cervarola dell'Appennino Settentrionale. *Bollettino della Società Geologica Italiana*, **94**, 157–176.

PLESI, G., BONANNI, G., BOTTI, F., DANIELE, G. & PALANDRI, S. 1998. Processi e tempi di costruzione della catena Appenninica nelle sue fasi Oligo-mioceniche: l'esempio della finestra di Pracchiola (Biostratigrafia, Petrografia e Analisi Strutturale con carta Geologica-Strutturale scala 1:20 000). *Bollettino della Società Geologica Italiana*, **117**, 841–894.

PLEUGER, J., FROITZHEIM, N. & JANSEN, E. 2005. Folded continental and oceanic nappes on the southern side of Monte Rosa (Western Alps, Italy): anatomy of a double collision suture. *Tectonics*, **24**, TC4013, doi: 10.1029/2004TC001737.

POLINO, R., DAL PIAZ, G. V. & GOSSO, G. 1990. Tectonic erosion at the Adria margin and accretionary process for the Cretaceous orogeny of the Alps. *In*: ROURE, F., HEITZMAN, P. & POLINO, R. (eds) *Deep Structure of the Alps*. Volume Speciale Società Geologica Italiana, Roma, **1**, 345–367.

PRINCIPI, G. & TREVES, B. 1985. Il sistema corso-appenninico come prisma d'accrezione. Riflessi sul problema generale del limite Alpi-Appennini. *Memorie della Società Geologica Italiana*, **28**, 549–576.

RAMPONE, B. & PICCARDO, G. B. 2000. The ophiolite-oceanic lithosphere analogue: new insights from Northern Apennine (Italy). *In*: DILEK, J., MOORES, E., ELTHON, D. & NICOLAS, A. (eds) *Ophiolites and Oceanic Crust: New Insights from Field Studies and Ocean Drilling Program*. Geological Society of America Special Paper, **349**, 21–34.

REHAULT, J. P., MASCLE, J. & BOILLOT, G. 1984. Evolution geodynamique de la Mediterranée depuis l'Oligocene. *Memorie della Società Geologica Italiana*, **27**, 85–96.

REUTTER, K. J., GUNTHER, K. & GROSCURTH, J. 1978. An approach to the Geodynamics of the Corsica-Northern Apennines double orogen. *In*: CLOSS, H., ROEDER, D. & SCHMIDT, K. (eds) *Alps, Apennines, Hellenides*. IUGG Scientific Report, 299–311.

ROCCHI, S., WESTERMAN, D. S., DINI, A., INNOCENTI, F. & TONARINI, S. 2002. Two-stage growth of laccoliths at Elba Island, Italy. *Geology*, **30**, 983–986.

ROSENBAUM, G., LISTER, G. S. & DUBOZ, C. 2002. Relative motion of Africa, Iberia and Europe during Alpine orogeny. *Tectonophysics*, **359**, 117–129.

ROSENBAUM, G. & LISTER, G. S. 2004. Neogene and Quaternary rollback evolution of the Tyrrhenian Sea, the Apennines and the Sicilian Maghrebides. *Tectonics*, **23**, TC1013, doi: 10.1029/2003TC001518.

ROSENBAUM, G. & LISTER, G. S. 2005. The western Alps from Jurassic to Oligocene: spatio-temporal constraints and evolutionary reconstruction. *Earth-Science Reviews*, **69**, 281–306.

ROSSETTI, F., FACCENNA, C., JOLIVET, L., GOFFÉ, B. & FUNICIELLO, R. 2002. Structural signature and exhumation P–T–t paths of the blueschist units exposed in the interior of the Northern Apennine chain, tectonic implication. *Bollettino della Società Geologica Italiana, volume speciale*, **1**, 829–842.

ROSSI, PH., LAHONDÉRE, J. C., LUCH, D., LOYE-PILOT, M. D. & JACQUET, M. 1994. Carte géologique de la France à 1/50.000, feuille Saint-Florent. BRGM.

ROSSI, PH., DURAND-DELGA, M., LAHONDÈRE, J. C. & LAHONDÈRE, D. 2003. Carte géologique de la France à 1/50.000, feuille Santo Pietro di Tenda. BRGM.

ROYDEN, L., PATACCA, E. & SCANDONE, P. 1987. Segmentation and configuration of subducted lithosphere in Italy: an important control on thrust-belt and foredeep-basin evolution. *Geology*, **15**, 714–717.

RUFFINI, R., COSCA, M. A, D'ATRI, A., HUNZIKER, J. C. & POLINO, R. 1995. The volcanic supply of the Taveyanne (Savoie, France): a riddle for Tertiary Alpine volcanism. *Accademia Nazionale delle Scienze*, **14**, 359–376.

SCANDONE, P. 1979. Origin of the Tyrrenian Sea and the Calabrian arc. *Bollettino della Società Geologica Italiana*, **98**, 27–34.

SCHMID, S. M., PFIFFER, O. A., FROITZHEIM, N., SCHÖNBORN, G. & KISSLING, E. 1996. Geophysical-geological transect and tectonic evolution of the Swiss-Italian Alps. *Tectonics*, **12**, 1036–1064.

SCHMID, S. M. & KISSLING, E. 2000. The arc of western Alps in the light of geophysical data on deep structure. *Tectonics*, **19**, 62–85.

SCHMID, S. M., FUGENSCHUH, B., KISSLING, E. & SCHUSTER, R. 2004. Tectonic map and overall architecture of the Alpine orogen. *Eclogae Geologique Helvetique*, **89**, 163–180.

SCHOLLE, P. A. 1970. The Sestri-Voltaggio line: a transform fault induced tectonic boundary between the Alps and the Apennines. *American Journal of Science*, **269**, 343–359.

SCHUMACHER, M. E. & LAUBSCHER, H. P. 1996. 3D crustal architecture of the Alps-Apennines join: a new view on seismic data. *Tectonophysics*, **260**, 349–363.

SELVAGGI, C. & AMATO, A. 1992. Intermediate-depth earthquakes in the Northern Apennines (Italy): evidence for a still active subduction? *Geophysics Research Letter*, **19**, 2127–2130.

SENO, S., DALLAGIOVANNA, G. & VANOSSI, M. 2005. Pre-Piedmont and Piedmont-Ligurian nappes in the central sector of the Ligurian Alps: a possible pathway for their superposition onto the inner Briançonnais units. *Bollettino della Società Geologica Italiana*, **124**, 455–464.

SÉRANNE, M. 1999. The Gulf of Lion continental margin (NW Mediterranean) revisited by IBS: an overview. In: DURAND, B., JOLIVET, L., HORVÁTH, F. & SÉRANNE, M. (eds) *Mediterranean Basins*. Geological Society of London, Special Publication, **156**, 15–36.

SERRI, G., INNOCENTI, F. & MANETTI, P. 1993. Geochemical and petrological evidence of the subduction of delaminated Adriatic continental lithosphere in the genesis of the Neogene-Quaternary magmatism of central Italy. *Tectonophysics*, **223**, 117–147.

SPAGNOLO, C. 2006. Late orogenic tectonics in the eastern sector of the Ligurian Alps. *Tesi di Dottorato, XVII ciclo, Università di Genova*.

SPAKMAN, W. & WORTEL, R. 2004. A tomographic view on Western Mediterranean Geodynamics. In: CAVAZZA, W., ROURE, F., SPAKMAN, W., STAMPFLI, G. M. & ZIGLER, P. A. (eds) *The Transmed Atlas—The Mediterranean Region from Crust to Mantle*. Springer, Berlin Heidelberg, 31–52.

SPERANZA, F., VILLA, I. M., SAGNOTTI, L., FLORINDO, F., COSENTINO, D., CIPOLLARI, P. & MATTEI, M. 2002. Age of the Corsica-Sardinia rotation and Ligure-Provencal Basin spreading: new paleomagnetic and Ar/Ar evidence. *Tectonophysics*, **347**, 231–251.

STAM, J. C. 1952. Gèologie de la région du Tenda Septentrional (Corse). Ph.D. Thesis, Universiteit van Amsterdam, 96pp.

STAMPFLI, G. M, MOSAR, J., MARQUER, D., MARCHANT, R., BAUDIN, T. & BOREL, G. 1998. Subduction and obduction process in the Swiss Alps. *Tectonophysics*, **296**, 159–204.

STAUB, R. 1924. Der Bau der Alpen. *Beitr geolog Karte Schweiz*, **46**.

STORTI, F. 1995. Tectonics of Punta Bianca promontory: Insights for the evolution of the Northern Apennines-Northern Tyrrhenian sea basin. *Tectonics*, **14**, 832–847.

STURANI, C. 1973. Considerazioni sui rapporti tra Appennino Settentrionale ed Alpi Occidentali. *Rendiconti Accademia Lincei*, **183**, 119–142.

TAVARNELLI, E. 1997. Structural evolution of a foreland fold-and-thrust belt: the Umbria-Marche Apennines, Italy. *Journal of Structural Geology*, **19**, 523–534.

THEYE, T., REINHARDT, J., GOFFÉ, B., JOLIVET, L. & BRUNET, C. 1997. Ferro and magnesiocarpholite from Mt. Argentario (Italy): first evidence for high-pressure metamorphism of the metasedimentary Verrucano sequence, and significance for P–T path reconstruction. *European Journal of Mineralogy*, **9**, 859–873.

TOSCANI, G., SENO, S., FANTONI, R. & ROGLEDI, S. 2006. Geometry and timing of deformation inside a structural arc: the case of the western Emilian folds (Northern Apennine front, Italy). *Bollettino della Società Geologica Italiana*, **125**, 59–65.

TREVES, B. 1984. Orogenic belts as accretionary prisms: the example of the Northern Apennines. *Ofioliti*, **9**, 577–618.

TRIBUZIO, R. & GIACOMINI, F. 2002. Blueschist facies metamorphism of peralkaline rhyolites from the Tenda crystalline massif (northern Corsica): evidence for involvement in the Alpine subduction event? *Journal of Metamorphic Geology*, **20**, 513–526.

VAN WAMEL, W. A. 1987. On the tectonics of the Ligurian Apennines (northern Italy). *Tectonophysics*, **142**, 87–98.

VANOSSI, M., PEROTTI, C. R. & SENO, S. 1994. The Maritimes Alps arc in the Ligurian and Tyrrhenian system. *Tectonophysics*, **230**, 75–89.

VANOSSI, M., CORTESOGNO, L., GALBIATI, B., MESSIGA, B., PICCARDO, G. & VANNUCCI, R. 1986. Geologia delle Alpi Liguri: dati, problemi, ipotesi. *Memorie Società Geologica Italiana*, **28**, 5–75.

VESCOVI, P. 1993. Schema evolutivo per le Liguridi dell'Appennino Settentrionale. *Atti Ticinensi di Scienze della Terra*, **36**, 89–112.

VESCOVI, P. 1998. Le unità subliguri dell'alta Val Parma (Provincia di Parma). *Atti Ticinensi di Scienze della Terra*, **40**, 215–231.

VESCOVI, P. 2005. The Middle Miocene Mt. Ventasso-Mt.Cimone arcuate structure of the Emilia Apennines. *Bollettino Società Geologica Italiana*, **124**, 53–67.

VITI, M., DE LUCA, J., BABBUCCI, D., MANTOVANI, E., ALBARELLO, D. & D'ONZA, F. 2004. Driving mechanism of tectonic activity in the northern Apennines: Quantitative insights from numerical modeling. *Tectonics*, **23**, TC4003, doi: 10.1029/2004TC0001623.

WARBOURTON, J. 1986. The ophiolite-bearing Schistes lustrés nappe in Alpine Corsica: A model for emplacement of ophiolites that have suffered HP/LT metamorphism. *Geological Society of America Memoir*, **164**, 313–331.

WATERS, C. N. 1990. The Cenozoic tectonic evolution of alpine Corsica. *Journal of the Geological Society, London*, **147**, 811–824.

WILDI, W. 1985. Heavy mineral distribution and dispersal pattern in penninic and ligurian flysch basins (Alps, northern Apennines). *Giornale di Geologia*, **47**, 77–99.

ZANZUCCHI, G. 1988. Ipotesi sulla posizione paleogeografica delle 'Liguridi Esterne' cretacico-eoceniche, nell'Appennino Settentrionale. *Atti Ticincensi Scienze della Terra*, **31**, 327–339.

ZIBRA, I. 2006. Late Hercynian granitoid plutons emplaced along a deep crustal shear zone. A case study from the S.Lucia Nappe (Alpine Corsica, France). *Tesi Dottorato, Università di Pisa*, 2003.

Index

Figures are shown in *italics*; tables in **bold**

accretion 307
accretion wedge 132, 189, 286, 330
    Apennine 421, 429, 430, *431*, 432, 433
actinolite 290
Adriatic (Apulian) Plate 5–36, 134, 146, 157, 403
Adula nappe 4, *264*, 266–271
    deformation 377–383
    lithology and structure 369–371
    petrology 371–377
    Tertiary subduction 365–387
age data 2, 8, 9, 70
    Austroalpine basement **14–17**, 27
    Carpathians 103–106
    Rieserferner pluton *29*
    Tauern Window 94–97
age of deformation
    Dinarides 353–354
    Monte Rosa nappe 265–266
    Valstrona di Omega 46, 48, 54, 59–61, 62
age of magmatic rock–suite 12–13
age of metamorphism 424
age, nannofossil **337–339**, 348, 350
age, petrology and isotopic age 117–136
age, Eocene–Miocene revised 335–359
Ahorn shear zone 199–204, 206–216
Ahrntal Fault 199, 214
ALCAPA *see* Alpine–Carpathian–Pannonian unit
Alpine Austroalpine–Penninic suture 8
Alpine deformation phases 211
Alpine Tethys 219
Alpine–Carpathian–Pannonian unit 169, *170, 282*
    convergence 317–332, *333*
amphibolite 373, 375, 382
amphibolite facies 23, 263, 265, 267, 397, 401
analysis, nannofossil 340–347
analytical data
    white mica **107–108**, *109*, **113**
analytical method
    $^{40}Ar/^{39}Ar$ dating 111–113
    fission-track 65, 172–173
    metamorphic minerals 286–287
anchizone 291–293, 297, 299, 301, 302, 308, 310
andalusite 28, 288, *289*, 396
anhydrite 89
anthracite 287, 296, 297
Antola nappe 425
apatite fission track
    age *29, 31*, 32, 36
    data 173, *175*, 177–181, *180, 182*
Apulian margin subduction 383–385
Apulian Plate *see* Adriatic Plate
$^{40}Ar/^{39}Ar$ dating 104–106
    analytical techniques 111–113
    Valstrona di Omega 46, 54–55, **56**, *57*, 60, 61, 62
Austroalpine basement 5–36
    age data **14–17**, 27
    cooling and exhumation 31–32
    cross-section *7, 19*

geochemistry 10–18
geodynamics 32–36
intrusive history 28–32
lithology *6*
magmatic evolution 9–18
metamorphic age 7–9
metamorphism 18–28
    age 27–28
    Alpine 35–36
    pre-Alpine 23–27
structural evolution 18–28
    post-collision 28–32
Avalonia 5

backthrust 425, 430
backthrusting 252, 253, 274, 365, *367*, 368
Barrovian metamorphism (HT metamorphism) 371, 373–377, 394, 396
basalt, sub-greenschist facies 298–300
biostratigraphy 335–359
bituminite reflectance 285, *294, 295*
blueschist 118, 120, 128, 129, 134, 424, 400
    Carpathians 101, 104, *105*, 106, 110
boudinage 50, *52, 53*
Brenner Fault 197, 214
Briançonnais
    basement 400, 401
    metamorphism 125, 129
    subduction 132, *136*, 383
brittle deformation 173
Bündnerschiefer nappes 220–221, *223*
Bucovinian nappes 171–172, 181, *183*
burial depth 187–189, *305*

Cadomia 5, *33*, 220
Cambrian 34
Canavese Zone 47
Carpathian embayment *170*, 317–318, 332
Carpathians *102, 103, 111*
cata-bituminite 294, 295
cataclastite 73
Celtic terrane 9
chemistry, metamorphic minerals 290–291
chlorite 290, 291
chlorite geothermometry 291, 306, 307
chloritoid 288, *289*, 396, *398*
chloritoid isograd 286, 298, 299–300
clastic wedge, Pienides 329–331
clay minerals 287
closure, Ligurian Tethys 432, 433
coal 285, 309
coalification map *294*
coesite 1, 393, 394
Collio Formation 70–73
collision 396, 397, 404
    Apennine *414*, 425, 430
    Variscan 5, 6–7, *33*, 35

collisional orogens 101–110
collision-related minerals 401–402
composite fault 234, 242, 244
continental convergence, Alpine 35
continental units
  isotope and petrological data *121*, 134
convergence
  Carpathian embayment 317, 332
cooling age, Adula nappe 382
cooling history, Getic detachment 310
cooling history, Maramures 169–195
cooling path, Rieserferner pluton *32*
cooling 302–311
Corsica, orogenic system 413–433
Cosustea mélange 286, 287, 290, 293, 298
Cosustea nappe *284*, 302
Cosustea unit 291, 293, 301, 306
crust, lower, deformation of 45–65
crust, lower, faulting in 215–216
crustal extension, southern Alps 80
crustal stacking 35
crustal thinning 48
crystallographic-preferred orientation (CPO) 256, 259–263

Danubian window, metamorphism 281–311
Defereggen Group 8, 9, *11*, **14–17**, 18, *20*, 21, 26, 33
Defereggen–Antholz–Vals line 8–9, 21, 30, 31, *33*, 36
deformation phases 211
  Adula nappe 377–383
  Dinarides 149–155
  Monte Rosa nappe 256–258, 270–271
  Tauern Window **204–205**
deformation
  brittle 9, 49, *53*, 54, 73, *78*, 79
  ductile 49, *51, 53*, 54, 73, 78
  pre-Alpine ductile 18–28
deformation, Stellihorn shear zone 263–275
Dent Blanche nappe 118–120
depositional environment 71, 79, 91–94
  Dinarides 336, 340, 350, 352–353
  Pienides 321–324, **328**
detrital mineral associations 287
diagenesis 287–291, 293, 296–298, 302
  syn-kinematic 309, 310
Dinarides 145–164
  deformation 149–155
  depositional environment 335–359
  palaeomagnetism 149, 155–157
  tectonic zones *147*
  thrusts 154
  topography *147*
diopside 290
dolomite 87, 93
duplex structure 90, 224–226
Durreck Muscoviteschist Group 8, 9, **14–17**, 26

earthquakes, Switzerland *232*, 234, 242, 244, 246
  focal mechanism *245*
eclogite 118, 120, 129, 373
eclogite facies 2, 4, 23, 28, 263, 264, 267
electron-microprobe analysis 285, 286

enriched mid ocean ridge basalt (N-MORB) 7, 10, 13, 32, 33
Eo-Alpine (Cretaceous) metamorphic event 9, 27, 35–36
Eo-Alpine thrusting *33*
epizone 291–293, 296–299, 301, 306
evaporite, Triassic 71
exhumation
  Adula nappe 383, 385
  Apennine 425, 426, 430
  central Alps 272, 273–275
  Ivrea Zone 45–65
  Rodna horst *187*
  Maramures 169–195
  Tauern Window 197–216
extension *50*, 70
  Monte Rosa nappe 251–276
External Dinarides 146
extrusion tectonics 3

fabric in mylonite 49–54
fabric in plutons 30–31
facies analysis 321–324, 329
fault orientation *243*, 244
faults, surface expression 234–244
  morphology *235, 237, 238*, 239
faults, post-glacial displacement 236, *237, 238*, 239
fission track age *29, 31*, 32, 34, 36
  Valstrona di Omega 46, 55, **59**, 60–62, *63*
fission track analysis 65, 172–173, *194, 195*
fission track data 174–177
flysch 417, *420*, 425
  in foreland basin 336
folding and shear 203–213
folds, Dinarides *153*
folds, recumbent *207–208*
foliation 204, *209*, 210
  mylonitic 256–259, *268*
  pre-Alpine 21
foraminifera, planktonic 335, **337–339**, 340
foreland basin, sedimentation 317–332

garnet *20*, 21, 22, 23, 24, 25, 27, *374*
geochemistry, Austroalpine basement 10–12
geochronology *see* age
geodynamic evolution
  Adriatic–Austroalpine plate 32–36
  Tertiary 393–404
geophysics
  Apennine–Corsica 415, 417
  Penninic nappes 234
  Salzach–Ennstal–Mariazell–Puchberg Fault (SEMP) 215–216
  Tauern Window 219–222, 224–226
geothermal gradient 186, 188
geothermobarometry 23–27
Getic detachment 4
  metamorphism 307–310
Getic-Supragetic units 282, 286, 301
Gondwana 5, 6, *33*, 220
  passive margin 34–35
goniometry 259–263
granite, Hercynian 219
granulite facies 48

graphite 88, 92, 220, 225, *296*
Grassi Detachment Fault 69–80
    age 70
    map *72*
    reconstruction 78–80
gravitational faults 234, 235–239
greenschist *152*, 298, 301, 373, 424
greenschist facies 189, 251, 265, 267, 401
    Danubian Window 281, 310–311
Greiner basin *225*
Greiner shear zone 199, 214
gypsum 222, 293

Helminthoid Flysch 417, *420*, 425
Helvetic nappes 244
Hercynian magmatic suite 219
high-pressure wedges 101–110
history of research
    Alps 1–4, 393–394
    Apennine–Corsica 413–415
Hochstegen marble 222
hornblende data **56**, 57
horst and graben 224

illite Kübler index 291–298, 307
imbricated crust 404
index minerals 394, 395
inertinite 294
inselberg, Croatia 146, 148, 162
Insubric Fault 3, 251–253, 256, 258, 266, 275
Insubric mylonite 365, *367*
Insubric Zone 47, 48
intermontane basin 83, *84*, 283
Internal Dinarides 145–164
intrusive history
    Austroalpine basement 28–32
isoclinal fold 18, *20*
isostatic rebound 242
isotope age data **122–124**
Ivrea Zone 1, 45–65

kaolinite 287, 298, 299, 302
K–Ar, Valstrona di Omega 55, **58**, *59*, 60, 61, 62, *63*
Kaserer Basin 87, 95–97
Katschberg Fault 197
Kübler index 285, 286–287
    illite 291–298, 307
kyanite 25, 26, 27, 396, 397
    Adula nappe 373, *374*, 375, 382

landscape and neofaulting 234–235
landslide *see* mass movement
Laurussia 5, *33*
Lepontine dome 3, 366, 368, 403
    deformation 253–254, 263–275,
lignite 309
Ligurian Knot 414, 432
lineament detection 234–235
liptinite 293
lithofacies, Pienides 321–324, 329
lithostratigraphy
    Adula nappe 369–371
    Austroalpine basement 6

    Danubian window *284–286*
    Northern Apennine *420*, 419–421
    Riffler–Schönach Basin **90**, 91–94
    Tauern Window *98*
loess 241

Mafic Complex 48
magmatic rock suite, pre-Variscan *11*
    zircon age 12–13
magnetic susceptibility 30–31
Malenco–Platta 134–136
mantle wedge 34
mantle-derived material 10, 13, 17, 18, 48
Maramures 3, 169–195
    burial 188–189
    exhumation 181–186
    fission track analysis 172–173, *194, 195*
        apatite data **178**
        zircon data **176**
    geology 170–171
    thermal modelling 177–181, *181, 182*
    thermochronological analysis 173–177
marble 92–94, *105*, 222
mass movement 235–239
Matreier Zone, cross–section *19*
maturation 297
Medvednica Mountains, tectonics 145–164
    palaeomagnetism 155–157
    reconstruction 157–162
    transport direction 149–155
mélange 286, 383, 429
Meliata Ocean 35, 131, 134
Meliata suture 101–110
Meliata unit 104, *105, 106, 111*
Meliata zone 2
metaconglomerate *86, 87*
metamorphic facies *396*, **397**, **399**
metamorphic grade *288*
metamorphic map, Alps *119*
metamorphic mineral associations 287–290
metamorphic petrology,
    metasediments 393–404
metamorphism 4
    Adula nappe 381–382
    Apennine–Corsica 423–425
    Austroalpine basement 18–28
    Stellihorn shear zone 263–265
metamorphism, low-grade 281–311
metamorphism, subduction-related 117–136
meta-porphyroid 8, *11*, 12
metasediments, petrology 393–404
mid ocean ridge basalt 7, *11*, 18, 33
mineralogy, metasediments 394–397
Moho 36, 428
mollusc **337**
monazite age 27, *29*, 35
    Valstrona di Omega 46, 48, 54, 59–62
Monte Rosa nappe 3, 251–276
    deformation 263–275
    goniometry 259–263
    metamorphism 263–265
    structure 256–259
moraine, faulted 239–242
MORB (mid ocean ridge basalt) 7, *11*, 18, 33

Morbegno Gneiss 70, 71–78
Mount Medvednica 3
mylonite 208, 214, 251, 256–263, 266, 275
    Morbegno Gneiss 71–73
        microstructure 73–78
mylonite belt 47, 49–54
mylonite zone 366, *367*, 382

nannoflora 47–348
nannofossil 335–359
    age **337–339**, 348, 350
    distribution **342–344**
nannoplankton 4, **349**, **351**
nappe stack 219–228, 282, 425
    Monte Rosa 253–254, 273
nappe stacking 149, *305*, 307, 308, 317, 319
    central Alps 365, 368, 369
    Cretaceous 170, 172, 179, *183*, *187*, 189
    and metamorphism 310
Neotectonic faulting 231–246
N-MORB (enriched mid ocean ridge basalt) 7, 10, 13, 32, 33
Noric Composite terrane 9
Northern Apennine, orogenic system 413–433
Northern–Defereggen–Petzeck Group 7, 18, 21, 26, *33*, 34, 35
    age data 9, **14–17**
    magmatism 9, 10, *11*
    thermobarometry *24*, 25

ocean floor metamorphism 283, 298, 302–306, 310
oceanic spreading 421
oceanic units
    isotope and petrological data *121*
olistolith 222
ophiolite 9, 117, 120, 134
    Apennine–Corsica 417, 425, 426, 429
    Dinaridic 146, 149, 152, 153, 162
ophiolitic mélange 254
optical microscopy 285, 286
organic matter reflectance 287, 293–298
organic maturation 297, 309
Orobic anticline 1, *70*
    cross-section 73
orogeny analyses 1–4
orthogneiss 8, 12

palaeocurrents, Pienides 324–*325*, 330
palaeodepth 186, 188
palaeogeographic reconstruction
    western Mediterranean 131, *131*, *133*
palaeogeography 132–135, *136*
    Dinaride 353
palaeomagnetic direction **157**, *159*, *160*
palaeomagnetism
    Dinarides 149, 155–157
    Medvednica Mountains 157–162, 163
palaeontology, micro-, Pienides 327, **328**, 329
palaeotectonic reconstruction *429*
Palaeotethys 35
palynomorphs **337–339**, **349**, **351**
Pangea break-up 84, 219
paragonite 291

pegmatite, Permian 8, 21, 28, *33*, 35
Penninic nappe stack 253–254, 273
Penninic nappes 226, 244, 368
    cross-section *223*, *366*
    lithology 221–222
    seismicity 234
Penninic Ocean Basin 220
Periadriatic fault 157, *161*, 162, 163, 251
    see also Insubric Fault
Permian thermal event 28, 35
petrography, Pienides 326–327
petrological data and subduction *121*, *128*
petrology and isotopic age 117–136
petrology, Adula nappe 371–377
Petrosani basin 283, 288, 309
Pfitsch–Mörchner Basin 85–87, 94–95, 97
phengite 396
photomicrographs
    boudinage *52–53*
    garnet *20*
    greenschist *152*
    index minerals *289*
    Meliata unit *105*
    metapelite *374*
    mylonite *52*, *53*, *75–78*, *258*, *263*
Piemonte–Liguria Ocean 120–125, 132, 135
Pienide nappe emplacement 317–332
Pienides *171*, 174, 179, 181, 189
plate tectonics 394
    Austroalpine 5–6
playa-lake deposits 94
Pogallo Line 47–48, 49, 61, 62
Ponteranica Conglomerate 71, 79
post-glacial faulting 236–242
post-Variscan basins *84–85*
prehnite 290, 298, 302
Preluca massif 173, 179, 188–189, *194*, *195*
pressure-temperature conditions 48, 394–404, 432
    Adula nappe 369, 371–377
    Austroalpine basement 23–27, 35
    Danubian window 302–303
pressure-temperature and subduction 118–130
    data for subduction metamorphism **126–127**
projection of lineations, calculations used 275–276
Proto-Tethys 33
pumpellyite 290, 298
pyrophyllite 298, 299, 396, 397

quartz fibres in faults 236
quartz recrystallisation *213*, 214
quartz texture in mylonite *76–78*
quartz-c-textures 21–23
quartzite 92–94

radiolaria 93
radiolarite *105*
rate of convergence 226, 231
rate of exhumation *186*, 186, 188–189
rate of opening 80
rate of uplift 226–227, 232, *233*, 234
reflectance 285
reflector, mylonite 215–216
reflectors, Tauern Window 224

ridge-top depressions 235, 236
riebeckite 301
Rieserferner pluton 28, 29, 30, 31, 36
Riffler Schonach Basin 85, 89–94, 97, 222
    cross-sections *223, 225*
Rodna horst 170, 172, 174, 181–190
    exhumation *187*
    fission track analysis, results *194, 195*
Rosarolo shear zone 49–54

Salzach–Ennstal–Mariazell–Puchberg Fault (SEMP) 197–216
    folding and shearing 203–211
Sava zone 146
sedimentary basin, Tauern Window 83–98
sedimentary sequence 149
    foreland basin 4, 317–332
    post-Variscan basins 84–94
    thickness 189
seismic data 224
seismicity, Swiss Alps 231–246
SEMP (Salzach–Ennstal–Mariazell–Puchberg Fault) 197–216
sensitive high resolution ion microprobe (SHRIMP) 13, 14
    age 265–266
    U–Pb data 48, 60, 63
    zircon age 120, 129, 130
serpentinite 86
Sesia Zone 47, 48, 118–120
Severin nappe 283, *284*, 286, 293, 298, 302
Severin Ocean 282
Severin unit 285–287, 290–291, 301, 306, 310
shear bands 18, *19*, 35
shear zone 49, 54, 424
    fabric *51, 52, 53*
    Tauern Window 197, 199–216
    Stellihorn 253, 255–266, 271
shearing, Adula nappe 385–387
sheath fold 18, *20*, 23, 54
SHRIMP *see* sensitive high resolution ion microprobe
sillimanite 26, 28, 375
Simplon fault 254, 266, 273, 274
slab break-off 28, 430, *431*, 432, 433
slab retreat 317–318
slab roll back 35
slickenfibre 236
slope stability 234–239
southern Alps 46
    subsidence history *62*
Southern Steep Belt 251–276, *366*
staurolite 25, 26, 27, 28, *29*, 396
    Adula nappe 373, *374*, 375, 382
Stellihorn shear zone 253, 255–266, 271
stretching lineation *73*, *152*, 258, 259, 426, *427*
stretching, Adula nappe 385–387
Strona–Ceneri Zone 47–49, *61*
structural cross-sections
    Adula nappe *371*
    Alps *227*
    Alps, central *366, 367, 384*
    Alps, eastern *198*
    Alps, metamorphic structure *403*
    Austroalpine basement *7, 19, 33, 223*
    Bucovinian nappe *183*
    Carpathians, Inner *111*
    Corsica–Northern Apennine *418, 431*
    Danubian Window *284*
    Grassi Detachment Fault *73, 79*
    Lepontine dome *253*
    Medvednica Mountains *150*
    neotectonic fault *238*
    Pienide nappe *330–331*
    Tauern Window *88, 89, 221, 222, 223, 225*
structural evolution
    Corsica 421–425, *431*
    Northern Apennine 423–433, *431*
subduction
    Adula nappe 365, 366, 368, 383–385
    Briançonnais 402–403
    Carpathians 110, 111
    continental *414*, 424, *431*, 432, 433
subduction depth *128*
subduction rate 134
subduction-related metamorphism 117–136, 306–307
    Apennine–Corsica 425, 426, 429
    lithologies **126–127**
subduction-related minerals 398–401
subduction, age of 129–130
sub-greenschist facies 298–300
suture 146
    Alpine–Apennine 8, *418*
    Austroalpine–Penninic 8
    Meliata 101–110
    Valaisan 254

Tauern Window 83–98, 129, 197–216
    deformation **205**, 211
    folding, age of 211
    metamorphism **205**
    Salzach–Ennstal–Mariazell–Puchberg Fault (SEMP) 197, *198*, 214
    shear zone 199–211
taxa cited **356–359**
tectonic sections *see* structural cross-sections
tectonic maps
    Adula nappe *264, 268, 369, 376, 378*
    Ahorn shear zone *213*
    Alpine–Carpathian–Pannonian unit *102, 103, 170, 282, 318, 333*
    Alps *198, 220, 274, 368*
    Apennine–Corsica *415, 416*
    central Alps *368*
    Corsica–Northern Apennine *415, 416, 422*
    Danubian Window *283*
    Dinarides *147*
    Maramures area *171, 174*
    Medvednica Mountains *150, 154, 156*
    Monte Rosa nappe *252, 260, 261*
    Pienides *320*
    Stellihorn shear zone *260*
    Tauern Window *221*
tectonic model, Apennine–Corsica *414*
tectonic stacking *33*
tectonics, review 1–4
tension cracks, open 235, 236
Tethys, palaeogeographic reconstruction *133*
thermal modelling apatite data 177–181, *181, 182*

thermobarometry *24*, 25
thermochronology, Maramures 173–177
thin-skinned tectonics 69
    Pienides 317–318
thrust wedge 425
thrusts, Dinarides *154*
Thurntaler Phyllite Group 8, *11*, 12, **14–17**, 18, 33
Tisza–Dacia blocks 169–170, 172
Tisza–Dacia unit 317–332
topography, Alps *2*
topography, Dinarides *147*
TRANSALP seismic line 216, 219, *221*, 224–226
transcurrent shearing 275
transport direction, Dinarides 149–155
tschermakite 26
Tuscan metamorphic units 419, 421, 426
Tyrrhenian Sea 421

ultramylonite 49, *51*
underplating *50*, 61, 401
underthrusting 424, 426, 432, 433
uplift path 184, *185*
uplift, Neotectonic 231–246

Val Biandino Quartz Diorite 70, 71–73, *74*, 78
Valais Ocean 129, 132, 134, 267, 403
Valaisan suture 254
Valle Biagio Granite 70, 71–73

Valseisia 255
Valstrona di Omega 45–65
    geochronology 59–61
Variscan collision 5, 6–7, *33*, 35
vibroseis data *224, 227*
Vienna Basin 197
vitrinite reflectance 285, 293–296, 307
volcanic arc basalt 7, 10, 13
volcanism in hanging wall 1
volcanism, Permian 69
volcanism, Tertiary 419, 421, 432

white mica *121*, 291, *374*, 382
    analytical data **107–108**, *109*, **113**
whole rock isotope characteristics 13–18
within-plate basalt 7, 10–12, 13, 17, 18, 33

xenoliths *74*
X-ray powder diffraction 285, 286–287

Zinsnock stock 29, 30
zircon age 12–13
zircon analysis 95–97
zircon fission track age *29, 31*, 34, 36
    Valstrona di Omega 46, 55,
        59, 60–62, *63*
zircon fission track data 174–177
zoisite 290, 396